GENETIC ASPECTS OF PLANT MINERAL NUTRITION

Developments in Plant and Soil Sciences

1. J. Monteith and C. Webb, eds.,
    Soil Water and Nitrogen in Mediterranean-type Environments. 1981. ISBN 90-247-2406-6
2. J.C. Brogan, ed.,
    Nitrogen Losses and Surface Run-off from Landspreading of Manures. 1981. ISBN 90-247-2471-6
3. J.D. Bewley, ed.,
    Nitrogen and Carbon Metabolism. 1981. ISBN 90-247-2472-4
4. R. Brouwer, I. Gašparíková, J. Kolek and B.C. Loughman, eds.,
    Structure and Function of Plant Roots. 1981. ISBN 90-247-2510-0
5. Y.R. Dommergues and H.G. Diem, eds.,
    Microbiology of Tropical Soils and Plant Productivity. 1982. ISBN 90-247-2624-7
6. G.P. Robertson, R. Herrera and T. Rosswall, eds.,
    Nitrogen Cycling in Ecosystems of Latin America and the Caribbean. 1982. ISBN 90-247-2719-7
7. D. Atkinson et al., eds.,
    Tree Root Systems and their Mycorrhizas. 1983. ISBN 90-247-2821-5
8. M.R. Sarić and B.C. Loughman, eds.,
    Genetic Aspects of Plant Nutrition. 1983. ISBN 90-247-2822-3
9. J.R. Freney and J.R. Simpson, eds.,
    Gaseous Loss of Nitrogen from Plant-Soil Systems. 1983. ISBN 90-247-2820-7
10. United Nations Economic Commission for Europe.
    Efficient Use of Fertilizers in Agriculture. 1983. ISBN 90-247-2866-5
11. J. Tinsley and J.F. Darbyshire, eds.,
    Biological Processes and Soil Fertility. 1984. ISBN 90-247-2902-5
12. A.D.L. Akkermans, D. Baker, K. Huss-Danell and J.D. Tjepkema, eds.,
    Frankia Symbioses. 1984. ISBN 90-247-2967-X
13. W.S. Silver and E.C. Schröder, eds.,
    Practical Application of Azolla for Rice Production. 1984. ISBN 90-247-3068-6
14. P.G.L. Vlek, ed.,
    Micronutrients in Tropical Food Crop Production. 1985. ISBN 90-247-3085-6
15. T.P. Hignett, ed.,
    Fertilizer Manual. 1985. ISBN 90-247-3122-4
16. D. Vaughan and R.E. Malcolm, eds.,
    Soil Organic Matter and Biological Activity. 1985. ISBN 90-247-3154-2
17. D. Pasternak and A. San Pietro, eds.,
    Biosalinity in Action: Bioproduction with Saline Water. 1985. ISBN 90-247-3159-3
18. M. Lalonde, C. Camiré and J.O. Dawson, eds.,
    Frankia and Actinorhizal Plants. 1985. ISBN 90-247-3214-X
19. H. Lambers, J.J. Neeteson and I. Stulen, eds.,
    Fundamental, Ecological and Agricultural Aspects of Nitrogen Metabolism in Higher Plants. 1986.
    ISBN 90-247-3258-1
20. M.B. Jackson, ed.
    New Root Formation in Plants and Cuttings. 1986. ISBN 90-247-3260-3
21. F.A. Skinner and P. Uomala, eds.,
    Nitrogen Fixation with Non-Legumes. 1986. ISBN 90-247-3283-2
22. A. Alexander, ed.
    Foliar Fertilization. 1986. ISBN 90-247-3288-3
23. H.G. v.d. Meer, J.C. Ryden and G.C. Ennik, eds.,
    Nitrogen Fluxes in Intensive Grassland Systems. 1986. ISBN 90-247-3309-X
24. A.U. Mokwunye and P.L.G. Vlek, eds.,
    Management of Nitrogen and Phosporus Fertilizers in Sub-Saharan Africa. 1986.
    ISBN 90-247-3312-X
25. Y. Chen and Y. Avnimelech, eds.,
    The Role of Organic Matter in Modern Agriculture. 1986. ISBN 90-247-3360-X
26. S.K. De Datta and W.H. Patrick Jr., eds.,
    Nitrogen Economy of Flooded Rice Soils. 1986. ISBN 90-247-3361-8
27. W.H. Gabelman and B.C. Loughman, eds.,
    Genetic Aspects of Plant Mineral Nutrition. 1987. ISBN 90-247-3494-0
28. A. van Diest, ed.,
    Plant and Soil: Interfaces and Interactions. 1987. ISBN 90-247-3535-1
29. United Nations, ed.,
    The Utilization of Secondary and Trace Elements in Agriculture. 1987. ISBN 90-247-3546-7

Genetic Aspects of Plant Mineral Nutrition

Proceedings of the Second International Symposium on
Genetic Aspects of Plant Mineral Nutrition, organized by
the University of Wisconsin, Madison, June 16–20, 1985

Edited by

W.H. GABELMAN
Professor of Horticulture, University of Wisconsin
Madison, Wis., USA

and

B.C. LOUGHMAN
University Lecturer in Biological Sciences and
Fellow of University College Oxford, Oxford, England

Chapters indicated with an asteriks in the table of contents were
first published in *Plant and Soil*, Volume 99:1 (1987)

1987 **MARTINUS NIJHOFF PUBLISHERS**
a member of the KLUWER ACADEMIC PUBLISHERS GROUP
DORDRECHT / BOSTON / LANCASTER

Distributors

for the United States and Canada: Kluwer Academic Publishers, P.O. Box 358, Accord Station, Hingham, MA 02018-0358, USA
for the UK and Ireland: Kluwer Academic Publishers, MTP Press Limited, Falcon House, Queen Square, Lancaster LA1 1RN, UK
for all other countries: Kluwer Academic Publishers Group, Distribution Center, P.O. Box 322, 3300 AH Dordrecht, The Netherlands

Library of Congress Cataloging in Publication Data

```
International Symposium on Genetic Aspects of Plant
   Mineral Nutrition (2nd : 1985 : Madison, Wis.)
   Genetic aspects of plant mineral nutrition.

   (Developments in plant and soil sciences)
   1. Plants--Nutrition--Genetic aspects--Congresses.
2. Plants, Effect of minerals on--Congresses.
I. Gabelman, W. H.   II. Loughman, B. C.   III. University
of Wisconsin--Madison.   IV. Title.   V. Series.
QK867.I424   1985        581.1'3         87-3885
```

ISBN-13: 978-94-010-8102-3   e-ISBN-13: 978-94-009-3581-5
DOI: 10.1007/978-94-009-3581-5

Copyright

© 1987 by Martinus Nijhoff Publishers, Dordrecht.
Softcover reprint of the hardcover 1st edition 1987

All rights reserved. No part of this publication may be reproduced, stored in a retrieval system, or transmitted in any form or by any means, mechanical, photocopying, recording, or otherwise, without the prior written permission of the publishers,
Martinus Nijhoff Publishers, P.O. Box 163, 3300 AD Dordrecht,
The Netherlands.

CONTENTS

Preface   IX

Section 1: Responses of wild plant ecotypes to nutrient deficiency stress

P. B. Vose, Genetical aspects of mineral nutrition — Progress to date   3
F. Stuart Chapin, III, Adaptations and physiological responses of wild plants to nutrient stress   15
A. Ostrowska, Application of ANE value and of shares of individual elements in this value for determining the difference between various plant species   27
C. J. Rosen and J. J. Luby, Variation in foliar elemental composition in Vaccinium crosses   45

Section 2: Screening techniques to detect nutritional differences under genetic control

* G. C. Gerloff, Intact-plant screening for tolerance of nutrient-deficient stress   55
A. J. Conner and C. P. Meredith, Somatic cell selection of mutants resistant to mineral stress   69
T. M. Andersen, E. Polle and C. F. Konzak, Screening spring wheat for drought tolerance   79
L. M. Gourley, Identifying aluminum tolerance in sorghum genotypes grown on tropical acid soils   89
R. R. Duncan and J. D. Sutton, Influence of field sampling techniques on the Al, Mn, Mg, and Ca nutritional profiles for acid soil tolerant susceptible sorghum genotypes   99

Section 3: Tolerance to salinity and to metal toxicities

* E. Epstein and D. W. Rains, Advances in salt tolerance   113
J. M. Stassart and J. Bogemans, Intervarietal ionic composition changes in barley under salt stress   127
B. Cartwright, A. J. Rathjen, D. H. B. Sparrow, J. G. Paul and B. A. Zarcinas, Boron tolerance in Australian varieties of wheat and barley   139
B. J. Scott, D. G. Burke and T. E. Bostrom, Australian research on tolerance to toxic manganese   153
R. H. Merry, Tolerance of plants to heavy metals   165

* F. P. C. Blamey, C. J. Asher and D. G. Edwards, Hydrogen and aluminium tolerance 173
* C. D. Foy, W. A. Berg and C. L. Dewald, Tolerances of Old World bluestems to an acid soil high in exchangeable aluminum 181

R. Magnavaca, C. O. Gardner and R. B. Clark, Comparisons of maize populations for aluminum tolerance in nutrient solution 189

R. Magnavaca, C. O. Gardner and R. B. Clark, Inheritance of aluminum tolerance in maize 201

R. A. Borgonovi, R. E. Schaffert, G. V. E. Pitta, R. Magnavaca and V. M. C. Alves, Aluminum tolerance in sorghum 213

* Ch. Hecht-Buchholz and J. Schuster, Responses of Al-tolerant and Al-sensitive Kearney barley cultivars to calcium and magnesium during Al stress 223

J. Bogemans and J. M. Stassart, Ion segregation in different plant parts within different barley cultivars under salt stress 239

P. R. Furlani and R. B. Clark, Plant traits for evaluation of responses of sorghum genotypes to aluminum 247

R. Magnavaca, C. O. Gardner and R. B. Clark, Elevation of inbred lines for aluminum tolerance in nutrient solution 255

Section 4A: Genotypic response to nutrient deficiency: Macronutrients

* B. C. Loughman, The application of *in vivo* techniques in the study of metabolic aspects of ion absorption in crop plants 269

G. Alagarswamy and F. R. Bidinger, Genotypic variation in biomass production and nitrogen use efficiency in pearl millet [*Pennisetum americanum* (L.) Leeke] 281

A. M. C. Furlani, R. B. Clark, W. M. Ross and J. W. Maranville, Differential phosphorus uptake, distribution, and efficiency by sorghum inbred parents and their hybrids 287

*A. J. Fist, F. W. Smith and D. G. Edwards, External phosphorus requirements of five tropical grain legumes in flowing-solution culture 299

R. R. Coltman, W. H. Gabelman, G. C. Gerloff and S. Barta, Genetics and physiology of low-phosphorus tolerance in a family derived from two differentially adapted strains of tomato (*Lycopersicon esculentum* Mill.) 309

T. Zaharieva, Genotypic differences in the chemical composition of maize plants grown on a calcareous chernozem 317

A. R. Memon and A. D. M. Glass, Genotypic differences in subcellular compartmentation of K^+: Implications for protein synthesis, growth and yield 323

J. F. Pedersen, J. H. Edwards and H. A. Torbert, Root morphological effects on Mg uptake in five tall fescue lines 331

B. Bochev, E. Neikova-Bocheva, N. Mitreva and G. Ganeva, Influence of different *Triticum aestivum* L. genomes and chromosomes on the assimilation of the main nutrient elements 343

* S. S. Figdore, W. H. Gabelman and G. C. Gerloff, The accumulation and distribution of sodium in tomato strains differing in potassium efficiency when grown under low-K stress 353
H. Perby and P. Jensen, Vegetative adaptation to N stress regimes in two barley cultivars with different N requirement 361
* M. Y. Siddiqi, A. D. M. Glass, A. I. Hsiao and A. N. Minjas, Genetic differences among wild oat lines in potassium uptake and growth in relation to potassium supply 369
J. J. Woodend, A. D. M. Glass and C. O. Person, Genetic variation in the uptake and utilization of potassium in wheat (*Triticum aestivum*) varieties grown under potassium stress 383
E. Alcantara and M. D. de la Guardia, Inheritance of response of sunflower inbreds to a low calcium/magnesium ratio 393
V. Kovacevic, L. J. Radic and N. Vekic, Genetic differences in the ear-leaf nutrient content of inbred lines of corn (*Zea mays* L.) 399

Section 4B: Genotypic response to nutrient deficiency: Trace elements

* R. D. Graham, J. S. Ascher, P. A. E. Ellis and K. W. Shepherd, Transfer to wheat of the copper efficiency factor carried on rye chromosome arm 5RL 405
* J. E. Bowen, Physiology of genotypic differences in zinc and copper uptake in rice and tomato 413
* E. P. Williams, R. B. Clark, W. M. Ross, G. M. Herron and M. D. Wit, Variability and correlation of iron-deficiency symptoms in a sorghum population evaluated in the field and growth chamber 425
N. Seetharama, R. B. Clark and J. W. Marranville, Sorghum genotype differences in uptake and use efficiency of mineral elements 437
J. G. Coors, Resistance to the European corn borer, *Ostrinia nubilalis* (Hubner), in maize, *Zea mays* L., as affected by soil silica, plant silica, structural carbohydrates, and lignin 445
E. Alcantara and M. D. de la Guardia, Differential response of sunflower genotypes to iron deficiency 457
* E. O. Leidi, M. Gomez and M. D. de la Guardia, Soybean genetic differences in response to Fe and Mn: Activity of metalloenzymes 463
S. M. Fatalieva, Ultrastructure of mesophyll cells grown on different levels of selenium of two pea genotypes 471

Section 5: Genetic variation in microorganism host interactions in mineral nutrition

F. A. Bliss, Host plant control of symbiotic N_2 fixation in grain legumes 479
* M. R. Sarić, Z. Sarić and M. Govedarica, Specific relations between some strains of diazotrophs and corn hybrids 495
* S. K. A. Danso, C. Hera and C. Douka, Nitrogen fixation in soybean as influenced by cultivar and Rhizobium strain 511

S. Rajapakse and J. C. Miller Jr, Intraspecific variability for VA mycorrhizal symbiosis in Cowpea (*Vigna unguiculata* [L.] Walp.)   523

Section 6: Germplasm resources and modification

W. H. Gabelman, Sources of germplasm for research on mineral nutrition   539

* T. M. Schettini, W. H. Gabelman and G. C. Gerloff, Incorporation of phosphorus efficiency from exotic germplasm into agriculturally adapted germplasm of common bean (*Phaseolus vulgaris* L.)   559

M. D. Casler and J. M. Reich, Genetic variability for mineral element concentrations in smooth bromegrass related to dairy cattle nutritional requirements   569

* T. K. Surowy and M. R. Sussman, Molecular cloning of the plant plasma membrane H^+-ATPase   579

R. B. Corey and S. M. Combs, Control of nutrient concentrations in plant growth media   591

S. M. Schwab, Considerations of vesicular-arbuscular mycorrhiza physiology in breeding for enhanced mineral uptake by plants   603

* M. R. Sarić, Progress since the first international symposium: Genetic aspects of plant mineral nutrition, Beograd, 1982, and perspectives of future research   617

Preface

This volume presents the proceedings of the Second International Symposium on Genetic Aspects of Plant Mineral Nutrition, held in Madison, Wisconsin in 1985.

The mechanisms by which plants acquire, transport and utilize essential mineral nutrients are highly complex. The means by which plants either exclude or tolerate ions of metals toxic to plants are equally complex.

The first symposium attempted to convene research scientists concerned with mineral nutrition for the purpose of exploring the kinds of mineral nutrition phenomena identified as being under genetic control. The first symposium also placed much emphasis on research to which genetic intervention might be applied.

At the second symposium more papers were presented on genetic and breeding research, a long-term objective of the first symposium. The second symposium also included biotic interactions under genetic control that either enhanced or impeded ion uptake, *e.g.* mycorrhizae and nitrogen fixing bacteria. This continuing dialogue is essential for a research area the complexity of which is due to its interdisciplinary nature.

It is particularly important that the fundamental mechanisms involved in absorption and transport of ions be understood before attempts at modifying them by genetic manipulation are embarked upon. A valuable aspect of the Symposium was the opportunity for physiologists and geneticists to understand each other's problems and the volume reflects this important interaction. We look forward to advances in the experimental approach to these problems involving recently developed techniques. It is our hope that at least some of our difficulties will be solved in papers to be given at the third Symposium to be held in June 1988 at the Institute of Crop Sciences and Breeding, Federal Research Centre of Agriculture, Braunschweig, Federal Republic of Germany.

W.H. Gabelman
B.C. Loughman

Section 1

Responses of wild plant ecotypes to nutrient deficiency stress

Genetical aspects of mineral nutrition — Progress to date

P.B. VOSE
92 Kentsford Road, Grange-over-Sands, Cumbria LA11 7BB, UK

Key words Al- and Mn-toxicity Genotype Low-input Nitrogen utilization Nutrient efficiency Salinity tolerance Selection techniques

Summary Recent years have seen greatly changed attitudes to the utilization of genetic differences in plant nutrition. There is now wide acceptance of the fact that cultivars may respond differently to nutritional factors, and that this may be applied to solving specific problems of soil fertility/stress. This response is under genetic control, and therefore improved response is accessible via screening, selection and normal plant breeding procedures. At the present time, due to perceived need, most work and current advances are taking place in regard to selecting for salinity tolerance, acid soil (Al and Mn) tolerance, and resistance to bicarbonate induced iron deficiency. There are still wide differences concerning selection methods, and clearly quite a number of approaches are possible including plant tissue culture. In some cases we lack adequate field validation of selection procedures. There is still little knowledge of the factors which make some genotypes 'responsive' and others 'non-responsive'. The latter should not necessarily be discarded as frequently they seem adapted to persistence under low fertility conditions, and may thus be a valuable gene source for low-input farming systems.

Introduction

This paper does not purport to be a complete review, for one thing it is nor very long since a rather detailed one[46] and secondly other participants are making specialised presentations. It is therefore an attempt to sketch in some highlights of progress, with some indications as to where the general field may be going.

There are various means of measuring progress, and the fact that the Second International Meeting assembled with more participants than previously is in itself a significant fact. However, it is surely in general attitudes over the last ten years that changes have most occurred. At one time it was necessary to explain that different genotypes or cultivars of crop plants might respond differently to various types of nutritional situation or soil stress, or that some cultivars were more successful than others in obtaining or utilising certain essential elements. Hitherto, plant breeders have attempted to get the maximum potential adaptability into their breeding material, and the approach has been largely successful[7], but it is now recognized that specifically adapted plant material may be necessary for extremely adverse soil conditions. Moreover, although the possibility of soil amendments including fertilizer is well understood, the use of specially adapted varieties can increase both effectiveness and economy.

H.W. Gabelman and B.C. Loughman (Eds.), Genetic aspects of plant mineral nutrition.
ISBN-13: 978-94-010-8102-3
© *1987 Martinus Nijhoff Publishers, Dordrecht/Boston/Lancaster.*

It seems rather surprising in retrospect that it took so long for this to be appreciated, when one considers all the plant ecological work that took place over many years relating plant species to soil type, pH and nutritional status. Probably it was not realised that so much variability in nutritional response was available in cultivated plants, as plant physiologists traditionally did not greatly favour comparative studies except between widely differing species. It is now much better appreciated that there can be wide genotypic differences in physiology and biochemistry. Additionally, in recent years there has been a much greater tendency to seek more widespread sources of germplasm, either through systematic testing of material from world collections or else on an *ad hoc* basis.

A recent authoritative report on plant and soil research priorities for developing countries[30] strongly emphasised the role that cultivar screening and selection might play in developing varieties with tolerance to major soil stresses. The need is not solely to improve cultivars for growing on soils with inherent problems, but also to improve the efficiency of already good cultivars under systems of intensive agriculture. Thus a Royal Society (UK) report[36] concerning the nitrogen cycle stated in connection with the losses of applied N-fertilizer, 'We identify a need for the production and introduction into agricultural practice of plant varieties with sustained or enhanced productivity but with the ability to use nitrogen more efficiently than many of those currently used.'

At the present time there seem to be four main areas attracting most work: tolerance to salinity; tolerance to soil acidity complex, especially aluminium toxicity; tolerance to bicarbonate induced iron deficiency chlorosis; and basic comparative studies of the mechanisms of genotypic differences. The reason for especial interest in these areas is of course perceived need, and the possibility, in the first three, of researchable approaches. The difficulty of defining researchable approaches for general efficiency of N and P utilization and productivity will be discussed in some detail later. The problems of breeding for improved dinitrogen fixation, and the possibility of using tissue culture methods to accelerate the selection process for salinity and acid tolerance, are of increasing interest.

The areas of problem soils are sufficiently large as to warrant specialised plant breeding programmes for this purpose, *e.g.* acid soils comprise 18 percent of total world soil area, or over 2.4 billion hectares, while in the case of salinity about 1000 million ha are affected worldwide[27] including at least 30 million ha in eastern Europe[41]. A third of irrigated soils, or about 77 million ha are sufficiently affected by salinity as to affect crop yields[15]. Salinity problems affect as much as 40 million ha of potential rice soils in the humid tropics[32]. About 25% of world soils are calcareous and liable to Fe-deficiency problems either on a regular basis or as a result of mismanagement or restricted water supply[45].

Selection for specific factors

Considerable current research is connected with the investigation and selection of salinity tolerant cultivars. It has been known for a long time that certain cultivars were more adapted to saline soils than were others, but little was made of this fact until Epstein and co-workers[16] showed that positive selection could produce tomato and barley lines of improved salt tolerance. At present, selection for, or investigation of, salt tolerant cultivars is taking place with such widely divergent subjects as turf grasses able to tolerate salt accumulation or low quality irrigation water[14,42], roadside use grasses able to tolerate the salinity that comes from winter de-icing salt[22], wheat able to grow under chronically saline conditions[24,26,39], and similarly rice[25,48], alfalfa[2] and tomato[37] to select just a few recent references.

It is generally agreed that the mechanism of salinity tolerance is very complex and is conferred by a number of factors, amongst which Na-exclusion from the leaves seems to be very important. Nevertheless, it was found in rice[48] that no single factor confers resistance and that varieties showing greatest tolerance to salt within the leaves were not necessarily those showing greatest overall resistance to salinity. This seems typical, and leads to a consideration of tissue culture as a means of selecting for tolerance to various soil factors[1]. The selection *via* callus culture of mutant cell lines and regenerated plants with higher NaCl tolerance has been summarised by Nabors and Dykes[29]. How are we to understand the higher tolerance for a factor selected through cell culture when such tolerance seems to be based in great part on structural/Na-transport modifications? The answer probably lies in the fact that there are almost certainly two major types of NaCl tolerance — one depending on Na-exclusion, or the whole plant transport factor, and the other a tolerance factor at the cellular level.

Cell culture selection might clearly be effective for the latter, but any selection for the former would seem to be fortuitous. Presumably in an ideal situation one should first find genotypes which have the Na-exclusion, or structural type of tolerance, and use this as basic material for cell culture selection to achieve further tolerance at the cellular level.

The situation is somewhat different in the case of selection for Al-toxicity tolerance in cell culture, where it is well-accepted that the mechanism of resistance to Al has a fundamental cellular basis[17]. Therefore callus culture selection should present no theoretical problem. The practical problem of keeping a known amount of aluminium available in the medium can apparently be overcome[12]. Similarly, other toxicity tolerances which are essentially cell-based should be amenable to selection through cell culture. Thus Bennetzen and Adams[6] have selected *Lycoper-*

sicum peruvianum for high levels of cadmium tolerance by means of suspension culture. The progeny cultures were found to be ten times more tolerant than the original.

Although tissue culture has possibilities for screening certain characters, subsequently regenerated plants will still have to undergo normal agronomic trials. The real significance of tissue culture methods lies in the possibility of selecting rare mutants from extremely large numbers. The feasibility of using mutation breeding techniques to obtain cultivars better adapted to specific soil stresses was discussed at length by Vose[44]. A major negative factor is that useful 'nutritional' characters seem mostly to be dominant, and mutants for dominant alleles are extremely rare. Nabors and Dykes[29] reported NaCl tolerant mutations, behaving as dominant alleles, arising in about one in every 500,000 cells.

Clearly, cell culture can be the means of both inducing mutations and screening them from the very large populations that are necessary.

Search for efficient cultivars

When one talks about the productivity of crops one is basically concerned with the uptake and utilisation of nitrogen and phosphorus, coupled of course with photoassimilation. The question of general efficiency is extremely complex, and although some progress is being made the patchwork nature of the work inevitably makes it easier to see the problems rather than the solutions: The definition of efficiency is itself a matter for discussion, for we have 'responding' and 'non-responding' varieties and genotypes according to the degree of response to increased nutrients. There are varieties which are 'efficient' or 'non-efficient' converters of nutrients into dry matter, some being specifically efficient when grown at lower levels of a particular nutrient. Varieties may be 'efficient' or 'inefficient' for uptake or translocation, or 'accumulators' or 'non-accumulators' of certain elements.

Therefore it depends on what is meant by 'efficiency'. I and some others have related efficiency mainly to the production of dry matter per unit of N (or other element) involved. This is valid for forage crops where all the above-ground production is utilised, but is true in only the most general manner for cereals and pulses, where the prime interest is in the grain and the harvest index

$$\left(\frac{\text{grain}}{\text{total yield}} \times 100\right), \text{ or the Nitrogen Index } \left(\frac{\text{grain N}}{\text{total N}} \times 100\right)$$

are more appropriate measures. Asana[4] pointed out that the 'most efficient' wheat, as measured by production of dry matter per unit N is not necessarily the most productive, because yield in cereals is determined by tillering, number of fertile tillers (ears), number of grains per ear, size of grain *etc*. Brinkman and Rho[8] found in a comparative test of oat cultivars at different N-levels, the cultivar showing greatest grain yield response to nitrogen did so because of better response of spikelets per plant and weight per kernel.

The question of method of cultivation must also be considered. This has been very well put by Dambroth and Bassam[13] in connection with root growth and plant density in sugar beet: 'maximum yield achieved in field experiments does not represent a physiological limitation of presently available cultivars but demonstrates that portion of the genetic potential which is realisable by the optimal utilization of the present means of cultivation'.

As was noted by Mengel[28] modern varieties of wheat and rice are very efficient in their use of P and K, because the harvest index is greatly increased compared with the older varieties, and less P and K is required as the result of fewer leaves requiring support. Nevertheless, on a 'per ha' basis the requirement for N, P, and K fertilizer may increase because the planting density must be increased both to achieve maximum yield and also for weed control. Modern varieties tend to be faster growing than the ones they replace, thus setting up increased demand for nutrients, which they will receive inadequately under natural soil fertility, and hence show great responsiveness to fertilizer application. The converse of this is that the older varieties are slower growing but can produce with a lower rate of nutrient supply. Frequently too, landraces or undeveloped ecotypes require little fertilization, but likewise have poor inherent physiological capacity for yield response.

Does this mean therefore that it will not be possible to improve the fundamental efficiency of utilisation of nutrients by crop plants, as opposed to merely moving around dry matter from unwanted parts, such as foliage, stem, etc. to desired components such as grain or fuit or tubers? Do low input cultivars necessarily imply low yield cultivars?

The work of the Madison group, as summarised by Gabelman and Gerloff[18], suggests that the answer is 'no' at least to the first of these questions. Growing genotypes of tomatoes and Phaseolus beans under nutrient stress conditions of K, P, N and Ca revealed some to be much more efficient than others and that broad sense heritability estimates of nutrient use efficiency were found to be high and that selection for efficiency would be effective. Of particular interest is the fact that in the case of progeny of breeding experiments, some genetic recombinants were much more efficient than either parent, suggesting that genotypes more efficient than the most efficient parents might be developed. This seems a rather clear case of hybrid vigour — indeed may be the basis of this phenomenon.

Similarly, Ramirez[34] compared 60 corn (*Zea mays* L.) inbreds for their capacity to produce dry matter per unit of N, P or K and found that there was considerable variation in the efficient use of nutrients. Although dry matter production was in general well correlated with total N, P and K accumulation *i.e.* larger plants had greater total nutrient content, nevertheless efficient use of nutrients appeared independent of plant size, which offers the possibility of developing genuinely low-input cultivars of moderate yield. These may not have the capacity for the very highest possible yields required for highly developed farming systems with large fertilizer inputs, but could be especially suitable for developing countries where cropping is often under constraints of little or no fertilizer and uncertain water. Under these conditions what is essential is reliable yield rather than ultimate yield.

At the present time we therefore need not only to develop cultivars appropriate to the conditions of the high-input industrialised countries but also for low-input conditions. Relatively non responsive genotypes should therefore not be necessarily discarded, as they may be a valuable gene source for persistence under low fertility conditions. In the future, as the economic situation of many developing countries improves, then quite likely their need will shift from low-input to high-input types.

The genetics of the efficiency of utilisation of major nutrient ions continues in the main to elude us, as regards real understanding, despite what theories are advanced. The fact is that there must be more than one mechanism operating to include uptake, transport and utilisation, and clearly the phenotypic expression is a summing of the separate mechanisms, all of which have their genetic control. This is not to take credit away from praiseworthy efforts to understand the genetics of nutritional response, but to point out that the net understanding is that of the main mechanism, which is confounded by the secondary processes that are operational.

Improved nitrogen utilisation efficiency must everywhere remain the main target. Can we use less N fertilizer and how can we better understand the mechanisms of nitrogen response? Can more efficient cultivars result in less contamination of groundwaters, as recently emphasized[36]. We know that some genotypes have better uptake capacity for nitrate or ammonium nitrogen; some cultivars produce more dry matter per unit N applied; other genotypes have better capacity for mobilization of nitrogen from the leaves and distributing it to the grain; some genotypes may be adapted to either high or low nutrition.

Most attention has been focussed on the nitrate reduction step of nitrogen assimilation, which is logical if it is accepted that in fertilized soils the activity of nitrifying bacteria normally results in nitrate being the dominant form of nitrogen available to the plant, but the obvious exceptions are when grassland is heavily fertilized with ammonium compounds and the normal situation is rice soils. Indeed, we now know that many acid soils keep applied ammonium fertilizer in the ammonium form for weeks. As it is now known that NO_3 and NH_4 ions are utilised equally well by plants, with certain differences in uptake and assimilation, it makes it important that the ammonium incorporation step should receive as much attention as has been applied to the reduction of nitrate, as it is in any event a following step in nitrate assimilation following nitrate reduction. As NH_4 ions are normally taken up by plants more rapidly than NO_3 ions, it is possible that the amination step is rarely rate-limiting, but the possibility exists.

The complexity of the factors influencing grain protein was shown by Huffaker and Rains[23] who found that the breakdown and retranslocation of protein from the foliage to the grain during seed filling was a critical physiological marker for grain protein content. Anza and UC44-111 showed marked differences in nitrate content of leaves, and although Anza contained considerably more nitrate it absorbed less nitrate in short term studies[39].

It is by the stepwise limiting factors in nitrogen acquisition and protein assimilation that we shall make progress, but in general this is an area which has not advanced much in the last few years. As noted above it seems almost inevitable that, due to the large number of separate but linked processes, any general study of nitrogen utilization efficiency will indicate a complex genetic situation incapable of simple resolution. Thus although it was shown that two classes of tomatoes could be recognized: N-responders and non-responders, inheritance of efficiency for nitrogen utilization was complicated with both dominance and additive gene effects[31]. At a more specific level, the level of nitrate reductase in two inbred lines of maize was controlled by two loci[47]. The answer must surely be to build up a composite picture of nitrogen efficiency concerning the components of uptake, transport, assimilation and remobilization.

In herbage grasses one does not have the complication of redistribution, as the product is harvested primarily in vegetative form. Early work with ryegrass, *Lolium perenne*[43], demonstrated that one could have high dry matter — low N content plants, while Antonovics et al.[3] found wild populations of *Lolium perenne* adapted to either high or low levels of habitat N. Subsequently Goodman[19,20,21] has demonstrated the possibility of selection for increased nitrogen uptake and yield in ryegrass. Although Goodman[20] found that nitrogen response is linked to nitrate reductase activity in ryegrass he also noted that ion (nitrate) uptake has two components: uptake velocity per unit size of root, and root size, and these two characters may vary at different stages of maturity. This suggests that if grass and cereal cultivars are to be developed which are more efficient in recovery of applied fertilizer nitrogen, with less nitrate being lost to groundwater pollution, then probably selection for increased root size will be an essential component.

Phosphorus is the second almost universal limiting factor for good yields. Variation in phosphorus nutrition can be important through: adaptability to low P; ability to gather P from low-P soils; response to P; tolerance to high P as in band fertilization and to P-associated Zn and Cu-deficiency susceptibility[40]. Metabolic differences in phosphorus uptake and assimilation have been described for *Phaseolus*, for maize cultivars, for tomatoes, and for soybeans. For example, Coltman et al.[11] have described differences in both growth, phosphorus acquisition and utilization for tomatoes grown under P-deficiency stress.

Root morphology is particularly relevant to the uptake of P and elements such as Fe, Mn and Zn where root interception and contact are important supply mechanisms. ^{32}P uptake by inbred lines of maize was related to the size of the root system, and the efficiency of P acquisition in maize can be increased by development and selection of hybrids with more fibrous root systems[38]. The soybean cultivar Aoda has twice the capacity for nutrient absorption at high nutrient levels, compared with Harosoy 63, despite the latter having a larger root system[35]. However, at low nutrient levels the capacity of each variety to absorb nutrients is about the same, as the more extensive Harosoy 63 root system then provides an advantage.

P-uptake prediction models[5] coupled with actual experiments with corn and soybeans have indicated that root size as measured by length and radius had the greatest effect on P-uptake. In other species root hairs were important for uptake[5], as for example with white clover[9].

Barley and wheat have been shown to have considerable cultivar differences in root depth and form, which can be significant for fertilizer use efficiency, drought resistance and response to minor element deficiencies. It is clear that plant breeding programmes could usefully provide a wider base for root selection, possibly aiming to establish more effective root systems on otherwise satisfactory cultivars. There is no evidence that photosynthetic systems cannot support a larger root system, indeed in most crop plants there is probably a reserve of photosynthetic capacity. Similarly a number of workers have found that there is no close relationship between rooting capability and size of tops. Therefore, selection for root-size should be possible, within limits, without influencing the above-ground parts.

A number of techniques for selecting larger root systems have been summarised by Clarke and Townley-Smith[10] in connection with drought tolerance, but are equally applicable to nutrient acquisition. Larger root systems do not, of course, remove the need for continued fertilizer application, but they do offer the possibility of much better fertilizer utilization and more resistance to marginal minor element deficiencies.

Conclusions

The quantity and variety of work now being carried out indicates clearly that this area of crop improvement has 'come of age'. The study of genetic differences in plant nutrition is no longer confined to a relatively small number of crop physiologists but is contributing to plant breeding and the development of improved varieties, especially in the case of acid soil and salinity tolerance, and resistance to ion deficiency chlorosis.

Nevertheless, considerable work remains to be done on the mechanisms that are responsible for yield response in relation to nitrogen and phosphorus fertilization. These are fundamental to the basic improvement of efficiency in crop plants, and to our understanding of potential low input cultivars.

References

1 Abrigo W M, Novero A V, Coronel V P, Cabuslay G S, Blanco L C, Parao F T and Yoshida S 1985 Somatic cell culture at IRRI. In Biotechnology in International Agricultural Research. Proc. Int. Center Seminar on 'International Agric. Res. Centers and Biotechnology': pp 149–167. IRRI, Manila.
2 Allen S G, Dobrenz A K, Schonhorst M and Stoner J E 1985 Heritability of NaCl tolerance in germinating alfalfa seedlings. Agron. J. 77, 99–101.
3 Antonovics J, Lovett J and Bradshaw A D 1966 The evolution of adaptation to nutritional factors in populations of herbage grasses. In Isotopes in Plant Nutrition and Physiology. Proc. IAEA Symposium, Vienna 5 6 Sept. 1966, pp 549–567. IAEA (Vienna).
4 Asana R D, Ramaiah P K and Rao M V K 1968 The uptake of nitrogen, phosphorus and potassium by the cultivars of wheat in relation to growth and development. Indian J. Plant Physiol. 9, 85–107.
5 Barber S A 1982 Soil-plant root relationships determining phosphorus uptake. In Proc. 9th Int. Plant Nutrition Colloquium Vol. 1. Ed. A Scaife, pp 39 44. Commonwealth Agricultural Bureaux Farnham Royal.
6 Bennetzen J L and Adams T C 1984 Suspension cultures of Lycopersicum peruvianum selected for high levels of cadmium tolerance. Plant Cell Reports 3, 258.
7 Borlaug N 1983 Feeding the World during the next doubling of the world population. In Chemistry and World Food Supplies, the New Frontiers. Chemrawn II, Perspectives and Recommendations. Eds. G Bixler and L W Shermilt. pp 133–158, IRRI, Manila.
8 Brinkman M A and Rho Y D 1984 Response of three oat varieties to N fertilizer. Crop Sci. 24, 973–977.
9 Caradus J R 1982 Genetic differences in the length of root hairs in white clover and their effect on phosphorus uptake. In Proc. 9th Int. Plant Nutrition Colloquium Vol. 1. Ed. A Scaife. pp 84–88, Commonwealth Agric. Bureaux, Farnham Royal.
10 Clarke J M and Townley-Smith T F 1984 Screening and selection techniques for improving drought resistance. In Crop Breeding, A Contemporary Basis. Eds. P B Vose and S G Blixt. Ch. 6, pp 137–162. Pergamon Press, Oxford. 443 p.
11 Coltman R, Gerloff G and Gabelman W 1982 Intraspecific variation in growth, phosphorus acquisition and phosphorus utilization in tomatoes under phosphorus deficiency stress. In Proc. 9th Int. Plant Nutrition Colloquium Vol. I. Ed. A Scaife. pp 117–122. Commonwealth Agric. Bureaux, Farnham Royal.

12 Connor A J and Meredith C P 1985 Simulating the mineral environment of aluminium toxic soils in plant cell culture. J. Expt. Bot. 36, 870-880.
13 Dambroth M and Bassam N EL 1982 Low input varieties: definition, ecological requirements and selection. In Genetic Specificity of Mineral Nutrition of Plants. Ed. M R Saric. Serb. Acad. Sci. and Arts, Beograd, 1982, pp 325-336.
14 Dudeck A E and Peacock C H 1985 Effects of salinity on seashore *Paspalum* turfgrass. Agron. J. 77, 47-50.
15 Eckholm E P 1975 Salting the earth. Environ. 17, 9-15.
16 Epstein E, Norlyn J D, Rush D W, Kingsbury R W, Kelley D B, Cunningham G A and Wrona A F 1980 Saline culture of crops: a genetic approach. Science 210, 399-404.
17 Foy C D, Chaney R L and White M C 1978 The physiology of metal toxicity in plants. Annu. Rev. Plant Physiol. 29, 511-566.
18 Gabelman W H and Gerloff G C 1982 The search for, and interpretation of, genetic controls that enhance plant growth under deficiency levels of a macronutrient. In Genetic Specificity of Mineral Nutrition of Plants. Ed. M Saric. Serbian Acad. Sci and Arts, Beograd 1982. pp 301-312.
19 Goodman P J 1977 Selection for nitrogen response in *Lolium*. Ann. Bot. 41, 243-256.
20 Goodman P J 1981 Genetic control of nitrogen response in *Lolium* species. In Plant Physiology and Herbage Production. Ed. C E Wright. Occasional Symp. No. 13, Brit. Grassl. Soc. pp 131-136.
21 Goodman P J 1982 Genetic variation in nitrogen nutrition of grasses and cereals and possibilities of selection. In Genetic Specificity of Mineral Nutrition of Plants. Ed. M R Saric. Serb. Acad. Sci. and Arts, Beograd 1982. pp 357-361.
22 Greub L J, Drolson P N and Rohweder D A 1985 Salt tolerance of grasses and legumes for roadside use. Agron. J. 77, 76-80.
23 Huffaker R C and Rains D W 1978 Factors influencing nitrate acquisition by plants; assimilation and fate of reduced nitrogen. In Nitrogen in the Environment. Vol. 2. Eds. D R Nielsen and J G MacDonald. pp 1-43. Academic Press, New York.
24 Kingsbury R W and Epstein E 1984 Selection for salt-resistant spring wheat. Crop Sci. 24, 310-315.
25 Lehman W F, Rutger J N, Robinson F E and Kaddah M 1984 Value of rice characteristics in selection for resistance to salinity in an arid environment. Agron. J. 76, 366-370.
26 Mashady A S, Heakal M S, Abdel-Aziz M H and Sayed H I 1985 Nutritional effects of non-steady state soil salinity on a salt-tolerance wheat cultivar. Plant and Soil 83, 223-231.
27 Massoud F I 1974 Salinity and alkalinity as soil degradation hazards. Rept. FAO/UNEP Expert Committee on Soil Degradation, FAO, Rome.
28 Mengel K 1982 Responses of various crop species and cultivars to mineral nutrition and fertilizer application. In Genetic Specificity of Mineral Nutrition of Plants. Ed. M Saric. Serbian Acad. Sci. and Arts, Beograd 1982. pp 233-245.
29 Nabors M W and Dykes T A 1985 Tissue culture of cereal cultivars with increased salt, drought, and acid tolerance. In Biotechnology in International Agricultural Research. Proc. Inter-Center Seminar on 'International Agric. Res. Centers and Biotechnology'. pp 121-138, IRRI, Manila.
30 National Research Council (U.S.) 1983 Soil Fertility and Plant Nutrition. In Chemistry and World Food Supplies: Research Priorities for Development; Report of a Workshop, Los Banos, Philippines, December 1982, pp 15-43. National Academy Press, Washington, DC.
31 O'Sullivan J, Gabelman W H and Gerloff G C 1974 Variations in efficiency of nitrogen utilization in tomato, *Lycopersicum esculentum*, grown under nitrogen stress. J. Am. Soc. Hort. Sci. 99, 543-547.
32 Ponnamperuma F N 1977 Screening rice for tolerance to mineral stresses. In Proc. Workshop. Beltsville, Ed. M J Wright. Cornell Univ. Agric. Exp. Sta. Special Pub., Ithaca (New York). pp 341-353.
33 Rao K P, Rains D W, Qualset C O and Huffaker R C 1976 Quoted in Huffaker R C and Rains D W, Ref. 23.
34 Ramirez R 1982 Efficient use of nitrogen, phosphorus and potassium by corn (*Zea mays* L.) inbreds. In Proc. 9th Int. Plant Nutrition Colloquium, Vol I. Ed. A Scaife. pp 515-520. Commonwealth Agric. Bureaux, Farnham Royal.

35   Raper C D and Barber S A 1970 Rooting systems of soybeans, II. Physiological effectiveness as nutrient absorption surfaces. Agron. J. 62, 585–588.
36   Royal Society (U.K.) 1983 The Nitrogen Cycle of the United Kingdom, A Study Group Report. Ed. W D Stewart. pp 1–33.
37   Sacher R F, Staples R C and Robinson R W 1983 Ion regulation and response of tomato to sodium chloride: A homeostatic system. J. Am. Soc. Hort. Sci. 108, 566–569.
38   Schenk M K and Barber S A 1979 Root characteristics of corn genotypes as related to P-uptake. Agron. J. 71, 921–924.
39   Storey R, Graham R D and Shepherd K W 1985 Modification of the salinity reponse of wheat by the genome of *Elytrigia elongatum*. Plant and Soil 83, 327–330.
40   Sumner M E, Boerma H R and Isaac R 1982 Differential genotypic sensitivity of soybeans to P-Zn-Cu imbalances. *In* Proc. 9th Int. Plant Nutrition Colloquium Vol. II, Ed. A Scaife. pp 652–657. Commonwealth Agric. Bureaux, Farnham Royal.
41   Szabolcs I 1971 European Solonetz soils and their reclamation. Akademia Kiado (Budapest).
42   Torello W A and Symington A G 1984 Screening of turfgrass species and cultivars from NaCl tolerance. Plant and Soil 82, 155–161.
43   Vose P B and Breese E L 1964 Genetic variation in the utilization of nitrogen by ryegrass species *Lolium perenne* and *L. multiflorum*. Ann. Bot. 28, 251.
44   Vose P B 1981 Potential use of induced mutants in crop plant physiology studies. *In* Proc. Symp. Induced Mutations, a Tool in Plant Research, IAEA (Vienna). pp 159–181.
45   Vose P B 1982 Iron nutrition in plants: a World Overview. J. Plant Nutr. 5, 233–249.
46   Vose P B 1983 Effects of genetic factors on nutritional requirements of plants. Chapter 4, *In* Crop Breeding: a Contemporary Basis. Eds. P B Vose and S Blixt. pp 67–114 Pergamon Press, Oxford, 443 p.
47   Warner R L, Hageman R H, Dudley J W and Lambert R J 1969 Inheritance of nitrate reductase activity in *Zea mays*. Proc. Nat., Acad. Sci. 62, 785–792.
48   Yeo A R and Flowers T J 1983 Varietal differences in the toxicity of sodium ions in rice leaves. Physiol. Plant. 59, 189–195.

Adaptations and physiological responses of wild plants to nutrient stress

F. STUART CHAPIN, III
Institute of Arctic Biology, University of Alaska, Fairbanks, AK 99775, USA

Key words Leaching Nutrient use efficiency Retranslocation Root Storage Uptake Wild plants

Summary All plants respond in a qualitatively similar way to low availability of major nutrients by reduced acquisition, lower tissue nutrient concentrations (high efficiency of nutrient use), reduced growth, and effective retranslocation of nutrients from senescing leaves. Plants compensate for reduced nutrient status by increasing their physiological potential to acquire the limiting nutrients. Plants adapted to high-nutrient habitats show the above responses to the greatest degree. Plants adapted to low-nutrient habitats have a high capacity to acquire those nutrients that are mobile in the soil (*e.g.* potassium, nitrate) and a low capacity to acquire less mobile ions (phosphate, ammonium). In response to pulses of nutrient availability, such plants exhibit luxury consumption of nutrients; these nutrient stores then support a slow growth rate over a long period of time. Although all plants growing under conditions of nutrient stress typically have high efficiency of nutrient use in producing biomass, there is currently little evidence that those plants adapted to infertile soils have a genetic potential for high rates of carbon or nutrient gain per unit nutrient. Such plants maximize their efficiency of nutrient use primarily by prolonging tissue life so that each unit of nutrient provides a maximum return before being lost from the plant.

Introduction

Food shortage is an inevitable consequence of the continued increase in the world's human population. Demands for increased food are particularly severe in third-world countries; in most cases these countries are already farming the best agricultural lands or have converted them to urban areas. The extent to which imports can meet rising food demands is limited by problems of international monetary exchange. A major consequence of this dilemma is that agriculture in these countries has expanded into areas that are marginal for crop production with respect to water or nutrient supply. To some extent the soils can be improved to meet the high nutrient demands of current high-yield crop varieties; however, this option is limited by the cost and availability of fertilizers. It is logical then to breed crops that can grow and produce an acceptable yield in infertile soils, and substantial progress has been made in this respect[19,33,43].

There are certain native plant species that typically occupy infertile soils and successfully grow and reproduce under these conditions. Are there consistent patterns of physiological traits exhibited by these species? If so, these might provide clues as to traits required for success in marginal soils; such traits might be selected for in breeding programs. In

H.W. Gabelman and B.C. Loughman (Eds.), Genetic aspects of plant mineral nutrition.
ISBN-13: 978-94-010-8102-3
© *1987 Martinus Nijhoff Publishers, Dordrecht/Boston/Lancaster.*

Table 1. Comparison of an agricultural crop (maize) and a tundra graminoid (*Eriophorum vaginatum*) in terms of tissue nutrient concentration, nutrient requirement (assuming mass flow meets entire plant requirement and plant transpires 500 g water per g tissue produced), observed soil solution concentrations, and percentage of the plant requirement that can be met by mass flow. Data for agricultural crop are from Barber (1963), data for tundra graminoid are from Chapin et al. (1980) and Chaplin (unpublished).

Element	Plant nutrient content (mg g^{-1})	Required soil soln. conc. (mg l^{-1})	Observed soil soln.conc. (mg l^{-1})	Percent of requirement met by mass flow
Agricultural crop				
N	20	40	4	10
P	2	4	0.05	1.2
K	20	40	4	1.2
Ca	2.2	4.4	33	750
Mg	1.8	3.6	28	778
Tundra sedge				
N	20	40	0.1	0.3
P	2.2	4.4	0.02	0.5
K	7.9	15.8	1	6.3
Ca	1.0	2.0	5	250
Mg	1.8	3.6	3	83

this paper I review the physiological adaptations of wild plants to nutrient stress. To a large degree these traits are associated with a syndrome of slow growth and conservative use of resources that may be incompatible with the agricultural goal of maximizing yield. The optimal patterns of agriculture in such areas may be a compromise between improving soils to meet the demands of crop plants and accepting a lower yield than would be possible on more fertile soils.

Nutrient uptake

The pathway of nutrients from the soil to the plant involves a series of processes: release into the soil solution (*via* mineralization of organic matter, weathering, or soil exchange processes), diffusion or mass flow to the root surface, and active uptake by the root. In infertile soils the quantity of nutrients brought to the root surface by mass flow meets only a small proportion of the plant demand[2,31], because (1) the soil solution contains low concentrations of nutrients (Table 1) and (2) species characteristic of these habitats have low rates of water flux to each unit of root. The slow water flux results from the high root:leaf ratio of plants grown in infertile soils[7,8,10] and the low rates of transpiration characteristic of slowly growing plants. These plants have low transpiration rates because they have low photosynthetic potentials, and photosynthetic potential is closely correlated with stomatal conductance[17]. Stomatal conductance, in turn, is the major control over transpiration. Because of the slow flux of soil solution to the root surface in infertile soils, diffusion becomes progressively more important as a mechanism of nutrient supply as soil fertility declines.

Plants can maximize nutrient diffusion to the root surface (1) by increasing root surface area (high root:shoot ratio, small root diameter, presence of root hairs and mycorrhizae), (2) by reducing the concentration at the root surface (by active uptake), or (3) by increasing the rate of supply from the bulk soil to the soil solution (by promoting decomposition of organic matter, secreting hydrolytic enzymes, weathering of primary minerals, or exchange reactions on the soil exchange complex). Most plants promote diffusion to the root surface under conditions of nutrient stress, but the relative importance of the above mechanisms differs among species. For example, most plants increase root:shoot ratio in response to nutrient stress[7,16]; however, wild plants adapted to infertile soils exhibit a consistently high root:shoot ratio that permits them to exploit maximally any nutrient flush, whereas plants adapted to high-resource availability reduce root growth and promote shoot growth to a greater degree in response to a nutrient flush[8,11,21]. There is no clear relationship between soil fertility and root diameter or presence of root hairs. The degree of mycorrhizal colonization increases as soil fertility declines in most species; only a few plant families, such as those characteristic of very wet soils (*e.g.* Cyperaceae) or very fertile soils (*e.g.* Chenopodiaceae and Cruciferae), characteristically lack mycorrhizae[8]. Presence of mycorrhizae and soil infertility both act to increase root longevity and further contribute to the high root:shoot ratio in infertile soils[8]. In summary, a high and relatively inflexible root:shoot ratio appears to be the major adaptation by which plants from infertile soils maximize their root surface area.

Roots of all plants increase their capacity (g^{-1} root) to absorb a limiting nutrient as reserves of that nutrient decline in the plant[8,23,24]; plants thereby maintain a low concentration of the nutrient at the root surface and maximize diffusion from the bulk soil to the root surface. However, species differ considerably in their capacity to absorb nutrients. Species characteristic of infertile soils have a high capacity to absorb ions that are mobile in the soil such as potassium[13,20,41] (but see[26]) and perhaps nitrate[13], but a low capacity to absorb less mobile ions such as phosphate[8,11,13,14,34,44] (but see[1,3]) and perhaps ammonium[13]. In the case of mobile ions, diffusion is relatively rapid, and diffusion shells extend a substantial distance from the root surface, so that a high capacity for uptake enables a root to compete effectively with roots of neighboring individuals. Immobile ions have smaller diffusion shells around each root[13,31], so roots of neighboring individuals are less likely to compete for nutrients; under these circumstances even a low nutrient uptake capacity is adequate to absorb those nutrients that reach the root surface and maintain a strong diffusion gradient to the root.

It is presently unclear to what extent roots under natural conditions can increase nutrient flux from unavailable soil pools to the soil solution and from there to the root. This could result from solubilization of rock phosphate[27], hydrolysis of organic phosphates with root surface phosphatases[45], stimulation of decomposition in the rhizosphere through root exudation of soluble organic compounds[35], or exchange reactions by which root exudates release nutrients from the soil exchange complex[18].

Growth rate

Perhaps the most striking characteristic of plants adapted to infertile soils is their slow growth rate[8,22]. Although there is probably selection for plants to maintain the maximum growth rate possible in a given environment, plants from infertile soils have adjusted their maximum potential growth rate to the rate of resource supply that they encounter. A slow growth rate may enable the plant to survive between occasional pulses of nutrient supply, whereas a more rapidly growing plant might exhaust its nutrient reserves, leading to a complex of nutrient deficiency symptons; such plants may be more vulnerable to herbivores or pathogens[5,8]. Frequently a plant with a slow growth rate will produce more growth over an annual cycle under conditions of nutrient limitation than will a plant with a higher growth potential[39] and will maintain a competitive advantage under these conditions.

The importance of slow growth in infertile environments has been emphasized by Grime[21,22] who points out that this trait is found in plants that tolerate most environmental stresses, whether they be inadequate nutrients, water, or light. In contrast, ruderal and competitive plants that occur in environments of higher resource availability generally have higher potential growth rates[21].

Growth rate has important implications for all aspects of plant nutrition. A slow growth rate reduces the demand for nutrients in new tissue, and nutrient demand is one of the primary determinants of the capacity of a plant to absorb nutrients[15]. Moreover, rates of leaf and root mortality are closely correlated with the rate of production of these organs, so slowly growing plants have a slower rate of turnover and loss of nutrients contained in these tissues (see below). The importance of growth rate *per se* in determining nutritional characteristics is seen in an experiment with barley in which removal of the endosperm to reduce seed reserves and growth rate resulted in the same syndrome of nutritional traits as was observed in a related slowly growing species: slow growth, low nutrient uptake capacity, and relatively high tissue nutrient concentrations[9].

Storage and luxury consumption

Plants store nutrients whenever supply exceeds immediate demand. Storage during the dormant season is enforced by the onset of unfavorable conditions or some cue (*e.g.* photoperiod) that predicts these conditions. As noted above, slow growth and increased longevity are important adaptations to infertile soils; annual plants are less common in these habitats[21]. Consequently, nutrient storage during the dormant season is important to survival in these environments. Nitrogen is stored primarily as amino acids and specialized storage proteins, whereas phosphorus is stored in various organic forms such as polyphosphate or phospholipid[11,12,25].

Storage during the period of active growth has been termed luxury consumption and occurs whenever uptake exceeds growth demands. When grown under a similar nutrient regime, low-nutrient-adapted plants exhibit more luxury consumption than high-nutrient-adapted plants, because absorption rate under high-nutrient conditions differs less among species than does maximum relative growth rate[8,11]: under high-nutrient conditions rapidly growing species reduce their capacity for nutrient uptake to a greater extent than do slowly growing species (see above) but show a much greater growth response to the high nutrient availability (Fig. 1). Consequently, the rapidly growing species dilutes the nutrient pool over a larger biomass, whereas the more slowly growing species accumulates nutrients as luxury consumption. Luxury consumption is extremely important ecologically, because this nutrient store enables the slowly growing species to continue to grow even during times when there is negligible uptake from soil[8]. Nutrients accumulated by luxury consumption during the period of active growth are stored primarily in vacuoles as inorganic phosphate, amino acids, *etc.*[4].

Nutrient use efficiency

Nutrient use efficiency (NUE) has been defined as the amount of dry matter produced per unit tissue nutrient and is the inverse of tissue nutrient concentration[19,42,44]. All plants exhibit an increase in NUE under conditions of nutrient stress (Fig. 1), in part because nutrient storage reserves in vacuoles decline[4], and fiber and carbohydrate concentrations increase[6,8,38]. Furthermore, on a whole plant basis, NUE increases under nutrient stress, because a larger proportion of plant biomass is allocated to tissues with low nutrient concentrations (*e.g.* roots as contrasted with

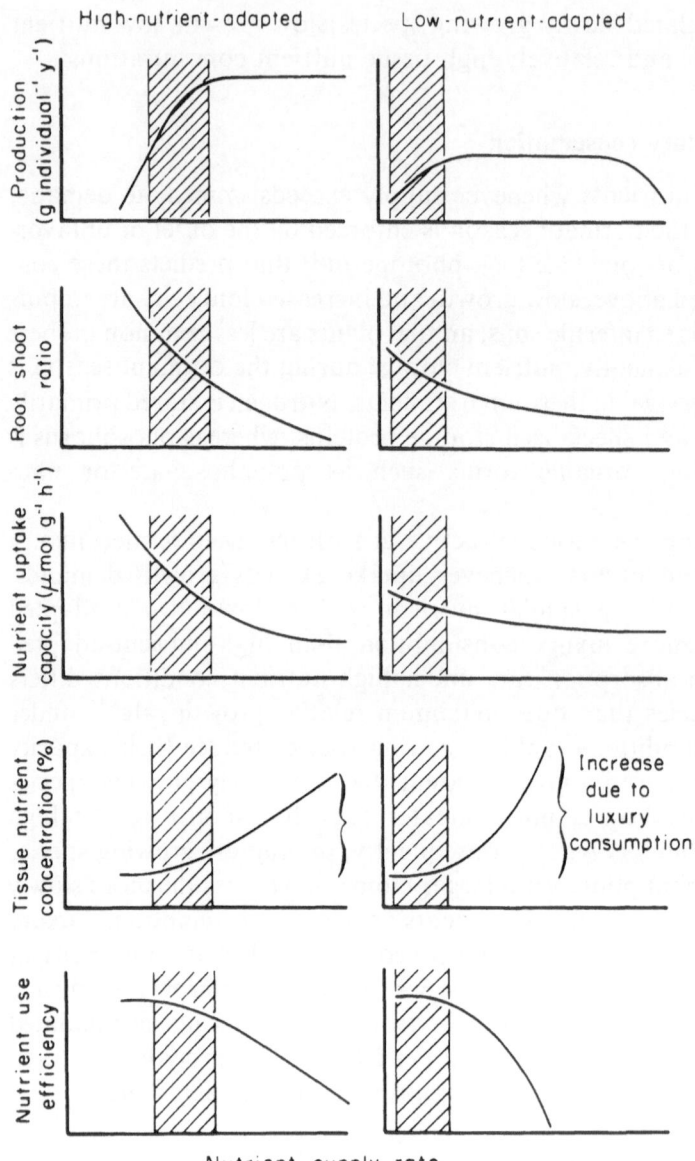

Fig. 1. Effect of nutrient supply upon growth and parameters related to nutrient uptake and use by high-nutrient-adapted and low-nutrient-adapted plants. *Hatched area* indicates normal range of nutrient supply where each plant type normally grows.

leaves or reproductive parts) under these conditions. When different species of wild plants are compared under identical growth conditions, slowly growing species generally have higher tissue nutrient concentrations (lower NUE) than do rapidly growing species, because their tissue nutrient reserves are diluted over a smaller biomass (Fig. 1). These differences among species may be minimal at low nutrient availability but become progressively more pronounced as nutrient availability increases, due to luxury consumption by the more slowly growing species[8].

A high NUE is advantageous, because it enables the plant to produce substantial biomass under a given rate of supply and thereby to accumulate (1) more tissue (*e.g.* roots and leaves) that can generate a positive feedback loop to acquire resources from the environment and (2) more nodes at which reproduction can occur. The disadvantages of a high NUE are that low leaf nitrogen concentrations lead to (1) low photosynthetic potential[17,30] and (2) low levels of reserves that can be mobilized to support further growth and reproduction. The relative costs and benefits of a high NUE depend upon the environment. A high NUE may be advantageous under conditions where there is relatively constant nutrient availability, so that nutrients will continue to be available to support future growth and reproduction; this is the common situation in agriculture. A high NUE (particularly for nitrogen) may also be advantageous where photosynthetic rate of the plant is not strongly limited by the photosynthetic potential per unit leaf weight, as under conditions of combined drought and nutrient stress or combined shade and nutrient stress. In fact, chaparral shrubs are notable for their high nitrogen use efficiency[17]. In the dense canopies of most agricultural crops, shade and nutrient stress may interact to produce a situation where a high NUE is advantageous. NUE has also be calculated for plant litter[42] and involves the factors controling NUE in mature leaves plus the process of retranslocation (see below).

NUE can also be studied with respect to particular processes such as photosynthesis or nutrient uptake. Photosynthetic nitrogen use efficiency (*i.e.* photosynthetic rate per unit nitrogen) increases with increasing tissue nitrogen concentration both within and among species[17], because leaves with a high nitrogen concentration have a larger proportion of leaf nitrogen allocated to photosynthetic proteins. Thus when considered in this way, plants adapted to or growing on infertile soils will have a low photosynthetic nitrogen use efficiency. Nitrogen use efficiency can also be considered as the total carbon gain per unit N over the life of the leaf. In this measure, photosynthetic nitrogen use efficiency is as great or greater in infertile habitats because such plants retain their leaves and photosynthesize for a longer period of time than do plants from fertile soils[10,32,37]

Nutrient loss

In perennial plants the rate of nutrient loss is just as important as rate of nutrient gain in determining plant nutrient budgets. Such plants could improve nutrient status by reducing the quantity of nutrients lost in leaching or litterfall. Plants with low tissue nutrient concentrations generally experience less nutrient loss due to leaching[29,40]. On average, about 50% of the nitrogen and phosphorus is retranslocated from a leaf before it is shed[12]. The proportion of nutrients retranslocated from a leaf ranges from 0–90% among species. Plants grown at low nutrient availability have low nutrient concentrations in mature leaves and litter; they generally retranslocate a smaller amount but a larger proportion of their maximum nutrient pool from senescing leaves prior to leaf abscission as compared to plants grown on more fertile soils[36]. However, there is conflicting evidence as to whether there are important differences among species adapted to different soil fertilities in terms of proportion of nutrients retranslocated[12].

Implications for agriculture

Comparisons of wild plants adapted to different soil fertility regimes suggests that the major adaptation to nutrient stress is a slow growth rate that entails a low annual demand of nutrients from the environment. Such species have a high root:shoot ratio and a high capacity to absorb mobile ions such as potassium from the soil but have a low capacity to absorb immobile ions like phosphate (Fig. 1). Low-nutrient-adapted plants generally have a similar or lower nutrient use efficiency compared to high-nutrient-adapted plants and do not show any clear adaptations to reduce nutrient loss except by retaining leaves and roots for a longer period of time. Such plants have low reproductive allocation. With the exception of the high capacity of low-nutrient-adapted plants to extract potassium from soil, all of these nutritional characteristics would be considered undesirable from an agricultural perspective.

However, there are important areas of ecological reseach that so far have received little attention but which could provide insight into development of crops for marginal lands. To date, most low-nutrient-adapted plants that have been studied by ecologists are perennial plants that occur on infertile soils. Many of the characteristics of these plants (*e.g.*, slow growth rate and luxury consumption) are important because they enable the plant to persist from one growing season to the next. In contrast, most crop plants are annuals, and the marginal agricultural lands on which they are currently being tested are considerably more fertile than the areas that have been studied by most nutritional ecologists.

An important area of future ecological research concerns the nutritional characteristics of annual plants on moderately infertile soils. Such plants may differ considerably from perennials that serve as the basis for most of the generalizations discussed in this paper. It remains to be determined which traits are important in annuals growing in marginal agricultural environments. The genetic work that has been done with crop plants[19,28] indicates that many nutritional characteristics are independently inherited and could be selected for in a breeding program. The most constructive approach to the problem of agriculture in marginal lands may require a combination of soil improvement, crop breeding for characteristics that more effectively exploit infertile soils (*e.g.* large fibrous root system), and a lowered expectation of crop yield.

Acknowledgement Research leading to these conclusions was supported by National Science Foundation grant BSR-8300397.

References

1 Andrew C S 1966 A kinetic study of phosphate absorption by excised roots of *Stylosanthes humilis, Phaseolus lathyroides, Desmodium uncinatum, Medicago sativa* and *Hordeum vulgare*. Aust. J. Agric. Res. 17, 611–624.
2 Barber S A, Walker J M and Vasey E H 1962 Principles of ion movement through the soil to the plant root. *In* Trans. Internat. Soc. Soil Sci., Commissions IV and V. Internat. Soil Conf: Soil Bureau, P.B. Lower Hutt, New Zealand. pp 121–124.
3 Barrow N J 1977 Phosphorus uptake and utilization by tree seedlings. Aust. J. Bot. 25, 571–584.
4 Bieleski R L 1973 Phosphate pools, phosphate transport and phosphate availability. Annu. Rev. Plant Physiol. 24, 225–252.
5 Bradshaw A D 1969 An ecologist's viewpoint. *In* Ecological Aspects of the Mineral Nutrition of Plants. Ed. I H Rorison. Blackwell, Oxford, pp 415–427.
6 Brady C J 1973 Changes accompanying growth and senescence and effect of physiological stress. *In* Chemistry and Biochemistry of Herbage. vol 2. Eds. G W Butler and R W Bailey. Academic Press, London. pp 317–351.
7 Brouwer R 1966 Root growth of grasses and cereals. *In* The Growth of Cereals and Grasses. Eds. F L Milthorpe and D J Ivins. Butterworths, London, pp 153–166.
8 Chapin F S, III 1980 The mineral nutrition of wild plants. Annu. Rev. Ecol. Syst. 11, 233–260.
9 Chapin F S, III and Bieleski R L 1982 Mild phosphorus stress in barley and a related low-phosphorus-adapted barleygrass: phosphorus fractions and phosphate absorption in relation to growth. Physiol. Plant. 54, 309–317.
10 Chapin F S, III, Johnson D A and McKendrick J D 1980 Seasonal movement of nutrients in plants of differing growth form in an Alaskan tundra ecosystem: Implications for herbivory. J. Ecol. 68, 189–209.
11 Chapin F S, III, Follett J M and O'Connor K F 1982 Growth, phosphate absorption and phosphorus chemical fractions in two *Chionochloa* species. J. Ecol. 70, 305–321.
12 Chapin F S, III and Kedrowski R A 1983 Seasonal changes in nitrogen and phosphorus fractions and autumn retranslocation in evergreen and deciduous taiga trees. Ecology 64, 376–391.
13 Chapin F S, III, Van Cleve K and Tryon P R 1986. Relationship of ion absorption to growth rate in taiga trees. Oecologia (Berl.) 69, 238–242.
14 Christie E K and Moorby J 1975 Physiological responses of semiarid grasses. I. The influence of phosphorus supply on growth and phosphorus absorption. Aust. J. Agric. Res. 26, 423–436.
15 Clarkson D T and Hanson J B 1980 The mineral nutrition of higher plants. Annu. Rev. Plant Physiol. 31, 239–298.
16 Davidson R L 1969 Effects of soil nutrients and moisture on root/shoot ratios in *Lolium perenne* L. and *Trifolium repens* L. Ann. Bot. N.S. 33, 571–577.

17 Field C and Mooney H A 1986 The photosynthesis-nitrogen relationship in wild plants. *In* On the Economy of Plant Form and Function. Ed. T J Givnish. Cambridge Univ. Press, Cambridge pp 25–55.
18 Gardner W K, Barber D A and Parbery D G 1983 The acquisition of phosphorus by *Lupinus albus* L. III. The probable mechanism by which phosphorus movement in the soil root interface is enhanced. Plant and Soil 70, 107–124.
19 Gerloff G C 1976 Plant efficiencies in the use of nitrogen, phosphorus and potassium. *In* Plant Adaptation to Mineral Stress in Problem Soils. Ed. M J Wright. Cornell Univ. Agric. Exptl. Stn. Ithaca, N.Y., pp 161–169.
20 Glass A D M and Perley J E 1980 Varietal differences in potassium uptake by barley. Plant Physiol. 65, 160–164.
21 Grime J P 1977 Evidence for the existence of three primary strategies in plants and its relevance to ecological and evolutionary theory. Am. Nat. 111, 1169–1194.
22 Grime J P and Hunt R 1975 Relative growth rate: its range and adaptive significance in a local flora. J. Ecol. 63, 393–422.
23 Harrison A F and Helliwell D R 1979 A bioassay for comparing phosphorus availability in soils. J. Appl. Ecol. 16, 497–505.
24 Hoagland D R and Broyer T C 1936 General nature of the process of salt accumulation by roots with description of experimental methods. Plant Physiol. 11, 471–507.
25 Jeffrey D W 1968 Phosphate nutrition of Australian heath plants. II. The formation of polyphosphate by five heath species. Aust. J. Bot. 16, 603–613.
26 Jensen P and Pettersson S 1978 Allosteric regulation of potassium uptake in plant roots. Physiol. Plant. 42, 207–213.
27 Kepert D G, Robson A D and Posner A M 1979 The effect of inorganic root products on the availability of phosphorus to plants. *In* The Soil-Root Interface. Eds J L Harley and R S Russell. Academic Press, London. pp 115–124.
28 Lindgren D T, Gabelman W H and Gerloff G C 1977 Variability of phosphorus uptake and translocation in *Phaseolus vulgaris* L. under phosphorus stress. J. Am. Soc. Hort. 102, 674–677.
29 Miller H G, Cooper J M and Miller J D 1976 Effect of nitrogen supply on nutrients in litter fall and crown leaching in a stand of Corsican pine. J. Appl. Ecol. 13, 233–248.
30 Natr L 1975 Influence of mineral nutrition on photosynthesis and use of assimilates. *In* Photosynthesis and Productivity in Different Environments. Ed. J P Cooper. Cambridge Univ Press, Cambridge. pp 537–555.
31 Nye P H and Tinker P B 1977 Solute movement in the soil-root system. Univ of California Press, Berkeley, 343 p.
32 Orians G H and Solbrig O T 1977 A cost-income model of leaves and roots with special reference to arid and semiarid areas. Am. Nat. 111, 677–690.
33 Ponnamperuma F N 1982 Genotypic adaptability as a substitute for amendments on toxic and nutrient-deficient soils. *In* Plant Nutrition 1982, vol 2. Ed. A Scaife. Commonwealth Agricultural Bureaux, Farnham House, U.K. pp 467–473.
34 Rorison I H 1968 The response to phosphorus of some ecologically distinct plant species. I. Growth rates and phosphorus absorption. New Phytol. 67, 913–923.
35 Rovira A D and Davey C B 1974 Biology of the Rhizosphere. *In* The Plant Root and Its Environment. Ed. E W Carson. University of Virginia, Charlottesville, VA. pp 153–204.
36 Shaver G R and Melillo J M 1984 Nutrient budgets of marsh plants: efficiency concepts and relation to availability. Ecology 65, 1491–1510.
37 Small E 1972 Photosynthetic rates in relation to nitrogen recycling as an adaptation to nutrient deficiency in peat bog plants. Can. J. Bot. 50, 2227–2233.
38 Smith D 1973 The nonstructural carbohydrates. *In* Chemistry and Biochemistry of Herbage. vol 1. Eds G W Butler and R W Bailey. Academic Press, London pp 105–155.
39 Specht R L and Groves R H 1966 A comparison of the phosphorus nutrition of Australian heath plants and introduced economic plants. Aust. J. Bot. 14, 201–221.
40 Tukey H B, Jr 1970 The leaching of substances from plants. Annu. Rev. Plant Physiol. 21, 305–324.

41 Veerkamp M T and Kuiper P J C 1982 The uptake of potassium by *Carex* species from swamp habitats varying from oligotrophic to eutrophic. Physiol. Plant. 55, 237–241.
42 Vitousek P 1982 Nutrient cycling and nutrient use efficiency. Am. Nat. 119, 553–572.
43 Vose P B 1982 Effects of genetic factors on nutritional requirements of plants. *In* Contemporary Bases for Crop Breeding. Eds. P B Vose and S Blixt. Pergamon Press, Oxford.
44 White R E 1972 Studies on mineral ion absorption by plants. I. The absorption and utilization of phosphate by *Stylosanthes humilis*, *Phaseolus atropurpureus* and *Desmodium intortum*. Plant and Soil 36, 427–447.
45 Woolhouse H W 1969 Differences in the properties of the acid phosphatases of plant roots and their significance in the evolution of edaphic ecotypes. *In* Ecological Aspects of the Mineral Nutrition of Plants. Ed. I H Rorison. Blackwell, Oxford, 357–380.

Application of ANE value and of shares of individual elements in this value for determining the difference between various plant species

A. OSTROWSKA
Department of Soil and Fertilization, Forest Research Institute, 3 Wery Kostrzewy Str., PL-00-362 Warszawa, Poland

Key words ANE   Nutrients   Plant species

Summary The content of nutrients in the leaves in various plant species was defined as a sum of many elements in the mass (ANE-Accumulation Nutritional Elements in meq per kg d.m.) and as the share of individual elements in ANE. Attention is also paid to some factors that determine the element accumulation in plant leaves. It has been found that ANE values for several plant species under study range from about 1500–2000 in pine needles, 2000–3000 in most deciduous trees, 3000–5000 in cultivated plants and are higher for vegetables. The nitrogen share in ANE is 60–70%, while of the remaining elements, the share of individual elements differs between the Mono- and Dicotyledoneous plants. The ANE value turned out to be characteristic of the individual plant species or groups of species as well as of their nutrition status and leaf age. In general, the difference between plant species appear mainly in the ANE value, while the distribution of elements within this value in the leaves of various plant species are similar

Introduction

The nutrient demands of a plant, accumulation of elements in its respective organs, response to deficiency or excess of elements as well as the ability to mobilize adaptative mechanisms at different nutritional stresses have been linked to the genetically conditioned properties of the plants.

Many papers reviewing the subject deal with the differences in responses to deficiency or excess of individual nutrients between species, varieties and cultivars.

Mechanisms of ion uptake and their transport within the plant as well as the morphological changes of tissues caused by deficiency or excess of the ion along with the biochemical transformations have also been studied[7,13,17,18,23,25,29,31,35,40,41,42].

Production of the biomass unit by plants requires a definite amount of specific elements, this amount being dependent upon both properties of the plant and upon conditions of its growth. The magnitude of production is determined directly by each of the elements interacting parallelly with the remaining ones.

The complexity of the nutrition factor resulting from the combined action of ions in the soil solution, during the transport within the plant as well as from the metabolic transformations complicate the determination of plant response to one given element separately. At the same time, the fact that many agents influence the relationships between the elements in the soil and their utilization by plants and the magnitude of yield may considerably affect the accumulation of specific elements in the mass of plants, even those genetically homogenic.

Different papers on accumulation of elements in plants define the interdependence of various elements by relating 1 or 2 components to each other or to the sum of the remaining ones[23,32,41]. However, interpretation of the results is difficult and, in the majority of cases, each element has been treated separately. The relationships between the ion content of the plant and the soil system can be represented by 2 values, *i.e.* as a sum of all elements and as a share of respective elements in this sum. Both values indicate the magnitude of yield as well as the quantity of accumulated elements per unit mass produced. Nutritional stresses resulting from changes in ionic systems of the growth environment manifest itself at first in the plant roots, whereas in the leaves, stability of accumulation is maintained even under considerable variation in nutrition conditions[38]. Therefore, the accumulation of elements in plant leaves seems to be a better indicator of differences resulting from the genetic variability than the accumulation in whole plants or in roots.

The purpose of this paper is to analyse the nutrient accumulation, defined as ANE (ANE — Accumulation of nutritional elements meq per kg d.m.), in the leaves of some plants, and to determine the contribution of individual elements to this accumulation. Attention is also paid to some factors that determine the element accumulation in plant leaves.

Materials and methods

The total content of elements in the plants was expressed in the form of miliequivalent values. As a basis for calculation the atomic mass was used as well as the valency of the ions taken up by the plants, namely: K^+, Na^+, Ca^{2+}, Mg^{2+}, Mn^{2+}, Fe^3, Al^{3+}, Al^{3+}, $P-H_2PO_4^-$, $S-SO_4^{2-}$, Cl^-, $N-NH_4^+$, NO_3^-.

The ANE unit was adopted in order to define the sum of elements while the per cent share of individual elements in the ANE helps to determine the qualitative-quantitative systems of elements.

Results are presented in the form of both the ANE as well as the per cent share of elements in this value (Table 1.)

Results and discussion

Values for several plant species have been compared using forest tree species as a basis. Element accumulation in leaves reflects their genetically determined nutrient demand since they have grown in their natural habitats[45].

Table 1. Contents of nutrients in leaves of different species

Plant	ANE	Share in ANE (%)										Characteristic of material	Author Year
		N	P	K	Ca	Mg	Na	Al	Fe	Mn			
Tilia spp	3913	55.8	1.8	7.7	27.8	6.7	–	–	–	–	1975 Leaves were sampled in July and	Insley 1981
	4292	46.6	1.6	7.1	38.7	6.4	–	–	–	–	1976 August from 24 trees, T. europea	
	3859	57.0	2.1	6.4	27.1	7.3	–	–	–	–	1978 12 trees, T. platyphylios — 5 trees,	
	3939	54.9	2.2	6.9	28.9	7.0	–	–	–	–	1979 T. euchlora — 5 trees, T. cordata — 2 trees. Leaves were sampled from the non shaded branches, slightly above the widest part of crown, in town, age — 30–50.	
Populus deltoides	2872	43.2	2.1	9.0	34.8	10.3	–	–	–	–	Leaves sampled from centre of crown in the beginning of September	Harvey 1981
Carya illionoensis	2580	50.9	2.1	6.1	29.8	10.9	–	–	–	–	Leaves sampled from centre of crown in the beginning of September	
Platanus acidentatis	2265	50.0	2.3	6.4	32.6	8.5	–	–	–	–	Leaves sampled from centre of crown in the beginning of September	
Liquidambar	2179	55.7	4.5	7.8	20.4	11.4	–	–	–	–	Leaves sampled from centre of crown in the beginning of September	
Fraxinus pensylvanica	2163	55.1	4.5	9.4	24.0	6.9	–	–	–	–	Leaves sampled from centre of crown in the beginning of September	
Quercus mutuallii Palmer	1956	60.2	2.7	8.7	19.4	8.9	–	–	–	–	Leaves sampled from centre of crown in the beginning of September	
Fraxinus pensylvanica	2254	45.2	5.4	13.7	28.5	7.0	–	–	–	–	Nursery — leaves sampled at the end of the fourth vegetation period	Wittwer 1980
Betula alleghaniensis	2706	64.7	1.4	7.4	17.0	7.1	0.1	0.4	0.1	1.4	70-year old stand, all species originate from one plantation, leaves sampled in autumn from centre of crown, before discoloration	Lea 1980
Fagus grandifolia	2540	69.4	1.4	7.0	14.3	5.9	0.1	0.2	0.0	1.3		
Acer saccharum	2170	65.2	1.5	6.4	19.8	5.4	0.1	0.2	0.1	1.3		
Acer rubrum	1949	63.0	1.3	6.9	20.5	6.0	0.2	0.2	0.2	1.6	70-year old stand, all species originate from one plantation, leaves sampled in autumn from centre of crown, before discoloration	

(continued)

Plant	ANE	Share in ANE (%)									Characteristic of material	Author Year
		N	P	K	Ca	Mg	Na	Al	Fe	Mn		
Betula nigra	3080	51.3	1.8	6.8	32.4	7.5	–	–	–	–	Nursery — leaves sampled at the end of the fourth vegetation period	Wittwer 1980
Populus tremuloides	2231	55.6	3.2	9.7	26.0	5.4	–	–	–	–	20 year old stand/Alaska/leaves sampled in August	Coyne 1977
Populus tremuloides	2842	60.9	1.9	8.3	21.3	7.6	0.1	–	0.2	–	47 year old stand, leaves sampled in August	Bartos 1978
Populus tremuloides	3130	62.0	2.3	12.8	16.5	7.0	0.0	–	0.1	–	116 year old stand, leaves sampled in August	Bartos 1978
Populus tremuloides	2707	59.6	2.9	13.5	14.0	9.2	–	–	–	–	Leaves sampled at the beginning of September from 1 year old plantation	Bowersox 1977
Populus deltoides	2658	66.3	2.9	13.2	12.8	4.8	–	–	–	–	Mature stand, leaves sampled in August	Leech 1981
Populus tremula tremuloides	2493	60.4	3.0	10.5	15.6	9.0	–	–	–	–	4 year old plants, plantation on a clearing, leaves sampled at the end of September	Gladysz 1977
Malus Mill. Jonatan	2882	60.5	2.0	8.9	17.1	10.5	0.2	–	–	–	Mature trees, leaves sampled from centre of crown in August, acid brown soil	Czarnowska 1983
M. Mc.Intosh	2800	62.8	2.1	7.8	17.5	10.2	0.2	–	–	–		not published
M. Lansberska	2922	58.2	2.1	8.6	18.4	11.9	0.2	–	–	–	Mature trees, leaves sampled from centre of crown in August, grey-brown soil	
M. Jonatan	2700	61.3	2.2	8.9	18.0	9.0	0.3	–	–	–		Czarnowska 1983
M. Lansberska	2627	61.7	2.3	10.9	16.2	8.6	0.3	–	–	–		
M. Mc.Intosh	2470	58.4	2.2	10.7	18.8	9.1	0.3	–	–	–	not published	
Pinus silvestris L.	1687	72.1	2.6	8.5	6.6	6.6	–	1.7	0.4	1.3	Average of 12 stands of III–V age class, fresh conifer forest situated in different regions, needles sampled from top of crown in October	Szczubialka 1981

Species									Description	Reference
Pinus silvestris L.	1560	71.3	3.2	8.0	7.2	6.9	—	1.9	0.8	0.8	Average of 5 stands, III–V age class, dry conifer forests situated in different regions, needles sampled from top of crown in October	Szczubiałka 1981
Pinus silvestris L.	1950	74.5	3.1	6.7	7.2	5.2	—	1.2	0.8	1.6	Average of 6 stands of IV–VIII age class, fresh mixed deciduous-conifer forests situated in different regions, needles sampled from top of crown in October	Szczubiałka 1981
Pinus silvestris L.	2161	71.0	3.0	9.9	4.2	4.6	—	4.3	2.1	0.9	Average of 15 nurseries situated in different regions, above ground part sampled at the end of the first vegetation period	Ostrowska 1982
Pinus silvestris L.	1857	75.0	3.4	12.3	2.4	5.1	0.3	—	0.9	0.5	Pot experiment — water culture, above ground part, sampled at the end of first vegetation period	Ostrowska 1982
Larix europ.	2250	67.9	2.4	7.0	12.0	4.8	—	1.0	0.4	4.4	11 year plantation, needles sampled in August	Nebe 1982
Larix europ.	2989	67.3	2.1	6.4	10.7	5.2	—	1.5	0.4	6.1	Values for various provenience	Nebe 1982
Larix europ.	1829	68.6	2.5	6.6	13.6	7.7	—	0.5	0.3	0.6	Values for various provenience	Nebe 1982

The ANE values for most of the deciduous tree species studied are contained within the limits of 2000–3000 (Table 1), ranging from 2000 in maple, oak and ash leaves, approaching 2500 in beech leaves and attaining 3000 in the leaves of birch and poplar.

The ANE value in the leaves of poplar varies between 2300–3000. It is of interest that the ANE values attain 2300 in the Alaskan species of poplar while in those growing in Central Poland it reaches 2,500.

The highest value of ANE — about 4000 — was found for the leaves of linden, whereas for apple it varied within the limits of 2500–3000 (Table 1). It should be stressed that the data under study reflect not only varying conditions of plant growth but also different varieties as well as the age of plants and, possibly, different stages of leaf growth, though the sampling methods may seem uniform, *i.e.* the samples were taken from the centre of the crown in August and September.

The ANE value in pine needles as well as the share of individual elements in this value have been broadly discussed in another paper[38]. It was shown, among others, that ANE in needles of the mature, older stands varies between 1400 and 2100 and depends upon nutrient resources of the site. Likewise, ANE in the above ground parts of annual plants varies within broad limits that vary with the nutrition conditions. It is noteworthy that at the optimal nutrient supply ANE oscillates within a narrow range of values between 1700–2000, in the needles of one-year old and of mature pines.

ANE oscillates within the limits of 2000–3000 in larch needles in a manner similar to that of deciduous trees. This relatively wide range of oscillation may result, apart from varying environmental conditions, from the time of sampling since larch needles undergo a full growth cycle during one vegetative season.

In Table 2, the values of ANE are compared in some cultivated species including vegetables. Consequently ANE amounts to about 2500–3000 in grasses, about 3300 in wheat, about 4000 in papilionacecus plants and up to 5000–6000 in leaves of vegetables. It is not easy to interpret these data in view of the fertilization factor, which enhances, as a rule, the value of ANE.

It should be noted that nutrient deficiency, as estimated by the analysis of a flag leaf of corn, was reported when ANE amounted to 1800.

ANE values of different plant species are contained within similar limits, however, among species under consideration, linden was shown to have a ANE value far higher than those of other deciduous species. On the other hand, ANE in pine needles is, as a rule, lower than in other tree species. Among cultivated plants, vegetables show a considerably higher values.

An attempt to relate the ANE values to the productivity of plants was undertaken in the course of a model experiment with pine. From the results it can be concluded that at the highest yields the ANE values oscillate in a relatively narrow range. At the lowest yields, only very low values of ANE were found, low yields, however, were sometimes reported at a very high value of ANE[38].

Calculations made by the author on the basis of ANE values obtained for different Alaskan tree species indicate that higher productivity of stands is usually associated with higher values of ANE[10]. This problem, however requires further study in respect of both the intraspecific variation of ANE due to the nutrient supply conditions as well as of the variation between species, that depends upon the production capacity of plants.

The share in ANE in the leaves of the majority of tree species is as follows: nitrogen 50–60%, calcium 15–30%, phosphorus 2–3% and subsequently, potassium and magnesium — within the limits of 7–12%. The most pronounced differences have been noted in the distribution of calcium and nitrogen within the given species and between species. It should be stressed that the systems of elements show a reiterative pattern, independent of the values of ANE, *e.g.* in leaves of linden, at ANE equal to 4000, the share of individual elements differs insignificantly from the share of elements in leaves of poplar, where ANE is about 2300.

In the needles of pine differing in age and growth conditions the components of ANE can be represented as follows: nitrogen 70–75%, phosphorus 2–3%, potassium up to 10%, while calcium and magnesium about 6–10% each. Similar distribution was reported in ANE of larch needles though the share of nitrogen is lower by several per cent while that of calcium higher than in pine needles. As it has been already mentioned, larch shows a share of elements in ANE similar to that of deciduous trees since its needles develop in the course of one year. In wheat, during the spreading period, the distribution of elements in ANE resembles that of pine needles.

Generally, the proportion of nitrogen in ANE in the plants under study varies around 50–60%. The remaining elements vary *e.g.* in grasses, in decreasing order: potassium, magnesium and calcium, in the papilionaceous plants: calcium, potassium and magnesium, in vegetables: potassium and calcium.

In the flag leaf of corn, assumed as optimal for this species, the share of elements in ANE was calculated to be as follows: nitrogen about 64%, phosphorus 4%, potassium 19%, magnesium and calcium — about 6% each (Table 2).

The process of leaf ageing is widely thought to be associated with the decrease in content of nitrogen, phosphorus, potassium, copper and zinc,

Table 2. Nutrient content in leaves in some species in different phases of development

Plant	Development phase	ANE	Share in ANE (%) N	P	K	Ca	Mg	Na	Fe	Mn	S	Characteristic of material	Author year
Triticum vulgare Vill.	spreading	3379	71.2	3.5	13.6	5.9	2.4	3.4	–	–	–	F experiment, non-fertilized	Brogowski 1983
Lolium perenne L.	60 days after germination	5760	65.4	2.1	12.8	10.0	4.9	0.7	–	–	4.0	Pot experiment heavily fertilized, above ground part	Smith 1982
Gramineae	blooming	274	51.8	4.6	15.9	10.2	13.6	3.7	–	–	–	Field experiment, average of 3 years and of 3 swath for 15 species	Moraczewski 1979
Gramineae	blooming	2484	46.9	5.1	15.2	11.2	12.0	9.4	–	–	–	Average of 3 years and of 3 swath for 1 species	Moraczewski 1979
Gramineae	blooming	3032	54.0	4.9	15.6	11.4	12.5	1.4	–	–	–	Average of 3 years and of 3 swath for 1 species	Moraczewski 1979
Papilionaceae	blooming	4392	51.9	3.2	12.9	19.3	10.7	2.0	–	–	–	Average of 3 years and of 3 swath for different species	Moraczewski 1979
Brassica	full vegetation	6699	57.5	1.7	21.2	17.5	1.4	–	0.2	0.2	–	Field experiment variant with highest yield	Kurczewski 1981
B. oleracea	budding	4735	63.3	3.5	25.5	4.9	2.3	–	0.1	0.1	–	Field experiment variant with highest yield	Kurczewski 1981
Allium cepa L.	6–7 leaf phase	5778	44.7	0.8	33.7	19.2	1.2	–	0.1	0.1	–	Field experiment variant with highest yield	Kurczewski 1981
Daucus carota L.	full vegetation leaves	5797	48.0	2.1	30.1	15.5	3.6	–	0.4	0.2	–	Field experiment variant with highest yield	Kurczewski 1981

												Field experiment variant with highest yield	
D. carota	crop leaves	5193	45.3	2.0	25.9	22.1	3.8	–	0.3	0.2	–	Field experiment variant with highest yield	Kurczewski 1981
Zea mays	flag leaf	1762	81.0	1.8	14.2	1.4	1.4	–			–	Nutrient deficite	Mengel
Zea mays	flag leaf	2327	76.8	2.6	16.6	2.1	2.1	–	–	–	–	low supply	Mengel
Zea mays	flag leaf	3937	63.5	4.0	19.1	6.3	6.3	–	–	–	–	optimal supply	
Zea mays	leaves	3534	55.7	2.4	23.6	8.3	6.6	–	–	3.2	–	Pot experiment, 4 weeks of growth highest yield	Summer 1981

and with a simultaneous increase in the amount of calcium, magnesium, manganese, iron, aluminium and boron[30].

In view of the countinuity of the ageing process that may begin soon after the leaf develops[7], time of leaf sampling is crucial for the ion system in the leaves.

Changes in the nutrient content in the leaves during the vegetative season were examined on the basis of results of several studies[2,14,20,37,45].

Analysis of the element content of leaves of trembling aspen performed at the beginning of August and October, separately for leaves from the top and the lower part of the shoot, showed that ANE in August was about 4000 while in October — about 2000. Lower values of ANE in leaves from the lower part, when compared to those from the upper part, were observed only in August. The proportions of individual elements in ANE were different at both the sampling times. Nitrogen and phosphorus in ANE decreased by about 30–40% in October, while that of calcium and magnesium increased, respectively, by about 30–40% and 15%, when compared to these elements in ANE in August. The proportion of potassium remained at an almost unchanged level at both times.

In the leaves of linden sampled from May to October the value of ANE decreased from about 5300 to about 3700, the proportion of nitrogen decreasing from about 70 to 40%, and that of calcium increasing from about 12 to 50%.

Among the remaining elements the share of magnesium increased slightly, while that of phosphorus and potassium decreased.

The value of ANE in needles of 30 year old pine, sampled from January to October, varied from 1300 to 1800. The lowest value of ANE was found in the oldest needles, developed during the preceding season and the highest — in the newly developed needles of current growth. In the samples taken at other times only slight differences in ANE have been encountered. A few per cent growth in calcium in the older needles and in potassium in the youngest ones was reported as far as the element systems in ANE are concerned. In general, however, during the period mentioned above, the proportion of elements in ANE is maintained at a similar level.

A comparison of current needles (needles I) with the one year old ones (needles II), sampled from many stands in October, showed only small differences in both ANE and in the share of elements in this value.

In ANE of the needles II a slight decrease was noted in nitrogen, phosphorus and potassium and an increase in calcium, manganese, iron and aluminium when compared with these elements in ANE in the needles I (Table 3).

In the leaves of bird-cherry the value of ANE decreases from 3100 to

2800 during the period from June to September, but the distribution of elements in ANE is similar during this time. A comparison of old and young shoots of Caragana, sampled with leaves in August, has shown that the difference in ANE value amounts to about 1500, to the advantage of young shoots. Furthermore, it has been found that old shoots contain more calcium and less potassium and magnesium than the young ones but differences in nitrogen were small (Table 3).

The chemical content of leaves changes downwards in the shoot and crown, and from the crown peripheries to its centre, as well as during the vegetative season[2,14,20,21,37]. The changes manifest themselves by a decrease of the ANE value, the decrease with age being, as a rule, more pronounced than that resulting from the position of leaves in the crown. At the same time, there occurs a shift in the element share in ANE, which consists, primarily, in a decrease of nitrogen and increase of calcium.

The influence of nitrogen supply on calcium accumulation in pine needles was discussed in earlier papers[38]. It should be added that different authors have also noted an increased accumulation of calcium under nitrogen deficit in various tree species.

According to Camp *et al.*[6] a decrease in photosynthetic activity of leaves, as the plants aged, was preceded by decrease in the amount of soluble proteins as well as by the lowering of the activity of some enzymes that probably accounts for a decrease in foliar nitrogen with age.

According to other authors[19] a relationship exists between lowering of photosynthetic activity and a decrease in the phosphorus and nitrogen content in spruce needles. Terry *et al.*[46] observed, however, a higher photosynthetic activity at a low calcium content. In general, the lower the nitrogen content in the leaves the higher the calcium content as plants get older, but an increase in the content of other elements, especially in magnesium and potassium[20,22] may also occur. Generally, Monocotyledoneous plants are thought to have more univalent cations, while Dicotyledoneous have more bivalent cations[4,5] related to different mechanisms of accumulation and of translocation of potassium and calcium in plants under study[8].

Differentiation of chemical content of shoots as, *e.g.* in Caragana[24], or of the whole above-ground biomass as, *e.g.*, in cereals[4] might result also from a differentiation of tissues in the course of time.

The contents of elements reported by different authors[27,33,36] in young leaves, that indicate the optimal nutrient supply are expressed, as a rule, by a nitrogen share in ANE equal to 60–70%, the remaining elements having 30–40%.

It is of interest that nutrient demands, calculated for the conifer and

Table 3. Effect of time of leaf sampling/of leaf age/on nutrient content in leaves

Plant	Sampling time, successive number of leaf	ANE	Share in ANE (%)									Characteristics of plant material	Author Year
			N	P	K	Ca	Mg	Mn	Fe	Al	S		
Populus tremula tremuloides hvbr.	8.08. leave 1–10	4114	73.8	5.9	9.1	6.4	4.3	0.05	–	–	–	Pot experiments, leaves were analysed successively from top down to basis of shoot	Gladysz 1983
	leave 21–26	3778	59.4	4.5	9.2	19.6	6.1	0.10	–	–	–		
Populus tremula tremuloides hvbr.	1.10											Pot experiments, leaves were analysed successively from top down to basis of shoot	Gladysz 1983
P. tremula	leave 1–10	2086	33.4	2.9	10.9	38.8	14.8	0.0	–	–	–		
	21–26	2056	38.2	2.7	11.4	33.8	13.8	0.0	–	–	–		
Prunus serotina	27.06	3118	68.7	1.7	13.2	9.1	7.2	–	–	–	–	Mature leaves were sampled, always from the top of shoots (growing, without discoloration)	Bowersox 1977
	24.07	2985	68.2	1.7	12.5	10.1	7.5	–	–	–	–		
	13.08	2803	68.5	1.7	12.1	10.2	7.4	–	–	–	–		
	4.09	2766	67.4	1.7	12.6	11.2	7.1	–	–	–	–		
Tilia spp	05	5330	69.7	4.2	7.9	12.2	6.0	–	–	–	–	Leaves were taken from crown of 15 trees at one-week and two-week intervals during 2 years. Results were calculated monthly only in 1977, in 1978 similar results were obtained	Insley 1981
	06	5013	62.6	2.9	6.5	22.0	5.9	–	–	–	–		
	07	4756	56.3	2.0	6.3	29.4	6.0	–	–	–	–		
	08	4730	52.7	1.6	5.0	34.9	5.8	–	–	–	–		
	09	4070	43.8	1.7	4.9	42.9	6.5	–	–	–	–		
	10	3751	38.9	1.9	4.0	47.9	7.0	–	–	–	–		
Caragana pygmaea	07												
	shoots young	3926	62.6	1.3	9.7	19.2	6.9	–	0.2	–	–	Shoots with leaves were sampled in dry-steppe in Mongolia	Kowalkowski 1980
	shoots old	2475	60.4	1.2	3.8	29.7	3.5	–	1.3	–	–		
Pinus silvestris L.	01/02	1472	70.8	3.3	9.9	5.6	6.1	–	–	–	4.3	Non fertilized, 25–35 year old stand fresh conifer, current needle (needles I) were sampled from tops of corn, 3 scales from 20–30 trees	Ostrowska 1980
	03	1326	70.7	3.6	9.4	5.0	5.8	–	–	–	5.5		
	04/05	1665	71.4	2.7	8.3	7.8	5.7	–	–	–	4.1		
	06	1842	69.2	3.6	13.0	5.7	4.0	–	–	–	4.5		
	07	1533	69.6	3.2	12.5	5.7	4.4	–	–	–	4.6		

Pinus silvestris L.												
08/09		1415	70.3	3.4	10.7	5.9	5.9	–	–	–	3.7	
10		1553	68.9	3.5	10.9	5.9	6.1	–	–	–	4.7	
10 needles I		1592	70.8	3.1	9.0	7.1	6.3	1.4	0.5	1.8	–	Average of 149 samples of pine stand in III–V classes of age growing on different sites, needles were sampled from top of crown, 1–3 whorl. Szozubiałka 1981
needles II		1644	67.8	2.7	7.5	11.8	5.3	2.0	1.1	2.1	–	

deciduous stands in Alaska, are expressed in the following shares in the sum of elements per mass unit: nitrogen about 50%, phosphorus 2–3%, potassium 10–15%, calcium 24–26% and magnesium 8–10%. In the leaves of the species in question the accumulation of nitrogen dominates, calcium accumulation being much lower. Under conditions, however, of nitrogen deficit a high accumulation was also reported[10]. According to the above data, differences between tree species in nutrient demand result only from the amount of mass units produced during a given time.

Other data suggest that the qualitative distribution of elements per mass unit produced approach the rule formulated for ion uptake.

It can be assumed that the plant tends to attain certain ratio of nitrogen to the remaining elements in areas where intensive photosynthesis takes place, mainly in the leaves, by a translocation of ions.

Special attention should be drawn to the qualitative similarities and quantitative differences in the nutrient demands of different plant species, which are probably linked with the intensity of photosynthesis as well as with the quality of produced metabolites but the problem requires further study.

Concluding remarks

Application of ANE values for assessing the total accumulation of nutrients in the leaves permits the establishment of a general state of plant nutrition as the factor which, in connection with other factors, influences the biomass production.

The ANE value is characteristic of the individual plant species or groups of species as well as of their nutrition state and leaf age. A comparison of ANE and of the shares of individual elements in this value in the leaves indicated that difference between species appear mainly in the ANE value. On the other hand, distribution of elements within ANE in the leaves of various plant species, as different as, *e.g.* young pines and wheat at the stage of spreading, are similar. The nitrogen share in ANE amounts to 60–70%, while 40–30% fall in the remaining elements, the proportions of individual elements differing between the Mono- and Dicotyledoneous plants.

The nutrition state of plants manifests itself in both the ANE values and the distribution of individual elements in this value. The ANE value in pine needles, under conditions of natural conifer-forest sites, was found to increase with the increase in the resources of the site, the share of elements being only slightly differentiated.

The processes of development and ageing of the leaves can be seen in the changes in ANE, the process of ageing being accompanied by a diminishing of the ANE and by a shift of elements within this value to the advantage of accumulated cations in relation to nitrogen, on the other hand, changes in the distribution within ANE, as a rule, were not encountered at the decrease of the ANE value resulting from the development of the leaves.

The analysis of plant material differing in respect of genetics, nutrition conditions, age of plants and methods of study has indicated that application of ANE value and the distribution of separate elements in this value may considerably facilitate the interpretation of genetical aspects of plant nutrition.

References

1 Bartos D L and Johnston R S 1978 Biomass and nutrient content of quaking aspen at two sites in the western United States. Q. Forest. Sci. 24, 273-280.
2 Bowersox T W and Ward W W 1977 Seasonal variation in foliar nutrient concentration of black cherry. Q. Forest. Sci.23, 429–432.
3 Bowersox T W and Ward W W 1977 Soil fertility, growth and yield of young hybrid poplar plantation in central Pennsylvania. Q. Forest. Sci. 23. 163–169.
4 Brogowski Z i Czarnowska K 1985 Stan jonowy 2 odmian pszenicy w różnych fazach rozwojowych na tle nawożenia azotowego. Rocz. Nauk Roln. in publish.
5 Brogowski Z, Czerwiński Z i Pracz J 1977 Stan równowagi jonowej a adporność drzew i krzewów parkowych na NaCl. Rocz. Naukl Roln. A, 102.
6 Camp P J, Huber S C, Burke J J and Moreland D E 1982 Biochemical changes of occur during senescence of wheat. Plant Physiol. 70 1641–1646.
7 Chapin F S 1983 Adaptation of selected trees and grasses to low availability of phosphorus. Plant and Soil 72, 283–287.
8 Chino M 1981 Species differences in calcium and potassium distribution within plants roots. Soil Sci. Plant Nutr. 27, 487–503.
9 Clark R B 1983 Plant genotype differences in uptake, translocation, accumulation, and use of mineral elements required for plant growth. Plant and Soil 72, 175–196.
10 Cleve K, Oliver L, Schlentner R, Viereck L A and Dyrnes C T 1983. Productivity and nutrient cycling in taiga forest ecosystems. Can. J. Forest Res. 13, 747–772.
11 Coyne P J and Cleve K V 1977 Fertilizer inducer morphological and chemical responses of a quaking aspen stand in interior Alaska. Q. Forest. Sci. 23, 92-192.
12 Dejaegere R, Neirinckx L, Stassart J M and Delegher V 1981 Mechanism of ion uptake across barley roots. Plant and Soil 63, 19–24.
13 Gabelman W H and Gerloff G C 1983 The search for and interpretation of genetic controls that enhance plant growth under deficiency levels of macronutrient. Plant and Soil 72, 335–350.
14 Gładysz A 1985 Zmienność zawartości składników pokarmowych w liściach sadzonek osiki w zależności od miejsca ich wystepowania na pedzie. Prace IBL *In press.*
15 Gładysz A, Janson L i Szczubiałka Z 1977 Zawartość składników pokarmowych w liściach dwuletniej osiki. Sylwan 3, 21-32.
16 Harvey E and Kennedy J R 1981 Foliar nutrient concentration and hardwood growth influenced by cultural treatments. Plant and Soil 63, 307–316.
17 Hecht-Buchholz Ch 1983 Light and electron microscopic investigation of the reactions of various genotypes to nutritional disorders. Plant and Soil 72, 151–165.
18 Horst W J 1983 Factors responsible for genotypic manganese tolerance in cowpea (*Vigna unguiculata*). Plant and Soil 72, 213–218.
19 Hom J L and Oechel W C 1983 The photosynthetic capacity, nutrient content, and nutrient use efficiency of different needle age classes of black spruce (*Picea mariana*) found in interior Alaska. Can. J. Forest. 13, 834–839.

20 Insley H, Boswell R C and Gardiner J B H 1981 Foliar macronutrients (N, P, K, Ca, Mg) in lime (*Tilia* spp). II. Seasonal variation. Plant and Soil 61, 391–401.
21 Insley H, Boswell R C and Gardiner J B H 1981 Foliar macronutrients (N, P, K, Ca, Mg) in lime (*Tilia* spp). I. Sampling techniques. Plant and Soil 61, 377 389.
22 Jain R K, Grag V K and Khanduja S D 1981 Macronutrient element composition of leaves from some ornamental shrubs grown in normal and alcali soil. J. Hortic. Sci. 56, 169–171.
23 Jeschke W D 1983 Cation fluxes in excised and intact roots in relation to specific and varietal differences. Plant and Soil 72, 197–212.
24 Kowalkowski A 1980 Rola karagany w kształtowaniu pokrywy glebowej suchego stepu centralnej Mongolii. Rocz. glebozn. XXXI, 133–150.
25 Kramer D 1983 Genetically determined adaptations in roots to nutritional stress: correlation of structure and function. Plant and Soil 72, 167–173.
26 Kurczewski H 1981 Wpływ wysokich dawek nawozów mineralnych na plony i skład chemiczny wybranych roślin warzywnych w zależności od nawożenia organicznego. Pr. doktorska SGGW-AR, Warszawa.
27 Lea R, Tierson W C, Bickelhaupt D H and Leaf A L 1980 Differential foliar responses of northern hardwoods to fertilization. Plant and Soil 54, 419–439.
28 Leech R T and Kim Y T 1981 Foliar analysis and DRIS as a guide to fertilizer amendments. Forest. Chronic. 57, 17 21.
29 Loughman B C, Roberts S C and Goodwin-Bailey 1983 Varietal differences in physiological and biochemical responses to change in ionic environment. Plant and Soil 72, 245–259.
30 Mengel K and Kirby E A 1983 Podstawy Żywienia Roślin. PWRiL, Warszawa, 80–83.
31 Mengel K 1983 Response of various crop species and cultivars to fertilizer application. Plant and Soil 72, 305–319.
32 Moszyńska B and Tatur A 1978 Characteristics of the process of decaying of grass leaves. Pol. Ecol. Stud. 4, 143–150.
33 Moraczewski R i Niczyporuk A 1979 Przydatność rolnicza niektórych gatunków traw i roślin motylkowych do produkcji pasz na łące umiarkowanie mokrej. Rocz. Nauk Roln. A-104, 95 112.
34 Nebe W und Rzeźnik Z 1982 Zur Ernährung der Lärche im 11-jährigen Provenienzversuch Siemianice. Forstwirtschaf. 1.
35 Nielsen N E and Schjorring J K 1983 Efficiency and kinetics of phosphorus uptake from soil by various genotypes. Plant and Soil 72, 255–230.
36 Nowosielski O 1972 Zasady Opracowywania Zaleceń Nawozowych w Ogrodnictwie. PWN, Warszawa.
37 Ostrowska A 1980 Dynamika składników w igłach sosny zwyczajnej Pinus silvestris. Roczn. glebozn. XXXI, 43–51.
38 Ostrowska A 1985 Wpływ warunków żywienia na układy jonowe w roślinach na przykładzie sosny zwyczajnej (*Pinus silvestris* L.) — Effect of nutritional conditions on the ionic systems in plants on the base of pine. Skierniewice 25/85.
39 Ostrowska A, Szczubiałka Z i Gawliński S 1982 Pinivit — wieloskładnikowy płynny nawóz do nawożenia dolistnego roślin iglastych. IBL, Warszawa, 1–58.
40 Petersson S and Jensen P 1983 Variation among species and varieties in uptake and utilization of potassium. Plant and Soil 72, 231–237.
41 Popp M 1983 Genotypic differences in the mineral metabolism of plants adapted to extreme habitats. Plant and Soil 72, 261–263.
42 Sarić M R 1983 Theoretical and practical approaches to the genetic specifity of mineral nutrition of plants. Plant and Soil 72, 137–150.
43 Smith G S, Goold G J, Johnston C M and Upsdell M 1982 Water use and chemical composition of ryegrass (Lolium) cultivars. Plant and Soil 69, 21 29.
44 Summer M E 1981 Diagnosing the sulfur requirement of corn and wheat using foliar analysis. Soil Sci. Soc. Am. Proc. 45, 87–90.
45 Sžcubialka Z Zawartość azotu i składników mineralnych w igłach jako podstawa oceny stanu zaopatrezenia sosny zwyczajnej (*Pinus silvestris* L.) w składniki pokarmowe. Pr. dokt. SGGW-AR, Warszawa.

46 Terry N and Huston R P 1975 Effects of calcium on the photosynthesis of intact leaves and isolated chloroplasts of sugar beets. Plant Physiol. 55, 923–927.
47 Wittwer R F and Immel M J 1980 Chemical composition of five deciduous tree species in four-year-old, closely spaced plantations. Plant and Soil 54, 461–467.

Variation in foliar elemental composition in Vaccinium crosses*

C.J. ROSEN and J.J. LUBY
Department of Soil Science and Department of Horticultural Science and Landscape Architecture, University of Minnesota, St. Paul, MN 55108, USA

Key words Blueberry nutrition Calcium Highbush Lowbush Manganese *Vaccinium* species

Summary Highbush blueberry culture is limited primarily to low pH coarse-textured soils high in organic matter. Lowbush blueberry species are used in U.S. breeding programs for the improvement of several characters including adaptation to a broader range of soil physical and chemical properties. The objective of this study was to determine the influence of various levels of lowbush ancestry on the mineral nutrition of progeny plants. The crosses had lowbush ancestry levels of 0%, 25%, 50%, and 100%. Recently matured leaves on 8-year-old plants were sampled during the middle of the fruiting season.

Significant differences in foliar composition were detected for many nutrients; however, leaf total N, P, Ca, Cu, and Mn appeared to show the strongest relationship to lowbush ancestry. High concentrations of these nutrients were associated with a higher percentage of lowbush ancestry. Except for leaf Mn, similar trends were also observed for the parental clones. Certain lowbush parents were also associated with high leaf levels of B. Magnesium, K, Zn, Fe and water-soluble NH_4-N and NO_3-N, showed no obvious relationship to lowbush background.

Introduction

Highbush blueberry (*Vaccinium corymbosum* L.) culture is generally limited to low pH, coarse-textured soils high in organic matter[6,8]. Use of the lowbush blueberry species, *V. angustifolium* Ait., in breeding programs has been proposed to improve several characters including adaptation to a broader range of soil physical and chemical properties[6].

Variation in nutrient composition of different plant species may be one indication that differences in nutrient absorption mechanisms exist[5]. Such a characteristic would be important for Fe, Mn, and Zn nutrition when breeding for tolerance to high pH soils. Although extensive comparisons have not been made, foliar nutrient composition of highbush and lowbush species has been shown to differ[1,4,8]. Generally, leaf Mn concentrations have been found to be higher in the lowbush species. Our objective was to characterize the influence of low/highbush blueberry ancestry on the nutritional status of progeny plants.

Materials and methods

The parental material (Table 1) consisted of 12 clones representing highbush (*V. corymbosum*), lowbush (*V. angustifolium* and *V. angustifolium* var. *nigrum* [Wood] Dole) and species hybrids. The

---

* Published as Paper No. 14507, Journal Series, Minnesota Agricultural Experiment Station.

H.W. Gabelman and B.C. Loughman (Eds.), Genetic aspects of plant mineral nutrition.
ISBN-13: 978-94-010-8102-3
© *1987 Martinus Nijhoff Publishers, Dordrecht/Boston/Lancaster.*

Table 1. Parents of the progenies in which foliar nutrient levels were examined

Parent[1]	Ancestry	% Lowbush[2]
MN84	*Vaccinium angustifolium*	100
N70249	*V. angustifolium*	100
GRVa	*V. angustifolium*	100
R2P4	*V. angustifolium* × *V. corymbosum*	50
GR 1	*V. angustifolium* × *V. corymbosum*	50
GR 2	*V. angustifolium* × *V. corymbosum*	50
MN61	*V. corymbosum* USDA 1193 × *V. angustifolium* var. *nigrum*	50
B 16	*V. corymbosum* (G65 × 'Ashworth')	0
B 10	*V. corymbosum* (G65 × 'Ashworth')	0
B 6	*V. corymbosum* (G65 × 'Ashworth')	0
B 11	*V. corymbosum* (G65 × 'Ashworth')	0
B 11	*V. corymbosum* (G65 × 'Ashworth')	0

[1] MN84, N70249, GRVa were collected from a wild stand in northern Minnesota. R2P4, GR 1, GR 2, are clones resulting from open-pollination of a half-high plant growing in the Harvard Forest (Massachusetts)-selected at Grand Rapids, Minnesota
[2] Approximate percentage of lowbush ancestry of clone.

parents were crossed in combinations that resulted in progenies with approximately 0, 25, 50 or 100% lowbush ancestry. The progenies were planted in 1976 at Becker, Minnesota on a Hubbard loamy sand (Udorthentic Haploboroll, sandy mixed). Several important soil chemical properties are presented in Table 2. The experiment consisted of a randomized complete block design with 4 replications. Each plot consisted of 12 seedlings spaced 1.2 m apart in rows spaced 2.4 m apart. From 1 to 4 plants of each parental clone were grown in an unreplicated adjacent block. Fertilizer applications each season included 27 kg/ha, N, 17 kg/ha P and 94 kg/ha K applied in mid-April and 24 kg/ha N applied in early June. Herbicide applications and irrigation were provided as necessary in accordance with commercial standards[7].

Recently matured leaves from each progeny and parent plant were sampled on July 18, 1984, during the middle of the fruiting season. Leaves from plants in each plot were bulked in approximately equal proportions by weight (100 g fresh weight per plot). Leaf samples were rinsed three times in deionized water, dried at 60°C, and ground in a Wiley mill to pass through a 40 mesh screen. Total N from Kjeldahl digests and water soluble NO_3-N and NH_4-N were determined conductrimetrically[3]. Other elements were determined using an inductively coupled plasma spectrometer[9]. Data were analyzed using analysis of variance procedures and means were separated using the least significant difference (LSD) test. Linear correlation was used to determine the relationship between lowbush ancestry and nutrient composition of the leaves.

Results and discussion

Significant differences among progenies for leaf nutrient composition were detected for all elements except Al and water soluble NO_3-N (Table

Table 2. Chemical properties of the field soil (means of 4 replications)

Soil depth (cm)	Organic matter (%)	pH	Elements extracted (ppm)									
			NH_4-N	NO_3N	P	K	Ca	Mg	Fe	Mn	Zn	Cu
0–15	2.0	5.0	1.4	3.9	83	165	450	70	103	34	1.2	0.5
15–30	1.6	5.0	0.5	0.5	33	83	468	68	70	21	0.6	0.5

OM – by wet oxidation; pH – 1:1 soil-water; P – Bray (0.025 N HCl, 0.03 N NH_4F); K, Ca, Mg – 1N neutral ammonium acetate; Fe, Mn, Zn, Cu – DTPA (diethylenetriamine pentaacetic acid pH 7.3); NH_4, NO_3 – 2N KCl (moist soil).

Table 3. Influence of lowbush ancestry on leaf nutrient concentrations in blueberry crosses (means of 4 replications)

Lowbush ancestry (%)	Cross	Leaf nutrient concentration (dry weight basis)											
		N (%)	NH_4-N (ppm)	NO_3N	P (%)	K	Ca	Mg	Fe (ppm)	Al	Mn	Zn	Cu	B
0	B6 × B11	1.87	163	15	0.11	0.62	0.48	0.19	64	136	837	11	3	39
0	B6 × B1-1	1.64	121	21	0.10	0.69	0.48	0.17	65	133	652	10	3	30
0	B10 × B11	1.66	139	19	0.10	0.62	0.45	0.19	71	161	732	12	3	38
25	GR1 × MN61	1.80	165	12	0.11	0.57	0.59	0.20	61	146	544	11	4	42
25	B6 × MN61	1.78	154	14	0.11	0.56	0.57	0.18	71	159	699	11	3	37
25	B10 × GR1	1.79	151	14	0.11	0.62	0.58	0.20	75	173	530	13	3	40
25	B10 × GR2	1.69	142	17	0.11	0.58	0.65	0.22	60	161	567	15	4	38
25	B16 × GR1	1.76	127	14	0.10	0.65	0.50	0.17	71	149	468	10	3	32
25	B16 × GR2	1.69	180	18	0.10	0.53	0.63	0.20	62	143	482	11	3	37
50	GR1 × R2P4	1.74	136	16	0.10	0.62	0.55	0.18	68	160	512	11	4	37
50	N70249 × B11	1.80	149	14	0.11	0.55	0.69	0.20	77	176	1302	14	3	50
50	N70249 × B1-1	1.89	146	17	0.11	0.68	0.57	0.19	74	167	1338	12	3	54
50	GRVa × B1-1	1.81	131	14	0.11	0.58	0.60	0.18	69	151	1208	11	3	41
50	MN84 × B11	1.80	120	18	0.11	0.58	0.60	0.19	70	141	504	10	4	36
50	GR2 × R2P4	1.76	159	17	0.11	0.57	0.63	0.20	62	152	553	11	4	40
50	B10 × GRVa	1.76	153	16	0.11	0.61	0.58	0.19	63	160	843	13	3	39
50	B16 × GRVa	1.77	146	16	0.11	0.54	0.59	0.17	63	150	1200	11	3	41
75	N70249 × GR2	1.77	194	19	0.11	0.56	0.69	0.20	66	159	1131	14	3	49
75	N70249 × MN61	1.88	167	16	0.12	0.53	0.69	0.22	70	170	1169	13	4	65
75	GRVa × R2P4	1.72	166	17	0.11	0.53	0.66	0.17	63	160	1119	10	3	37
75	MN84 × MN61	1.87	146	18	0.12	0.58	0.58	0.18	69	143	498	11	5	34
75	MN84 × GR2	1.86	141	12	0.12	0.58	0.73	0.21	68	161	634	12	4	40
100	MN84 × GRVa	1.85	143	16	0.12	0.62	0.68	0.17	73	168	1237	11	4	39
Significance[1]		**	**	ns	**	**	**	**	*	ns	**	**	**	**
LSD		0.12	34	—	0.01	0.07	0.08	0.03	10	—	240	2	1	8
Corr. coeff.[2]		0.51	0.19	—	0.71	−0.37	0.77	0.02	0.17	—	0.45	0.09	0.44	0.37
Significance[3]		**	ns	—	**	ns	**	ns	ns	—	*	ns	*	ns

[1] F test for differences among crosses; *, $P < 0.05$; **, $P < 0.01$; ns, not significant.
[2] correlation (r) between % lowbush ancestry and leaf nutrient concentration.
[3] test for absolute value of $r > 0$; *, $P < 0.05$; **, $P < 0.01$; ns, not significant.

Table 4. Influence of lowbush ancestry on leaf nutrient concentrations in blueberry parent plants (bulked samples from 1 to 4 plants in an unreplicated plot)

Lowbush ancestry (%)	Parent	Leaf nutrient concentration (dry weight basis)											
		N (%)	NH$_4$-N (ppm)	NO$_3$-N	P (%)	K	Ca	Mg	Fe (ppm)	Al	Mn	Zn	Cu	B
0	B10	1.51	116	14	0.14	0.5	0.65	0.20	53	138	518	11	4	41
0	B16	1.57	74	7	0.09	0.64	0.51	0.16	52	118	454	10	4	25
0	B6	1.78	81	9	0.10	0.55	0.54	0.18	52	110	609	8	4	28
0	B1-1	1.78	78	10	0.10	0.89	0.37	0.37	48	118	552	9	3	28
0	B11	1.66	104	12	0.10	0.56	0.52	0.19	61	141	617	10	3	37
50	R2P4	1.74	131	18	0.10	0.54	0.58	0.19	64	138	686	8	3	43
50	GR2	1.86	114	12	0.10	0.60	0.66	0.24	54	149	691	14	4	40
50	GR1	1.97	114	14	0.12	0.81	0.60	0.21	79	201	590	10	4	51
50	MN61	1.63	116	12	0.10	0.54	0.62	0.21	57	172	172	12	4	29
100	N70249	2.12	93	11	0.13	0.79	0.74	0.19	60	153	1095	12	5	64
100	GRVa	1.82	174	17	0.10	0.66	0.66	0.17	72	146	1289	10	4	36
100	MN84	1.96	81	10	0.11	0.59	0.60	0.16	57	121	461	8	5	32
Corr. coeff.[1]		0.71	0.41	0.35	0.63	0.12	0.67	−0.04	0.50	0.27	0.51	0.14	0.62	0.47
Significance[2]		**	ns	ns	*	ns	*	ns	ns	ns	ns	ns	*	ns

[1] correlation (r) between % lowbush ancestry and leaf nutrient concentration.
[2] test for absolute value of r > 0; *, $P < 0.05$; **, $P < 0.01$; ns, not significant.

3). Because of the large number of progeny examined and the diverse genetic background of the parents, significant differences were expected. Of importance to the present study, however, was to determine whether these differences had any relationship to low/highbush ancestry. Leaf nutrient concentrations of progeny and parents grouped according to lowbush background are presented in Tables 3 and 4, respectively. Significant positive correlations between % lowbush ancestry and leaf concentrations of N, P, Ca and Cu were detected for both parents and progeny. For the progeny only, leaf Mn was positively correlated with lowbush ancestry.

Although total N in leaves was found to be correlated with lowbush background, water soluble leaf NH_4-N and NO_3-N were not. Concentrations of leaf NO_3-N were relatively low in all crosses and parents. Soil tests indicated that NO_3-N was the dominant N form available to the blueberry roots (Table 2). Because blueberry species have been shown to absorb and utilize NO_3-N[10], these low levels of leaf NO_3-N suggest that an efficient nitrate reductase system exists within the blueberry plant. For all crosses and parents, water soluble NO_3-N and NH_4-N were less than 1% of the total N. Phosphorus leaf concentrations tended to increase with lowbush ancestry, a trend similar to that of N. For both elements, variation from 0% lowbush to 100% lowbush was less than 20%.

Leaf Ca showed the greatest correlation with % lowbush compared to the other major cations, Mg and K (leaf Na was less than 200 ppm — data not presented). For both parents and progeny, the leaf Ca/Mg ratio had a higher correlation (r = 0.73 in parents and r = 0.84 in progeny) than Ca alone. This relationship suggests a preference for Ca over Mg by lowbush compared to highbush roots. Other ratios, K/Ca + Mg, K/Ca, K/Mg and Ca/K + Mg, were no more highly correlated with lowbush ancestry than Ca alone.

Leaf Fe and Zn showed no significant relationship to low/highbush background. Other research has demonstrated differences in Fe accumulation within highbush progenies at both low and high solution pH levels[2]. Variation within blueberry species for Fe accumulation is apparently as important as variation between species.

In contrast to Fe, leaf Mn in progeny plants was significantly correlated to lowbush ancestry. This association has been observed in a number of other studies[1,4,8]. For parent plants, the association between Mn and lowbush ancestry was not significant. However, the parents MN 61 and MN 84 appeared to be atypical in that leaf Mn concentrations were relatively low. All progeny with MN 84 as a parent (with the exception of MN 84 × GRVa) were relatively low in Mn. Although records indicate that MN 84 is a wild lowbush selection, its stature is more upright than is typical of *V. angustifolium*. This suggests that MN 84 is a result of some introgression from highbush species. MN 61 contained lowest leaf Mn concentrations, but when used in crosses did not appear to influence Mn levels in its progeny. The results indicate that high Mn accumulation tended to be dominant in crosses involving a high Mn accumulating parent. The physiological reasons for high Mn accumulation by *V. angustifolium* and its derivatives cannot be determined from this study. However, high accumulation of Mn may be advantageous on higher pH soils where Mn availability tends to be reduced.

Leaf B was not significantly correlated with lowbush background, however, certain lowbush parents tended to concentrate B. Of particular interest is the fact that N70249 appeared to be a high B accumulator. As with Mn, all progeny in which N70249 was used as a parent had relatively high leaf concentrations of B.

Leaf Cu was significantly correlated with lowbush background, although concentrations were generally low for all parents and progeny.

The large number of positive correlations indicates that lowbush genotypes tended to concentrate nutrients to a greater extent than highbush genotypes. A number of possibilities may account for this effect. Lowbush species generally have smaller leaves and shorter stature than highbush species. Although dry matter accumulation measurements have not been made, it is conceivable that highbush species have lower nutrient concentrations in part due to a dilution effect. That is, for a given amount of nutrient absorbed, more dry matter is produced. Differences in root morphology and distribution may also play a role[5]. The magnitude of difference observed with Mn, however, suggests that more than these effects are involved for this element.

This study has demonstrated that the proportion of lowbush or highbush ancestry has a marked effect on leaf elemental composition of the blueberry plant. It appears that leaf analysis of blueberries to determine nutritional status may be useful only with knowledge of ancestry of the sampled plants. Further studies are needed to determine the mechanisms involved and whether similar nutritional relationships occur on higher pH soils.

Acknowledgement The assistance of S Vest and B Fischbacher during the course of this study is greatly appreciated.

References

1 Ballinger W E 1966 Soil management, nutrition and fertilizer practices. pp 132–178. *In* Blueberry Culture. Eds. P Eck and N F Childers. Rutgers Univ. Press, New Brunswick, NJ.

2 Brown J C and Draper A D 1980 Differential response of blueberry (*Vaccinium*) progenies to pH and subsequent use of iron. J. Am. Soc. Hortic. Sci. 105, 20–24.
3 Carlson R M 1978 Automatic separation and conductimetric determination of ammonia and dissolved carbon dioxide. Anal. Chem. 50, 1528–1532.
4 Chandler C K, Korcak R F and Draper A D 1984 Analysis of leaf and soil samples from a planting of blueberry seedlings growing on an unmulched, upland soil. Fruit Crops 1984. The Ohio State University. Research Circular 283, 69–71.
5 Epstein E 1972 Mineral Nutrition of Plants: Principles and Perspectives. Chap. 12. John Wiley and Sons, New York.
6 Galletta G J 1975 Blueberries and cranberries. *In* Advances in Fruit Breeding. Eds. J Janick and J N Moore. pp 154–196. Purdue Univ. Press, West Lafayette, Ind.
7 Hoover E, Rosen C, Wildung D, Luby J, Hertz L, Heaps J and Stienstra W 1984 Blueberry production in Minnesota Agric. Ext. Ser. Univ. of MN AG-FO2241.
8 Korcak R F, Galletta G J and Draper A 1982 Response of blueberry seedlings to a range of soil types. J. Am. Soc. Hortic. Sci. 107, 1153–1160.
9 Munter R C, Halverson T L and Anderson R D 1984 Quality assurance for plant tissue analysis by ICP-AES. Commun. Soil Sci. Plant Anal. 15, 1285–1322.
10 Rosen C J, Luby J J and Buchite H J 1885 Nitrogen utilization and solution pH responses in half-high blueberries. (Abstract) Hort Science 20, 563.

Section 2

Screening techniques to detect nutritional differences under genetic control

Intact-plant screening for tolerance of nutrient-deficiency stress

G.C. GERLOFF
Department of Botany, University of Wisconsin, Madison, WI 53706, USA

Key words Deficiency tolerance Equivalent stress Nutrient efficiency Nutritional genotypes Nutrient stress P deficiency Screening procedures

Summary Critical factors in the selection of appropriate screening procedures to detect different phenotypic responses to nutrient-deficiency stress are discussed. Various morphological, anatomical, and physiological plant factors responsible for adaptations to nutrient deficiency, particularly low-P stress, are reviewed. Also, the relative effectiveness of various screening culture techniques for detecting phenotypic efficiencies based on specific plant features are considered.

The relative ineffectiveness of liquid culture media in detecting plant factors critical in P acquisition from low-P natural environments is recognized, and a culture medium that is effective under these conditions is described. P adsorbed onto alumina, after mixing with coarse sand, serves as a P source in nutrient cultures. Buffered P concentrations approximating soil solution concentrations are maintained in this system, and P availability at the root surface seems diffusion-limited. With this system, significant differences in the growth of tomato strains under P stress were detected.

The desirability of screening phenotypes at the same degree of depression from maximum yield (equivalent deficiency stress) is discussed. The need for evaluations at equivalent stress is associated with the capacities of plants in general to respond to deficiency stress with morphological and physiological changes that may not be under genetic control, for example an increase in root:shoot ratio. Additional capacity to adjust the same plant factors often are characteristic of specific phenotypes. The relative growth of the same tomato strains under equivalent and non-equivalent P-deficiency stress is compared. Significant strain differences were observed under both conditions. However, the relative responses among strains for several efficiency parameters were very different under the two types of stress.

Introduction

A critical aspect of most projects in nutritional plant genetics is the isolation of intraspecific differences in capacity for growth under specific nutritional conditions, particularly under nutritional stress. A wide range of morphological, anatomical, and physiological plant features can be responsible for intraspecific variations in response to nutrient stress and some screening procedures obviously will be more effective than others in identifying phenotypic responses based on specific plant features. Selecting the most appropriate and effective screening procedure for detecting intraspecific differences based on specific physiological-morphological factors is an easily neglected aspect of the genetic approach to plant nutrition.

This discussion will review some of the critical factors in selecting the most effective procedures to employ in screening for tolerance of nutrient-deficiency, particularly P-deficiency stress.

H.W. Gabelman and B.C. Loughman (Eds.), Genetic aspects of plant mineral nutrition.
ISBN-13: 978-94-010-8102-3
© *1987 Martinus Nijhoff Publishers, Dordrecht/Boston/Lancaster.*

Table 1. Physiological and morphological plant features associated with genotypic adaptations to nutrient deficiency stress.

A. **Nutrient acquisition from the environment**
1. Morphological factors
   a. Root-shoot ratio, particularly the increases that characteristically occur under deficiency stress.
   b. Extent of the root system, both vertical and lateral.
   c. Density of the root system within a unit of soil as related to —
      (1) Diameter of lateral roots.
      (2) Frequency and length of root hairs.
2. Root modification of the microenvironment, for example by proton efflux and organic exudate production.
3. Root-microorganism interactions, for example with mycorrhizal fungi and N-fixing bacteria.
4. Biochemical ion-uptake mechanisms.

B. **Nutrient movement across root to xylem**
1. Transfer across endodermis.
2. Release to xylem.
3. Control of rate of nutrient uptake and distribution.

C. **Nutrient distribution and remobilization in shoot**
1. Retransport from older to younger leaves and from vegetative to reproductive parts.
2. Chelates involved in xylem transport of trace element cations, for example Fe.
3. Variations in nutrient distribution among subcellular storage compounds and compartments, for example in vacuoles.

D. **Utilization in metabolism and growth**
1. Capacity for normal metabolism at reduced tissue concentration of a nutrient.
2. Lower element concentration in supporting structures, particularly the stem.
3. Element substitution, for example partial substitution of Na for K functions.

Specific tolerance mechanisms vs screening procedures

As an initial point, plant geneticists and plant physiologists involved in genetic studies of deficiency-stress tolerance must continually be aware of (1) the various morphological, anatomical, and physiological factors that might contribute to intraspecific variations in deficiency stress response and (2) the availability of screening media and procedures that will effectively detect intraspecific differences based on those specific plant features.

To illustrate the above point, a number of the plant factors that could be responsible for adaptations to nutrient deficiency are summarized in Table 1. The various possibilities are divided into four categories associated with (1) nutrient acquisition from the environment, (2) nutrient movement across the root to the xylem, (3) nutrient distribution and remobilization in the shoot, and (4) nutrient utilization in metabolism and growth. Phenotypic differences in most of the listed possibilities have been established. Modifications of the list undoubtedly are possible.

The primary point of Table 1 for this discussion is to recognize that some screening procedures are more effective than others in detecting tolerance of nutrient deficiency stress based on specific plant factors.

Screening procedures

Before considering specific screening procedures, two general aspects of screening techniques that are of major concern in nutritional plant breeding will be mentioned.

First, inheritance experiments require large numbers of plants. To be attractive to plant breeders, a screening procedure should be simple enough to permit evaluations of hundreds and even thousands of plants with reasonable effort and expense. For the same reason, tests that permit evaluations during the seedling stage are highly desirable.

A second general factor is that usually priority should be given to screening procedures that detect morphological and physiological factors of major significance in successful plant growth under field conditions. This is well illustrated by the factors critical in P availability to plants in soil systems[3,22,23]. Average soils are characterized by very low soil solution P concentrations[3], perhaps $1-5 \mu M$ compared to the 500 × higher $1-2 \, mM$ concentrations in synthetic nutrient media. In the soil system, P availability at root surfaces also is diffusion limited[4]. This is in sharp contrast to the conditions of availability in liquid nutrient cultures in which agitation continually brings a fresh supply of P to root surfaces.

Because of the nature of the factors that control P availability in the soil, the root:shoot ratio and the production of root hairs and mycorrhizae can be of critical importance under soil P-deficiency stress, even though they are of limited importance in growth in synthetic nutrient media. In contrast, it has been suggested that in diffusion-limited systems variations in the activity of the root P-uptake mechanism may be of relatively little importance in adaptation to P-deficiency stress[6,7].

Several of the most common culture media and procedures for detecting nutritional phenotypes of interest are indicated below. Some of the advantages and limitations of each approach for detecting specific types of efficiency adaptations also are indicated, with the emphasis on plant features critical for growth in P-deficient soils.

1. Small-volume nutrient cultures

Most plant nutritionists probably would consider the culture of single plants in individual, relatively small containers providing 1–2 l of one of the accepted synthetic nutrient media as the standard nutrient culture technique. The nutrient solution, of course, must be modified to provide an amount of the nutrient under study that will result in the desired degree of nutrient-deficiency stress.

These cultures are relatively easy to prepare, and they effectively screen for efficiencies based on physiological or anatomical factors other than nutrient acquisition, that is they detect efficiencies based on nutrient movement in the plant and utilization in metabolism.

Some of our first efforts were with this approach[24,25,26]. It effectively distinguished strain differences in efficiency of K utilization that resulted in oven-dry vegetative yield differences as high as 47 per cent.

The small-volume culture technique, unfortunately, is of limited effectiveness in screening for root morphological factors critical in the acquisition of P and several other nutrients from the soil. One reason is that in solution cultures, nutrients continually are brought to root surfaces by agitation. Also, during most of a growth period the P concentration in the nutrient solution is much higher than in the soil solution. As a result, several adaptive features induced by low P, such as root hair initiation and growth, may not be detected[17,19].

2. Soil

Because soil is the medium in which tolerant selections eventually must produce superior yield, an argument can be made for carrying out the required screening in that medium.

Soil may be the best choice in some cases. However, there are disadvantages, for example in duplicating in different experiments the nutrient conditions provided by a specific soil. Even in storage the availability of some nutrients may change over a period of time. The extraction of intact root systems when plants are grown in soil also can be difficult, and yet root characteristics often are critical in the success of efficient strains. Undoubtedly, whenever possible, it is desirable to check the performance of selections made in nutrient cultures by comparing their growth in a deficient soil[9].

3. Large-volume nutrient cultures

A second possibility for screening in nutrient culture is to grow plants representing a number of phenotypes in a large-volume tank. This approach has been used satisfactorily in screening for tolerance of high salt levels[15] and should be equally satisfactory in screening for tolerance to heavy metals. In these situations, the plants probably do not significantly modify the initial salt or metal levels even over extended time periods. This is not true in screening for tolerance of nutrient-deficiency stress at concentrations of P and K approximating the low concentrations in soil solutions. Even with frequent adjustments, it is difficult to maintain the desired P or K nutrient levels.

Another disadvantage of this system is that, in common with small-volume cultures, it is ineffective in detecting root morphological factors critical in P and K acquisition under soil conditions.

4. Continuous-flow cultures, instrument-maintained concentrations

Continuous-flow cultures in which instrumentation is used to automatically maintain a soil-solution concentration of the element under study is a useful modification of the large-volume culture system. The apparatus developed at the Grassland Research Institute in England is an example[5,8].

An obvious advantage of such a system is that very low nutrient concentrations can be maintained. With the Grassland Research Institute apparatus, plants were grown at P concentrations as low as $0.05\,\mu M$[5]. The cost of the equipment for such a setup is not insignificant and may limit its application. Also, for diffusion-limited nutrients, availability at root surfaces continues to be by different mechanisms than in soils.

5. Cell and tissue culture

Finally, recognition that widespread interest in cell and tissue culture means that this technique will be seriously considered for detecting the nutritional differences of interest in this discussion. Cell and tissue culture will be the subject of other presentations, and because of possible duplication, this discussion will not elaborate on that approach in reference to detecting responses to nutrient-deficiency stress.

Cell and tissue cultures in general would seem useful for detecting differences that involve nutrient utilization in metabolism. Those factors relating to root systems and requiring whole plants for expression would be missed.

For the reason indicated, intact plants have been used in almost all of our studies. It was considered desirable to evaluate not only root factors important in nutrient uptake under natural conditions but also shoot factors that may contribute to uptake and movement: for example, transpiration, an energy source for root functions, and hormones that may regulate uptake and transport.

Development of the P-alumina culture system

Because the techniques considered are inadequate for detecting some plant factors important in P acquisition from low-P natural environ-

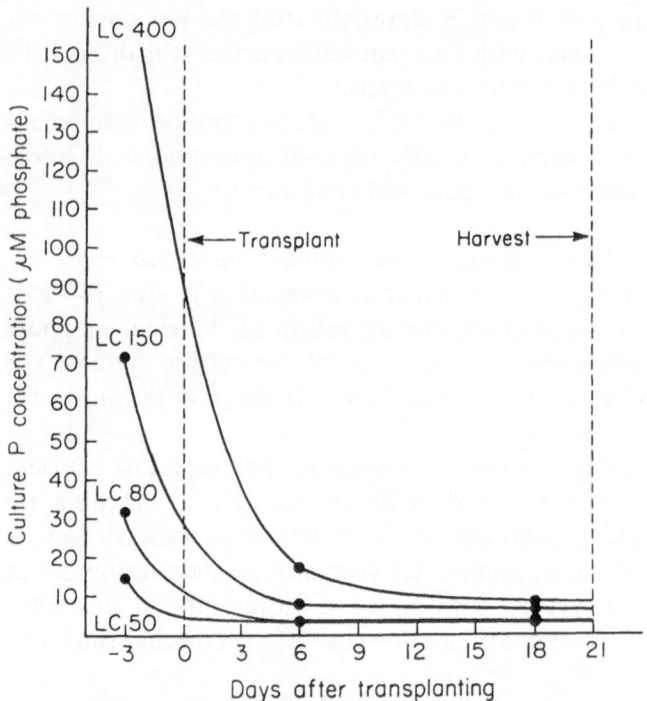

Fig. 1. Culture P concentrations at various times before and after transplanting for treatments containing 50 g alumina/pot loaded at several P concentrations (LC). Cultures were leached for 15 days before transplanting.

ments, we have given considerable attention to developing a screening system that would more closely approximate conditions of P availability in soils. The approach was to develop a culture system in which: (1) the P concentration would approximate P concentrations in average soils, (2) the P concentration could be buffered at a desired level, and (3) the availability of P at the root surface would be diffusion-limited.

A system was developed in which P adsorbed onto alumina, after mixing with coarse sand, served as a P source in nutrient cultures[10]. The P concentration maintained in the cultures was determined by the P concentration in the solution with which the alumina was loaded over an extended period. Applications of Hoagland's solution minus P provided the other essential nutrients.

Figure 1 shows the concentration of P in solutions extracted by applying suction to the culture pots three days prior to transplanting and at two intervals during a 21-day growth period[10]. As indicted, longer periods of washing of the alumina after P loading are required to stabilize the P concentration in cultures containing alumina loaded with high concentrations of P. At loading concentrations of 50 and 80 mM,

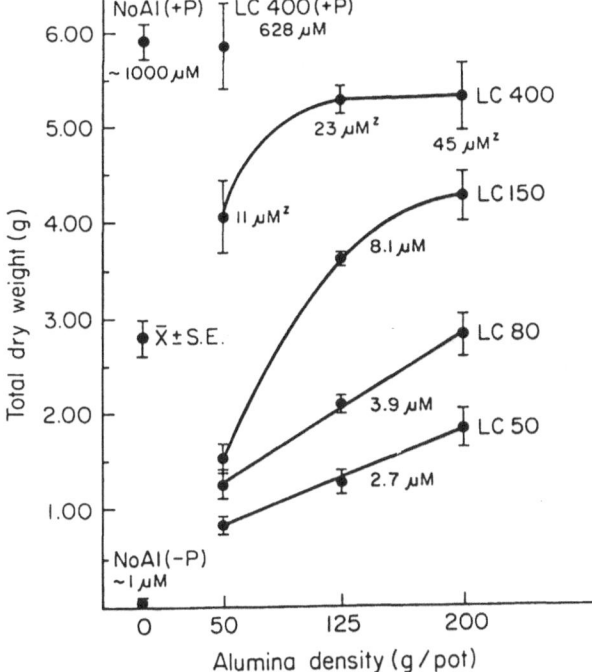

Fig. 2. Relationship between alumina density in sand-alumina cultures and plant total dry weight at several solution P concentrations. Stable average concentrations are noted; where footnoted[z] treatment concentrations were not stable during the entire period of plant growth.

the P concentration in the culture solution seemed reasonably stable at approximately 5 μM throughout the culture period.

Increasing the amount of alumina loaded at a specific P concentration in the sand-alumina cultures had no effect on the P concentration in the culture solution. However, as indicated in Figure 2, in which alumina density as g/culture pot is plotted against plant dry weight produced, an increase in the amount of alumina per pot at the lower P concentrations resulted in substantial increases in plant yield at a constant P concentration[10]. This indicates that, as in soil, diffusion limited P availability at the root surface.

Figure 3 shows the relationship between the concentration of P in culture extracts and plant dry weight produced[10]. All cultures were provided with 50 g of alumina but the alumina had been loaded at varying P concentrations. The extraction procedure clearly measured a useful concentration parameter to which the tomato plants were highly responsive.

The sand-alumina system was used to compare the tolerance of tomato strains to P-deficiency stress[11]. Extremes in tolerance were selected from the mass screening data and were reevaluated in several experi-

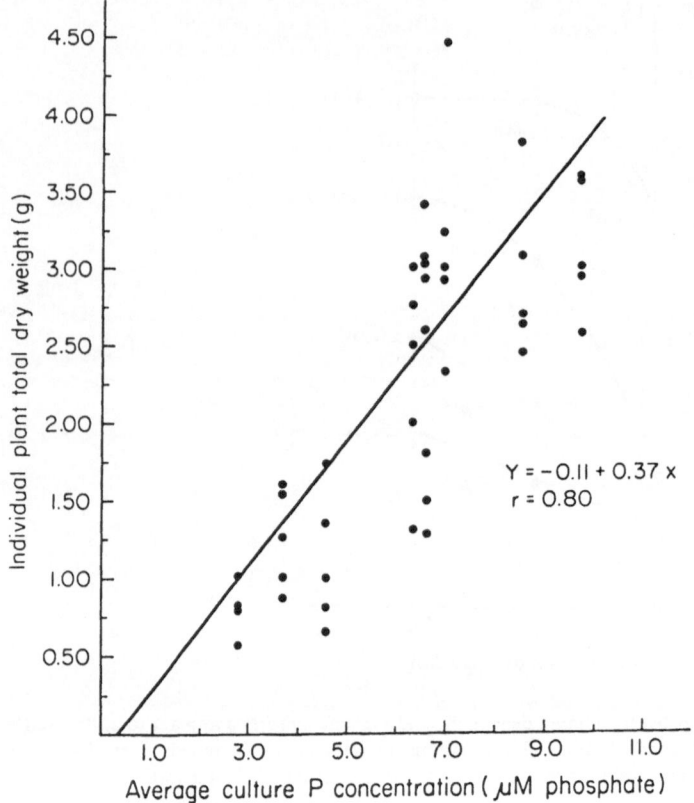

Fig. 3. Linear regression of individual plant total dry weights on average culture P concentrations from all treatments containing 50 g/pot alumina. Data are omitted from treatments with unstable concentrations. Individual points are replicate total dry weights. A significant portion of the TDW variability at a given treatment concentration is due to block effects ($P < 0.05$). The slope of the regression line is significant at $P \leq 0.01$.

ments. The average dry weights from four experiments are presented in Table 2. The sand-alumina system successfully divided the strains into tolerant and intolerant groups with strains 127 and 214 in the intolerant category. The strain with the highest average dry weight (479) produced 8.03 g dry wt or 73% more growth than the 4.58 g from the strain with the lowest dry weight (214). There was no significant difference in the yield of the strains when all were grown under adequate P.

Detailed comparisons of the morphological and physiological characteristics of the 7 strains demonstrated that three factors were responsible for the observed growth differences: the root:shoot ratio, an aspect of P acquisition other than root morphology, and efficiency of P utilization following absorption. As shown in Table 3, these tolerance factors were expressed to varying degrees in the 7 strains.

Table 2. Mean dry weights of tomato strains grown at low P and high P in sand-alumina cultures

Strain	Low P total dry wt (g)^{X,Y}	Tolerance rating^X	High P shoot dry wt (g)^{X,Y}
55	6.89 A^U	(T)	3.07 A
127	4.85 B	(I)	3.17 A
134	6.77 A	(T)	3.66 A
159	6.46 A	(T)	3.34 A
212	7.45 A	(T)	3.80 A
214	4.58 B	(I)	3.56 A
479	8.03 A	(T)	3.51 A

^{Y} Overall means of values from four experiments
^{X} Overall mean separation and low-P tolerance ratings
(T = Tolerant, I = Intolerant) based on viewing "experiment" as a random variable in overall ANOVA.
^{U} Means followed by the same letter do not differ significantly according to a 5% LSD

Screening at equivalent P stress

A screening problem that does not involve the culture medium has been of recent concern in our studies. The point of interest is referred to as the need to screen under equivalent deficiency stress, that is to make growth comparisons among strains when all strains are at the same degree of depression from maximum yield.

Several factors contribute to the importance of screening under equivalent stress. One factor is that plants in general respond to deficiency stress with specific morphological and physiological changes, for example an increased root:shoot ratio, increased rate of nutrient uptake, and decreased tissue nutrient concentration. In addition, there may be strain-specific responses to the same morphological and physiological factors that are under genetic control. A second reason for screening at equivalent stress is that when grown at a single level of a nutrient, differentially adapted strains usually are under different degrees of deficiency stress at harvest. As a result, it is difficult to determine the degree to which physiological and morphological differences are strain-specific responses under genetic control rather than general responses.

Data on two of the strains studied (Tables 2 and 3) serve as an example of the point of interest. Inefficient strain 214 produced much less growth than efficient strain 55 when grown at a single P concentration in the sand-alumina system, 4.58 vs 6.89 g (Table 2). Nevertheless, the root systems of the two strains were approximately the same size (Table 3). Therefore, strain 214 had a significantly higher root:shoot ratio than 55 as shown by relative root extension ratios (RER) of 1.38 vs 0.97 (Table 3). RER values are expressed as meters of root per gram of shoot. The relative P utilization ratio (PUR) also was significantly higher for 214, 1.07 vs 0.91 mg plant dry wt. per mg of P absorbed.

Table 3. Relative means for traits associated with tolerance to P deficiency in tomato strains producing different total dry weight (TDW) in three sand-alumina culture experiments

Strain	Root diameter	Root length	RER[z]	PARG[y]	PARM[x]	Total P uptake	PUR[w]	TDW
479(t)[v]	1.01 b[u]	1.22 a	0.93 b	1.03 b	1.05 abc	1.27 a	0.98 bc	1.25 a
212(t)	0.95 c	1.10 b	0.93 b	0.98 b	1.05 abc	1.11 bc	1.04 ab	1.16 ab
55(t)	0.93 c	1.06 bc	0.97 b	1.21 a	1.09 ab	1.18 ab	0.91 d	1.08 bc
134(t)	1.00 b	0.95 d	0.89 b	1.02 b	1.15 a	1.08 bc	0.96 cd	1.07 bc
159(t)	0.94 c	0.99 cd	0.94	1.07 b	1.02 bc	1.00 c	1.00 bc	1.01 c
127(i)	1.15 a	0.74 e	0.97	0.86 c	0.97 c	0.76 d	1.04 ab	0.77 d
214(i)	1.02 b	0.99 cd	1.38 a	0.84 c	0.67 d	0.67 d	1.07 a	0.72 d
Absolute means	0.320	187.3	35.7	8.08	0.28	14.56	446.8	6.38
Units	mm	m	m/g	mg/da/g	mg/da/m	mg	mg/mg	g

[z] RER = root extension ratio (m total root length/g shoot dry weight).
[y] PARG = net P assimilation rate per day per g root dry weight.
[x] PARM = net P assimilation rate per day per m root length.
[w] PUR = P utilization ratio (mg TDW produced per mg P absorbed).
[v] Overall tolerance ratings of tolerant (t) and intolerant (i).
[u] Means followed by the same letter are not statistically different according to Fisher's "protected" LSD 0.05.

Nevertheless, the overall rating for strain 214 was intolerant apparently due to a very inefficient P-acquisition mechanism as suggested by the much lower relative P assimilation rate (PARM) of 0.67 vs 1.09 for 55, expressed as mg P uptake per meter of root (Table 3).

To determine if the varying strain responses represented true physiological-morphological strain differences, tomato strains were evaluated by a procedure designed to make yield comparisons at an equal degree of nutrient-deficiency stress[12].

Table 4. Means of per cent shoot dry weight (SDW) reduction due to P deficiency stress and corresponding means of root extension ratio (RER) and P utilization ratio (PUR) for tomato strains grown at various levels of P selected to reduce SDW equivalently and at a single level of low P

Strain	Equivalent stress multi-P-level study				Single low-P-level screening studies		
	Ave. soln. P conc. (μM)	% RED	RER[x]	PUR[w]	% RED	RER[x]	PUR[w]
55(T)	3.2	49 AB[u]	63 A	439 A	10–30	35 B[u]	407 D
127(I)	6.0	43 AB	45 C	366 B	40–60	35 B	465 AB
134(T)	3.7	44 AB	57 AB	347 B	10–30	32 B	429 CD
212(T)	5.1	37 B	44 C	354 B	10–30	34 B	465 AB
214(I)	6.6	53 A	66 A	360 B	40–60	49 A	478 A
479(T)	3.7	55 A	47 BC	364 B	10–30	33 B	438 BC
MEAN		47	54	372		36	447

[x] RER = m total root length/g sdw
[w] PUR = mg dry weight/mg P
[u] Means followed by the same letter do not differ significantly according to a 50% LSD

Six tomato strains previously rated as tolerant or intolerant of P-deficiency stress were cultured in the sand-alumina system under a range of P concentrations. The resulting data were used to select for each strain a P-loading concentration and a sand-alumina P concentration[12] that would produce 50% of the growth in a high-P control.

Table 4 compares the responses of the six tolerant and intolerant strains in terms of two morphological-physiological parameters when the strains were grown at one P-level and showed differing stress at harvest and when grown to equivalent stress under the indicated P concentrations. Of the six strains, all but 212 were considered to have been under equivalent stress as indicated by the per cent reduction in yield.

The RER and PUR data under equivalent stress show that significant differences in both parameters continued to be observed under equivalent stress. However, a comparison of the RER and PUR data under the two conditions of P supply makes it obvious that the relationships between strains for both root extension ratio and P utilization ratio changed significantly under equivalent stress. For example, under equivalent stress, the root:shoot ratio for strain 214 was no longer significantly higher than for strain 55 as shown by RER values of 66 and 63, respectively. Likewise the relative PUR values had reversed. Under differing stress, the value for 214 had been significantly higher; under equivalent stress the reverse was true with a value of 439 for strain 55 compared to only 360 for 214.

The conclusion from this study is that the most reliable evaluations of genetically-controlled intraspecific responses to deficiency stress are obtained under equivalent stress. However, the considerable effort required for equivalent stress evaluations is hardly consistent with the earlier suggestion that to be attractive to plant geneticists screening procedures should be relatively simple. A suitable compromise would seem to be to carry out initial screening at a single nutrient level and with a single harvest. Final evaluations of phenotypic response and selections of strains for detailed physiological or genetic studies then should be made under equivalent stress.

Screening for other nutrient efficiencies

The discussion thus far has been directed primarily to P. It is obvious that factors that are unique in the availability and utilization of other elements mean that the most effective P-stress screening procedures may not be applicable to those elements. This point can be illustrated with Ca and K.

1. Calcium

We have observed, as have others[1,2], that H^+ and other cations effectively interfere with Ca uptake from a synthetic nutrient solution. This is critical in determining the degree of plant response to a specific stress Ca concentration[18]. It also suggests that when screening for response to Ca stress, pH control is critical and that the concentrations of cations other than Ca should be comparable to their concentrations in soil solutions rather than in common synthetic nutrient media.

As a second point on Ca, recognition that, in contrast to P, mass flow of soil water to the root surface is of major importance in bringing Ca to root absorbing surfaces. As a result, morphological root factors that increase root surface area would seem to be less critical in Ca acquisition than in P availability and screening procedures that involve liquid nutrient culture may be satisfactory.

2. Potassium

Two unique features of K utilization seem worth mentioning.

Na partially substitutes for K in its metabolic functions[21]. Therefore, the possibility that strains differ in capacity for this substitution must be considered and should be a part of a screening program. Such differences do occur in tomato[20].

Secondly, a major physiological role of K is as a non-specific activator of enzymes involved in many different aspects of metabolism[16]. As a result, K seems to offer unique opportunities for phenotypic differences based on K utilization in metabolism. Screening procedures involving liquid nutrient cultures or cell and tissue cultures should be effective in detecting these differences.

Additional needs in screening procedures

It should not be concluded that the procedures discussed adequately screen for all plant factors that may be the basis of phenotypic responses to deficiency stress. There is need for further improvements.

One additional need is for procedures that fully recognize the possible roles of microorganisms in influencing nutrient availability in the root environment. The formation of mycorrhizae is one example. In regard to mycorrhizae, the sand-alumina system was shown to be very satisfactory for establishing mycorrhizae on tomatoes grown in the sand-alumina system. This undoubtedly is not the only suitable synthetic system in which mycorrhizae will form. However, the ease of maintaining a very low but controlled P concentration makes the sand-alumina system attractive for mycorrhizae studies.

A second need in screening procedures is for improved media for evaluating differing capacities of root systems to modify the root environment under conditions comparable to those in the field. The Fe-deficiency tolerance of certain strains of dicots associated with the efflux of protons and/or Fe-reducing capacity is an obvious example. It seems advantageous to make such strain evaluations not in conventional nutrient media but in media that approximate the conditions in calcareous soils in which Fe-chlorosis is a widespread practical problem[13,14].

References

1 Arnon D I and Johnson C M 1942 Influence of hydrogen ion concentration on the growth of higher plants under controlled conditions. Plant Physiol. 17, 525–539.
2 Arnon D I, Fratzke W E and Johnson C M 1942 Hydrogen ion concentration in relation to absorption of inorganic nutrients by higher plants. Plant Physiol. 17, 515–524.
3 Barber S A 1984 Soil Bionutrient Availability. John Wiley and Sons, New York.
4 Bhat K K S and Nye P H 1973 Diffusion of phosphate to plant roots in soil. I. Quantitive autoradiography of the depletion zone. Plant and Soil 38, 161–175.
5 Breeze V G, Canway R J, Wild A, Hopper M J and Jones L H P 1982 The uptake of phosphate by plants from flowing nutrient solution. I. Control of phosphate concentration in solution. J. Exp. Bot. 33, 183–189.
6 Chapin F S III 1980 The mineral nutrition of wild plants. Annu. Rev. Ecol. Syst. 11, 233–260.
7 Chapin F S III and Bieleski R L 1982 Mild phosphorus stress in barley and a related low phosphorus-adapted barley grass: Phosphorus fractions and phosphate absorption in relation to growth. Physiol. Plant. 54, 309–317.
8 Clement, C R, Hopper M J, Canway R J and Jones L H P 1974 A system for measuring the uptake of ions by plants from flowing solutions of controlled composition. J. Exp. Bot. 25, 81–99.
9 Coltman R, Gerloff G and Gabelman W 1982 Intraspecific variation in growth, phosphorus acquisition and phosphorus utilization in tomatoes under phosphorus-deficiency stress. In Proc. of Ninth Int. Plant Nutrition Colloq. Ed. A Scaife, pp. 117–122.
10 Coltman R R, Gerloff G C and Gabelman W H 1982 A sand culture system for simulating plant responses to phosphorus in soil. J. Am. Soc. Hort. Sci. 107, 938–942.
11 Coltman R R, Gerloff G C and Gabelman W H 1985 Differential tolerance of tomato strains to maintained and deficient levels of phosphorus. J. Am. Soc. Hort. Sci. 110, 140–144.
12 Coltman R R, Gerloff G C and Gabelman W H 1986 Equivalent stress comparisons for evaluating physiological and morphological differences among tomato strains differentially tolerant to P deficiency. J. Am. Soc. Hort. Sci. Submitted.
13 Coulombe B A, Chaney R L and Wiebold W J 1984 Use of bicarbonate in screening soybeans for resistance to iron chlorosis. J. Plant Nutr. 7, 411–425.
14 Coulombe B A, Chaney R L and Wiebold W J 1984 Bicarbonate directly induces iron chlorosis in susceptible soybean cultivars. Soil Sci. Soc. Am. J. 48, 1297–1301.
15 Epstein E and Norlyn J D 1977 Seawater-based crop production: A feasibility study. Science 197, 249–251.
16 Evans H J and Sorger G J 1966 Role of mineral elements with emphasis on the univalent cations. Annu. Rev. Plant Physiol. 17, 47–76.
17 Foehse, D and Jungk A 1983 Influence of phosphate and nitrate supply on root hair formation of rape, spinach and tomato plants. Plant and Soil 74, 359–368.
18 Giordano de B L, Gabelman W H and Gerloff G C 1982 Inheritance of differences in calcium utilization by tomatoes under low-calcium stress. J. Am. Soc. Hort. Sci. 107, 664–669.
19 Itoh S and Barber S A 1983 Phosphorus uptake by six plant species as related to root hairs. Agron. J. 75, 457–461.

20 Makmur A, Gerloff G C and Gabelman W H 1978 Physiology and inheritance of efficiency in potassium utilization in tomatoes grown under potassium stress. J. Am. Soc. Hort. Sci. 103, 545 549.
21 Mengel K and Kirkby E A 1982 Principles of Plant Nutrition. Int. Potash Institute, Worblaufen-Bern, Switzerland.
22 Nye P H 1977 The rate-limiting step in plant nutrient absorption from soil. Soil Sci. 125, 292-297.
23 Nye P H and Tinker P B 1977 Solute Movement in the Soil-Root System. Studies in Ecology. Vol. 4. Blackwell Scientific Publishers, Oxford, England.
24 O'Sullivan J, Gabelman W H and Gerloff G C 1974 Variations in efficiency of nitrogen utilization in tomatoes (*Lycopersicon esculentum* Mill.) grown under nitrogen stress. J. Am. Soc. Hort. Sci. 99, 543-547.
25 Shea P F, Gerloff G C and Gabelman W H 1968 Differing efficiencies of potassium utilization in strains of snapbeans, *Phaseolus vulgaris* L. Plant and Soil 28, 337-346.
26 Whiteaker G, Gerloff G C, Gabelman W H and Lindgren D 1976 Intraspecific differences in growth of beans at stress levels of phosphorus. J. Am. Soc. Hort. Sci. 101, 472-475.

Somatic cell selection of mutants resistant to mineral stress

A.J. CONNER* and C.P. MEREDITH
Department of Viticulture and Enology, University of California, Davis, CA 95616, USA

Key words Aluminum resistance Aluminum toxicity Cell culture Mineral stress Mutant *Nicotiana plumbaginifolia* Somatic cell selection

Summary Genotypes with enhanced resistance to mineral stress are particularly amenable to selection in plant cell cultures. Both ion toxicities and nutrient deficiencies have severe effects at the cellular level and can thus be expected to inhibit cells in culture. Many of the mechanisms by which plants adapt to mineral stresses are also of a cellular nature. Indeed, known genotypic differences in resistance to mineral stress have been shown to be expressed in cultured cells in several cases. Thus, it is reasonable to expect that genotypes differing in resistance to mineral stress can be identified and isolated in cultured cells and that their resistance may be expressed in regenerated plants and their progeny. Using a selection medium designed to simulate the mineral environment of acid soils, we have successfully isolated aluminum-resistant mutants from cultured cells of *Nicotiana plumbaginifolia* Viv. A total of 246 aluminum-resistant variants were isolated from non-mutagenized homozygous diploid cell cultures. Resistance was retained in 119 of these variants after each was individually cloned and reselected from single cells. After 6–9 months of growth in the absence of aluminum, resistance was retained in callus cultures of 67 of the reselected variants. Complete plants were regenerated from 40 of these stable aluminum-resistant variants and all transmit aluminum resistance to their seedling progeny (selfed and backcrossed) in segregation ratios expected for a single dominant mutation.

Introduction

That new plant genotypes can be obtained by selection in cell cultures has been clearly established[1,18]. However, most of the mutants obtained to date have been selected for resistance to simple chemical stresses such as antibiotics. Mutants of agricultural significance have been rare. For agriculturally important traits, there are very few reports that unequivocally demonstrate the expression of the selected phenotype in regenerated plants and genetic transmission to their progeny.

Cell selection is based upon the application of a selection pressure that kills or inhibits wild-type cells but permits the preferential growth of rare cells with the desired phenotype. The design of a selection strategy is straightforward for phenoytpes such as drug resistance. The cells are simply subjected to an inhibitory drug concentration in the culture medium. Most agriculturally significant phenotypes, however, are more complex and their selection requires more sophisticated selection strategies. Since selection pressure applied to cultured cells acts only on cellular phenotypes, an agricultural trait must have a recognized cellular

---

* Present address: Crop Research Division, DSIR, Private Bag, Christchurch, New Zealand.

manifestation that is expressed in cultured cells if it is to be selectable. This requirement is a major limitation to the application of cell selection for crop improvement[2,3,21].

Many agriculturally important traits are not cellular; they reside at levels of organization above the cell and may require the integration of many plant functions (*e.g.* plant architecture). Others involve specialized processes that are cellular but may not operate normally in cultured cells (*e.g.* flavor). Of those traits that are expressed in cultured cells, many are still so poorly understood that we may be unable to recognize their cellular manifestation.

If the desired cellular phenotype is known, it may then be possible to devise a selection strategy that will permit its isolation. There may be cases in which the selection strategy initially seems obvious, but care must be taken to avoid over-simplification. It is essential to fully consider the agricultural setting for which the new genotype is destined. Oversimplification may lead to the selection of genotypes that succeed in the laboratory but are not adapted to the complexities of the agricultural environment.

We believe that resistance to mineral stress, both deficiencies and toxicities, is a class of agriculturally important phenotypes that is among the most amenable to cell selection. In this paper we review the basis for this contention and illustrate its potential by discussing our recent work on the selection of aluminum-resistant mutants.

Resistance to mineral stress as a selectable trait in cell culture

Cellular basis of mineral stress

Mineral stresses include both deficiencies of essential nutrient elements and excesses of toxic elements. Mineral nutrients generally play fundamental cellular roles (*e.g.* enzyme activation, components of fundamental molecules) and so are just as critical to cultured cells as to whole plants. Likewise, toxic ions effect injury via the disruption of fundamental cellular mechanisms (*e.g.* competition with essential elements for uptake, inactivation of enzymes, displacement of essential elements from functional sites)[12]. Consequently, cultured plant cells are sensitive to artcificially applied mineral stresses, both nutrient deficiencies and mineral toxicities.

Mechanisms by which plants resist mineral stresses may also be of a cellular nature. For example, more efficient uptake of a deficient nutrient element may result from an altered membrane transport mechanism, or more efficient utilization can be achieved via higher enzyme affinity for an element[15]. Resistance to ion toxicities might be achieved by exclusion at the plasmalemma, detoxification via binding to an organic molecule, or altered target sites[14].

Salinity is sometimes included in discussions of mineral stress. However, salinity involves osmotic stress in addition to the toxicity produced by excesses of mineral elements such as Na^+ and Cl^-. The additional complexity of salt stress has significant implications for cell selection strategies[21]. Because selection for salt resistance requires a different approach than does that for simple mineral stress, it falls outside the realm of this discussion.

Plant response to mineral stress is under genetic control[12] and genotypic differences have been attributed to a single major gene in a number of cases[11]. That these genetic differences may have a cellular basis is indicated by observations that some genotypic differences in whole-plant response to mineral stress are expressed in cultured cells. Differences between *Agrostis* genotypes in resistance to zinc and copper toxicity were maintained in callus culture[31]. Qureshi *et al.*[27] reported similar findings for Anthoxanthum genotypes differing in resistance to zinc and lead, as did Sain and Johnson[28] for soybean genotypes differing in resistance to iron deficiency. Christianson[4], however, did not observe the expected differences in zinc sensitivity in Phaseolus genotypes. In tobacco, callus cultures of five cultivars differed significantly in manganese resistance, but the differences did not correspond to those observed in seedlings[25].

Creating toxicities and deficiencies in cell culture

Selection strategies for the isolation of variants resistant to particular mineral stresses are relatively simple in concept. The cells can simply be subjected to the deficiency or toxicity by modifying the mineral composition of the nutrient medium in which they are grown. However, the chemical milieu of standard nutrient media is quite different from that of soil and it is important to consider this in designing the selection medium. Not only is the concentration of the critical element important, but factors such as pH and other interacting elements must be taken into account and perhaps modified. While a selection strategy consisting of the simple addition or deletion of the critical element may be appealing and may even lead to resistant plants, these plants may fail in the agricultural setting for which they are intended. To maximize the probability of success, the selection strategy employed should aim to simulate the soil environment.

The selection of variants resistant to mineral stress

Reports of selection for resistance to mineral stress in plant cell cultures have so far been few in number and their success has been limited. Several attempts have been thwarted by difficulties in regenerating plants from the selected cell lines, *e.g.* aluminum-resistant tomato cells[20], manganese-resistant tobacco cells[25]. In other cases, details regarding plant regeneration or expression in plants have not been reported, *e.g.* phosphorus efficient red clover cells[26], mercury-resistant petunia cells[5], manganese- and aluminum-plus-manganese-resistant carrot cells[24], cadmium-resistant *Datura* cells[17]. A preliminary report suggested that plants regenerated from aluminum-resistant carrot cells both express and transmit the resistance[23].

We have recently selected aluminum-resistant mutants from cell cultures of *Nicotiana plumbaginifolia* Viv. Not only is the aluminum resistance expressed in whole plants derived from the selected cells, it is also transmitted to progeny in a defined genetic pattern. We believe this represents the first unequivocally successful use of somatic cell selection to genetically modify plant response to mineral stress. Because the purpose of our study was to determine whether genetic adaptation to mineral stress could be manipulated at the cellular level, and not to improve any particular crop, we chose to work with a model species. While a number of crop species are now amenable to cell culture manipulations, for the most part they do not have the combined features of ease of plant regeneration from established cultures and amenability to genetic analysis. *N. plumbaginifolia* is considered a model species for plant somatic cell genetics because cell cultures are easily established and maintained, plant regeneration from established cultures is routine, the plants are relatively small and have a short generation time, controlled pollinations are easy to make and yield large numbers of seeds, and the species is a true diploid, thus facilitating genetic analysis. We thus considered it ideal for our study.

Aluminum-resistant mutants from cell cultures: a case study

The importance of aluminum resistance

Aluminum toxicity is one of the most serious environmental stresses affecting crop productivity[22]. Widespread in acid soils throughout the world, it is especially severe in developing countries in tropical regions[13,29]. Although breeding and selection for aluminum resistance is highly effective[11,13], limitations exist for some crops, either because sufficiently resistant germplasm has not been identified or because the crop is not easily manipulated genetically. The selection of aluminum-resistant genotypes at the cellular level may circumvent some of these difficulties. In addition, genotypes known to possess cellular resistance mechanisms, by virtue of their having been isolated at the cellular level, may serve as useful material for physiological and biochemical studies of mechanisms of aluminum resistance.

The design of an aluminum toxic medium

A major problem in studying aluminum toxicity in plant cell culture is the rapid precipitation of aluminum ions in standard culture media. Although this problem was recognized by Ojima and Ohira[24], no measures were taken to prevent it and excessively high aluminum concentrations were required to inhibit cell growth of *Daucus carota* and to select an aluminum-resistant cell line. Other studies have circumvented aluminum precipitation in culture media by chelating aluminum with EDTA[19,20,21]. There are several reasons why we no longer favor this approach[8]. Plants growing in aluminum toxic soils are not exposed to Al-EDTA. In addition, chelation with EDTA may not only alleviate aluminum toxicity, but observed toxicity may be due to the EDTA rather than aluminum.

To overcome the problems asociated with aluminum precipitation, we have modified the inorganic composition of the culture medium to simulate conditions existing in aluminum toxic soils[8]. To prevent the precipitation of aluminum, the phosphate concentration was reduced from $1250 \mu M$ to $10 \mu M$ and the pH from 5.8 to 4.0. Both of these changes are absolutely essential for the expression of aluminum toxicity in cell culture. Two changes that further increase aluminum toxicity are the reduction of the calcium concentration from 3.0 mM to 01.1 mM and the use of unchelated iron. Since the gelling properties of agar are inhibited at pH 4.0, cells were cultured on filter paper supported by polyurethane foam saturated with liquid medium[7]. The only limitation to the growth of *N. plumbaginifolia* cells on this modified medium was the reduced phosphate concentration. This was partially overcome by 'pre-loading' the cells with phosphate and renewing the culture medium every second day.

With this modified medium, dose-response experiments established that plated cells, callus cultures, *in vitro* grown shoots, and seedlings all showed a similar sensitivity to aluminum, with total inhibition of growth occurring at or above $600 \mu M$[9]. This consistent expression of toxicity in developmentally different plant systems reflects the fundamental cellular basis of aluminum toxicity[16].

Selection strategies

Two strategies were employed in the selection of aluminum-resistant variants from *N. plumbaginifolia* cell cultures:
(1) A one-step *direct method* in which cells were plated onto the modified medium supplemented with $600 \mu M$ aluminum, and

Table 1. Summary of the selection and characterization of aluminum-resistant variants from cell cultures of *Nicotiana plumbaginifolia*. (Adapted from 10)

Total number of variants selected	246
Cloning and reselection unsuccessful	127
Cloned and reselected from single cells	119
Lost through contamination or senescence	4
Primary callus aluminum-sensitive*	48
Fertile plants not regenerated	23
Fertile plants regenerated	25
Seedling progeny aluminum-sensitive	24
Seedling progeny segregate for resistance	1
Primary callus aluminum-resistant*	67
Fertile plants not regenerated	27
Fertile plants regenerated	40
Seedling progeny aluminum-sensitive	0
Seedling progeny segregate for resistance	40

* The aluminum resistance of the primary callus was determined after 6–9 months of growth in the absence of aluminum.

(2) A two-step *rescue-method* in which cell suspensions were cultured in the modified medium supplemented with 600 μM aluminum for 10 days and then plated onto the standard medium (no aluminum) for the recovery of survivors.

With each selection strategy the strict control of both cell density and aggregate size was critical for the total inhibition of background growth of wild-type cells and thus the stringency of the selection[9].

The rescue selection method proved to be vastly superior to the direct method for selecting aluminum-resistant variants. Not only was it far less laborious and simpler to use, but it was more stringent and variants could be recovered in less time and at a higher frequency[9].

Selection and characterization of variants

A total of 246 aluminum-resistant variants were isolated from non-mutagenized homozygous diploid cell cultures of *N. plumbaginifolia*; 29 *via* the direct method and 217 *via* the rescue method. The stabilty of aluminum resistance was carefully characterized in both callus cultures and in the seedling progeny of plants regenerated from the variant cell lines (Table 1). To eliminate the possibility that the originally selected variants may have been chimeras of aluminum-resistant and wild-type cells, each variant was individually cloned and re-selected from single cells[9]. Only 119 (48%) of the variants could be reselected in this manner. They were then maintained as callus in the absence of aluminum for 6–9 months and then retested for growth on 600 μM aluminum. Resistance was lost in 48 of these variants, but retained in 67 (Table 1). There was no association between the degree of aluminum resistance and callus vigor in the absence of aluminum[10], suggesting that the resistance is (1) not a consequence of increased overall vigor, and (2) not detrimental in the absence of aluminum challenge. Fertile plants could be regenerated from 25 of the 48 unstable variants and, as expected, they failed (with one exception) to transmit aluminum resistance to their seedling progeny (Table 1). The non-conformity of the one variant was attributed to a somaclonal genetic event, independent of the initial cell selection, that occurred during the plant regeneration phase[10]. Fertile plants were regenerated from 40 of the 67 stable variants. Their seedling progeny were scored for resistance (survival in the presence of 600 μM aluminum). All 40 of these variants transmitted aluminum resistance to their sexual progeny (Table 1), and were confirmed as true mutants[10]. In all instances, the selfed and backcrossed seedling progeny segregated for aluminum resistance in the ratios expected for a single dominant mutation[6].

The genetic and physiological basis of aluminum resistance is currently under further study in our laboratories. We intend to construct a series of isogenic lines, each with a different aluminum resistance gene that might govern a different physiological or biochemical mechanism of resistance to aluminum.

Agricultural implications

We have clearly shown that mutants with increased resistance to aluminum can be selected from plant cell cultures. We recognize, however, that improved performance of these mutants in aluminum toxic acid soils must be demonstrated before definite statements can be made regarding the applications of cell selection in generating aluminum-resistant germplasm for crop improvement programs. These critical experiments will be performed as soon as plants homozygous for aluminum resistance are identified and sufficient seed can be harvested. Nevertheless, since the selection medium simulated aluminum toxic soils and aluminum resistance was expressed in both cultured cells and seedlings, we are optimistic that at least some component of this resistance will be expressed in a field situation.

Conclusion

On the basis of the evidence it is reasonable to predict that new crop genotypes with resistance to important mineral stresses can be produced by somatic cell selection. Cultured cells are sensitive to mineral stresses, known resistance mechanisms operate in cultured cells, selection strategies can be readily devised, and resistant cell lines have been isolated.

It is also encouraging that responses to mineral stress are often under simple genetic control.

Our success in selecting aluminum-resistant mutants from cell cultures can be largely attributed to the care taken in designing the culture system[7], the modification of the inorganic composition of the culture medium[8], and the development of stringent selection strategies[9] prior to the initiation of selection experiments. For other mineral stresses, careful consideration of environmental and physiological factors may suggest other modifications to culture procedures that will elicit responses similar to those in whole plants. The subsequent design of effective and appropriate selection strategies will increase the likelihood of recovering agriculturally useful mutants from cultured plant cells.

References

1. Chaleff R S 1981 Genetics of Higher Plants: Applications of Cell Culture. Cambridge University Press, Cambridge, 184 p.
2. Chaleff R S 1983 Considerations of developmental biology for the plant cell geneticist. *In* Genetic Engineering of Plants: An Agricultural Perspective. Eds. T Kosuge, C P Meredith and A Hollaender. Plenum Press, New York, 499 p.
3. Chaleff R S 1983 The isolation of agronomically useful mutants from plant cell cultures. Science 219, 676–682.
4. Christianson M L 1979 Zinc sensitivity in *Phaseolus*: expression in cell culture. Env. Exp. Bot. 19, 217–221.
5. Colijn C M, Kool A J and Nijkamp H J J 1979 An effective chemical mutagenesis procedure for *Petunia hybrida* cell suspension cultures. Theor. Appl. Genet. 55, 101–106.
6. Conner A C 1985 Selection of aluminum-resistant mutants from plant cell culture. Ph.D. Diss. University of California, Davis.
7. Conner A J and Meredith C P 1984 An improved polyurethane support system for monitoring growth in plant cell cultures. Plant Cell Tissue Organ Culture 3, 59–68.
8. Conner A J and Meredith C P 1985a Simulating the mineral environment of aluminum toxic soils in plant cell culture. J. Exp. Bot. 36, 870–880.
9. Conner A J and Meredith C P 1985b Strategies for the selection and characterization of aluminum-resistant variants from plant cell culture. Planta 166, 466–473.
10. Conner A J and Meredith C P 1985c Large scale selection of aluminum-resistant mutants from plant cell culture: expression and inheritance in seedlings. Theor. Appl. Genet. 71, 159–165.
11. Devine T E 1982 Genetic fitting of crops to problem soils. *In* Breeding Plants for Less Favorable Environments. Eds. M N Christiansen and C F Lewis. John Wiley and Sons, New York, 459 p.
12. Epstein E 1972 Mineral Nutrition of Plants: Principles and Perspectives. John Wiley and Sons, New York, 412 p.
13. Foy C D 1984 Adaptation of plants to mineral stress in problem soils. *In* Origins and Development of Adaptations. Eds. D Evered and G M Collins. Pitman Press, Bath.
14. Foy C D, Chaney R L and White M C 1978 The physiology of metal toxicity in plants. Annu. Rev. Plant Physiol. 29, 511–566.
15. Gerloff G C 1976 Plant efficiencies in the use of nitrogen, phosphorus, and potassium., *In* Plant Adaptation to Mineral Stress in Problem Soils. Ed. M J Wright. Cornell University Agricultural Experiment Station, Ithaca, 420 p.
16. Haug A 1984 Molecular aspects of aluminum toxicity. CRC Critical Reviews Plant Sciences 1, 345–373.

17 Jackson P J, Roth E J, McClure P R and Naranjo C M 1984 Selection, isolation and characterization of cadmium-resistant *Datura innoxia* suspension cultures. Plant Physiol. 75, 914–918.
18 Maliga P 1984 Isolation and characterization of mutants in plant cell culture. Annu. Rev. Plant Physiol. 35, 519–542.
19 Meredith C P 1978 Response of cultured tomato cells to aluminum. Plant Sci. Lett. 12, 17–24.
20 Meredith C P 1978 Selection and characterization of aluminum-resistant variants from cell cultures. Plant Sci. Lett. 12, 25–34.
21 Meredith C P 1984 Selecting better crops from cultured cells. *In* Gene Manipulation in Plant Improvement. Ed. J P Gustafson. Plenum Press, New York, 668 p.
22 National Academy of Sciences, USA 1977 World Food and Nutrition Study: the Potential Contributions of Research. National Academy of Sciences, USA, Washington.
23 Ojima K and Ohira K 1982 Characterization and regeneration of an aluminum-tolerant variant from carrot cell cultures. *In* Plant Tissue Culture 1982. Ed. A Fujiwara. The Japanese Association for Plant Tissue Culture, Tokyo, 839 p.
24 Ojima K and Ohira K 1983 Characterization of aluminum and manganese tolerant cell lines selected from carrot cell cultures. Plant Cell Physiol. 24, 789–797.
25 Petolino J F and G B Collins 1985 Manganese toxicity in tobacco (*Nicotiana tabacum* L.) callus and seedlings. J. Plant Physiol. 118, 139–144.
26 Phillips G C and Collins G B 1981 Growth and selection of red clover (*Trifolium pratense* L.) cells on low levels of phosphate. (Abstract) Agron. Abstr. 1981, 187.
27 Qureshi J A, Collin H A, Hardwick K and Thurman D A 1981 Metal tolerance in tissue cultures of *Anthoxanthum odoratum*. Plant Cell Reports 1, 80–82.
28 Sain S L and Johnson G V 1984 Selection of iron-efficient soybean cell lines using plant cell suspension culture. J. Plant Nutr. 7, 389–398.
29 Sanchez P A, Bandy D E, Villachica J H and Nicholaides J J 1982 Amazon basin soils: management for continuous crop production. Science 216, 821–827.
30 Smith R H, Bhaskaran S and Schertz K 1983 Sorghum plant regeneration from aluminum selection media. Plant Cell Reports 2, 129—132.
31 Wu L and Antonovics J 1978 Zinc and copper tolerance of *Agrostis stolonifera* L. in tissue culture. Am. J. Bot. 65, 268–271.

Screening spring wheat for drought tolerance

T.M. ANDERSEN, E. POLLE and C.F. KONZAK
Department of Agronomy and Soils, Washington State University, Pullman, WA 99164-6420, USA

Key words Drought Polyethylene glycol Wheat

Summary An improved method to screen wheat seedlings for drought tolerance using polyethylene glycol-8000 (PEG) was developed. Plants were started in nutrient solution and transferred to PEG after stable growth was achieved. Plants were held in a plexiglass framework that allowed the repeated weighing of a single holder with plants from a single genotype. The same set of plants was used to plot a fresh weight growth curve through the entire experiment. This procedure has allowed us to screen larger numbers of genotypes than otherwise would be possible.

The response curve of genotypes differed in reaction to PEG treatments, allowing us to rank genotypes by tolerance to PEG-induced stress. Results may be influenced by environmental conditions such as temperature, humidity, and light intensity. Low relative humidity or high temperature reduced the amount of fresh weight, bringing the growth curves closer to horizontal. Tests across a range of environmental conditions produced different growth curves among genotypes, but the ranking of the genotypes remained constant. The method is a simple procedure to screen for tolerance to drought stress during the vegetative period. One locally adapted, high yielding spring wheat line was found to perform as well as the low yielding drought tolerant standard.

Introduction

Some plant species are better able to tolerate drought stress than others. Genetic variability for drought tolerance also exists within many plant species[1,5,6,7,9,11]. The efficient exploitation of that variability for crop improvement would be facilitated by a fast and accurate screening method for identifying drought tolerant genotypes[3].

The breeding problem is made complex by the fact that plants may experience drought stress at any stage in their growth cycle, and plant responses to drought stress may vary with the physiological stage of development[1,10]. Moreover, no correlation has been found between seedling reaction to drought stress and germination under moisture deficient conditions[1], indicating the independent nature of the physiological processes responsible for the plant responses at these stages. Further plant responses to drought stress during grain filling appears under a still different genetic control (Blum, Konzak, unpublished observation) Thus, it is unlikely that a single criterion screening method can be used to select a genotype able to respond favorably to drought stress. However, it is possible to partition the stress circumstances into at least three main plant growth periods: (1) germination, (2) vegetative growth and development, and (3) grain filling. Seedlings are convenient materials for screening from the standpoint of ease of evaluating, space, and time.

H.W. Gabelman and B.C. Loughman (Eds.), Genetic aspects of plant mineral nutrition.
ISBN-13: 978-94-010-8102-3
© *1987 Martinus Nijhoff Publishers, Dordrecht/Boston/Lancaster.*

Several methods utilizing seedlings to screen for drought tolerance have been developed using an osmoticum such as polyethylene glycol (PEG)[1,12]. For some genotypes, the seedling response apparently is a good indicator of the plant's reaction to drought stress later in its life cycle. However, early screening of seed plants via seedlings tests must still be considered as a method for identifying genotypes that will tolerate drought stress in the vegetative stage.

Many seedling drought tolerance screening methods have used dry plant weights to measure response to drought stress[1]. This paper describes an improved method for evaluating the growth response of plant genotypes to induced drought stress using fresh weight of spring wheat seedlings grown in PEG-8000.

Materials and methods

Plant materials

Four reference genotypes (three spring wheats and one spring durum) were used to develop the method. The spring wheats were Tobari 66 (International Maize and Wheat Improvement Center), having low drought tolerance[1], K8005223 and WA6921 (advanced lines from the spring wheat breeding program at Washington State University) having respectively high and low drought tolerance based on preliminary tests. The durum, Inbar (Israel), also has a high tolerance to drought as shown earlier by Blum[1]. In addition, spring wheat genotypes of previously unknown response were evaluated.

Plant growth conditions and weighing method

To shorten screening period and obtain uniform seedlings for the tests, seeds were soaked for two hours in deionized water on day 1. Seeds were placed without light in a germination chamber consisting of a 15 × 25 × 30 cm plastic tub with a plexiglass framework supporting the seeds (approximately 0.5 cm) over deionized water. Seeds were placed on germination blotter paper extending down into the water to maintain a constant exposure to moisture. High chamber humidity was maintained by misting the inside and covering the container with plastic wrap. On day 3, when roots were 2–4 cm and coleoptiles 0.5 to 1.5 cm, depending on the genotype, 10 to 14 seeds were transferred to each plexiglass holder (Figure 1). The holders were held in a plexiglass frame floated in nutrient solution (1/4 Hoaglands) at a constant level in a 13 × 13 × 30 cm plastic tub using styrofoam blocks on each end of the frame. On day 7, the number of plants was thinned to 10 uniform plants, the first weight was taken and the nutrient solution was changed. For weighing, each holder plus plants was removed from the framework; the holder and roots were blotted with tissue paper, weighed, and returned to the 'tubs' with nutrient solution. Plant weights are recorded directly into a data collector. On day 10, a second weight was taken and the nutrient solution was replaced with fresh nutrient solution with PEG-8000. On day 14, a third weight was taken and the PEG and nutrient solution changed. The final weight was taken on the 18th day of growth (Figure 2).

Plant growth conditions

On day 3, the plant growth tubs were placed in the environment to be maintained for the remaining growth period. The standard environment for the experiments was a growth chamber with 15/20 °C night/day temperatures with a daylength of 16 hours. Humidity fluctuated from 50 to 100% and light intensity was $400\,\mu E\,cm^{-2}\,sec^{-1}$. Alternate environments were a different growth chamber in which light intensity was $72\,\mu E\,cm^{-2}\,sec^{-1}$ and a laboratory bench on which the humidity was 18 to 20%, the light intensity was $45\,\mu E\,cm^{-2}\,sec^{-1}$, and temperature was 20–24 °C.

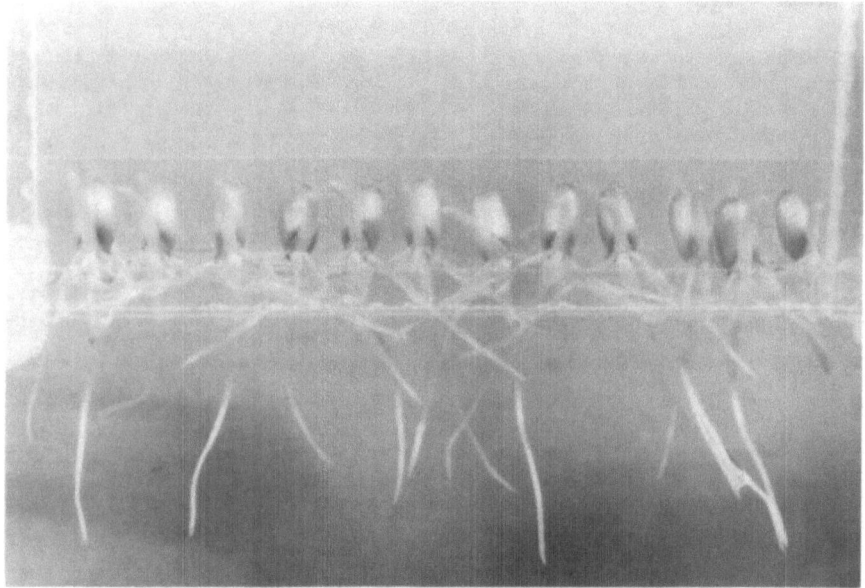

Fig. 1. Two-day-old germinated seeds after placement on holder.

Fig. 2. Seventeen-day-old seedlings grown in nutrient solution, left; or exposed to PEG-8000 treatment (20%) center, or 22% right.

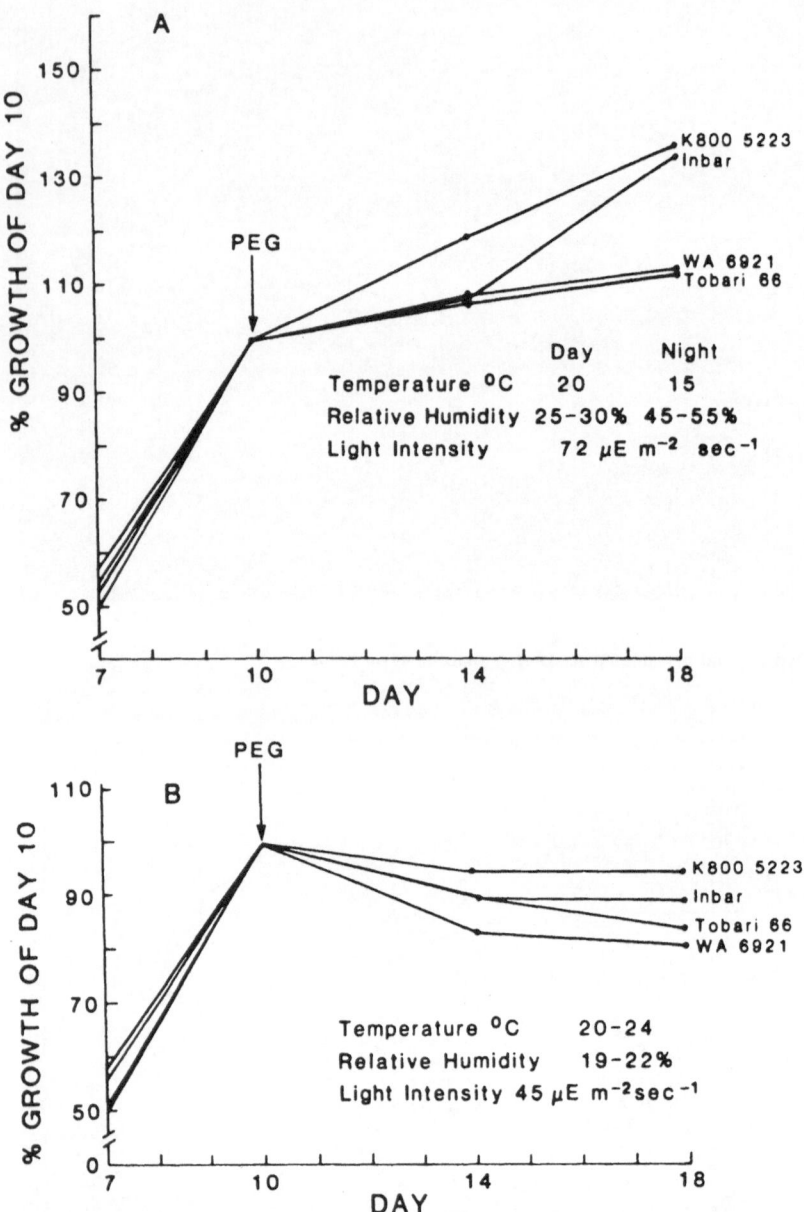

Fig. 3. Growth response of wheat genotypes to PEG-8000 treatments at various ambients (A, B, C).

Water potential determinations

On day 10, whole leaf water potentials and leaf sap osmotic potentials were measured on five plants per genotype. On day 18 similar measurements were taken on PEG-treated and control plants. The first leaf was cut off with a razor blade and placed in a PMS Instrument Co. pressure chamber for whole leaf water potential determination. The leaf was then frozen in liquid nitrogen, thawed, and the sap expressd out of the leaf. Osmotic potential of sap was measured psychrometrically.

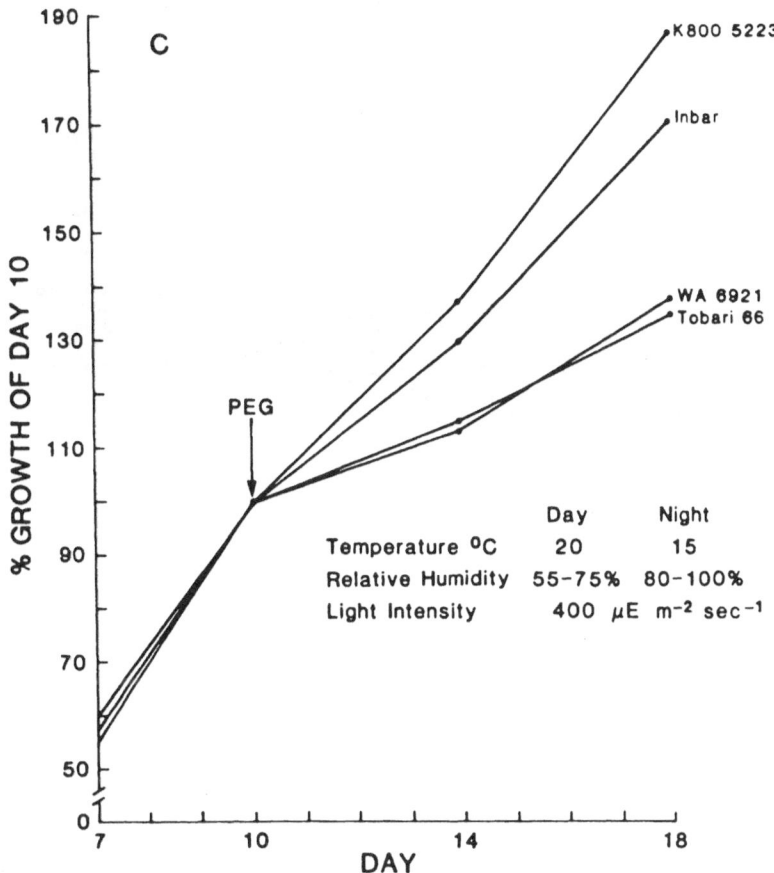

Results

Growth curves for the four reference genotypes are presented in Figure 3A, B and C and show consistent varietal differences in response to PEG solutions. Under increasingly severe conditions of increased T° and reduced R.H. the percent growth of the genotypes after exposure to PEG was greatly reduced, while the rank remained constant. Under conditions of increased light intensity, growth after PEG treatment was improved. This result indicated that the ranking of genotypes for response to PEG was similar over a wide range of environments and showed that the method may be effective for screening genotypes, when reference testers are included.

Growth curves for 15 genotypes are given in Figure 4.

Fresh weight percentage (Table 1) ranged from 84 to 87% for the reference genotypes and also several other genotypes measured. The

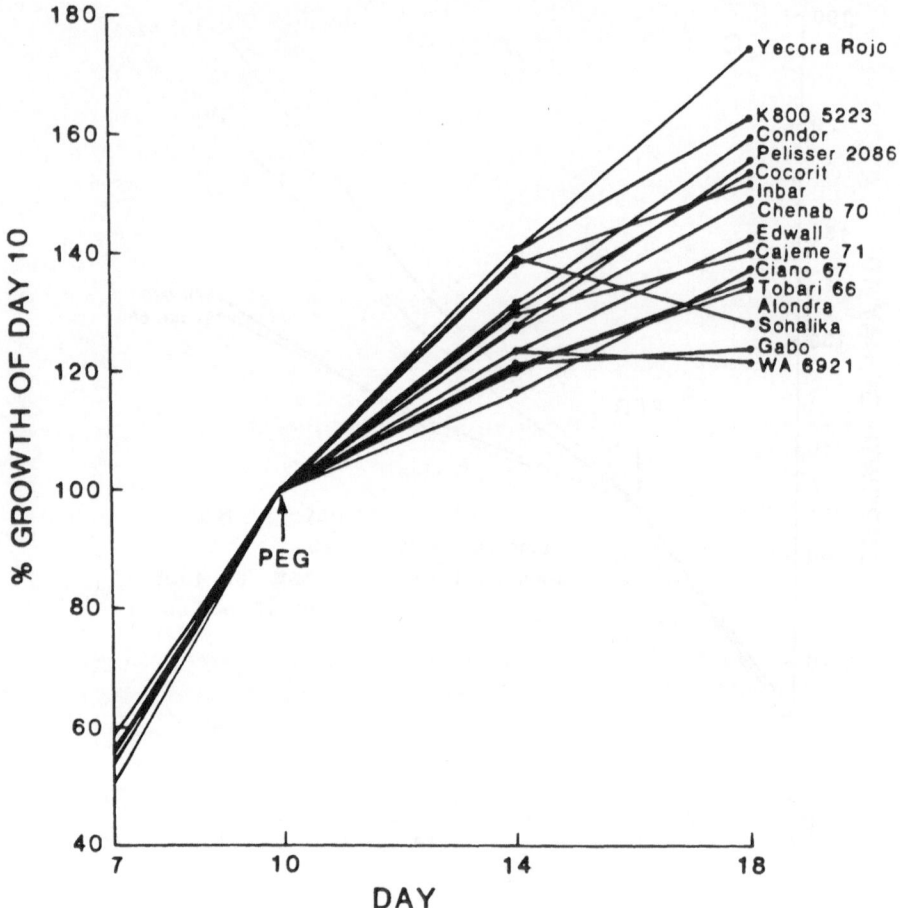

Fig. 4. Growth response of wheat genotypes to PEG-8000 treatment.

narrow percentage range indicates a good correlation of the fresh weight to the dry weight measurements.

As shown in Table 2, total water potentials of the first leaf were the same for three of the genotypes in control nutrient solutions. However, the total water potential of plants grown in 20 or 22% PEG solutions for

Table 1. Percentage moisture of wheat genotypes at day 18 after 8 days in PEG-8000 [+]

	$\dfrac{\text{(Fresh weight } - \text{ dry weight)}}{\text{Fresh weight}}$
WA6921	84
K8005223	86
Yecora Rojo	87
Tobari 66	86

[+] There were no significant differences.

Table 2. Whole leaf water potentials (megapascals) of three spring wheat genotypes pre and post PEG treatment

	Control		PEG treatment (day 18)	
	Day 10	Day 18	20%	22%
Yecora Rojo	−0.20a[+]	−0.25a	−0.73b	−0.72b
K8005223	−0.22a	−0.23a	−0.77b	−0.83b
WA6921	−0.21a	−0.19a	−0.67b	−0.72b

[+] Means followed by the same letter are not significantly different ($p = 0.05$).

8 days dropped significantly. There were no significant differences between genotypes or between the two PEG solutions. Fewer differences were measured in leaf sap osmotic potential (Table 3).

Discussion

The rating of the four reference genotypes remained constant over the concentration range from 18 to 22% PEG, although this is a relatively more narrow range of concentrations than used by Blum et al.[1] and Johnson and Asay[8]. However, we found that very small changes in PEG concentration over the 18–22% range could result in large changes in plant growth. Moreover, we did not find that direct transfer of seedings from Hoagland to 20% PEG always caused a cessation of growth, with regrowth occurring after adjustments or conditioning. For most genotypes, the response to PEG was a slowed growth rate, although for the more severe conditions there was a net loss of wet weight after exposure to PEG. Even under these conditions the ranking of the genotypes remained constant. Conditioning the plants to the PEG by a step-wise progression to 20% PEG (*i.e.*, 5% → 10% → 15% → 20%) (data not shown) resulted in a similar ranking of the genotypes and only extended the experiment.

It was observed that Yecora Rojo and K8005223 have shorter root systems than WA6921. No correlation was found between root length and drought tolerance. This suggests, as expected, that the response to

Table 3. Leaf sap osmotic potentials (megapasals) of three spring wheat genotypes pre and post PEG treatment[+]

	Control		PEG treatment (day 18)	
	Day 10	Day 18	20%	22%
Yecora Rojo	−1.20	−1.03	−1.25	−1.09
K8005223	−1.06	−1.06	−1.17	−1.39
WA6921	−1.20	−0.94	−1.33	−1.10

[+] There are no significant differences.

PEG-8000 observed here may measure true drought stress tolerance, rather than drought escape.

In all parameters measured, differences between PEG treatment and controls were greater than differences between genotypes. This finding was consistent with observations by Fischer and Sanchez[5]. The PEG method apparently measured differences that were unrelated to osmotic adjustment. However, differences in osmotic adjustment apparently are not closely correlated with drought stress tolerance[10].

Determining the growth pattern of genotypes in an osmotic solution by fresh weight is a fast and simple method to determine genotype reaction.

We believe the method has potential for wide application for screening plants for vegetative period tolerance to drought stress. The method may not evaluate genotype potential for drought stress tolerance during germination or grain filling periods. Drought stress tolerance at germination or grain filling probably is influenced by separate genetic systems and must be evaluated by procedures tailored to produce drought stress during these growth periods.

A PEG method likely could be developed for screening genotypes for germination at low water potential. However, a test such as that developed by Blum *et al.*[2] to measure the ability of plants to translocate metabolites into grain will likely be needed to screen for tolerance to grain fill period drought stress.

Acknowledgements Scientific Paper No. *7250*. College of Agriculture Research Center, Project 1568. Supported in part of USDA-S&E STEEP Grant No. 83–CRSR–2–2313.

References

1 Blum A, Sinmena B and Ziv 0 1980 An evaluation of seed and seedling drought tolerance screening tests in wheat. Euphytica 29, 727–736.
2 Blum A, Poiarkova H, Golan G and Mayer J 1983 Chemical desiccation of wheat plants as a simulator of post-anthesis stress. I. Effects on translocation and kernel growth. Field Crops Res. 6, 51–59.
3 Clark J M and McCraig T N 1982 Evaluation of techniques for screening for drought resistance in wheat. Crop Sci. 22, 503–506.
4 Fischer R A 1981 Optimizing the use of water and nitrogen through breeding of crops. Plant and Soil 58, 249- 278.
5 Fischer R A and Sanchez M 1979 Drought resistance in spring wheat cultivars. II. Effect on plant water relations. Aust. J. Agric. Res. 30, 801–804.
6 Fischer R A and Wood J T 1979 Drought resistance in spring wheat cultivars. III. Yield associations with morpho-physiological traits. Aust. J. Agric. Res. 30, 1000–1020.
7 Hurd E A 1974 Phenotype and drought tolerance in wheat. Agr. Met. 14, 39–55.
8 Johnson D A and Asay K H 1978 A technique for assessing seedling emergence under drought stress. Crop Sci. 18, 520–522.
9 Kilen T C and Andrew R H 1969 Measurement of drought resistance in corn. Agron. J. 61, 669–672.

10 Morgan J M 1980 Osmotic adjustment in the spikelets and leaves of wheat. J. Exp. Bot. 31, 655–665.
11 O'Toole J C, Aquino R S and Alluri K 1978 Seedling stage drought response in rice. Agron. J. 70, 1101–1103.
12 Singh P N, Prasad R, Salim M and Sharg A 1984 An improved system for subjecting plants to water stress. Biol. Plant. 26, 16–21.

Identifying aluminum tolerance in sorghum genotypes grown on tropical acid soils

LYNN M. GOURLEY
Department of Agronomy, Mississippi State, MS 39762, USA

Key words   Acid soil   Aluminum   Field screening   *Sorghum bicolor*   Ultisol

Summary   A field screening procedure to evaluate a large portion of the sorghum [*Sorghum bicolor* (L.) Moench] world collection for tolerance to Al-toxic, low base status acid soils was developed that is now being employed in tropical South America. An Ultisol at Quilichao, Colombia, South America, with 63% Al-saturation, had Al-toxicity levels sufficient to kill sensitive sorghum genotypes, but not too high to prevent tolerant genotypes from producing reasonable yields of grain. Using soil classification maps of Africa to trace locations where particular genotypes were originally collected, genotypes were systematically selected for testing from areas with highly acid soils. The greatest genetic diversity in sorghum for Al tolerance might be expected to be found in the tropical areas with high percentages of acid soils. Higher percentages of sorghum lines in the tolerant and moderately tolerant category were selected from the acid soil areas in Kenya and Uganda than from the other African countries.

Grain yield trials of the more agronomically desirable Al-tolerant genotypes grown at three Al-saturation levels appeared to have acceptable qualities necessary for commercial low-input production strategies.

Introduction

The vast underutilized areas of acid soils in the tropics are one of the last frontiers for agriculture land expansion. Traditional agricultural practices have not been applied successfully in these areas because high-input technology has been impractical for resource-poor farmers. National research agencies in many countries are searching for low-cost production technology to help alleviate their food- and feed-grain deficits.

The greatest potential for expanding agricultural production in some of the potentially arable lands of the world lies in the tropical rain forest and savanna regions dominated by acid soils classified as Oxisols and Ultisols[5,6,14]. Approximately 43% of the land area in the tropics consists of the soil orders Oxisols and Ultisols with 55% of these in tropical America and 39% in tropical Africa[14]. These acid and leached tropical soils are characterized by low inherent fertility, exemplified by low pH, low cation exchange capacity, high levels of soluble Al and sometimes Mn, and high P fixation. A low or moderate-input technology, compared to traditional high-input technology, must rely on the use of Al-tolerant sorghum lines and a moderate input of lime, P, and other fertilizer to alleviate some of the serious soil constraints.

H.W. Gabelman and B.C. Loughman (Eds.), Genetic aspects of plant mineral nutrition.
ISBN-13: 978-94-010-8102-3
© *1987 Martinus Nijhoff Publishers, Dordrecht/Boston/Lancaster.*

The introduction of sorghum ecotypes from Africa, the origin of sorghum, could possibly provide Al-tolerant germplasm for use in Tropical America and other areas[6]. Successful evaluation of sorghum for tolerance to toxic effects of Al were conducted in field tests on acid Cerrado soils in Brazil[11,12,15,16]. Scientists from the Brazilian National Sorghum Program have screened sorghum germplasm from Uganda and converted exotic lines from the Texas A & M University/USDA Sorghum Conversion program. Less than 1% of the sorghum world collection[1] has been screened for Al toxicity tolerance in Latin America because of local quarantine restrictions. Several studies using nutrient culture[3,4,7,10] and greenhouse soil[8,15] techniques to screen sorghum for tolerance to Al toxicity have been reported. Sorghum grown in a greenhouse in acid soil with 64% Al saturation showed concordant responses to Al toxicity compared to field grown plants. However, coefficients of determination were only 52% for shoot dry weight × grain yield and 41% for root dry weight × grain yield. Colombian field validation studies at 63% Al saturation for sorghum[9] and 85% Al saturation for rice[5] showed that most genotypes rated as Al tolerant by nutrient culture techniques would be rated as susceptible under field conditions.

The author spent two years at the International Center for Tropical Agriculture (CIAT), Cali, Colombia, initiating a program to screen sorghum lines for adaptation to the 'acid soil complex' of the tropics. The purpose of this paper was to report methods developed for field screening and some of the genetic variability in the sorghum world collection for tolerance to acid soils in Tropical America.

Materials and methods

Site selection and field preparation

The primary screening site was an area on the CIAT-Quilichao substation (80 km southeast of Cali) which had an Ultisol classified as a clayey, oxidic, isohyperthermic, Typic Palehumult. Only Al-tolerant genotypes and advanced breeding lines were evaluated on a secondary site at the CIAT/ICA-Carimagua substation (eastern Colombia on the Llanos) a savanna Oxisol classified as a clayey, kaolinitic, isohyperthermic, Tropeptic Haplustox. The site at the CIAT-Quilichao substation was best suited for testing large numbers of genotypes. A virgin soil at this site had a topsoil (0–20cm) Al-saturation of 80%, organic matter concentration of about 7%, and a Mn concentration of less than $20 \mu g\ g^{-1}$. Other chemical characteristics of both Colombian tropical soils are shown in Table 1.

In 1983, a 1.5 ha plot in permanent pasture at CIAT-Quilichao was divided into 20 grid squares and composite soil samples of each grid were taken at 0–20 and 20–40 cm depths for analysis (at the CIAT soil testing laboratory). The analyses are shown in Table 2.

The test field was divided into two blocks of 0.6 ha each (Blocks I and II), and one of 0.3 ha (Block III). Each block was separated by a 2 m alley. Each block received broadcast applications equivalent to 500 kg $CaCO_3$ as dolomitic limestone (12% Mg), 100 kg N as urea, 131 kg P as concentrated super phosphate, 83 kg K as KCl, 4.0 kg Zn as $ZnSO_4$, and 1.0 kg B as borax ha^{-1}. Additional applications of calcitic limestone equivalent to 1000 kg $CaCO_3$ ha^{-1} on Block II and 3500 kg $CaCO_3$ ha^{-1} on Block III were broadcast. After application, each plot was tilled several times during the course of the next several weeks to thoroughly mix the amendments into the upper 20 cm of soil and to kill remnant pasture grasses. Periodic rains promoted equilibration of the amendments.

Germplasm selection

From about 3000 sorghum lines in the portion of the world collection maintained at Purdue University (Dr. John Axtell), 1000 were chosen initially for seed increase. Using soil classification maps of Africa to note the location where particular lines were originally collected, lines were systematically selected which originated from acid soil areas. Seed of these first 1000 lines was increased at CIAT during the winter of 1982–83. The plants were grown on soils under relatively optimum growing conditions. Many exotic lines must be increased for seed in the tropics because about 60% of the world collection is photoperiod sensitive. Self-pollinated seed of 730 pure lines that produced sufficient quantities for testing were used in this study (Table 3).

Screening and yield trials

Single row plots (3 m long) of the selected entries were planted in each of the Blocks I, II, and III, in October 1983. In addition, field validation trials of sorghum genotypes reported to be Al tolerant in nutrient culture studies were planted in a randomized complete block design of three replications in each of the blocks. Visual ratings of plants (1 = most Al tolerant to 4 = least tolerant) were made for entries in Block I. Visual differentiation for Al tolerance among genotypes in Blocks II and III was not possible. Visual ratings were described as 1 = good plant color, well filled panicles, little stress or Al-toxicity symptoms; 2 = some yellowing of leaves, reduced panicle size, some stress and Al-toxicity symptoms; 3 = stunted plants, yellowing and dead leaves, small panicles with little grain, many stress symptoms; and 4 = severely stunted or dead plants 2 to 3 weeks after emergence.

Since two plantings per year were possible at this location (preferably planted in March and September) a second screening trial of additional world collection lines and the Al-tolerant entries from the first screening trial were planted in Blocks I and II in April 1984. Replicated yield trials of the 22 most agronomically desirable Al-tolerant lines from the first screening were also planted in each of the treatment blocks. One meter sections from each replication of the single row yield trial entries were harvested for grain yield.

Results and discussion

Both Ultisols and Oxisols in Colombia have high levels of soluble Al and low effective cation exchange capacity which yield Al-saturation levels of 90% or above (Table 1). Commercially available grain sorghum cultivars presently used have been susceptible to the toxic effects of Al at saturated levels above about 40%. During periods of drought these cultivars have been unable to obtain water in the unlimed subsoil because of their poor rooting characteristics. The 70 million ha of savanna Oxisols with less than 8% slope in Colombia comprise the main target area for increasing grain sorghum production.

Field preparation

Applications of 500, 1500, and 4000 kg ha^{-1} CaCO$_3$ reduced the 80% Al-saturation level of the virgin soil to 63, 45, and 32% Al saturation, respectively, in the three blocks (Table 2). Dolomitic limestone was added to each block to supply fertilizer quantities of Ca and Mg. Calcitic limestone was used to supply CaCO$_3$ at the two higher lime rates to

Table 1. Characteristics of two tropical Colombian soils used to screen sorghum for Al tolerance[+]

Soil	Depth (cm)	Clay (%)	O.M. (%)	P (μg g^{-1})	pH	Exchangeable cations (c mol kg^{-1})				Effective CEC	Al sat. (%)
						Al	Ca	Mg	K		
Ultisol	(CIAT-Quilichao)										
	0–20	71	7.1	1.8	4.1	2.7	0.65	0.49	0.36	4.21	64
	20–35	77	4.0	1.1	4.0	2.7	0.31	0.04	0.13	3.25	83
	35–62	84	1.9	0.9	4.3	3.2	0.24	0.02	0.09	3.65	88
	62–91	88	0.7	0.9	4.4	1.1	0.15	0.02	0.06	1.43	77
	91–105	89	1.5	1.2	4.4	2.0	0.22	0.01	0.04	2.34	85
Oxisol	(CIAT/ICA-Carimagua)										
	0–12	38	4.0	1.0	4.5	3.8	0.20	0.20	0.10	4.40	86
	12–32	41	2.0	1.0	4.6	2.8	0.10	0.10	0.10	3.10	89
	32–58	43	1.7	Trace	4.8	2.1	0.10	0.10	0.10	2.30	91
	58–88	45	0.9	Trace	5.2	0.7	0.10	0.10	0.10	0.90	78
	88–148	45	0.6	Trace	5.1	0.6	0.10	0.10	0.10	0.80	75

[+] Data taken from CIAT information. Methods for extraction and/or determination were: P — Bray II; pH — 1:1 soil:water; exchangeable cations — 1N KCl from 100 g of soil; effective cation exchange capacity (ECEC) — sum of exchangeable cations; and Al saturation — exchangeable Al divided by ECEC times 100.

eliminate Mg as an additional variable. Soluble phosphate fertilizer was applied at the relatively high rate of 131 kg P ha^{-1} to allow for the measure of Al tolerance and not P deficiency. Rock phosphate fertilizer was not used because its solubility normally increases with decreasing soil pH. Zinc and B were also added to insure they would not be limiting.

One of the objectives of this study was to establish Al-toxicity levels sufficiently high to kill sensitive genotypes, but not too high to prevent

Table 2. Mean topsoil (0 to 20 cm) chemical characteristics of the CIAT-Quilichao soil used to screen sorghum for Al tolerance (chemical properties were before and after amendment with lime and fertilizer)

Soil characteristics	Virgin	CaCO$_3$ added (kg ha^{-1})[+]		
		500 Block I	1500 Block II	4000 Block III
pH (H$_2$O)	4.5	4.4	4.6	5.0
P (μg g^{-1})	2.3	17.9	16.2	17.8
Ca (c mol kg^{-1})	0.68	1.24	2.44	3.33
Mg (c mol kg^{-1})	0.18	0.52	0.53	0.51
K (c mol kg^{-1})	0.15	0.24	0.26	0.23
Al (c mol kg^{-1})	3.90	3.40	2.65	1.90
Effective CEC (c mol kg^{-1})	4.91	5.40	5.88	5.97
Al saturation (%)	80.4	63.0	45.1	31.8

[+] The first 500 kg ha^{-1} CaCO$_3$ was dolomitic limestone. The remainder of the CaCO$_3$ was calcitic limestone. Methods for extraction and/or determination were: pH — 1:1 soil:water; P — Bray II; exchangeable cations — 1N KCl from 100 g of soil; effective cation exchange capacity (ECEC) — sum of exchangeable cations; and Al saturation — exchangeable Al divided by ECEC times 100.

tolerant genotypes from producing reasonable grain yields. The high level of Al saturation selected (63%) was able to do this. This high level of Al saturation in the initial screening described most of the sorghum lines as less than tolerant to Al, but identified a low percentage of lines with good tolerance to the acid soil complex. To have economical potential, genotypes with the highest level of Al tolerance would allow resource-poor farmers to apply minimum quantities of lime for commercial production. Since only the topsoil is usually modified with soil amendments, the roots of genotypes with high levels of Al tolerance would be more likely to penetrate the higher Al-saturation levels in the subsoil.

Germplasm selection

The International Crops Research Institute for the Semi-Arid Tropics (ICRISAT) has a worldwide mandate for sorghum improvement research, which includes germplasm collection and conservation. A world collection of some 22 000 sorghum lines is maintained by ICRISAT at Hyderabad, India. More than 15 000 of these lines are in long-term storage at the U.S. National Seed Storage Laboratory at Fort Collins, Colorado. An additional 9000 lines are maintained in the U.S. by the USDA and other research agencies[1]. Purdue University has about one half of the world collection in medium-term storage. Evaluation of world collections ranks as the weakest link in the germplasm management system[2].

The first step in utilizing this exotic germplasm was nonrandom selection of those lines that might have a higher probability of being Al tolerant and that were obtainable in the Western Hemisphere, thus avoiding quarantine delays.

Field screening observations

A visual rating scale was designed to evaluate sorghum genotypes for Al tolerance at physiological maturity. Visual rating scales with more categories (1 to 9) proved too complicated for the wide range of maturities and plant heights in the material evaluated. Only minor Al-toxicity symptoms of the most Al-sensitive genotypes were observed at the 1500 $CaCO_3$ rate (Block II). Continued discussion of ratings and observations is limited to the 500 $CaCO_3$ rate (Block I). Initial attempts to rate the genotypes at four-week intervals were impractical. All of the genotypes germinated and had good plant stands at the 63% Al-saturation level. However, most Al-susceptible genotypes died within two to three weeks. The other genotypes differed in their responses to Al-toxicity stress at the different stages of plant growth. Possible reasons for this change in rating is discussed later.

Table 3. Aluminum-tolerance ratings for 730 world collection lines of sorghum by country of origin

Country	Number of lines	Al-tolerance rating[+] (%)			
		1	2	3	4
Ethiopia	158	13	21	37	29
Kenya	116	15	37	30	18
Nigeria	161	6	25	38	31
Tanganyika	14	14	36	29	21
Uganda	104	14	48	22	16
Upper Volta	82	4	36	32	28
Zaire	16	12	37	31	19
Misc.	224	10	23	35	32
Total	730	11	31	32	26

[+] 1 = Al tolerant, 4 = Al susceptible; see text for description of visual ratings.

Table 3 shows the ratings of the first 730 world collection genotypes evaluated by country of origin. Acid soils areas in Kenya and Uganda had higher percentages of entries in categories 1 and 2 (more Al tolerant) than those evaluated from the other countries listed. Conclusions concerning the country(s) in Africa yielding the best sources of Al tolerance will not be discussed until the 3000 genotypes originally selected have been tested.

Field evaluations for a stress response are not controlled as easily by the researcher as those conducted in the laboratory. Therefore, the exact cause-and-effect relationships in the field must, in many instances, be obtained by deduction or await further information. Other disadvantages of field tests include weather related variability, insect and disease damage, and the inability to easily assess the quantity and quality of roots affected by Al toxicity. Advantages of field evaluation over laboratory methods are that the agronomic information like grain yield, yield components, plant height, maturity, and resistance to foliar diseases and insects can be collected. Field evaluations of sorghums to the 'tropical acid-soil complex' could be affected by many factors. Several of these factors are subsequently discussed.

Aluminum. Tolerance to Al is usually the foremost factor causing problems to plants grown in the acid soils of Tropical America. Without adequately high levels of Al in the test media, susceptible genotypes with no practical degree of Al tolerance could be selected and the measure of other genetic or agronomic factors would be negated. The degree of Al saturation in the upper 20 cm of the topsoil can be altered to the desired level with relatively good accuracy. This level of Al saturation usually remains quite stable for a year or more after the first one or two crops have been grown.

Phosphorus. Although not intended as a variable in the screening procedure, the association of P availability and Al saturation must be considered. Some genotypes screened showed typical purple-leaf symptom of P deficiency for four weeks after planting even though the high rate of P (131 kg ha^{-1}) had been applied when the test plot was prepared. Tropical American soils readily fix added phosphates, thus, added soluble phosphates are only partially recovered by plants under optimum conditions. The high amounts of hydrous oxides (Al and Fe) of many tropical soils can result in enormous phosphate-fixing capacities.

Calcium and magnesium. Calcium and magnesium in liming amendments can reduce the level of Al saturation when applied to tropical acid soils. In these leached, low base status soils, Ca and Mg in fertilizer quantities are important to plant nutrition. Calcium is essential to root elongation and is not translocated basipetally in roots, thus soil Ca must be located at the root site. The quantity of Ca in some tropical subsoils is insufficient to assure appropriate root growth. Calcium ions of some compounds, such as $CaSO_4$, $CaCl_2$, and $Ca(NO_4)_2$, will move downward in the soil profile (up to 100 cm in a year) in association with its anion[13]. This is not the case with $CaCO_3$. Critical Ca concentrations for sorghum root growth in these soils is not known. However, without sufficient Al tolerance, roots will not grow anyway.

Even though Mg is essential for root development, it is translocated basipetally. Therefore, Mg does not necessarily have to be present at all soil sites where root development takes place. Lower amounts of Mg in soils are therefore needed compared to Ca.

Root mass. The relative rate of root and shoot growth of a particular sorghum genotype can affect its visual Al-tolerance field rating. Some genotypes produce rapid shoot growth in the juvenile development stage while others appear stunted and stressed. Many of the lines that produced good early shoot growth died before or during the grain-filling period, probably because more vegetative growth was produced than could be supported by the root system. Some slow growing lines recovered and produced grain yields nearly equal to their genetic potential on a soil with a lower Al-saturation level. This phenomenon appeared to be because of a priority for partitioning of photosynthate in different tissue during various stages of growth in the different genotypes. Those genotypes tolerant to high levels of Al apparently partitioned a greater portion of their photosynthetic resources in developing roots. Since grain production is the ultimate goal under the prescribed conditions, visual field rating for Al tolerance was of little value before the genotypes reached physiological maturity.

Drought. Drought can be a production constraint on acid soils in the tropics, even during favorable rainfall periods. Despite clay contents of 40 to 70%, Oxisols and Ultisols in the tropics respond more like sandy loams of the temperate zones relative to water infiltration. Water holding capacity in the upper soil profile of the tropical soils is low. During short periods without rain, plants with only sufficient Al tolerance to develop roots in the topsoil frequently fail because of drought. Therefore, it is important to have sufficiently high Al-saturation levels in the topsoil for adequate Al-tolerance screening and to use irrigation only to 'rescue' or prevent the loss of the screening trials. When plants are in full foliage, wilting during the relative short periods of drought provided an additional measure of the degree of Al tolerance. Those genotypes not wilting under full radiation were obtaining sufficient water from the subsoil. Genotypes that wilted under these conditions did not have sufficient Al tolerance for roots to penetrate the higher Al-saturated subsoil.

Table 4. Effect of Al saturation on sorghum grain yields (t ha^{-1})

Cultivar	Visual[‡] rating 63%	Al saturation level[+]		
		63%	45%	32%
IS 8577	1.3	5.0	6.2	5.7
IS 7151	1.7	4.8	5.7	5.2
M-91057-117	2.0	4.6	4.4	4.9
3DX57/1/1/910	1.3	4.5	4.6	5.1
(IS 12573CXSC108-3)-7-3-5-1-1-1	1.7	4.4	4.5	5.5
(GPR 168 XCS-170-6-17)-1-1	2.0	4.4	4.7	5.1
IS 7132	1.0	4.3	4.9	6.9
IS 8933	1.0	4.1	4.9	6.7
IS 2765	1.4	4.0	4.6	7.1
79 SEPON 54	2.0	3.8	3.7	5.0
IS 7173C	1.0	3.6	4.3	4.8
IS 8612	2.0	3.2	3.3	5.6
IS 8860	2.0	3.1	3.6	5.4
M-90378	3.0	2.9	3.1	3.2
(F3B554XF3B441)−2	2.3	2.8	3.4	3.9
(F3B554XF3B441)-1	2.0	2.5	2.8	3.9
79 SEPON 8	2.7	2.3	3.2	6.3
79 SEPON 11	3.0	2.0	3.2	3.6
TX 415	4.0	0.5	1.2	2.2
Mean	2.0	3.5	4.0	5.1
S.E. (mean)	0.2**	0.4**	0.4**	0.6**
C.V. (%)	18	19	19	20

[+] Aluminum saturation levels 63, 45 and 32% were produced by 0.5, 1.5, and 4.0 t ha^{-1} CaCO$_3$ rates, respectively.
[‡] 1 = Al tolerant, 4 = Al susceptible for plants grown with 63% saturation. Visual ratings for cultivars in the 45 and 32% Al-saturation trials were each 1.0.
** Significant at $P \leq 0.01$.

Interactions. The total of all factors and their interactions should affect the visual rating for Al tolerance. Consistently high ratings for Al tolerance under the conditions prescribed in the screening process should identify the upper range of genetic variability for the overall 'Al-tolerance complex'.

Under the 63% Al-saturation conditions of the screening trial, about 10% of the genotypes exhibited high degrees of tolerance to Al toxicity. Most of the genotypes, however, were agronomically unacceptable as grain sorghum varieties in their present form.

Yield trials

A few of the more agronomically desirable Al-tolerant genotypes are now being evaluated for grain yield. Preliminary results of grain yields at the three levels of Al saturation at the CIAT-Quilichao substation are given in Table 4. After further testing, the release of some of these genotypes for direct commercial use may be recommended. Parental material for breeding programs has been identified.

Through cooperation and collaboration with international agricultural research centers, national programs, universities, and commercial seed companies, this screening method has identified and the research program has made available Al-tolerant exotic sorghum germplasm to scientists conducting research on the tropical acid soil frontiers in developing countries.

Acknowledgement This is a report of research conducted by the author while assigned at CIAT, Cali, Colombia. The author wishes to thank USAID for funding through the International Sorghum/Millet (INTSORMIL) Collaborative Research Support Program and CIAT for the use of facilities. Published as journal contribution No. 6145 the Mississippi Agricultural and Forestry Experiment Station, Mississippi State, Mississippi 39762.

References

1. Acheampong E, Anishetty N M and Williams J T 1984 A world survey of sorghum and millets germplasm. AGPG:IBPGR/83/5 . International Board for Plant Genetic Resources (IBPGR) Secretariat, Rome, 45 p.
2. Anonymous 1984 Conservation and utilization of exotic germplasm to improve varieties. *In* The 1983 plant breeding research forum (Executive Summary). Pioneer Hi-Bred International, Inc., Des Moines, Iowa, 63 p.
3. Bastos C R 1981 Toxicity of aluminum in nutrient solution to sorghum seedlings. M.S. Thesis, Mississippi State University, Mississippi State, MS, 57 p.
4. Bastos C R 1982 Inheritance study of aluminum tolerance in sorghum in nutrient culture. Ph.D. Thesis, Mississippi State University, Mississippi State, MS, 80 p.
5. CIAT (International Center for Tropical Agriculture) Annual Report 1984 Rice program. CIAT, Cali, Colombia, 59 p.
6. Doggett H 1970 Sorghum. Longsman, Green and Co., London, 403 p.
7. Furlani P R and Clark R B 1981 Screening sorghum for aluminum tolerance in nutrient solutions. Agron. J. 73, 587–594.
8. Gourley L M 1983 Breeding sorghum for tolerance to aluminum-toxic tropical soils. pp 299 313. *In* Proc. Plant Breeding Methods and Approaches in Sorghum Workshop for Latin

America. Eds. F R Miller, V Guiragossian and A Betancourt. International Center for Maize and Wheat Improvement (CIMMYT), El Batan, Mexico.
9 Gourley L M 1985 Finding and utilizing exotic aluminum-tolerant sorghum germplasm. *In* Evaluating Sorghum for Tolerance to Aluminum-Toxic Tropical Soils in Latin America. Eds. L M Gourley and J G Salinas. International Center for Tropical Agriculture (CIAT), Cali, Colombia. (*In press*).
10 Malavolta E, Nogueira F D, Oliveira I P, Nakayama L and Eimori I 1981 Aluminum tolerance in sorghum and bean-methods and results. J. Plant Nutr. 3, 687–694.
11 Pitta G V E, Schaffert R E, Borgonovi R A, Vasconcellos C A, Bahia Jr. A F C and Oliveira A C 1979 Evaluation of sorghum lines to high soil acidity conditions. p 128. *In* Proc. 12th Brazilian Corn Sorghum Res. Conf. Goiania, GO, Brazil (*In Portuguese*).
12 Pitta G V E, Trevisan W L, Schaffert R E, de Franca G E and Bahia Jr. A F C 1976 Evaluations of sorghum lines under high acidity conditions. pp 553–557. *In* Ann. XI Brazilian Maize Sorghum Rev. Piracicaba, SP, Brazil (*In Portuguese*).
13 Ritchey K D, Souza D M G, Lobanta E and Correa O 1980 Calcium leaching to increase rooting depth in a Brazilian savanna oxisol. Agron. J. 72, 40–44.
14 Sanchez P A and Salinas J G 1981 Low-input technology for managing oxisols and ultisols in tropical America. Adv. Agron. 34, 279–406.
15 Santos Jr. H L, Baligar V C and Vasconcellos C A 1980 Screening sorghum genotypes for aluminum tolerance. p 24. *In* Ann. XIII Brazilian Maize Sorghum Rev., Londrina, P, Brazil (*In Portuguese*).
16 Schaffert R E, McGrate A J, Trevisan W L, Bueno A, Meira J L and Rhykerd C L 1975 Genetic variation in *Sorgum bicolor* (L.) Moench for tolerance to high levels of exchangeable aluminum in acid soils of Brazil. pp 151–160. *In* Proc. Sorghum Workshop, Univ. of Puerto Rico, Mayaguez, Puerto Rico.

Influence of field sampling techniques on the Al, Mn, Mg, and Ca nutritional profiles for acid soil tolerant and susceptible sorghum genotypes

R.R. DUNCAN and J.D. SUTTON
Department of Agronomy, Georgia Experiment Station, Griffin, GA 30212-5099, USA

Key words Acid soil stress Growth stages *Sorghum bicolor* (L.) Moench

Summary Several sampling techniques (variable number of plants per sample, specific leaf sampling, sequential growth stage sampling, and specific plant part sampling) were investigated for sorghum [*Sorghum bicolor* (L.) Moench] under acid soil pH levels > 6.0 and < 4.8. Plant Al, Mn, Mg, and Ca concentrations were determined for samples of three genotypes which varied in their reaction to acid soil stress conditions. Five plants provided a sufficient sample for sorghum grown under variable acid soil stress conditions, although a low quantity of plant material of the most susceptible genotype was a problem. Nutrient concentrations were generally highest for GS_1 (emergence to panicle initiation) samples and decreased with subsequent growth stages, regardless of soil pH. Sampling leaves 2, 3 or 4 (from the top of the plant) during GS_2 (panicle initiation to anthesis) and GS_3 (anthesis to physiological grain maturity) provided representative concentrations for the four nutrients. In comparison with other plant parts for whole plants, leaf samples provided good representative plant samples for nutrient analyses, regardless of soil pH. Differential Ca uptake by genotypes may be a key component in the overall acid soil tolerance mechanism.

Introduction

The amount of a nutrient that a plant may need for growth and reproduction varies among plant species and/or varieties[8]. The technique by which plant samples are collected has a major effect on the results obtained from various analyses[23]. Due to the complex and heterogenetic nature of most plants and their parts, several factors such as the nutrients concerned, their concentration and distribution in the particular plant tissue, and the precision of analytical results are essential in interpreting research data. These factors can affect sampling frequency and technique as well as quantity and quality of samples needed for analyses. The specific tissue to be sampled as well as the time of sampling are related to the concentration of the nutrients and to the physiological age and yield of the plant[3].

Great care must be taken when selecting plant parts to be sampled due to variations in the nutrient concentrations of different plant parts from the same plant[18,40]. For generalized sampling, leaves were determined the best tissues to sample for analyses[36,46]. Sampling of leaves for such crops as corn (*Zea mays* L.), soybean (*Glycine max* L. Merr.), sorghum, tobacco (*Nicotiana tabacum* L.), and cotton (*Gossypium hirsutum* L.) has been suggested[26]. Leaf sampling has also been recommended for crops such as apples (*Malus domestica* Borkh.), strawberries (Fragaria × ananassa Duch.)[5] and oranges (*Citrus sinensis* L. Osb.)[37].

In some cases, tissue samples other than leaves can be used and may be superior[14]. For sugarcane (*Saccharum* spp.), the best tissue to analyze for K, Ca, Mg and P was the sheath tissue of leaves 3, 4, 5 and 6 counting from the top of the plant downward[10]. For N, the chlorenchyma of the same leaves was considered the best sample for analysis. For determination of K in alfalfa (*Medicago sativa* L.), the middle stem of the plant was the best for sampling[30]. Sampling whole corn plants was determined to be better than sampling leaves[33]. Several researchers have recommended the sampling of petioles of such crops as celery (*Apium graveolens* L.), grapes (*Vitis vinifera* L.)[26], cotton[22], sugar beet (*Beta vulgaris* L.)[43] and lima bean (*Phaseolus limensis* Macf.)[44].

The physiological age of the plant tissue and the amount of available nutrients in the soil are the most important factors affecting nutrient concentration in plants[36]. Although the nutrient content of plants varies with age[3], the amount of variation depends on the plant species[17] and the particular nutrient concerned[19]. When expressed on a dry weight basis, most plant nutrient concentrations tend to be higher during early vegetative growth and decrease as the plant matures[42].

A significant change in the nutrient concentration of the vegetative portions of plants has been reported when fruit or seed development begins[25]. Consequently, vegetative samples (leaves, *etc.*) should not be taken after pollination for many grain crops such as barley (*Hordeum vulgare* L.), corn, oats (*Avena sativa* L.), rice (*Oryza sativa* L.), rye (*Secale cereale* L.), sorghum and wheat (*Triticum aestivum* L. em. Thell.). Usually, samples should be taken during or prior to the reproductive stage, but other stages of growth (depending on the crop) may also be beneficial for sampling.

Plant tissue considered to be the same physiological age has shown variation in nutrient concentration as the growing season progresses[6,36,45]. If the nutrient concentrations in plant samples are to be interpreted correctly, the growth stage in which the samples were taken must be defined[3]. The sampling of the last fully expanded leaf would probably provide tissue that is equivalent to the same physiological age as any other tissue[3].

The amount of plant material needed for analysis depends on several factors. Plant type, size of land area being studied[42], physical condition of plants, diversity of soil and purpose for sampling should be considered before determining the number of plants to be analyzed[25]. The proper sample size should include enough tissue for analyses as well as provide a good representation of the plants in the area[42]. The best way to achieve

Table 1. Various sampling techniques used for nutrient analyses of grain sorghum

Plant part	Nutrients monitored	Number of plants sampled	Time of sampling	Reference
Leaves, stems, peduncle, head	NPK	10	Weekly starting 23 days after planting	28
4th leaf from top	Mg	NA*	Late pollination	16
Blades, sheaths, stems, heads	Mg, Ca	5	Weekly starting when collar of 3rd leaf visible	20
Leaf 1, 2, 3, 4, 5; roots; upper, middle and lower stem	Mg, Zn, Cu, Fe	5	Boot stage	32
Top 2 leaves, bottom leaves	P, K, Ca, Mg, Mn, Fe, Cu, Al, Mo	12	18 days after planting	7
3rd leaf from top	Al, Fe, Mn, Cu, Zn, Mg, K, Ca, P	10	Physiological grain maturity	12
Grain	N, P	NA	Grain maturity	31
Upper 16 leaves, lower leaves	K, Ca, Mg, Mn, Cu, Zn, P, Fe, B, Al, S, Mo, Cr, Co, Ni, Pb, Se, Hg, Sr	5	24 days after germination	9
2nd leaf from top	P, Ca, Mg, K, Cu, Mn, Fe, Zn, Al	10	Physiological grain maturity	11
Shoots, roots	Fe, Mn, P	NA	4 weeks after germination	27
Dead leaves, live leaves	P	5	40, 60, 80 and 100 days after planting	35

* Not available

a proper sample size may be collection of as many plants as practical[38].

The number of plants per sample has been suggested for several crops such as corn, sorghum, soybeans, alfalfa, tobacco, cotton, tomato (*Lycopersicon esculentum* Mill.), strawberry, pecan (*Carya illinoensis* Wang K. Koch), and grapes[26]. A recommended system for sampling sorghum for nutrient concentration has been reported[24]: the sampling of the above-ground portion of 20 to 30 whole plants during the seedling stage of growth, the fully developed leaf below the whorl of 12 to 25 plants prior to the heading stage and the entire second leaf from the top of 15 to 25 plants at heading. Whole-plant seedlings of sorghum collected 25 to 36 days after planting were found to be the most valuable for diagnosing nutrient deficiencies[29].

Several factors may cause a reduction in the availability of a nutrient. One such factor is soil acidity[1,4]. Acid soils can also be responsible for poor root development and the toxic or deficient reaction to some elements[15,21]. The extent of the detrimental effects of acid soils on plant response varies depending on soil types and plant species or varieties[2,15].

Table 2. Type, pH and concentration of Al, Mn, MG and Ca in soils on which sorghum was grown, 1982 to 1983 at Blairsville, Calhoun and Pike County, Georgia USA

Location	Year	Soil type	H$_2$O pH	Mean soil analysis – postemergence + postharvest samples (μg/g)			
				Al	Mn	Mg	Ca
Blairsville	1982	Congaree sandy loam	6.2	0.3	41	84	1609
		(Typic Udifluvent)	4.6	7.6	38	24	279
Blairsville	1983	Congaree sandy loam	6.5	0.1	36	111	2267
		(Typic Udifluvent)	4.8	2.9	36	22	264
Calhoun	1982	CedarBluff silt loam	6.1	0.3	69	106	821
		(Fragiaquic Paludults)	4.3	9.6	97	16	172
Calhoun	1983	CedarBluff silt loam	6.5	0.1	47	114	869
		(Fragiaquic Paludults)	4.4	7.6	57	17	193
Pike County	1982	Cecil sandy loam	6.3	0.5	50	71	1241
		(Typic Hapludults)	4.8	2.2	50	13.6	236
Pike County	1983	Cecil sandy loam	6.5	0.3	49	81.9	1453
		(Typic Hapludults)	4.5	5.0	63	14.6	225

Several sampling techniques which have been used for grain sorghum are summarized in Table 1. Sampling techniques ranging from whole plants to individual plant parts have been used for sorghum.

Materials and methods

Three grain sorghum lines each having different acid soil stress tolerances [SC283 — acid soil tolerant, SC599 — moderately acid soil tolerant, and NB9040 — acid soil susceptible][13] were planted in a randomized complete block design with 4 replications under 2 different soil treatments [pH > 6.0 (high pH) and pH < 4.8 (low pH)]. Specific soil types as well as soil pH values and extractable nutrient concentrations are presented in Table 2. Locations and specific agronomic data have been described elsewhere[39]. Soil analyses were conducted using the Mehlich I method[34].

Plots were sampled during various growth stages[13]. The first sampling occurred at the end of GS$_1$ (planting to panicle initiation) stage of growth, approximately 35 days after emergence of the seedling and coinciding with panicle initiation. The second sampling period took place at the end of the GS$_2$ (panicle initiation to bloom) stage of growth, approximately 70 to 80 days after planting and immediately prior to or during anthesis. The third and final sampling occurred 110 to 115 days after planting at the end of the GS$_3$ (bloom to physiological maturity) stage of growth.

After plant samples had been collected, they were dried at 70°C in a forced-draft oven for at least one week. All samples were ground to pass a 40-mesh stainless steel screen using a Wiley mill. Each sample was wet-ashed with a HNO$_3$/HClO$_4$ (5:2 v/v) mixture and analyzed for Mn, Mg and Ca by atomic absorption spectrophotometry and for Al by a Catechol Violet procedure[41].

Samples for analysis were collected as follows:

Plant number sampling

During each of the 3 growth stage sampling periods, 36 random whole plants (excluding roots) were cut at approximately 3 cm from the soil surface. These plants were then divided into groups consisting of 1, 5, 10 and 20 whole plant samples.

Leaf sampling

During the GS$_1$ growth stage, 10 randomly selected plants from each plot were sampled for individual leaves by collecting the fifth leaf from the bottom and moving up the plant. During the

Table 3. Genotype Al, Mn, Mg, and Ca concentrations involving whole plant samples over all sampling periods[§]

Genotype	Al	Mn	Mg	Ca
	(µg/g)		(%)	
a. High soil pH[+]				
SC283	303	56	0.31	0.56
SC599	306	48	0.27	0.42
NB9040	269	49	0.29	0.37
LSD (0.05)	NS	3	0.01	0.02
b. Low soil pH[*]				
SC283	320	269	0.19	0.42
SC599	356	203	0.17	0.37
NB9040	350	169	0.18	0.35
LSD(0.05)	NS	20	0.01	0.10

[+] $N = 620$ [*] $N = 416$ [§] (on dry weight basis)

GS_2 and the GS_3 growth stages, similar samples were taken except that the flag leaf was used as leaf-1 and the other leaves were collected moving down the plant. During each of the sampling periods, 10 additional plants were randomly selected from each plot and all of their leaves were collected for composite (bulk) leaf samples.

Growth stage sampling

During the GS_1, GS_2 and GS_3 stages of growth, 10 randomly selected whole plant samples (excluding roots) were cut at approximately 3 cm from the soil surface.

Plant part sampling

During the GS_2 and GS_3 stages of growth, 20 randomly selected whole plants (excluding roots) were cut at approximately 3 cm from the soil surface. From these 20 plants, 10 were composited for whole plant samples. The other 10 plants were separated into heads, leaves and stems which were then bulked into specific plant part samples.

Results and discussion

Plant number sampling

The nutrient concentrations of whole plant samples from specific genotypes grown at two acid soil pH levels revealed no significant differences among genotypes for Al (Table 3). All genotypes contained substantially higher Mn concentration in the samples from soils with low pH values. The highest Mn concentrations were found in SC283 samples, regardless of soil pH. The acid soil susceptible genotype NB9040 contained the lowest concentration of Mn in samples from the low soil pH plots. As expected, concentrations of Mg and Ca were lower in samples collected from low soil pH than from high soil pH plots. SC283 had the highest concentrations for both elements, regardless of acid soil pH stress level. Differences among genotypes for Mg concentrations were quite small on the low soil pH plots. However, Ca concentrations were more distinctly separated and may be directly involved in the acid soil tolerance mechanism.

Table 4. Interaction of genotype by number of plants per sample at two soil pH levels

Genotype	No. plts	Al (μg/g)	Mn	Mg (%)	Ca
High soil pH					
SC283	1	316	55	0.29	0.53
	5	291	56	0.30	0.56
	10	337	58	0.30	0.54
	20	279	58	0.32	0.58
SC599	1	292	46	0.26	0.40
	5	320	48	0.27	0.42
	10	310	49	0.26	0.41
	20	305	50	0.29	0.45
NB9040	1	280	47	0.28	0.38
	5	260	50	0.28	0.37
	10	274	48	0.27	0.36
	20	262	52	0.30	0.38
LSD (0.05)		NS	7	0.04	0.07
Low soil pH					
SC283	1	324	248	0.18	0.41
	5	356	284	0.19	0.43
	10	353	239	0.18	0.40
	20	333	290	0.20	0.43
SC599	1	325	209	0.17	0.37
	5	345	204	0.17	0.37
	10	415	188	0.16	0.35
	20	392	204	0.17	0.38
NB9040	1	337	152	0.17	0.31
	5	383	191	0.18	0.37
	10	465	161	0.18	0.34
	20	308	170	0.19	0.37
LSD (0.05)		NS	53	NS	0.06

When comparing the genotype by whole-plant sample number interactions, no differences were noted for Al concentrations in samples from either soil pH level and for Mg concentrations in samples from the low soil pH plots (Table 4). Within genotypes, the number of plants sampled provided equivalent concentrations of all nutrients, regardless of soil pH. Consequently, we concluded that 5 plants were sufficient for sampling sorghum under variable acid soil stress conditions. We had problems with the small quantity of sample material for subsequent analyses of the most susceptible genotypes, especially during the reproductive growth stage.

Leaf sampling

Comparisons of genotypes by growth stage for leaf nutrient concentrations revealed no signficant differences for Al and the values were extremely high which could have been due to soil contamination on the

Table 5. Specific genotype leaf-sample nutrient concentrations by growth stage

	Al	Mn	Mg	Ca
	(μg/g)		(%)	
GS$_1$[+]				
SC283	887	77	0.49	1.14
SC599	864	70	0.36	0.79
NB9040	758	80	0.44	0.79
LSD (0.05)	NS	5	0.03	0.05
GS$_2$[*]				
SC283	676	50	0.36	0.77
SC599	605	54	0.35	0.58
NB9040	340	55	0.37	0.51
LSD (0.05)	73	3	0.01	0.02
GS$_3$[§]				
SC283	862	193	0.39	0.85
SC599	518	72	0.38	0.70
NB9040	677	85	0.37	0.60
LSD (0.05)	100	5	0.02	0.02

[+] N = 450 [*] N = 553 [§] N = 429

unwashed juvenile plant leaves (Table 5). SC283 and NB9040 were higher in leaf Mn concentration than SC599 during GS$_1$. SC283 had the highest leaf Mg and Ca concentrations among the genotypes during juvenile growth.

Table 6. Specific leaf number by growth stage interaction for nutrient concentrations across soil pH levels and genotypes

Leaf number[+]	Growth stage											
	GS$_1$[†]				GS$_2$				GS$_3$			
	Al	Mn	Mg	Ca	Al	Mn	Mg	Ca	Al	Mn	Mg	Ca
	μg/g		%		μg/g		%		μg/g		%	
1	47	38	0.21	0.23	265	37	0.16	0.27	267	61	0.26	0.60
2	72	48	0.27	0.48	292	44	0.21	0.37	281	62	0.29	0.68
3	161	35	0.23	0.36	359	47	0.24	0.43	417	70	0.33	0.69
4	269	41	0.26	0.42	390	47	0.27	0.47	537	77	0.35	0.68
5	376	45	0.28	0.49	495	51	0.33	0.55	656	90	0.40	0.73
6	529	57	0.32	0.68	555	57	0.38	0.65	1319	116	0.49	0.81
7	685	74	0.42	0.95	774	62	0.46	0.80	1371	130	0.54	0.85
8	946	89	0.51	1.19	903	71	0.52	0.91	1904	147	0.56	0.90
9	1530	111	0.63	1.38	979	77	0.59	1.03	1062	138	0.65	1.03
10	1772	130	0.68	1.58	1106	90	0.61	1.11	925	136	0.73	1.05
BK[§]	716	62	0.33	0.67	328	49	0.31	0.57	450	77	0.37	0.70
LSD (0.05)	243	21	0.07	0.07	204	5	0.06	0.08	255	11	0.06	0.04

[+] Leaf 10 corresponded to actual leaf number 5 in chronological age. The bottom 5 leaves had senesced or died completely by the first sampling
[†] Counting from top downward
[§] Bulk sample of 10 leaves

Table 7. Interaction of growth stage by genotype on nutrient concentration of whole plant samples by soil pH level

Genotype	GS	Al	Mn	Mg	Ca
		µg/g		%	
High soil pH					
SC283	GS_1	481	62	0.41	0.80
	GS_2	158	45	0.28	0.47
	GS_3	272	62	0.24	0.39
SC599	GS_1	537	56	0.35	0.60
	GS_2	183	42	0.25	0.36
	GS_3	201	47	0.21	0.31
NB9040	GS_1	507	62	0.38	0.55
	GS_2	111	41	0.27	0.31
	GS_3	183	46	0.21	0.25
LSD (0.05)		116	5	0.03	0.04
Low soil pH					
SC283	GS_1	443	301	0.21	0.52
	GS_2	222	264	0.16	0.34
	GS_3	243	233	0.17	0.36
SC599	GS_1	457	253	0.19	0.45
	GS_2	321	185	0.17	0.32
	GS_3	231	140	0.15	0.29
NB9040	GS_1	452	194	0.19	0.41
	GS_2	310	166	0.18	0.31
	GS_3	189	118	0.16	0.27
LSD (0.05)		137	46	0.02	0.04

For leaf samples collected during GS_2, SC283 was significantly higher in Mn and Ca concentrations than the other two genotypes. NB9040 was substantially lower in Al concentration than either SC283 or SC599. Very small differences were detected for leaf Mg concentrations among the genotypes.

During GS_3, SC283 leaf nutrient concentrations were highest for Al, Mn, and Ca. NB9040 leaf samples had higher Al and Mn but lower Ca concentrations than SC599. Since the leaf samples collected during this late reproductive growth stage reflect movement of these nutrients into this plant part, very distinctive differences were noted among the genotypes. SC283, the most acid soil tolerant genotype, had high leaf values for the four nutrients. SC599. a moderately tolerant genotype, had the lowest leaf concentrations for Al and Mn, but was significantly higher in Ca concentration than NB9040, the most susceptible genotype. Consequently, because of the complexity of the acid soil tolerance mechanism, we concluded that the ability of genotypes to tolerate toxic levels of Al and Mn were important components in the tolerance mechanism. However, differential Ca uptake by specific genotypes may be a key component in the overall mechanism. The mechanisms by which these three genotypes withstand acid soil stress conditions may also be different.

Table 8. Interaction of genotype by plant part nutrient accumulation averaged over GS_2 and GS_3 sampling periods

Genotype	Plant part	Al (μg/g)	Mn	Mg (%)	Ca
High soil pH					
SC283	Lvs	459	71	0.36	0.77
	Stms	041	57	0.24	0.35
	Hds	107	44	0.19	0.14
	10-p[†]	194	55	0.25	0.43
SC599	Lvs	378	55	0.33	0.58
	Stms	34	40	0.20	0.27
	Hds	96	37	0.17	0.13
	10-p	179	45	0.22	0.33
NB9040	Lvs	317	62	0.34	0.52
	Stms	31	39	0.23	0.25
	Hds	55	28	0.17	0.07
	10-p	120	42	0.23	0.28
LSD (0.05)		79	8	0.03	0.04
Low soil pH					
SC283	Lvs	437	279	0.21	0.63
	Stms	43	282	0.15	0.29
	Hds	136	134	0.18	0.13
	10-p	240	205	0.17	0.35
SC599	Lvs	502	193	0.16	0.49
	Stms	91	161	0.14	0.24
	Hds	119	101	0.18	0.14
	10-p	286	144	0.15	0.30
NB9040	Lvs	376	152	0.17	0.46
	Stms	68	160	0.17	0.27
	Hds	91	92	0.19	0.10
	10-p	290	141	0.16	0.28
LSD (0.05)		144	61	0.03	0.05

[†] Ten whole plants bulked together

The pattern of nutrient accumulation for specific leaf samples at each growth stage revealed that the bottom leaves on the plant should be avoided for sampling Al concentrations due to soil contamination (Table 6). Sampling leaves during GS_1 should exclude the bottom leaves to avoid soil contamination or this plant material will need to be thoroughly washed. We concluded that sampling leaves 2, 3, or 4 (from the top of the plant) during GS_2 and GS_3 would provide representative samples and did not exhibit a wide variation in nutrient concentration. Those leaves are known to be directly involved in grain filling during the reproductive growth stage.

Growth stage sampling

Interaction of growth stage with genotype showed that nutrient concentrations were generally highest for GS_1 samples and decreased thereafter, regardless of soil pH (Table 7). Al concentrations were consistently highest for all genotypes in GS_1 samples. Concentrations of Mn were higher and Mg and Ca lower for samples from low pH soils than from high pH soils. GS_2 and GS_3 samples provided similar nutrient concentrations within genotypes, particularly for those samples from low pH soils.

Plant part sampling

When each genotype was divided into specific plant parts, a pattern of nutrient accumulation could be compared according to soil pH level (Table 8). The highest accumulation of the four nutrients for SC283 occurred in the leaves, regardless of soil pH. SC283 also had a high accumulation of Mn in the stems under low soil pH stress conditions. SC599 had high accumulations of Al and Ca in the leaves, but comparable concentrations of Mn and Mg distributed among the leaves, stems, and heads in samples from low soil pH plots. Under high soil pH conditions, the highest concentrations of the four elements were detected in the leaf samples. For NB9040, high Al and Ca values were found in leaf samples from low pH soils, while Mn and Mg values were comparable among leaves, stems, and heads. NB9040 leaf samples had the highest nutrient concentrations on high pH soils. We concluded that leaf samples provide a good representative plant sample for nutrient analyses, regardless of soil pH stress level.

References

1 Baker D 1976 Soil chemical constraints in tailoring plants to fit problem soils. I. Acid soils. *In* Proc. Workshop on Plant Adaptation to Mineral Stress in Problem Soils. Ed. M J Wright. Beltsville MD USA. pp 127–140.
2 Bartuska A M and Unger I A 1980 Elemental concentrations in plant tissues as influenced by low pH soils. Plant and Soil 55, 157–161.
3 Bates T E 1971 Factors affecting critical nutrient concentrations in plants and their evaluation: A review. Soil Sci. 112, 116–130.
4 Black C A 1957 Soil and Plant Relationships. John Wiley & Sons, Inc. New York, New York USA.
5 Bould C, Bradfield F G and Clarke G M 1960 Leaf analysis as a guide to the nutrition of fruit crops. J. Sci. Food Agric. 11, 229–242.
6 Bradley G A and Fleming J W 1960 The effect of position of leaf and time of sampling on the relationship of leaf phosphorus and potassium to yield of cucumbers, tomatoes and watermelons. Proc. Am. Soc. Hort. Sci. 75, 617–624.
7 Brown J C, Clark R B and Jones W E 1977 Efficient and inefficient use of phosphorus by sorghum. Soil Sci. Soc. Am. J. 41, 747–750.
8 Clark R B 1983 Plant genotype differences in the uptake, translocation, accumulation, and use of mineral elements required for plant growth. Plant and Soil 72, 175–196.
9 Clark R B, Pier P A, Knudsen D and Maranville J D 1981 Effect of trace element deficiencies and excesses on mineral nutrients in sorghum. J. Plant Nutr. 3, 357–374.

10 Clements H F 1964 Interactions of factors affecting yield. Annu. Rev. Plant Physiol. 15, 409–442.
11 Duncan R R 1981 Variability among sorghum genotypes for uptake of elements under acid soil field conditions. J. Plant Nutr. 4, 21–32.
12 Duncan R R, Dobson Jr. J W and Fisher C D 1980 Leaf elemental concentrations and grain yield of sorghum grown on an acid soil. Commun. Soil Sci. Plant Anal. 11, 699–707.
13 Eastin J D 1972 Efficiency of grain dry matter accumulation in grain sorghum. *In* Proc. 27th Annual Corn and Sorghum Res. Conf., Ed. D Wilkinson. Chicago, IL USA, 7–17 p.
14 Emmert F H 1959 Chemical analysis of tissue as a means of determining nutrient requirements of deciduous fruit plants. Proc. Am. Soc. Hort. Sci. 73, 521–547.
15 Foy C D 1975 Toxic factors in acid soil. Hort. 53, 38–43.
16 Gallaher R N, Harris H B, Anderson O E and Dobson Jr. J W 1975 Hybrid grain sorghum response to magnesium fertilization. Agron. J. 67, 297–300.
17 Guha M M and Mitchell R L 1966 The trace and major element composition of the leaves of some deciduous trees. Plant and Soil 24, 90–112.
18 Harrington J F 1944 Some factors influencing the reliability of plant tissue testing. Proc. Am. Soc. Hort. Sci. 45, 313–317.
19 Howlett F S 1961 Variation patterns established by foliar analysis of vegetable plants. *In* Plant Analysis and Fertilizer Problems. Pub. No. 8. Ed. W Reuther. Amer. Inst. Biol. Sci. The Lord Baltimore Press. USA. pp 355–388.
20 Jacques G L, Vanderlip R L and Whitney D A 1975 Growth and nutrient accumulation and distribution in grain sorghum. Agron. J. 67, 607–611.
21 Jackson W A 1967 Physiological effects of soil acidity. *In* Soil Acidity and Liming. Eds. R W Pearson and F A Adams. Amer. Soc. Agron. Madison, WI USA. pp 43–124.
22 Joham H E 1951 The nutritional status of the cotton plant as indicated by tissue tests. Plant Physiol. 26, 76–89.
23 Jones Jr. J B 1981 Analytical techniques for trace element determinations in plant tissues. J. Plant Nutr. 3, 77–92.
24 Jones Jr. J B and Eck H V 1973 Plant analysis as an aid in fertilizer corn and grain sorghum. *In* Soil testing and Plant Analysis. Eds. L M Walsh and J D Beaton. Soil Sci. Soc. Amer. Inc. Madison, WI USA. pp 349–364.
25 Jones Jr. J B and Steyn W J A 1973 Sampling, handling, and analyzing plant tissue samples. *In* Soil Testing and Plant Analysis. Eds. L M Walsh and J D Beaton. Soil Sci. Soc. Amer. Inc. Madison, WI USA. pp 249–270.
26 Jones Jr. J B, Large R L, Pfleiderer D B and Klosky H S 1971 How to properly sample for a plant analysis. Crops Soils 23, 15–18.
27 Kuo S and Mikkelsen D S 1981 Effect of P and Mn on growth response and uptake of Fe, Mn, and P by sorghum. Plant and Soil 62, 15–22.
28 Lane H C and Walker H J 1961 Mineral accumulation and distributon in grain sorghum. Texas Agric. Exp. Stn. MP-533.
29 Lockman R B 1972 Mineral composition of grain sorghum plant samples. III. Suggested nutrient sufficiency limits at various stages of growth. Comm. Soil Sci. Plant Anal. 3, 295–303.
30 McGlashan M L 1961 Terminology in plant- and soil-water relations. Nature 189, 207–209.
31 Mathers A C, Thomas J D, Stewart B A and Herring J E 1980 Manure and inorganic fertilizer effects on sorghum and sunflower growth on iron-deficient soil. Agron. J. 72, 1025–1029.
32 Ohki K 1975 Manganese supply, growth, and micronutrient concentration in grain sorghum. Agron. J. 67, 30–32.
33 Phillips M W and Barber S A 1959 Estimation of available soil potassium from plant analysis. Agron. J. 51, 403–406.
34 Procedures used by state soil testing laboratories in the southern region of the United States 1984 Southern Cooperative Series Bulletin 190. 16 p.
35 Sharpley A and Reed L W 1982 Effect of environmental stress on the growth and amounts and forms of phosphorus in plants. Agron. J. 74, 19–22.
36 Smith P F 1962 Mineral analysis of plant tissues. Annu. Rev. Plant Physiol. 13, 81–108.

37 Smith P F, Reuther W, Specht A W and Hrnicar G 1954 Effect of differential nitrogen, potassium, and magnesium supply to young Valencia orange trees in sand culture on mineral composition especially of leaves and fibrous roots. Plant Physiol. 29, 349–355.
38 Steyn W J A 1961 The errors involved in the sampling of citrus and pineapple plants for leaf analysis purposes. *In* Plant Analysis and Fertilizer Problems. Ed. W Reuther. Amer. Inst Biol. Sci. Washington, D.C. pp 409–430.
39 Sutton J D 1984 Sampling grain sorghum [*Sorghum bicolor* (L.) Moench] for chemical analyses. Master of Science Thesis, University of Georgia, 228 p.
40 Thomas W 1937 Foliar diagnosis: principles and practice. Plant Physiol. 12, 571–599.
41 Thomas L C and Chamberlain G J 1980 Colorimetric Chemical Analytical Methods. The Tintometer Ltd. Salisbury, England. pp 85–87.
42 Ulrich A 1952 Physiological basis for assessing the nutritional requirements of plants. Annu. Rev. Plant Physiol. 3, 207–228.
43 Ulrich A 1955 Influence of night temperature and nitrogen nutrition on the growth, sucrose accumulation and leaf minerals of sugar beet plants. Plant Physiol. 30, 250–296.
44 Ulrich A and Berry W L 1961 Critical phosphorus levels for lima bean growth. Plant Physiol. 36, 626–632.
45 Ulrich A and Hills F J 1967 Principles and practices of plant analysis. *In* Soil Testing and Plant Analysis. Eds. G W Hardy *et al*. Part II. Soil Sci. Soc. Am. Spec. Pub. No. 2. pp 11–24.
46 Wallace T 1961 The Diagnosis of Mineral Deficiencies in Plants. Chemical Publ. Co., Inc. New York, New York USA.

Section 3

Tolerances to salinity and to metal toxicities

Advances in salt tolerance

EMANUEL EPSTEIN and D.W. RAINS
Departments of Land, Air and Water Resources, and of Agronomy and Range Science, University of California, Davis, CA 95616, USA

Key words   Cell culture   Molecular biology   Plant breeding   Salinity   Seawater   Selection

Summary   Advances in and prospects for the development of salt tolerant crops are discussed. The genetic approach to the salinity problem is fairly new, but research has become quite active in a short span of time. Difficulties and opportunities are outlined. Salinity varies spatially, temporally, qualitatively, and quantitatively. In addition, the responses of plants to salt stress vary during their life cycle. Selection and breeding, including the use of wide crosses, are considered the best short-term approaches to the development of salt tolerant crops, but the new biotechnological and molecular biological techniques will make increasingly important contributions. Cooperation is called for among soil and water scientists, agronomists, plant physiologists and biochemists, cytologists, and plant geneticists, breeders, and biotechnologists. Given such cooperation and adequate support for these endeavors, the potential for increasing productivity in salt-affected areas can be realized.

Introduction

The general topic that this symposium is devoted to, genetic aspects of plant mineral nutrition, is quite new. In discussing, in 1939, 'hereditary variation in plant nutrition,' Harvey[10] referred to a total of nine publications on the subject, all original research papers. When, for a 1976 workshop, one of us tried to collect all the information then available on the same topic it was possible to list nine review articles[4]. No original research papers were included. Finally, when the same job was repeated in 1984[6], again omitting individual research papers, the listing contained 35 items, about a third of which were edited volumes, proceedings, and the like. So the rise of interest in this field has been remarkable.

The development of interest in a genetic approach to the specific problem of salinity and salt tolerance has paralleled that of the broader context in which it belongs[6]. It is by now recognized in many parts of the world that there exists genetic variation in salt tolerance within and among species, and that this variation may be used to develop crops specifically tailored to salt-affected soils. This recognition is most welcome; without it no progress can be made. At the same time it must be admitted that progress has not been as rapid as was initially hoped. It is worthwhile to consider briefly why this is so; there are some lessons to be learned from such a consideration. Some of these lessons are considered in this paper.

H.W. Gabelman and B.C. Loughman (Eds.), Genetic aspects of plant mineral nutrition.
ISBN-13: 978-94-010-8102-3
© *1987 Martinus Nijhoff Publishers, Dordrecht/Boston/Lancaster.*

Table 1. Summary of physical properties and major ions in shallow ground water from observation wells, farm drain sumps, and collector drains in the area served by the San Luis Drain in the southwestern part of the San Joaquin Valley of California. May 5-21, 1984. From reference 2

Properties and constituents	All physiographic zones (130 samples)		
	Minimum	Median	Maximum
Physical properties			
Specific conductance (μmho/cm at 25 C)	431	3,655	68,000
pH (standard units)	6.6	7.5	8.5
Temperature (C)	14	19.0	24.5
Cations			
Calcium (mg/L)	25	310	630
Magnesium (mg/L)	10	92	4,000
Sodium (mg/L)	33	460	30,000
Potassium (mg/L)	0.6	3.0	36
Anions			
Bicarbonate plus carbonate (as HCO_3) (mg/L)	112	290	1,150
Sulfate (mg/L)	39	1,700	65,000
Chloride (mg/L)	29	290	16,000

(Data from U.S. Geological Survey. There were no data for some constituents in five samples)

Salinities — not 'Salinity'

First, what do we mean by salinity and salt-affected soils? To judge by many papers devoted to the physiology of the salt relations of plants, salinity is to be equated with NaCl, or mixtures of it and $CaCl_2$. That does not, however, correspond to the facts of life.

Table 1 gives a summary of physical and chemical properties of shallow groundwater from observation wells, sumps, and collector drains in the area served by the San Luis Drain in the southwestern part of the San Joaquin Valley of California, an agriculturally very important area. The Coast Range to the west of that area consists of marine sedimentary rocks; the soils of the area are alluvial, the result of the erosion of the Coast Range. Some of the chemical characteristics listed in Table 1 reflect that fact. The predominant anion often is SO_4^{2-}, not Cl^-, and Mg^{2+} concentrations may exceed those of Ca^{2+} by large factors. Magnesium and Ca^{2+} combined may exceed the concentration of Na^+.

That, then, is just one salt-affected area that fails to conform to the stereotype of 'Salinity.' But another aspect of Table 1 needs comment: the tremendous range of the values. For Na^+, the range spans a factor of over 900/1, even in this limited area of 4.8×10^5 hectares. For other major ions the corresponding factors are not quite so spectacular, but are nevertheless very large in the context of plant responses.

The variability referred to is often evident on a very small scale[21,22]. Such observations led to the conclusion that since the best yields will be achieved in the least saline areas, selections should be made for high yield on non-saline soils[21]. This argument, however, ignores that genotypic differences in respect specifically to salt tolerance have been found in barley, wheat, and other crops, and that wild, salt tolerant germplasm is available for introgression of that trait into economic crops; see the literature assembled by Epstein[6]. This subject will be discussed below. Meanwhile it is certainly true that spatial variability in salinity, both quantitative and qualitative, is a very real difficulty, and one that more attention will have to be paid to in the future[18,22].

Salinity over time

We have discussed spatial variation in both qualitative and quantitative terms. Another problem is variation over time, both in the soil and water the crops are exposed to, and in the responses of the plants to this variability.

Salinity varies not only in space but also in time. In areas with mediterranean climates, winter rains may leach salt below the root zone, enabling growers to get a good stand of grain in early spring. With the advent of the dry summer, however, saline well water often has to be used, while in more severely arid regions, the seeds may have to cope with salinity from the beginning.

Even under a steady salinity regime, however, time enters into the salt responses of plants because the salinity tolerance of many plants varies markedly during the course of their life cycle[27]. We are thus faced with changes over time in the salinity to which the plants are exposed, and with genetically controlled changes within the plant in its responses during ontogeny.

Selection and breeding

The conventional approach of the plant breeder has been to select and breed varieties for high yield, disease resistance, quality, and other desirable traits. For the most part breeders have conducted this selection and breeding under environmentally favorable conditions. There have been two principal reasons for this. First, breeders did not want environmentally stressful conditions to confound the results of breeding for the specific traits they were aiming for, and second, most modern plant breeding has been conducted in areas and for areas that have relatively benign environments, especially in respect to soil and water, or could be rendered so by fertilization, reclamation, and irrigation.

Soils afflicted with minerally stressful conditions, however, occupy an estimated 30% of the land area of the world[5]. Marginal land will increasingly have to be used for crop production. The best lands are already farmed, and population pressures are highest in many areas with problem soils, including saline ones.

It is therefore essential that a genetic dimension be added to the more traditional approach of reclamation, drainage, use of excess irrigation water to leach salt below the root zone, and other such management practices, essential as they are and will continue to be.

Progress in this enterprise has so far been modest. To this date we are unaware that certified seed of a single variety of any crop specifically selected and bred for saline conditions has been released to growers. There are a number of reasons for this. (a) There is the problem of spatial and temporal variation in salinity and its qualitative variability, already mentioned. (b) Easily recognized markers for salt tolerance, especially early in ontogeny, but indicative of subsequent performance, have not been identified. (c) Screening throughout the life cycle of large numbers of entries, while reliable, is extremely time consuming and costly. (d) Our basic understanding of mechanisms of salt tolerance is as yet scant. (e) Salt tolerance is a polygenic trait, governed at levels of organization ranging from subcellular to organismic. (f) Above all, compared with the large numbers of breeders aiming for more conventional desiderata, the number of breeders addressing the problem of salt tolerance is as yet minuscule.

Just the same, selection and breeding, not necessarily altogether conventional, will probably for some years at least offer the best prospects for success. There are several reasons for this appraisal.

First, very little has as yet been done, and is being done, along this line. To the best of our knowledge not one of the world collections of major crops has yet been systematically screened for salt tolerance.

Second, what limited attempts have been made have given evidence of the possibility of advances in salt tolerance. Lines of barley and wheat have been selected in solution cultures salinized with a sea salt mix to the equivalent of 100% seawater salinity for barley and 50% for wheat[7]. Seawater salinity was used to explore the possibility of actually using seawater for irrigation along coastal deserts. There are, however, at least two more reasons for choosing media with seawater-type salinity for selection. One is that where conventional irrigation is practiced in areas near the coast, seawater intrusion is common, making this particular salinity a matter of concern to growers. And secondly, as already pointed out, in important agricultural areas such as the west side of the San Joaquin Valley, the salinity of soils and water bears considerable resemblance to seawater salinity. For such areas, a selection medium with seawater-type salinity is therefore indicated. A number of wheat selections resulting from screening 5,000 wheat accessions in the world collection in a seawater-type solution culture have done well in field trials in the San Joaquin Valley[11].

For other areas where qualitatively different salinities prevail, appropriate selection media need to be used. It is not to be expected that we can develop universally 'salt tolerant' cultivars that would do well in every salt affected area.

Selection and breeding can be greatly expanded in their effectiveness by going beyond the target species; that is, by using germplasm of closely and not so closely related salt tolerant wild species to introgress salt tolerance into the economic target species We have done this with some success for the tomato[23], and presently Dvorák and collaborators[3] are attempting it with wheat. All this experience and that of others around the world[6], shows the potential of a genetic approach to the problems posed by salt-affected soils. Because use of exotic, salt tolerant germplasm can enlarge the potential genetic variability available, we believe that this approach of using wide crosses holds out great promise of success. Jones and Qualset[11] have provided a comprehensive discussion of breeding for environmental stress tolerance, including salt tolerance in particular. The conservation of wild germplasm[33] is as essential for the development of this trait as it is for that of so many other desirable traits.

Cell and tissue culture

Cell and tissue culture techniques offer a number of advantages not found in the conventional selection and breeding procedures currently used to enhance tolerance of plants to saline environments. These advantages include: (a) Characterization, at the cellular level, of physiological markers associated with salt tolerance. (b) A relative lack of differentiated tissues which reduces complications arising from morphological variability found in differentiated plant tissues. (c) Ability precisely to control environmental and nutrient conditions for manipulation of plant cell populations. (d) Manipulation of large numbers of cells, increasing the probability of obtaining variants with the desired characters. (e) Utilization of procedures developed with microbial systems such as haploid production, somatic hybridization, gene transfer and mutant induction and selection.

There are, however, a number of difficulties in applying cell culture procedures to the selection and genetic manipulation of plant cells for salt tolerance.

The character under selection isolated at the cellular level will have to be expressed at the whole plant level. This will require regeneration of the cells to whole plants. Although the list of plant species capable of regeneration continues to increase not all species under all conditions can be regenerated.

Another difficulty is the effect of culture time on regeneration frequency. The longer cells are maintained in culture the more difficult it is to regenerate these cells to plants. Concomitantly the long-term cultures show enhanced variability in characters unrelated to the selected character resulting in a very heterogeneous population of regeneration plantlets with the target trait potentially masked by the non-targeted traits[29].

It is essential that the regenerated plants maintain the characteristics of the original plant genotype while incorporating the desired genetic character without incorporating inimical genes. This is not always the situation using the cell culture techniques. The critical issue is to understand the limitations of the technique and to apply the technology in the appropriate situation[14].

Plant cell lines exposed to saline environments have been selected for enhanced tolerance to salinity[19,20]. In most of the reported examples the enhanced tolerance was documented at the cellular level but not at the whole plant level.

Plant cell lines from a number of species have been selected for tolerance to salinity. These include *Nicotiana sylvestris*, *N. tabacum*, *Capsicum annuum*, *Saccharum officinarum*, *Medicago sativum*, *Oryza sativa*, *Coffea arabica*, *Datura inoxia*, *Citrus sinensis*, *C. aurantium*, *Crepis capillaris*, *Hordeum vulgare*, *Lycopersicon esculentum* and *Colocasica esculenta*. The list is not exhaustive but it does demonstrate the many species to which this technique has been applied successfully[20,30].

Minimal success has been achieved in demonstrating the expression of salt tolerance by plants regenerated from salt selected cell lines. Tobacco plants regenerated from selected cell lines showed greater survival rate when watered with saline water[17]. Rice cells selected for tolerance to salinity have been regenerated but plants were partially sterile[32]. Sodium chloride selected alfalfa cells have been regenerated to plants but no improvement in tolerance to salinity was demonstrated[1]. A more comprehensive review of the application of selection techniques to improving plant performance in saline environments is found in a paper by Stavarek and Rains[30]. In that paper details of the technique and its utility are discussed and the conclusion is drawn that the techniques developed to select plant cells for resistance to salt stress are feasible, but that further studies will be required on media refinement and on the basic genetic and physiological processes related to salt stress.

Molecular biology

The application of molecular biology to the genetic improvement of plant tolerance to salt stress offers exciting possibilities, particularly if coupled with a long-term commitment to developing basic information on the biochemical and physiological mechanisms related to salt tolerance.

The tools of molecular biology have been primarily developed from prokaryotes, organisms which are genetically and structurally relatively simple. The application of these techniques to higher plants will require extension of molecular tools to these more complex organisms and the further development of techniques not found necessary for manipulation of lower organisms.

Currently, genetic manipulation using molecular techniques requires that the alteration be controlled by one gene. Frequently it has been demonstrated that single gene traits are reflected in a single alteration in a particular biochemical pathway. This type of response, however, is very unlikely in a complex process such as salt tolerance. Plants possess multigenic regulation of stress tolerance and respond at both the cellular level and the organismic level. It will be necessary to manipulate a number of genetic processes simultaneously; they constitute a very complex system currently not readily attacked with the emerging technologies.

A number of approaches could be used to enhance the utility of these very powerful tools. These approaches will combine cell culture techniques with molecular manipulation of plant genomes.

Somaclonal variation

Regeneration of plants from plant cells is not a guarantee that the regenerated plant will be an exact copy or a clone of the parental line used to culture the cell lines. These phenotypic variants provide a potential source of genetic variability and are very important to selection systems and as biological materials for the molecular biologist. The variability can be increased through mutagenic agents; the variation has been expressed phenotypically in a number of regenerated plant species[26].

A number of mechanisms have been proposed as being responsible for somaclonal variation. These include: chromosome breakage and rearrangement, asymmetric exchange of sister chromatids and their segregation upon mitosis, transposable genetic elements and gene amplification and diminution. The basic cause of somaclonal variation may be mutation of an allele or expression of existing genetic traits which are not expressed at the organismal level.

Mutagenesis

Directed mutation through chemical mutagens or through physical factors such as radiation provides opportunities to enhance genetic variability. The mutagenic agent can be applied to the plant, to individual tissues, to organs (seeds, ovaries) or to cells. After exposure of the target material a selection screen is necessary if the desired character is to be isolated from the treated population. Identification of mutants will provide markers for cellular genetic manipulations and these selected lines will be available for physiological research and germplasm development[9].

A number of mutants have been isolated from plant tissues and cells challenged with mutagenic agents. Chlorate, an analog of nitrate, has been used to select for cells deficient in nitrate reductase[15,16]. A similar approach has been used for barley seedlings[12]. A number of auxotrophs have been recovered from mutagenized plant cells. These cells require various amino acids, nucleic acids and vitamins for growth. The significance of these mutants is the isolation of cell lines with markers that can be used to identify genetic traits. A linkage between the nutrient requirement and a desired trait that is expressed at the cellular level provides an effective screen for selection of whole plant characteristics at the cellular level[13].

Somatic hybridization

Protoplast fusion with formation of somatic hybrids provides an opportunity to transfer genetic information from one species to another[25]. The fusion of protoplasts from two different plant species and recovery of hybrid cell clones requires a selection system to isolate hybrid cells. For example the imposition of a saline screen should enrich a population of fused hybrid cells containing a combination of genetic traits which enhance tolerance of that environmental stress. The transfer of genetic traits through somatic hybridization involves a number of processes or mechanisms. The fusion of protoplasts of sexually incompatible species may produce fertile amphidiploid somatic hybrids. It is also possible to produce heterozygous lines within a species that is normally vegetatively propagated. Chromosome manipulation and elimination permit the transfer of only part of the nuclear genetic information from one species to another. This is very important if the species containing genetic traits for salt or drought tolerance is not an agronomic crop. The other, non-essential traits do not have to be introduced into the agronomic species and therefore do not have to be removed through backcrossing or recurrent selection so as to retain only the desired character.

The role of somatic hybridization in plant improvement is becoming more and more significant. Shepard *et al.*[28] have demonstrated the feasibility of using this procedure in a breeding scheme for potatoes. As with all of the 'new technologies' there are limitations. Currently very few important agronomic crops have produced protoplasts, whether single or fused, which have been successfully regenerated. Good success has been had with tobacco, potato, carrot, petunia and others but the major food crops are only now being incorporated into the list of plant species amenable to somatic manipulation[31].

Recombinant DNA

The concept of genetic engineering of plants has commonly focused on the use of recombinant DNA technology. In fact most of the previous discussion could be defined as genetic engineering since we have focused on the manipulation of the plant genome to enhance variability in genetic characters to increase the potential of selecting for the desired trait

Recombinant DNA technology also provides an opportunity to increase variability by a directed introduction of a gene into a plant cell. A number of methods are available for the transfer of genes[24].

Plant cells may take up DNA directly when protoplasts are coprecipitated with calcium phosphate. To date no conclusive evidence that plant cells have been transformed using this approach has appeared.

Vectors, such as plant viruses, are capable of transferring DNA segments into plant cells. In this situation, a biological organism is used as a vehicle to introduce into the target cell the gene or a portion of a gene which carries the genetic information necessary to enhance stress tolerance of the target call. This technique has still to be applied successfully.

One of the major difficulties is the identification of markers for transformed genes. It will be of little use to transform cells with insertion of desired genes if these cells cannot be isolated. The current approach is to use resistance to such drugs as kanamycin or neomycin. If this resistance can be linked to a desired gene, then the transformed cells treated with these drugs will be the only cells isolated.

A very valuable technique, commonly used in animal systems, is the direct injection of DNA from the nucleus of one species into the nucleus of another species. The potential power of this approach is obvious but it will be limited by our ability to screen and isolate cell lines genetically transformed by these injections of DNA. One of the most successful techniques currently available to transfer genes or DNA segments of genes is the use of a naturally occurring vector called the T_i plasmid. This plasmid is a circular strand of DNA present in a genus of soil bacteria called *Agrobacterium*. These bacteria induce 'crown galls' on plants by transferring their DNA plasmid to plant cells. The DNA is incorporated into the plant chromosomal DNA and the cell is transformed. Recombinant DNA technology could be applied to this system through 'splicing' a gene with specific characteristics into the plasmid. This plasmid could then be introduced into the plant cell via the T_i vector and, if expressed, provide a means to incorporate genetically a desired character. Theoretically this approach will permit manipulation of any gene from any species and result in a directed genetic change so as to obtain a desired characteristic.

A general program which applies genetic engineering techniques to improve stress tolerance and water use efficiency of plants might involve a sequence of steps as outlined below:

Molecular cloning of stress tolerance genes from beneficial microorganisms.

Recombinant DNA analysis of the structure of genes regulating tolerance to water or salt stress.

Development of plant gene vectors for genetic engineering of stress tolerant plants.

Application of recombinant DNA technology for molecular cloning of plant stress tolerance genes.

Use of cell culture techniques for the selection of stress tolerant mutant plants.

Transfer of these genetic traits for stress tolerance into a range of crop plants for agriculture.

The applications of these techniques have great potential and this area should be actively pursued. It is apparent, however, that our understanding of mechanisms related to stress tolerance and water use efficiency coupled with the very preliminary status of our ability to apply genetic engineering techniques to higher plants will limit the usefulness of this approach. Success will not be realized without substantial investments in time and resources devoted to basic studies of plant processes and plant genetic systems as well as to transferring genetic engineering technology from the prokaryotes to the eukaryotes.

Research on mechanisms

Research on mechanisms of salt tolerance needs to be carried out with the understanding that this trait is governed at both the organismic and cellular levels. Most crop species are not markedly salt tolerant; what salt tolerance they possess depends more often than not on the degree to which the salt can be excluded, especially from the shoot. This means that such phenomena as absorption and long-distance transport of ions need to be studied — processes that cannot be investigated at the purely cellular level. At the same time, even these studies, while transcending the cellular level, involve that level as well. For example, effective sequestration of sodium in the vacuoles of the root cortex is a cellular phenomenon, but it also is instrumental in keeping the ions so sequestered from moving readily into the xylem and toward the photosynthetic tissues. Such sequestration — a cellular phenomenon — is thus one mechanism that participates in the regulation of long-distance transport.

On the other hand, the research with cell and tissue culture clearly shows that there are purely cellular mechanisms of salt tolerance. The mechanisms that lend themselves to an approach at the cellular level include selectivity in ion transport and compartmentation, energy costs of salt resistance mechanisms, and organic osmotica and their role in salt tolerance. Research at both the organismic and the cellular levels of organization, however basic in nature, has practical implications, for it may lead to the discovery of markers for salt tolerance traits and thus speed up the development of salt tolerant crops.

The ultimate test

By whatever means a genotype resistant to a certain type and severity of salinity is developed it is essential that it be tested in the field, under appropriate agronomic management. It is therefore necessary to establish close cooperation among soil and water scientists, plant physiologists and biochemists, cytologists, and plant geneticists and breeders — including the new 'breed' of molecular geneticist[8]. It is the agronomist in the field, and then the grower, who will be the ultimate judges of success or failure. Given close cooperation among the specialists mentioned and support for their endeavors, the potential for increasing productivity in salt-affected areas can be realized.

References

1   Croughan T P 1981 The application of cell culture techniques to the selection and study of salt tolerance. Ph.D. Diss. University of California, Davis.
2   Deverel S J, Gilliom R J, Fujii R, Izbicki J A, Fields J C 1984 Areal Distribution of Selenium and Other Inorganic Constituents in Shallow Groundwater of the San Luis Drain Service Area, San Joaquin Valley, California: A Preliminary Study. U. S. Geological Survey Water Resources Investigations Report 84 4319. Sacramento, California.
3   Dvorák J, Ross K, Mendlinger S 1985 Transfer of salt tolerance from *Elytrigia pontica* (Podp.) Holub to wheat by the addition of an incomplete *Elytrigia* genome. Crop Sci. 25, 306 309.
4   Epstein E 1977 Adaptation of crops to salinity. *In* Plant Adaptation to Mineral Stress in Problem Soils. Ed M J Wright. A Special Publication of Cornell University Agricultural Experiment Station, Ithaca, N.Y., pp 73 82.
5   Epstein E 1981 Impact of Applied Genetics on Agriculturally Important Plants: Mineral Metabolism. Report to the Office of Technology Assessment, United States Congress. Department of Land, Air and Water Resources, University of California, Davis. 42 p.
6   Epstein E 1985 Salt-tolerant crops: origins, development, and prospects of the concept. Plant and Soil 89, 187 198.
7   Epstein E, Norlyn J D, Rush D W, Kingsbury R W, Kelley D B, Cunningham G A, Wrona A F 1980 Saline culture of crops: a genetic approach. Science 210, 399 404.

8 Frey K J 1985 The unifying force in agronomy — biotechnology. Agron. J. 77, 187–189.
9 Gottschalk W 1981 Mutation: higher plants. Progress in Botany 43, 139 152.
10 Harvey P H 1939 Hereditary variation in plant nutrition. Genetics 24, 437 461.
11 Jones R A, Qualset C O 1984 Breeding crops for environmental tolerance. *In* Application of Genetic Engineering to Crop Improvement. Eds. G B Collins and J G Petolino. Martinus Nijhoff/Dr W Junk, Publishers, Dordrecht, pp. 305 340.
12 Kleinhofs A, Warner R L, Muehlbauer R J, Nilan R A 1978 Induction and selection of specific gene mutations in *Hordeum* and *Pisum*. Mutation Res 51, 29 42.
13 Maliga P 1984 Isolation and characterization of mutants in plant cell cultures. Annu. Rev. Plant Physiol. 35, 519 542.
14 Meins F Jr. 1983 Heritable variation in plant cell culture. Annu. Rev. Plant Physiol. 34, 327–346.
15 Muller A J, Grafe R 1978 Isolation and characterization of cell lines of *Nicotiana tabacum* lacking nitrate reductase. Molec. Gen. 161, 67–76.
16 Murphy T M 1982 Analysis of distributions of mutants in clones of plant-cell aggregates. Theor. Appl. Genet. 61, 367 372.
17 Nabors M W, Gibbs S E, Bernstein C S, Meis, M E 1980 NaCl-tolerant tobacco plants from cultured cells. Z. Pflanzenphysiol. 97, 13–17.
18 Papadopoulos I Rendig V V 1983 Tomato plant response to soil salinity. Agron. J. 75, 696–700.
19 Rains D W, Croughan T P, Stavarek S J 1980 Selection of salt-tolerant plants using tissue culture. *In* Genetic Engineering of Osmoregulation. Impact on Plant Productivity for Food, Chemicals, and Energy. Eds. D W Rains, R C Valentine and A Hollaender. Plenum Press, New York, pp 279- 292.
20 Ram N V R, Nabors M W 1986 Salinity tolerance. *In* Biotechnology Application and Research. Eds. P N Cheremissinoss and R P Ouellette. pp 623 642. Technomic Publications, Inc., Lancaster.
21 Richards R A 1983 Should selection for yield in saline regions be made on saline or non-saline soils? Euphytica 32, 431 -438.
22 Richards R A, Dennett C W, Qualset C O, Epstein E, Norlyn J D, Winslow M D 1986 Variation in yield of grain and biomass in wheat, barley, and triticale in a salt-affected field. Field Crops Res. *In press*.
23 Rush D W, Epstein E 1981 Breeding and selection for salt tolerance by the incorporation of wild germplasm into a domestic tomato. J. Am. Soc. Hort. Sci. 106, 699 704.
24 Schell J, Van Montagu M, Holsters M, Depicker A, Zambryski P, Dhaese P, Hernalsteens J-P, Leemans J, De Greve H, Willmitzer L, Otten L, Schröder J and G 1982 Plant cell transformations and genetic engineering. *In* Plant Improvement and Somatic Cell Genetics. Eds. I K Vasil, W R Scowcroft, K J Frey. Academic Press, New York, pp 255 276.
25 Schieder O 1982 Somatic hybridization: A new method for plant improvement. *In* Plant Improvement and Somatic Cell Genetics. Eds. I K Vasil, W R Scowcroft, K J Frey. Academic Press, New York, pp 150–178.
26 Scowcroft W R, Larkin, P J 1982 Somaclonal variation: A new option for plant improvement. *In* Plant Improvement and Somatic Cell Genetics. Eds. I K Vasil, W R Scowcroft, K J Frey. Academic Press, New York, pp 159 178.
27 Shannon M C 1984 Breeding, selection, and the genetics of salt tolerance. *In* Salinity Tolerance in Plants. Strategies for Crop Improvement. Eds. R C Staples and G H Toeniessen. John Wiley and Sons, New York, pp 231–254.
28 Shephard J F, Bidney D, Shahin E 1980 Potato protoplasts in crop improvement. Science 208, 17 24.
29 Stavarek S J, Croughan T P, Rains D W 1980 Regeneration of plants from long-term cultures of alfalfa cells. Plant Sci. Letters 19, 253–261.
30 Stavarek S J, D W Rains 1984 The development of tolerance to mineral stress. Hort Sci. 19, 377 382.
31 Sybena J 1983 Genetic manipulation in plant breeding: somatic versus generative. Theor. Appl. Genet. 66, 179–201.

32 Wong C, Ko S, Woo S 1983 Regeneration of rice plantlets on NaCl-stressed medium by anther culture. Bot. Bull. Acad. Sinica 24, 59–64.
33 Yeatman C W, Kafton D, Wilkes G, Eds 1984 Plant Genetic Resources. A Conservation Imperative. American Association for the Advancement of Science Selected Symposium Series. Vol. 87. Westview Press, Boulder, Colorado.

Intervarietal ionic composition changes in barley under salt stress

J.M. STASSART and J. BOGEMANS
Laboratorium Plantenfysiologie, Vrije Universiteit Brussel, Paardestraat 65, B-1640 St. Genesius — Rode, Belgium

Key words Accumulation Barley Calcium Chloride Hoagland Ions Leaves Potassium Salt stress Sodium Vermiculite

Summary The effect of salt stress at 1% NaCl (w/v = 171 mM) on different barley cultivars was tested over a growth period of 9 weeks. It was found that the 14 cultivars tested reacted quite differently to the stress situation. Growth and ionic content were followed in three to four plant parts, such as old leaves, young leaves, sheath and the ear as it appeared. The best growing plants also showed the most efficient salt segregation between their different parts. It is our aim to analyse in further detail (metabolism) how these plants differ from each other to explain their better salt resistance. There were some genetic links between these cultivars, but the genetic background was rather confused, due to uncontrolled intercrossing with other unknown types.

Introduction

Barley and most other world wide economic crops have never been selected for salt tolerance as such. After working for some years on two summer barley cultivars, Union and Menuet, we intended to make a survey over 12 more commercial cultivars (Table 1). Our objective was to identify cultivars seemingly tolerant or sensitive to high salt concentration and to try and find out what if any, physiological differences were involved in such differential response. Initally, our attention was focused on plant growth and its reduction due to salt treatment.

Materials and methods

Seeds were soaked in deionized water for some hours before being transferred to moist sand for germination overnight. Two row malting barley has a very strong germination power and seedlings come out simultaneously, which has the advantage of producing very equal populations of plants. (Stassart J.M. and Bogemans J., unpublished data).

Sand was washed away and the germinated seeds were transferred to square pots 13 cm a side. These pots were filled to 2/3 with moisted vermiculite, this artificial soil is only used for anchoring the plants. About seven seedlings were set per pot and covered with some more vermiculite, then watered again. Three pots were used as a replicate for one treatment or one cultivar. One week after germination all plants were given Hoagland's solution at half strength, together with the usual micro-elements, with the following ionic composition KNO_3, 5mM; $Ca(NO_3)_2$, 1.5 mM; $MgSO_4$, 1 mM and $NH_4H_2PO_4$, 1 mM.

The micro-nutrients: B, 46 μM; Mn, 9.1 μM; Cu, 0.31 μM; Zn, 0.76 μM; Mo, 0.20 μM and Fe, 53.7 μM.

Twice a week 100 ml of this nutrient solution was given per pot. Plants to be salinized were given salt solution in addition to the Hoagland's solution. Salt solution was also given at 100 ml volume

H.W. Gabelman and B.C. Loughman (Eds.), Genetic aspects of plant mineral nutrition.
ISBN-13: 978-94-010-8102-3
© *1987 Martinus Nijhoff Publishers, Dordrecht/Boston/Lancaster.*

Table 1. List of barley varieties used in the presented experiment

Number	Cultivar	Type
1	Union	summer barley
2	Menuet	summer barley
3	Igri	winter barley
4	Sonja	winter barley
5	Auriga	winter barley
6	Egmont	summer barley
7	Isaura	winter barley
8	Candida	winter barley
9	Hebe	summer barley
10	Herta	summer barley
11	Regent	summer barley
12	Iban	summer barley
13	Atem	summer barley
14	Fripone	summer barley
15	Legia	summer barley

* All cultivars used are two row barley.

and the reference plants received the same volume of water. The total volume of solutions given per week was enough to keep the vermiculite substrate moist so drought could be avoided. Previous experiments showed us that this was also the right amount of nutrient needed for the number of plants used per pot. During the winter, artificial lighting was used when daylight requirements seemed insufficient. Two weeks after adding the solutions, plants were thinned to five plants per pot. The redundant plants thinned out were analysed for ion content, to see if any indication could already be found at this early stage of development concerning the later evolution of the plant. The second harvest was carried out at five weeks, just before earshooting and the third harvest at the end of week nine. At these two harvest times the plants were cut into different parts, described as follows:

Old leaves: leaves 1 to 4
Young leaves: leaves 5 to 6 at week five and leaves 5 to 9 at week nine.
Stem or sheath: left over as leaves had been cut off.
Ear: as it appeared at the week-nine harvest.

In previous experiments all leaves were harvested separately, but it was found that a group of certain leaves had a very similar behavior with respect to ion uptake and accumulation, so the leaves with the same physiological reaction were put together in this experiment and divided into parts such as old leaves and young leaves.

The fresh weight of samples was measured immediately after harvest and after drying over night in an oven, dry weight was measured. Samples were then dry-ashed in a muffle furnace at 560 C for about three hours. Ashes were dissolved in 1.5 M HNO_3 and brought to volume by water before being measured by flame photometry for sodium, potassium and calcium chloride was measured on a chloro-counter. Results are expressed in micro-equivalents per gram dry weight (μeq/g dr wt).

Results and discussion

Cultivar 9, 'Hebe' did not germinate succesfully and was discarded from further experimentation.

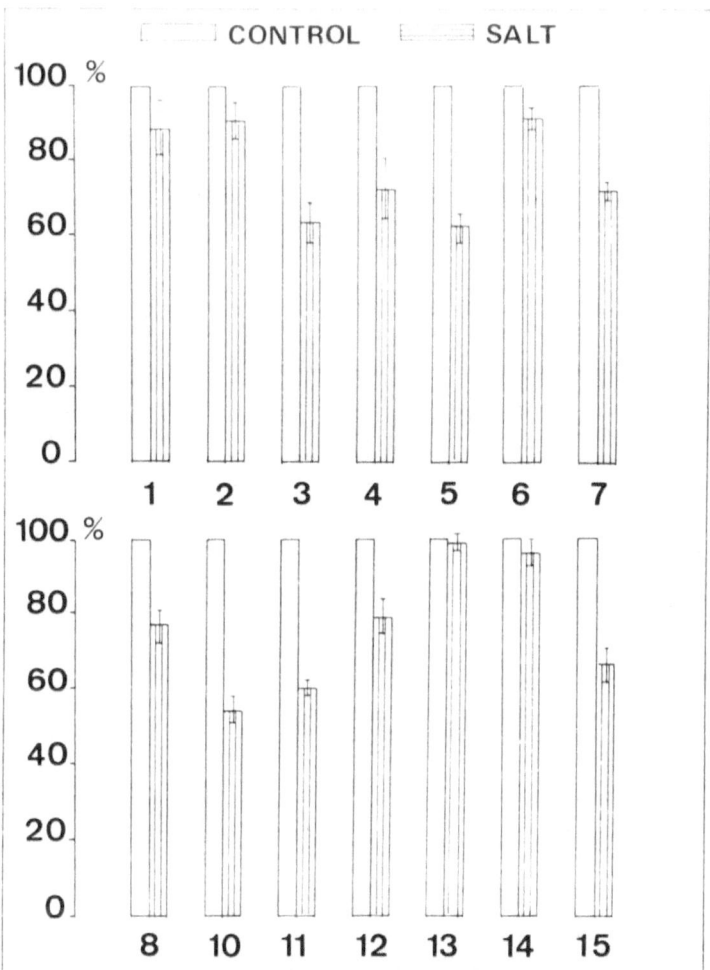

Fig. 1. Dry matter production after 2 weeks. Percentual expression of the salt treatment against the control plant, being set as 100% for each variety. Not taking into account the differences in growth between control plants.

Measurements after two weeks

Growth. It was clear from the start by comparing the plants which has been on salt, with control plants that the winter barley type was in general sensitive to salt application. For summer barley it was found that there were some cultivars which grew well under salt stress, but some others that grew poorly. These were:

Herta 46% growth reduction
Regent 40% growth reduction
Iban 21% growth reduction

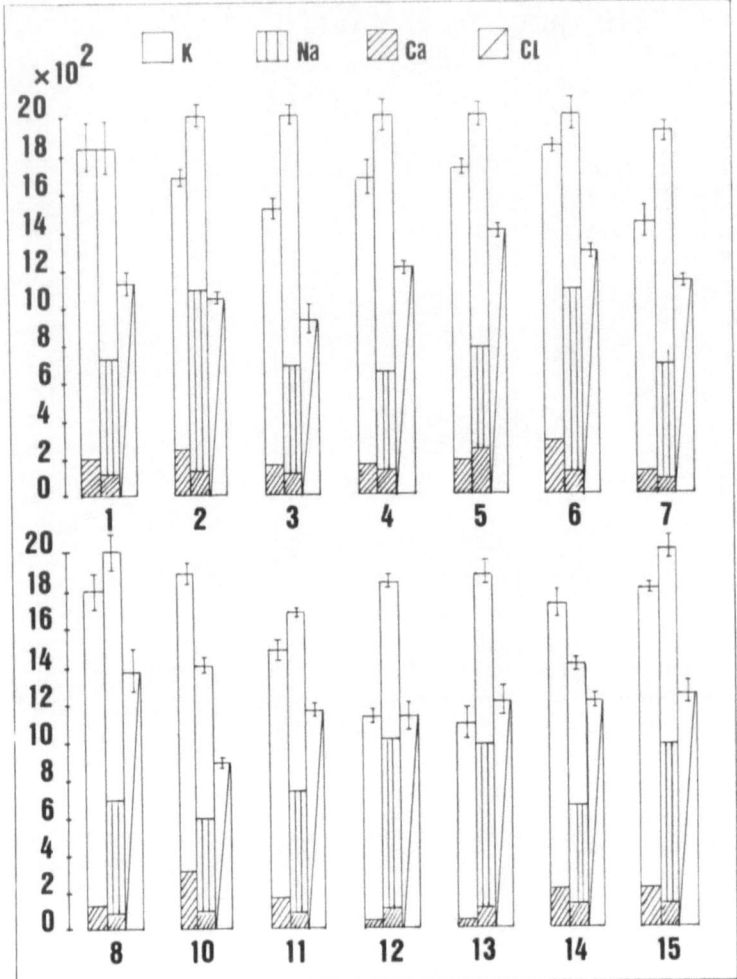

Fig. 2. Ionic composition of control and salt treatment plants after 2 weeks. Expressed in μeq/g dr wt.

Growth reduction expressed as a reduction in dry matter production can be clearly seen in Figure 1. The barley cultivars which showed the strongest reduction after only two weeks came out as being the most sensitive later on (Figs. 1 and 3).

Ionic content. After two weeks the development of the plant reached only the stage of the first two leaves and no indication could be found (Fig. 2) that the plant would later tend to become tolerant or sensitive to salt. However summer barley has a tendency to accumulate more sodium than winter barley.

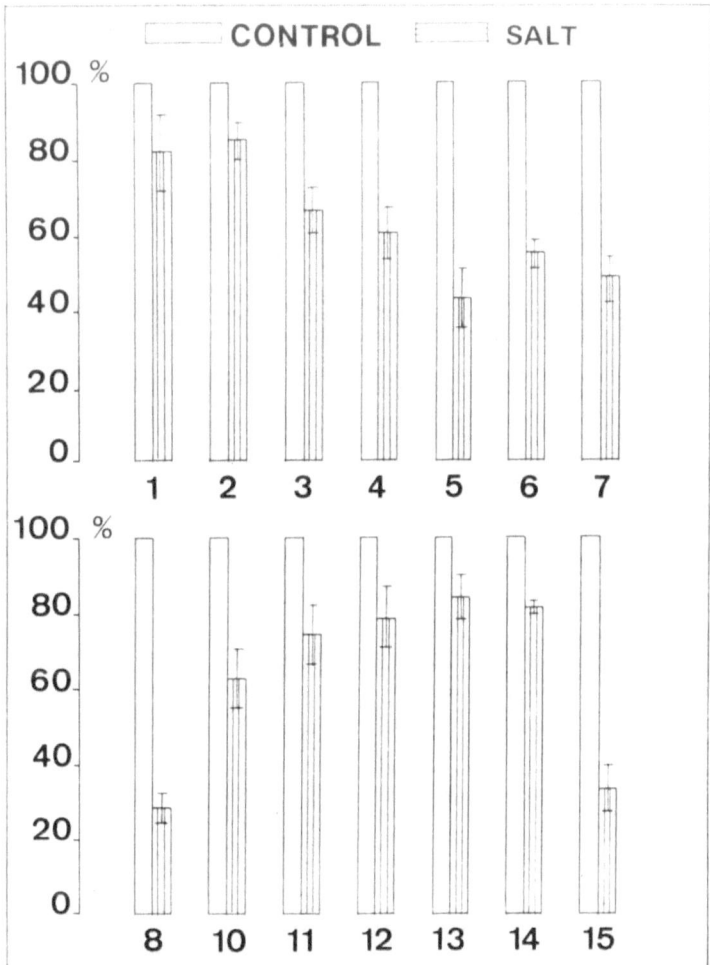

Fig. 3. Dry matter production after 5 weeks. Percentual expression of the salt treatment against the control plant, being set at 100% for each variety. Not taking into account the differences in growth between control plants.

Measurements after five weeks

Growth. At this stage of the development of the plants one could distinguish three main groups of plants, depending on the growth reduction afflicted by a saline treatment (Fig. 3).

Within 20% reduction = rather tolerant
Upto 40% reduction = semi-tolerant
Over 40% reduction = sensitive

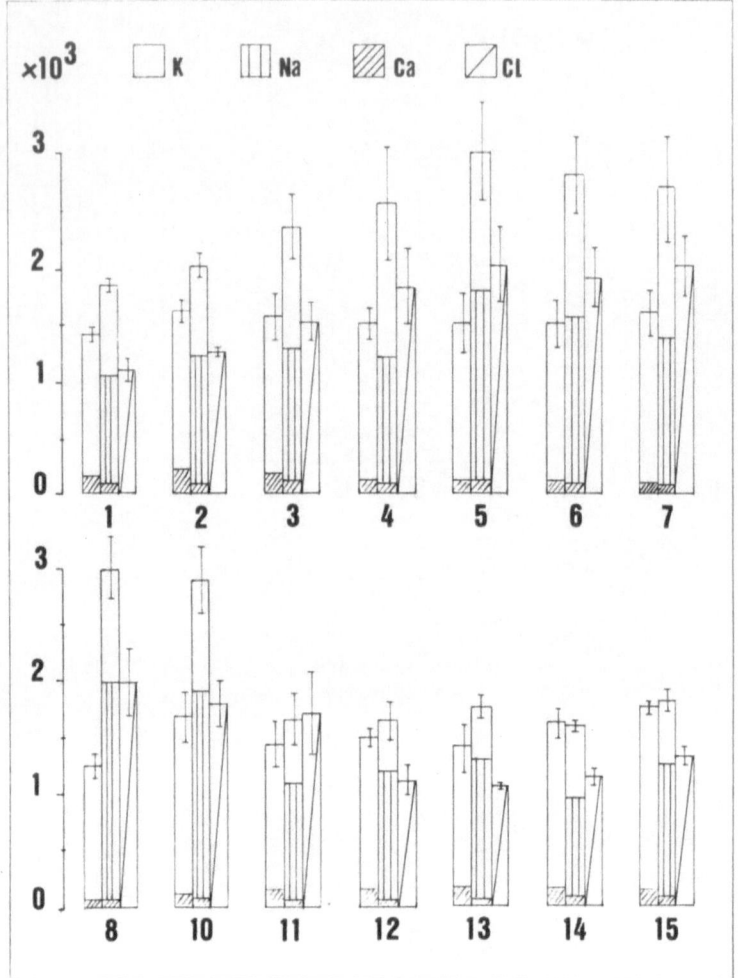

Fig. 4. Ionic composition of control and salt treatment plants after 5 weeks. Expressed in µeq/g dr wt.

From the data one sees that the cultivars 1, 2, 11, 12, 13 and 14 show only a limited growth reduction and seem not to be very much affected by the salt treatment as much as growth is concerned. *i.e.* they are rather tolerant to salt. In a second group the cultivars 3, 4 and 10 show an intermediate growth reduction and since they do not belong to either group we called them semi-tolerant. The third group with cultivars 5, 6, 7, 8, and 15 show a very large reduction in growth and are more likely to be sensitive to a salt treatment especially where growth and development is concerned.

Ion content. Here too, some differences could be found within the cultivars (Fig. 4). The differences became clear at this stage of the plants

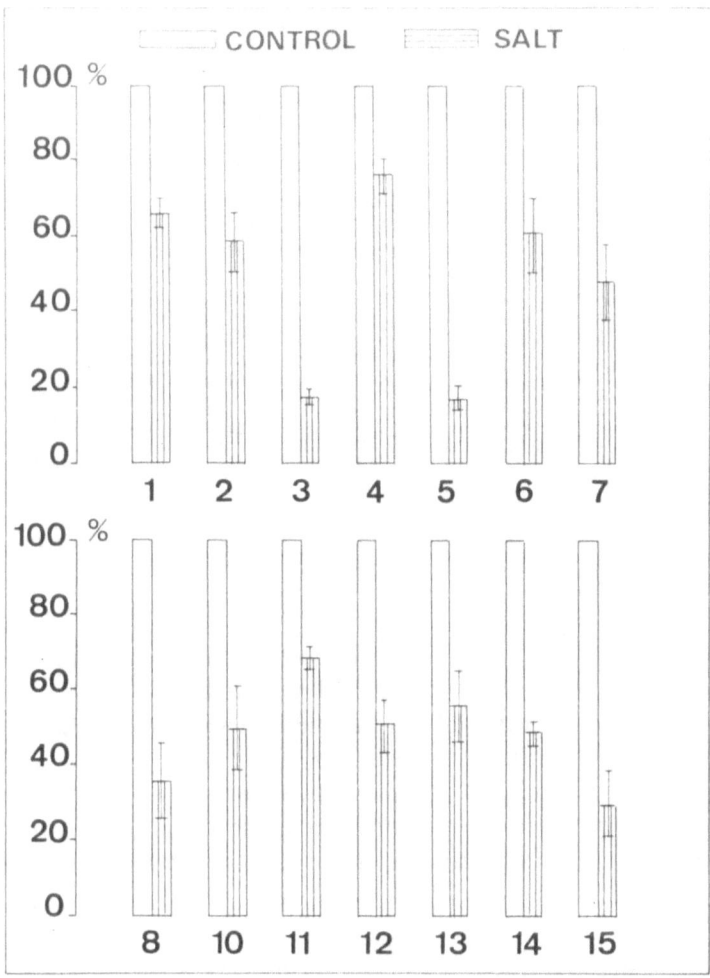

Fig. 5. Dry matter production after 9 weeks. Precentual expression of the salt treatment against the control plants, being set as 100% for each variety. Not taking into account the differences in growth between control plants.

development especially in the sodium and chloride content. These findings on ion content tended to confirm the classification made by measuring growth. On the other hand potassium did not show any correlation on a whole plant basis. Only by looking into more detail at the different plant parts could one find strong indications about the reaction to salinity, as will be discussed later on.

Measurements after nine weeks

Growth. The tendencies found in week five seemed to be confirmed but due to the season in which we grew our barley, the winter barley stayed

Table 2. Dry matter production of the ear, expressed in gram per plant and percent of the total plant dry weight, for salt and control plants. Also the percentual reduction of dry matter between ear of salt treated plants compared with these of control plants

Variety	Salt treatment		Control		Reduction
	g/pl.	% plant	g/pl.	% plant	%
1	0.1802	30%	0.1891	21%	−4%
2	0.0899	15%	0.1397	14%	−36%
3	–		0.0960	13%	−100%
4		–	–	–	
5	–		–	–	–
6	0.0747	12%	0.1597	15%	−53%
7	–	–	–		–
8		–	–	–	
9*					
10	–		0.1361	14%	−100%
11	0.0280	6%	0.1537	21%	−82%
12	0.0545	11%	0.1689	19%	−68%
13	0.0931	13%	0.2150	19%	−57%
14	0.1048	17%	0.2269	24%	−54%
15	–	–	0.1204	14%	−100%

in the vegetative state, resulting in appreciable reduction of dry matter production compared with the summer barley cultivars. One cultivar (Egmont) showed good growth and development at week nine where it seemed to lag behind the others at week five, (Fig. 5). Of course dry matter production is not the best criterion of salt resistance for a crop like barley where the production of seeds is the most important feature. In this respect we obtain similar results in expressing the dry matter of the ear as an absolute value of weight or relative to the plants own dry matter (Table 2).

Ion content. The results obtained at week five were confirmed. After nine weeks the ion composition in the ear is the most important feature, since the seeds are economically the main part and are important aswell

Table 3. Ionic composition of ear of salt treated plants. Total cation and Cl expressed in μeq/g dr wt. and the percentual distribution of ions within one treatment

Cultivar number	Cations	Chloride	K	Na	Ca
			(%)		
1	272	72	76.	20	4
2	612	245	61	35	4
6	736	268	86	12	2
11	714	286	60	35	5
12	954	92	58	40	2
13	655	193	75	23	2
14	601	162	81	15	4

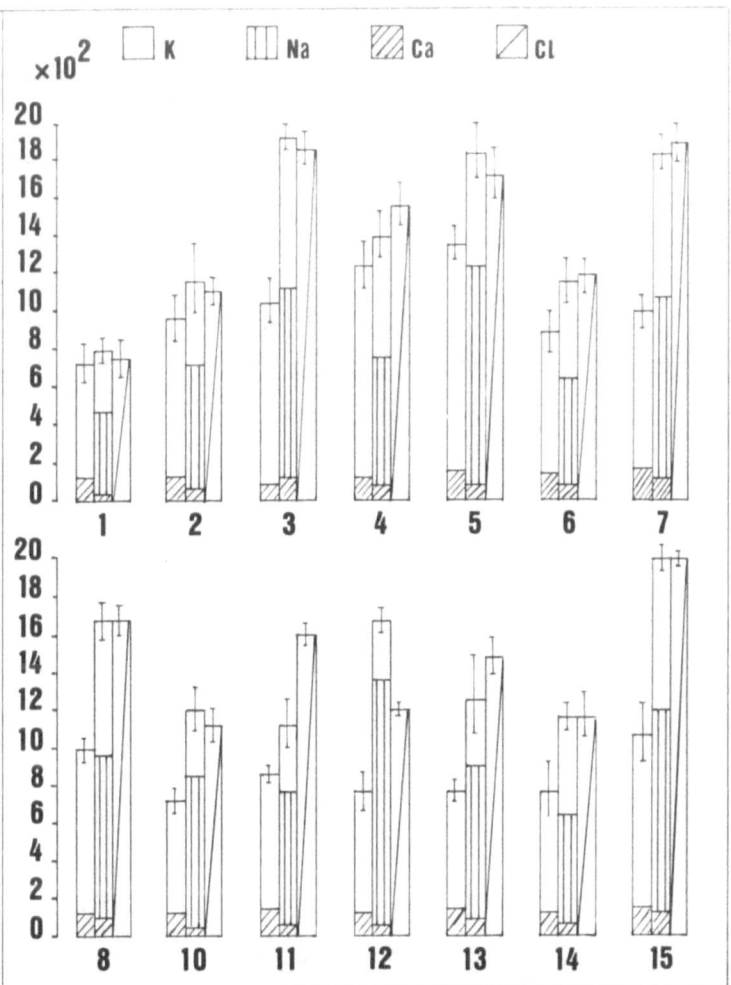

Fig. 6. Ionic composition of control and salt treatment plants after 9 weeks. Expressed in µeq/g dr wt.

for the following generations. Table 3 shows clearly that even for the more tolerant cultivars there can be large differences in the total ion content of the ear as well as in their relative importance in the pool. Evidently some cultivars can exclude sodium more efficiently than others. These are more specificly: Union (1), Menuet (2), Egmont (6) and Fripone (14). Intermediate would be the cultivars Herta (10) and Regent (11). Poor performers are Iban (12) and Legia (15). To check this classification of salt tolerance of barley cultivars, we checked their ability to segregate ions in the ear (Table 3) and found that the cultivars Union (1), Egmont (6) and Fripone (14) had the best K/Na ratios in the developing ear, and grew best under saline conditions.

Table 4. Ionic composition of whole plants versus young leaves for the salt treated plants. Only the plants coming to full development (shooting ear) were presented here. Results expressed in µeq/g dr wt

Cultivar number	Plant			Young leaves		
	K	Na	K/Na	K	Na	K/Na
1	326	434	0.75	372	460	0.81
2	358	700	0.51	420	645	0.65
6	499	632	0.79	532	508	1.05
11	325	712	0.46	391	610	0.64
12	282	1320	0.21	273	971	0.28
13	324	844	0.38	322	780	0.41
14	507	577	0.88	579	524	1.10

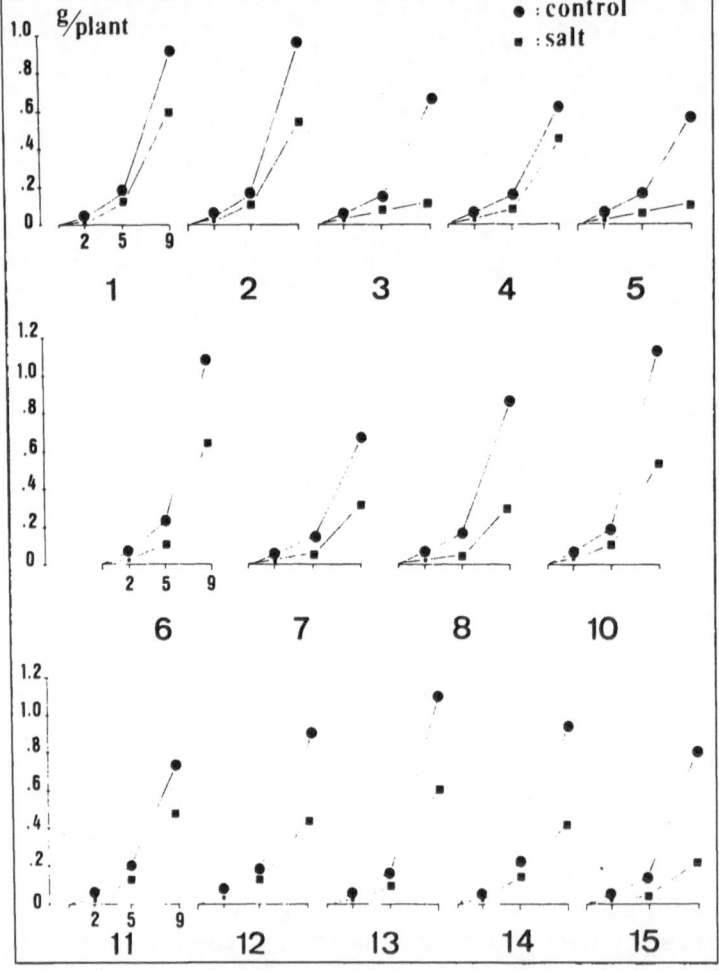

Fig. 7. Growth of whole plants over 9 weeks, in absolute value. Expressed in gram dry matter per plant for each variety.

By comparing ion content and K/Na ratio in the young leaves with that of whole plants one can find which are the most succesfull cultivars on this basis. As shown in Table 4, it seems clear that young leaves selectively exclude sodium. Most cultivars show a better K/Na ratio in young leaves than in whole plants, but again Union (1), Egmont (6) and Fripone (14) show the best ratios. So one finds a consistent trend in the most tolerant plants, their effectiveness in avoiding sodium from their young developing plant parts.

It is our aim now to get further research done on metabolic implications which would explain such differences in behavior under salt stress. Of course the more we look into the details of different plan parts the more we hope to reach an understanding of the processes involved in salt tolerance. Only then can we think of having a genetical approach of this problem.

Boron tolerance in Australian varieties of wheat and barley

B. CARTWRIGHT[*], A.J. RATHJEN[**], D.H.B. SPARROW[**], J.G. PAULL[**], and B.A. ZARCINAS[*]
[*] *CSIRO Division of Soils, PMB 2, Glen Osmond, South Australia 5064* and [**] *Waite Agricultural Research Institute, PMB 1, Glen Osmond, South Australia 5064*

Key words Barley Boron toxicity Tolerance Wheat

Summary Tolerance of boron among varieties of wheat and barley was studied in field trials on high-boron and normal soils in South Australia. Grain yield differed significantly between varieties, as did boron concentration in tops, and in grain at maturity. The relative order of yields of barley and wheat varieties varied between the sites of different soil boron status.

In wheat, both specific tolerance (relatively high yield despite accumulation of high concentrations of boron in tissues) and resistance (good yield with below-average tissue concentrations of boron) were identified in individual varieties.

Identification was made of groups of wheat varieties with common ancestries which showed tolerance or sensitivity to boron, respectively. These groups had significantly different yields and concentrations of boron in grain. Pedigree did not guarantee tolerance however, since boron-tolerant and sensitive varieties of wheat were identified as derivatives of a single parental cross.

The wide variation of nutrient absorption among varieties produced no general statistical relationship between yield and concentrations of boron, or associated elements such as calcium or sodium. Yield was not related to vigour of vegetative growth in the high-boron soil.

The results call into question the general applicability of 'critical' boron concentrations as indicators of toxicity status for genotypes of wheat and barley.

Introduction

Boron toxicity in semi-tolerant crops such as barley[6] growing in alkaline sodic soils has recently been recognised in Australia[3]. The only other documented occurrences of boron toxicity were in citrus and vine (boron-sensitive[6]) grown under irrigation in saline calcareous earths near Mildura, northwestern Victoria[15,19].

In the state of South Australia, boron toxicity was found to be more widespread than previously suspected[4], and consideration is therefore being given to selecting varieties of barley and wheat with improved tolerance of boron.

As there is presently no information on the sensitivity of Australian cereal varieties to boron toxicity, we report here comparisons of boron tolerance under field conditions among current commercial varieties and advanced selections of barley (*Hordeum vulgare* L.) and wheat (*Triticum aestivum* L.).

H.W. Gabelman and B.C. Loughman (Eds.), Genetic aspects of plant mineral nutrition.
ISBN-13: 978-94-010-8102-3
© *1987 Martinus Nijhoff Publishers, Dordrecht/Boston/Lancaster.*

Table 1. Some properties of a high-boron Natrixeralf at Two Wells, South Australia

Depth (cm)	pH	EC	Cl	CaCO$_3$ (%)	ESP (%)	Silt + Clay (%)	Mannitol extr. B (mg/kg)	Saturation extr. B (mg/kg)
		(1:5 suspension)						
		(mS/cm)	(mg/kg)					
0–10	7.8	0.18	20	0	12	34	3.6	0.6
10–20	8.6	0.52	320	0	28	64	52.8	16.0
20–30	9.2	0.97	450	1	31	68	104.0	18.9
30–40	9.4	1.47	620	8	32	65	101.0	17.5
40–50	9.3	1.37	410	19	33	68	82.5	16.1
50–60	9.3	1.39	490	14	34	63	67.4	13.9
60–70	9.3	1.38	500	11	35	59	57.3	12.4
70–80	9.5	1.64	760	8	34	59	51.2	11.5
80–90	9.5	1.64	760	8	34	60	45.3	9.6
90–100	9.5	1.47	520	7	34	59	42.9	7.9

Location: Lat. 34 35'S, Lon. 138 30' E; Two Wells, South Australia.
Elevation: 40 m, on flat coastal plain.
Rainfall: 396 mm.
Land use: cereal production and sheep grazing.
Soil: Typic Natrixeralf, clayey, mixed (calcareous), thermic.

Materials and methods

(a) Field trial sites

A trial site with soil containing a high concentration of soluble boron of natural origin was located at Two Wells, 25 km N of Adelaide, South Australia. The terrain was a flat coastal plain in a region which has a Mediterranean-type climate. The mean annual rainfall is 396 mm (winter maximum). The soil was uniform over this trial area, with a potentially toxic concentration of boron in the subsoil (Table 1). The soil was classified as Typic Natrixeralf, clayey, mixed (calcareous), thermic[20].

Trials serving as controls were carried out with barley and wheat at the Waite Agricultural Research Institute (WARI). This site is 30 km south of Two Wells, on the gently sloping Adelaide plain. The site has an annual rainfall of 627 mm. Extractable boron concentration at this site was 'normal' (Table 2). This soil was classified as a Typic Rhodoxeralf, clayey, thermic.

(b) Field trial design

The trials were carried out in association with the Interstate Barley and Wheat Variety Trial Programmes[10,14] (IBVT, IWVT) run by Australian plant breeding centres and state Departments of Agriculture. A trial in 1983 at Two Wells compared 25 varieties of barley in a nearest-neighbour design[22] with 4 replications. The barley lines included current commercial varieties and advanced breeding selections. A similar trial at WARI compared the performances of the same varieties.

In 1984 two trials using randomized complete block designs were carried out with wheat at Two Wells. The first of these compared boron tolerances among 10 current commercial varieties (with 4 replications). The second experiment used 150 lines (with 2 replications) to assess the variation in performance among a wide range of wheat genotypes under conditions of high boron supply. This trial included wheat selections from Australia and many other countries. Control data was obtained from IWVT and other selection trials at WARI.

Trial plots were 4.2 m length by 4 drill-row width. In each case the plots were sown with superphosphate at 100 kg/ha in accordance with local farming practice.

(c) Sample collection and analysis

At Two Wells a sample of 5 plants (whole tops) was selected at random from each plot of barley at boot stage in 1983, and near anthesis from wheat in 1984. Grain yields were measured at maturity.

Table 2. Some properties of a Rhodoxeralf soil at the Waite Agricultural Research Institute, South Australia

Depth (cm)	pH	EC (mS/cm)	Cl (mg/kg)	CaCO₃ (%)	ESP (%)	Silt + Clay (%)	Mannitol extr. B (mg/kg)
0–10	6.8	0.28	110	0	2.2	37	1.1
20–30	7.1	0.19	70	0	1.4	49	1.3
30–40	7.4	0.20	60	0	1.5	62	1.2
40–50	7.8	0.27	100	0.2	1.4	79	1.3
50–60	7.9	0.24	60	0.2	1.4	78	1.2
60–70	8.0	0.28	70	0.2	1.1	76	1.2
70–80	8.0	0.30	70	0.2	1.1	54	1.2
90–100	8.6	0.32	70	6.6	1.2	49	1.3
110–120	8.8	0.35	60	14.0	2.1	46	1.3

Location: Lat. 34°58′S Lon. 138°38′E; Urrbrae, South Australia.
Elevation: 122 m, on coastal plain adjacent Mt Lofty Ranges.
Land use: Cereal production and sheep grazing.
Rainfall: 627 mm
Soil: Typic Rhodoxeralf, clayey, thermic.

and a samples of grain reserved for analysis. Samples of grain (but not tops) were available from the IBV and IWV trials carried out at WARI.

Plant samples were carefully washed with distilled water and dried overnight at 80°C in a forced-draught oven. A sub-sample was ground in a stainless-steel mill to pass a 1 mm screen. Digests of plant material were prepared from 1 g oven-dried material for analysis of boron and 13 other elements simultaneously by inductively coupled plasma spectrometry[23]. Analytical results were checked against certified values for NBS Standard Reference Material No. 1571 (orchard leaves) and standard samples of barley and wheat.

Soil profile cores were collected at the trial sites using a hydraulically driven auger (5 cm diam.). Soil samples were taken at 10 cm depth intervals for chemical analysis. Soil samples were spread on aluminium trays and dried at 40°C in a forced-draught oven for several days. The samples were crushed and sieved to collect the fraction passing through a 2 mm screen. Determinations of pH, EC and chloride (in a 1:5 water suspension), exchangeable cations, and particle size fractions were carried out according to standard procedures[13]. Boron was extracted from soils at pH 8.5 with 0.1 M CaCl₂ and 0.05 M mannitol[2]. Saturation extracts of soils[17] were filtered over a period of 1 hour[11].

(d) Statistical analysis

Analysis of variance of yields and chemical analyses was carried out using a Nearest Neighbour routine[22], or with the MINITAB statistical software[18].

Results

(a) Soil properties

The soils at Two Wells and WARI were of similar texture, although the maximum concentration of silt + clay was greater, and occurred deeper in the WARI soil. At both sites the soils were highly alkaline, and pH increased with depth. Carbonates were present throughout the subsoil at Two Wells, but at WARI were only substantial below one metre depth. There was no evidence of salinity at WARI, and only minor accumulation of salts at Two Wells.

The soils differed markedly in sodicity and in concentrations of extractable boron. The WARI soil showed no evidence of sodicity and mannitol-extractable boron was within the 'normal' range throughout the profile. At Two Wells however, the profile was sodic throughout, especially in the subsoil. Extractable boron concentration increased sharply with depth, reaching a maximum where silt + clay was maximal. The concentration of soluble boron in the saturation extract was high, and within the range of potential toxicity even for boron-tolerant species of plants[17]. This soluble boron had a similar distribution in the soil profile to that of mannitol-extractable boron.

(b) Toxicity symptoms

Symptoms of boron toxicity on barley at Two Wells were generally as described by Eaton[6], but varied among varieties with respect to the size of brown lesions, and their density. Some varieties (*e.g.* Akka) were affected to the extent that stems were weakened and plants appeared lodged or storm-damaged. Other varieties (*e.g.* Clermont) remained strongly erect although showing characteristic foliar symptoms of boron toxicity. These effects have not previously been described in relation to boron toxicity, but have since been reproduced in a glasshouse (M.A. Kausar and B. Cartwright, unpublished). There were no symptoms of boron toxicity on plants grown at WARI.

Wheat grown at Two Wells showed different symptoms from those observed on barley. Specifically, the only visual symptom on wheat was gradual yellowing and necrosis from the tips of the older leaves until, in the most sensitive varieties, all leaves were affected. There was no occurrence of brown spotting or stem weakening.

(c) Tolerance of boron in barley

Yield and analytical data for barley grain at maturity and whole barley tops at boot-stage are shown in Tables 3 and 4, respectively. The tabulated data are for five commercial varieties widely grown in Australia at present, and an advanced selection (WI-2584) which was identified from these trials to have good tolerance of boron. The varieties are ranked in order of their grain yields at Two Wells. The tables also show the maximum, minimum and least significant differences (LSD) found for each parameter among the 25 varieties compared at Two Wells and WARI. As it is not possible to tabulate here all the data for every variety, these values allow the performance of the named varieties to be assessed relative to the complete sets.

There were highly significant differences in yields among varieties at both trial sites. Relative comparisons in yields were made by standardising the data as percentages of the mean yield of all varieties at each site.

Table 3. Grain yield and grain properties of some barley varieties from the 1983 trials

Site:	Two Wells				WARI			
Barley variety	Yield (g/plot)	% Site mean	1000-GM (g)	Grain boron (mg/kg)	Yield# (t/a)	% Site mean	1000 GM* (g)	Grain+ boron (mg/kg)
WI-2584	1287	147	35.4	9.8	5.81	127	39.8	1.3
Forrest	1041	119	41.5	9.1	5.02	109	46.2	1.5
Schooner	744	85	31.1	11.5	4.32	94	40.7	1.8
Galleon	696	80	28.9	11.1	4.80	105	42.1	1.3
Clipper	638	73	30.5	10.4	3.90	85	41.3	1.9
Stirling	586	67	27.9	14.2	4.55	99	40.6	1.3
Site maximum:	1287	147	43.8	14.2	5.81	127		
Site minimum:	439	50	21.3	7.2	3.03	66		
LSD: $P < 0.05$	60		4.0	4.3	0.26			
$P < 0.01$	110		6.1		0.40			
$P < 0.001$	242		9.8		0.66			

data from Nitschke[14].
* Means for 4 IBVT sites (Nitschke[14]).
+ Mean conc. boron in grain grown in WARI in 1984 and 1985 (S.D. = ±0.52).

Pedigrees:
WI-2584: ((Abed Deba*WI-2335)/22*(CD28*WI-2231)/165
Forrest: Atlas 57//(A16)/(Prior*Ymer)
Schooner: Proctor*4/Ethiopian line CI 3208-1
Galleon: ((Clipper*Hiproly)//(Proctor*CI-3576)
Clipper: Proctor/Prior A/Proctor/CI-3576
Stirling: Dampier//(A14)*Prior/Ymer/3/Piroline

Barley yields at WARI (Table 3) are from IBV-trial data reported by Nitschke[14]. At both sites variety WI-2584 gave the highest yield, and of the varieties considered here, Forrest gave the second highest yield. The relative yield advantage of these two varieties above the site mean yield was greater at Two Wells than at WARI.

The remaining varieties gave significantly lower yields than WI-2584 and Forrest, and their relative order differed between the two field sites. At WARI, Stirling was similar in performance to Schooner and Galleon, but gave significantly lower yield at Two Wells, and its relative performance declined compared to the WARI site. It was also observed that toxicity symptoms were especially pronounced on Stirling at Two Wells.

Boron concentrations in tops and grain at Two Wells were generally high, and within a range considered toxic[3]. At Two Wells, WI-2584 and Forrest had significantly lower boron concentrations in grain than Stirling. Grain of these varieties grown at WARI in 1984 and 1985 had boron concentrations in the range 1.3–1.9 mg/kg (Table 3), which is considered normal[3,12].

Differences in 1000-grain mass among the varieties were highly signifi-

Table 4. Yield and analysis of some barley varieties at boot-stage in the 1983 trial at Two Wells

Barley variety	DM/100 tillers (g)	Boron (mg/kg)	Ca/B	Sodium (mg/kg)
WI-2584	87.5	72.3	24.3	5630
Forrest	136.8	88.0	20.2	5325
Schooner	85.3	109.9	21.4	6790
Galleon	85.9	100.5	20.4	6250
Clipper	86.4	132.3	17.3	7050
Stirling	77.5	94.8	19.8	9147
Maximum:	220.3	156.1	26.9	10945
Minimum:	60.3	63.3	11.0	4250
LSD: $P < 0.05$	32.0		9.7	2530
$P < 0.01$	48.5			3835

cant. In general, the varieties which showed the most severe toxicity symptoms (e.g. Stirling) had markedly pinched grain and the lowest yields. Comparison with data for 1000-grain mass for the same varieties grown at other sites in the region (Table 3) shows that all the varieties suffered some reduction in grain filling at Two Wells, and the effect was severe in the case of Schooner, Galleon, Clipper and Stirling. However, there was no statistical relationship between grain boron concentration and 1000-grain mass which applied across all varieties.

In the highly sodic soil at Two Wells, the highest yielding varieties (e.g. WI-2584, Forrest) had significantly lower sodium concentration in tops than low-yielding varieties such as Stirling, though none of the sodium concentrations were considered toxic. Sodium concentrations were not statistically related to boron concentrations or to yield.

Significant differences were found among the population of 25 varieties for Ca/B ratio in tops and in grain, but among the varieties shown in Tables 3 and 4 this factor was not related to yield or boron concentrations. No general relationship was found between the intensity of toxicity symptoms or dry mass/100 tillers (Table 4) at boot-stage with either yield or absorption of boron.

(d) Tolerance of boron in wheat

The commercial varieties of wheat showed significant differences in yields at both field sites, though yields at Two Wells were generally lower than at WARI (Table 5). At Two Wells the yields of the highest yielding variety (Aroona) and the lowest yielding variety (Warigal) differed significantly (407 and 236 g/plot, respectively). However, the concentrations of boron in the tops of these varieties were not significantly different (76.6 and 77.1 mg/kg, respectively).

Table 5. Grain yield and boron concentration in plant tops of wheat varieties in the 1984 trials

Site:	Two Wells				WARI		
Wheat variety	Yield (g/plot)	% Site mean	Tops boron (mg/kg)	Grain boron (mg/kg)	Yield (g/plot)	% Site mean	Grain boron (mg/kg)
Aroona	407	123	76.6	4.9	630	97	(1.5 ± 1.2)*
Spear	398	120	66.1	4.0	715	110	
Bayonet	380	115	94.0	8.0	593	91	
Halberd	363	110	47.6	3.5	556	86	
Bindawarra	361	110	125.7	7.0	636	98	
Egret	298	90	88.9	6.4	NA	NA	
Oxley	296	89	100.6	7.8	633	98	
Kite	282	85	89.2	8.2	593	91	
Condor	271	82	100.2	7.3	729	112	
Warigal	236	71	77.1	5.0	753	116	
LSD: $P < 0.05$	122		39.7	1.1	59		
$P < 0.01$			57.7	1.7	109		
$P < 0.001$				2.7			

* Mean conc. boron in grain of 13 varieties grown in 1983.
NA not available.

Pedigrees:
Aroona: (WW-15*Raven)/24/43; WW-15 synonym Anza
Spear: (Sabre*MEC3*Insignia)
Bayonet: (Pitic-62*Glaive)
Halberd: Scimitar*Kenya C6042*Bobin//Insignia 49
Bindawarra: ((Mexico-120*Koda)*Raven)/122/16
Egret: (WW-15*Heron)
Oxley: (WW-80/2*WW-15)
Kite: Norin 10B/Eureka//5*Falcon
Condor: (WW-80/2*WW-15)
Warigal: (WW-15*Raven)/101/17

In contrast, the varieties Halberd and Bindawarra had boron concentrations in tops (47.6 and 125.7 mg/kg, respectively) and in grain (3.5 and 7.0 mg/kg, respectively) that were significantly different. Despite these differences, the varieties gave similar yields at Two Wells (363 and 361 g/plot, respectively).

Boron concentrations in grain grown at WARI were not obtained for the set of varieties grown at Two Wells. A normal[12] boron concentration of 1.5 ± 1.2 mg/kg was found in grain of 13 varieties in an adjacent trial. This data compares with a mean concentration of 6.1 ± 1.9 mg/kg in grain at Two Wells. Boron concentrations in tops near anthesis at Two Wells were more variable than found previously in barley, but again did not show any overall statistical relationship to final yield.

Several pairs of varieties showed contrasting relative yields at the two sites. For example, Warigal gave the lowest yield at Two Wells, but at

Table 6. Relative yields and boron concentrations in grain of some high-yielding and low-yielding lines selected from 150 lines

Highest yielding varieties[*]	% Site mean	Grain boron (mg/kg)	Lowest yielding varieties[s]	% Site mean	Grain boron (mg/kg)
(TJB*MKR)/TW1/10	136	6.6	(MMC*WIQ)/29/9	78	7.5
((WIQ*KP)*WIMH)/6/12	133	4.4	(Kinsman*CP)/TS19/7	77	10.0
(Bindawarra*MKR)/111	131	6.8	(MM*WIGL)*MKR)/24	74	8.4
Spear	130	4.8	(CP*Warimba)/GW7/15	70	8.9
WIQ*KP)*WIMH)/6	126	4.8	(MM*MMC)/47/S15	68	7.7
(PF*Bindawarra)/7/3	126	5.8	(Kinsman*CP)/TS19/8	60	7.7
(TJB*MKR)/32/17	125	7.8	(WIGL*MMC)/S1/10	58	9.1
High-yield mean		5.9	Low-yield mean		8.5
Grain boron LSD: $P < 0.01$		2.2			

[*] selections include Insignia-Halberd family, or ((Mexico-120*Koda)*Raven) [=MKR] in ancestry.
[s] selections include Warigal, (Champlein*Pitic-62) [=CP], or ((Mexico-22A*Mengavi)*Crim) [=MMC] in ancestry.

Pedigrees (see also Table 5):
Bindawarra: ((Mexico-120*Koda)*MKR)/122/16
CP: ((Champlein*Pitic-62)
Kinsman: U.K. variety
KP: (Kloka*Pitic-62)
MM: (Mexico-22A*Mengavi); Mexico-22A = Siete Cerros 66 (8156)
MMC: ((Mexico-22A*Mengavi)*Crim)
PF: (Pitic-62*Festiguay)
Spear: Sabre*MEC3*Insignia
TJB: Unreleased U.K. variety
Warimba: (Mexico-22A*Mengavi)/25/4
WIGL: Warigal = (WW-15*Raven)/101/17
WIMH: (Warimek*Halberd); Warimek = (Mexico-120*Koda)
WIQ: Wariquam = (Mexico-120*Quadrat)

WARI gave the highest yield, and yielded significantly better than Aroona. Also at WARI, Bindawarra yielded significantly better than Halberd; Spear yielded more than Bayonet; and Condor yielded more than Kite. In each of these cases at Two Wells, the yields of these varieties were similar or in the reverse order.

A second trial at Two Wells in 1984 sought to identify extremes of performance among a wide range of wheat genotypes under conditions of high boron supply. Among 150 lines compared, 'high-yield' and 'low-yield' groups were arbitrarily selected to include varieties yielding at least 20% above or below the site mean yield, respectively. Within these groups, some lines with pedigree relationships were identified (Table 6).

The pedigrees of these selections included (a) in the high-yield group: members of the Insignia-Halberd family (*e.g.* Spear), or the line ([Mexico-120*Koda]*Raven), *e.g.* Bindawarra; and (b) in the low-yield group:

Warigal, CP = (Champlein*Pitic-62), or members of the (Mexico-22A*Mengavi) family (MM, MMC; see Table 6).

The varieties in the highest-yielding group yielded 25 to 36% (site maximum) above the mean yield of all varieties, while in the lowest-yielding group the varieties yielded 22 to 42% (site minimum) below the site mean. The pooled yields of these two groups were significantly different at $P < 0.01$. The mean concentration of boron in grain of the high-yield group was significantly lower than that in the low-yield group ($P < 0.001$), probably due to the diluting effect of the higher yields.

Discussion

Recognition that boron toxicity may affect dryland cereal production in parts of South Australia[3,4] has stimulated interest in selecting varieties better adapted to potentially toxic soils. We consider here three aspects of the problem.

(a) Soil properties

The soils at Two Wells and WARI are representative of large areas used for cereal production near Adelaide. The soils are formed in colluvium derived from weathering of Precambrain tillites, shales and impure limestones of the adjacent Mt. Lofty Ranges. Sumartojo and Gostin[21] have shown that marine sediments of the Adelaide geosyncline may contain concentrations of total boron exceeding 200 mg/kg. This accounts for sporadic accumulations of toxic concentrations of boron in soils throughout this region. In affected soils, boron accumulation is usually found in the sodic, clayey subsoil, as at Two Wells (Table 1) or the Gladstone soil described by Cartwright et al.[3]. The A-horizons are of variable depth, and generally contain 'normal' concentrations[2] of mannitol-extractable boron, i.e. 0.5–5.0 mg/kg.

Where surface soils are sufficiently deep, a cereal variety with a shallow rooting habit would be able to avoid an underlying subsoil containing a toxic concentration of soluble boron. Seasonal conditions would then determine whether a toxicity problem would become apparent by controlling moisture availability in the topsoil. On the other hand, it is observed that erosion of the topsoil may lead to exposure of toxic subsoils, so that 'scald' patches become apparent in most seasons.

(b) Resistance and tolerance of boron

Tolerance is regarded here specifically as 'having potential for relatively high yield despite accumulation of above-average tissue concentrations of boron'. In contrast, resistant varieties are able to maintain yield potential by excluding boron (at least from tops), and thereby maintain below-average tissue concentrations of boron.

An illustration of these two types of response was provided by comparing boron concentrations in tops in wheat varieties of similar yield, for example in the resistant variety Halberd (47.6 mg/kg), and the tolerant variety Bindawarra (125.7 mg/kg; Table 5). Such genotypic variation in nutrient uptake is of course already known[5,7], and in some cases the mechanism of exclusion of the toxic element may not necessarily reside within the plant itself, but in symbiotically associated myccorhizae[1], for example.

The barley varieties grown at Two Wells showed a similar range in boron concentrations at the boot-stage to that found in wheat near anthesis, although the concentrations were more variable (Tables 4 and 5). Forrest and Stirling had numerically similar concentrations of boron at boot-stage (88.0 and 94.8 mg/kg, respectively), but gave highly significant differences in tiller dry mass and final yield. Stirling barley seems to be highly susceptible to boron toxicity, while WI-2584 and Forrest are relatively tolerant.

Sodium concentrations were within the normal range, as found previously[3], and so it is reasonable to interpret effects on yield as mainly due boron toxicity rather than soil sodicity. However, the lower accumulation of sodium by WI-2584 and Forrest may indicate a generally better adaptation to growth in sodic soils.

Final yield in barley was not closely related to the vigour of vegetative growth (Table 4). Although Forrest had greater tiller mass than WI-2584, it yielded significantly less grain. Moreover, the vegetative growth of WI-2584 was not different from that of much lower yielding Stirling at Two Wells.

The concentrations of boron we find[3] in cereal grain on high-boron soils are commonly much higher than those observed by Gupta[8,9] in barley affected by boron toxicity, and may indicate a real difference in adaptation by Australian varieties.

(c) Selection for tolerance of boron

Our experimental programme aims to obtain an understanding of the mechanisms of boron tolerance in cereals. It is also intended that the work should lead to production of boron-tolerant commercial varieties. The most suitable varieties by definition, must be capable of yielding well under field conditions. To establish the range of boron tolerance, a wide range of genotypes must first be compared in the field before detailed studies are carried out with the most promising selections grown in pot or solution culture.

The data presented here show that screening of varieties on the basis of low concentrations of boron in vegetative tissues would not necessarily identify the most useful lines. For example, the final yield of Bindawarra wheat was not significantly different from that of more highly ranked varieties, although it had a significant higher concentration of boron in tops (Table 5). Vegetative growth alone also was not a reliable indicator of final yield in the high-boron soil at Two Wells, as shown by the comparison between WI-2584 and Forrest varieties of barley (Table 4).

Varietal pedigree was potentially a useful indicator of boron tolerance, as shown by some common ancestries among respective groups of high-yielding and low-yielding wheats (Table 6). Even here caution is required since varieties derived from a common parental cross exhibited different performances at WARI and Two Wells. For example, the wheat varieties Aroona and Warigal are both derivatives from the cross (WW-15*Raven). Both these varieties had among the lowest concentrations of boron in tops and grain, but Warigal showed much greater sensitivity to boron than Aroona when their relative yields at the two sites are compared (Table 5).

Aroona tends to reach anthesis more rapidly than Warigal (or Halberd), and this earliness is accentuated when water supply is restricted (Rathjen[16]). These varieties had similar concentrations of boron in tops and grain at Two Wells, but Aroona yielded significantly more than Warigal (407 and 236 g/plot, respectively). In contrast, Warigal yielded more than Aroona at WARI (753 and 630 g/plot, respectively). The earliness of Aroona may serve as an 'avoidance' mechanism enabling development to maturity before it becomes necessary to seek moisture from a toxic subsoil.

In varieties which are later maturing and also deep-rooted, a toxic concentration of boron in the subsoil may act differentially on more sensitive varieties to limit access by roots to subsoil moisture, especially at the end of the growing season (November) when crops in southern Australia normally experience moisture-stress. We presently have no data on this aspect of the growth of these varieties.

The wheat varieties Condor and Oxley were also derivatives of a common cross, *i.e.* (WW-15*WW-80/2*). Condor gave the second highest yield (after Warigal) at WARI, and Oxley gave a significantly lower yield (729 and 633 g/plot, respectively). At Two Wells, Condor and Oxley produced similar yields (271 and 296 g/plot, respectively), which were only slightly above the lowest yielding variety (*i.e.* Warigal, 236 g/plot). The relatively greater decline in yield by Condor confirms that progeny of different sensitivities to boron may be derived from a single cross of parent lines.

These results call into question the general use of so-called 'critical' values for defining toxicity status of plants. It is clear that varieties may show degrees of tolerance or resistance to boron, and may give different yield responses relative to each other when exposed to high concentrations of soluble boron in soils. There seems little possibility therefore that a given critical value would have general applicability for the broad range of genotypic variation evident within barley or wheat.

Acknowledgements J W Chigwidden and M G Howe assisted with field work, and Ms L R Spouncer carried out chemical analyses.

References

1 Bradley R, Burt A J and Read D J 1981 Mycorrhizal infection and resistance to heavy metal toxicity in *Calluna vulgaris*. Nature 929, 335–337.
2 Cartwright B, Tiller K G, Zarcinas B A and Spouncer L R 1983 The chemical assessment of the boron status of soils. Aust. J. Soil Res. 21, 321–232.
3 Cartwright B, Zarcinas B A and Mayfield A H 1984 Toxic concentrations of boron in a red-brown earth at Gladstone, South Australia. Aust. J. Soil Res. 22, 261–272.
4 Cartwright B, Zarcinas B A and Spouncer L R 1986 Boron toxicity in South Australian barley crops. Aust. J. Agric. Sci. 37, (In press).
5 Clarkson D T and Hanson J B 1980 The mineral nutrition of higher plants. Annu. Rev. Plant Physiol. 31, 239–298.
6 Eaton F M 1944 Deficiency, toxicity and accumulation of boron in plants. J. Agric. Res. 69, 237–277.
7 Foy C D, Chaney R L and White M C 1978 The physiology of metal toxicity in plants. Annu. Rev. Plant Physiol. 29, 511–566.
8 Gupta U C 1971 Boron and molybdenum nutrition of wheat, barley and oats in Prince Edward Island soils. Can. J. Soil Sci. 51, 415–422.
9 Gupta U C 1972 Interaction effects of boron and lime on barley. Soil Sci. Soc. Am. Proc. 36, 332–334.
10 Hancock T W, Mayo O and Puckridge R J 1985 Interstate Wheat Variety Trials. Annual Report (1983 Programme). South Australian Dept. Agric., Adelaide.
11 Holmgren G G S, Juve R L and Geschwender R C 1977 A mechanically controlled variable rate leaching device. Soil Sci. Soc. Am. J. 41, 1207–1208.
12 Kleese R A, Rasmusson D C and Smith L H 1968 Genetic and environmental variation in mineral element accumulation in barley, wheat and soybeans. Crop Sci. 8, 591–593.
13 Loveday J (ED.) 1974 Methods of analysis of irrigated soils. Tech. Comm. No. 54, Commonw. Bur. Soils, Farnham Royal, U.K.
14 Nitschke A 1985 Interstate Barley Variety Trials. Annu Report (1983 Programme). South Australian Dept. Agric., Adelaide.
15 Penman F and McAlpin D M 1949 Boron poisoning in citrus. Vict. J. Dep. Agric. (Aust.) 47, 181–189.
16 Rathjen A J 1981 'Aroona'. Aust. Inst. Agric. Sci. J. 47, 234–235.
17 Richards L A (Ed.) 1954 Diagnosis and improvement of saline and alkaline soils. Agric. Handb. U.S. Dep. Agric. No. 60. Washington D.C.
18 Ryan T A, Joiner B L and Ryan B F 1982 Minitab reference manual. Pennsylvania State Univ., Philadelphia.
19 Sauer M R 1958 Boron content of sultana vines in the Mildura district. Aust. J. Soil Res. 9, 123–129.
20 Soil Survey Staff 1975 Soil Taxonomy. Agric. Handbook No. 436. U.S. Dept. Agric., Washington, D.C.

21 Sumartojo J and Gostin V A 1976 Geochemistry of the late Precambrian Sturt Tillite, Flinders Ranges, South Australia. Precamb. Res. 3, 243–252.
22 Wilkinson G N, Eckert S R, Hancock T W and Mayo O 1983 Nearest neighbour (NN) analysis of field experiments. J. Roy. Statist. Soc. B 45, 151–211.
23 Zarcinas B A and Cartwright B 1983 Analysis of soil and plant material by inductively coupled plasma-optical emission spectrometry. Optimisation of operating parameters, calibration of the spectrometer and quantification of inter-element interferences. CSIRO Aust., Div. Soils Tech. Pap. No. 45.

Australian research on tolerance to toxic manganese

B.J. SCOTT,
Agricultural Research Institute, P.M.B., Wagga Wagga, New South Wales, Australia 2650

D.G. BURKE
State Chemistry Laboratory, 5 Macarthur St., Melbourne, Victoria, Australia 3002

and T.E. BOSTROM
Electron Microscope Unit, University of Sydney, New South Wales, Australia 2006

Key words Australia *Brassica* spp. Distribution Manganese Organic acids Toxicity Tolerance *Triticum aestivum* Wheat X-ray analysis

Summary The occurrence of manganese toxicity in Australia is outlined and recent research on species and cultivar tolerance is briefly described. Manganese tolerance in *Brassica campestris, B. napus* and wheat is reported with emphasis on the characteristic of manganese tolerance in wheat.

X-ray analysis of manganese distribution in a tolerant wheat indicated that the manganese was not highly localised within the plant, although highest manganese levels occurred in the leaf epidermis and vascular bundles, and in the root cortex.

Accumulation of organic acids was associated with manganese damage to a sensitive wheat variety (Teal). The high aconitate, malate and citrate was interpreted as indicating a disruption in the Krebs cycle. The tolerant variety (Carazinho) was characterised by slight symptoms and limited organic acid changes, although the tissue levels of manganese were higher than those in the sensitive variety.

Introduction

Manganese toxicity has been reported on the north coast of New South Wales in French beans (*Phaseolus vulgaris*)[48,50], and soybeans (*Glycine max*; P.J. Desborough pers. com.) and toxicity in beans also occurs in southern Queensland (G.E. Rayment pers. com.). Manganese toxicity may also be responsible for the low legume component in some Queensland pastures[44]. Soils high in manganese have been reported on the southern tablelands of N.S.W. and manganese toxicity problems occur in lucerne, (*Medicago sativa*)[17], subterranean clover (*Trifolium subterraneum*)[8,9,31,32] and rape (*Brassica campestris*)[9]. On the south west slopes of N.S.W. manganese toxicity has been identified in subterranean clover[49] and lucerne[53]. This problem extends into north east Victoria as elevated levels of manganese have been recorded in wheat (*Triticum aestivum*)[52] and subterranean clover (J.R. Hirth pers. com.). Toxicity has been reported in Tasmania in a range of manganese sensitive species[35,36,56], but there appear to be no reports from South Australia and Western Australia[41,42].

H.W. Gabelman and B.C. Loughman (Eds.), Genetic aspects of plant mineral nutrition.
ISBN-13: 978-94-010-8102-3
© *1987 Martinus Nijhoff Publishers, Dordrecht/Boston/Lancaster.*

Interest in Australia in the use of plant tolerance to high manganese is due to several factors. The use of lime in correcting manganese toxicity is frequently limited, the problem is often sporadic and the cost of lime treatment in this country is relatively high. The effectiveness of lime in correcting manganese toxicity is limited in the following way:-

(a) Extreme climatic conditions *e.g.* waterlogging and dry, hot conditions can lead to toxicity even on limed soils[49]. Toxicity has been reported on near neutral soils in some species and circumstances[21]. This climatic effect makes for variability in toxicity both between and within a growing season.

(b) It has been shown[8] that soil high manganese can be present at depths greater than practical liming depths (10 to 15 cm in most Australian situations). Thus the source of high soil manganese may be from soil deeper than the practicable depth of lime use.

While lime is recommended in some cropping situation (*e.g.* vegetable crops in Tasmania and beans on the north coast of N.S.W.) many of the affected areas are used for extensive, low input grazing enterprises (*e.g.* the southern tablelands and slopes of N.S.W.). In these relatively low input and low return operations, lime use is difficult to justify economically.

Tolerance research

Tolerance in pasture grasses

Tropical grass species have been compared and found to vary widely in their tolerance to high manganese in solution culture[51]. Rayment and Verrall[44] found greater tolerance in kikuyu than white clover. Culvenor[12] found that *Phalaris aquatica* L. was relatively tolerant of high manganese, although differences did exist between lines. However, he believes that aluminium toxicity was likely to be more limiting to production on the southern tablelands of N.S.W.

Tolerance in pasture legumes

Tolerance in pasture legumes has been extensively studied in Australia[3,14,46,54]. Tolerance of manganese in legumes has been reviewed by Helyar[32] and a list of relative tolerances of a wide range of species is reported. A major problem with legume studies is the separation of rhizobial tolerance, host plant tolerance and the tolerance of the symbiotic system. Hutton, Williams and Andrew[33] assessed tolerance in *Macroptilium atropupureum* both with and without supplied mineral nitrogen. They concluded that lines varied in their reaction to high manganese and that their tolerance did not vary with the source of nitrogen. Presumably then, host plant tolerance was the dominant factor controlling nodulated plant performance.

Osborne, Pratley and Stewart[40] tested seven subterranean clover cultivars for tolerance to high solution manganese with mineral nitrogen and found a range of tolerance. A subsequent study[15] has identified a wider range of tolerance in seventy six lines tested. A sub set of eighteen cultivars was retested for tolerance with mineral and symbiotic nitrogen. Broadly, the results indicated that, provided the cultivar formed an effective symbiosis with the rhizobia used (in a non-Mn stress control), the tolerance with mineral nitrogen was related to that with symbiotic nitrogen.

Salisbury and Downes[47] tested lucerne with mineral nitrogen for manganese tolerance and found tolerant plants within all varieties tested in this outcrossing species. At Wagga we have confirmed these results using similar methods.

There is some evidence that the health of grazing animals may be damaged by high intakes of dietary manganese but the situation is not well underestood[20,55]. Also, tolerance to manganese in some plant species and cultivars is associated with tolerance of high tissue levels of manganese[14]. It is possible that selection or breeding for tolerance to manganese in some pasture species may increase plant tissue levels and cause animal health problems. Selection for low tissue manganese levels in a tolerant species (*Phalaris aquatica* L.) has been studied[13].

Tolerance in crop species

A range of crop (and some pasture) species has been tested for their manganese tolerance in solution culture experiments[12,14,54]. All these studies have identified a wide range of tolerance in the material tested. Considerable tolerance has been identified in centro (*Centrosema pubescens*), sunflower (*Helianthus annuus*), safflower (*Carthamus tinctorius*), sweet potato (*Ipomoea batatas*) and in oats (*Avena sativa*), sugar beet (*Beta vulgaris*) and lupin (*Lupinus angustifolius*).

With oilseed rape (*Brassica campestris*) one of us (B.J.S.) and N. Wratten, Wagga tested 42 lines and found some manganese tolerant plants in only four cultivars. These four tolerant groups have been both inter and intra crossed and reselected over one or two additional cycles and more uniformly tolerant populations have been obtained. Manganese tolerance in soybean has been extensively examined by an Australian group working at the University of Sydney. This group has identified a range of tolerance[11,28], and studied the effects of pH and calcium supply[27] and temperature[29] on tolerance. More recent studies have investigated growth and seed yield[22] and transport and distribution of manganese in two cultivars contrasting in tolerance[23]. Interactions with other plant nutrients have also been investigated[24,25]. The characteristic of manganese tolerance in soybean was found to be heritable[26].

At Wagga our main emphasis has been on cereals, particularly wheat, and we have tested for aluminium and manganese tolerances using sub-irrigated gravel beds[16,45]. This method is similar to that used by Hutton, Williams and Andrew[33] and described by Andrew[2]. Within wheat lines there is a wide range of tolerance to high levels of manganese in solution culture.

The cultivars most tolerant to manganese come from a range of countries, but South America is a major source. Examples are Cotipora and Carazinho (Brazil) and Collafen and Mexifen (Chile). The most sensitive cultivar tested by us was Teal (Australia). The basis of tolerance appeared to be the ability of tolerant cultivars to accumulate higher levels of manganese, in both tops and roots, than the sensitive cultivars but they were undamaged by these elevated tissue levels (Scott, Fisher and Spohr — unpublished). This has been noted by others[19].

The manganese tolerance character is clearly heritable and preliminary experiments using a Carazinho/Teal cross indicate that two recessive genes may control tolerance but this is not confirmed (J.A. Fisher pers. comm.). Breeding of tolerant lines has been undertaken[16] but in the absence of a rapid test for manganese tolerance in wheat a semi-commercial scale sub-irrigated gravel bed was used to test the breeding lines. Rapid testing of breeding material may be possible if a greater understanding of the plant reaction to manganese toxicity stress could be developed.

Distribution of manganese in a tolerant wheat

Carazinho (manganese tolerant) and Teal (manganese sensitive) wheats were grown in the sub-irrigated gravel bed under high manganese (180 mg Mm/l). The plants (tops and roots) were then tested by x-ray microanalysis to determine the distribution of manganese in various plant tissues.

Pieces of root about 4 mm long and small pieces from about the mid-region of the leaf were excised, placed in small copper stubs so as to present a transverse section (TS) upwards, and frozen in liquid nitrogen (LN_2). These were stored in LN_2 until required; the remainder of the plants were stored in a refrigerator. Only the leaf specimen from the Mn-tolerant variety was recovered from the LN_2 undamaged. It was therefore necessary to take further specimens from the refrigerated plants, which appeared to be still in good condition. Before analysis, the frozen specimens were transferred to an ultramicrotome equipped with a cryostat attachment, and the exposed surface was planed at −90°C to provide a flat transverse surface for morphological identification and analysis.

Microanalysis

The specimen stubs were transferred (without allowing significant warming or atmospheric moisture condensation) to the cold stage of a JEOL JSM-U3 SEM, equipped with an Edax energy-dispersive X-ray detector and an Ino-Tech Ultima II multichannel analyser combined with a Data General Nova 3 computer. Because of hardware problems, no quantitative processing of the X-ray spectra was possible, so Mn was determined in terms of the apparent presence or absence of a manganese peak in the spectrum. Specimens were not coated, and an accelerating voltage of 15 kV was used throughout.

Initially the cold stage was maintained at between −140°C and −120°C (at which temperatures the sublimation of ice from the specimen was very small), and the specimen was first analysed in the fully hydrated state. The temperature was then raised to about −60°C to etch (partially freeze-dry) the transverse surface. The temperature was lowered again when sufficient ice had been removed. This procedure increases the effective concentration of solutes from fresh-weight to approaching dry-weight percentages. Analyses were then carried out on different areas and cell types in the exposed transverse surface.

No Mn was detected in any of the specimens in the fully hydrated frozen state. In the Mn-sensitive plant (var. Teal), where lower Mn levels were expected, Mn was never found at more than barely detectable levels.

Mn could be detected after etching both in the leaf and root specimens from the Mn-tolerant plant (var. Carazinho) (Tables 1 and 2). In the leaf, Mn was mainly associated with the vascular bundles, and was also found in the epidermal cells. In the root, Mn was present in the cortex, but was not detected in the stele. It is possible that the Mn in the cortex is intercellular (apoplastic) as well as intracellular; it was not possible on etched material to distinguish analytically between these two locations.

The main problem encountered during these experiments was that of detection sensitivity. It is not surprising that Mn was not detected in the fully hydrated frozen material. Even after etching, the surface tissue may not be fully dehydrated, so Mn concentrations may still be below dry-weight values. Of course if the Mn is highly localised then the detection is greatly improved, but no such areas were found. In all analyses of etched samples, the peak for K was very prominent. Some analyses also showed the presence of smaller amounts of Cl, P, S, and Ca.

Table 1. Analysis for Mn in leaf sample of a tolerant variety (Carazinho).

Area analysed	Mn*
Epidermis (cells on adaxial and abaxial surfaces)	+ +
Mesophyll cells	+
Small vascular bundle	+ +
Large vascular bundles:	
Sheath cells (some cells showed no Mn)	
Phloem	+ +
Xylem	O to + + (variable)

* + +: present
+: trace
O: not detected

The presence of a large K peak in the spectra was good evidence that a large fraction (if not all) of the diffusible solutes was retained by the experimental procedure. The results from the leaf sample from the Mn-tolerant plant are perhaps the most reliable as this sample was frozen when fresh. The other samples were taken from refrigerated material, and it is possible that some redistribution of solutes could have occurred during storage.

Unfortunately it was not possible to establish whether there was any difference in the distribution pattern of manganese between the tolerant and sensitive varieties as no distribution could be obtained for Teal. The distribution in Carazinho was similar to that described by Memon et al.[38] for the manganese accumulator *Acanthopanax sciadophylloides* in that manganese was observed in epidermis and mesophyll. Memon et al. also observed high manganese in the bundle sheath cells in the petiole.

The lack of apparent strong localization of manganese in Carazinho contrasts with the observations of Blamey, Joyce, Edwards and Asher[5]. These authors found very strong localization of manganese in the trichomes of sunflower when exposed to high concentrations of manganese. The sunflower appeared to be tolerant of high manganese and may even excrete manganese to the exterior through these leaf hairs.

Table 2. Analysis for Mn in root sample of a tolerant variety (Carazinho).

Area analysed	Mn*
Cortex	
Overall areas (some areas showed only traces)	+ +
Inner cell just exterior to endodermis	+ +
Endodermis	O, + (variable)
Stele:	
Entire area during etching	O
phloem	O
Xylem	O

* as for Table 1.

Metabolic effects of high manganese in wheat

Manganese is specifically required for the photolytic activity in chloroplasts, for isocitrate dehydrogenase and malic enzyme activity and is a non-specific activator of some aerobic respiratory enzymes[6,39].

Little is known of the metabolic effects of high Mn levels in wheat, but in cotton toxicity is associated with destruction of auxin through increased activity of indole acetic acid oxidase, elevated activities of peroxidase and polyphenol oxidase, lower activities of catalase, ascorbic acid oxidase and glutathione oxidase, lower ATP content and lower respiration rate[18]. Manganese can also act as an error-producing factor during mitochondrial DNA replication and inhibits protein synthesis[4,43].

It is not yet known whether any of these biochemical activities is the primary target of manganese toxicity. However, the involvement of oxygenases in the plant response to elevated Mn levels and the strong oxidising power of Mn^{3+} indicates that Mn toxicity involves interference in biological oxidation-reduction systems. So to determine whether elevated levels of Mn had a more fundamental metabolic effect than that described for cotton a technique known as metabolic profiling[34] was applied to wheat grown at high Mn levels in solution culture. Since involvement in oxidation-reduction processes were indicated the class of metabolites selected for profiling were the organic acids as these are intermediates in the Krebs cycle providing reducing equivalents for subsequent oxidative phosphorylation through the electron transport chain.

Two wheat cultivars (Carazinho and Teal) were grown for 4 weeks on coarse gravel with nutrient solution as described earlier. The basal solution contained 0.5 mg Mn/l and additional Mn was added as $MnSO_4$ to supply increases of 0, 30, 90 and 180 mg Mn/l. After harvesting, organic acids were extracted and quantified using capillary chromatography[10]. Duplicates were dried, weighed and analysed for Mn.

Teal began to show some toxicity symptoms at 30 mg Mn/l and extensive marginal chlorosis was apparent at 90 mg Mn/l. At 180 mg Mn/l the dry matter yield of the tops of Teal was 38% of the control yield. Carazinho only displayed very slight toxicity symptoms when subjected to 180 mg Mn/l and there was no apparent reduction in dry matter yield. As $MnSO_4$ levels increased, both varieties increased their tissue levels of Mn. The tolerant variety (Carazinho) continued to accumulate Mn to much higher levels than the susceptible variety (Teal). At 180 mg Mn/l dry tops of Carazinho had 3867 ppm Mn while Teal had 1725 ppm. Clear toxicity symptoms seen in Teal at 30 mg Mn/l were associated with tissue levels of 632 ppm Mn.

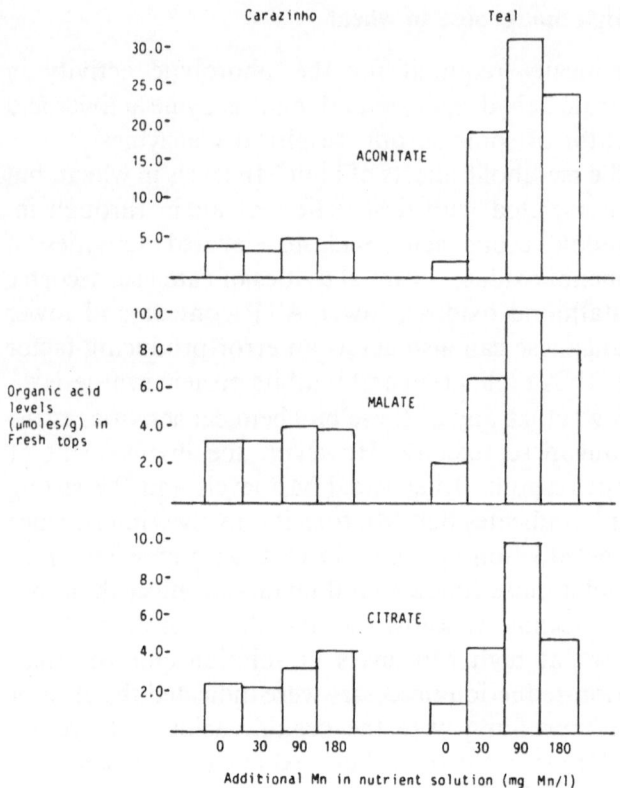

Fig. 1. Organic acid levels in the fresh tops of manganese tolerant (Carazinho) and sensitive (Teal) wheat cultivars after 4 weeks growth in nutrient culture with various solution manganese additions.

All three acids which were detected when Teal was grown under normal conditions began to increase at 30 mg Mn/l, but the rise in organic acid levels was most dramatic at 90 mg Mn/l (Fig. 1). At this level of Mn aconitate accumulated to 15 times the normal value and malate and citrate levels increased by a factor of 10. At 180 mg Mn/l the organic acid accumulation was not as marked, possibly because at this stage toxicity had reduced overall metabolic activity which resulted in decreased levels of Krebs cycle precursors and intermediates. Excluding the 180 mg Mn/l data, increased levels of organic acids correlated well with reduced dry matter yield and increased marginal leaf chlorosis. Carazinho showed only minor toxicity symptoms and a small increase in organic acid levels at the highest Mn level studied.

The accumulation of these acids can be explained by inhibition of one or more of the enzymes for which they are substrates. The block can be due to either a direct inhibitory effect of Mn on the activity of these few specific enzymes (malate dehydrogenase, aconitase, isocitrate dehydrogenase) or it can be an indirect effect of a secondary inhibitor or reduced availability of oxidation cofactors such as NAD^+. It is not yet clear whether the aconitate which accumulated in the sensitive variety was in the trans or cis configuration. Trans-aconitate is a product of both aconitate isomerase[1] and citrate dehydrase[7] and thus may be a detoxication product for removal of either isocitrate (*via* cis-aconitate) or of citrate from metabolically active sites in the cell to inactive vacuolar storage[37]. Resolution of the configuration of accumulating aconitate and determination of the proportions of citrate and isocitrate accumulated will be needed before a comprehensive explanation of metabolite accumulation through enzymatic inhibition can be proposed. The phenomenon of organic acid accumulation, though, provides a metabolic marker for future research into the primary biochemical lesion of manganese toxicity in wheat.

The data outlined above show that a significant metabolic disturbance occurred at a more fundamental level of cellular metabolism than has been previously reported. Though this disturbance may not be the primary site of manganese toxicity, this metabolic lesion would be sufficient to cause cellular disruption and ultimately plant death.

The tolerance of the variety Carazinho to high internal levels of Mn may be a function of: 1) altered sensitivity to enzymatic inhibition by Mn and 2) its ability to effectively remove Mn to storage pools where it is metabolically unavailable unless required.

The first hypothesis can be directly tested by comparing the activity of the appropriate enzyme from tolerant and sensitive plants under increased levels of Mn *in vitro*. However, if a negative result for this experiment is to be a valid indication of the latter hypothesis, further work on the biochemistry of toxicity is required to determine the exact enzymatic activity or activities blocked as a result of high cellular Mn levels.

Acknowledgement The authors wish to thank Dr. G C Cox for his cooperation in the x-ray analysis work reported in this paper.

References

1 Altekar W, Bhattacharyya P, Rangachari P, Maskati F and Rao M 1965 Aconitate isomerase in sugarcane. Indian J. Biochem. 2, 132-133.
2 Andrew C S 1974 Automatic sub-irrigation sand culture technique for comparative studies in plant nutrition Lab. Pract. 23, 20—21.
3 Andrew C S and Hegarty M P 1969 Comparative responses to manganese excess of eight tropical and four temperate legume species. Aust. J. Agric. Res. 20, 687–696.
4 Baranowska H, Ejchart A and Putrament A 1977 Manganese mutagenesis in yeast. ∛. On mutation and conversion induction in nuclear DNA. Mutat. Res. 42, 343- 345.
5 Blamey F P C, Joyce D C, Edwards D G and Asher C J 1986 Role of trichomes in sunflower tolerance to manganese toxicity. Plant and Soil 91, 171-180.

6   Bould C, Hewitt E J and Needham P 1983 Diagnosis of Mineral Disorders in Plants. Her Majesty's Stationary Office, London.
7   Brauer D and Teel M 1981 Metabolism of trans-aconitic acid in maize. I. Purification of two molecular forms of citrate dehydrase. Plant Physiol. 68, 1406–08.
8   Bromfield S M, Cumming R W, David D J and Williams C H 1983 Change in soil pH, manganese and aluminium under subterranean clover pasture. Aust. J. Exp. Agric. Anim. Husb. 23, 181–191.
9   Bromfield S M, Cumming R W, David D J and Williams C H 1983 The assessment of available manganese and aluminium status in acid soils under subterranean clover pastures of various ages. Aust. J. Exp. Agric. Anim. Husb. 23, 192–200.
10 Burke D G, Hilliard E P Watkins K, Russ V and Scott B J 1985 Determination of organic acids in seven wheat varieties by capillary gas chromatography. Anal. Biochem. (Submitted).
11 Carter O G, Rose I A and Reading P F 1975 Variation insusceptibility to manganese toxicity in 30 soybean genotypes. Crop Sci. 15, 730–732.
12 Culvenor R A 1986 Tolerance of *Phalaris aquatica* L. lines and some other agricultural species to excess manganese, and the effect of aluminium on manganese tolerance in *P. aquatica*. Aust. J. Agric. Res. 36, 695–708.
13 Culvenor R A, Oram R N and David D J 1986 Genetic variability for manganese concentration in *Phalaris aquatica*. growing in acid soils. Aust. J. Agric. Res. 37, 409–416.
14 Edwards D G and Asher C J 1982 Tolerance of crop and pasture species to manganese toxicity. Proceedings of the 9th International Plant Nutrition Colloquium. Warmick, UK: 145–150.
15 Evans J, Scott B J and Lill W J 1987 Manganese tolerance in subterranean clover (*Trifolium subterranean* L.) genotypes grown with nitrate or symbiotic nitrogen. Plant and Soil 97, 207–215.
16 Fisher J A and Scott B J 1984 Breeding wheat for acid soils. Proceedings of the Australian Plant Breeding Conference, Adelaide Feb. 1983, 74–75.
17 Flemons K and Siman A 1970 Goulburn lucerne failures linked with induced manganese toxicity. Agric. Gaz. N.S.W. 89, 662–663.
18 Foy C D 1984 Physiological effects of hydrogen, aluminium and manganese toxicities in acid soil. Agronomy: A Series of Monographs, No. 12. Soil Acidity and Liming, 2nd Edition. Am Soc. Agron. Inc; Madison, Wisconsin. pp 57–98.
19 Foy C D, Fleming A L and Schwartz J W 1973 Opposite aluminium and manganese tolerances in two wheat varieties. Agron. J. 65, 123–126.
20 Grace N D 1973 Effect of high dietary Mn levels on the growth rate and the level of mineral elements in the plasma and soft tissues of sheep. N.Z. J. Agric. Res. 16, 177–180.
21 Grasmanis V O and Leeper G W 1966 Toxic manganese in near-neutral soils. Plant and Soil 25, 41–48.
22 Heenan D P and Campbell I C 1980 Growth, yield components and seed composition of two cultivars as affected by manganese supply. Aust. J. Agric. Res. 31, 471–476.
23 Heenan D P and Campbell L C 1980 Transport and distribution of manganese in two cultivars of soybean (*Glycine max* (L.) Merr.). Aust. J. Agric. Res. 31, 943–949.
24 Heenan D P and Campbell L C 1981 Influence of potassium and manganese on growth and uptake of magnesium of soybeans (*Glycine max* (L.) Merr. cv. Bragg). Plant and Soil 61, 447–456.
25 Heenan D P and Campbell L C 1983 Manganese and iron interactions on their uptake and distribution in soybean (*Glycine max* (L.) Merr.) Plant and Soil 70, 317–26.
26 Heenan D P, Campbell L C and Carter O G 1981 Inheritance of tolerance to high Mn supply in soybeans. Crop Sci. 21, 626–627.
27 Heenan D P and Carter O G 1975 Response of two soya bean cultivars to manganese toxicity as effected by pH and calcium levels. Aust. J. Agric. Res. 26, 967–974.
28 Heenan D P and Carter O G 1976 Tolerance of soybean cultivars to manganese toxicity. Crop Sci. 16, 389–391.
29 Heenan D P and Carter O G 1977 Influence of temperature on the expression of manganese toxicity in two soybean varieties. Plant and Soil 47, 219–227.
30 Helyar K R 1970 Aluminium and manganese in acid soils. Agric. Gaz. N.S.W. 81, 572.

31 Helyar K R 1970 Excessive levels of aluminium and manganese in acid soils. Aust. Plant Nut. Conf. Mt. Gambrier, South Australia, Sept., 1970.
32 Helyar K R 1978 Effects of aluminium and manganese toxicity on legume growth. pp xx xx. In Mineral Nutrition of Legumes in Tropical and Sub-Tropical Soils. Eds. C S Andrew and E J Kamprath. Melbourne, Australia.
33 Hutton E M, Williams W T and Andrew C S 1978 Differential tolerance to manganese in introduced and bred lines of *Macroptilium atropurpureum*. Aust. J. Agric. Res. 29, 67 79.
34 Jellum E 1977 Profiling of human body fluids in healthy and diseased states using gas chromatography and mass spectrometry with special reference to organic acids. J. Chromatogr. 143, 427–462.
35 Lamp C A 1965 Manganese toxicity of marrowstemmed kale (*Brassica oleracea* L. var *acephela* D.C.) and its relation to some other plant nutrients. M. Agric. Sci. Thesis, University of Melbourne.
36 Lamp C A 1966 Root and related forage crops in Tasmania. Bulletin 41, New Series, Tasmanian Dept. of Agric.
37 MacLennan D and Beevers H 1964 Trans-aconitate in plant tissues. Phytochem. 3, 109–113.
38 Memon A R, Chino M, Takeoka Y, Hara K and Yatazawa M 1980 Distribution of manganese in leaf tissues of manganese accumulator: *Acanthopanax sciadophylloides* as revealed by electronprobe X-ray microanalyzer. J. Plant. Nutr. 2, 457–476.
39 Mengel K and Kirkby E A 1981 (3rd Edition) Principles of Plant Nutrition. International Potash Institute, Bern.
40 Osborne G J, Pratley J E and Stewart W D P 1981 The tolerance of subterranean clover (*Trifolium subterranean* L.) to aluminium and manganese. Field Crops Research 3, 347–358.
41 Porter W M, Cox W J and Wilson I 1980. Soil acidity — Is it a problem in Western Australia? J. Agric. West. Aust. 21, 126 133.
42 Porter W M and Yeates J S 1984 Effects on plant growth. J. Agric. West. Aust. 25, 123–124.
43 Putrament A, Baranowska H, Ejchart A and Jachymczykw W 1977 Manganese mutagenesis in yeast. VI Mn uptake, mitochondrial DNA replication and E^r induction comparison with other divalent cations. Mol. Gen. Genet. 151, 69–76.
44 Rayment G E and Verrall K A 1980 Soil manganese tests and the comparative tolerance of kikuyu and white clover to manganese toxicity. Trop. Grassl. 14, 101-114.
45 Read B J and Scott B J 1983 Tolerance of barleys to aluminium and manganese. Proceedings of the Australian Plant Breeding Conference, Adelaide Feb., 1983, 333–334.
46 Robson A D and Loneragan J F 1970 Sensitivity of annual *Medicago* species to manganese toxicity as affected by calcium and pH. Aust. J. Agric. Res. 21, 223–232.
47 Salisbury P A and Downes R W 1982 Breeding lucerne for tolerance to acid soils. Proceedings of the Second Australian Agronomy Conference. Wagga, New South Wales; 339.
48 Siman A, Cradock F and Autry-Hall F 1966 Manganese toxicity in beans at Terranora. Agric. Gaz. N.S.W. 77, 168-172.
49 Siman A, Cradock F W and Hudson A W 1974 The development of manganese toxicity in pasture legumes under extreme climatic conditions. Plant and Soil 41, 129–140.
50 Siman A, Cradock F W, Nicholls P J and Kirton H C 1971. Effects of calcium carbonate and ammonium sulphate on manganese toxicity in an acid soil. Aust. J. Agric. Res. 22, 201–214.
51 Smith F W 1979 Tolerance of seven tropical pasture grasses to excess manganese. Comm. Soil Sci. Pl. Anal. 10, 853–867.
52 Sparrow L A and Uren N C 1984 Monitoring manganese in acid soils and wheat in the field. Proceedings of the National Soils Conference. Brisbane, Australia. p 349.
53 Spencer K and Moye D V 1964 Excess manganese as a factor in lucerne establishment. Aust. Plant Nut. Conf., Perth, Western Australia, August 1964.
54 Temple-Smith M G and Koen 1982 Comparative response of poppy (*Papaver somniferum* L.) and eight crop and vegetable species to manganese excess in solution culture. J. Pl. Nutr. 5, 1153–1169.
55 Underwood E J 1977 In Trace Elements in Human and Animal Nutrition. Academic Press, New York. pp 189 190.
56 Wade G C 1961 Manganese toxicity of vegetables with particular reference to cauliflowers. Tasm. J. Agric. 32, 285-287.

Tolerance of plants to 'heavy metals'

R.H. MERRY
CSIRO Division of Soils, PMB 2, Glen Osmond, South Australia 5064

Key words Ectomycorrhizae Environmental factors Heavy metals Indicator species

Summary Australian research on the tolerance of plants to metal toxicity is summarized. Aspects of the effects of interactions with environmental factors such as temperature are discussed as are the effects of symbionts on uptake mechanisms, and the effects of metal tolerance on the distribution of indicator plants used in orebody location.

Introduction

The occurrence and physiology of 'heavy metal' tolerance and toxicity in plants have been extensively reviewed[1,7,12]. This paper presents some aspects of research carried out in Australia. Toxic metals are of concern in Australia as they affect both urban and rural environments, and there has been interest in the responses of plants to copper, lead and zinc as an aid to biogeochemical prospecting for orebodies. Recently, an account of sources of copper pollution of agricultural soils in Australia has been presented[13] which can be more broadly extended to other toxic metals and metalloids. Despite the general interest in metal uptake and toxicity in plants, there have apparently been no systematic attempts in Australia to study genetic factors and varietal differences, and their relationship with heavy metal tolerance. Although the specific details of the genetics of the tolerance of plants to heavy metals have not been extensively researched, inferences can be drawn about responses and tolerance to heavy metals. Some of these will be discussed.

What is heavy metal toxicity?

As has been observed previously[7], visual evidence of specific metal toxicity is not common (although, for example, the visual symptoms of manganese toxicity in cruciferous plants are fairly distinctive). The most striking visual effects usually results from release of excessive amounts of toxic metals in situations where, because of chemical form, the metals are readily available for uptake by plants. Examples are the effects on plant growth of materials released in tailings dam failures, mine spoil heaps, and of copper/chromium/arsenic compounds released following the rupture of tanks containing wood preservatives. In fact, some of the visual effects may not be directly due to the toxic heavy elements but to side

H.W. Gabelman and B.C. Loughman (Eds.), Genetic aspects of plant mineral nutrition.
ISBN-13: 978-94-010-8102-3
© *1987 Martinus Nijhoff Publishers, Dordrecht/Boston/Lancaster.*

effects resulting from increased soil acidity (aluminium and manganese toxicity), salt effects, or even deficiency of a major nutrient, induced or natural. Plants prone to copper toxicity are also usually prone to iron deficiency as, for example, citrus and some cultivars of soybean[3]. Perhaps the most important symptom of heavy metal toxicity is stunted roots which are not normally visible so that the main effects of toxicity are seen only as decreased top growth. For example, apple trees and clover plants have been observed growing in a soil (3000 mg kg^{-1} of copper extracted by EDTA, soil pH 5.5) contaminated with material from an abandoned copper mine. No visual symptoms were evident apart from stunted growth. Apparent heavy metal toxicities may not always be a simple response to excessive amounts of toxic metal and adequate characterisation should be carried out on both the growing medium, be it soil or nutrient solution, and critical element concentrations within the plant.

Effects of growing environment

In studies of toxic metals in soils it is difficult to obtain control soils that differ from contaminated soils only in the concentration of a specific element, and it is somewhat unusual for toxic elements to occur singly. It is in these respects, and the generally longer equilibration times in the field environment that experimentation with soils from the field differ most from glasshouse and nutrient culture studies where elements may be examined singly (by addition of salts) and may have unrealistically high availability to plants. Factors such as temperature, light intensity, confinement of the rooting zone and water availability have large effects on the response of a plant to nutrient elements[11] as well as toxic metals from sewage sludge[6].

Genotypic differences in the capacity for absoption of toxic metals in response to temperature can be illustrated by an experiment[8] where silver beet (*Beta vulgaris* L.), radish (*Raphanus sativus* L.) and subterranean clover (*Trifolium subterraneum* L.) were grown in an orchard soil contaminated with copper, lead and arsenic and maintained at soil temperatures of 12°C and 22°C. At both soil temperatures, the concentration of copper in silver beet was much greater than those in radish or subterranean clover. Concentrations of arsenic in subterranean clover, radish and silver beet were low at 12°C (Table 1) although significantly different from each other. At 22°C however, the arsenic concentration in subterranean clover was much greater than in both radish and silver beet (Table 1). The reason for this result for subterranean clover cannot be explained in terms of increased availability of soil arsenic since all three species would then be expected to show similar increases with tem-

Table 1. Effect of soil temperature on growth and composition of three plant species

Species	Subclover tops		Silver beet tops		Radish tops		Radish roots	
Soil temp.	12°C	22°C	12°C	22°C	12°C	22°C	12°C	22°C
Dry wt.+	14.4	25.5***	4.8	11.1***	5.2	7.7***	9.5	9.9
Copper	21	30***	77	134***	19	26**	11	16*
Lead	4.8	11.9***	3.3	9.6***	3.2	5.0**	8.4	13.2*
Arsenic	3.9	19.8***	1.4	2.4**	2.5	3.9*	2.5	2.1

+ Dry weight in g per pot, concentrations in mg/kg.
*, **, *** different at P less than 0.05, 0.01 and 0.001, respectively.
For experimental details see Merry et al.[8]

perature. Nor was it likely to be due to large differenes in root morphology since the root systems of subterranean clover and silver beet appear to be similar.

Effects of symbiotic organisms

Under normal growing conditions, mycorrhizal fungi infect the roots of most plants but their effects on heavy metals are usually not known and are rarely considered. Mycorrhizal infection by races of fungi isolated from contaminated spoil have been found to be involved in the exclusion mechanisms for copper and zinc in *Calluna vulgaris*[2]. The roots of *Pinus radiata* seedlings grown in a soil developed over a natural orebody (800 mg Pb kg^{-1}) were infected with ectomycorrhizal fungi (Plate 1, courtesy R.C. Foster). Under the electron microscope, electron-dense particles (arrowed) were found in the vacuoles (V) and encrusting gel (G) of mycorrhizal fungi which are not found in similar hyphae grown in soils with low lead concentration. These particles are too small to be analysed by the electron-probe microanalyser but have an electron density characteristic of lead deposits. Since symbiotic organisms also adapt to contaminated soils[2], they can be expected to be complicating factors in investigations of the genetics of metal tolerance in plants.

Toxicity and essential trace elements

The trace elements copper, zinc, manganese, *etc* are essential for plants for growth and reproduction. At low concentrations in the growing medium a plant must secure enough of each element for its requirements, but as concentrations in the growing medium approach toxic levels, the plant must have mechanisms to avoid the uptake or translocation of toxic amounts to vegetative and reproductive parts of the plant if it is to survive. These effects can be illustrated by wheat grain grown on soils contaminated by lead, zinc and cadmium from a smelter (Fig. 1).

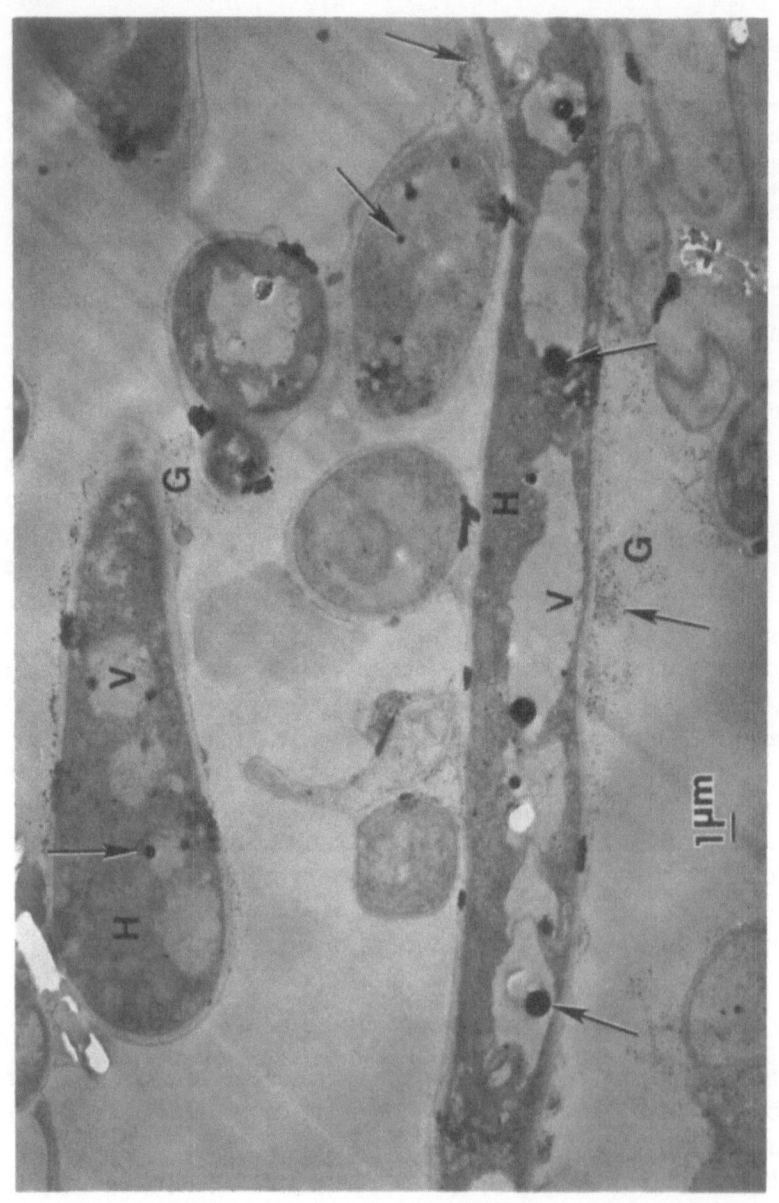

Plate 1. Particles (arrows) characteristic of lead deposits in the vacuoles (V) and gel (G) of ectomycorrhizal hyphae of *Pinus radiata* grown in soil containing 800 mg Pb kg^{-1} (courtesy R.C. Foster, CSIRO Division of Soils, Glen Osmond).

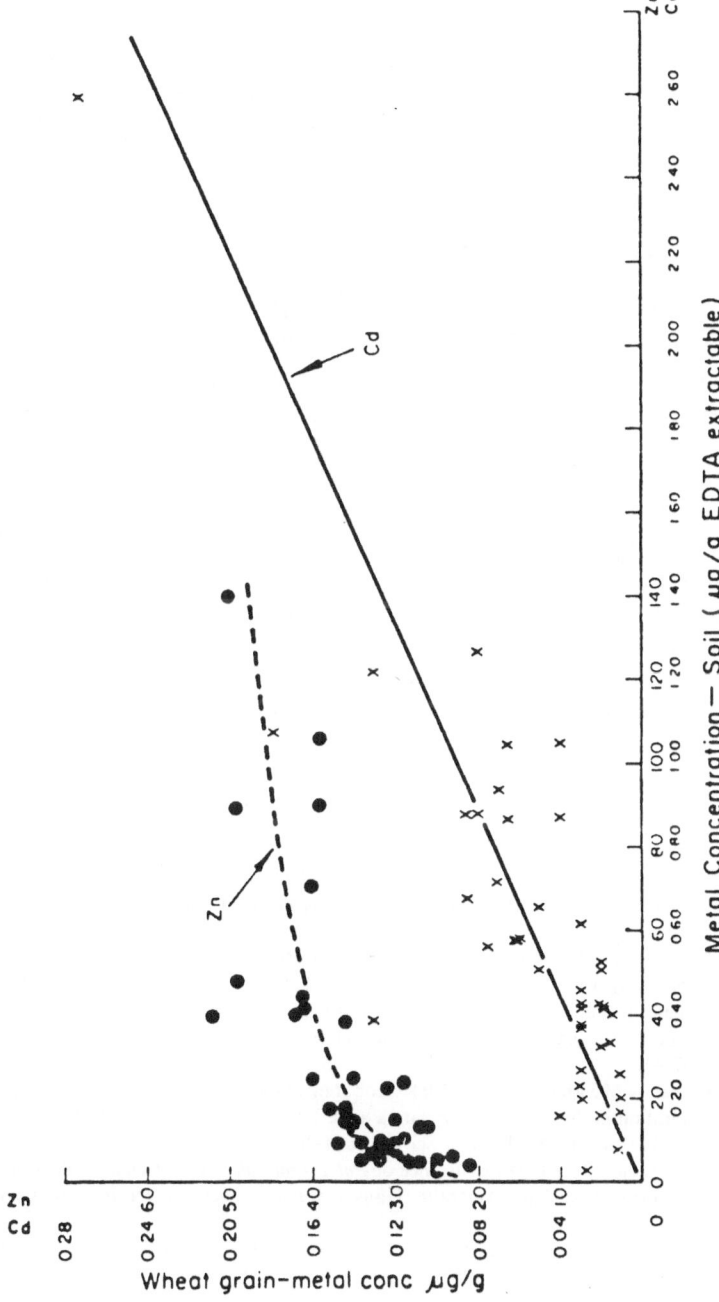

Fig. 1. Relationship of soil Cd and Zn to wheat grain Cd and Zn (from Merry et al.)[9].

Whereas the cadmium and lead (not shown) concentrations in the grain increased linearly with soil concentrations, zinc in the grain, which was initially near deficiency concentrations, increased logarithmically with soil concentration but did not increase above about 50 mg Zn kg^{-1} even when the supply of zinc in the soil reached much higher levels. The mechanism controlling zinc uptake under these conditions must operate at the root surface or during transport through the root since the same logarithimic form of uptake was observed for plant tops.

Indicator species and metal tolerance

A number of plant species, including *Polycarpaea* sp., have been used successfully by prospectors in northern Australia as indicators for copper, lead and zinc orebodies[4,10]. High concentrations of metals in the soil apparently control vegetation associations. *Polycarpaea glabra* accumulates zinc but is able to exclude copper (and lead), maintaining a leaf concentration of about 20 mg Cu kg^{-1} in plants growing on soils containing up to 1 percent of copper[10]. It is apparently the copper or lead that eventually proves toxic to the plant even though zinc concentrations in the plants may be much higher. *Polycarpaea spirostylis* also shows a strong association with high copper concentrations in the soil[5] although it was pointed out that other ecological factors which were not measured, such as microclimatic variation, rock cover, interspecific competition, inefficient seed dispersal, *etc.*, could equally well be influencing the distribution of the plants.

References

1. Antonovics J, Bradshaw A D and Turner R G 1971 Heavy metal tolerance in plants. Adv. Ecol. Res. 7, 1–85.
2. Bradley R, Burt A J and Read D J 1981 Mycorrhizal infection and resistance to heavy metal toxicity in *Calluna vulgaris*. Nature (London) 292, 335–337.
3. Brown J C, Holmes R S and Specht A W 1955 Iron, the limiting element in chlorosis: Part II Copper-phosphorus induced chlorosis dependent upon plant species and varieties. Plant Physiol. 30, 457-462.
4. Cole M M, Provan D M J and Tooms J S 1968 Geobotany, biogeochemistry and geochemistry in mineral exploration in the Bulman — Waimua area, Northern Territory, Australia. Trans. Instn. Min. Metall. (Sect. B: Appl. Earth Sci.) 77, B81–B104.
5. Correll R L and Taylor R G 1974 Occurrence of *Polycarpae spirostylis* at Daisy Bell copper mine, Emuford, north Queensland, Australia. Trans. Instn. Min. Metall. (Sect. B: Appl. Earth sci.) 83, B30- B34.
6. de Vries M P C and Tiller K G 1978 Sewage sludge as a soil amendment, with special reference to Cd, Cu, Mn, Ni, Pb and Zn — comparison of results from experiments conducted inside and outside a glasshouse. Environ. Pollut. 16, 231-240.
7. Foy C D, Chaney R L and White M C 1978 The physiology of metal toxicity in plants. Annu. Rev. Plant Physiol. 29, 511–566.

8   Merry R H, Tiller K G and Alston A M 1986 The effects of contamination of soil with copper, lead and arsenic on the growth and composition of plants. 1. Effects of season, genotype, soil temperature and fertilisers. Plant and Soil 91, 115–128.
9   Merry R H, Tiller K G, de Vries M P C and Cartwright B 1981 Contamination of wheat crops around a lead-zinc smelter. Environ. Pollut. (Series B) 2, 37 48.
10 Nicolls O W, Provan D M J, Cole M M and Tooms J S 1965 Geobotany and geochemistry in mineral exploration in the Dugald River area, Cloncurry district, Australia. Trans. Instn. Min. Metall. 74, 695–799.
11 Stephanson R C and Collis-George N 1974 The importance of environmental factors in soil fertility assessments. II Nutrient concentration and uptake. Aust. J. Agric. Res. 25, 309 316.
12 Thurman D A 1981 Mechanism of metal tolerance in higher plants. *In* Effect of Heavy Metal Pollution on Plants. Volume 2. Ed. N W Lepp. Applied Science Publishers, London pp 239 249.
13 Tiller K G and Merry R H 1981 Copper pollution of agricultural soils. *In* Copper in Soils and Plants. Eds. J F Loneragan, A D Robson and R D Graham. Academic Press, Sydney pp 119–137.

Hydrogen and aluminium tolerance

F.P.C. BLAMEY, C.J. ASHER and D.G. EDWARDS
Department of Agriculture, University of Queensland, St. Lucia, Queensland 4067, Australia

Key words Aluminium Hydrogen Soil acidity

Summary Reduced productivity due to soil acidity has been demonstrated with subterranean clover and wheat in many parts of Australia. Nodulation in clover appears to be more sensitive to low pH than growth of the host plant in the presence of adequate mineral nitrogen. Low pH is associated with aluminium toxicity in a number of species and nodulation in clover is more sensitive than growth of the host plant to Al. Decrease in soil pH is associated with significant increases in exchangeable Al. The breeding of Al tolerant wheats in Australia involves a backcross programme utilizing the transfer of tolerance from the Brazilian cultivar Carazinho.

Introduction

Although soil acidity was identified as a problem in the early 1960s in Australia, it has been only recently that the extent, or probable extent, of the problem has been recognized[9,10,18,32,41,43]. Poor plant productivity due to soil acidity factors has been reported in southern and eastern Australia[9,10,18,19,22,29,32,40,41], particularly in the subterranean clover (*Trifolium subterraneum*)/wheat (*Triticum aestivum*) belt, in Tasmania[34,39], in the south west of Western Australia[42], and in the high rainfall areas of northern Australia[13,35,38]. In addition to the relatively late recognition of the extent of soil acidity problems in Australia, many of the earlier studies overlooked the importance of toxic aluminium as a major factor contributing to poor plant growth on acid soils.

Effects of low pH on plant growth

For many years it was accepted that poor crop performance on acid soils was due to hydrogen ion toxicity. While nowadays this is not a widely-held hypothesis, soil pH does play an important role in nutrient availability and nutrient uptake by plant roots. Also, there is often a close relationship between soil pH and Al status in individual soils.

While pH has an important role in soil fertility, the effects of soil pH *per se* on plant growth are difficult to study. This results from the interrelationships between pH, Al concentrations in the soil solution, and the availability of essential nutrients, as well as differences between rhizosphere pH and the pH of the bulk of the soil.

To overcome these difficulties, flowing solution culture experiments have been used in which solution pH has been maintained within narrow

H.W. Gabelman and B.C. Loughman (Eds.), Genetic aspects of plant mineral nutrition.
ISBN-13: 978-94-010-8102-3
© *1987 Martinus Nijhoff Publishers, Dordrecht/Boston/Lancaster.*

limits[8,23,25]. Islam et al.[23] reported marked differences among plant species in response to solution pH. Ginger (*Zingiber officinale*) and tomato (*Lycopersicon esculentum*) produced maximum or near maximum dry matter yields at pH 7.5 to 8.5, whereas cassava (*Manihot esculenta*), French bean (*Phaseolus vulgaris*), wheat, and maize (*Zea mays*) produced maximum yields at pH 5.5 to 6.5. Within-species differences in response to pH have been reported for sunflower (*Helianthus annuus*)[8]. Critical solution pH (90% of maximum total dry matter yield) for four sunflower cultivars ranged from pH 3.9 for Hysun 11, through pH 4.3 for Hysun 32, to pH 4.5 for Hysun 21 and Hysun 31 (Fig. 1). A solution pH of 3.5 was lethal to all four cultivars tested.

Legumes are likely to be subject to the same effects of acidity as other plant species, and in addition there is a possiblity of specific effects on the legume-*Rhizobium* symbiosis. Recent research on subterranean clover has confirmed and extended earlier research[28] on the effects of pH on nodulation in this species. In a flowing solution culture experiment in which adequate nitrogen was supplied, growth of 12 subterranean clover

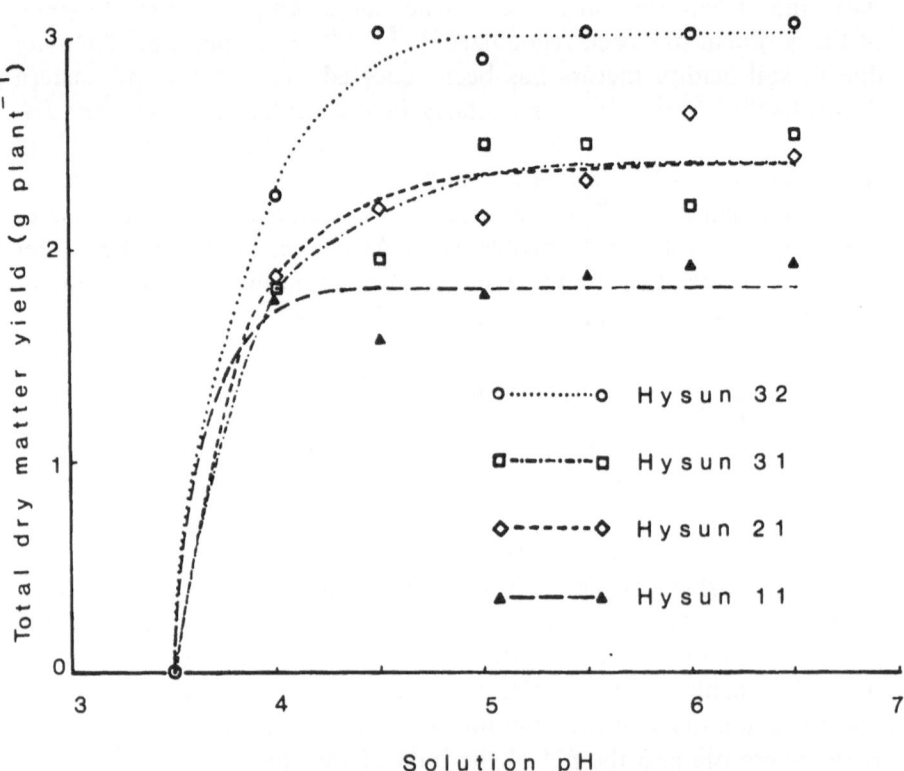

Fig. 1. Response of four sunflower cultivars to solution pH in flowing solution culture[8].

cultivars was vigorous at pH 4.0, and there was no response to pH over the range 4.0 to 6.5[26]. However, in a subsequent flowing culture experiment[27] in which no mineral nitrogen was added, nodulation and dry matter yields were severely depressed at pH 4.0 in all 11 cultivars studied. In 10 out of the 11 cultivars, there was substantial and statistically significant depressions in nodulation and dry matter yield also at pH 4.5. Hence, in the subterranean clover cultivars tested, nodulation was much more sensitive to low pH than was growth of the host plant in the presence of adequate mineral nitrogen.

Toxic Al effects in solution culture

Solution culture offers a suitable technique for the study of toxic Al concentrations on plant growth. However, the chemistry of Al in solution is complex, and Al may be precipitated from nutrient solutions through the addition of alkali to adjust solution pH or through the addition of phosphorus. Indeed, many studies have used pH levels and P concentrations that would result in considerable loss of Al from solution. In addition to precipitation, Al may be present in solution in polymeric forms. Recent research[7] strongly suggests that only monomeric Al is toxic to root growth.

In addition to the conversion of Al into biologically inert forms, the activity of monomeric Al is reduced by high the concentrations of nutrients used in many solution culture studies. This led Blamey *et al.*[7] to use the sum of the activities of the monomeric Al species (Al^{3+}, $Al(OH)^{2+}$, $Al(OH)_2^+$, $Al(OH)_3^0$ and $Al(SO_4)^+$) to predict the effect of Al toxicity on soybean (*Glycine max*) root elongation. Alva *et al.*[1] have further shown the advantage of using this approach over the use of total Al added, or the measured concentrations of total or monomeric Al in solution.

While these studies with whole plants[1,7] have been rather short term (c. 4 days), the principles established have provided a sound basis for the study of the longer term effects on plant growth of low Al concentrations ($< 100 \mu M$) as found in soil solutions. Nutrient concentrations, especially phosphorus, must be kept low. The techniques that have permitted the study of Al toxicity at realistic concentrations have been the flowing solution culture technique[2,4] and the programmed nutrient addition technique[3]. Plants that have been studied using these techniques include sunflower and sorghum (*Sorghum bicolor*) (Blamey *et al.*, unpublished), subterranean clover[25], and grain legumes (soybean, cowpea (*Vigna unguiculata*), mungbean (*Vigna radiata*), and peanut (*Arachis hypogaea*)) (Suthipradit *et al.*, unpublished).

In the case of aluminium toxicity there is again substantial evidence to suggest the existence of special effects on the legume-*Rhizobium* symbiosis. In six *Stylosanthes* spp., nodulation was delayed and depressed, and dry matter yields reduced when plants were grown at 25 μM Al in solution[16]. However, in plants supplied with adequate mineral nitrogen, the same Al concentration caused a significant yield reduction in only one of these species (*S. guianensis* cv. Cook). Subsequently, the nodulation of other tropical pasture legumes was found to be more sensitive to Al in solution than was growth of the host plant[30]. However, in *S. guianensis* cv. Oxley, host plant growth and nodulation appeared to be about equally sensitive to Al toxicity. In subterranean clover, host plant growth was significantly depressed only at Al concentrations $\geqslant 24 \mu M$, but nodulation was severely reduced at a concentration of 6 μM Al[24]. At 12 μM Al, nodulation was reduced to zero in six out of the 13 cultivars tested.

The exact nature of these effects of Al on nodulation is at present not well understood. In *Stylosanthes*, it was suggested that Al acted at the infection or nodule initiation stage[17]. There was no effect of Al concentrations up to 100 μM on the functioning of nodules once established[14] or on the survival of free-living Rhizobium (CB756) at Al concentrations up to 100 μM[15]. However, 6 μM Al has been found to markedly reduce the survival of an acid-tolerant strain of *Rhizobium trifolii* (RSP-24) both in nutrient solution and in the subterranean clover rhizosphere[24].

Toxic Al effects in soils

Reports of soil acidity problems have come from all states of Australia. In the south west of Western Australia[5,6] and in south eastern Australia[9,10,41], increasing soil acidity has been associated with the use of legume-based pastures and leys, particularly on the coarser textured soils. A decline in pasture persistence and productivity has been associated with decreases of approximately one pH unit in the surface 100 mm over a 50 year period[41]. The decrease in soil pH has resulted in appreciable increases in exchangeable Al. Exchangeable Al levels in the surface 100 mm increased from a mean of 0.05 meq 100 g^{-1} (2% Al saturation) in 0 to 10 year old pastures to 0.59 meq 100 g^{-1} (25% Al saturation) in pastures that were 35 to 51 years old[41]. However, in acid, Al-toxic soils low in available P from north-eastern New South Wales, liming increased the yield of white clover (*Trifolium repens* cv. Ladino) only where P fertilizer was applied[22].

In Queensland, the area of acid soils is small compared with the total potentially arable area in the state[12]. However, the acid soils largely occur along the better-watered eastern coast where soil acidity factors may severely limit crop and pasture production[11,38]. In particular, the extremely high exchangeable Al in the subsoil (up to 21.3 meq 100 g^{-1} or 84% Al saturation)[11] may severely limit root growth.

Sugarcane (*Saccharum* spp.) production has a long hisory on the acid soils of the Queensland coastal belt. The production of sugarcane, with the high rates of nitrogen fertilizer used, has led to further acidification of the soils, and there is concern regarding this trend despite the tolerance of sugarcane to Al toxicity.

In Tasmania, agricultural production is, with few exceptions, confined to soils which were extremely to moderately acid prior to development[31]. Pastures on krasnozems have responded to the application of lime[34]. Although the reasons for the pasture yield responses were not identified, yield responses in poppy (*Papaver somniferum*) were attributed primarily to the alleviation of Al toxicity in soils with 5 to 11% Al saturation (0 to 150 mm)[39]. High Al saturation in the subsoil caused poor root growth even in limed plots.

Most existing soil tests for Al can be calibrated satisfactorily within a particular soil type, but the calibrations often do not hold across a diversity of soils. In three Al toxic Queensland soils, soil solution Al concentration ranked the soils correctly in terms of Al toxicity to *Stylosanthes* spp, whereas exchangeable Al and Al saturation did not[13]. However, total Al in the soil solution may contain variable amounts of biologically inactive Al, including polymers and organic complexes. In view of the finding that monomeric Al was closely associated with soybean root growth in solution culture[7], it may be that the concentration, or better still the activity, of monomeric Al in the soil solution would provide a superior measure of Al toxicity. Research is being undertaken at the University of Queensland to evaluate this hypothesis.

Breeding for Al tolerance

It has long been recognized that genetic differences in tolerance to Al toxicity exist among plant species. For example, the growth of *Stylosanthes humilis* on an acid soil was better than that of *Desmodium intortum* because of better fine root development of the former species in acid soil layers[33]. The genetic differences in tolerance to Al toxicity within species has led to the breeding of cultivars with greater tolerance to Al toxicity. In Australia, emphasis as been placed on wheat breeding for Al tolerance[21]. A backcross breeding program is in progress to transfer the Al tolerance of the Brazilian cultivar, Carazinho, to Australian wheats. Included in the program is selection for resistance to stripe rust and *Septoria tritici* blotch. A mathematical model has been developed to provide two tolerance indices, *viz*. the slope of the initial yield decline with increasing Al concentration, and the Al concentration at which maximal yield depression is first reached[37]. Highest grain yields in cereals on acid soils (35 to 59% Al saturation) have been attained using a combination of liming and the planting of Al tolerant cultivars of wheat (cv. Olympic), triticale (cvv. Satu and Tyalla), and oats (*Hordeum vulgare* cv. West)[20,36].

References

1 Alva A K, Edwards D G, Asher C J and Blamey F P C 1985 Effects of P:Al molar ratio and calcium concentration on plant response to aluminum toxicity. Soil Sci. Soc. Am. J. 50, 133–137.
2 Asher C J 1981 External concentrations of trace elements for plant growth: use of flowing solution culture techniques. J. Plant Nutr. 3, 163 180.
3 Asher C J and Cowie A M 1970 Programmed nutrient addition — A simple method for controlling the nutrient status of plants. Proc. Aust. Plant Nutr. Conf., Mount Gambier, S.A. 1(b), 28 32.
4 Asher C J and Edwards D G 1983 Modern solution culture techniques. *In* Inorganic Plant Nutrition. Ed. A Lauchli and R L Bieleski. Encyclopedia of Plant Physiology. New Series 1, 94–119.
5 Barrow N J 1964 Some responses to lime on established pastures. Aust. J. Exp. Agric. Anim. Husb. 4, 30–33.
6 Barrow N J 1965 Further investigations of the use of lime on established pastures. Aust. J. Exp. Agric. Anim. Husb. 5, 442 449.
7 Blamey F P C, Edwards D G and Asher C J 1983 Effects of aluminum, OH:Al and P:Al molar ratios, and ionic strength on soybean root elongation in solution culture. Soil Sci. 136, 197 207.
8 Blamey F P C, Edwards D G, Asher C J and Kim M-K 1982 Response of sunflower to low pH. Proc. 9th Int. Plant Nutr. Coll., Warwick, U.K. 1, 66 71.
9 Bromfield S M, Cumming R W, David D J and Williams C H 1983 Change in soil pH, manganese and aluminium under subterranean clover pasture. Aust. J. Exp. Agric. Anim. Husb. 23, 189 191.
10 Bromfield S M, Cumming R W, David D J and Williams C H 1983 The assessment of available manganese and aluminium status in acid soils under subterranean clover pastures of various ages. Aust. J. Exp. Agric. Anim. Husb. 23, 192 200.
11 Bruce R C 1984 Chemical properties of acid B horizons of some Queensland soils. Proc. Natn. Soils Conf., Brisbane p. 314.
12 Bruce R C and Crack B J 1978 Chemical attributes of some Australian acid tropical and subtropical soils. *In* Mineral Nutrition of Legumes in Tropical and Subtropical Soils. Ed. C S Andrew and E J Kamprath. CSIRO, Melbourne pp 59–73.
13 Carvalho M M de, Andrew C S, Edwards D G and Asher C J 1980 Comparative performance of six *Stylosanthes* species in three acid soils. Aust. J. Agric. Res. 31, 61 76.
14 Carvalho M M de, Asher C J, Edwards D G and Andrew C S 1982 Lack of effect of toxic aluminium concentrations on nitrogen fixation by nodulated *Stylosanthes* species. Plant and Soil 66, 225 231.
15 Carvalho M M de, Bushby H V A and Edwards D G 1981 Survival of rhizobium in nutrient solutions containing aluminium. Soil Biol. Biochem. 13, 541 542.
16 Carvalho M M de, Edwards D G, Andrew C S and Asher C J 1981 Aluminum toxicity, nodulation, and growth of *Stylosanthes* species. Agron. J. 73, 261 265.
17 Carvalho M M de, Edwards D G, Asher C J and Andrew C S 1982 Effects of aluminum on nodulation of two *Stylosanthes* species grown in nutrient solution. Plant and Soil 64, 141 152.

18 Cregan P D, Sykes J A and Dymock A J 1979 Pasture improvement and soil acidification. Agric. Gaz. N.S.W. 90(5), 33 35.
19 Cumming R W, Bromfield S M and David D J 1984 Soil profile changes and longevity of incorporated and surface applied lime. Proc. Natn. Soils Conf., Brisbane p. 312.
20 Doyle A D and Bradley J 1982 Lime for cereals on acid soils in northern New South Wales. Proc. 2nd Aust. Agron. Conf., Wagga Wagga. Aust. Soc. Agron., Melbourne p. 260.
21 Fisher J A and Scott B J 1984 Breeding wheats for acid soils. In Annual Res. Rept. Agric. Res. Inst., Wagga Wagga and Agric. Res. Adv. Stn., Temora. Dept. Agric. N.S.W. p. 14.
22 Holford I C R 1985 Effects of lime on yields and phosphate uptake by clover in relation to changes in soil phosphate and related characteristics. Aust. J. Soil Res. 23, 75–83.
23 Islam A K M S, Edwards D G and Asher C J 1980 pH optima for plant growth. Results of a flowing solution culture experiment with six species. Plant and Soil 54, 339 357.
24 Kim M-K, Asher C J, Edwards D G and Date R A 1985 Aluminium toxicity: Effects on growth and nodulation of subterranean clover. Proc. XV Int. Grasslands Congr., Kyoto (In Press).
25 Kim M-K, Edwards D G and Asher C J 1985 Aluminium limitations on growth of subterranean clover cultivars supplied with inorganic nitrogen. Aust. J. Agric. Res. (In Press).
26 Kim M-K, Edwards D G and Asher C J 1985 Tolerance of Trifolium subterraneum cultivars to low pH. Aust. J. Agric. Res. 36, 569–578.
28 Loneragan J F and Dowling E J 1958 The interaction of calcium and hydrogen ions in the nodulation of subterranean clover. Aust. J. Agric. Res. 9, 464–472.
29 McLachlan K D 1980 Nutrient problems in sown pasture on an acid soil. 2. Role of lime and superphosphate. Aust. J. Exp. Agric. Anim. Husb. 20, 568–575.
30 Murphy H E, Edwards D G and Asher C J 1984 Effects of aluminium on nodulation and early growth of four tropical pasture legumes. Aust. J. Agric. Res. 35, 663–673.
31 Nicolls K D and Dimmock G M 1965 Soils. In Atlas of Tasmania. Ed. J L Davies. Lands and Surveys Dept., Hobart pp. 26 29.
32 Osborne G J and Wright W A 1978 Increasing soil acidity threatens farming system. Agric. Gaz. N.S.W. 89(1), 21.
33 Pinkerton A and Simpson J R 1983 Effects of subsoil acidity and phosphorus placement on growth, root development and phosphorus uptake by Stylosanthes humilis and Desmodium intortum. Aust. J. Agric. Res. 34, 109 118.
34 Rowe B A 1982 Effects of limestone on pasture yields and the pH of two krasnozems in north-western Tasmania. Aust. J. Exp. Agric. Anim. Husb. 22, 100 105.
35 Russell J S 1966 Plant growth on a low calcium status solodic soil in a subtropical environment. I. Legume species, calcium carbonate, zinc and other minor element interactions. Aust. J. Agric. Res. 17, 673–686.
36 Scott B J 1982 Resposnes to lime by cereals. Proc. 2nd Aust. Agron. Conf., Wagga Wagga. Aust. Soc. Agron., Melbourne p. 260.
37 Spohr L J 1984 Analysis of aluminium tolerance data. In Annual Res. Rept. Agric. Res. Inst., Wagga Wagga and Agric. Res. Adv. Stn., Temora. Dept. Agric. N.S.W. pp. 84–85.
38 Teitzel J K and Bruce R C 1972 Fertility studies of pasture soils in the wet tropical coast of Queensland. 4. Soils derived from metamorphic rocks. Aust. J. Exp. Agric. Anim. Husb. 12, 281 287.
39 Temple-Smith M G, Wright D N, Laughlin J C and Hoare B J 1983 Field response of poppies (Papaver somniferum L.) to lime application on acid krasnozems in Tasmania. J. Agric. Sci., Camb. 100, 485–492.
40 Vimpany I and Bradley J 1984 Lime requirement of acid soils. Proc. Natn. Soils Conf., Brisbane 309 p.
41 Williams C H 1980 Soil acidification under clover pasture. Aust. J. Exp. Agric. Anim. Husb. 20, 561–567.
42 Yeates J S and McGhie D A 1984 The effects of soil acidity on pasture production in the high rainfall areas of south west Western Australia. Proc. Natn. Soils Conf., Brisbane p. 310.
43 Yeates J S, Porter W M, Robson A D and Glencross R N 1984 Proc. Natn. Soil Acidity Workshop, Western Australia, September 2 7, 1984 pp. 76.

Tolerances of Old World bluestems to an acid soil high in exchangeable aluminum

C.D. FOY,
Plant Stress Laboratory, Plant Physiology Institute, USDA-ARS, Beltsville, MA 20705 USA

W.A. BERG and C.L. DEWALD
Southern Plains Range Research Station, USDA-ARS, Woodward, OK 73801, USA

Key words Al toxicity *Bothriochloa caucasica* *Bothriochloa intermedia* *Bothriochloa ischaemum* Low imput agriculture Marginal soils Warm season grasses

Summary Twenty-nine genotypes of Old World bluestems (*Bothriochloa intermedia, B. ischaemum* and *B. caucasica*) were screened for Al tolerance in greenhouse pots of acid Tatum subsoil which was unlimed at pH 4.1 and limed at pH 5.3. Three strains of weeping lovegrass (*Eragrostis curvula*) and El Reno side-oats grama grass (*Bouteloua curtipendula*) were also included as indicators of acid and alkaline soil tolerance, respectively. At pH 4.1 only 5 of the 29 bluestems and the 3 weeping lovegrasses produced measurable yields of tops or roots. The remaining 24 bluestems and side-oats grama either died or barely remained alive (due to frequent watering) with no appreciable growth. Weeping lovegrass was significantly more tolerant to the acid soil than any of the bluestems; relative top yields (pH 4.1/pH 5.3) were 101, 94 and 79% for the FQ71, common and FQ22 strains, respectively. Among the 5 bluestems that survived at pH 4.1, relative top yields ranged from 19 to 46%. Bluestems PI 300860 and PI 300857 (both *B. intermedia*) appeared more tolerant than PI 300886 (*B. intermedia*) and PI 312442 (*B. caucasica*) with PI 300858 (*B. intermedia*) being intermediate; however, all 5 showed promise for use on acid soils that are high in exchangeable Al. Genotypes that failed to grow at pH 4.1 included members of *B. intermedia, B. ischaemum* and *B. caucasica*. Some of these, such as PI 300825 (*B. intermedia*) and PI 300765 (*B. caucasica*), were among the highest yielders at pH 5.3. None of the 10 genotypes of *B. ischaemum* survived at pH 4.1.

Introduction

Old World bluestems of the genus *Bothriochloa* are warm season grasses with considerable potential for use in beef production and soil conservation in areas where tall fescue, a cool season grass, is now the principal pasture species[13]. The Old World bluestems were introduced primarily to reclaim marginal cropland and to complement native pastures in the southern Great Plains, USA. To date, a wide range of germplasm has been evaluated for forage quality, tolerance to low temperature and drought and overall forage and beef production potential. However, this work has been conducted primarily on near-neutral or alkaline soils of reasonably high fertility[1,2,11]. For best use of such germplasm, additional information is needed concerning tolerances to mineral stresses of toxicity or deficiency.

Aluminum toxicity, associated with soil acidity, is a major growth-limiting factor for plants on many marginal soils[3]. Even where surface soils are limed, excess soluble or exchangeable Al in acid subsoils may

H.W. Gabelman and B.C. Loughman (Eds.), Genetic aspects of plant mineral nutrition.
ISBN-13: 978-94-010-8102-3
© *1987 Martinus Nijhoff Publishers, Dordrecht/Boston/Lancaster.*

Table 1. Characteristics of Tatum subsoil at two lime levels

CaCO$_3$ added (g kg^{-1})	Final pH*	Exchangeable cations** c mol (+) kg^{-1}					Sum of cations	Al sat. (%)***	Exchangeable Mn^{2+} (µg g^{-1})
		Ca^{2+}	Mg^{2+}	K$^+$	Na$^+$	Al^{3+}			
0	4.1	0.15	0.52	0.68	0.52	4.93	6.79	72.6	4.1
3	5.3	7.74	0.27	0.52	0.55	1.07	10.15	10.5	1.6

* pH was determined on 1:1 soil to water suspensions of composite samples from each lime level.
** Ca, Mg, K, Mn and Na were extracted with molar NH$_4$OAC at pH 7.0; Al was extracted with molar KCl.
*** $\dfrac{\text{KCl extr. Al}}{\text{KCl extr. Al} + \text{NH}_4\text{OAc extr. Ca} + \text{Mg} + \text{K} + \text{Na}} \%$

Table 2. Description of Old World bluestems and other grasses used in the experiment

Entry	PI	Species	Origin
442	312442	*Bothriochloa caucasica*	USSR
765	300765	*B. caucasica*	USSR
477	301477	*B. ischaemum*	Unknown
481	301481	*B. ischaemum*	Unknown
505	301505	*B. ischaemum*	Burma
506	301506	*B. ischaemum*	Hong Kong
516	501516	*B. ischaemum*	India
517	301549	*B. ischaemum*	Pakistan
535	301535	*B. ischaemum*	Afghanistan
573 (WW Spar)	301573	*B. ischaemum*	Pakistan
599	301599	*B. ischaemum*	Pakistan
602	301602	*B. ischaemum*	Pakistan
604	301604	*B. ischaemum*	Pakistan
607	301607	*B. ischaemum*	Pakistan
Ganada	107017	*B. ischaemum*	Turkestan
776	300776	*B. ischaemum*	Japan
807	300807	*B. intermedia*	Pakistan
811	300811	*B. intermedia* var. *indica*	Pakistan
814	300814	*B. intermedia* var. *indica*	India
815	300815	*B. intermedia* var. *indica*	Pakistan
820	300820	*B. intermedia* var. *Montana*	Pakistan
822	300822	*B. intermedia* var. *indica*	Pakistan
825	300825	*B. intermedia* var. *Montana*	Pakistan
857	300857	*B. intermedia* var. *Montana*	Pakistan
858	300858	*B. intermedia* var. *Montana*	Pakistan
860	300860	*B. intermedia* var. *Montana*	India
886	300886	*B. intermedia* var. *Montana*	India
893	300893	*B. intermedia*	India
908	300908	*B. intermedia*	Oklahoma, USA
El Reno	Side oats grama	*Bouteloua curtipendula*	
Common WL	Weeping lovegrass	*Eragrostis curvula*	Africa
FQ 22WL	Weeping lovegrass	*Eragrostis curvula*	Africa
FQ 71 WL	Weeping lovegrass	*Eragrostis curvula*	Africa

restrict root development and thereby reduce the use of subsoil water and nutrients by plants. Hence, Al-tolerant genotypes would be of great value in a low cost forage production system on such sites. The objective of our study was to screen a range of bluestem germplasm for tolerance to Al in an acid, Tatum subsoil.

Materials and methods

Tatum clay loam subsoil (clayey, mixed, thermic, Typic Hapludult) was collected from a wooded site near Orange, Virginia. This soil is used routinely to screen plants for tolerance to Al[4,5,7]. Half of the soil was left unlimed (final pH 4.1), and the remainder was limed to a final pH of 5.3 (Table 1). Before planting, both batches of soil were fertilized with 100, 109 and 137 μg g^{-1} of N, P and K, respectively, added as NH_4NO_3 and KH_2PO_4 in solution and mixed throughout the pots, along with the lime treatments. The experimental design was a randomized block consisting of 33 genotypes, two lime levels and three replications. The study was conducted in a greenhouse at Beltsville, Maryland, during March and April of 1984. Supplemental incandescent lights were used to extend day length to 16 hours.

Plant genotypes used in the experiment are shown in Table 2. Old World bluestem entries were from three species: *Bothriochloa intermedia*, (R. Br.) A. Camus; *B. caucasica* (Trin.) C.E. Hubb; and *B. ischaemum* (L) Keng var. ischaemum. The 29 bluestem genotypes used were selected from an original group of 800 introductions chosen for desirable forage traits (C.L. Dewald-personal communication). Three genotypes of weeping lovegrass, (*Eragrostis curvula* (Schrader) Nees (Common, FQ 22 and FQ71) and El Reno side oats grama grass, *Bouteloua curtipendula* (Michx) Torr., were also included as indicators of acid or alkaline soil sensitivity. As a group weeping lovegrasses are quite tolerant to acid soils or mine spoils[8], but FQ 71 and common are more tolerant than FQ 22[10]. Conversely, FQ 22 is more resistant to Fe-related chlorosis on calcareous soil than FQ 71 or common[9,14]. The FQ 22 strain of weeping lovegrass is also more Fe-efficient than FQ 71 in nutrient solutions[6]. Side oats grama is generally resistant to Fe-related chlorosis on calcareous soils (W.A. Berg — personal communication).

Fifty-three days after seeding, the plants were photographed, harvested, dried and weighed. The tops of selected genotypes were analyzed for P by vanadomolybdophosphate[12], and for Al, Ca, Mg, K, Fe and Mn by atomic absorption spectrophotometry. Roots of selected genotypes were also washed out of the soil, photographed, dried and weighed.

Results

Both species and genotypes within species differed widely in tolerance to the unlimed soil at pH 4.1 (Table 3). Figure 1 shows a striking difference between two genotypes (860 and 822) from the species *B. intermedia*. On unlimed soil at pH 4.1, only 5 of the 29 bluestems and the 3 weeping lovegrasses produced measurable yields of tops or roots (Table 3). The remaining 24 bluestems and side-oats grama grass either died or barely remained alive (due to frequent watering) with no appreciable growth. The weeping lovegrass genotypes were significantly more tolerant to the acid soil than any of the bluestems tested; relative top yields (pH 4.1/pH 5.3) were 101, 94 and 79% for FQ 71, common and FQ 72 genotypes, respectively.

Fig. 1. Differential tolerances of two Old World bluestems to an acid, Al-toxic Tatum subsoil. *Left*: PI 300860 and PI 300822 with no lime at pH 4.1. *Right*: PI 300860 and PI 300822 with 3 g kg $^{-1}$ CaCO$_3$ at pH 5.3. Both entries are from the species *Bothriochloa intermedia*.

Among the 5 bluestems that grew at pH 4.1, relative top yields (pH 4.1/pH 5.3) ranged from 19 to 46%. Bluestem genotypes 860 and 857 (both *B. intermedia*) were more tolerant to the acid soil than 886 (*B. intermedia*) and 442 (*B. caucasica*) with 858 (*B. intermedia*) being intermediate. On unlimed soil at pH 4.1, genotypes 860 and 857 produced

Table 3. Top and root growth of weeping lovegrasses and Old World bluestems on acid, Al toxic Tatum subsoil at two lime levels

Entry	Top dry wt.* (g pot^{-1})		Relative top wt.*	Root dry wt. (g pot-l*)		Relative root wt.*
	pH 4.1	pH 5.3	pH 4.1/pH 5.3%	pH 4.1	pH 5.3	pH 4.1/pH 5.3%
FQ 71 WL	5.41 a	5.35 ab	101 a	1.98 a		
Common WL	5.16 a	5.49 a	94 a			
FQ 22 WL	3.27 b	4.14 cd	79 a			
Bluestem 860	1.87 c	4.04 cdef	46 b	1.59 ab		
857	1.82 c	4.13 cde	44 b	1.58 abc	3.20 a	49 a
858	1.66 cd	4.35 bcd	38 b	1.32 bc	2.89 a	46 a
886	0.87 d	3.80 c–h	23 b	1.03 c	2.67 ab	39 a
442	0.85 d	4.48 abc	19 b	0.51 d	1.96 b	26 a
477	0**	0.43 d				
481	0	2.26 jk				
505	0	3.14 e–k				
506	0	3.03 e–k				
576	0	2.01 k				
517	0	2.93 f–k				
535	0	2.45 jk				
573	0	2.57 jk				
599	0	2.89 g k				
602	0	2.83 g–k				
604	0	2.40 jk				
607	0	3.01 e–k				
765	0	4.11 cde		0**	1.94 b	
776	0	2.60 jk				
807	0	2.41 jk		0	2.01 b	
811	0	3.84 c–h		0	2.89 a	
814	0	3.28 d–j		0	2.62 ab	
815	0	3.83 c–h				
820	0	2.66 h–k				
822	0	3.95 c–g		0	2.48 ab	
825	0	4.74 abc				
893	0	3.71 c–i				
908	0	2.58 jk				
Ganada	0	2.94 f–k				
El Reno	0	3.11 e k		0	0.96 c	

* Within a column, any two means having a letter in common are not significantly different at the 5% level by Duncan's Multiple Range Test.
** No measureable growth. Poor germination at both lime levels.

significantly higher top yields than 886 and 442, but on limed soil at pH 5.3, these differences were equalized or reversed (Table 3). Bluestem genotypes that failed to grow at pH 4.1 included members of all 3 species tested. None of the 10 genotypes of *B. ischaemum* made any measurable growth of pH 4.1; however, genotype 477 did not receive a fair test because of poorly germinating seeds. Two bluestem genotypes that failed to grow at pH 4.1, 765 (*B. caucasica*) and 825 (*B. intermedia*), were among the highest yielders at pH 5.3.

Table 4. Concentrations of mineral elements in the tops of weeping lovegrass and Old World bluestem genotypes grown on Tatum subsoil at two lime levels

Entry	Mineral element concentrations in plant tops*						
	Ca	Mg	K	P	Al	Mn	Fe
	$g\,kg^{-1}$				$\mu g\,g^{-1}$		
No lime pH 4.1							
FQ 71 WL	0.67 a	0.60 a	13.9 cd	1.43 d	41 ab	54 bc	28 ab
Common WL	0.30 c	0.43 a	7.9 d	1.60 cd	49 ab	33 d	15 b
FQ 22 WL	0.67 a	0.83 a	28.4 bc	1.83 bc	71 a	45 cd	50 a
Bluestem 860	0.27 c	1.20 a	35.6 ab	1.93 ab	49 ab	53 bc	24 ab
Bluestem 857	0.40 bc	1.47 a	44.2 a	1.90 abc	50 ab	60 bc	24 ab
Bluestem 858	0.50 b	2.93 a	47.6 a	1.77 bc	56 ab	82 a	39 ab
Bluestem 886	0.43 b	1.83 a	47.3 a	1.96 ab	33 b	66 ab	53 a
Bluestem 442	0.43 bc	1.80 a	44.9 a	2.17 a	55 ab	53 bc	38 ab
Lime pH 5.3							
FQ 71 WL	3.97 a	0.53 bc	15.3 bc	1.47 b	32 a	71 a	21 a
Common WL	2.03 a	0.30 c	9.0 c	1.43 b	32 a	52 a	10 a
FQ 22 WL	3.30 a	0.43 bc	14.3 bc	1.93 a	28 a	80 a	36 a
Bluestem 860	3.77 a	1.30 ab	35.6 a	1.47 b	40 a	69 a	22 a
Bluestem 857	3.23 a	1.10 abc	28.0 ab	1.43 b	39 a	55 a	12 a
Bluestem 858	5.17 a	1.83 a	40.8 a	1.43 b	23 a	98 a	21 a
Bluestem 886	3.10 a	1.03 abc	29.4 ab	1.70 ab	42 a	55 a	21 a
Bluestem 442	2.23 a	0.93 bc	25.0 ab	1.63 ab	34 a	49 a	16 a

* Within a lime level, and within a column, any two means having a letter in common are not significantly different at the 5% level by Duncan's Multiple Range Test.

Plant genotypes showing greatest sensitivity to the unlimed soil at pH 4.1 could not be chemically analyzed because of insufficient growth. At pH 4.1, the tops of the 5 surviving bluestems tended to contain lower concentrations of Ca and higher concentrations of Mg, K, P and Mn than did those of the (more tolerant) 3 weeping lovegrasses (Table 4). The two groups did not differ appreciably in concentrations of Al and Fe. None of the genotypes accumulated excessive Al or Mn in plant tops. At pH 4.1, common weeping lovegrass tops tended to contain lower concentrations of Ca, Mg, K, Mn and Fe than did FQ 71 and FQ 22 weeping lovegrasses or the 5 surviving bluestem genotypes. The tops of FQ 22 weeping lovegrass tended to contain higher concentrations of Fe than did those of FQ 71, but the difference was not significant at the 5% level. At pH 4.1, the tops of the more acid soil tolerant 860 and 857 bluestems tended to contain lower concentrations of Ca, Mg, K and Fe than did those of the less tolerant 886 and 442 genotypes. The Al and Mn concentrations were not appreciably different in these two groups.

At pH 5.3 the tops of bluestems which grew at pH 4.1 tended to contain higher concentrations of Ca, Mg and K than did those of the 3 weeping lovegrasses. At pH 5.3, common weeping lovegrass tops contained lower concentrations of Ca, Mg, K and Fe than did those of FQ

71 and FQ 22. The FQ 22 weeping lovegrass genotype tended to contain higher concentrations of both Fe and Mn in its tops than did those of FQ 71.

Discussion

This preliminary study showed that Old World bluestem species, and genotypes within species, differed significantly in tolerance to an acid (pH 4.1) Tatum subsoil having 72.6% of its cation exchange capacity saturated with Al. At pH 4.1 the Al stress was so high that only 5 of the 29 bluestems survived and made measurable growth. However, if the soil pH had been slightly higher (pH 4.5–5.0), other genotypes might have also produced acceptable growth. Additional work is planned to describe the lime response curves of selected tolerant and sensitive genotypes in greater detail. Acid soil tolerance seemed more prevalant within *B. intermedia* than *B. ischaemum*. Of the two *B. caucasica* genotypes tested, one was moderately tolerant and the other sensitive to the acid soil. Although no bluestem genotype tested approached the weeping lovegrasses in acid soil tolerance, the 5 genotypes that survived at pH 4.1 show promise for use on acid soils, particularly if the pH is above 4.1.

Bluestem genotypes noted for outstanding forage production on neutral soils (477, 765, 573)[1] were extremely sensitive to the acid soil at pH 4.1. Also sensitive were the Fe-efficient 535 and 822 genotypes (W.A. Berg and C.L. Dewald, unpublished data). The most acid soil tolerant genotypes (860 and 857) originated in India and Pakistan, respectively, but we have no specific knowledge of whether or not they have acid soil backgrounds.

Within the 5 bluestems that survived at pH 4.1, differential tolerance to the acid soil was not associated with striking differences in the mineral element contents of plant tops. None of the plants accumulated excesses of Al or Mn. Concentrations of Fe appeared low in some cases (both weeping lovegrasses and bluestems), but there were no obvious Fe deficiency symptons.

Bluestems that survived at pH 4.1 contained higher concentrations of Mg, K, Mn and Fe than did the weeping lovegrasses when both were grown on the unlimed soil. The FQ 71 and FQ 22 weeping lovegrasses tended to accumulate higher concentrations of Ca, Mg, K, Mn and Fe than did common. Such compositional differences could be of significance in determining forage quality and nutrient balance in grazing animals.

Acknowledgements We thank M L McCloud of the Plant Stress Laboratory for chemical analyses of soil and plant samples and David Starner, superintendent of the Virginia Polytechnic Institute and State University Piedmont Research Station at Orange, Virginia for help in obtaining the Tatum soil.

References

1. Coyne P I, Bradford J A and Dewald C L 1982 Leaf water relations and gas exchange in relation to forage production in four Asiatic bluestems. Crop Sci. 22, 1036-1040.
2. Faix J J, Kaiser C J and Hinds F E 1980 Quality, yield and survival of Asiatic bluestems and an Eastern gama grass in Southern Illinois. J. Range Mgt. 33, 388-390.
3. Foy C D 1984 Physiological effects of hydrogen, aluminum and manganese toxicities in acid soils. *In* Soil Acidity and Liming. Ed. F Adams. (Agronomy 12). Am. Soc. Agron., Madison, WI, pp 57-98.
4. Foy C D, Armiger W H, Briggle L W and Reid D A 1965 Differential aluminum tolerances of wheat and barley varieties in acid soils. Agron. J. 57, 413-417.
5. Foy C D and Fleming A L 1982 Aluminum tolerances of two wheat genotypes related to nitrate reductase activities. J. Plant Nutr. 5, 1313-1333.
6. Foy C D, Fleming A L and Schwartz J W 1981 Differential resistance of weeping lovegrass genotypes to an iron-related chlorosis. J. Plant Nutr. 3, 537-550.
7. Foy C D, Jones J E and Webb H W 1980 Adaptation of cotton genotypes to an acid, Al-toxic soil. Agron. J. 72, 833-839.
8. Foy C D, Oakes A J and Schwartz J W 1979 Adaptation of some *Eragrostis* species to calcareous soil and acid mine spoil. Commun. Soil Sci, Plant Anal. 10, 953-968.
9. Foy C D, Voigt P W and Schwartz J W 1977 Differential susceptibilities of weeping lovegrass strains to an iron-related chlorosis on calcareous soils. Agron. J. 69, 491-496.
10. Foy C D, Voigt P W and Schwartz J W 1980 Differential tolerance of weeping lovegrass genotypes to an acid coal mine spoil. Agron J. 72, 859-862.
11. Henry D S 1983 Forage yield and quality on Kentucky surface mine spoils. J. Soil Water Conservation. 38, 56-58.
12. Jackson M L 1958 Soil Chemical Analysis. Prentice-Hall, Inc., Englewood Cliffs, N.J.
13. Sims P L and Dewald C L 1982 Old World bluestems and their forage potential for the Southern Great Plains — A review of early studies. USDA-ARS, ARM-S-28/October, 15 pp.
14. Voigt P W, Dewald C L, Matocha J E and Foy C D 1982 Adaptation of iron-efficient and inefficient lovegrass strains to calcareous soils. Crop Sci. 22, 672-676.

Comparisons of maize populations for aluminum tolerance in nutrient solution

R. MAGNAVACA,
National Corn and Sorghum Research Center, EMBRAPA, C.P. 151, 35 700 Sete Lagoas, M.G., Brazil

C.O. GARDNER and R.B. CLARK
Department of Agronomy and U.S. Department of Agriculture, Agricultural Research Service, University of Nebraska, Lincoln, NE 68583, USA

Key words Al effects on roots Al Ca Cl Cu Fe K Mg Mn P S Zn (*Zea mays*)

Summary The original Brazilian maize (*Zea mays* L.) base population 'Composto Amplo' and a fourth cycle of selection from 'Composto Amplo' were compared with the Hays Golden vareity, Nebraska B Synthetic, Corn Belt × Brazilian, and Corn Belt × Caribbean populations for Al tolerance when grown in nurtient solutions (241 μmol Al L^{-1}). Root lengths of the original 'Composto Amplo' population were slightly longer than those of the selected 'Composto Amplo' population when grown with Al, and these were considerably longer than the other populations. The selection of 'Composto Amplo' maize population for Al tolerance when grown in an acid Brazilian Oxisol soil, based on grain yield, decreased the frequency of the more Al tolerant plants in favor of intermediate Al tolerant plants. When both of the 'Composto Amplo' populations were compared with the temperate Corn Belt (Hays Golden and Nebraska B Synthetic) and Corn Belt × - tropical (Brazilian and Caribbean) populations, the 'Composto Amplo' populations had a higher frequency of genes for Al tolerance than the other populations. The population derived from the cross of Corn Belt × Brazilian materials consistently ranked high in Al tolerance relative to Hays Golden and Nebraska B Synthetic. The introgression of Brazilian germplasm into other populations increased the tolerance of these populations to Al toxicity.
 The tropical populations had lower root Al concentrations, but similar Al contents as the temperate populations. Shoot concentrations of Cl and Fe were higher and P, S, and Zn were lower and shoot contents of Ca, K, Cl, Mn, and Cu were higher in the tropical compared to the temperate populations. Root concentrations of K, Cl, and Mn were higher and P, S, and Cu lower and root contents of Mg, Ca, K, Cl, Mn, and Zn were higher and S and Cu were lower in the tropical compared to the temperate populations. The tropical × temperate populations were intermediate to the tropical and temperate populations for element concentrations and contents.

Introduction

A major constraint to maize (*Zea mays* L.) production on tropical soils such as those found in Brazil is the problem created by excess acidity. Many mineral element deficiencies and toxicities occur in acid soils. Aluminum toxicity is one of the most serious problems[4]; hence, strategies to alleviate this disorder by adding lime or P or by developing more tolerant germplasm have been pursued. A promising strategy for overcoming Al toxicity has been to develop breeding procedures for the identification, selection, and improvement of Al-tolerant genotypes.
 Maize population improvement for acid soil tolerance has received

H.W. Gabelman and B.C. Loughman (Eds.), Genetic aspects of plant mineral nutrition.
ISBN-13: 978-94-010-8102-3
© *1987 Martinus Nijhoff Publishers, Dordrecht/Boston/Lancaster.*

considerable attention in recent years. Bahia et al.[1] screened 195 populations in a Brazilian acid Oxisol soil and found considerable variation among populations for adaptation to such soils. Fewer populations were tested in subsequent years and CMS 14 and CMS 30 ('Composto Amplo') showed the greatest potential for population improvement by recurrent selection[10]. Beginning in 1975, the 'Composto Amplo' population was used to select for Al tolerance by measuring grain yield when grown on a Brazilian acid Oxisol with 45% Al saturation[11]. Each year, 500 half-sib families were tested, and the selected families were recombined at Sete Lagoas, Brazil, until four cycles of selection were completed.

The objectives of the study reported here were to (i) obtain information about changes in Al tolerance that occurred from selection for high grain yield in a population grown on acid soils of Brazil, (ii) compare non-tolerant and Al-tolerant populations for differences in mineral element uptake, (iii) examine the effect of tropical germplasm introgressed into Corn Belt populations on Al tolerance, and (iv) determine how this information might be used in breeding for Al tolerance in maize.

Materials and methods

Germplasm used

Six maize populations chosen for this study included (1) the original 'Composto Amplo', (2) a fourth cycle selection from 'Composto Amplo', (3) Hays Golden, (4) Nebraska B Synthetic, (5) Corn Belt × Brazilian Composite, and (6) Corn Belt × Caribbean Composite. The original 'Composto Amplo' is a composite with a broad genetic base developed at the Institute of Genetics, Luiz de Queiroz College of Agriculture, University of Sao Paulo, Piracicaba, S.P., Brazil. This population has been used extensively in the maize breeding program at the National Corn and Sorghum Research Center, Sete Lagoas, M.G., Brazil. Four cycles of selection between and within half-sib families of the original 'Composto Amplo' population were obtained by growing the plants of each cycle on a Brazilian acid Oxisol soil containing 45% Al saturation. The Hays Golden variety, developed at the Fort Hays (Kansas) Experiment Station, was collected in southwestern Nebraska and has been maintained at the Nebraska Agricultural Experiment Station. The Nebraska B Synthetic was developed from 32 inbred lines by Dr. J. H. Lonnquist at the Nebraska Agricultural Experiment Station where it has been maintained. A Corn Belt × Brazilian Composite was developed by Dr. J. H. Lonnquist by compositing two crosses: Corn Belt Composite × Brazilian #5 and Corn Belt Composite × Brazilian Composite 2 #1. The Corn Belt Composite consisted of the varieties Hays Golden, Barber Reid, Krug Yellow Dent, Lancaster Surecrop, and Golden Republic. A Corn Belt × Caribbean Composite was also developed by Dr. J. H. Lonnquist compositing R_{III} × Caribe #4, SSS_{III} × Caribe #4, $Krug_{III}$ × Caribe #4, and CBC × Colombian #6. R_{III}, SSS_{III}, and $Krug_{III}$ are improved synthetic varieties developed by three cycles of recurrent selection for general combining ability from Nubold Reid, Stiff Stalk Synthetic (Iowa), and Krug Yellow Dent, respectively[18]. CBC is a Corn Belt Composite, and US342 is a double cross hybrid (C15 × KYS) × (WF9LH × 3811LH).

Growth of plants

Captan [N-(trichloromethylthio)-4-cyclohexene-1, 2-dicarboximide] treated maize seeds were germinated between rolled paper towels kept moist with aerated water. Seven-day-old uniform sized seedlings without visual root injury were transferred to a plastic plate (42 plants per plate) and grown in 6.5 L of aerated nutrient solution containing 241 μmol Al L^{-1} as KAl(SO$_4$)$_2$. The nutrient solution and techniques used for growing plants have been described[3,14]. The composition of the nutrient solution (μmol element L^{-1}) was 10,900 NO$_3$-N, 3500 Ca, 2300 K, 1300 NH$_4$-N, 850 Mg, 590 Cl, 580 S, 45 P, 25 B, 9.1 Mn, 2.29 Zn, 0.63 Cu, 0.83 Mo, and 77 Fe as FeHEDTA (ferrichydroxyethylethylenediaminetriacetate). The pH of nutrient solutions was adjusted initially to 4.0, monitored daily, and adjusted to 4.0 when necessary. Water was added to maintain solution volumes.

Plants were grown in a controlled environment room with 16 h light at 27 ± 1°C and 8 h darkness at 19 ± 1°C. The photosynthetic photon flux density was 150 μE m^{-2} s^{-1} at plant height (40 cm below the lamps) provided by fluorescent lamps (Agro-Lite, cool white, F40)*. The experimental design was a randomized complete block with three replications. Each plot (plate) had 42 plants.

Handling of plants and traits measured to assess Al tolerance

When seedlings were initially transferred to treatment solutions, the lengths of seminal roots were measured, and adventitious roots that had started to grow were removed. Plants were grown in treatment solutions for 10 days. When experiments were terminated, the final seminal root length and length of the longest adventitious root were measured. Roots and lower leaves were thoroughly water rinsed, blotted dry, separated into shoots and roots (residual kernel pieces removed), dried in a forced-air oven at 70°C for a minimum of four days, weighed, and ground to pass a 0.5 mm screen.

The traits used to assess plants for tolerance to Al were (a) initial seminal root length, (b) final seminal root length, (c) longest adventitious root length, (d) relative seminal root length (the final seminal root length divided by the initial seminal root length), (e) net seminal root length (the final seminal root length minus the initial seminal root length), (f) shoot dry weight, and (g) root dry weight. Mineral element concentrations and contents were also determined to evaluate the populations for relationship of mineral elements to Al tolerance. Concentrations of Mg, Al, P, S, Cl, K, Ca, Mn, Fe, Cu, and Zn were determined in 13 mm diameter pellets (100 mg) of dried shoot and root tissue by energy-dispersive x-ray fluorescence spectrometry[17].

Results and discussions

Means for four of the traits measured in the six populations are presented in Table 1. The results were relatively consistent from one trait to another. The original and the selected (four cycles) 'Composto Amplo' populations both showed higher levels of Al tolerance than the other populations. A high frequency of favorable genes for Al tolerance existed in the original 'Composto Amplo' population as well as in the selected 'Composto Amplo'. The Corn Belt × Brazilian population which has been grown annually in Nebraska on non-acid soils for many years displayed more Al tolerance than either of the Corn Belt populations and the Corn Belt × Caribbean population. The latter three populations differed little from each other in Al tolerance. The introgression of Brazilian germplasm into a Corn Belt population apparently increased the Corn Belt population tolerance to Al toxicity. Root development of the six populations is illustrated in Figure 1.

---

* Mention of a company, trademark, or proprietary product does not constitute a guarantee or warranty of the product by the University of Nebraska or the U.S. Department of Agriculture, and does not imply its approval to the exclusion of other products that may also be suitable.

Table 1. Means for relative seminal root length (RSRL), longest adventitious root length (LARL), final seminal root length (FSRL), and net seminal root length (NSRL) of six maize populations grown in nutrient solution with Al (mm root^{-1})

Populations	RSRL	LARL	FSRL	NSRL
'Composto Amplo' original	2.09 a[†]	176 a	341 cd	178 a
'Composto Amplo' selected (4th cycle)	1.94 b	148 b	310 b	149 b
Corn Belt × Brazilian	1.27 c	115 c	158 c	33 c
Corn Belt × Caribe	1.08 d	97 d	150 cd	11 d
Hays Golden	1.12 d	71 e	167 c	18 cd
Nebraska B Synthetic	1.20 cd	81 de	135 d	22 cd

[†] Means followed by a common letter are not significantly different at $\alpha = 0.05$ according to Duncan's New Multiple Range Test.

Estimates of variability among individual plants within each population for four Al tolerance traits are reported in Table 2. The variance in the original 'Composto Amplo' population was higher than that of the selected 'Composto Amplo' for all measured traits, and both had greater plant-to-plant variances than the other populations. The Corn Belt × Brazilian population had more useful variability than the other three non-Brazilian populations. The Corn Belt × Caribbean and Hays Golden populations showed relatively little variability for Al tolerance.

The decrease in variability of the selected 'Composto Amplo' population compared to the original 'Composto Amplo' was best demonstrated

Fig. 1. Root development of maize populations grown in nutrient solutions with 241 μmol Al L^{-1}. Left to right: 1 = original 'Composto Amplo', 2 = selected 'Composto Amplo', 3 = Corn Belt × Brazilian, 4 = Corn Belt × Caribbean, 5 = Hays Golden, and 6 = Nebraska B Synthetic.

Table 2. Analyses of variance for relative seminal root length (RSRL), longest adventitious root length (LARL), final seminal root length (FSRL), and net seminal root length (NSRL) of six maize populations grown in nutrient solution with Al

Source of variation	df	Mean square values			
		RSRL	LARL	FSRL	NSRL
Replications	2	0.0005	264.89	30.11	35.65
Populations	5	0.5927**	4905.47**	24663.36**	16644.69**
Error	10	0.0051	90.82	136.77	90.98
CV (%)		4.9	8.3	5.6	13.9
Among plants	738	0.1055	2540.05	3213.06	2531.69
'Composto Amplo' original	123	0.2904	7513.12	9459.39	8219.02
'Composto Amplo' selected	123	0.2023	4112.17	5257.98	4840.76
Corn Belt × Brazilian	123	0.0752	1164.51	1900.91	1222.07
Corn Belt × Caribe	123	0.0046	984.95	606.13	88.97
Hays Golden	123	0.0105	697.96	827.21	235.47
Nebraska B Synthetic	123	0.0503	767.61	1226.72	583.84

** Statistical significance at $\alpha = 0.01$.

using relative seminal root lengths (RSRL). Values above 1.0 represent greater tolerance to Al (Fig. 2). When frequency distributions of the two populations were compared, the frequency of the highly tolerant genotypes in the selected population decrease. The selection of germplasm grown in soils with relatively high Al saturation favoured the selection of germplasm with intermediate Al tolerance rather than germplasm with high Al tolerance. A possible explanation for the selection of intermediate types would be a negative correlation between a high level of Al tolerance and grain yield but information about this is limited. However, Martini et al.[19] and Silva[22] reported that Brazilian Al tolerant wheat (*Triticum aestivum* L.) varieties produced as well as other high yielding varieties when grown on soils without Al toxicity problems.

The use of nutrient solutions with Al could be an effective method to screen maize germplasm for Al tolerance. Large numbers of progenies could be evaluated for Al tolerance in nutrient solution, and the more tolerant progenies could than be evaluated for grain yield in the field on

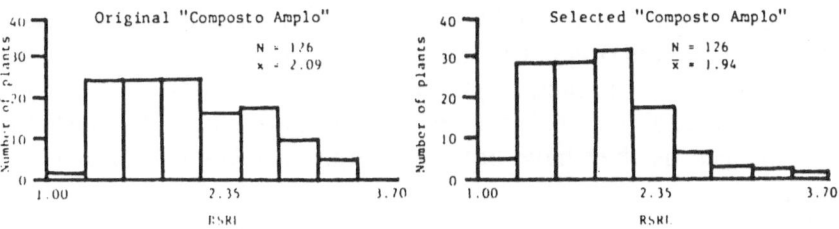

Fig. 2. Frequency distributions of relative seminal root length (RSRL) of plants from the original and 'Composto Amplo' populations grown in nutrient solution with 241 μmol Al L^{-1}.

soils with Al toxicity stresses. This two-stage selection program, which corresponds to the Hazel and Lush[15] use of independent culling levels for two traits, has promise for simultaneously improving both traits with minimum work. As Hazel and Lush[15] pointed out, the use of a selection index involving two traits is generally more efficient in animal breeding, but screening for Al tolerance among maize families could be accomplished on a large scale at minimum cost. Yield testing, on the other hand, is very expensive and should be limited to those families with relatively high Al tolerance. If any negative correlation between yield and Al tolerance is caused by linkage, genetic recombination should occur and simultaneous selection for two traits will be effective for both. To the extent that negative correlation is caused by pleiotropy, little can be done to improve both traits simultaneously. Additional information on the nature of the correlation is needed.

Because of the large differences among the six populations for Al tolerance, the roots and shoots of the plants were analyzed for mineral elements to determine if mineral elements could be related to genotype differences to Al tolerance. Dry matter yields and mineral element concentrations and contents in roots and shoots are presented in Tables 3 and 4.

Shoot and root dry matter yields (Table 3) followed trends similar to root length measurements for Al tolerance. That is, higher dry matter yields and longer roots were both associated with greater Al tolerance. Mineral element concentrations did not follow the same trend. Total dry matter yields of the populations were: original 'Composto Amplo' > selected 'Composto Amplo' ≥ Corn Belt × Caribbean = - Corn Belt × Brazilian > Hays Golden = Nebraska B Synthetic. The two temperate (Corn Belt) populations has the lowest dry matter yields. The Corn Belt × Caribbean population which was affected as badly as the temperate populations by Al had higher shoot and root dry matter yields.

Origin of the populations should be considered in the interpretation of the mineral element data (Tables 3 and 4). Since concentrations of mineral elements were not determined for these same populations grown in nutrient solutions without Al, the effect of Al is partially confounded by the effect of plant origin. Critical plant nutrient requirements were not determined for these maize populations according to origin, however, and nutrient requirements for these different maize populations need to be established as they have been for sorghum (*Sorghum bicolor* (L.) Moench)[9,13]

The Al concentrations in roots could be used to separate the populations by origin (Table 3); the tropical (Brazilian and Caribbean) populations had lower Al concentrations than the temperate populations (Table

Table 3. Dry matter yields and mineral element concentrations in shoots and roots of six maize populations grown in nutrient solution with Al

Population	Dry matter yield (mg 42 plants^{-1})	Mineral element concentration										
		Mg	Ca	K	P	S	Cl	Al[†]	Mn	Fe	Cu	Zn
		(mg g^{-1} dry wt.)							(μg g^{-1} dry wt.)			
Shoots												
'Composto Amplo' original	8670a[‡]	2.9 d	7.3 a	65.4 b	2.6 c	3.2 c	8.8 a	—	47.5 a	103 b	9.2 ab	32.6 b
'Composto Amplo' selected	7000 b	3.0 cd	7.0 ab	70.7 a	3.0 c	3.5 c	9.2 a	—	47.3 a	108 b	9.0 ab	35.4 b
Corn Belt × Brazilian	6670 b	3.1 cd	6.4 c	62.2 bcd	4.6 ab	5.1 b	7.5 cd	—	39.1 ab	141 a	9.5 ab	53.2 a
Corn Belt × Caribbean	6800 b	3.4 bc	6.5 bc	60.5 cd	4.4 b	5.2 ab	7.9 bc	—	37.3 ab	129 a	7.7 b	50.6 a
Hays Golden	4930 c	3.5 ab	6.8 abc	59.3 d	4.9 ab	5.3 ab	7.1 d	—	41.9 ab	126 a	9.6 a	48.2 a
Nebraska B Synthetic	4570 c	3.9 a	6.3 c	64.8 bc	5.2 a	5.5 a	8.0 b	—	34.9 b	132 a	10.6 a	51.7 a
Roots												
'Composto Amplo' original	3170 a	2.5 a	6.0 a	45.6 b	2.8 c	5.5 d	4.6 a	3.7 d	28.6 a	416 d	13.7 c	44.8 a
'Composto Amplo' selected	2430 b	2.4 a	4.8 bc	57.5 a	3.0 c	5.9 d	3.4 b	4.0 cd	24.3 a	496 cd	12.0 c	47.0 a
Corn Belt × Brazilian	2030 c	2.4 a	5.7 a	42.9 b	4.9 b	16.9 c	3.0 b	5.3 bc	14.2 b	554 bc	60.5 b	46.9 a
Corn Belt × Caribbean	2100 bc	2.2 ab	5.4 ab	38.1 c	4.8 b	18.1 bc	3.2 b	5.3 bc	15.9 b	571 bc	51.6 b	44.8 a
Hays Golden	1200 d	1.8 b	4.9 bc	33.1 d	5.6 a	22.2 a	2.4 c	8.8 a	12.3 b	705 a	83.1 a	45.2 a
Nebraska B Synthetic	1330 d	2.2 ab	4.6 c	38.2 c	5.2 ab	18.8 b	3.0 b	6.3 b	16.2 b	666 ab	78.1 a	42.8 a
Statistical significance of F test for treatment												
Shoots	**	**	*	**	**	**	**		ns	**	ns	**
Roots	**	*	**	**	**	**	**	**	**	**	**	ns

[†] The aluminum concentrations in shoots were below the minimum detectable limit of the instrument ($<100\,\mu$g g^{-1}).
[‡] Means followed by a common letter are not significantly different at $\alpha = 0.05$, according to Duncan's New Multiple Range Test.
*, ** Statistical significance at $\alpha = 0.05$ and $\alpha = 0.01$, respectively; ns = nonsignificant.

3). These differences were not significant for Al content (total amount in plant parts), which would indicate a possible dilution effect from larger roots (Table 4). Since Al in the shoots was very low, Al apparently was not translocated from roots to shoots, which has also been noted in other studies with maize[2,20]. The Al may have become bound to the root surfaces[20] and root cell walls[7,16].

Concentrations of Mg, Ca, and K in the plants were dependent on the element and the plant part (Table 3). Concentrations of Mg in shoots were higher for both of the temperate populations, but these populations differed only slightly in root Mg concentration. The discrimination among populations for Ca in the shoots and roots did not appear to be related to plant origin. Both of the 'Composto Amplo' populations had higher K concentrations in shoots and roots than the other populations. The contents of Mg, Ca, and K followed approximately the same trend as the dry matter yield differences (Table 4). Detrimental effects of Al on Mg, Ca, and K uptake and translocation, as well as differences in critical levels of these elements in each population, may help to explain the results obtained. Effects of Al on element uptake have been reported for maize[2,21].

Concentrations of P and S were different from those of Mg, Ca, and K (Table 3). The two Al tolerant 'Composto Amplo' populations had lower P and S concentrations in both the shoots and roots than did the non-tolerant populations. The temperate × tropical populations consistently had higher P and S contents in both shoots and roots than the other populations (Table 4). The tropical and temperate populations did not differ in P and S in shoots, but the temperate populations had lower contents of P and higher contents of S in roots. Mineral element uptake and distribution were not related to root length measurements not to dry matter yields of the different populations. Origin of population may be important for element concentration and content. Chlorine concentrations and contents were similar to Mg, Ca, and K.

The P concentrations and contents noted in the Al tolerant and non-tolerant populations used in this study did not agree with results reported by others. The contents of P noted (Table 4) did not indicate that Al interfered with P absorption and translocation. The lower P concentrations in the Al tolerant populations may have been due to plant growth dilution. A relationship between P nutrition and Al tolerance in maize plants has been reported[5,6,12,20].

Although mineral element data obtained in most studies have been experiments involving homogeneous maize genotypes, the data obtained in this study were obtained from heterogenous populations. A higher level of Al in nutrient solutions would probably be required for Al to cause detrimental effects on P uptake. Although Al decreased root length

Table 4. Mineral element contents in shoots and roots of six maize populations grown in nutrient solution with Al

Genotype	Mineral element content										
	Mg	Ca	K	P	S	Cl	Al[†]	Mn	Fe	Cu	Zn
	(mg 42 plants[-1])							(μg 42 plants[-1])			
Shoots											
'Composto Amplo' original	25.1 a[‡]	63.7 a	565 a	22.7 b	28.0 b	76.2 a	—	413 a	896 ab	80.3 a	283 b
'Composto Amplo' selected	20.8 bc	48.9 b	495 b	21.0 b	24.8 c	64.1 b	—	332 ab	759 bc	63.1 b	247 c
Corn Belt × Brazilian	20.9 bc	42.7 b	415 c	30.9 a	34.0 a	49.8 d	—	260 bc	940 a	63.5 b	354 a
Corn Belt × Caribbean	22.9 ab	44.2 b	411 c	30.2 a	35.0 a	53.9 c	—	254 d	880 ab	52.4 b	344 a
Nebraska Hays Golden	17.2 d	33.7 c	293 d	24.3 b	26.1 bc	35.0 e	—	207 cd	623 c	43.4 b	238 c
Nebraska B Synthetic	17.7 cd	28.9 c	296 d	23.5 b	25.3 c	36.4 e	—	160 d	603 c	48.0 b	235 c
Roots											
'Composto Amplo' original	8.00 a	19.1 a	144 a	8.8 ab	17.2 c	14.5 a	11.7 a	91 a	1307 a	43.5 c	142 a
'Composto Amplo' selected	5.80 b	11.8 b	139 a	7.3 bc	14.5 c	8.2 b	9.7 ab	60 b	1227 ab	29.6 c	115 b
Corn Belt × Brazilian	4.90 b	11.6 b	87 b	10.0 a	34.4 a	6.2 c	10.7 ab	29 c	1128 ab	122.8 a	95 c
Corn Belt × Caribbean	4.50 b	11.3 b	80 b	10.1 a	38.1 a	6.6 bc	11.1 ab	33 c	1199 ab	108.5 ab	94 c
Nebraska Hays Golden	2.20 c	5.9 c	40 c	6.8 c	26.7 b	2.9 d	10.5 ab	15 c	846 b	99.8 b	54 d
Nebraska B Synthetic	3.00 c	6.1 c	50 c	6.9 c	25.0 b	4.0 d	8.4 b	21 c	885 b	102.3 b	57 d
Statistical significance of F test for treatment											
Shoots	**	**	**	**	**	**	ns	**	**	**	**
Roots	**	**	**	**	**	**		**	ns	**	**

[†] It was not possible to estimate the content of Al in shoots since the concentrations were usually below the minimum detectable limit of the instrument (<100 μg g[-1]).
[‡] Means followed by a common letter are not significantly different at α = 0.05, according to Duncan's New Multiple Range Test.
*, ** Statistical significance at α = 0.05 and α = 0.01, respectively, ns — nonsignificant.

dramatically, no P deficiency symptoms were noted on any of the plants in the populations.

The results for Mn, Fe, Cu, and Zn indicated different responses for each element. The more Al tolerant tropical populations did not show differences for Mn and Cu in shoots compared to the non-tolerant populations (Table 3). Iron and Zn concentrations were lower in the tolerant populations than in the non-tolerant populations. Aluminum tolerant populations had root concentrations of Zn similar to non-tolerant populations, but had lower root Fe and Cu and higher root Mn.

Distinctions between tolerant and non-tolerant populations were noted for Mn contents in the shoots which were higher in the Al tolerant populations than in the non-tolerant populations (Table 4). No differences among populations were noted for Fe contents in the roots, but tolerant populations had higher Mn and Zn contents and lower Cu contents. These results agree with comparisons made between Al tolerant and non-tolerant inbred maize lines for Mn, but not for Fe, Zn and Cu[2]. The concentrations and contents of Cu in roots were extremely low in the Al tolerant populations compared with the non-tolerant populations. Duncan[8] and Furlani[13] reported that differences in concentrations of Fe, Mn, and Cu might be used to distinguish Al tolerant and non-tolerant genotypes of sorghum.

References

1. Bahia A F C, Franca G E, Pitta G V E, Magnavac R, Mendes J F, Bahia F G F T C and Pereira P 1978 Evaluation of corn inbred lines and populations in soil acidity conditions. pp 51–58. *In* XI Annu. Brazilian Maize Sorghum Conf., Piracicaba, S.P., Brazil. (*In Portuguese*)
2. Clark R B 1977 Effect of aluminum on growth and mineral elements of Al-tolerant and Al-intolerant corn. Plant and Soil 47, 653–662.
3. Clark R B 1982 Nutrient solution growth of sorghum and corn in mineral nutrition studies. J. Plant Nutr. 5, 1039–1057.
4. Clark R B 1982 Plant response to mineral element toxicity and deficiency. pp 71–142. *In* Breeding Plants for Less Favorable Environments. Eds. M N Christiansen and C F Lewis John Wiley & Sons, New York.
5. Clark R B and Brown J C 1974 Differential mineral uptake by maize inbreds. Commun. Soil Sci. Plant Anal. 5, 213–227.
6. Clark R B and Brown J C 1974 Differential phosphorus uptake by phosphorus-stressed corn inbreds. Crop Sci. 14, 505–508.
7. Clarkson D T 1967 Interactions between aluminum and phosphorus on root surfaces and cell wall material. Plant and Soil 27, 347–356.
8. Duncan R R 1981 Variability among sorghum genotypes for uptake of elements under soil field conditions. J. Plant Nutr. 4, 21–32.
9. Duncan R R, Dobson J W Jr and Fisher C D 1980 Leaf elemental concentrations and grain yield of sorghum grown on an acid soil. Commun. Soil Sci. Plant Anal. 11, 699–707.
10. EMBRAPA (Brazilian Department of Agricultural Research) 1980 National Maize and Sorghum Research Center Tech. Report 1979. EMBRAPA, Brasilia, D.F., Brazil. (*In Portuguese*)
11. EMBRAPA (Brazilian Department of Agricultural Research) 1981 National Maize and

Sorghum Research Center Tech. Report 1980. EMBRAPA, Brasilia, D.F., Brazil. (*In Portuguese*)

12 Foy C D and Brown F C 1964 Toxic factors in acid soils. II. Differential aluminum tolerance of plant species. Soil Sci. Soc. Am. Proc. 28, 27 32.
13 Furlani P R 1981 Effects of aluminum on growth and mineral nutrition of sorghum genotypes. Ph.D. Diss. Univ. of Nebraska, Lincoln. Diss. Abstr. 42, 1260B.
14 Furlani P R and Clark R B 1981 Screening sorghum for aluminum tolerance in nutrient solution. Agron. J. 73, 587–594.
15 Hazel L N and Lush J L 1943 The efficiency of three methods of selection. J. Heredity 33, 393 399.
16 Klimashevsky E L, Markova Y A, Bernatzkaya M L and Malysheva A S 1972 Physiological responses to aluminum toxicity in root zones of pea varieties. Agrochimica 26, 487–496.
17 Knudsen D, Clark R B, Denning J L and Pier P A 1981 Plant analysis of trace elements by x-ray. J. Plant Nutr. 3, 61–75.
18 Lonnquist J H 1961 Progress from recurrent selection procedures for the improvement of corn populations. Nebraska Agric. Expt. Stn. Res. Bull. 197.
19 Martini J A, Kochhann R A, Gomes E P and Langer F 1977 Response of wheat cultivars to liming in some high Al Oxisols of Rio Grande do Sul, Brazil. Agron. J. 69, 612–616.
20 Rasmussen H P 1968 The mode of entry and distribution of aluminum in *Zea mays*: Electron microprobe x-ray analysis. Planta 81, 28–37.
21 Rhue R D and Grogan C O 1977 Screening corn for Al tolerance using different Ca and Mg concentrations. Agron. J. 69, 755–760.
22 Silva A R 1976 Application of the plant genetic approach to wheat culture in Brazil. pp 223-231. *In* Plant Adaptation to Mineral Stress in Problem Soils. Ed. M J Wright. Cornell Univ. Agric. Exp. Stn., Ithaca, NY.

Inheritance of aluminum tolerance in maize

R. MAGNAVACA,
National Corn and Sorghum Research Center, EMBRAPA, C.P. 151, 35,700 Sete Lagoas, M.G., Brazil

C.O. GARDNER and R.B. CLARK
Department of Agronomy and U.S. Department of Agriculture, Agricultural Research Service, University of Nebraska, Lincoln, NE 68583, USA

Key words Additive effects Backcrosses Dominance effects Epistasis F_2 generations General and specific combining abilities Heterosis Relative seminal root length *Zea mays*

Summary The inheritance of Al tolerance in maize (*Zea mays* L.) was studied in nutrient solution. Analysis of relative seminal root lengths of six generations (P_1, P_2, F_1, F_2, BC_1, and BC_2) derived from crosses between tolerant and non-tolerant inbred lines showed that additive gene effects contributed most to genetic variation for Al tolerance of the materials included in this study. Dominance effects accounted for only half as much variation as did additive effects. Effects of epistasis contributed little compared to other gene effects. The frequency distributions of plants within the F_2 generations were continuous, unimodal, and typical for quantitatively inherited traits. There was some tendency for non-tolerance to be dominant over tolerance, but it was not consistent. In a diallel cross among inbred lines, the analysis of F_1 crosses indicated that the variance for general combining ability explained most of the variation, but specific combining ability was statistically significant in each case.

Introduction

The inheritance of Al tolerance in maize (*Zea Mays* L.) has been studied by several authors using different techniques to induce and assess Al toxicity[8,9,14,19,22,24,25]. Rhue *et al.*[22] grew plants of F_2 generations and the backcrosses of more sensitive parental lines in nutrient solutions with 250 μmol Al L^{-1} (126 μmol P L^{-1}) at different concentrations of Ca. Because of the tendency for a 3:1 segregation in the F_2 generations and a 1:1 segregation in the backcross generations among the maize inbred lines, they concluded that Al tolerance was controlled by a single locus. However, additional evidence by these authors indicated control of Al tolerance to be by a multiple allelic series[21].

Three tolerant and two non-tolerant maize lines were used to develop F_1, F_2, and backcross generations studied by Garcia *et al.*[9]. Seeds from these generations were planted in sand and irrigated with nutrient solution containing 0 and 2780 μmol Al L^{-1}. Seven days after germination, root lengths of seedlings grown with and without Al were measured to determine relative root lengths among the genotypes. The three F_1 hybrids studied and their backcrosses to the Al tolerant parents were relatively tolerant to Al. Backcrosses to the non-tolerant parents showed an approximate 1:1 segregation. In the F_2 generations, a bimodal distribution was observed with approximately a 3:1 segregation. These investigators concluded that Al tolerance in maize was controlled by a single dominant gene with possible alterations by modifiers.

Variance component estimates for Al tolerance were determined for the maize variety 'Piranao'[8]. A total of 144 progenies were produced according to the Comstock and Robinson[3] mating design I. Progenies were evaluated in a pot experiment with an acid soil. Based on shoot and root dry weights, it was reported that the most important component of genetic variability was dominance variance.

Naspolini et al.[19] obtained estimates of general and specific combining abilities from a diallel cross of 10 inbred maize lines previously selected for Al tolerance. The 45 possible single crosses were tested in a field experiment on acid soil with three levels of Al saturation (0, 45, and 64%). The estimates of the combining ability variances for grain yield were signficant even though they were variable among Al saturation levels. The magnitudes of general combining ability variances were higher than those of the specific combining ability variances, indicating the importance of additive gene effects.

Maize populations were tested for Al tolerance using nutrient solutions in a complete diallel cross involving the parental populations, F_1 crosses, and reciprocals[14]. The variance for general combining ability explained most of the variation, but specific combining ability was also significant.

Since information is somewhat limited, a better understanding of the genetic control of Al tolerance in maize is needed. Such information would be useful to help evaluate the potential of improving maize through selection and to greatly improve the efficiency of breeding programs aimed at helping to solve maize production problems in Al toxic acid soils. The objective of the research reported here was to obtain additional information on the inheritance of Al tolerance in maize and to assess the relative importance of additive and dominance effects and epistasis on Al tolerance.

Materials and methods

Growth of plants

Captan [N-(trichloromethylthio)-4-cyclohexene-1,2-dicarboximide] treated maize seeds were germinated between rolled paper towels kept moist with aerated water. Seven-day-old uniform sized seedlings without visual root injury were transferred to a plastic plate (42 plants per plate) and grown in 6.5 L of aerated nutrient solution containing 185 μmol Al L^{-1} as KAl(SO$_4$)$_2$. The nutrient solution and techniques used for growing plants have been described[1,15]. The composition of the nutrient solution (μmol element L^{-1}) was 10 900 NO$_3$-N, 3500 Ca, 2300 K, 1300 NH$_4$-N, 850 Mg, 590 Cl, 580 S, 45 P, 25 B, 9.1 Mn, 2.29 Zn, 0.83 Mo, 0.63 Cu, and 77 Fe as FeHEDTA (ferric hydroxyethyl-ethylenediaminetriacetate). The pH of nutrient solutions was adjusted initially to

4.0 ± 0.1 and maintained at this pH throughout the experiment. Water was added daily to maintain solution volumes.

Plants were grown in a controlled environment room with 16 h light at 27 ± 1 C and 8 h darkness at 19 ± 1 C. The photosynthetic photon flux density was 150 μE m^{-2} s^{-1} at plant height (40 cm below the lamps) provided by fluorescent lamps (Agro-Life cool white, F40)*.

Handling of plants and traits measured to assess Al tolerance

When seedlings were transferred to treatment solutions, the initial lengths of seminal roots were measured. Plant were grown in Al treatment solutions for 10 days, when the experiments were terminated and the final seminal root lengths measured. Relative seminal root length (RSRL) was used to evaluate plants for Al tolerance. RSRL values were determined by dividing the final seminal root length by the initial length. This trait was chosen to assess Al tolerance because it has been found to be one of the better traits to assess Al toxicity[6,16]. The greater the RSRL value, the greater the Al tolerance.

Germplasm, experimental design, and statistical analysis of experiments

Experiment 1. Six generations (P_1, P_2, F_1, F_2, BC_1, BC_2) from each of six sets of crosses (B37 × A635, B37 × C103, Mo17 × B37, Mo17 × H84, W117 × A554, and W117 × A635) were used. The experimental design was a modified randomized complete block with three replications. Mean RSRL values were determined using six plants per plot, except for the F_2 generation in which 12 plants per plot were used.

Subdivision of the degrees of freedom and sums of squares for generations within each set were handled according to Mather and Jinks[17] generation means model. The following coefficients for parameters of the complete model were used to obtain the sums of squares:

Generation	Coefficients for parameters					
	m	[d]	[h]	[i]	[j]	[l]
P_1	1	1	0	1	0	0
P_2	1	−1	0	1	0	0
F_1	1	0	1	0	0	1
F_2	1	0	1/2	0	0	1/4
BC_1	1	1/2	1/2	1/4	1/4	1/4
BC_2	1	−1/2	1/2	1/4	−1/4	1/4

where m = the mean of the two parents; [d] = the cumulative additive gene effects; [h] = the cumulative dominance gene effects; [i] = the cumulative additive × additive epistatic effects; [j] = the cumulative additive × dominance epistatic effects and [l] = the cumulative dominance × dominance epistatic effects.

The estimates of the six parameters were obtained by using the equations:

$$m = 1/2\bar{P}_1 + 1/2\bar{P}_2 + 4\bar{F}_2 - 2\overline{BC}_1 - 2\overline{BC}_2$$

$$[d] = 1/2\bar{P}_1 - 1/2\bar{P}_2$$

$$[h] = 6\overline{BC}_1 + 6\overline{BC}_2 - 8\bar{F}_2 - \bar{F}_1 - 3/2\bar{P}_1 - 3/2\bar{P}_2$$

$$[i] = 2\overline{BC}_1 + 2\overline{BC}_2 - 4\bar{F}_2$$

$$[j] = 2\overline{BC}_1 - \bar{P}_1 - 2\overline{BC}_2 + \bar{P}_2$$

$$[l] = \bar{P}_1 + \bar{P}_2 + 2\bar{F}_1 + 4\bar{F}_2 - 4\overline{BC}_1 - 4\overline{BC}_2.$$

* Mention of a company, trademark, or proprietary product does not constitute a guarantee or warranty of the product by the University of Nebraska or the U.S. Department of Agriculture, and does not imply its approval to the exclusion of other products that may also be suitable.

Experiment 2. Four sets of F_2 progenies involving maize parental lines in the crosses A554 × W117, B37 × A635, B37 × C103, and Mo17 × H84 were used. The experimental design was a randomized complete block with four replications. The statistical analysis of RSRL values was conducted using indidivual plant data.

Experiment 3. Eight parental inbred maize lines (A554, A635, B37, C103, C164, H84, Mo17, and W117) and progenies of their diallel crosses (28 F_1 crosses without reciprocals) were tested for Al tolerance in nutrient solutions. The experimental design was a 6 × 6 triple lattice (three replications). The lattice analysis for RSRL was performed according to procedures outlined by Cochran and Cox[2]. Based on the effective error variance from the lattice analysis and the experimental error from the randomized complete block analysis, the relative efficiency of the lattice was determined. Where the relative efficiency of the lattice was less than 10%, the data were analyzed as a randomized complete block design and means were not adjusted.

Since the 36 treatments consisted of 28 F_1 progenies and eight parental lines, the sum of squares for treatments was subdivided into parents, parents *vs.* crosses, and among crosses according to procedures of Gardner and Eberhart[10]. The parents were excluded from the diallel cross part of the analysis which was then analyzed according to experimental method No. 4 of Griffing[11], involving one set of F_1 progenies with no parents and excluding reciprocal crosses. Only the $p(p-1)/2$ (p = number of parental inbred maize lines) different F_1 mean values were used in estimating general and speciifc combining ability variances.

Results and discussion

Experiment 1

Means of RSRL values measured in the different generations of six crosses are presented in Table 1. In Sets 1 and 2, F_1, F_2, and backcross generations were similar to the susceptible parents, but in Sets 3, 4, 5, and 6, the F_1, F_2, and backcross generations were similar to the more Al tolerant parents. The greatest amount of heterosis was exhibited in Sets 3 and 5.

The analysis of variance for the RSRL values measured in different generations of the six crosses (Table 2) indicated statistical significance among generation means in all six sets. Subdivision of the sums of squares for generations within sets into those attributable to each of the different kinds of gene effects is presented in Table 3. Estimates of genetic parameters in each set are shown in Table 4.

Table 1. Mean relative seminal root lengths (RSRL) measured in different generations derived from six maize crosses and grown in nutrient solutions (Experiment 1)

Set	Generation					
	P_1	P_2	F_1	F_2	BC_1	BC_2
1	B37 / 1.34	A635 / 1.62	1.20	1.32	1.26	1.37
2	C103 / 1.65	B37 / 1.21	1.24	1.24	1.25	1.19
3	W117 / 1.42	A635 / 1.69	1.74	1.59	1.55	1.63
4	Mo17 / 1.33	B37 / 1.25	1.39	1.35	1.44	1.43
5	A554 / 1.24	W117 / 1.35	1.48	1.39	1.41	1.55
6	H84 / 1.64	Mo17 / 1.29	1.54	1.52	1.46	1.39

Table 2. Analysis of variance of relative seminal root length (RSRL) obtained from six maize generations derived from six different crosses grown in nutrient solutions (Experiment 1)

Source of variation	df	Mean squares
Replications	2	0.0124
Sets	5	0.2188**
Replications × sets	10	0.0032
Generations within sets	30	0.0486**
Generations within Set 1	5	0.0688**
Generations within Set 2	5	0.0914**
Generations within Set 3	5	0.0373**
Generations within Set 4	5	0.0163**
Generations within Set 5	5	0.0341**
Generations within Set 6	5	0.0437**
Error	60	0.0030
Total	107	
CV (%)		3.9

** Statistical significance at $\alpha = 0.01$.

Sets 1, 2, 3, and 6 involved generations derived from crosses between tolerant and non-tolerant lines. In these sets, additive gene effects explained most of the genetic variation, but dominance contributed a significant amount of variance except in Set 6. Dominance accounted for about half as much of the total genetic variance as did additive gene effects in Sets 1, 2, and 3. Epistasis was significant in Sets 2 and 6. Each of the three kinds of epistasis variance in Set 2 was significant, although small compared to the variances due to additive and dominance effects. Dominance × dominance epistasis was significant in Set 6, but it was small compared to the variance due to additive effects. The model with epistasis effects included substantially increased the amount of genetic variance that could be explained compared to the model with only additive and dominance effects, except for Set 1 (see multiple r^2 values in Table 4).

Table 3. Mean squares of relative seminal root length (RSRL) obtained from six maize generations derived from six different sets (Experiment 1)

Source of variation[†]	df	Mean squares of set nos.					
		1	2	3	4	5	6
Generations within set	5	0.06880**	0.0914**	0.0373**	0.0163**	0.0341**	0.0437**
Additive	1	0.22188**	0.2688**	0.1128**	0.0994	0.0382**	0.1733**
Dominance	1	0.1114**	0.1107**	0.0568**	0.0351**	0.0904**	0.0080
Epistasis	3	0.00270	0.0258**	0.0056	0.0123*	0.0140**	0.0124**
Additive × additive	1	0.00001	0.0319**	0.0110	0.0040	0.0041	0.0008
Additive × dominance	1	0.00800	0.0314**	0.0034	0.0012	0.0097	0.0116
Dominance × dominance	1	0.00007	0.0141*	0.0024	0.0318**	0.0283**	0.0249**

*, ** Statistical significance at $\alpha = 0.05$, and $\alpha = 0.01$, respectively.
[†] Parameters defined by Mather and Jinks[17].

Table 4. Estimates of the additive, dominance, and epistasis effects for relative seminal root length (RSRL) obtained from six maize generations derived from six different sets (Experiment 1)

Set model[†]	Additive	Dominance	Additive × additive	Additive × dominance	Dominance × dominance	Multiple R^2 100 (%)
Set 1						
3 Parameters[‡]	−0.172	−0.232	–	–	–	92.7
6 Parameters[‡]	−0.188	−0.295	−0.027	0.163	0.037	94.9
Set 2						
3 Parameters	0.189	−0.228	–	–	–	80.5
6 Parameters	0.222	0.762	−0.067	−0.323	0.503	96.9
Set 3						
3 Parameters	−0.123	0.163	–	–	–	61.3
6 Parameters	−0.133	−0.010	0.013	0.107	0.207	88.5
Set 4						
3 Parameters	0.035	0.129	–	–	–	34.7
6 Parameters	0.042	1.228	0.367	−0.063	−0.757	62.2
Set 5						
3 Parameters	−0.073	0.206	–	–	–	67.3
6 Parameters	−0.053	1.237	0.340	−0.180	−0.713	89.0
Set 6						
3 Parameters	0.152	0.061	–	–	–	70.9
6 Parameters	0.172	−0.952	−0.360	−0.197	0.670	83.6

[†] Parameters defined by Mather and Jinks[17].
[‡] The 3 parameter model excludes epistasis; the six parameter model includes epistasis.

In crosses between lines susceptible to Al (sets 4 and 5), dominance gene effects and dominance × dominance epistasis explained most of the genetic variation. Additive gene effects were of some importance in Set 5, but not in Set 4.

Experiment 2

The four F_2 progenies of single crosses A554 × W117, B37 × A635, B37 × C103, and Mo17 × H84 (Sets 1, 2, 5, and 6 of Experiment 1) differed in their mean Al tolerance when grown in nutrient solutions (Table 5). They also differed in the amount of plant-to-plant variation within the F_2 generations; Mo17 × H84 had the greatest variance and B37 × C103 had the least. The magnitude of the variances among plants within F_2 generations was not related to the magnitude of the differences between their parents in Al tolerance. A554 and W117 differed the least in RSRL values (0.11, Table 1), but had next to the highest variance in their F_2 generation; and B37 and C103 differed the most (0.44), but had the lowest F_2 generation variance. The two crosses involving B37 had the lowest variances among F_2 generation plants.

The frequency distributions of the F_2 generations are presented in

Table 5. Analysis of variance for relative seminal root length (RSRL) in F_2 maize generations of four different crosses grown in nutrient solutions (Experiment 2)

Source of variation	df	Mean squares
Replications	3	0.0088
F_2 generations	3	0.0420**
Error	9	0.0017
Total	15	
CV (%)		3.1
Among plants	656	0.0259
F_2 generation 1 (B37 × A635)	164	0.0185
F_2 generation 2 (B37 × C103)	164	0.0163
F_2 generation 3 (A554 × W117)	164	0.0294
F_2 generation 4 (Mo17 × H84)	164	0.0397

** Statistical significance at $\alpha = 0.01$.

Figure 1. The figure shows susceptible plants were more frequent than tolerant ones. The highest frequencies of Al-tolerant plants occurred in the cross involving Mo17 × H84. All frequency distributions were continuous, unimodal, and typical for a quantitatively inherited trait. Higher frequencies in the susceptible ranges for three crosses (numbers closer to 1.0), indicated a preponderance of genes dominant for susceptibility to Al tolerance. Plants in the F_2 generation of Mo17 × H84 showed the most nearly normal distribution.

Experiment 3

The analysis of variance for RSRL of eight parental lines and their 28 single-cross hybrids (diallel crosses) grown in nutrient solution are

Table 6. Analysis of variance for relative seminal root length (RSRL) in eight maize parents and their 28 F_1 crosses (diallel) grown in nutrient solution (Experiment 3)

Source of variation	df	Mean squares
Replications	2	0.0121
Treatments (unadj.)	35	0.3301**
Parents	7	0.1778**
Parents vs crosses	1	1.5699**
Crosses	27	0.3237**
General combining ability	7	0.8911**
Specific combining ability	20	0.1251**
RCBD error	70	0.0481
Blocks within replications (adj.)	15	0.0544
Intra-block error	55	0.0464
Total	107	
Average effective error		0.0479
Relative efficiency of lattice (%)		100.4
CV (%)		11.8

** Statistical significance at $\alpha = 0.01$.

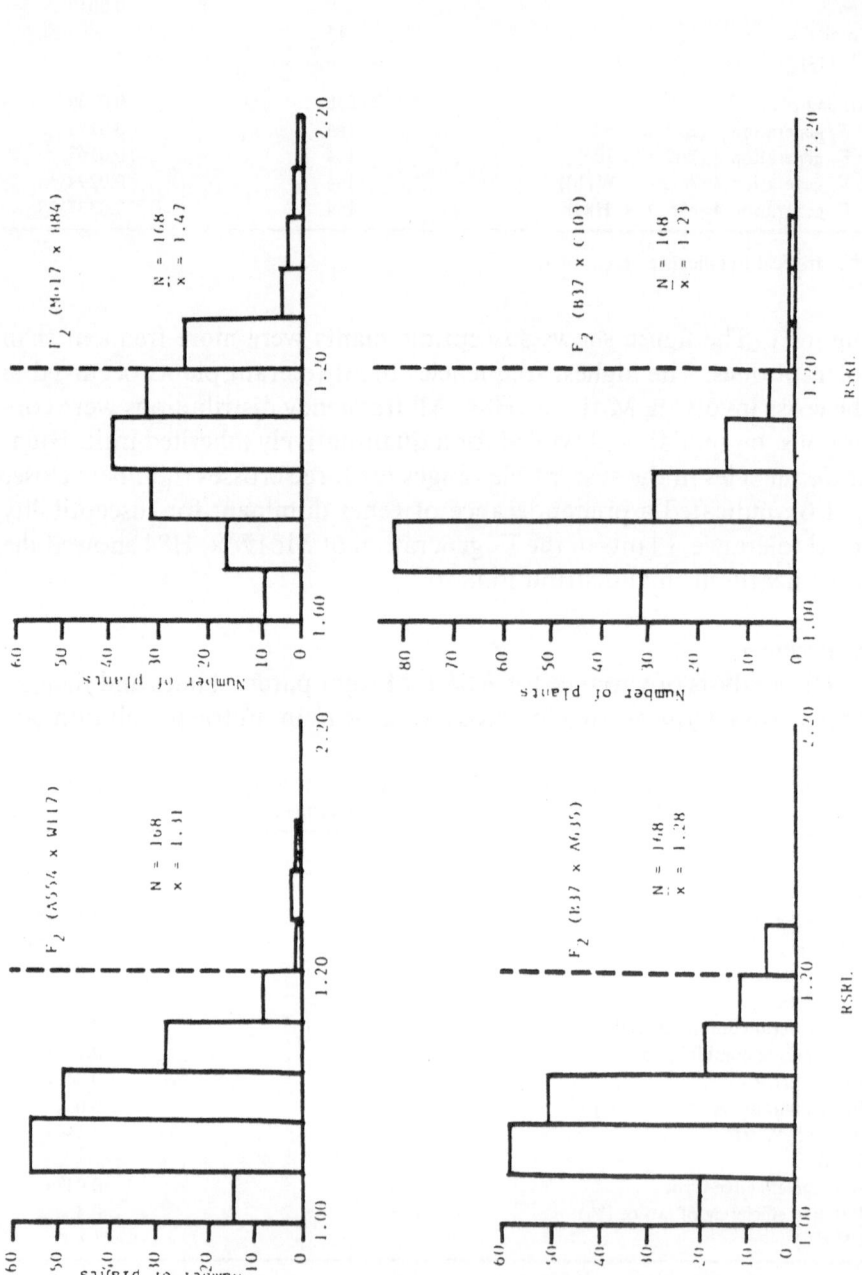

Fig. 1. Frequency distributions for relative seminal root length (RSRL) of plants from four different F_2 maize generations grown in nutrient solution (Experiment 2). (The dotted line represents genotypes considered to have good Al tolerance).

Table 7. Parental and F₁ means for relative seminal root length (RSRL) of maize grown in nutrient solution (Experiment 3)

Inbred lines	F₁ means									Parental† means
	Mo17	C103	H84	C164	B37	W117	A635	A554	Cross mean	
Mo17		1.75	1.67	1.75	1.64	1.46	1.78	1.99	1.72	1.57 bcde
C103			1.81	1.82	1.72	1.63	1.85	2.37	1.85	2.06 a
H84				2.23	1.55	1.92	1.85	2.67	1.96	1.79 ab
C164					1.79	2.13	2.01	2.83	2.08	1.74 abc
B37						1.78	1.49	1.86	1.69	1.71 abcd
W117							2.27	2.03	1.89	1.39 cde
A635								2.19	1.92	1.50 bcde
A554									2.28	1.30 e

† For parents, means followed by a common letter are not significantly different at = 0.05, according to Duncan's New Multiple Range Test.

presented in Table 6. Statistical significance was noted for differences among parental lines, for parental lines *vs.* crosses, and for general and specific combining abilities. The parental lines *vs.* crosses comparison was an indication of heterosis in the F_1 generation crosses, which suggested a preponderance of genes dominant for tolerance in this set of lines. The variance for general combining ability explained most of the variation for each of the six variables studied, but specific combining ability was also important. As in Experiment 1, additive gene effects were the most important for controlling tolerance but dominance was also important.

The parental and F_1 generation means and estimates of general and specific combining ability effects are presented in Tables 7 and 8. Line C103 possessed the greatest tolerance to Al toxicity, adn B37, C164, and

Table 8. Estimates of general combining ability (GCA) and specific combining ability (SCA) effects for relative seminal root length (RSRL) of maize generations grown in nutrient solution (Experiment 3)

Inbred lines	SCA effects								GCA effects
	Mo17	C103	H84	C164	B37	W117	A635	A554	
Mo17		0.1484	−0.0585	−0.1158	0.2232	−0.1877	0.0965	−0.1063	−0.2373
C103			−0.0694	−0.1999	0.1523	−0.1699	0.0199	0.1195	−0.0847
H84				0.0865	−0.1380	−0.0022	−0.1113	0.2926	0.0403
C164					−0.0485	0.0639	−0.0952	0.2087	0.1843
B37						0.1727	−0.1563	−0.2057	−0.2713
W117							0.3895	−0.2666	−0.0417
A635								−0.1424	−0.0047
A554									0.4151

Standard errors of differences: Between two GCA effects = 0.0730
  Between two SCA effects with one common parent = 0.1633
  Between two SCA effects with no parents in common = 0.1461

H84 tended to be intermediate in Al tolerance. Results of Experiment 3 were somewhat in agreement with those of Experiment 1 in that higher Al tolerance was noted for C103 and H84 and lower Al tolerance was noted for A554, Mo17, and W117. However, the results for A635 and B37 were not consistent. Differences in seed quality between the two experiments may have been important. A635 seedlings with greater vigor were used in Experiment 3 than in Experiment 1. Rhue and Grogan[21] and Rhue et al.[22] reported low Al tolerance for A554, B37, and Mo17 tested in their nutrient solution studies.

A554 had the highest general combining ability and consistently performed well in crosses with the other lines. As a line, A554 had a low level of Al tolerance. Similar types of responses were noted for A554 when it was used in crosses by Rhue et al.[22], who suggested that specific factors existed in the Al non-tolerant A554 which enhanced the expression of Al tolerance in the F_1 generation, but their nature was not understood. In our study, the better than expected performance of A554 for Al tolerance was expressed particualrly in crosses with the more Al-tolerant lines A635, C103, C16, and H84.

Specific combining ability effects were highest for the crosses C164 × A554, H84 × A554, and W117 × A635, and lowest for B37 × A554, C103 × C164, and W117 × A554. The results described for the inheritance of Al tolerance by many authors were not in complete agreement with the results of the studies reported here. The differences may be explained by differences in genotypes and techniques used. Rhue[20], Rhue et al.[22], and Stockmeyer et al.[25] used 250 μmol Al L^{-1} and 126 μmol P L^{-1} in their Al tolerance studies. Thus, the concentrations of Al and P used were different from the concentrations used in our inheritance studies (185 μmol Al L^{-1} and 45 μmol P L^{-1}). The methods Rhue and colleagues[20,21,22,25] used to assess Al tolerance were also different because they usually separated plants in the F_2 and backcross generations into classes as well as using root growth as the criterion for separation. Also, the seedlings were transplanted to treatments earlier (2- to 3-day old seedlings) compared to our study (7-day-old seedlings).

Silva[24] used techniques in which the seeds were germinated in sand and irrigated with nutrient solution containing 2780 μmol Al L^{-1}. Maize seedlings were stressed with Al at an early stage of germination or were germinated with Al in some experiments. After 7 days, the root length was measured and relative root lengths were determined by comparing measurements on plants grown with and without Al.

In a study with barley (*Hordeum vulgare* L.) seedlings, May et al.[18] concluded that the response of plants to variations in the environment decreased rapidly with the age of the plants. Thawornwong and Van

Diest[26] pointed out that young rice (*Oryza sativa* L.) seedlings were more susceptible to Al toxicity than older plants. Sartain and Kamprath[23] noted that short-term root elongation studies with soybean cultivars [*Glycine max* (L.) Merr.] only took into account the effects of Al on cell elongation and cell division. However, in another study with soybeans in which selection was made in a population, Hanson and Kamprath[12] concluded that Al tolerance based on growth in nutrient solutions apparently identified genetic differences for tolerance in established plants as well as tolerance that appeared unique in the seedling stage. This may account for some of the differences between the results reported in our studies compared to those of the other studies described.

Galvao and Silva[8] based their evaluations of Al tolerance in maize on shoot and root dry matter yields, which are not always related to Al tolerance[6,17,14]. Lopes *et al.*[14] and Naspolini *et al.*[19] found that the magnitude of general combining ability variances was higher than that of the specific combining ability variances. These results agree with those noted in our study.

It is evident that Al tolerance in maize is quantitatively inherited in the lines studies. However, the differences in level of Al tolerance of Brazilian maize inbred lines compared to USA inbred lines[16] do not eliminate the possibility of a major gene for Al tolerance as described by Silva[24]. Further investigation should be conducted using the more tolerant Brazilian lines in crosses with the less tolerant USA lines.

Several different mechanisms explaining plant Al tolerance have been described in the literature[4,5,6,13]. It is not clear how these mechanisms are related and which may be the most important in attempting to explain Al tolerance. Evidence does not support the concept that a single major gene controls Al tolerance; a more complex genetic system is more probable. Perhaps at specific stages of plant development, one mechanism could be more important than another and simple genetic control of Al tolerance might exist at specific growth stages.

References

1 Clark R B 1982 Nutrient solution growth of sorghum and corn in mineral nutrition studies. J. Plant Nutr. 5, 1039 1057.
2 Cochran W G and Cox G M 1957 Experimental Designs, 2nd edition. John Wiley & Sons, New York.
3 Comstock R E and Robinson H F 1948 The components of genetic variance in populations of biparental progenies and their use in estimating the average degree of dominance. Biometrics 4, 254 266.
4 Foy C D, Chaney R L and White M C 1978 The physiology of metal toxicity in plants. Annu. Rev. Plant Physiol. 29, 511 566.
5 Foy C D and Fleming A L 1978 The physiology of plant tolerance to excess available

aluminum and manganese in acid soils. pp 301–328. *In* Crop Tolerance to Suboptimal Land Conditions. Ed. G A Jung. Am. Soc. Agron., Madison, WI.

6. Furlani P R 1981 Effects of aluminum on growth and mineral nutrition of sorghum genotypes. Ph.D. Diss., Univ. of Nebraska, Lincoln. Diss. Abstr. 42, 1260B.
7. Furlani P R and Clark R B 1981 Screening sorghum for aluminum tolerance in nutrient solution. Agron. J. 73, 587–594.
8. Galvao J D and Silva J C 1978 Inheritance of aluminum tolerance in the maize variety Piranao. Ceres 25, 71–78. (*In Portuguese*)
9. Garcia O Jr, da Silva W J and Massei M A S 1979 An efficient method for screening maize inbreds for aluminum tolerance. Maydica 24, 75–82.
10. Gardner C O and Eberhart S A 1966 Analysis and interpretation of the variety cross diallel and related populations. Biometrics 22, 439–452.
11. Griffing B 1965 Concept of general and specific combining ability in relation to diallel crossing systems. Aust. J. Biol. Sci. 9, 463–493.
12. Hanson W D and Kamprath E J 1979 Selection for aluminum tolerance in soybean based on seedling-root growth. Agron J. 71, 581–586.
13. Lafever H N 1981 Genetic differences in plant response to soil nutrient stress. J. Plant Nutr. 4, 89–109.
14. Lopes A L, Magnavaca R, Bahia A F C and Gama E E G 1985 Performance of maize populations for aluminum tolerance in nutrient solution. Brazilian Agric. Res. (*In press*) (*In Portuguese*)
15. Magnavaca R 1982 Genetic variability and the inheritance of aluminum tolerance in maize (*Zea mays* L.). Ph.D. Diss., Univ. of Nebraska, Lincoln. Diss. Abstr. 43, 2073B (1983).
16. Magnavaca R, Gardner C O and Clark R B 1986 Evaluation of inbred maize lines for aluminum tolerance in nutrient solution. pp 189–199. *In* Genetic Aspects of Plant Mineral Nutrition. Eds. W H Gabelman and B C Loughman. Martinus Nijhoff/Dr. W. Junk Publ., The Hague, The Netherlands.
17. Mather K and Jinks J L 1971 Biometrical Genetics. Cornell Univ. Press, Ithaca, NY.
18. May L H, Chapman F H and Aspinall D 1964 Quantitative studies of root development. Aust. J. Biol. Sci. 18, 25–35.
19. Naspolini V, Bahia A F C, Viana R T, Gama E E G, Vasconcellos C A and Magnavaca R 1981 Performance of inbreds and single crosses of corn in soils under 'cerrado' vegetation. Ciencia Cultura 33, 722–727. (*In Portuguese*)
20. Rhue R D 1979 Differential aluminum tolerance in crop plants. pp 61–80. *In* Stress Physiology in Crop Plants. Eds. H Mussell and R C Staples. John Wiley & Sons, New York.
21. Rhue R D and Grogan C O 1977 Screening Corn for Al tolerance using different Ca and Mg concentrations. Agron. J. 69, 75–760.
22. Rhue R D, Grogan C O, Stockmeyer E W and Everett H L 1978 Genetic control of aluminum tolerance in corn. Crop. Sci. 18, 1063–1067.
23. Sartain J B and Kamprath E J 1978 Aluminum tolerance of soybean cultivars based on root elongation in solution culture compared with growth in acid soil. Agron. J. 70, 17–20.
24. Silva W J da 1979 Selection of tolerant corn to aluminum toxicity. pp 107–113. *In* XIII Annu. Conf. Genetic Society, Jaboticabal, S.P., Brazil. (*In Portuguese*)
25. Stockmeyer E W, Everett H L and Rhue R H 1978 Aluminum tolerance in maize seedlings as measured by primary root length in nutrient solutions. Maize Genet. Coop. Newsl. 52, 15–16.
26. Thawornwong N and Van Diest A 1974 Influences of high acidity and aluminum on the growth of lowland rice. Plant and Soil 41, 141–159.

Aluminum tolerance in sorghum

R.A. BORGONOVI*, R.E. SCHAFFERT, G.V.E. PITTA, R. MAGNAVACA and V.M.C. ALVES
EMBRAPA, National Corn and Sorghum Research Center. Caixa Postal 151. 35700 - Sete Lagoas, MG, Brazil

Key words Additive effects Aluminum tolerance Dominance effects General combining ability Relative seminal root length *Sorghum bicolor* (L.) Moench Specific combining ability

Sumary Aluminum (Al) tolerance in sorghum (*Sorghum bicolor* (L.) Moench) was studied in nutrient solution. The relative seminal root lengths of lines from the World Collection (IS Numbers) and other sources suggested that a reasonable amount of genetic variation apparently exists in sorghum for tolerance to Al toxicity. The estimates of general and specific combining ability, based on F_1 crosses, indicated that the variances for general combining ability of male and female (R and A lines) sorghum lines accounted for most of the variation (52%) for the trait, but specific combining ability was also important. This high ratio of general combining ability emphasizes the importance of the additive gene effects over non-additive gene effects for aluminum tolerance in sorghum.

Introduction

The 'cerrado' region of Brazil occupies approximately 180 million hectares or about 20% of the total area of the country. The great majority of the soils under 'cerrado' vegetation are unfertile due to high phosphorus adsorption capacity, low pH, high aluminum saturation, low cation exchange capacity; and thus a generalized deficiency of nutrients, principally phosphorus, calcium, magnesium, potassium and zinc[9,15].

The latosol soils which comprise 56% of this area[22] are normally high in clay content but due to their structure, the water infiltration rate is high and the water holding capacity is low[16]. Water stress that occurs during periods of one to four weeks without rain, called 'veranicos', are common in the 'cerrado' region and according to several authors may cause limitations to sensitive annual crops grown without supplemental irrigation[14,16,26].

Assuming the great potential of the 'cerrado' region for grain sorghum production[4,14], and the technical and economical constraints for subsoil acidity correction, the importance of developing Al tolerant sorghum cultivars is evident.

Genetic variability of sorghum germplasm and breeding material for aluminum tolerance in field and in greenhouse conditions have been demonstrated[3,11,13,17,19,21,23,24,25]. Schaffert *et al.*[25] evaluated 30 grain sorghum hybrids in acid soils. The results showed an association of alumi-

* Deceased March 1986.

H.W. Gabelman and B.C. Loughman (Eds.), *Genetic aspects of plant mineral nutrition.*
ISBN-13: 978-94-010-8102-3
© 1987 Martinus Nijhoff Publishers, Dordrecht/Boston/Lancaster.

num tolerance with differential root growth and drought tolerance among the hybrids.

Pitta et al.[19,21] evaluated 1200 sorghum lines from different origins in acid soils with 50% aluminum saturation. The results showed that some germplasm from Uganda and Tanzania were more Al tolerant than germplasm from other sources. Salinas and Sanchez[23] using a nutrient solution with two Al levels (0 and 8 ppm) and two P levels (0.05 and 0.20 ppm) quantified the relative growth rate (RGR) and relative root extension rate (RER) as variables to characterize Al tolerance in five sorghum genotypes: two commercial hybrids (TEY 101 and RS610) and three lines (SC 112-14, SC 334-9 and TX 7078). The hybrid, TEY 101 was the most efficient cultivar under P and Al stresses. Santos et al.[24] obtained significant correlations between soil screening in greenhouse conditions and field experiments for Al tolerance. The correlation coefficients for the estimates at 64% Al saturation were 0.72** for dry weight of tops × grain yield and 0.64** for dry weight of roots × grain yield. Malavolta et al.[17] evaluated 30 grain sorghum hybrids in nutrient solution at five Al levels (0, 3, 6, 12 and 24 ppm) and concluded that total dry weight of the seedlings after three weeks growth was a better estimate of Al tolerance than seedling shoot weight, root weight, seedling height or root length. The Al concentration that best differentiated the genotypes was 12 ppm.

Bastos and Gourley[3] using a modified Steinberg nutrient solution with 6 ppm Al at pH 4.0, concluded that the rate of seminal root elongation after a 6-day treatment period was the best index to measure Al tolerance. They also found that 27 of 158 genotypes were equal to or exceeded the tolerant check IS 12666 C (SC 165-14). According to Furlani[11] the response of sorghum genotypes to Al in nutrient solutions varied, depending on the germplasm as well as the Al concentration. At an Al concentration of 43 μM (1.2 ppm), very few genotypes were considered nontolerant to Al. However, at a concentration of 96 μM (2.7 ppm) aluminum, the genotypes SC 283 and SC 112-14, previously classified as Al tolerant, confirmed their behavior. In general the methods used to assess Al tolerance in sorghum have involved different techniques, Al concentrations and cultivars, leading to conflicting results.

A few studies dealing with inheritance of Al tolerance in sorghum have been reported in the literature[2,5,11,20]. The results of Pitta et al.[20] concerning the field evaluation of two female lines (BR 007 and Wheatland), three pollinator lines (TX 2536, SC 112-14 and TAM 428) and their hybrids, suggested the presence of a small number of genes controlling Al tolerance. Furlani[11] considered the inheritance of this trait to be complex.

Bastos[1], evaluating the relative root length of sorghum lines and their F_1 hybrids in nutrient solutions with two Al levels (0 and 4 ppm), indicated that the genetic control of tolerance was complex. Later, Bastos[2] utilizing five F_1 crosses, two sensitive parental lines (TX 415 and 7B113), two tolerant parental lines (SC 175-14 and SC 237-14) and their F_2 populations, observed transgressive segregations of the F_2, and concluded that different genes were probably involved in Al tolerance.

The results of Borgonovi et al.[5] indicated differences between hybrids involving the Al tolerant line SC 283 (IS 7173 C) and several male sterile Al-sensitive lines. The relative seminal root growth (RSRG) of the hybrids between this line and male sterile lines Wheatland A, CMSXS 168 A and Redlan A was superior to the hybrids made with BR 007 A and TX 623 A. Apparently there is a specific combining ability effect for Al tolerance, indicating that the inheritance of this trait may be complex.

Considering that information for screening sorghum cultivars for Al tolerance and the inheritance of this trait is incomplete and somewhat conflicting, this research was conducted to obtain additional information on the behavior of the known sources of Al tolerance in sorghum and on the relative importance of additive and non-additive effects in the control of this trait.

Materials and methods

Seed germination, seedling handling and growth in nutrient solution

The germplasm screening experiments were conducted in greenhouse conditions at CNPMS/EMBRAPA to identify sorghum lines with different degrees of Al tolerance and used the techniques proposed by Furlani and Clark[12], Furlani[11] and Magnavaca[18].

Sorghum (*Sorghum bicolor* (L.) Moench) seeds treated with CAPTAN* [N-(trichlorometyl)-thio-4-cyclohexane-1,2-dicarboximide] were germinated in paper towels rolled into tubes and kept moist with aerated water for seven days. On the eighth day the seedlings were examined for possible root damage; the initial seminal root length (ISRL) was measured and recorded and the seedlings were transferred to plastic containers with 8.5 l of nutrient solution. The nutrient solution used an Al concentration of 180 μM Al (4.8 ppm) in the form of $KAl(SO_4)_2 \cdot 12H_2O^{18}$. The composition of the nutrient solution was (μM element l^{-1}): 3,500 Ca; 2,400 K; 850 Mg; 10,300 NO_3-N; 1,300 NH_4-N; 45 P; 850 S; 25 B; 590 Cl; 0.63 Cu; 9.1 Mn; 0.83 Mo, 2.29 Zn and 77 Fe as FeHEDTA (ferric hydroxyethylenediaminetriacetate). The pH of the solution was initially adjusted to 4.0 and maintained throughout the period of evaluation.

Each container had a suspended plexiglass top with 20 mm holes for placement of the seedlings and two holes of 10 mm diameter for aeration and system monitoring. The seedlings were fixed in the holes with sponge rubber so that development of adventitious roots was not impeded. The nutrient solution was maintained with constant aeration during the growth period (normally 10 to 12 days).

At the end of the growth period, the seedlings were removed and the roots were examined for visual symptoms of Al toxicity. The final seminal root length (FSRL) and longest adventitious root

* Mention of a company, trademark, or proprietary product does not constitute a guarantee or warranty of the product by the authors, and does not imply its approval to the exclusion of other products that may also be suitable.

length (LARL) were recorded for each plant. The relative seminal root growth was calculated as follows (Furlani, 1981):

$$RSRG(\%) = \left(\frac{ISRL}{FSRL} - 1\right) \times 100$$

Genotypes, experimental design and statistical analyses of experiments

Experiment 1. Thirty genotypes were evaluated in nutrient solution in a 5 × 6 rectangular lattice design with three replications. The RSRG and LARL mean values were estimated using seven plants per plot. The statistical analyses was performed according Cockran and Cox[6].

The genotypes were chosen based on previous reports of their performance in Al tolerance screenings conducted at CNPMS/EMBRAPA and in other programs.

Experiment 2. Three male-sterile lines (BR 007, Wheatland and Redlan), eight restorer lines (SC 048, SC 418, SC 112-14, CMSXS 154, 3DX57/1/1/910, V 20-1-1-1, 156-P-5-Serere-1 and 5DX61/612) and their 24 F_1 hybrids were evaluated for Al tolerance in nutrient solution. The experimental design was a 6 × 6 triple lattice. The RSRG mean values were estimated using seven plants per plot. The statistical analyses were performed according to Cochran and Cox[6]. For the estimation of combining ability effects, the parental lines were excluded and the data were analyzed using a fixed effects model according to Design II of Comstock and Robinson[7].

A coefficient comparable to heritability was used for the estimation of the genetic proportion of the phenotypic variability coefficient called 'coefficient of genotypic determination' according to Fonseca[10]. Two ratios were obtained as follows (see Table 3 for definition of terms):

(a) $\hat{b}_R = \dfrac{\hat{V}_{g(R)}}{\hat{V}_{g(R)} + \hat{V}_{s(AR)} + \dfrac{\hat{V}_e}{k.a}}$ for R lines

(b) $\hat{b}_A = \dfrac{\hat{V}_{g(A)}}{\hat{V}_{g(A)} + \hat{V}_{s(AR)} + \dfrac{\hat{V}_e}{k.r}}$ for A lines

Results and discussion

Experiment 1

The origin, group and the mean RSRG and LARL of the genotypes evaluated in experiment 1 are shown in Table 1. The analyses of variance for these two traits are presented in Table 2. Although some consistency of results for the two traits is evident, the RSRG showed a better differentiation of tolerant lines than LARL. Among the germplasm evaluated, the following showed the best performance: 9DX9/11, 5DX61/612, IS 7173 C, 156-P-5-Serere-1, IS 3625 C, IS 12666C, IS 7254 C, CMSXS 154, V 20-1-1-1, CMSXS 604, IS 12564 C and IS 1335 C. The performance of these lines in this experiment agreed with the results obtained in acid soils with high Al saturation[4,8,19,21] and in nutrient solution[4,13], confirming the effectiveness of this technique at this Al level. Although only a few of the lines for the sorghum germplasm collection have been screened for Al tolerance, the relatively high frequency of lines of Uganda, Nigeria and Tanzania classified as Al tolerant, should be

Table 1. Mean relative seminal root growth (RSRG) and longest adventitious root length (LARL) of genotypes evaluated for Al tolerance in nutrient solution (Experiment 1)

Identification	Origin	Group	RSRG (%)	LARL (cm)
9DX9/11	Uganda	–	45.9	20.2
5DX61/612	Uganda		43.0	18.9
IS7173C (SC283B)	Tanzania	Conspicuum	40.9	15.8
156-P-5-Serere-1	Uganda	–	36.4	24.6
IS3625 C (SC549)	Nigeria	Conspicuum	35.8	20.3
IS12666 C (SC 175-14)	Ethiopia	Zera-Zera	34.8	13.3
IS7254 C (SC 566-14 B)	Nigeria	Caudatum	34.4	17.2
CMSXS 154	Brazil	–	34.0	19.2
V20-1-1-1	Uganda	–	33.9	22.1
CMSXS 604	Brazil	–	33.9	18.5
IS12564 C (SC 048)	Sudan	Zera-Zera	32.9	23.2
IS1335 C (SC 418)	Tanzania	Caudatum-Kafir	31.3	22.4
3DX57/1/1/910	Uganda	–	29.6	20.4
IS2744	Africa	Caudatum-Kafir	25.5	20.8
IS7542 C (SC 408)	Nigeria	Caudatum-Guineense	24.5	14.7
IS1309 C (SC 322)	Tanzania	Nigricans	17.4	16.9
IS2677	Africa	Caffrorum-Durra	16.8	25.0
IS502	Mexico	Caffrorum	15.8	13.3
IS11612 (SC 112-14)	Ethiopia	Zera-Zera	11.9	13.8
IS3674	–	–	9.9	11.2
IS3802		Feterita	8.5	15.7
IS2368	–		7.5	6.2
IS3676	–	Kafir	7.5	3.9
TX2536	USA	–	6.9	2.6
IS1143 C (SC 208)	India	Durra	6.2	2.2
TX415	USA	–	4.8	1.8
IS8361 (Wheatland B)		–	4.1	3.3
TX623 B	USA	–	3.9	2.3
BR 300 (hybrid)	Brazil	–	8.1	4.9
CMSXS 348 (hybrid)	Brazil	–	9.0	5.2

considered as an indication that lines from these regions may be more tolerant to Al toxicity and should be a source of germplasm for additional screening studies.

Table 2. Analyses of variance for relative seminal root growth (RSRG) and longest adventitious root length (LARL) of sorghum genotypes grown in nutrient solution with Al (Experiment 1)

Source of variation	Df	Mean squares	
		RSRG	LARL
Replications	2	1.160	0.035
Treatments (unadj)	29	908.950**	281.968**
RCBD error	58	5.708	2.470
Blocks within replications (adj)	15	5.484	3.031
Intra-block error	43	5.787	2.274
Total	89		
CV%		8.79	8.95

** Statistical significance at $\alpha = 0.01$

Table 3. Analysis of variance for relative seminal root growth (RSRG) of 24 F_1 hybrids grown in nutrient solution (Experiment 2)

Source of variation	Df	Mean squares	Expectations of mean squares[1]
Replications	(k-1) 2		
General combining ability (A lines)	(a-1) 2	404.157**	$\hat{V}_e + k\hat{V}_{s(AR)} + kr\hat{V}_{g(A)}$
General combining ability (R lines)	(r-1) 7	166.156**	$\hat{V}_e + k\hat{V}_{s(AR)} + ka\hat{V}_{g(R)}$
Specific combining ability (A × B)	(a-1)(r-1) 14	18.738**	$\hat{V}_e + k\hat{V}_{s(AR)}$
ERROR	(ar-1)(k-1) 46	1.308	$\hat{V}_e$

** Statistical significance at $\alpha = 0.01$

$1/\hat{V}_e$ = environmental variance

$\hat{V}_{s(AR)}$ = variance due to the specific combining ability arising from interaction of female (A) and male (R) parents

$\hat{V}_{g(R)}$ = variance arising from genetic differences in the combining ability among male (R) parents, and

$V_{g(A)}$ = variance arising from genetic differences in the combining ability among female (A) parents.

Experiment 2

The analysis of variance for RSRG of 24 F_1 hybrids grown in nutrient solution is presented in Table 3. Statistical significance was noted for general combining ability (GCA) of the A and R lines and for specific combining ability (SCA). The estimates of the components of variance for RSRG, the ratios of male (R lines) and female (A lines), GCA effects and the ratio of GCA and SCA effects are presented in Table 4. All R lines and A lines showed similar contributions for GCA, but the variance of GCA effects were about three times greater than the variances of SCA effects. This difference indicates the relative importance of additive gene effects over the non-additive gene effects on the genetic variation for Al tolerance in the lines used in this study. This result is similar to the results described by Bastos[2] who evaluated the relative root length of five crosses among four sorghum lines and concluded that additive gene action was probably involved in Al tolerance.

Table 4. Estimates of components of variance obtained from the analysis of variance for relative seminal root growth (RSRG), ratios of male (R) over female (A), general combining ability (GCA) effects and GCA over specific combining ability (SCA) effects for RSRG. (Experiment 2)

	Components of variance			$GCA_R\ GCA_A$	GCA SCA
	$\hat{V}_{S(AR)}$	$\hat{V}_{g(A)}$	$\hat{V}_{g(R)}$	$V_{g(R)}\ \hat{V}_{g(A)}$	$(\hat{V}_{g(R)} + \hat{V}_{g(A)}) \cdot 2\hat{V}_{S(AR)}$
RSRG	5.810** + 2.210	16.059** + 12.437	16.380** + 8.734	1.020	2.792

** Statistical significance at $\alpha = 0.01$

Table 5. Estimates of general combining ability (GCA) effects of restorer (R) lines and of male-sterile (A) lines for relative seminal root growth (RSRG) (Experiment 2)

RSRG %	R lines							A lines		
	SC048	SC418	SC112-14	CMSXS154	3DX57/1/1/910	V20-1-1-1-156	P-5-Serere-1	5DX61/612	BR007	Wheatland	Redlan
GCA	−4.92**	−4.77**	−4.15**	−5.32**	9.58**	3.94**	−6.45**	12.09**	−7.49**	6.65**	0.85*

*, ** Statistical significance at $\alpha = 0.05$ and $\alpha = 0.01$, respectively.

The estimated values of the coefficient of genotypic determination ($\hat{b}_R = \hat{b}_A = 0.73$) indicated that a large portion of the variability observed for RSRG may be explained by genetic factors. This coefficient, which is a measure of the genetic proportion of the phenotypic variability, was used because of the fixed effects model adopted. These results agreed with those obtained by Bastos[2] who reported heritability (h_2) values of relative root length in sorghum between 0.58 and 0.87.

Since the variance for GCA effects explained most of the variation (52%) of RSRG, the SCA effects which accounted for 12% of RSRG variation were omitted from this paper. The SCA effects varied from −5.64 to 5.86.

The estimates of GCA effects of R and A lines based on the F_1 crosses are presented in Table 5. The R lines 5DX61/612, 3DX57/1/1/910 and V-20-1-1-1 showed a positive effect in crosses with the male sterile lines. The analyses of GCA effects of A lines indicated differences between them. Wheatland had a low level of Al tolerance, but its GCA did not indicate this tolerance. These results are in agreement with those reported by Borgonovi et al.[4] and should be considered as an indication of a possible existence of specific genetic factors in the Al non-tolerant line Wheatland that enhanced the expression of Al tolerance in the F_1 hybrids.

References

1 Bastos C R 1981 Toxicity of aluminum in nutrient solution to sorghum seedlings. M.S. Thesis, Mississippi State University, Mississippi State, MS.
2 Bastos C R 1982 Inheritance study of aluminum tolerance in sorghum in nutrient culture. Ph.D. Thesis, Mississippi State University, Mississippi State, MS.
3 Bastos C R and Gourley L M 1982 Rapid screening of sorghum seedlings for tolerance to low pH and aluminum. In ICRISAT 1982 Sorghum in the Eighties. Proceedings of the international symposium on sorghum, 2–7 November 1981, Patancheru AP, India.
4 Borgonovi R A, Santos F G and Schaffert R E 1982 Population breeding in sorghum I development of BRP5BR with tolerance to aluminum toxicity. p 34. In Ann. XIV Brazilian Maize and Sorghum Congress, Empasc, Florianópolis, SC, Brazil (In Portuguese).
5 Borgonovi R A, Schaffert R E and Pitta G V E 1984 Breeding aluminum tolerant sorghums. In Evaluating Sorghum for Tolerance to Al Toxic Tropical Soil in Latin America. Proceedings of the international workshop, May 28–June 02 1984 (In press).
6 Cochran W G and Cox G M 1957 Experimental Designs, 2nd ed. John Wiley Sons, New York.
7 Comstock R E and Robinson H F 1948 The components of genetic variance in populations of biparental progenies and their use in estimating the average degree of dominance. Biometrics 4, 254–266.
8 Duncan R R 1981 Registration of GP1R acid soil tolerant sorghum germplasm population. Crop Sci. 21, 637.
9 EMBRAPA 1978 CPAC annual report for 1976–77. Planaltina, DF, Brazil (In Portuguese).
10 Fonseca T C 1978 Estimation of coefficients for selecting mulberry (Monus alba L.) artificial hybrids. M.S. Thesis, University of São Paulo, Piracicaba, SP.
11 Furlani P R 1981 Effects of aluminum on growth and mineral nutrition of sorghum genotypes. Ph.D. Thesis, University of Nebraska, Lincoln.

12 Furlani P R and Clark R B 1981 Screening sorghum for aluminum tolerance in nutrient solution. Agron. J. 73, 587–594.
13 Furlani P R, Bastos C R, Borgonovi R A and Schaffert R E 1984 Differential response of sorghum genotypes for tolerance to aluminum in nutrient solutions *In* Ann XV Brazilian Maize and Sorghum Congress, Epeal, Maceio, Al, Brazil (*In Portuguese*).
14 Goedert WJ, Lobato E and Wagner E 1980 Agricultural potencial of brazilian cerrado. Pesq. Agrop. Bras. 15, 1–17 (*In Portuguese*).
15 Lopes A S and Cox F R 1977 A survey of the fertility status of surface soils under 'cerrado' vegetation in Brazil. Soil Sci. Am. J. 41, 743–747.
16 Lopes A S 1983 'Cerrado' soils: characteristics, properties and management. International Potash and Phosphate Institute. International Potash Institute. Piracicaba, SP, Brazil (*In Portuguese*).
17 Malavolta E, Nogueira F D, Oliveira I P, Nakayama L and Eimori I 1981 Aluminum tolerance in sorghum and bean — methods and results. J.Plant Nutr. 3, 687–694.
18 Magnavaca R 1982 Genetic variability and the inheritance of aluminum tolerance in maize (*Zea mays* L.). Ph.D. Thesis, University of Nebraska, Lincoln.
19 Pitta G V E, Trevisan W L, Schaffert TR E, França G E and Bahia A F C 1976 Evaluation of sorghum lines under soil acidity conditions I. pp 553–557. *In* Ann. XI Brazilian Maize and Sorghum Research Conference, Piracicaba, SP, Brazil (*In Portuguese*).
20 Pitta G V E, Schaffert R E and Borgonovi R A 1979a Evaluations of sorghum parents and hybrids to high soil acidity conditions. p 127 *In* Proc. XII Brazilian Corn and Sorghum Research Conference, Emgopa, Goiania, GO, Brazil (*In Portuguese*).
21 Pitta G V E, Schaffert R E, Borgonovi R A, Vasconcellos C A, Bahia A F C and Oliveira A C 1979b Evaluation of sorghum lines. II. p 128. *In* Proc. XII Brazilian Corn and Sorghum Research Conference, Emgopa, Goiania, GO, Brazil (*In Portuguese*).
22 Sanchez P A, Lopes A S and Buol S W 1974 Cerrado research center: preliminary project proposal. Raleigh, NC (mim.).
23 Salinas J G and Sanchez P A 1978 Tolerance to Al toxicity and low available P. pp 115–137. *In* Agronomic-Economic Research on Soils of the Tropics. Annual Report for 1976–77. North Caroline State University, Raleigh, NC.
24 Santos H L, Baligar V C and Vasconcellos C A 1980 Screening sorghum genotypes for aluminum tolerance. p 24 *In* Ann. XIII Brazilian Maize and Sorghum Research Conference, Iapar, Londrina, PR, Brazil (*In Portuguese*).
25 Schaffert R E, McCrate A, Tervisan W L, Bueno A, Meira J L and Rhykerd C L 1975 Genetic variation in *Sorghum bicolor* (L.) Moench for tolerance to high levels of exchangeable aluminum in acid soils of Brazil. pp 151–160 *In* Proc. Sorghum Workshop, University of Puerto Rico, Mayaguez, Puerto Rico.
26 Wolf J M 1977 Probabilities of wet season dry spells for Brasilia, DF. Pesq. Agrop. Bras. 12, 141–150 (*In Portuguese*).

Responses of Al-tolerant Dayton and Al-sensitive Kearney barley cultivars to calcium and magnesium during Al stress

Ch. HECHT-BUCHHOLZ and J. SCHUSTER
Institut für Nutzpflanzenforschung — Pflanzenernährung, Technische Universität Berlin, D-1000 Berlin 33

Key words Aluminium stress Aluminium tolerance Barley Calcium Magnesium Meristem Mitotic activity Regeneration capacity Root growth

Summary Two barley cultivars differing in Al tolerance, Kearney (Al-sensitive) and Dayton (Al-tolerant) were exposed to Al stress with varied Ca and Mg concentrations in the nutrient solution. Increase in calcium and magnesium supply protected root meristems and root growth from Al toxicity more effectively in the Al-tolerant cultivar than in the Al-sensitive one. Lateral roots were much more sensitive to Al than adventitious roots. Exposure to 0.33 mM Al with low concentrations of Ca (1.3 mM) and Mg (0.3 mM) caused damage to root tips in both cultivars. Increasing the Ca concentration to 4.3 and 6.3 mM prevented root tip damage in Dayton but not in Kearney. In the Al-tolerant cultivar Dayton, however, the root tips regenerated even at the low Ca concentration of 1.3 mM, whereas 6.3 mM Ca was necessary for this to occur in Kearney. This difference was due to the fact that Dayton's root meristem cells were more resistant to damage. Magnesium responses also varied between the two cultivars. At the lowest Ca concentration an increase in Mg to 6.3 mM permitted regeneration of damaged Kearney root tips and completely prevented any damage in Dayton. It is to be assumed that the different responses of the two cultivars are due to differences in plasma membrane properties.

Introduction

Plant species and even cultivars within a species, vary enormously in their susceptibility to Al stress[13,16,17], but the mechanisms for differing Al-tolerance are far from being understood. Tolerance of Al has been associated with an increase of pH in the rhizosphere (making toxic Al species less available), a lower root cation exchange capacity[10], greater resistance to Al induced nutritional disorders[13], detoxification by forming Al-complexes with organic acids[22] or phenolic compounds[13] or binding with proteins[2] and a greater resistance of the plasmalemma to the entry of Al into the cytoplasm[18].

Reduction of Ca concentration in plants exposed to Al has been reported for various crops[13]. Furthermore, there have been numerous reports demonstrating the vital role of Ca ions as well as Mg ions in enabling plants to overcome an excess of Al and other ions[14,15,23,25,30] Because it is well documented that inhibition of root growth by Al is due to impairment of root meristems[9,18], the object of the present experiment was to study whether the deleterious effects of Al on root tips and developing laterals could be overcome by increasing the Ca and Mg concentration in the nutrient solution. Two variously Al-tolerant barley

Table 1. Ca- and Mg-concentration as used in the screening solution for Al tolerance

Treatment	Al mM	Ca mM	Mg mM	Ca/Al mM
1	0	1.3	0.3	
2	0.33	1.3	0.3	4
3	0.33	4.3[+]	0.3	13
4	0.33	6.3[+]	0.3	19
5	0.33	1.3	6.3[++]	4

[+] Ca was added as $CaSO_4$ to obviate an increase in nitrate concentration
[++] Mg was added as $MgSO_4$

cultivars were used: Dayton (Al-tolerant), developed on acid soils in Ohio and Kearney (Al-sensitive), developed on less acid soils in Kansas and Nebraska.

Materials and methods

Barley (*Hordeum vulgare* L.) cultivars were obtained from C.D. Foy, Beltsville, USA. The cultivar Kearney was known to be Al-sensitive, whereas Dayton was Al-tolerant. Plants were cultivated in a growth room at 24°C, a relative humidity of 60–80%, a light intensity of 10 000 lux and a day length of 16 h. For germination seeds were placed on moistened filter paper and kept in darkness for 3 days. The seedlings by then had developed 5–7 primary roots. The plants were then transferred for 3 days to a 1/10 strength nutrient solution and thereafter to the full medium for 7 days. Constant aeration of the nutrient solution was provided and the pH maintained at 4.8 by daily adjustment. Experiments were started with 14 day old plants by applying 0.33 mM Al and varied Ca and Mg concentrations to the nutrient solution. Each pot (1.5 l) was planted with 5 plants and the nutrient solution was renewed every 3 days. The nutrient solution was essentially similar to that used by Foy *et al.*[10]. The composition of the nutrient solution was in mM: 1.27 $Ca(NO_3)_2 \cdot 4H_2O$; 0.27 $Mg(NO_3)_2 \cdot 6H_2O$; 0.38 KNO_3; 0.08 K_2SO_4; 0.1 K_2HPO_4; 0.21 NH_4NO_3; 0.07 $(NH_4)_2SO_4$; 1.5×10^{-3} $MnCl_2 \cdot 4H_2O$; 7.9×10^{-3} H_3BO_3; $0.7 \cdot 10^{-3}$ $ZnSO_4 \cdot 7H_2O$; 0.2×10^{-3} $CuSO_4 \cdot 5H_2O$; 0.1×10^{-3} $Na_2MoO_4 \cdot 2H_2O$; 2 ppm Fe as $FeSO_4 \cdot 7H_2O$ and 3 ppm Fe as Fetrilon (BASF). Al was applied as $K Al(SO_4)_2$.

The following treatments were used (Table 1):
Examination of plants: One adventitious root of every plant was measured every 2 days. Visual symptoms of root tips and shoots were recorded.
Root tip preparation: Fixation: glutaraldehyde, OsO_4; dehydration: acetone; embedding: Spurr's resin; cutting: Ultramicrotome-Ultrotome III (LKB).
Staining procedure: Semi-thin sections (10 μm) were mounted in a drop of water on a slide and left to airdry. Then the longitudinal sections were stained with 1% toluidine blue in 1% borax solution.
Photographs: Stained sections as well as fresh root tips were photographed with a Zeiss photomicroscope.

Results

Effects of Al exposure (low concentrations of Ca and Mg)

Root tip meristem. After 6 h of exposure to 0.33 mM Al with low concentrations of Ca and Mg the root tips of barley cultivars displayed no external visual symptoms, the meristems appeared to be intact (Fig.

Table 2. Differential Al tolerance among barley cultivars in relation to mitotic activity in the root tip

Treatment	Number of mitotic figures visible in longitudinal sections of the root tip			
	Proximal meristem		Calyptrogen (distal meristem)	
	Dayton	Kearney	Dayton	Kearney
Control	18	24	2.3	2.1
6 h 0.33 mM Al	20	22	1.3	1.7
24 h 0.33 mM Al	6	5	0.0	0.0
4 d 0.33 mM Al	7+	0+	–	

+ only proximal meristem (pm), calyptrogen (d) is destroyed and no differentiation between pm and d is possible

1.2) in comparison with control (Fig. 1.1). In Al-sensitive Kearney, however, autolysed cells occurred in the rhizodermal cells of the proximal meristem at the boundary between root apex and root cap (Fig. 1.2). After 24 h exposure to Al the root tips of Kearney were swollen and had acquired a brownish colour. At that stage, however, root tips of Dayton retained an undisturbed root surface, whereas the roots of Kearney were severely damaged and displayed a disintegrated outer shape. The root cap meristem and the proximal meristem of the root tips were damaged in both cultivars (Figs. 1.3 and 1.4). In Al-tolerant Dayton, however, the proximal meristem appeared to be less impaired (Fig. 1.4). Of note was that the quiescent centre cells in Kearney were autolysed (Fig. 1.3) while they appeared to be still viable in Dayton (Fig. 1.4).

Mitotic activity. Mitotic figures (metaphase) in the proximal meristem and the root cap meristem (calyptrogen) were counted in the longitudinal sections (Table 2). Even though this method did not allow valid statistical analysis there was some indication that 6 h of exposure had already caused inhibition of mitotic activity in the distal meristem of both cultivars. After 24 h of Al exposure, the number of mitotic figures in the proximal meristem had markedly decreased and there were no mitotic figures visible any more in the calyptrogen. This was the case in both cultivars. After 4 days of Al exposure, however, differences between the two cultivars became evident: there was no meristematic activity left in Kearney, whereas in Dayton cells were still dividing.

Regeneration capacity. Within 4 days the severity of Al toxicity symptoms in Kearney roots increased, leading to necrotic, brown root tips and a total inhibition of root elongation. Microscopic examination revealed a total loss of plasma in all cells of the root tips. In contrast, Dayton

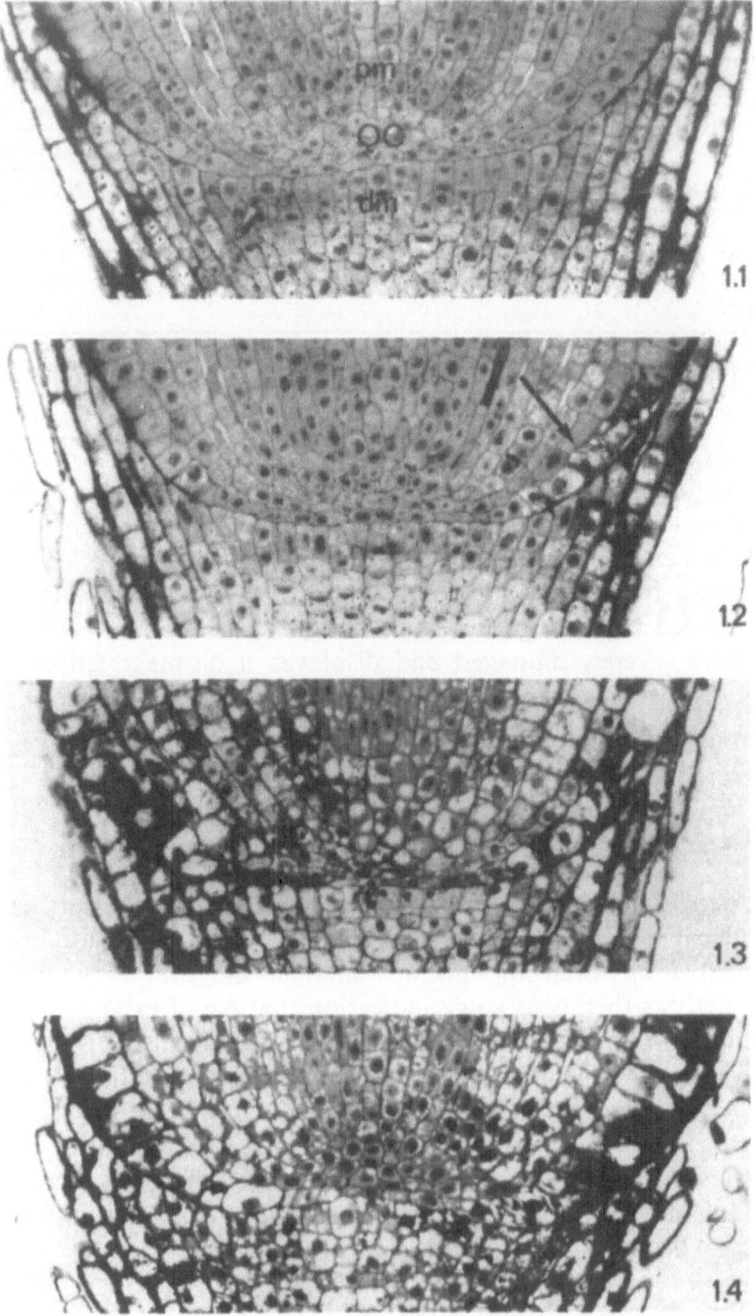

Figs. 1.1.–1.4. Longitudinal sections of root tips of barley cultivars. pm = proximal meristem, dm = distal meristem (calyptrogen), QC = quiescent centre

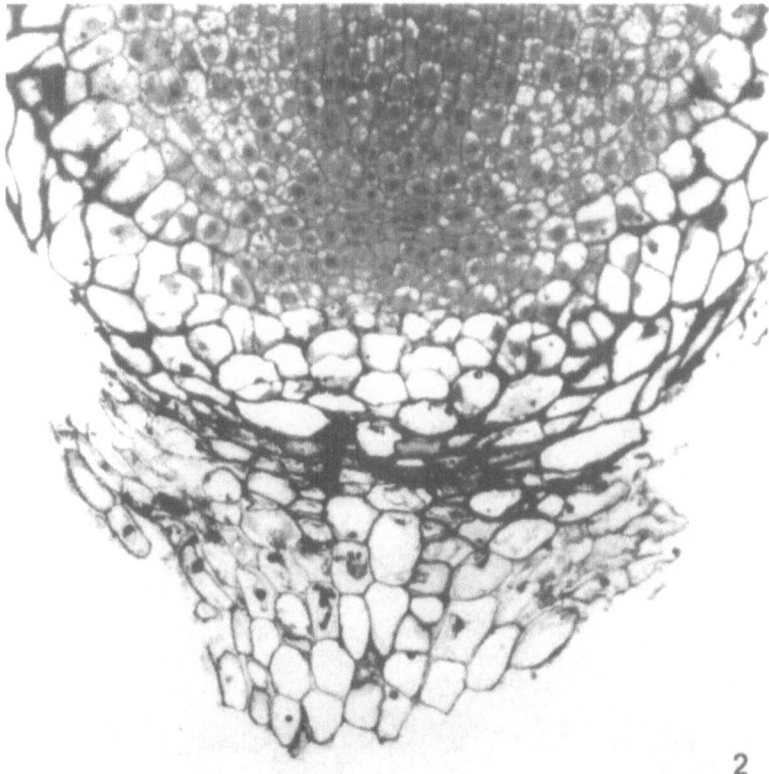

Fig. 2. Regenerating root tip of cultivar Dayton after 4 days exposure to 9 ppm Al. The proximal meristem is forming a new root cap.

retained viable root meristem cells; new root caps were formed and root growth continued.

New root caps, separated from the old ones by several layers of collapsed cells (Fig. 2) had their origin in the same group of initials which normally form root body cells in the proximal direction. This kind of regeneration activity was maintained during the whole 16 days of the

---

Fig. 1.1. Dayton without Al exposure.

Fig. 1.2. Kearney after 6 h exposure to 0.33 mM Al. Autolysed cells occur in the rhizodermal cells of the proximal meristem at the boundary between root apex and root cap (*arrow*).

Fig. 1.3. Kearney after 24 h exposure to 0.33 mM Al. The cells of the meristems including the quiescent centre are severely damaged.

Fig. 1.4. Cultivar Dayton after 24 h exposure to 0.33 mM Al. The cells of the root meristems are injured but the quiescent centre cells appear to be still intact.

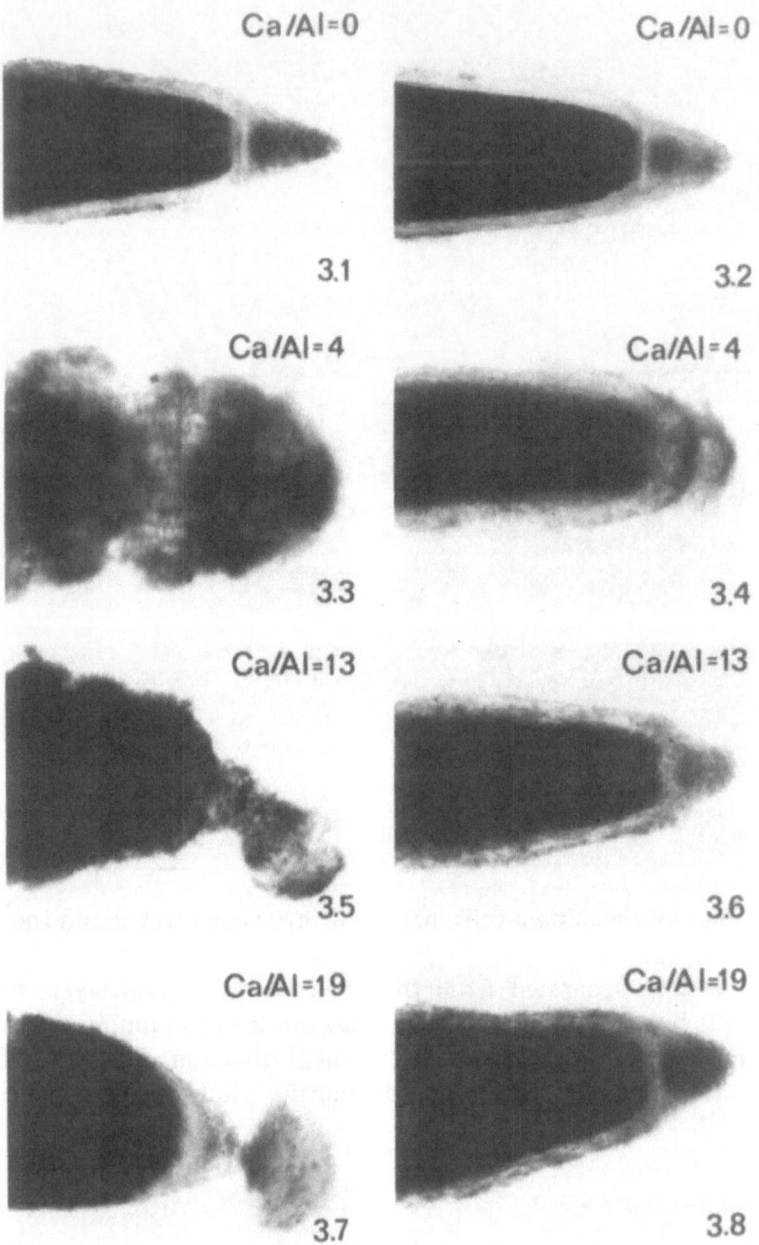

Figs. 3.1.-3.8. Whole root tips of barley cultivars. Left: Kearney, right: Dayton, without Al and after 5 days of exposure to Al with different Ca/Al ratios.

Fig. 3.1. Kearney, without Al.

Fig. 3.2. Dayton, without Al.

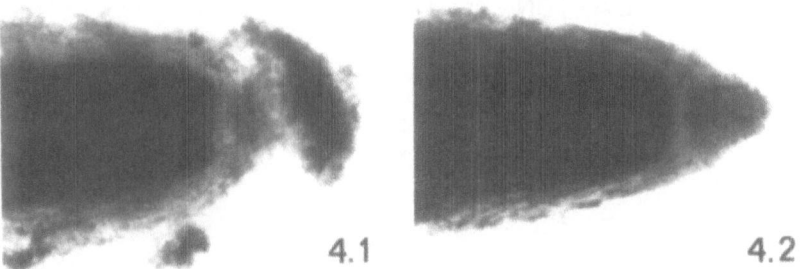

Figs. 4.1.–4.2. Whole root tips of barley cultivars after 5 days of exposure to Al with a high level of Mg.

Fig. 4.1. Kearney, 0.33 mM Al, 6.3 mM Mg, root tip is regenerating.

Fig. 4.2. Dayton, 0.33 mM Al, 6.3 mM Mg, root tip is intact.

experiment. The newly formed root cap cells were recognized as such by their amyloplasts.

Influence of increasing Ca and Mg concentration

Root tips. Light microscopic observation of fresh, intact root tips revealed that the barley cultivars Kearney and Dayton showed different responses to 0.33 mM Al with varied Ca and Mg concentrations, in comparison to the control (Figs. 3.1 to 3.8). Exposure to Al at the low Ca level (Ca/Al = 4) caused damage to root tips in both cultivars (Figs. 3.3 and 3.4) compared with Figs. 3.1 and 3.2 contents. Increasing the Ca concentration to 4.3 and 6.3 mM (Ca/Al = 13, Ca/Al = 19) prevented root tip damage in Dayton (Figs. 3.6 and 3.8), but not in Kearney (Figs. 3.5 and 3.7). In Dayton, however, the root tips regenerated even at the low Ca concentration of 1.3 mM (Fig. 3.4), whereas 6.3 mM Ca was necessary for this to occur in Kearney (Fig. 3.7). Mg responses also varied between the two cultivars. At the lowest Ca concentration (1.3 mM) increase in Mg to 6.3 mM permitted regeneration of damaged Kearney root tips (Fig. 4.1) and completely prevented damage in Dayton (Fig. 4.2).

Fig. 3.3. Kearney, 0.33 mM Al, Ca/Al = 4, root tip is severely damaged.

Fig. 3.4. Dayton, 0.33 mM Al, Ca/Al = 4, root tip is regenerating a new cap.

Fig. 3.5. Kearney, 0.33 mM Al, Ca/Al = 13, root tip is severely damaged.

Fig. 3.6. Dayton, 0.33 mM Al, Ca/Al 13, root tip is regenerating a new cap.

Fig. 3.7. Kearney, 0.33 mM Al, Ca/Al 19, root tip is regenerating a new cap.

Fig. 3.8. Dayton, 0.33 mM Al, Ca/Al 19, root tip remained intact.

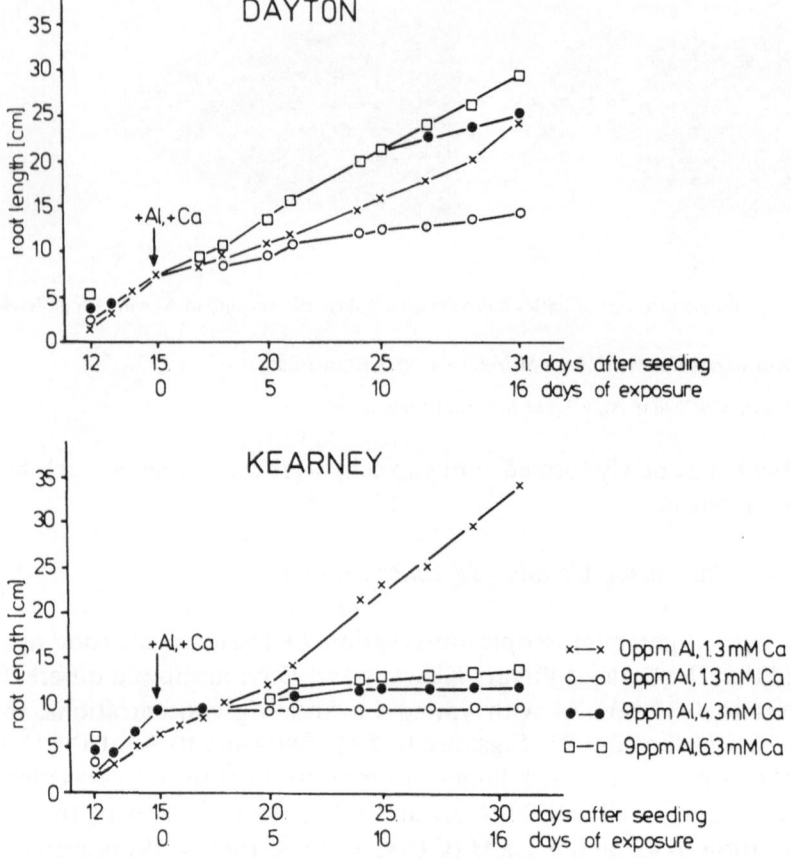

Fig. 5.1. Effect of calcium on adventitious root elongation of barley cultivars during exposure to excess Al (9 ppm = 0.33 mM).

Root length. The differential response of the root tips to Al stress was reflected in root growth. (Figs. 5.1 and 5.2 and 6). At the two higher Ca levels (4.3 mM and 6.3 mM Ca), the roots of Dayton subjected to Al maintained an undisturbed root surface (Figs. 3.6 and 3.8) as well as normal growth (Fig. 5.1). At the low Ca level, however, after 16 days of Al exposure the roots of Dayton were regenerating, but were significantly shorter than those of the control (Fig. 6). In Kearney Ca concentrations of 1.3 mM and 4.3 mM did not repress the lethal effect of Al. Root growth discontinued after 4 days (Fig. 5.1). But with 6.3 mM Ca roots of Kearney showed a similar response as those of Dayton at 1.3 mM Ca (Figs. 5.1 and 6). However, in all series of tests, root length of Kearney was always significantly less than that of the control. Mg responses in terms of regeneration capacity and root elongation also varied between the two cultivars. At the lowest Ca concentration an increase in Mg to

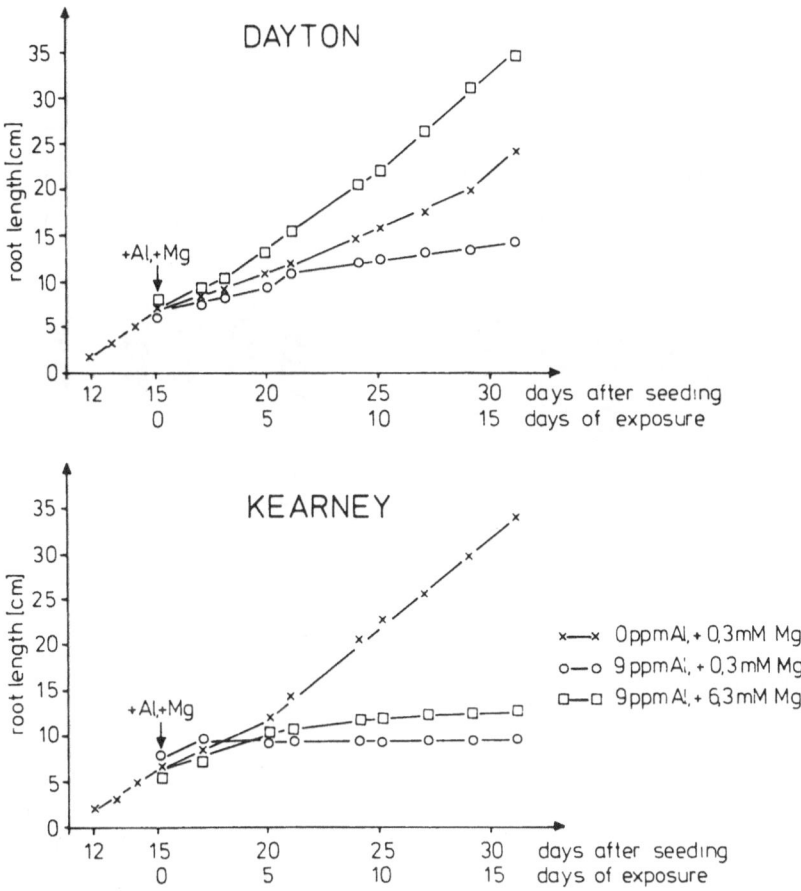

Fig. 5.2. Effect of magnesium on adventitious root elongation of barley cultivars during exposure to excess Al.

6.3 mM promoted root elongation of Al-exposed Dayton, but there was only a slight beneficial effect in the case of Kearney (Figs. 5.2 and 6).

Shoot length. Whereas there were clear responses of root length to Al stress and varied Ca and Mg concentrations in both cultivars, there was no obvious response of shoot length in Dayton. In Kearney shoot length decreased significantly due to Al stress and responded only slightly to the highest Ca level (Fig. 6).

Inhibition of apical dominance. Damage of the root meristem leads to an impairment of apical dominance. Depending on the severity of root tip damage due to Al, the distance between lateral root initiation and root apex was reduced (Table 3). In Kearney plants exposed to Al lateral

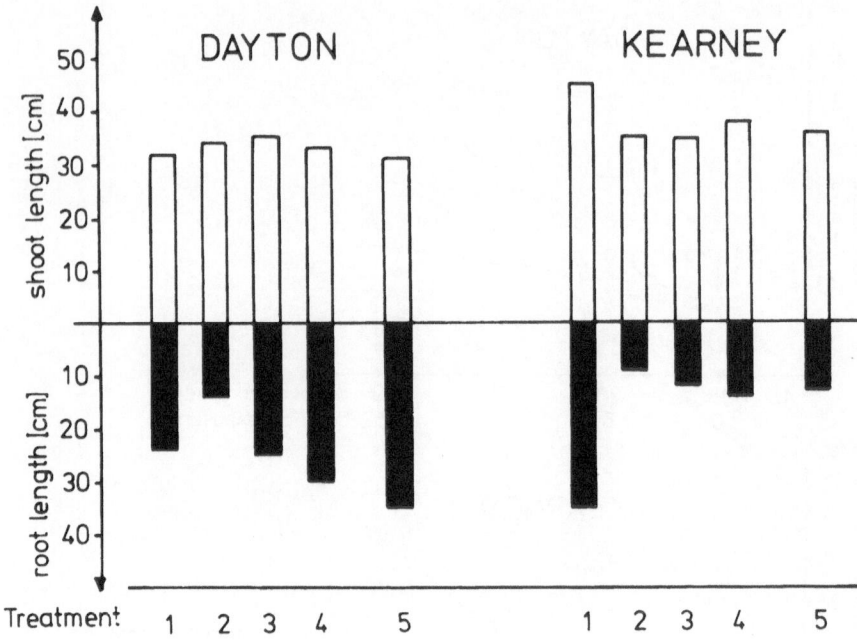

Fig. 6. Effect of calcium and magnesium on shoot and root length of barley cultivars exposed to 0.33 mM Al (pH 4,8). Time of exposure: 16 days. Treatment: (see table 1).

roots occured about 5 mm from the apex at 1.3 mM and 4.3 mM Ca. Under these conditions the root tips of Kearney were necrotic. When there were still viable cells in the meristematic zone, the distance between lateral root initiation and apex was at least 10 mm. This was the case in Dayton even at the low Ca level and in Kearney only at the high Ca level. In the presence of high Ca levels Dayton seemed to be unaffected by Al, *i.e.* the distance of 40 mm between laterals and apex was similar to that in the controls of both Dayton and Kearney.

The high Mg concentration had a similar protective effect on apical dominance as the high Ca concentration; this effect was greater in Dayton than in Kearney.

Table 3. Effect of Ca and Mg on apical dominance of barley cultivars under Al stress

Treatment			Distance between laterals and root apex (mm)	
Al (mM)	Ca (mM)	Mg (mM)	Dayton	Kearney
0	1.3	0.3	35	40
0.33	1.3	0.3	10	5
0.33	4.3	0.3	25	5
0.33	6.3	0.3	40	10
0.33	1.3	6.3	40	10

Lateral root development. Under Al stress the structure of lateral root cells was already affected before they broke through the cortex. In comparison with the control (Figs. 7.1 and 7.2) lateral root growth was totally inhibited in both cultivars (Figs. 7.3 and 7.4) at the low Ca level, and the laterals appeared on the surface of the adventitious root as callus like nodules. With 4.3 mM Ca lateral growth was only slightly inhibited in Dayton (Fig. 7.6); with 6.3 mM Ca normal lateral growth occurred (Fig. 7.8). In Kearney lateral growth was inhibited at all Ca levels (Figs. 7.3, 7.5, and 7.7). Increase in Mg concentration prevented inhibition of lateral root development in Dayton but not in Kearney (not shown).

Discussion

The reduction of mitotic figures in the root tip meristem is the earliest symptom of Al-toxicity being visible after 6 h[7,20]. Inhibition of mitosis by Al is well documented[3,7,20]. Our experiments corroborated these findings, but also gave evidence on the resumption of cell division under Al stress. That more resistant meristematic areas allow the emergence of new roots when the Al stress is removed has been emphasized in former reports[9]. Restoration of mitotic activity under continuous Al supply has been documented recently in experiments with cowpea genotypes[20]. In the present study the Al-tolerant Dayton, even at the low Ca level, retained viable meristem cells under Al stress allowing the formation of a new root cap. By contrast, higher Ca and Mg levels were necessary for this to occur in the Al-sensitive barley cultivar.

Controversial views exist as to whether or not the quiescent centre is essential for root cap regeneration[4,8,18]. After removal of Al, Henning[18] reported formation of new root apices in which the quiescent centre cells were necrotic. Our results, however, led to the assumption that only a limited area of the quiescent centre might be the sole source for regeneration. Increase in Ca or Mg supply protected root-tip integrity and the root meristems of adventitious roots and laterals more in the Al-tolerant cultivar Dayton than in the Al-sensitive Kearney. Different impairment of the root tip meristem indicated differing inhibition of apical dominance, characterized by the distance between lateral initiation and root apex. This might serve as a good screening parameter for Al tolerance.

In accordance with other investigators[26] we found that root elongation appeared to be a better parameter for screening Al tolerance than shoot growth.

In recent reports on the influence of Al-stress on forest trees it was claimed[28] that the ratio of Ca/Al is an important parameter for resistance toward Al-toxicity. In the Al-sensitive Kearney a Ca/Al ratio of 4 was already lethal, whereas in conifers such as *Picea abies*, the Ca/Al ratio

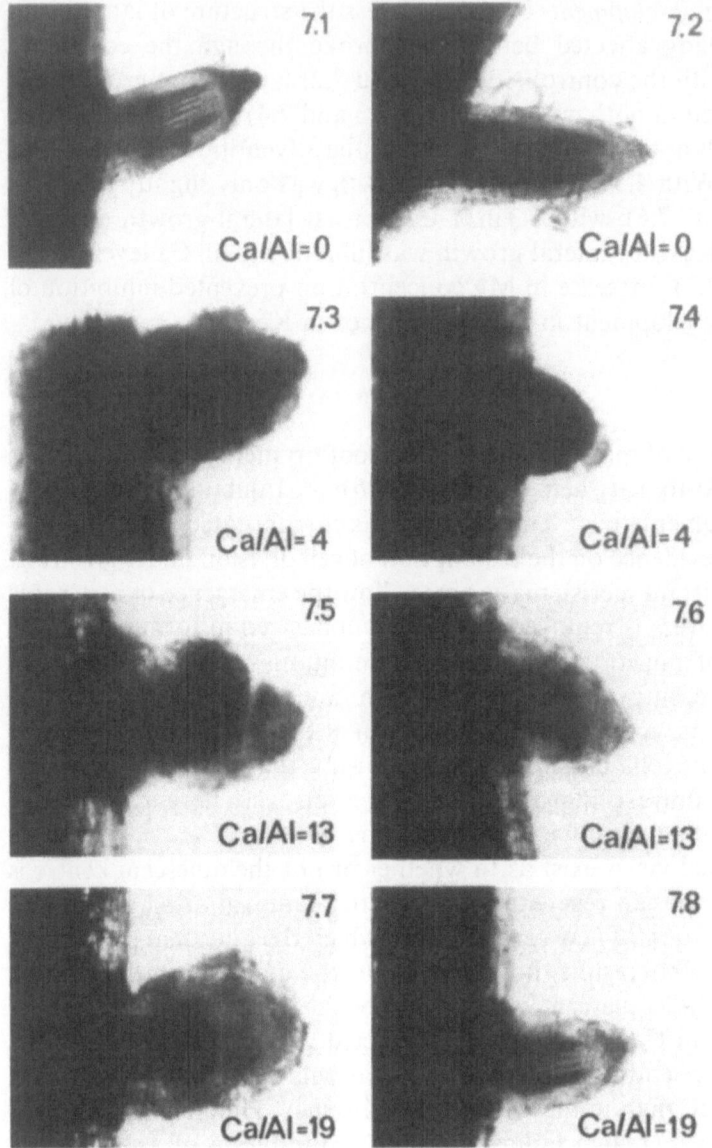

Figs. 7.1.–7.8. Developing laterals of barley cultivars. left: Kearney, right: Dayton, without Al and after 16 days of exposure to Al with different Ca/Al ratios.

Fig. 7.1. Kearney, 0 Al, Fig. 7.2. Dayton, 0 Al.

Fig. 7.3. Kearney, 0.33 mM Al, Ca/Al 4.

Fig. 7.4. Dayton, 0.33 mM Al, Ca/Al 4.

Fig. 7.5. Kearney, 0.33 mM Al, Ca/Al 13.

Fig. 7.6. Dayton, 0.33 mM Al, Ca/Al 13.

Fig. 7.7. Kearney, 0.33 mM Al, Ca/Al 19.

Fig. 7.8. Dayton, 0.33 mM Al, Ca/Al 19.

has to be under 1 before damage to the root tips occurs. This demonstrates that conifers are much more tolerant to Al than barley.

The present finding that Mg was as effective in protecting roots from Al as Ca is in accordance with results obtained with corn in breds (*Zea mays* L.)[27]. The differential response of Al-tolerant and sensitive genotypes in terms of root growth and visual symptoms of Al-toxicity to Ca and Mg under Al stress has been documented by several authors[1,24,27]. However, the reason for different response of the variously Al-tolerant cultivars to an increase of the Ca and Mg concentration under Al stress is not clear. It is widely accepted that Al substitutes for Ca at the exchange-absorption sites of the 'Free Space' including the cell walls[21] and the mucilage[19]. Roots of Al sensitive cultivars have been claimed to accumulate more Al because of a greater cation-exchange capacity[10]. In this case more Mg and Ca for the sensitive cultivar might be necessary to compete with Al.

The inhibitory effect of Al on Ca uptake and Ca transport is well known[22]. One of the assumed mechanisms of Al tolerance is a greater resistance to Al induced Ca- or Mg-deficiency[13]. This view is supported by results with various crops showing that reduction of Ca in the plants was more pronounced in the Al-sensitive cultivars than in the more tolerant ones[10,11,12]. There was a direct protective effect of Mg and Ca on the morphology of root tips and developing laterals. From experiments with variously Al-tolerant cowpeas Horst *et al.*[20] concluded that Al-tolerance was related to lower Al-uptake into the root tips rather than to less inhibition of Ca and Mg uptake.

Ca and Mg are considered to influence the molecular construction of cell membranes[5,6] and this might control their permeability to Al. Insufficient supply of Ca and Mg, especially in the case of rapidly dividing and elongating cells in the root tip, might lead to an increased Al uptake into root cells. If this explanation is valid, the fact that different amounts of Ca or Mg ions are necessary to protect variously Al-tolerant genotypes from Al stress could be due to differential properties of their plasma membranes. This would support the idea of Henning[18] 'that the only satisfactory explanation for varietal tolerance to Al lies in the molecular constructions of a cell's plasmalemma'. In recent investigations it has been stressed that conditions aimed at disturbing the plasma membrane's permeability such as unphysiologically high temperatures and application of metabolic inhibitors consequently lead to an increased Al uptake into root cells[21,29].

Acknowledgement We wish to thank Carmen Wolfram for expert technical assistance.

References

1 Ali S M E 1973 Influence of cations on aluminium toxicity in wheat (*Triticum aestivum* Vill, Host). Ph. D. Thesis. Oregon State University. Corvallis.
2 Aniol A 1984 Induction of aluminium tolerance in wheat seedlings by low doses of aluminium in the nutrient solution. Plant Physiol. 76, 551–555.
3 Baier R, Münnich H, Heinke F and Göring H 1976 Zytologische Untersuchungen zur Wirkung von Aluminiumionen auf Maiswurzeln. Wissenschaftliche Zeitschrift der Humboldt-Universität zu Berlin. Math.-Nat.R. XXV 6, 840–844.
4 Barlow P 1974 Regeneration of the root cap of primary roots of *Zea mays*. New. Phytol. 73, 937–954.
5 Caldwell C R and Haug A 1981 Temperature dependence of the barley root plasma membrane-bound Ca^{2+} — and Mg^{2+} — dependent ATPase. Physiol. Plant. 53, 117–124.
6 Changeux J P, Tung J and Kittel C 1967 On the cooperativity of biological membranes. Proc. Natl. Acad. Sci. USA 57, 335–341.
7 Clarkson D T 1966 The effect of aluminium and some other trivalent metal cations on cell division in the root apices of *Allium cepa*. Ann. of Bot. 29, 309–315.
8 Feldman L J 1976 The *de novo* origin of the quiescent center in regeneration root apices of *Zea mays*. Planta (Berl.) 128, 207–212.
9 Fleming A L and Foy C D 1968 Root structure reflects differential aluminium tolerance in wheat varieties. Agron. J. 60, 172–176.
10 Foy C D, Fleming A L, Burns G R and Armiger W H 1967 Characterization of differential aluminium tolerance among varieties of wheat and barley. Soil Sci. Soc. Am. Proc. 31, 513–521.
11 Foy C D, Fleming A L and Armiger W H 1969 Aluminium tolerance of soybean varieties in relation to calcium nutrition. Agron. J. 61, 505–511.
12 Foy C D, Fleming A L and Gerloff G C 1972 Differential aluminium tolerance in two snapbean varieties. Agron. J. 64, 815–818.
13 Foy C D, Chaney R L and White M C 1978 The physiology of metal toxicity in plants. Annu. Rev. Plant Physiol. 29, 511–566.
14 Furlani P R and Clark R B Screening sorghum for aluminium tolerance in nutrient solutions. Agron. J. 73, 587–592.
15 Grimme H 1983 Aluminium induced magnesium deficiency in oats. Z. Pflanzenernaehr. Bodenkd. 146, 666–676.
16 Hecht-Buchholz Ch and Foy C D 1981 Effect of aluminium toxicity on root morphology of barley. Plant and Soil 63, 93–95.
17 Hecht-Buchholz Ch 1983 Light and electron microscopic investigations of the reactions of various genotypes of nutritional disorders. Plant and Soil 72, 151–165.
18 Henning S J 1975 Aluminum toxicity in the primary meristem of wheat roots. Ph. D. Thesis. Oregon State University. Corvallis.
19 Horst W J, Wagner A and Marschner H 1982 Mucilage protects root meristems from aluminium injury. Z. Pflanzenphysiol. 105, 435–444.
20 Horst W J, Wagner A and Marschner H 1983 Effect of aluminum on root growth, cell division rate and mineral element contents in roots of *Vigna unguiculata* genotypes. Z. Pflanzenphysiol. 109, 95–103.
21 Huett D O and Menary R C 1979 Aluminum uptake by excised roots of cabbage, lettuce and Kikuyu grass. Aust. J. Plant Physiol. 6, 643–653.
22 Jones L H 1961 Aluminum uptake and toxicity in plants. Plant Soil 13, 297–310.
23 Johnson R E and Jackson W A 1964 Calcium uptake and transport by wheat seedlings as affected by aluminium concentrations. Soil Sci. Soc. Am. Proc. 28, 381–386.
24 Kerridge P C 1969 Aluminium toxicity in wheat (*Triticum aestivum* Vill., Host.). Ph. D. Thesis. Oregon State University. Corvallis.
25 Kirkby E A and Mengel K 1976 The role of magnesium in plant nutrition. Z. Pflanzenernaehr. Bodenkd. 139, 209–222.

26 Reid D A, Fleming A L and Foy C D 1971 A method of determining Al response of barley in nutrient solution. Agron. J. 63, 600–603.
27 Rhue R D and Grogan C O 1977 Screening corn for Al-tolerance using different Ca and Mg concentrations. Agron. J. 69, 755–760.
28 Rost-Siebert K 1983 Aluminium-Toxizität und -Toleranz an Keimpflanzen von Fichte (*Picea abies* Karst.) und Buche (*Fagus silvatica* L.). Allg. Forstzeitschr. 26/27, 686–689.
29 Wagatsuma T 1983 Effect of non-metabolic conditions on the uptake of aluminium by plant roots. Soil Sci. Nutr. 29 (3), 323–333.
30 Wallace A and Frolich E 1966 Calcium requirements of higher plants. Nature 209, 634.

Ion segregation in different plant parts within different barley cultivars under salt stress

J. BOGEMANS and J.M. STASSART
Laboratorium voor Plantenfysiologie, Vrije Universiteit Brussel, Paardestraat 65, B-1640 St. Genesius — Rode, Belgium

Key words Barley cultivars Ion transport Salt stress

Summary Fourteen barley varieties were screened for their salt tolerance when stressed with 171 mM NaCl. The barley varieties consisted of 9 summer and five winter cultivars. There was no direct genetic link between the cultivars. Plants were harvested after 2, 5 and 9 weeks and divided them into different plant parts, old leaves (leaves 1 to 4), young leaves (5 to 8), stem and ear. Growth and ion contents were measured. All plants could survive the first five weeks of salt stress, after that time almost all summer and one winter variety could withstand the stress. All sensitive varieties had a tendency to accumulate high amounts of Na and Cl in the developing younger leaves. It was suggested that tolerant plants were able to retranslocate K-ions more efficiently from older leaves to the developing leaves, exchanging them with Na-ions. They also had the tendency to maintain favorable K/Na ratios in younger leaves and stems.

Introduction

From earlier work on ion distribution during ontogenesis of the barley cultivar Union under salt stress[1,5], we could generally observe three aspects of the plant with respect to ion transport.

During the early developmental stage there was a selective uptake of K from root to shoot, together with an accumulation of Na in roots. In a second stage, when younger leaves (leaves 5 to 8), developed, Na accumulated in the older leaves, whereas potassium was retranslocated to the developing organs.

The final stage of development was characterised by the reduction of growth of the stem and younger leaves. Good K/Na ratios were observed in the ear and younger leaves, especially the flag leaf. The formation of the ear led us to suppose that the variety was salt tolerant and that the observed ion transport features were in a sense related to it. The aim of the present study was to test several barley cultivars for their salt tolerance and to relate it to their transport characteristics.

Materials and methods

Barley seeds were germinated on moistened sand overnight, transplanted in pots filled with vermiculite in the greenhouse and for the first week received only water. From the second week till the 9th week all plants received 100 ml of one half Hoagland solution plus microelements twice a week. Salt stressed plants were treated with 100 ml of 171 mM NaCl twice a week, whereas control plants received an additional 100 ml of distilled water per plant. Plants were harvested after 2, 5 and 9 weeks (3 replicates of 5 plants) and were divided into old leaves (leaf 1 to 4), young leaves (leaf

Table 1. Barley varieties used in the experiment

Number	Name	Seed row	Sort
1	Union	2	summer barley
2	Menuet	2	summer barley
3	Igri	2	winter barley
4	Sonja	2	winter barley
5	Auriga	2	winter barley
6	Egmont	2	summer barley
7	Isaura	2	winter barley
8	Candida	2	winter barley
10	Hebe	2	summer barley
11	Herta	2	summer barley
12	Iban	2	summer barley
13	Atem	2	summer barley
14	Friponne	2	summer barley
15	Legia	2	summer barley

5 to 8), stem and ear. After dry weight measurement all plant parts were dry ashed and analyzed for K, Na, Ca and Cl content. Cation content was measured by flame photometry and Cl colorimetrically by titration (Chlor-O-counter). Fifteen barley varieties were screened under the above mentioned conditions (Table 1).

Results

Growth

At the second week harvest some differences in growth of the older leaves could be observed (Fig. 1). Most of the varieties already showed

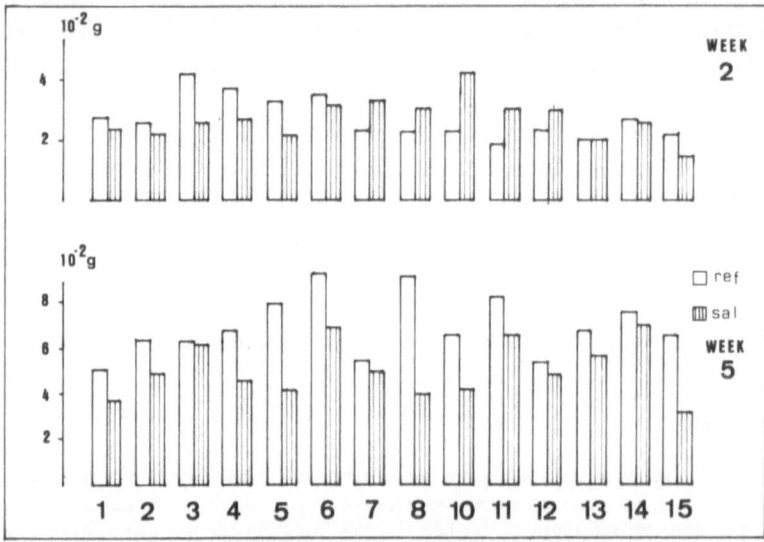

Fig. 1. Dry matter production (g dry weight per plant) after 2 and 5 weeks in old leaves of reference plants (ref) and salt stressed plants (sal).

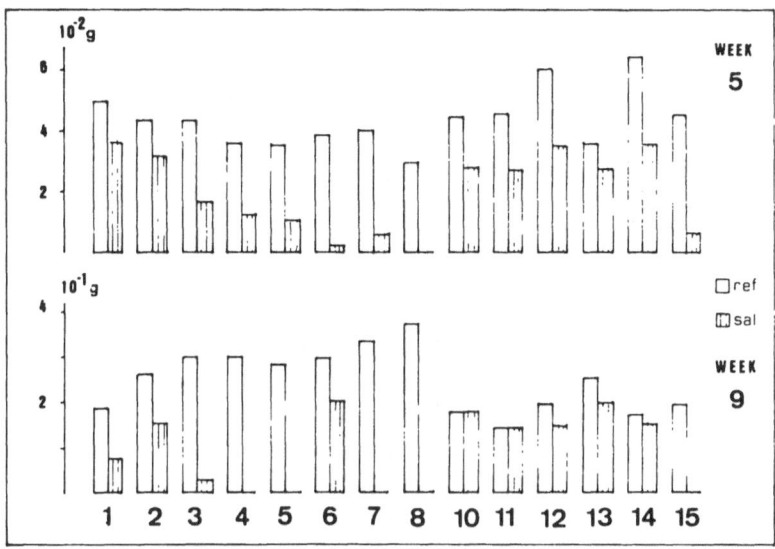

Fig. 2. Dry matter production (g dry weight per plant) after five and nine weeks in young leaves of reference plants (ref) and salt stressed plants (sal).

a reduction in growth of these leaves, but some of them (variety: 7, 8, 10, 11, 12) were stimulated under stressed conditions. After five weeks all salt treated plants showed a reduction in growth of their older leaves *e.g.* varieties 5, 8 and 15 were reduced by almost 50%. At the beginning of the fifth week (Fig. 2) not all of the varieties had formed younger leaves (leaf 5 and 6), but testing the cultivars for leaf development after 9 weeks, five varieties (4, 5, 7, 8 and 15) failed to reach full development. A special characteristic of these 5 varieties was their formation of tillers, so it was difficult to separate these tillers (three leaf stage) from previously developed plant. Comparing stem growth after 5 and 9 weeks (Fig. 3), stems were reduced only in the final developmental stage. For the above mentioned reasons the varieties 4, 5, 7, 8 and 15 were discarded for growth and ion content measurements. All other plants also tillered after 9 weeks, but their dry weights were negligible in comparison with the mature plants.

Ion content

After two weeks of salt stress (Fig. 4), ion contents of old leaves differ very little. Generally the potassium content is higher than sodium, providing a good K/Na ratio. Chloride seems to be equilibrated with potassium rather than with sodium. After 5 weeks the ion patterns are completely changed (Fig. 5). In most of the summer barley varieties potassium content was reduced by more than 50%. For the winter

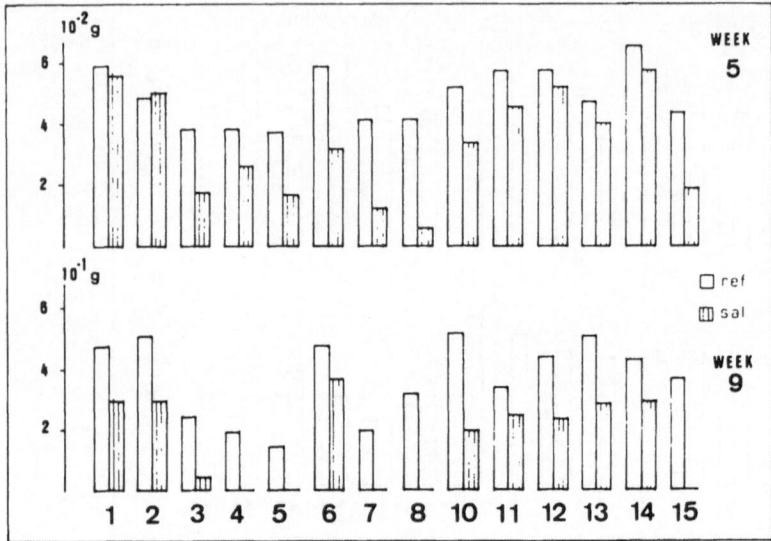

Fig. 3. Dry matter production (g dry weight per plant) after five and nine weeks in stems of reference plants (ref) and salt stressed plants (sal).

varieties (with the exception of var. Igri) the K content remained constant or the decline was less compared with summer barley. K content fell and sodium was accumulated in the older leaves. Whereas chloride increased very little, this element seemed to be equilibrated with the Na or (Na + K) concentration. The ion content in the younger leaves

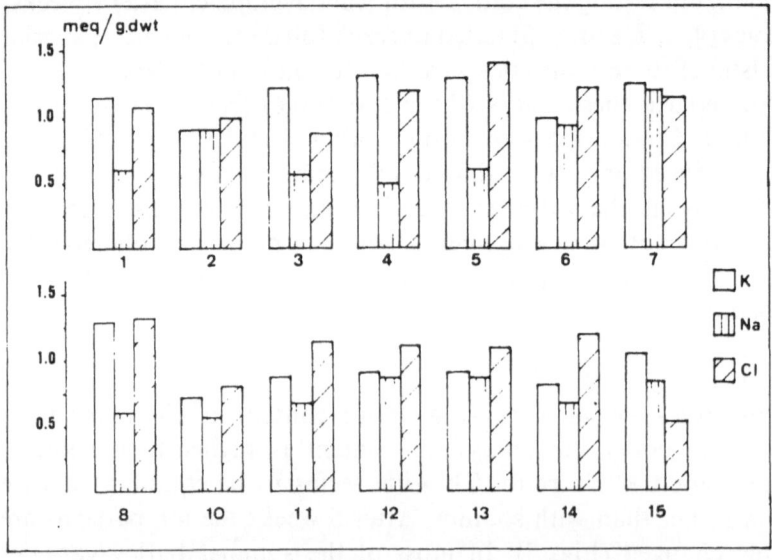

Fig. 4. Ionic composition in old leaves after two weeks of salt treatment.

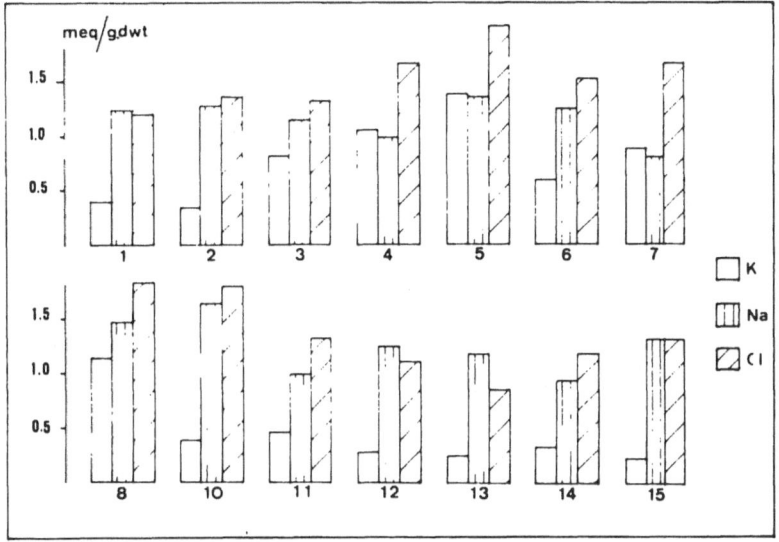

Fig. 5. Ionic composition in old leaves after five weeks of salt treatment.

showed two tendencies (Fig. 6). With the exception of the variety Legia (15) all summer varieties had a potassium content which was equal or higher than the chloride content and all ion contents were about at the same level but in winter barley and the var. Legia the reverse was true. The ion content in the varieties 4, 5 and 7 reached high levels. Very high levels of ions were observed in many of the stems of winter barley after

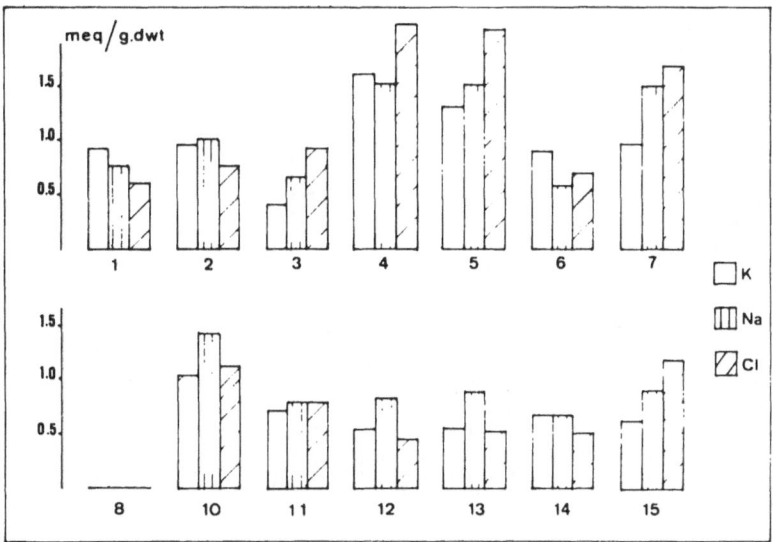

Fig. 6. Ionic composition in young leaves after five weeks of salt treatment.

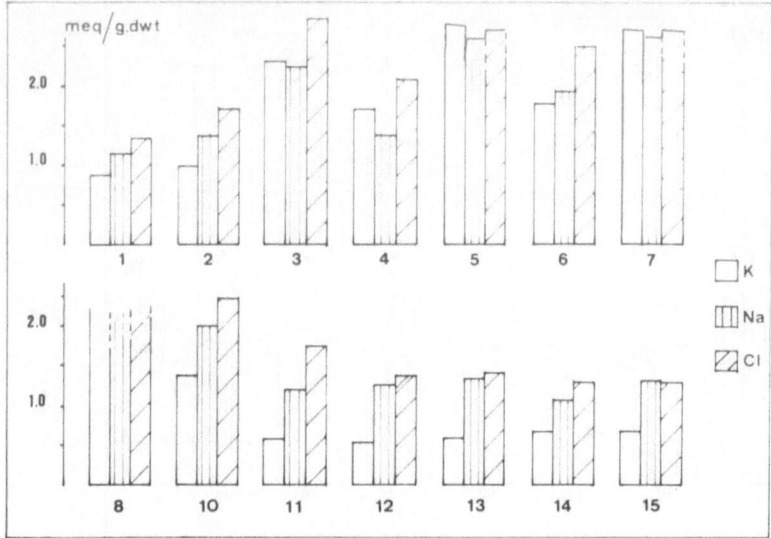

Fig. 7. Ionic composition in stem after five weeks of salt treatment.

5 weeks of salt stress (Fig. 7) and were difficult to compare with those concentrations found in summer barley. Among the summer varieties ion content patterns in stems are comparable. As already mentioned all winter varieties with the exception of var. Igri, and only one summer cultivar (Legia) ceased growing in the latest developing stage. The ion concentrations in the young leaves of the remaining plants (Fig. 8) are

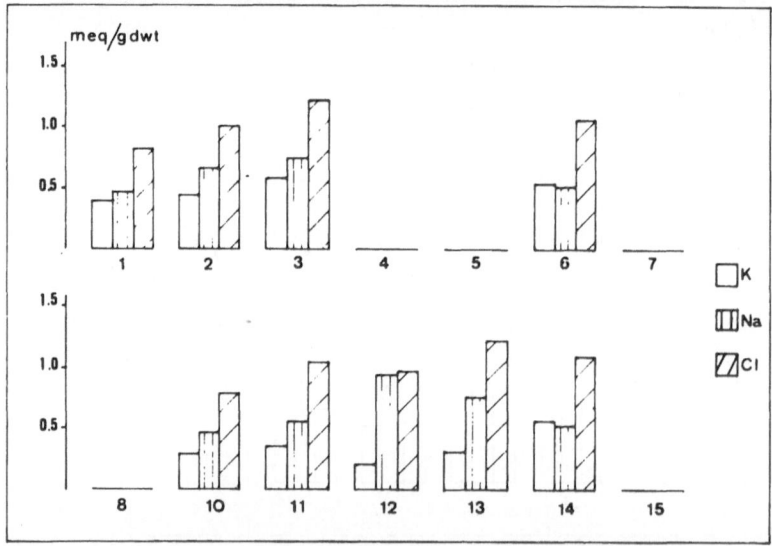

Fig. 8. Ionic composition in young leaves after nine weeks of salt treatment.

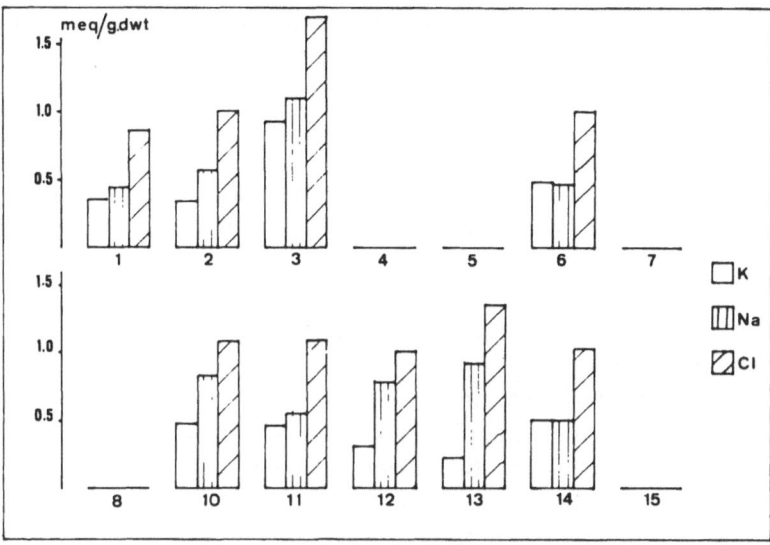

Fig. 9. Ionic composition in stem after nine weeks of salt treatments.

comparable with each other. As in the older leaves in the earlier developmental stages, potassium concentration decreased in the younger leaves of almost all varieties. Sodium content fell in practically all summer varieties, whereas chloride still accumulated to a level equal to (Na + K) concentration. The same tendency was observed for the ion concentrations in the stems after 9 weeks (Fig. 9).

Discussion

Due to the method of harvesting one should be careful in judging the varieties for salt tolerance or sensitivity on the basis of the dry weight of the different plant parts, since a reduction control plants could also mask a growth retardation. Therefore it is only after nine weeks when ears have emerged that one can estimate the differences in salt tolerance among the varieties. Generally it became clear that among the varieties tested, winter barley plants are most sensitive towards salt stress and the reverse is true for summer varieties but there are some exceptions. Among the winter varieties cv. Igri withstood salt stress, whereas the summer variety Legia was salt sensitive.

On the basis of the build up of the ion concentrations in the different plant parts, one can deduce some general ion transport characteristics. In the early tillering stage sodium was excluded from the shoot. Though the variety Legia was very poor in growth it had a tendency to exclude chloride from the shoot at this initial stage.

After five weeks when old leaves were mature, one could already distinguish some differences between summer and winter varieties. On the basis of the decline in the potassium concentration one could deduce that summer varieties are capable of retranslocating K^+ more efficiently. With the exception of var. Legia all summer barley varieties could survive the salt stress until the ninth week. In the final developing stage sodium concentrations in young leaves and stems tend to be lower than at their initial stage. Compared with our findings on the variety Union[1,5] total ion uptake at this stage is low and K^+ is merely provided by the retranslocation mechanism. Plants which could control their K/Na ratios in the younger leaves efficiently were also able to form ears in the final developing stage.

The data represented here confirms the earlier findings of Greenway[3,4], whereby ion changes in the older plant parts are related to the growth of barley on saline soils. However the variety Legia which also shows an identical ion pattern in the older leaves succumbed to the salt stress. In comparison with the other summer varieties this variety was already severely reduced in growth after two weeks. This variety could be a chloride excluder and could not maintain this type of regulation until full maturity. However it strengthens the implication that besides ion regulation in the developing leaves, probably under phloem control[4], the ion regulation between root and shoot, which is mainly under xylem transport control[2], is also of importance in relation to salt tolerance.

References

1 Bogemans J and Stassart J M 1984 Membrane Transport in Plants. Academia. Prague, 474 p.
2 Davis R F 1984 Membrane Transport in Plants. Academia. Prague, 489 p.
3 Greenway H 1962 Plant response to saline substrate Aust. J. Biol. Sci. 15, 39–57.
4 Greenway H, Gunn A, Pitman M G and Thomas D A 1965 Plant Plant response to saline substrate VI. Aust. J. Biol. Sci. 18, 525–540.
5 Stassart J M and Bogemans J 1984 Membrane Transport in Plants Academia. Prague, 510.

Plant traits for evaluation of responses of sorghum genotypes to aluminum

P.R. FURLANI
Instituto Agronomico, C.P. 28, 13 100 Campinas, Sao Paulo, Brazil

and R.B. CLARK
U.S. Department of Agriculture, Agriculture Research Service, Department of Agronomy, University of Nebraska, Lincoln, NE 68583, USA

Key words Aluminum tolerance Nutrient solution Root lengths Root weights *Sorghum bicolor* (L.) Moench

Summary Sorghum [*Sorghum bicolor* (L.) Moench] genotypes known to respond to Al toxicity in acid soils were grown in nutrient solutions with Al and measurements of roots related to differential responses to Al were made. Traits assessed were final seminal, net seminal, relative net seminal, total adventitious, mean adventitious, longest adventitious root lengths, number of adventitious roots, and shoot and root dry and fresh matter yields. Longest adventitious root and net seminal root lengths were the most sensitive plant traits for assessment of sorghum plant responses to Al. Longest adventitious root length required only one measurement at the end of the treatment and net seminal root length required two measurements at the time plants were treated and after treatment. The nondestructable root measurements for Al toxicity tolerance might be useful in genetic and plant breeding studies to allow better asessment of adaptation to the stresses of Al toxicity in acid soils.

Introduction

Differential tolerance to aluminum (Al) of sorghum [*Sorghum bicolor* (L.) Moench] genotypes has been demonstrated in potted soil[4,12], nutrient solutions[8], and in field plots[2,3,7]. If plant breeders are to make practical use of differential Al tolerance in breeding programs, rapid and inexpensive screening procedures need to be used to reduce costs and time for field and controlled environment experiments. Some of this methodology has already been developed[8,9,10,13].

Plant differences to Al toxicity have been studied using various traits like visual toxicity ratings, shoot and root dry matter yields, root staining, root regrowth after Al treatment, total root growth, and relative root lengths of plants grown with and without Al. In studies with sorghum[8], differences in Al toxicity among genotypes could not be easily detected from shoot and root dry matter yields. Visual Al toxicity symptom ratings and relative root lengths were found to be more helpful in assessing sorghum plants for Al toxicity tolerance. The correlation value between relative root length and visual al toxicity symptoms was significant but not high (-0.36). Better quantitative traits need to be used to separate tolerant and nontolerant germplasm for Al toxicity responses.

H.W. Gabelman and B.C. Loughman (Eds.), Genetic aspects of plant mineral nutrition.
ISBN-13: 978-94-010-8102-3
© 1987 Martinus Nijhoff Publishers, Dordrecht/Boston/Lancaster.

The objective of this study was to assess several root traits as quantitative measurements of Al toxicity in sorghum.

Materials and methods

Growth of plants

Two systems were used to grow sorghum plants in nutrient solutions with treatments. System I consisted of growing a large number of young seedlings (up to 252) for 7 days with Al. System II consisted of growing limited numbers of older seedlings (up to 66) for 14 days with Al.

Seed was germinated between moist rolled paper towels. Three-day-old seedlings of similar size were transferred to a plastic plate that held up to 252 plants and grown in 15 L of 0.5-strength nutrient solution containing different levels of Al (System I) or 0.3-strength nutrient solution with no Al. Eight-day-old seedlings grown without Al were transferred to plastic plates that held up to 66 plants (wrapped loosely with sponge rubber) and grown in 8 L of 0.5-strength nutrient solution containing different levels of Al (System II). Techniques for germination, seedling growth, nutrient solution composition, and handling of plants have been described[6,8].

Two experiments each were conducted in a controlled environment growth chamber and in a greenhouse. Growth chamber conditions were 17 h light at $28 \pm 1°C$ and 7 h darkness at $23 \pm 1°C$. Light was provided by incandescent plus metalarc lamps at a photosynthetic photon flux density of $350 \mu mol\, m^{-2} s^{-1}$ at plant height (50 cm below the lamps). Greenhouse conditions were (May to August) $33 \pm 4°C$ daytime and $25 \pm 3°C$ nighttime temperatures. The experimental design used in each experiment was a split-plot with Al level as the main plot and sorghum genotypes as the subplot. Experiments were arranged in randomized complete blocks with three (Experiments 1 and 2) or two (Experiments 3 and 4) replications.

Plant traits measured

The following traits were measured on individual plants: (i) initial seminal root length (ISRL) of the longest seminal root at the time seedlings were transferred to treatment solutions; (ii) final seminal root length (FSRL) of the longest seminal root at the time the experiment was terminated; (iii) net seminal root length (NSRL) calculated by FSRL — ISRL; (iv) relative net seminal root length (RNSRL) calculated by NSRL/ISRL; (v) total adventitious root length (TARL) determined by the addition of the length of each individual primary adventitious root; (vi) number of adventitious roots (NA); (vii) mean adventitious root length (MARL) calculated by TARL/NA; (viii) length of the longest adventitious root (LARL); (ix) shoot and root fresh weights; and (x) shoot, total root, seminal root, and adventitious root dry weights.

Experiments

Experiment 1. Plants were grown using System I in a growth chamber with six levels of Al (0, 19, 37, 56, 74, and 93 μM) and seven genotypes (CK60, 'Martin', NB9040, NB3494, SC33-9-8-E4, SC369-3-1JB, and TX415). Plant traits measured were ISRL, FSRL, NSRL, and RNSRL.

Experiment 2. Plants were grown using System II in a growth chamber with six levels of Al (0, 37, 46, 56, 65, and 74 μM) and six genotypes (CK60, Martin, NB9040, SC33-9-8-E4, SC369-3-1JB, and TX415). Plant traits measured were TARL, NA, MARL, and LARL.

Experiment 3. Plants were grown using System I in a greenhouse with four levels of Al (0, 74, 148, and 222 μM) and seven genotypes (MN1204, SC175-14, SC208, SC283, TX415, TX623A, and 156-P-5-2-1). The plant trait measured was NSRL.

Experiment 4. Plants were grown using System II in a greenhouse with two levels of Al (92 and 185 μM) and six genotypes (MN1204, SC175-14, SC208, SC283, TX415, and TX623A). The plant trait measured was LARL.

Results and discussion

Seminal root lengths

The length of seminal roots decreased as Al increased in nutrient solutions, and wide differences among sorghum genotypes were detected (Table 1). At 74 μM Al, NSRL and RNSRL gave the widest differences among genotypes tolerant and nontolerant to Al. Even though FSRL for Al nontolerant plants (Martin, SC33-9-8-E4, and TX415) grown with varied Al levels decreased more than for Al tolerant plants (NB9040 and SC369-3-15B), the differences did not easily separate genotypes for tolerance to Al. Correction for the initial length of the seminal roots should be included in the measurements to enhance differences of Al on seminal root growth. Both NSRL and RNSRL values included corrections for the initial seminal root length. NSRL and RNSRL values could be used to evaluate genotypes for Al tolerance. For example, at 74 μM

Table 1. Final seminal root length (FSRL), net seminal root length (NSRL), and relative net seminal root length (RNSRL) of seven sorghum genotypes grown with Al in nutrient solution (Experiment 1)

Al level (μM)	Genotype							
	TX415	Martin	SC33-9-8-E4	NB3494	CK60	SC369-3-1JB	NB9040	Mean
FSRL (mm plant^{-1})								
0	144	140	142	89	139	175	139	138 A[†]
19	144	160	151	85	147	163	148	143 A
37	106	131	135	68	133	123	149	121 AB
56	88	88	108	46	134	119	142	104 B
74	60	59	86	38	93	109	109	79 C
93	54	57	77	35	75	80	90	67 C
LSD (0.05) for genotype and Al level				17				
NSRL (mm plant^{-1})								
0	100	97	91	63	99	126	97	96 AB
19	102	115	100	58	106	119	104	100 A
37	61	89	82	42	90	76	100	77 ABC
56	41	45	57	20	91	77	95	61 C
74	18	17	34	10	51	64	65	37 D
93	13	14	22	8	32	36	41	24 D
LSD (0.05) for genotype and Al level				15				
RNSRL								
0	2.3	2.3	1.8	2.4	2.5	2.6	2.3	2.3 A
19	2.4	2.6	2.0	2.1	2.6	2.7	2.4	2.4 A
37	1.4	2.2	1.5	1.7	2.1	1.6	2.1	1.8 B
56	0.9	1.0	1.1	0.8	2.1	1.9	2.0	1.4 BC
74	0.4	0.4	0.6	0.4	1.2	1.4	1.5	0.9 CD
93	0.3	0.3	0.4	0.3	0.7	0.8	0.9	0.5 D
LSD (0.05) for genotype and Al level				0.5				

[†] For each plant trait, means followed by a common letter are not significantly different at $P = 0.05$ according to Duncan's new multiple range test (DNMRT).

Al the differences between the means of the three more Al nontolerant and the two more Al tolerant genotypes were 4.3- and 3.6-fold for NSRL and RNSRL, respectively compared to 2.1-fold for FSRL.

Adventitious root lengths

Seminal roots are important in the early stages of sorghum plant growth, but become less important as the adventitious (crown roots become the major portion of the root system[1]. Therefore, an evaluation of adventitious roots for Al toxicity tolerance should be importnt. The number of adventitious roots initiated by sorghum plants increased and TARL, MARL, and LARL, decreased as Al in solution increased (Table 2). The mean variability between 0 and 74 μM Al over all genotypes for these traits was 1.4-, 1.9-, 2.4-, and 2.2-fold for NA, TARL, MARL, and LARL, respectively. The mean variability between the three Al nontolerant and the two Al tolerant genotypes grown at 74 μM Al was 1.3-, 2.4-, 3.2-, and 3.7-fold for NA, TARL, MARL, and LARL respectively. The widest differences among genotypes were obtained at both 65 and 74 μM Al in these experiments.

Except for one of the more Al tolerant genotypes (SC369-3-1JB), the number of adventitious roots initiated increased for the genotypes when plants were grown with Al compared to plants grown without Al. This increase was generally greater in the Al nontolerant than in the Al tolerant genotypes. Similar results were obtained for barley (*Hordeum vulgare* L.)[11]. This increase in number of adventitious roots per plant with increased Al stress could be attributed to the response of plants to generate new roots when other roots were damged by Al.

The adventitious root traits which gave the largest differences due to Al treatment or genotype were MARL and LARL. Of these two traits, LARL took less time to measure and was relatively easy to obtain because only one measurement at the end of the treatment period was required.

Dry and fresh weights

Shoot dry weights for the genotypes grown with Al showed few differences and this trait was not good to assess plants for Al toxicity (Table 3). However, differences were noted among genotypes for root dry weights. Both seminal and adventitious root dry weights showed significant differences with the trend of seminal roots being more sensitive to Al than the adventitious roots. Fresh weights of the plant parts showed similar trends as dry weights (data not reported). Differences among Al nontolerant genotypes for fresh and dry weights did not appear in roots until plants had been in Al treatments 8 days; shoot

Table 2. Number of adventitious roots (NA), total adventitious root length (TARL), mean adventitious root length (MARL), and longest adventitious root length (LARL) of six sorghum genotypes grown with Al in nutrient solution (Experiment 2)

Al level (μM)	Genotye						
	TX415	Martin	SC33-9-8-E4	CK60	SC369-3-1JB	NB9040	Mean
NA							
0	6.1	6.5	5.6	5.9	6.0	5.5	5.9 C[†]
37	6.1	6.7	5.8	5.4	5.2	5.9	5.9 C
46	7.4	6.9	6.5	5.7	5.5	5.4	6.3 C
56	7.8	8.4	8.2	6.8	6.2	6.7	7.4 B
65	7.8	9.2	8.0	7.4	5.9	7.1	7.6 AB
74	8.8	9.5	8.1	8.2	6.3	6.8	8.0 A
LSD (0.05) for genotype and Al level				1.2			
TARL (mm plant^{-1})							
0	1283	1451	1017	1407	1355	1482	1333 A
37	682	980	734	1161	1196	1543	1049 B
46	576	662	644	978	1155	1405	903 C
56	509	558	583	1133	907	1493	863 C
65	343	463	460	904	821	1364	726 D
74	326	463	435	923	894	1095	689 D
LSD (0.05) for genotype and Al level					238		
MARL (mm plant^{-1})							
0	215	222	182	240	227	272	227 A
37	111	147	127	217	228	262	182 B
46	77	98	99	178	218	264	156 C
56	65	66	71	167	148	225	124 D
65	44	51	57	122	142	192	101 DE
74	37	49	54	114	142	161	93 E
LSD (0.05) for genotype and Al level					49		
LARL (mm plant^{-1})							
0	406	436	311	403	427	397	397 A
37	212	275	245	359	439	443	329 B
46	146	165	176	313	411	428	273 C
56	131	136	141	287	306	409	235 D
65	89	100	107	196	325	372	198 E
74	83	84	90	183	314	314	178 E
LSD (0.05) for genotype and Al level					61		

[†] For each plant trait, means followed by a common letter are not significantly different at $P = 0.05$ according to Duncan's new multiple range test (DNMRT).

weights did not change until plants had been in Al treatments for 12 days (data not reported).

The responses of the genotypes to Al in nutrient solution (Tables 1 and 2) agreed with the results noted for sorghum genotypes grown in acid soils under controlled environmental conditions[4,5]. Recent results showed NB9040 to be susceptible to Al when grown in the field[7], so additional experiments were conducted to determine the effect of higher Al levels on genotype responses to Al. The genotypes chosen for these studies had

Table 3. Shoot, total root, seminal root, and adventitious root dry matter yields of six sorghum genotypes grown with Al in nutrient solution (Experiment 2)

Al level (μM)	Genotype						
	TX415	Martin	SC33-9-8-E4	CK60	SC369-3-1JB	NB9040	Mean
Shoot (mg plant^{-1})							
0	137	146	127	138	173	218	156 B[†]
37	118	133	140	134	134	207	144 C
46	133	137	125	142	146	207	148 BC
56	151	157	141	141	152	202	157 B
65	134	164	155	135	148	198	157 B
74	148	201	161	141	151	218	170 A
LSD (0.05) for genotype and Al level				24			
Total root (mg plant^{-1})							
0	81	103	86	85	96	134	98 A
37	62	87	82	88	78	132	88 A
46	56	72	63	80	80	119	78 C
56	52	70	66	81	81	123	79 C
65	32	56	56	67	71	117	66 D
74	32	59	53	66	76	124	68 D
LSD (0.05) for genotype and Al level				16			
Seminal root (mg plant^{-1})							
0	39	49	42	40	44	57	45 AB
37	39	55	46	51	38	56	47 A
46	33	49	33	50	40	53	43 B
56	24	39	29	43	38	46	36 C
65	12	25	21	31	36	44	28 D
74	10	24	17	27	36	51	28 D
LSD (0.05) for genotype and Al level				8			
Adventitious root (mg plant^{-1})							
0	42	54	44	45	53	77	53 A
37	23	32	36	37	40	77	41 B
46	23	23	30	30	40	66	35 C
56	27	31	37	38	42	77	42 B
65	20	31	35	37	35	72	38 BC
74	22	35	36	39	40	74	41 B
LSD (0.05) for genotype and Al level				11			

[†] For each plant trait, means followed by a common letter are not significantly different at $P = 0.05$ according to Duncan's new multiple range test (DNMRT).

been grown in acid soils in Brazil[2] and the USA[7]. Only NSRL and LARL were used to assess Al toxicity tolerance of these genotypes. Significant differences among genotypes were noted for these traits (Tables 4 and 5), showing that NSRL anb LARL were useful traits to differentiate genotypes for responses to Al. If sufficient Al was added to solutions, even the Al tolerant genotypes (SC175-14, SC283, and MN1204) had little or no seminal root growth (Table 3). Adventitious roots did not appear to be affected as severely as seminal roots by the higher Al levels (Tables 3 and 4), and the LARL gave similar results as NSRL for Al tolerance. Both root traits related well to results found in the field[2,7].

Table 4. Net seminal root length (mm plant^{-1}) (NSRL) of Al tolerant and nontolerant sorghum genotypes grown with Al in nutrient solution (Experiment 3)

Al level (μM)	Al nontolerant					Al tolerant			
	SC208	TX415	TX623A	156-P-5-2-1	Mean	SC175-14	SC283	MN1204	Mean
0	182	188	146	133	162 A†	149	72	132	118 B
74	13	11	6	16	12 C	158	93	145	132 B
148	5	3	2	4	4 C	19	20	41	27 C
222	4	3	3	5	4 C	6	8	7	7 C
LSD (0.05)		9					18		

† Means followed by a common letter are not statistically different at $P = 0.05$ according to Duncan's new multiple range test (DNMRT).

Table 5. Longest adventitious root length (mm plant^{-1}) (LARL) of sorghum genotypes grown with Al in nutrient solution (Experiment 4)

Al level (μM)	Al nontolerant				Al tolerant			
	SC208	TX415	TX623A	Mean	SC175-14	SC283	MN1204	Mean
92	60	61	51	57 B†	321	255	290	289 A
185	27	19	20	22 C	279	212	293	261 A
LSD (0.05)		9				65		

† Means followed by a common letter are not statistically different at $P = 0.05$ according to Duncan's new multiple range test (DNMRT).

An advantage of the adventitious root measurements over the seminal root measurements was that a measurement of the former was only required at the termination of the experiment. This was because the adventitious roots had not yet initiated at the time plants were placed in Al treatments (8-day-old plants). However, plants assessed for adventitious root traits were older and had to be spaced further apart on the plates because of root tangling during subsequent growth. Advantages that seminal root traits had over adventitious root traits were that plants could be assessed for Al toxicity at a younger age and more plants could be tested per unit space. Care had to be used not to injure roots when initial seminal root length measurements were made.

Studies were also conducted to determine how soon plants could be assessed for Al toxicity after introduction to Al. Differences among genotypes were obtained 4 days after treatment with Al for seminal root traits, but only after 8 days in Al for adventitious root traits (data not reported).

Either seminal or adventitious root measurements should be meaningful quantitative traits to assess Al toxicity tolerance in sorghum genotypes. Considerations must be given to space, time, and labor when choosing traits to assess plants for Al toxicity. These traits, especially adventitious root measurements, have been effective for screening genotypes in genetic and plant breeding programs in Brazil for improving sorghum growth and production on Al toxic soils. Reduced inputs of lime or amendments to alleviate high Al toxicity problems have been encouraging from use of this new germplasm.

References

1. Blum A, Arkin G F and Jordan W R 1977 Sorghum root morphogenesis and growth. I. Effect of maturity genes. Crop Sci. 17, 149-153.
2. Borgonovi R A, Santos F G and Schaffert R E 1982 Population breeding in sorghum. I. Development of BRP5BR with tolerance to aluminum toxicity, p. 34. *In* XIV Brazilian Maize and Sorghum Congress, Florianopolis, S.C., Brazil (*In Portuguese*).
3. Borgonovi R A, Schaffert R E and Pitta G V E 1985 Breeding aluminum tolerant sorghums. *In* Evaluating Sorghum for Tolerance to Al-Toxic Tropical Soils in Latin America. Ed. L M Gourley. International Center for the Arid Tropics (CIAT), Cali, Colombia. (*In press*)
4. Brown J C and Jones W E 1977 Fitting plants nutritionally to soils. III. Sorghum. Agron. J. 69, 410-414.
5. Brown J C, Clark R B and Jones W E 1977 Efficient and inefficient use of phosphorus by sorghum. Soil Sci. Soc. Am. J. 41, 747-750.
6. Clark R B 1982 Nutrient solution growth of sorghum and corn in mineral nutrition studies. J. Plant Nutr. 5, 1039-1057.
7. Duncan R R, Clark R B and Furlani P R 1983 Laboratory and field evaluations of sorghum for response to aluminum and acid soil. Agron. J. 75, 1023-1026.
8. Furlani P R and Clark R B 1981 Screening sorghum for aluminum tolerance in nutrient solutions. Agron. J. 73, 587-594.
9. Polle E, Konzak C F and Kittrick J A 1978 Rapid screening of wheat for tolerance to aluminum in breeding varieties better adapted to acid soils. USAID Tech. Serv. Bull. 21, U.S. Agency for International Development, Washington, DC.
10. Polle E, Konzak C F and Kittrick J A 1978 Rapid screening of maize for tolerance to aluminum in breeding varieties better adapted to acid soils. USAID Tech. Serv. Bull. 22, U.S. Agency for International Development, Washington, DC.
11. Reid D A, Fleming A L and Foy C D 1971 A method for determining aluminum response of barley in nutrient solution in comparison to response in Al-toxic soil. Agron. J. 63, 600-603.
12. Santos, H L, Baligar V C and Vasconcellos C A 1980 Screening sorghum genotypes for aluminum tolerance. p. 24. *In* XIII Braxilian Maize and Sorghum Congress, Londrina, P.R., Brazil (*In Portuguese*).
13. Wright M J (Ed.) 1976 Plant Adaptation to Mineral Stress in Problem Soils. Cornell Univ. Agric. Exp. Stn., Ithaca, NY.

Evaluation of inbred maize lines for aluminum tolerance in nutrient solution

R. MAGNAVACA,
National Maize and Sorghum Research Center, EMBRAPA, C.P. 151, 35 700 Sete Lagoas, M.G. Brazil

C.O. GARDNER and R.B. CLARK
Department of Agronomy and U.S. Department of Agriculture, Agricultural Research Service, University of Nebraska, Lincoln, NE 68583, USA

Key words Al and P levels to assess Al tolerance Brazilian and USA inbred lines Plant measurements for Al tolerance *Zea mays*

Summary Maize (*Zea mays* L.) inbred lines were grown in nutrient solution with different levels of Al and P to study their genetic variability for Al tolerance. Plant measurements for determining inbred line responses to Al were also evaluated. The best traits to assess maize for Al tolerance were seminal and adventitious root lengths.

Brazilian maize inbred lines were more tolerant to Al for the traits measured than USA inbred lines when grown in nutrient solution. Most inbred lines tested showed decreased root lengths, but a few Brazilian lines were not affected by the Al levels used. At higher Al levels most Brazilian lines were not affected as severely as the USA lines. Tolerance of the inbred lines to Al was increased by inclusion of P in nutrient solutions. The greatest Al-induced decrease in root length generally occurred at the lowest P level. The combination of 185 μmol Al L^{-1} and 45 μmol P L^{-1} in nutrient solutions was the best combination of Al and P to evaluate maize genotypes for Al tolerance in this study.

Introduction

Acid soils appear frequently in tropical areas. Oxisols which are strongly weathered soils and have low cation exchange capacities, occupy 8.1% of the world land area[8]. In tropical areas, Oxisols exhibit major mineral element deficiencies and toxicities; deficiencies of P, Ca, Mg, and Zn are common, toxic exchangeable Al is usually high, and the fixation of P by soil particles is extensive. Oxisols, Ultisols, and Inceptisols are estimated to occupy approximately 1000 million hectares in the tropics[24]. This corresponds to 33% of the total potentially arable land area of the world useful for crop production without irrigation.

Kamprath[17] indicated that the Al saturation in soils should be less than 45% for maximum growth of maize (*Zea mays* L.), but Olmos and Camargo[20] found that 25% Al saturation reduced maize yields. On a Brazilian acid Oxisol soil that had 55% Al saturation, 363 inbred maize lines were evaluated for their response to acid soil toxicity[1]. The soil had received no lime applications but had received relatively high amounts of N and P. Fifteen days after planting 19% of the inbred lines died and 45 days later 70% of the inbred lines were dead. Decreases in maize yield due to Al toxicity have also been reported for Oxisols in Brazil[9,10], for ferrallitic and ferruginous soils in Madagascar[25]; for Oxisols and Ultisols of Puerto Rico[3]; for Andosols from northeastern Japan[23]; and for Ultisols of Pennsylvania in the USA[11].

Genetic variability in Al tolerance has been demonstrated among maize inbred lines, hybrids, and varieties for Al tolerance[1,2,6,10,16,22]. In these studies, evaluations of maize germplasm for tolerance to Al were made in field experiments where plants were grown on acid soils, in greenhouse experiments where plants were grown in pots of acid soil, in germination studies where seeds were irrigated with Al-containing nutrient solutions in sand, and in studies where seedlings were grown in nutrient solutions with Al. The different approaches to assess Al tolerance have been responsible for many of the conflicting results that have been reported for genotypic differences in tolerance to Al[15]. Screening techniques used for the assessment of Al tolerance must consider the appropriate combination of factors like concentrations of Al, Ca, Mg, P, and K, the solution pH, source of N (ammonium or nitrate), and temperature. The traits used to measure Al genotypic differences also need to be considered[14].

The objectives of this research were to: (i) develop relatively easy and inexpensive techniques to evaluate seedlings of maize genotypes for Al tolerance; (ii) study genetic variation among homogeneous maize inbred lines for Al tolerance; and (iii) compare responses to tropical (Brazilian) and temperate (USA) maize inbred lines to Al.

Material and methods

Growth of plants

Seeds treated with captan [N-(trichloromethylthio)-4-cyclohexene-1,2-dicarboximide] were germinated seven days in rolled paper towels kept moist with aerated distilled water. The temperature was 27 ± 1 C for 16h with fluorescent lights (Agro-Lite cool-white, F40)* at $150 \mu E\,m^{-2}s^{-1}$ and 8 h of darkness at 19 ± 1 C. Seven-day-old uniform-sized seedlings without visual root injury were transferred to plastic plates containing either 126 (Experiment 1) or 42 (Experiment 2) plants per plate and grown in 6.5 L nutrient solution with treatments. Distilled water was added regularly to maintain volume. The initial pH of nutrient solutions was adjusted to 4.0, monitored daily, and adjusted to 4.0 if needed with 1 N HCl or NaOH. Plants were grown in treatment solutions 10 days before experiments were terminated.

The composition of nutrient solutions used for growth of maize was (μmol element L^{-1}): 10 900 NO_3-N; 3500 Ca; 2300 K; 1300 NH_4-N; 850 Mg; 590 S; 590 Cl; 25 B; 9.1 Mn; 2.29 Zn; 0.88 Mo; 0.63 Cu; 77 Fe as FeHEDTA (ferric hydroxethylethylenediaminetriacetate). Phosphorus and Al concentrations varied with the treatment. Aluminum was added to the nutrient solutions as $KAl(SO_4)_2$ and P as KH_2PO_4. Details of procedures used for growing plants in nutrient solution have been described[5,15].

* Mention of a company, trademark, or proprietary product does not constitute a guarantee or warranty of the product by the University of Nebraska or the U.S. Department of Agriculture, and does not imply its approval to the exclusion of other products that may also be suitable.

Growth conditions for Experiment 1 were 17 h light at 28 $\pm$ 1 C and 7 h darkness at 23 $\pm$ 1 C in a controlled environment chamber. Photosynthetic photon flux density (PPFD) was 350 μE m^{-2} s^{-1} at plant level (60 cm below the lights). Lamps providing light were high pressure sodium (General Electric, Lucalux, LU4008)[3] and metal halide (General Electric, multivapor, MV400)[3]. For Experiment 2, plants were grown in a controlled environment room with 16 h light at 27 $\times$ 1 C and 8 h darkness at 19 $\pm$ 1 C. Lamps providing light were fluorescent (Agro-Light cool-white, F40)[3] yielding 150 μE m^{-2} s^{-1} PPFD at plant height (100 cm below the lamps).

Plant measurements to assess Al tolerance

In Experiment 1, only seminal roots were allowed to develop and the adventitious roots that had started to grow were removed. In Experiment 2, adventitious roots were allowed to develop. At the time seedlings in both experiments were put into treatment solutions, the initial seminal root lengths were measured for each plant. When experiments were terminated, the final seminal root length and the length of the longest adventitious root were measured on each plant. These and two other calculated root traits were used to assess plants for Al tolerance. Definition of each trait is:

Initial seminal root length (ISRL): The length of the primary seminal root when plants were transferred to treatment nutrient solutions;

Final seminal root length (FSRL): The length of the primary seminal root when experiments were terminated (plants grown 10 days in treatment solutions);

Relative seminal root length (RSRL): Calculated by dividing FSRL by ISRL;

Net seminal root length (NSRL): Calculated by subtracting ISRL from FSRL; and

Longest adventitious root length (LARL): The length of the longest adventitious (synonymous with crown) root when experiments were terminated (plants grown 10 days in treatment solutions).

Maize genotypes

The USA inbred lines used were obtained from the maize breeding program of the Department of Agronomy, University of Nebraska, Lincoln even though several of the lines were originally developed in other states. The inbred lines used were A554, A635, B37, B73, C103, CI64, H84, Mo17, N7A, N28, N132, N139, N152, N156, N168, N174, and W117.

The Brazilian inbred lines used were developed at and belong to the collection of the National Maize and Sorghum Research Center, Sete Lagoas, M.G., Brazil, and brought to Nebraska. Each of these lines had undergone at least six generations of selfing and are identified as L69, L153, L297, M1001, M1002, M1003, M1004, M1005, M1007, M1009, M1010.

Treatments, experimental design, and statistical analysis

Experiment 1. Treatments included four Al levels (0, 74, 148, and 222 μmol L^{-1}) and 21 maize inbred lines (B37, L69, L153, L297, M1001, M1002, M1003, M1004, M1005, M1007, M1009, M1010, Mo17, N7A, N28, N132, N139, N152, N156, N168, and N174). A split-plot design was used with Al levels as whole plots and inbred lines as subplots. Whole plots were arranged in a randomized complete block design with three replications and subplots consisted of six plants each. Each nutrient solution container, therefore, constituted a whole plot and included 21 subplots of six plants each. The P level in nutrient solutions was 64 μmol L^{-1}. Each plant of each subplot was evaluated for Al tolerance using ISRL, FSRL, NSRL, and RSRL, and subplot means were calculated. The analyses of variance were computed for each variable using subplot means of six plants.

Statistical analyses were performed using a fixed effects model. The Al levels sums of squares were partitioned into their linear and quadratic components. The Al level and genotype interactions sums of squares were likewise partitioned into genotype $\times$ Al levels linear and genotypes Al levels quadratic components.

To test for linear and quadratic effects of Al on each genotype, a different subdivision of degrees of freedom was allocated according to the following model:

$$Y_{ijn} = u + r_i + g_n + a_{jn} + e_{ijn}.$$

where Y_{ijn} = the measurement of nth genotype in the jth Al level in the ith replication, u = the overall mean effect, r_i = the effect of the ith replication, g_n = the effect of the nth genotype a_{jn} = the effect

Table 1. Mean relative seminal root length (RSRL), final seminal root length (FSRL), and net seminal root length (NSRL) of maize inbred lines grown in nutrient solution with Al (Experiment 1)

Inbred source	Inbred line	RSRL				FSRL (mm plant^{-1})				NSRL (mm plant^{-1})			
		0†	74	148	222	0	74	148	222	0	74	148	222
Brazil	L69	2.45	2.33	1.99	2.25	154	154	125	128	91	87	64	68
	L153	2.98	3.04	2.36	2.11	217	207	189	153	144	138	109	81
	L297	3.82	3.75	2.23	1.87	244	247	156	110	173	178	83	50
	M1001	3.44	3.23	2.33	1.80	297	242	198	163	210	164	111	71
	M1002	2.17	2.55	2.36	2.23	121	142	116	135	64	84	66	74
	M1003	2.54	2.76	2.40	1.88	191	198	177	147	116	125	105	69
	M1004	2.59	2.56	2.43	1.96	235	251	243	191	144	155	142	93
	M1005	3.89	3.79	3.16	2.88	239	241	213	165	177	175	144	106
	M1007	3.81	2.63	1.61	1.24	347	233	123	114	255	148	44	21
	M1009	3.45	2.62	1.58	1.32	247	187	136	107	174	113	49	27
	M1010	4.44	3.26	2.67	1.73	308	245	205	144	233	168	129	63
	Mean	3.23	2.96	2.28	1.93	236	213	171	142	162	140	95	66
USA	B73	1.97	1.65	1.35	1.32	194	175	131	121	95	69	33	29
	Mo17	2.99	2.59	1.99	1.64	167	143	103	96	108	85	51	37
	N7A	2.18	1.81	1.25	1.16	195	149	123	107	105	65	25	15
	N28	3.48	3.00	1.94	1.42	249	260	126	109	171	172	55	32
	N132	2.37	2.28	1.51	1.29	240	205	154	134	139	114	51	30
	N139	2.07	1.94	1.51	1.33	226	216	156	143	116	105	54	35
	N152	278	2.32	1.56	1.36	311	278	182	153	197	157	66	43
	N156	3.08	2.71	1.81	1.44	262	225	171	125	174	137	76	37
	N168	2.39	2.12	1.78	1.40	248	224	161	138	147	121	72	40
	N174	3.22	2.08	1.69	1.43	206	174	135	108	138	88	59	32
	Mean	2.65	2.25	1.64	1.38	230	205	144	123	139	111	54	33
Overall mean		2.96	2.62	1.98	1.67	233	209	158	133	151	126	74	50

† μmol Al L^{-1}.

of the jth Al level within the nth genotype, e_{ijn} = the random pooled component of error associated with the nth genotype at the jth Al level.

In this model, the pooled error mean squares provided only an approximated F-test for effects of Al levels within genotypes.

Experiment 2. Treatments included nine combinations of a 3^2 factorial for three Al levels (37, 111, and 185 μmol L^{-1}) and three P levels (22.5, 45, and 67.5 μmol L^{-1}). Seven maize inbred lines (A554, A635, CI64, L69, L153, L297, and Mo17) were used. Traits measured were ISRL, FSRL, RSRL, NSRL, and LARL. The individual measurement analyzed was the mean value of six plants per plot. The experimental design was a split-plot with Al and P combinations in the whole plots and genotypes allotted to the subplots. Two replications of whole plots were arranged in a randomized complete block design.

The statistical analyses were performed using a fixed effects model and dividing Al and P levels sums of squares into linear and quadratic components. The Al × P level interaction sums of squares were also subdivided into interactions involving the linear and quadratic effects of Al and P.

Results and discussion

Experiment 1

The means for RSRL, FSRL, and NSRL and the analyses of variance

Table 2. Analysis of variance for relative seminal root length (RSRL), final seminal root length (FSRL), and net seminal root length (NSRL) on inbred maize lines grown in nutrient solution with Al (Experiment 1)

Source of variation	df	Mean squares		
		RSRL	FSRL (mm plant^{-1})	NSRL (mm plant^{-1})
Replications	2	0.2201	2714.77	2044.46
Al levels	3	21.7961**	133091.67**	133612.82**
Al linear	1	64.0803**	390368.00**	392624.26**
Al quadratic	1	0.0157	32.14	5.73
Deviation	1	1.2922	8874.87	8208.46
Error a	6	0.5145	2207.64	2392.30
Lines	20	2.5982**	12620.94**	9940.54**
Lines × Al levels	60	0.3760**	2369.23**	2228.05**
Lines × Al linear	20	0.9152**	5408.34**	5385.76**
Lines × Al quadratic	20	0.1317	987.02*	776.56*
Deviation	20	0.0812	712.34	521.83
Error b	160	0.0972	519.32	459.37
Total	251			
CV (%)		13.5	12.4	21.3

*, ** Statistical significance at $\alpha = 0.05$ and $\alpha = 0.01$.

for the USA and Brazilian inbred maize lines evaluated at four Al levels in Experiment 1 are presented in Tables 1 and 2. A consistency of results for the three traits was evident from the analyses of variance in Table 2. The linear responses to Al levels, differences among inbred lines, and the interaction of Al levels × lines were all statistically significant. Because of the significant Al levels × lines interactions, different analyses of variance were calculated and the linear and quadratic responses of Al levels were evaluated for each inbred line[18].

The Brazilian inbred lines L69 and M1002 were not affected by Al level (Table 1). Each of the other lines decreased in seminal root length with increasing Al level in the nutrient solution. However, at the highest Al level (222 μmol L^{-1}) the Brazilian lines L153, L297, M1001, M1003, M1004, M1005, and M1010 were affected less than the other inbred lines, especially when the RSRL and NSRL traits were used. Although differences at low levels of Al were noted, each of the USA and two of the Brazilian lines (M1007 and M1009) grew poorly at high levels of Al. In addition to the inbred lines L69 and M1002, the other inbred lines that grew relatively well under high Al stress should be desirable for production of hybrids to be grown on soils with high Al saturation.

The Brazilian inbred lines were generally more tolerant of Al than the USA inbred lines. The Brazilian lines were not developed from plants grown on soils with high levels of Al saturation, but the apparent random fixation of genes for tolerance to Al in inbred lines is an indication of the variability for tolerance in populations from which inbred

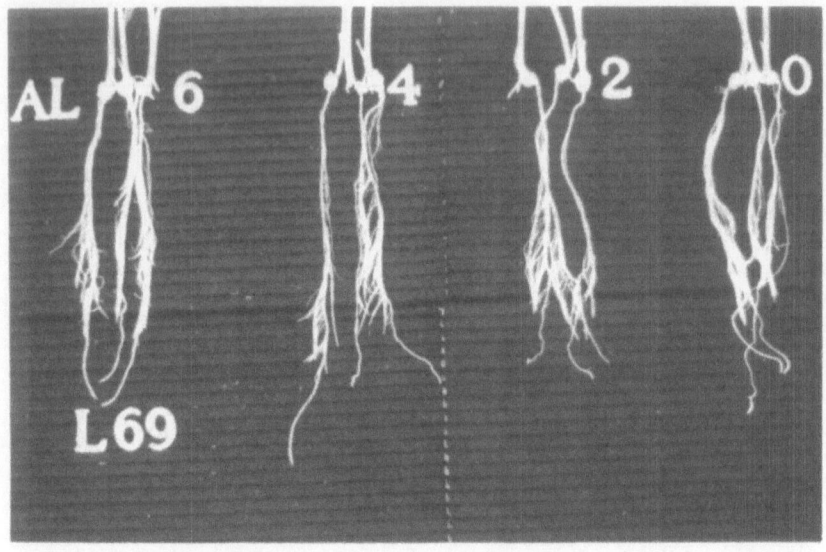

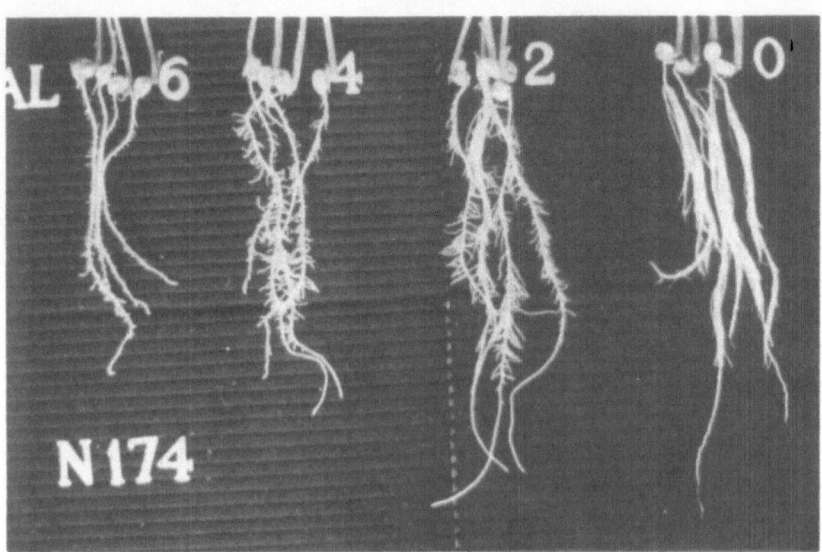

Fig. 1. Roots of the inbred maize lines L69 and N174 grown in nutrient solution at 222, 148, 74, and 0 μmol Al L^{-1} (6, 4, 2, and 0 mg Al L^{-1}) (left to right).

lines are developed. These lines had been tested for Al tolerance on acid soils in Brazil, and most of them showed high Al tolerance and grew well on the acid soils with high Al saturation[1].

The screening method used in this experiment showed considerable potential for large scale evaluations of inbred lines and progenies of

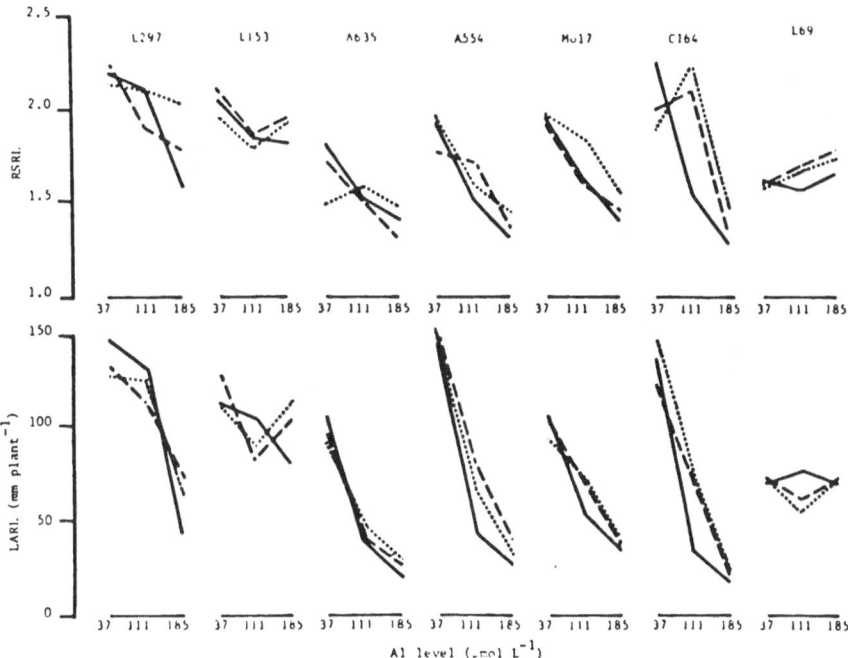

Fig. 2. Relative seminal root length (RSRL) and longest adventitious root length (LARL) of maize inbred lines grown at various levels of Al and P in nutrient solution (Experiment 2).

maize. By using six plants per plot, it was possible to evaluate 21 lines per container. Discrimination among inbred lines was relatively easy at 222 μmol Al L^{-1}, especially when using RSRL to assess Al tolerance. The utilization of the high Al level used overcame the necessity of calculating relative values from measurements of the same lines in solutions without Al. Thus, the size of the experiment could be decreased to half. Figure 1 shows visual symptoms of Al toxicity for the tolerant L69 and the non-tolerant N174 maize inbreds.

Experiment 2

Seven inbred maize lines were evaluated at different levels of Al and P in nutrient solutions; three Brazilian lines (L69, L153, and L297) and four USA lines (A554, A635, CI64, and Mo17) were tested. The means of four traits (RSRL, LARL, FSRL, and NSRL) for each inbred line grown at various levels of Al and P are presented as response curves (Figs. 2 and 3). The analyses of variance of these four traits are presented in Table 3. Even though results varied slightly among the traits, significant differences were consistent among main effects of Al levels, lines, and the triple interaction of Al $\times$ P $\times$ lines. Also the lack of significance of main effects of P and P $\times$ line interaction was consistent. Only for RSRL was the Al $\times$ P interaction significant.

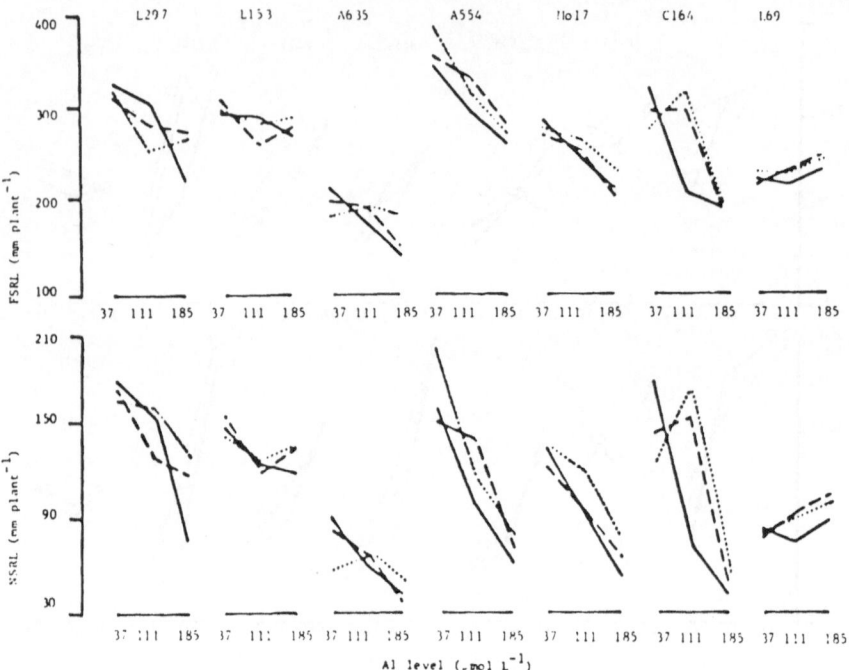

Fig. 3. Final seminal root length (FSRL) and net seminal root length (NSRL) of maize inbred lines grown at various levels of Al and P in nutrient solution (Experiment 2).

The inbred line L69 was not affected by Al level, and slight increases in root length at high levels of P were observed for each seminal root measurement. L69 was derived from a local Catete variety of the 'Cerrado' region in Brazil, and its response to Al level was similar in both Experiments 1 and 2.

The response of line L297 (derived from a cross of a tuxpeno line with the temperate single cross hybrid XL45) to Al levels depended on P level, especially for seminal root measurements. As the P level increased in the nutrient solution, the detrimental effects of Al decreased. The response of line L153, derived from a cross of a Catete population from Brazil and an Eto population from Colombia, for the four traits indicated some sensitivity to Al level, but somewhat independent of P level. However, L153 root lengths were greater than those of L69 at each Al and P level and greater than those of L297 at low P levels.

It was of interest to compare the results for these three Brazilian lines grown in nutrient solution with their grain yield response when grown in the field with three levels of Al saturation[10,19]. L69 was stable over each Al level in nutrient solution, but was intermediate in grain yield when grown in the field. L153 did not grow well at the high Al levels in nutrient solution, but two tons of lime ha^{-1} in the field was sufficient to promote

Table 3. Analysis of variance of relative seminal root length (RSRL), longest adventitious root length (LARL), final seminal root length (FSRL), and net seminal root length (NSRL) of maize inbred lines grown in nutrient solution with three levels each of Al and P (Experiment 2)

Source of variation	df	Mean squares			
		RSRL	LARL (mm plant^{-1})	FSRL (mm plant^{-1})	NSRL (mm plant^{-1})
Replications	1	0.0040	271.63	1225.78	257.14
Al levels	2	1.2314**	49749.11**	29056.89**	28618.88**
Al linear	1	2.4480**	96425.19**	57828.76**	56732.01**
Al quadratic	1	0.0149	3073.02**	285.02	505.75
P levels	2	0.0476	393.92	1230.05	1530.45
P linear	1	0.0953	624.30	2325.76	3012.01
P quadratic	1	0.0000	163.53	134.35	48.89
Al × P	4	0.0774*	185.08	856.26	1159.47
Al linear × P linear	1	0.2379**	565.79	2511.16	3001.78
Al linear × P quadratic	1	0.0012	172.02	190.72	172.02
Al quadratic × P linear	1	0.0668	1.17	323.15	1292.59
Al quadratic × P quadratic	1	0.0038	1.34	400.00	171.50
Error a	8	0.0182	254.81	1322.03	596.82
Lines	6	0.4775**	6518.90**	34879.68**	13590.75**
Al × lines	12	0.1396**	3703.32**	3382.19**	3349.94**
Al linear × lines	6	0.2231**	6176.50**	5972.96**	5753.90**
Al quadratic × lines	6	0.0561**	1230.15**	791.43	945.97**
P × lines	12	0.0171	136.16	241.08	246.09
P linear × lines	6	0.0199	213.55	305.96	335.37
P quadratic × lines	6	0.0142	58.78	176.21	156.81
Al × P × lines	24	0.0328**	232.83*	937.41*	680.98**
Al linear × P linear × lines	6	0.0201	193.99	627.74	426.28
Al quadratic × P linear × lines	6	0.0677**	313.96*	2370.95**	1434.57**
Al linear × P quadratic × lines	6	0.0139	90.34	330.11	205.80
Al quadratic × P quadratic × lines	6	0.0297	333.02*	420.83	657.25*
Error b	54	0.0152	126.47	514.38	280.34
Total	125				
CV (%)		7.03	13.98	8.77	15.07

*, ** Statistical significance at $\alpha = 0.05$ and $\alpha = 0.01$.

high grain yield. L297 responded adversely to Al levels in nutrient solution, but at high Al saturation in the field, grain yields were high. From these results, it was concluded that Al affected the high yielding L297 line by interfering with the uptake or utilization of P. This was not apparent for the other two Brazilian lines. At the highest level of Al, L29 and L153 were able to absorb and translocate P in a normal pattern. Relationships between P nutrition and Al tolerance have been reported[2,6,7,12,13,21].

The four USA inbred lines A554, A635, CI64, and Mo17 had decreased root lengths with increasing levels of Al. The largest decreases generally occurred at the lowest level of P. The inbred line CI64 was unique for seminal root measurements in that when grown with high levels of P, the addition of 111 μmol Al L^{-1} increased root length relative to the lowest level of Al. However, root development decreased sharply with the addition of the higher 185 μmol Al L^{-1}. Similar responses were reported for the inbred maize line B57[4]. Root growth may also be inhibited by high levels of P[5]. The addition of Al to the solution may have inactivated some of the P in solution. Because of the biological and statistical significance of the Al × P × line interaction, response curves for each trait of each line to the various levels of Al and P have been presented (Figs. 2 and 3).

Based on these results, 185 μmol Al L^{-1} and 45 μmol P L^{-1} in nutrient solutions were considered to be the best combination of Al and P to be used in genetic studies where 42 plants were grown per 6.5 L container for 10 days in treatment.

Although similarities among traits for assessing Al tolerance were noted in the maize inbreds, RSRL determined by the nutrient solution technique is recommended as the best of the traits to assess maize genotypes for Al tolerance. One concern was the effect of Al of root vigor during the seedling stage. RSRL was the only seminal root measurement that was corrected for ISRL. In Experiment 2, the correlations of ISRL with FSRL, NSRL, RSRL, and LARL were 0.68**, 0.31**, −0.02, and 0.21*, respectively. The NSRL and LARL analyses showed higher coefficients of variation compared to RSRL. NSRL and LARL would be recommended as second choices to RSRL as traits for assessing Al tolerance in nutrient solution. The FSRL could not be recommended as a suitable trait to assess Al tolerance.

References

1 Bahia A F C, Franca G E, Pitta G V E, Magnavaca R, Mendes J F, Bahia F G F T C and Pereira P 1978 Evaluation of corn inbred lines and populations in soil acidity conditions. pp 51–58. *In* XI Annu. Brazilian Maize Sorghum Conf., Piracicaba, S.P., Brazil. (*In Portuguese*)
2 Bahia A F C, Magnavaca R, Vasconcellos C A, Bahia F G F T C, Pitta G V E and Naspolini V 1979 Response curves to lime and phosphorus of corn hybrids in soil with high acidity. p 86. *In* XII Annu. Brazilian Maize Sorghum Conf., Goiania, GO., Brazil. (*In Portuguese*)
3 Brenes E and Pearson R W 1973 Root responses of three Gramineae species to soil acidity in an Oxisol and an Ultisol. Soil Sci. 116, 295–302.
4 Clark R B 1977 Effect of aluminum on growth and mineral elements of Al-tolerant and Al-intolerant corn. Plant and Soil 47, 653–662.
5 Clark R B 1982 Nutrient solution growth of sorghum and corn in mineral nutrition studies. J. Plant Nutr. 5, 1039–1057.
6 Clark R B and Brown J C 1974 Differential mineral uptake by maize inbreds. Commun. Soil Sci. Plant Anal. 5, 213–227.
7 Clark R B and Brown J C 1974 Differential phosphorus uptake by phosphorus-stressed corn inbreds. Crop Sci. 14, 505–508.
8 Dudal R 1976 Inventory of the major soils of the world with special reference to mineral stress hazards. pp 3–13. *In* Plant Adaptation to Mineral Stress in Problem Soils. Ed. M J Wright. Cornell Univ. Agric. Exp. Stn., Ithaca, NY.
9 EMBRAPA (Brazilian Department of Agriculture Research) 1978 Cerrado Agricultural Research Center Tech. Rep. 1976-1977. EMBRAPA, Brasilia, D.F. Brazil. (*In Portuguese*)
10 EMBRAPA (Brazilian Department of Agricultural Research) 1980 National Maize and

Sorghum Research Center Tech. Rep. 1979. EMBRAPA, Brasilia, D.F., Brazil. (*In Portuguese*)

11 Fox R H 1979 Soil pH, aluminum saturation, and corn grain yield. Soil Sci. 127, 330–334.

12 Foy C D and Brown J C 1964 Toxic factors in acid soils. II. Differential aluminum tolerance of plant species. Soil Sci. Soc. Am. Proc. 28, 27–32.

13 Foy C D, Chaney R L and White M C 1978 The physiology of metal toxicity in plants. Annu. Rev. Plant Physiol. 29, 511–566.

14 Furlani P R 1981 Effects of aluminum on growth and mineral nutrition of sorghum genotypes. Ph.D. Dissertation, Univ. of Nebraska, Lincoln. Diss. Abstr. 42, 1260B.

15 Furlani P R and Clark R B 1981 Screening sorghum for aluminum tolerance in nutrient solution. Agron. J. 73, 587–594.

16 Garcia Jr. O. da Silva W J and Massei M A S 1979 An efficient method for screening maize inbreds for aluminum tolerance. Maydica 24, 75–82.

17 Kamprath E J 1972 Soil acidity and liming. pp 136–149. *In* Soils of the Humid Tropics. Nat. Acad. Sci., Washington, DC.

18 Magnavaca R 1982 Genetic variability and the inheritance of aluminum tolerance in maize (*Zea mays* L.). Ph.D. Diss. Univ. of Nebraska, Lincoln. Diss. Abstr. 43, 2073B (1983)

19 Naspolini V, Bahia A F C, Viana R T, Gama E E G, Vasconcellos C A and Magnavaca R 1981 Performance of inbreds and single crosses of maize in soils under 'cerrado' vegetation. Ciencia Cultura 33, 722–727. (*In Portuguese*)

20 Olmos I L and Camargo M N 1976 Incidence of aluminum toxicity in Brazilian soils: Its characterization and distribution. Ciencia Cultura 28, 171–180. (*In Portuguese*)

21 Rasmussen H P 1968 The mode of entry and distribution of aluminum in *Zea mays*: Electron microprobe x-ray analysis. Planta 81, 28–37.

22 Rhue R D and Grogan C O 1977 Screening corn for Al tolerance using different Ca and Mg concentrations. Agron. J. 69, 755–760.

23 Saigusa M, Shoji S and Takahshi T 1980 Plant root growth in acid Andosols from northeastern Japan. 2. Exchange acidity Y_1 as a realistic measure of aluminum toxicity potential. Soil Sci. 130, 242–250.

24 Van Wambeke A 1976 Formation, distribution and consequences of acid soils in agricultural development. pp 15–24. *In* Plant Adaptation to Mineral Stress in Problem Soils. Ed. M J Wright. Cornell Univ. Agric. Exp. Stn., Ithaca, NY. 25 Velly J 1974 Observation on the acidification of some soils in Madagascar. J. Agron. Trop. 12, 1249–1262.

Section 4A

Genotypic response to nutrient deficiency: Macronutrients

The application of *in vivo* techniques in the study of metabolic aspects of ion absorption in crop plants

B.C. LOUGHMAN
Department of Agricultural Science, University of Oxford, Parks Road, Oxford, UK

Key words Maize NMR Phosphate Potato Varietal differences

Summary Little is known about the biochemical basis of the genotypic differences in the capacity for ion absorption and transport shown by many crop species. If these differences reflect the abundance of a specific membrane component or the activity of an enzyme we need to have some indication of the *in vivo* operation of these systems in whole plants. The *in vivo* assessment of glycolytic enzymes is illustrated by the effects of mannose on the transport of phosphate in maize varieties. The application of high resolution ^{31}P-NMR to the study of intermediary metabolism *in vivo* is also helpful in following transport capacity.

The five-fold rise in respiratory rate that occurs when freshly cut potato slices are maintained in aerated water for 24 hours is accompanied by the turning on of a wide range of biochemical systems. Major increases in the capacity for absorption of phosphate from low concentrations ($0.1\,\mu M$–$10\,\mu M$) and in the phosphorylative ability of the tissue are seen, indicating the synthesis of a carrier involved in phosphate transport. These capacities differ markedly between individual tissues of the tuber, *i.e.* pith, parenchyma, cortex and buds and large differences have been observed between comparable tissue from different varieties. Varieties grown under similar conditions have been compared and shown to exhibit different kinetics with respect to the development of the low concentration absorption site and in their sensitivity to the effects of uncouplers such as 2,4-dinitrophenol.

Introduction

We are concerned with the factors involved in the responses of plants under conditions of deficiency of major nutrients. My own interests have centred on phosphorus utilization under conditions of varied supply and on the effects of deficiency of other nutrients such as calcium, boron and zinc on the processes involved and I would like to pick out a few problems and indicate how recent experimental approaches may help us to understand some of the metabolic sequences that underlie the processes of absorption and transport of ions. Despite the weight of publications concerning ion uptake our understanding of the processes involved in whole plants is particularly sketchy and proposed mechanisms are often based on experimental results obtained with algal cells, storage tissue slices and pieces of root tissue.

Unless we are able to tackle those aspects of intermediary metabolism in roots that determine absorption and distribution of ions we will be confined to approaches based on analogies to enzyme kinetics. Clearly, these have contributed to our understanding but unless the successful combination of the physiological and biochemical approaches can be

H.W. Gabelman and B.C. Loughman (Eds.), Genetic aspects of plant mineral nutrition.
ISBN-13: 978-94-010-8102-3
© *1987 Martinus Nijhoff Publishers, Dordrecht/Boston/Lancaster.*

achieved it will be difficult to focus on the genetic implications of current research in plant nutrition.

A number of contributions in this volume concern the assessment of the absorption capacity of genotypes under conditions of deficiency of phosphorus or nitrogen and it is clear that selection of genotypes for efficient growth under conditions of low availability of particular nutrients is of great practical importance. The more we can examine the biochemical differences in the genotypes that underlie the variation in the absorption ability the more readily can we hope to identify the key components and use them as a possible aid to selection. Whether the differences lie in the activities of a single enzyme or a carrier protein or glycolipid in the plasmalemma, tonoplast or other membrane one must be in a position to accurately assess that particular component. Another difficulty is inherent in the way in which such assessments are made. If two varieties differ in a particular capacity for transport it is possible to extract the tissue and measure the activity of a particular enzyme and perhaps correlate this activity with the metabolic process involved in the transport. However the enzyme activity in the extract might bear little relation to that in the intact tissue for a number of reasons. Firstly, the tissues could differ in their content of metals, metabolites and enzyme inhibitors leading to inactivation during extraction. Secondly, the functional levels of the enzyme in the intact tissues could be similar even though wide differences were apparent in the extracts from the two sources. Some success is being achieved in animal systems in correlating *in vivo* enzyme levels with those obtained by extraction[4] but the presence of the vacuolar contents in plants causes extra problems.

The importance of some form of *in vivo* assessment of enzyme activity and metabolite levels in whole plants cannot be overstressed and the application of suitable NMR techniques to these problems is now bringing encouraging results[8,19]. A number of approaches to this problem of *in vivo* assessment are possible and in some cases it is clear that it is the only way of measuring specific enzymes that lose or alter their activities when the tissues are disrupted, however mildly. It is also clear that a number of biochemical systems can only be followed by an *in vivo* approach because of the complexity of the enzyme systems or their lability. A simple method of coping with very labile enzyme systems has been used to follow the metabolism of phenoxyacetic acid growth regulators, namely vacuum infiltration of small segments of tissue weighing a few mg with a few μl of the ^{14}C labelled compound followed at intervals by transfer of the tissue to the baseline of chromatogram[6]. Reliable kinetic data can be readily obtained by this technique and it has been considerable advantage in experiments designed to show that the rate of hydroxylation and glucosylation may be correlated with the degree of tolerance to a herbicidal compound. The activity of the enzymes concerned is assessed under conditions that normally occur within the plant even though the tissue has to be destroyed to make the analysis. A combination of this technique with that of ^{13}C-NMR promises to increase the sophistication of the experimental approaches open to those interested in correlating internal metabolic activities with specific aspects of plant nutrition. My own work in this area has been concerned primarily with phosphate nutrition under a wide range of supply and environmental conditions and I would like to concentrate on this ion to illustrate some of the difficulties facing those of us who are interested in detecting metabolic differences in varieties of crop plants that have very different responses on soils of low phosphorus availability or those where other essential nutrients are in short supply.

The phosphorus found in higher plants is absorbed from the soil solution where concentrations of inorganic phosphate of the order of 0.1–10 μM are in equilibrium with insoluble forms from which the ambient concentration of free ions is maintained. The process of absorption must be highly efficient if concentrations of > 20 mM are to be maintained within the cells of the plant and it is dependent on the availability of other ions particularly those of calcium, boron and zinc[16,17]. Some evidence is available to support the view that $H_2PO_4^-$ arrives in the cytoplasm whereas it is possible that the almost instantaneous incorporation of incoming phosphate occurs in the membrane itself[11]. If the initial incorporation into organic forms, occurring within 1 sec after exposure of roots to 1 μM KH$_2$^{32}PO$_4$ does in fact take place in the mitochondria it is clear that the processes involved in crossing the plasmalemma and moving through the cytoplasm are extremely rapid. It is now accepted that plants are able to absorb phosphate faster than it can diffuse in the soil solution to the root surface[2].

The process of entry probably results in the transfer of inorganic orthophosphate across the plasmalemma into the cytoplasm there to be incorporated into nucleoside phosphates, sugar phosphates, nucleoside diphospho sugars, *etc*. The rapidity with which this incorporation occurs causes difficulties in isolating the biochemical stages of the overall processes but some progress has been made. A further special feature of this metabolism is that although all the phosphate entering the root from the soil solution is incorporated into organic forms, dephosphorylation occurs prior to the transfer of P$_i$ to the vacuole or to the xylem for onward transfer to the leaf cells[15].

Assessment of enzyme activities *in vivo*

Whole plants

Although much of our knowledge of absorption, transport and utilization of phosphate has been gathered initially by analytical methods

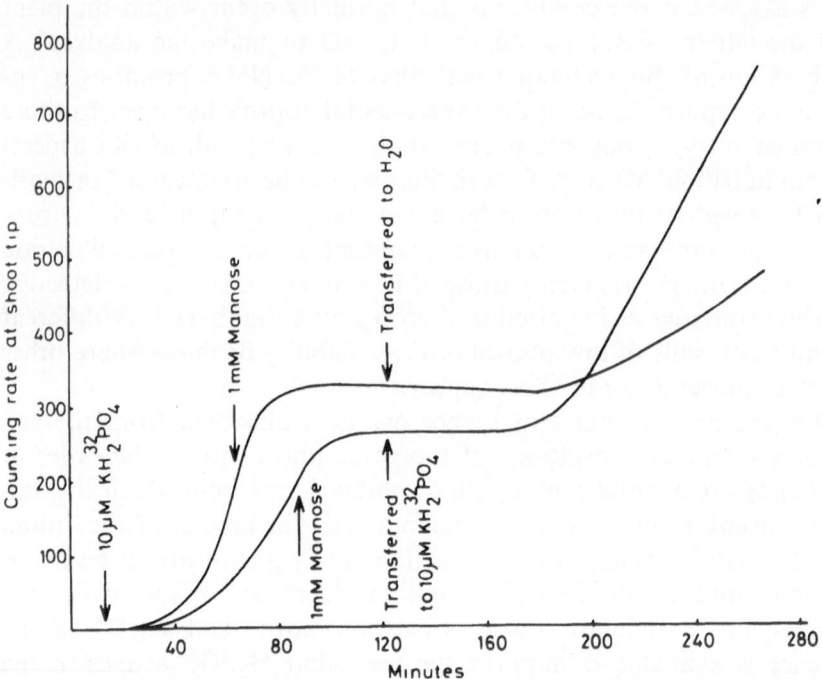

Fig. 1. The effect of 1 mM mannose on the transport of phosphate to the shoots of 10-day-old maize plants growing in culture solution containing 1 μM $KH_2{}^{32}PO_4$. Single plants are attached to a G.M tube. The detection area is 0.5 cm^2 at a distance of 0.5 cm from the top of the 1st leaf.

followed by the use of radioisotopes in short term experiments it has been difficult to examine specific metabolic problems without sacrificing the plant. Some degree of *in vivo* assessment has been possible by external monitoring of labelled whole plants in assessing internal metabolic changes. As an example, the selective effect of mannose between the absorption and transport of phosphate is well documented[12]. Figure 1 shows the transport of P_i to the leaves from roots in 1 μM solution by a 10 day old maize plant and the effect of 1 mM mannose added to the root environment on this transport. Cessation of transport occurs within minutes of addition of mannose and restarts after a time on removal of mannose. The incoming phosphate is sequestered as mannose-6-phosphate in the roots thus preventing onward transport of P_i by the normal mechanism and extraction of the phosphorylated forms shows this clear difference in the root metabolism[14]. After transfer to mannose-free solution the metabolic pattern reverts to that of the pre-mannose phase and as a result normal transport resumes. The key factor in these effects is the low activity of phosphomannoisomerase (PMI) in the roots of *Zea mays* and other cereals whereas species with high enzyme levels such as *Phaseolus aureus* are not affected by addition of mannose to the root environment. In sensitive species and within such species the rate of reversal of the mannose effect is variable and can be used as a measure of *in vivo* activity of the enzyme[14]. The important point is that an external monitoring of the transport rate can provide an assessment of the *in vivo* activity of the key enzyme and this can rarely be achieved by other experimental approaches. Some evidence for wide variations of PMI activity within species has been obtained with varieties of maize. If similar techniques were available for assay of other enzymes possibly involved in ion transport the assessment of their relevance would be made easier. It is of particular interest that the closely related enzyme, phosphoglucoisomerase is being extensively studied within a number of species because of its widespread use as a genetic marker in plant breeding programmes[7].

The availability of high resolution ^{31}P-NMR spectrometers has recently focussed attention on *in vivo* assay of metabolic sequences and this method promises to revolutionise our approach to problems of entry and efflux of ions, intracellular movement between organelles and the sequential transformation of metabolites. Among other factors affecting absorption and transport of phosphate are those related to the availability of other elements particularly those concerned with maintenance of membrane integrity such as calcium, boron and zinc. Deficiency of boron has been shown to reduce entry of phosphate in a wide range of species without affecting efflux rates whereas zinc deficiency increases efflux without affecting influx[17]. As boron deficiency develops, the ATPase activity of the isolated plasmalemma fraction falls in parallel with the capacity for ion transport and both can be restored very rapidly by addition of low concentrations of boron. The experimental evidence suggests that boron is essential for the maintenance of the conformation of the ATPase in the membrane rather than participating directly in the ATPase action[20]. Such experiments illustrate that the capacity for phosphate absorption is one of the first properties to be reduced when boron and zinc are in short supply but it would be so much easier to interpret these experiments if an *in vivo* assay of the ATPase was possible. The NMR technique of saturation transfer is currently being used to tackle this type of problem, particularly with respect to the assay of mitochondrial phosphorylation in maize roots[21].

Root systems

A typical *in vivo* ^{31}P-NMR spectrum of actively metabolising plant tissue, *e.g.* whole root systems or root tips of young maize seedlings shows, in addition to cytoplasmic hexose phosphates and P_i, vacuolar P_i and cytoplasmic nucleoside phosphates and nucleoside diphosphosugars, (Fig. 2.). The hexose phosphates in the cytoplasm reflect the

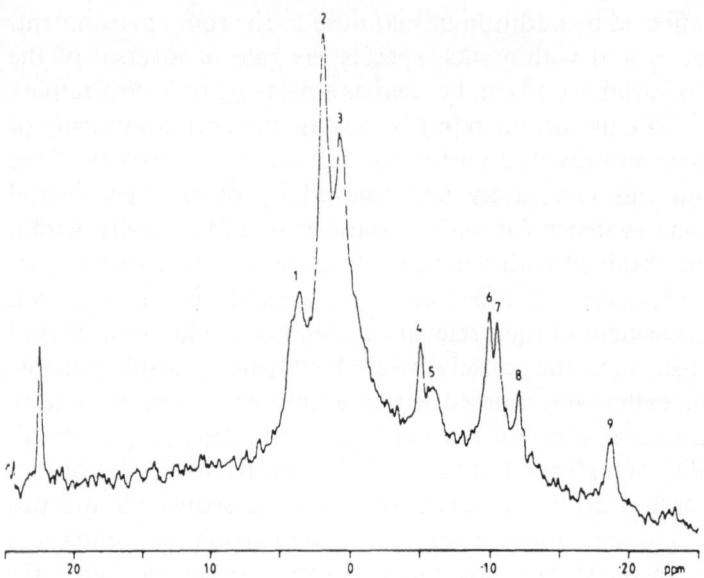

Fig. 2. 121.49 MHz ^{31}P spectrum of 160, 4.0 mm root tips from 54-hour-old maize seedlings (*Zea mays*, L. var LG11). This 16 000 scan spectrum was obtained with a fast acquisition cycle and a selective preamplifier, with the sample in a 10 mm tube. Peak assignments 1. Cytoplasmic hexose phosphate. 2. Cyt. P_i. 3. Vac. P_i. 4, 6 and 9. γ, α and β phosphates of nucleoside triphosphates. 7 and 8. Nucleoside diphosphosugars plus NAD in 7.

glycolytic activity of the tissue and considerable changes are observed as cells become deficient in essential elements such as boron. The nucleotides reflect the oxidative capacity of the cells and the levels respond very rapidly to the presence of uncouplers and ionophores that affect proton pumps. The nucleoside diphosphosugars are characteristic of tissue with a high capacity for polysaccharide synthesis and recent NMR evidence of the type shown in Figure 2 suggests that the *in vivo* levels of these compounds are much higher than those calculated from analysis after acid extraction of the root material and are similar to the level of ATP. ^{31}P-NMR experiments have shown that the concentration of inorganic phosphate in the cytoplasm of plant cells is very closely controlled[10].- Comparison of cells of starved plants with those of normal and high phosphate status indicates that the cytoplasmic level remains constant and that excess phosphate is normally deposited in the vacuole from which it can be readily transferred if the cytoplasm content is reduced by metabolic requirements. It is also clear that the greater part of the inorganic phosphate of mature cells is found in the vacuole. Even if 95% of the P_i is vacuolar, when the cytoplasm represents only 5% of the cell then the actual concentrations may be very similar. This is an important point when invoking the level of cytoplasmic P_i in the control of metabolic processes in organelles such as the control of starch synthesis by the P_i/triosephosphate translocator of the chloroplast membrane[5]. The use of mannose to sequester P_i as mannose-6-phosphate in an attempt to influence these processes by altering the P_i/triose P ratio can be questioned if, in fact, the P_i level does not actually vary, as is the case in root tips. Improved NMR methods might, in the future distinguish phosphorus compounds in mitochondria, chloroplasts and amyloplasts in addition to cytoplasm and vacuole and enable a new degree of awareness of the complex interactions involved during photosynthesis, respiration and photorespiration.

Storage tuber tissues

Another aspect of phosphate metabolism amenable to examination by ^{31}P-NMR is that of the mobilisation of nutrients during the release of buds of storage tubers from the dormant state. The technique has demonstrated that the phosphate content and distribution of its metabolites is very different in specialized tissues within the potato tuber[8]. Virtually all the soluble phytic acid in the potato tuber is confined to the cells of the pith. As a result of this localisation of the ester in a very small volume of the tuber the cellular concentration reaches more than 20 mM and the chemical shift indicates that it is confined to the vacuoles at pH 5.5 $\pm$ 0.1. The narrow line widths indicate that the level of paramagnetic ions in the vacuole must be very low and that the vacuoles observed must all have a pH close to 5.5. Some P_i is present and all the phytic acid resonances are lost when extracts are treated with phytase leaving a single large resonance attributable to P_i.

This major fraction of the total phytic acid existing in a freely soluble state in the vacuole has particular significance for the nutrition of the developing tuber. The pith arms connect the central pith to the developing buds and spectra obtained with pith samples during the course of plant development indicate that phytic acid is being broken down in the pith even after many weeks. The pith tissue of potato probably provides the highest concentration of soluble phosphorylated compound in any plant cell and focusses attention on the role of this tissue in redistribution of phosphate reserves after the breaking of dormancy. Although the vacuolar concentration of phytic acid in pith cells is very high compared to parenchyma cells which contain only low concentrations of P_i in the vacuole the cells absorb phosphate at similar rates in some varieties[8]. This surprising observation confirms the clear-cut separation of vacuolar and cytoplasmic components and shows that the vacuolar content of esterified phosphate has little or no influence on the rate of entry of P_i across the plasmalemma.

The problems encountered in comparing metabolic activities within species are exemplified in experiments with potato tuber slices, tissue which has been extensively used by Laties[9] and others for ion transport studies. The distribution of P between cytoplasm and vacuole differs enormously in pith and parenchyma cells. One aspect of this difference is that the development of the low concentration absorption site occurs at very different rates when fresh slices are incubated in aerated water for 24 hours (Fig. 3). An unusual discontinuity is often observed but the development of the capacity in pith proceeds at twice the rate of that in the parenchyma and the final capacity in the pith tissue is over twice that of the parenchyma. The discontinuity occurring between 6 and 10 hours (Fig. 3) possibly reflects a change in the transport capacity of the plasmalemma or tonoplast.

Thus differences occur in the pattern and rate of carrier site development, and in the final capacity achieved in varieties selected from those raised under identical conditions on the same soil. In general when comparing varieties of crop plants the whole plant or at least the root system is used and it is possible that the major differences in particular

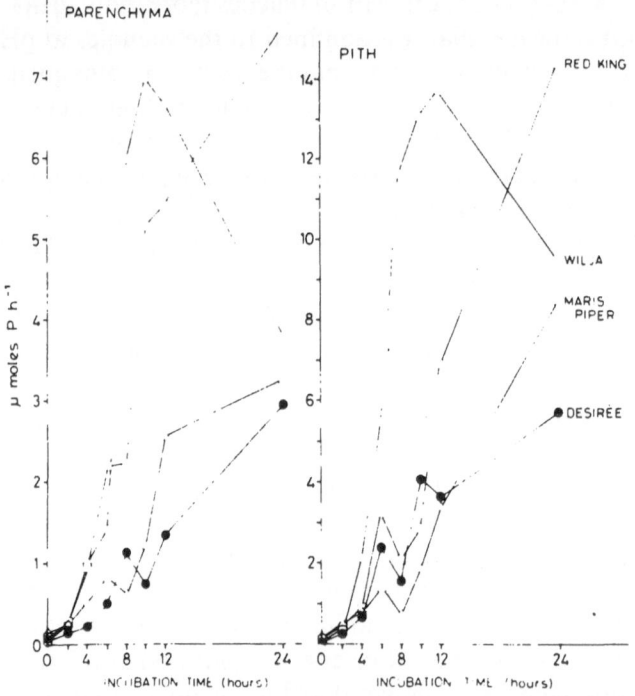

Fig. 3. The development of the capacity to absorb phosphate from $10\,\mu M$ KH_2PO_4 during incubation of freshly cut 1 mm slices of pith and parenchyma tissues from potato tubers of four varieties at 25 C. Absorption for 30 minutes at 25 C at pH 5.5.

capacities of individual cell types will be obscured by the overall absorption. Nevertheless the metabolic basis of the overall differences could be located in a particular cell type and the identification of such locations should help in the elucidation of the mechanisms concerned in the genetic differences observed.

Further differences can be observed if the responses of the tissues to metabolic inhibitors are examined. The potato varieties mentioned earlier clearly differ in their ability to absorb phosphate after incubation of the freshly cut slices in water. After the maximal capacity has been reached the slices rapidly metabolize incoming phosphate and this process is, as expected, inhibited by 2,4-dinitrophenol (DNP). However, at 0.1 mM DNP, a concentration that completely prevents phosphorylation in Wilja, only 60% inhibition occurs in Desirée and the young root systems of the potato varieties also show differences in sensitivity to inhibition. Young root systems of barley (6 days) are 50% uncoupled by 0.1 mM DNP whereas at 18 days 99% inhibition of phosphorylation occurs[13]. Again, the sensitivity to glycolytic inhibitors such as iodoacetate decreases with age and it is clear that responses to inhibitors must be examined under very clearly defined conditions before useful conclusions can be drawn about their relevance. Within the same tuber, pith and parenchyma tissues differ in their sensitivity to inhibitors as well as in the ability to develop an absorption site or carrier mechanism and this may well be true of different cell types between the epidermis and xylem parenchyma of the root, a fact which complicates the interpretation of metabolic studies associated with ion transport.

Bowling[3] has pointed out that a significant rise in vacuolar pH can be detected across the roots of sunflower and if the vacuolar pH is reflected in that of the cytoplasm one would expect differences in metabolic responses. The effect of mannose on phosphate metabolism and transport is reduced by rise in pH indicating that the key enzyme, PMI, known to have a relatively high pH optimum becomes more active as the pH rises. It appears therefore that external changes in pH might be reflected in the pH of the cytoplasm. Alternatively the absorption of mannose itself could be reduced as the external pH increased but whatever the cause it is important that such effects be recognised.

Too little attention has been paid to the biochemical events that mediate transport processes and a major effort must be made to identify carrier systems concerned in the absorption of nutrients by plant cells before the differences in the absorption capacity of individual genotypes can be explained at the molecular level. An examination of four varieties of maize selected for similar phosphate content of the seeds shows that absorption rates of 10 day old seedlings differ by a factor of three

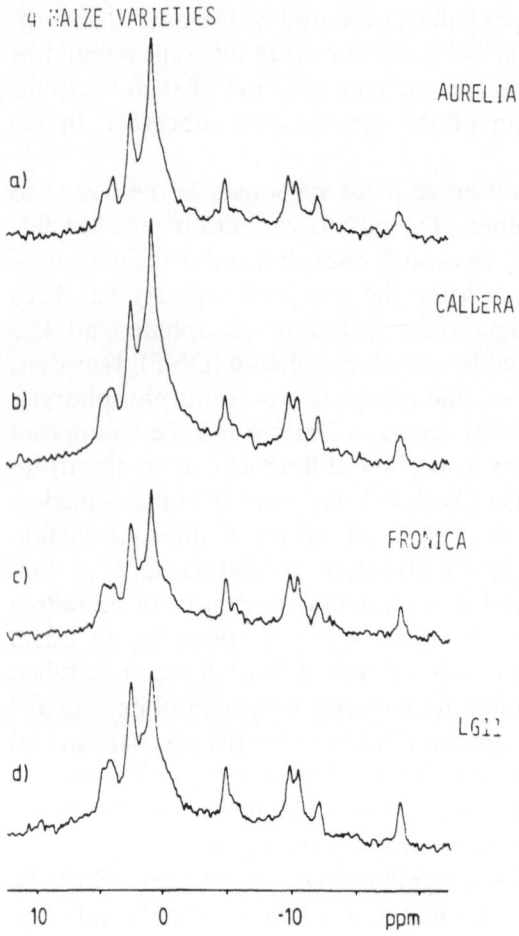

Fig. 4. The 121.49 MHz ^{31}P spectrum of 160, 4.0 mm root tips from 56-hour-old maize seedlings of four varieties. Assignments as in Fig. 2. Preparation of material and acquisition conditions were similar to Fig. 2.

whereas transport rates to the shoot are much less affected. The levels of phosphomannoisomerase as judged by the rate of reversal of transport after transfer to mannose free solution also differs considerably[18]. All four varieties are extremely sensitive to dinitrophenol being completely uncoupled at 20 µM. NMR studies of root tips of these varieties 48 hours after germination show that the levels of ATP and UDPG are identical but that major differences are seen in the distribution of inorganic phosphate between the cytoplasm and vacuole (Fig. 4). Such differences may have relevance to the absorption and transport capacities of the tissues and emphasize the role of *in vivo* assessment in elucidating the factors that underlie differences in ion utilization.

Conclusion

The current trend towards *in vivo* assessment of metabolic sequences should be of great value to plant nutritionists in their search for methods by which the genetic basis for differences in ion transport can be examined. Very significant differences are apparent in the capacities of cultivars to absorb ions from low concentrations or in their ability to exclude potentially toxic ions. Some progress is being made in elucidating the physiological and biochemical roles played by nutrients such as boron and zinc but we are a long way from establishing the primary site of action, if any, of such essential micronutrients.

Recent interest in calcium binding proteins such as calmodulin, known to be present in relatively high concentrations in leaves has caused a complete reassessment of the role played by calcium in plants because of the realisation that cytoplasmic levels of the free ion are likely to be around $0.1 \mu M$ with the greater portion of the total cell calcium bound or localised in the vacuole. Many aspects of plant physiology are open to reinterpretation in the light of such findings and it is likely that micronutrients such as manganese are present at cytoplasmic levels much lower than previously thought. The fact that manganese ions added to the bathing solution reduce the P_i resonance in the vacuoles of cells of maize root tips without affecting that of the cytoplasm implies that the Mn^{++} ions either move extremely rapidly into the vacuole or that binding proteins in the cytoplasm maintain the cytoplasmic Mn^{++} level at such a low level that the P_i resonance shows no broadening effect characteristic of the presence of paramagnetic ions. It is therefore likely that the level of free Mn^{++} ions in the cytoplasm is less than $10 \mu M$. Such data illustrate another important use of NMR of particular relevance to plant nutritionists, *i.e.* that it provides a new experimental approach to the direct measurement of absorption and transport rates of paramagnetic ions in tissues and whole plants. The possibilities of using other nuclei such as ^{15}N and ^{39}K open up exciting challenges and already progress is being made in the field of nitrate utilization by whole plants[1]. It is likely that the next few years will see considerable progress in our knowledge of the biochemical basis of mechanisms controlling the transport of ions and *in vivo* techniques will surely play a major part in these advances.

Acknowledgement This research was supported by the Cecil Pilkington Charitable Trust.

References

1 Belton P S, Lee R B and Ratcliffe R G 1985 A ^{14}N nuclear magnetic resonance study of inorganic nitrogen metabolism in barley, maize and pea roots. J. Exp. Bot. 36, 190–210.

2 Bhat K K S and Nye P H Diffusion of phosphate to plant roots in soil. Plant and Soil 38, 161–175.
3 Bowling D J F 1973 A pH gradient across the root. J. Exp. Bot. 24, 1041–1045.
4 Brindle K M, Brown F F, Campbell I D, Foxall D L and Simpson R J 1982 A ^{1}H n.m.r. study of isotope exchange catalysed by glycolytic enzymes in the human erythrocyte. Biochem. J. 202, 589–602.
5 Herold A, Lewis D H and Walker D H 1976 Sequestration of cytoplasmic orthophosphate by mannose and its differential effect on photosynthetic starch synthesis in C_3 and C_4 species. New Phytol. 76, 397–407.
6 Hutber G N, Lord E I and Loughman B C 1978 The metabolic fate of phenoxyacetic acids in higher plants. J. Exp. Bot. 29, 619–629.
7 Jones T W A 1985 The kinetics and thermal stability of phosphogluco-isomerase isozymes of ryegrasses. Physiol. Plant. 63, 365–369.
8 Kime M J, Loughman B C, Ratcliffe R G and Williams R J P 1982 The application of ^{31}P nuclear magnetic resonance to higher plant tissues. J. Exp. Bot. 33, 656–669.
9 Laties G G 1967 Metabolic and physiological development in plant tissues. Aust. J. Science 30, 193–203.
10 Lee R B and Ratcliffe R G 1983 Phosphorus nutrition and the intracellular distribution of inorganic phosphate in pea root tips. A quantitative study using ^{31}P-NMR. J. Exp. Bot. 34, 1222–1244.
11 Loughman B C 1960 Uptake and utilization of phosphate associated with respiration changes in potato tuber slices. Plant Physiol. 35, 418–424.
12 Loughman B C 1966 The mechanism of absorption and utilization of phosphate by barley plants in relation to subsequent transport to the shoot. New Phytol. 65, 388–397.
13 Loughman B C 1974 The metabolism of phosphate in young roots. In Structure and function of primary root tissues. Ed. J Kolek, Veda, Bratislava. pp 439–446.
14 Loughman B C 1978 Metabolic factors and the utilization of phosphorus by plants. In Phosphorus in the Environment: its Chemistry and Biochemistry. Eds. R Porter and D W Fitsimons. Elsevier, North Holland. pp 155–174.
15 Loughman B C 1981 Metabolic aspects of the transport of ions by cells and tissues of roots. In Structure and Function of Plant Roots. Eds. R Brouwer et al. Martinus Nijhoff, The Hague. pp 237–245.
16 Loughman B C and Russell R S 1957 The absorption and utilization of phosphate by young barley plants. J. Exp. Bot. 8, 280–293.
17 Loughman B C, Webb M J and Loneragan J F 1982 Zinc and the utilization of phosphate in wheat plants. In Plant Nutrition. Ed. A Scaife. CAB. pp 335–340.
18 Loughman B C, Roberts S C and Goodwin-Bailey C I 1983 Varietal differences in physiological and biochemical responses to changes in the ionic environment. Plant and Soil 72, 245–259.
19 Loughman B C and Ratcliffe R G 1984 Nuclear magnetic resonance and the study of plants. In Advances in Plant Nutrition Vol. 1. Eds. P B Tinker and A Lauchli. Praeger, New York. pp 241–283.
20 Parr A J, and Loughman B C 1983 Boron and membrane transport in plants. In Metals and Micronutrients. Ed D A Robb and W S Pierpoint. pp 87–107. New York: Academic Press.
21 Roberts J K M, Wemmer D and Jardetzky O 1984 Measurement of mitochondrial ATPase activity in maize root tips by saturation transfer ^{31}P nuclear magnetic resonance. Plant Physiol. 74, 632–639.

Genotypic variation in biomass production and nitrogen use efficiency in pearl millet [*Pennisetum americanum* (L.) Leeke]*

G. ALAGARSWAMY and F.R. BIDINGER
International Crops Research Institute for the Semi-arid Tropics (ICRISAT), Patancheru, A.P. 502 324, India

Key words   Field study   Genotypic variation   Nitrogen uptake   Nitrogen use efficiency   *Pennisetum americanum*

Summary   Twenty diverse pearl millet genotypes ranging from landraces to high yielding hybrids were studied for genotypic variation in nitrogen (N) use efficiency in high (100 kg N/ha) and low fertility (20 kg N/ha) over two years in the field.
   The combined data over years and fertility levels indicated that despite taking up similar amounts of N, genotypes differed significantly in biomass production and thus in N use efficiency. A West African genotype, Souna B, had N use efficiency values 32% higher than the less efficient Indian genotype BJ 104 even though both genotypes had similar N uptake. An increase in N fertility decreased N use efficiency since the percentage increase in biomass was smaller than the percentage increase in N uptake.

Introduction

Nitrogen (N) fertilizers are required to maximize cereal crop food production in some environments and global production and consumption of fertilizer N are expanding. In spite of an increase in the use of N fertilizers, the poor efficiency with which many crop plants recover fertilizer N from the soil has been generally recognized[2]. Low recovery of fertilizer N from the soil and growing concern over the nitrate pollution of ground waters, have increased the need to improve the plant's efficiency to recover and use the fertilizer N.
   Breeding plants for less favourable environments emphasizes developing plants to fit their environment rather than modifying the soil through fertilizer input. For plant breeding practices to be successful in alleviating soil stress problems in crop production, genetic variability must exist within the plant species. Many aspects of mineral nutrition are under genetic control[8]; the efficiency of N utilization in tomatoes has particularly been reported to be controlled by a few major dominant genes[15]. Genotypic differences in N uptake and/or utilization have been recognized in maize[3,6,12], sorghum[10], pearl millet[1] and wheat[7]. However, detailed information on genetic differences in N utilization has not been reported for pearl millet. Such information may be important since this crop is generally grown in infertile soils with little or no fertilizer input.

* Approved as Journal Article No. 555 by the International Crops Research Institute for the Semi-Arid Tropics (ICRISAT).

H.W. Gabelman and B.C. Loughman (Eds.), Genetic aspects of plant mineral nutrition.
ISBN-13: 978-94-010-8102-3
© *1987 Martinus Nijhoff Publishers, Dordrecht/Boston/Lancaster.*

Identification of intraspecific variation in N use efficiency is also necessary to investigate mechanisms controlling such variations in pearl millet. The specific objectives of this field investigation were to:

(i) identify and document within-species variation for the ability of pearl millet (*Pennisetum americanum* (L.) Leeke] genotypes to absorb and utilize N, and

(ii) study the effect of levels of N applied on N utilization.

Materials and methods

From a large diverse genotype set that was evaluated for performance under 20 and 100 kg N/ha, the top and bottom ten genotypes were selected for detailed investigation reported here. These 20 genotypes representing a range of materials including improved landraces, inbred lines, varieties and high yielding F_1 hybrids (Table 1) were grown on Alfisol at the International Crops Research Institute for the Semi-Arid Tropics (ICRISAT), Patancheru, India during 1978 and 1980 rainy season at high (100 kg N + 25 kg P/ha) and low (20 kg N + 9 kg P/ha) fertility, using di-ammonium phosphate and Urea. Available soil N was measured only during 1980. Nearly 50 kg N/ha of nitrate N accumulated in the profile (120 cm depth) at the beginning of the season. The experimental site has been designated as high and low fertility environments based on pearl millet productivity and N uptake by the plants (the maximum values: for biomass 1030 and 675 g/m^2; for N uptake 12.4 and 5.8 g/m^2 for HF and LF respectively). The rainfall during the cropping season (June-September) was 1007 in 1978 and 727 mm in 1980. The experiment was conducted as a split plot design with fertility as main plots and genotypes as subplots. Plots were 4 m (1978) or 9 m (1980) long and consisted of four rows at 75 cm spacing. Excess seeds were planted and thinned to a final stand of 13.3 plants/m^2.

Table 1. List of genotypes, genetic constitution, their origin and days to flowering

Genotype	Genetic constitution	Country of origin	Days to flowering HF	LF
Souna B	Open pollinated variety	Mali	60	62
P3 Kolo	Open pollinated variety	Niger	57	57
3/4 Hainei Khirei	Open pollinated variety	Niger	54	56
IP 2757*	Germplasm accession	Niger	48	49
Serere 17	Breeding line	Uganda	52	52
Serere Comp. 1(M)	Breeding composite	Uganda	49	50
Serere Comp. 2(M)	Breeding composite	Uganda	46	47
Super Serere Comp.	Breeding composite	Uganda	49	49
ICI 266	Inbred line	India	48	49
700112	Breeding line	Nigeria	52	53
700651	Breeding line	Nigeria	52	54
700440	Breeding line	Nigeria	53	56
700331	Breeding line	Nigeria	56	57
700772	Breeding line	Nigeria	50	50
700250	Breeding line	Nigeria	57	56
700441	Breeding line	Nigeria	51	51
700471	Breeding line	Nigeria	53	53
ICMS 7703	Open pollinated variety	India	47	49
BK 560	F_1 Hybrid	India	44	45
BJ 104	F_1 Hybrid	India	42	43

* IP = International *Pennisetum*

Table 2. Intraspecific variation in biomass, N uptake and N use efficiency in 20 pearl millet genotypes.*

Genotypes	Biomass (g/m²)	N uptake (g/m²)	N use efficiency (g/g)
Souna B	813 ab*	7.5 a c	123 a
P3 Kolo	740 a e	7.8 a c	106 b d
3/4 Hainei Khirei	731 a f	7.1 a c	112 bc
IP 2757	717 b g	7.4 a c	109 bc
Serere 17	751 a d	7.1 a c	105 b d
Serere Comp. 1(M)	769 a c	8.2 ab	105 b-d
Serere Comp. 2(M)	717 b g	7.9 a c	101 c e
Super Serere Comp.	714 b-g	7.1 a c	113 bc
ICI 266	612 h	6.8 a–c	102 c e
700112	727 a–f	7.0 a–c	114 ab
700651	711 c-g	7.6 a–c	105 b e
700440	696 c h	7.2 a c	109 bc
700331	761 a d	7.9 a c	111 bc
700772	673 c-h	7.7 a–c	102 c e
700250	663 d h	7.1 a c	101 c e
700441	632 f h	7.1 a c	102 c e
700471	598 h	6.5 c	103 b e
ICMS 7703	818 a	8.4 a	108 bc
BK 560	645 e h	7.7 a c	95 de
BJ 104	620 gh	7.7 a-c	93 e

* Data represent means over two years and two fertility levels.
* Any means within a column followed by the same letter are not significantly different at $P = 0.05$, according to Duncan's Multiple Range Test.

At physiological maturity (black layer at the hylar region of the seed) an 80 cm length of the two internal rows of the plot was cut 1 cm above the soil (16 plants). These samples were oven dried at 80 °C for 72 h, weighed and ground to pass through a 0.5 mm screen. Organic nitrogen was determined from 0.5 g acid (2 ml sulphuric acid + hydrogen peroxide) digested plant material using a colorimetric method in a Technicon Auto Analyser. From the above ground biomass and N concentration values N uptake and N use efficiency (above ground biomass production per unit of N in the plant) values were calculated. The term N use efficiency used here is equivalent to the inverse of N concentration.

Results and discussion

Genotypic variation in N use efficiency

Differences in N uptake between two fertility levels were substantial (genotypic mean N uptake: 10.5 and 4.5 g/m² for HF and LF respectively). The fertility X genotype and the year X fertility X genotype interactions were not significant. Hence for the sake of simplicity, the combined data for two fertility rates over two years were used to document genotypic variations for N uptake and N use efficiency. The genotype means were ranked for biomass, N uptake and N use efficiency and compared (Table 2). Genotypes differed widely in N use efficiency.

Table 3. Effects of high fertility (HF) and low fertility (LF) on N use efficiency of 20 pearl millet genotypes

Genotypes	Nitrogen use efficiency (g/g)[a]	
	HF	LF
Souna B	105.5	140.4
P3 Kolo	91.8	120.1
3/4 Hainei Khirei	97.2	126.9
IP 2757	86.6	131.6
Serere 17	91.5	118.6
Serere Comp. 1(M)	92.0	118.7
Serere Comp. 2(M)	85.0	116.2
Super Serere Comp.	93.3	131.6
ICI 266	81.2	122.1
700112	97.4	131.4
700651	93.2	116.5
700440	91.3	126.3
700331	94.0	128.1
700772	79.3	124.3
700250	89.5	113.3
700441	85.2	119.6
700471	87.1	118.5
ICMS 7703	90.1	125.7
BK 560	97.2	113.4
BJ 104	75.3	111.4
Mean	89.2	122.7

[a] Data represent mean over two years.

Except for genotype 700471, genotypes did not differ in N uptake. However, consistent and significant ($P < 0.001$) differences in N use efficiency between genotypes were measured. The N use efficiency values were highest in West African improved landrace Souna B and lowest in Indian hybrid BJ 104. Differences in the ability of genotypes to utilize N were not related to N uptake. Souna B had 30% higher N use efficiency than BJ 104, even though both genotypes had a similar N uptake values. The differences in N use efficiency between these genotypes were confirmed in nutrient culture solution experiments (Alagarswamy *et al.* unpublished data).

Genotypic differences for N use efficiency at similar N uptake levels are important for better utilization of fertilizer N. The improvement in plant N use efficiency combined with better root distribution and improved fertilizer management practices are likely to improve crop productivity and economy of fertilizer application. Such economic benefits have been anticipated[13]. In addition benefits could be achieved by reducing environmental pollution by nitrate in the ground water.

Significant correlations existed between N use efficiency and days to flowering among all the genotypes (r = 0.45 $P < 0.001$), indicating that more efficient genotypes tended to be later maturing. However, biomass production was not related to days to flowering.

Effects of N level on N use efficiency

The effect of applied N level on N use efficiency for the 20 genotypes is presented in Table 3. Increase in the amount of applied N decreased N use efficiency in all genotypes. The N use efficiency values decreased by 25% in an efficient genotype Souna B compared to a 32% reduction in an inefficient genotype BJ 104 with an increase in N supply. These results are in contrast to those reported for sorghum, where both efficient and inefficient genotypes showed similar decrease in N use efficiency as N was increased in nutrient solution[17]. The decline in N use efficiency at HF was due to differential effects of N supply on N uptake and biomass production. The overall mean N uptake for the genotypes increased by 130% while the biomass increased by only 60% in response to increased N level. N use efficiency was thus inversely related to N uptake ($r = -0.74$, $P < 0.001$). Increased N supply is generally known to reduce N use efficiency in maize[12,14], wheat, triticale and rye[9,16], and in sorghum[11,17].

Plant species differ in N use efficiency. Plants with C_4 metabolism utilize N more efficiently than C_3 species[5]. C_4 plants have been found to have higher CO_2 exchange rates per unit tissue N and a steeper linear increase in CO exchange rate with increased leaf N than C_3 plants[4]. However, the mechanisms for intraspecific variations in N use efficiency are not yet known. Hence documenting the existence of intraspecific variation in N use efficiency is a necessary preliminary prior to investigate the mechanisms for variation in N use efficiency among pearl millet genotypes.

Acknowledgements We wish to thank Drs. R B Clark, J W Maranville and N. Seetharama for reviewing the manuscript. The technical assistance of Mr. D S Raju is gratefully acknowledged.

References

1 Alagarswamy G and Bidinger F R 1982 Nitrogen uptake and utilization by pearl millet [*Pennisetum americanum* (L.) Leeke]. *In* Proc. of Ninth International Plant Nutrition Colloquium, Ed. A Scaife, pp 12–16, Warwick, England.
2 Allison F E 1966 The fate of nitrogen applied to soils. Adv. Agron. 18, 219–258.
3 Beauchamp E G, Kannenberg L W and Hunter R B 1976 Nitrogen accumulation and translocation in corn genotypes following silking. Agron. J. 68, 418–422.
4 Bolten J K and Brown R H 1980 Photosynthesis of grass species differing in carbon dioxide fixation pathways V. Response of *Panicum maximum*, *Panicum milioides* and tall fescue (*Festuca arundinacea*) to nitrogen nutrition. Plant Physiol. 66, 97–100.
5 Brown R H 1978 A difference in N use efficiency in C_3 and C_4 plants and its implications in adaptation and evolution. Crop Sci. 18, 93–98.

6. Chevalier P and Schrader L E 1977 Genotypic differences in nitrate absorption and partitioning of N among plant parts in maize. Crop Sci. 17, 897–901.
7. Cox M C, Qualset C O and Rains D W 1985 Genetic variation for nitrogen assimilation and translocation in wheat. 1. Dry matter and nitrogen accumulation. Crop Sci. 25, 430–435.
8. Epstein E 1972 Mineral Nutrition of Plants: Principles and Perspectives. John Wiley and Sons. Inc. NY, 412 p.
9. Gashaw L and Mugwira L M 1981 Response of triticale, wheat and rye to N in a greenhouse. Commun. Soil Sci. Plant Anal. 12, 289–297.
10. Maranville J W, Clark R B and Ross W M 1980 Nitrogen efficiency in grain sorghum. J. Plant Nutr. 2, 577–589.
11. Mirhadi M J, Yoshida S and Kobayashi Y 1979 Studies of the productivity of grain sorghum. 1. Nitrogen nutrition of grain sorghum. Japan J. Crop Sci. 48, 483–489.
12. Moll R H, Kamprath E J and Jackson W A 1982 Analysis and interpretation of factors which contribute to efficiency of nitrogen utilization. Agron. J. 74, 562–564.
13. Mudahar M and Highnett T 1982 Energy and fertilizer: Policy implications and options for developing countries. IFDC Publication No. T20, Muscle Shoals, Alabama.
14. Rhoads F M and Stanley Jr R L 1984 Yield and nutrient utilization efficiency of irrigated corn. Agron. J. 76, 219–223.
15. O'Sullivan J, Gabelman W H and Gerloff G C 1974 Variations in efficiency of nitrogen utilization in tomatoes (*Lycopersicon esculentum* Mill.) grown under nitrogen stress. J. Am. Soc. Hort. Sci. 99, 543–547.
16. Terman G L 1979 Yields and protein content of wheat grain as effected by cultivar, N and environmental growth factors. Agron. J. 437–440.
17. Zweifel T R 1982 An evaluation of grain sorghum [*Sorghum bicolor* (L) Moench] hybrids for efficiency of nitrogen use. Ph.D Thesis, University of Nebraska, Lincoln, 111 p.

Differential phosphorus uptake, distribution, and efficiency by sorghum inbred parents and their hybrids

A.M.C. FURLANI,
Instituto Agronomico, C.P. 28, 13 100 Campinas, Sao Paulo, Brazil

R.B. CLARK, W.M. ROSS and J.W. MARANVILLE
U.S. Department of Agriculture, Agricultural Research Service and Department of Agronomy, University of Nebraska, Lincoln, NE 68583, USA

Key words Heritability of phosphorus traits Heterosis Phosphorus nutrition Plant breeding *Sorghum bicolor*

Summary Four sorghum [*Sorghum bicolor* (L.) Moench] female inbred parents, four male inbred parents, and their 16 F_1 hybrids were grown in a greenhouse in nutrient solutions with 75 and 150 μmol P per plant. The objectives were to determine differences among genotypes for growth, P concentrations and contents, P distribution among plant parts, and P efficiency (dry matter produced per unit P in the plant), and to determine inheritance of these P traits.

Differences existed among the parents and hybrids for responses to P, and the largest differences appeared when plants were grown at the low P level. Two male parents (NB9040 and SC120-15) appeared to be more tolerant (more efficient) to low P than the other male parents ('Plainsman' and SC33-9-8-E4). Differences among female parents were smaller than among male parents. Hybrids from NB9040 and SC120-15 had higher dry matter yields, higher upper/lower leaf P ratios, and higher efficiency ratios than hybrids from Plainsman and SC-33-9-8-E4. Hybrids from the female parents 'Wheatland' and 'Redland' had higher dry matter yields and larger root systems than hybrids from the female parents 'Combine Kafir-60' and KS35. Heterosis for total dry matter yield and dry matter produced per unit P was observed in most hybrids.

The differences in P absorption, distribution, and efficiency indicated that these characteristics were genetically controlled. The better growth of the male parents under low P and the transfer of the trait to their hybrids indicated the importance of dominant genes, although additive genes appeared to be involved in the variability of P uptake and efficiency traits. The improvement of sorghum lines and hybrids for growth and performance at low P levels appears feasible.

Introduction

Even though variability among species for absorbing, translocating, and utilizing P has been reported for nearly 50 years[15], only recently has attention been given to differential responses of genotypes within species to low P levels. Most studies have been concerned with identifying plants which grow better at low P and accumulate different concentrations of P[1,2,4,5,7-12,14,15,20,21,24,26], but few studies have been concerned with the inheritance of these traits.

Sorghum [*Sorghum bicolor* (L.) Moench] genotypic responses to P deficiency stress have been observed in plants grown in nutrient solutions and soils; and large differences among genotypes for dry matter yield, P content, P distribution in plant parts, and efficiency ratios (dry matter production per unit P) have been noted[4,10,11,12].

H.W. Gabelman and B.C. Loughman (Eds.), Genetic aspects of plant mineral nutrition.
ISBN-13: 978-94-010-8102-3
© *1987 Martinus Nijhoff Publishers, Dordrecht/Boston/Lancaster.*

Multigenic differences between efficient and inefficient bean (*Phaseolus vulgaris* L.) strains were found for P utilization grown under P deficiency stress[26]. Narrow-sense heritability for P tolerance in beans grown under P deficiency stress was estimated to be about 40%[19]. Under high levels of P, inheritance of P efficiency in beans (measured as dry matter production per unit P) was found to be multigenic in hybrid crosses between efficient and inefficient parents. Dominance and heterosis were also observed in crosses between two efficient parents[22]. Maternal inheritance was not observed in any of these studies. Single-, multi-, and additive-gene inheritance traits were also noted for K and N efficiency in plants[20,21,22,23].

The purpose of this study was to obtain preliminary information on the inheritance of the traits responsible for higher efficiency in the uptake and use of P by sorghum genotypes. The F_1 hybrids and their parents were tested in nutrient solutions under low P conditions.

Materials and methods

Experiment 1

Four female parents ['Combine Kafir-60' (CK60), KS35, 'Wheatland', and 'Redland'], four male parents [NB9040, SC120-15 (IS2816), 'Plainsman', and SC33-9-8-E4 (IS12553)] and their 16 F_1 hybrids were grown separately in 7 L of nutrient solution under greenhouse conditions during the same summer season (June and July). Day lengths were 16 ± 1 h at temperatures of $36 \pm 4°C$. Nighttime temperatures were $24 \pm 3°C$.

Four-day-old uniform-sized seedlings (germinated between moist paper towels) were transferred to a plate holding 126 plants and grown in 0.5-strength nutrient solution for five days as described by Clark[6]. At seven days, the seedlings were suspended in full-strength nutrient solution from styrofoam lids holding six plants per lid. The composition of the full-strength nutrient solutions was (μmol element L^{-1}) 11 400 NO$_3$-N, 3700 Ca, 3620 K, 1390 NH$_4$-N, 970 Cl, 910 S, 780 Mg, 25 B, 9 Mn, 2.3 Zn, 0.8 Mo, 0.6 Cu, and 25 Fe as FeHEDTA [ferric N-hydroxyethylethylenediaminetriacetate]. Phosphorus was added initially at 226 μmol and another 226 μmol was added one week later to give 75 μmol P L^{-1} per plant as KH$_2$PO$_4$. Excessive concentrations of P added at one time can cause severe 'red-speckling' to appear on the lower leaves of many sorghum genotypes[13]. Since plants constantly change nutrient solution pH[3], the pH of solutions was monitored every other day but not changed. The initial solution pH was 5.2, followed by a decrease to 4.0 ± 0.2 in each container within 3 to 4 days. The solutions remained at this pH value throughout the experiment which was normal for plants grown in nutrient solutions[3]. The 24 genotypes with five replications of six plants per pot were arranged in randomized complete blocks. The experiment was terminated when plants were 26 days of age (19 days in treatment solution) and in the 8- to 9-leaf stage.

Experiment 2

This experiment used the same experimental design, genotypes, and procedures described for Experiment 1, except that nutrient solutions were changed 10 days after the seedlings were put in the treatment solutions. Phosphorus levels were increased 2-fold and added in two increments of 450 μmol each week to give 150 μmol per plant as KH$_2$PO$_4$. The experiment was terminated when plants were 28 days of age (19 days in treatment solutions) and in the 10- to 11-leaf growth stage.

When both experiments were terminated, roots were rinsed in distilled water; and plants were separated into lower leaves (leaves 1 to 5), upper leaves (leaf 6 and above), and roots in Experiment 1, into lower leaves (leaves 1 to 5), middle leaves (leaves 6 and 7), upper leaves (leaf 8 and above), and roots in Experiment 2. Plant samples were dried at 70 C in a forced-air oven, weighed, and

Table 1. Mean squares and coefficients of variation (CV) for dry matter yield, P content, upper/lower leaf P, and dry matter produced per unit P (DM/P) in sorghum hybrids and parents grown in nutrient solution with 75 µmol P L^{-1}

Source of variation	Degrees of freedom	Dry matter yield			P content					Upper/lower leaf ratio	DM/P
		Leaves	Roots	Total	Upper leaves	Lower leaves	Roots	Total			
		g 6 plants^{-1}			mg 6 plants^{-1}						mg DM mg^{-1} P (10^{-3})
Blocks	4	2.70*	0.51*	4.92*	6.37*	0.13	1.52	5.47		119.78*	0.094*
Genotypes	23	2.26*	3.36*	10.15*	3.53*	0.35*	1.34*	7.11*		119.83*	0.181*
Hybrids (H)	15	1.51*	1.42*	5.12*	1.91	0.21*	1.41*	5.42		120.40*	0.125
Males (M)	3	5.59*	3.67*	17.61*	2.38*	0.48*	3.26*	12.47*		300.88*	0.38
Females (F)	3	0.63*	1.73*	2.76*	3.14	0.02	0.68	1.18		20.73	0.034
F × M	9	0.44*	0.57*	1.75*	1.34	0.19*	1.03	4.49		93.47*	0.067*
Parents (P)	7	1.89*	2.10*	6.74*	4.80*	0.45*	1.38*	6.87		87.23*	0.08
M	3	2.82*	2.76*	10.19*	3.93	0.75*	1.06	12.14*		116.68*	0.152*
F	3	0.84*	2.14*	4.85*	3.43	0.23*	0.99	3.50		62.11	0.034
F vs. M	1	2.23*	0.00	2.07*	11.52*	0.19	3.50*	1.18		74.21*	0.000
H vs. P	1	16.24*	14.34*	109.40*	18.97*	1.70*	0.03	34.00*		339.54*	1.706*
Error	92	0.21	0.16	0.54	1.61	0.05	0.64	3.50		29.56	0.026
Total	119										
CV, %		8.1	8.1	6.9	18.8	35.9	17.8	15.7		40.1	17.3

* Significant at $P = 0.05$.

Table 2. Mean squares and coefficients of variability (CV) for dry matter yield, P content, upper/lower leaf P, and dry matter produced per unit P (DM/P) in sorghum hybrids and parents grown in nutrient solution with 150 μmol P L⁻¹

Source of variation	Degrees of freedom	Dry matter yield			P content				Upper/lower leaf	DM/P	
		Leaves	Roots	Total	Upper leaves	Lower leaves	Roots	Total		ratio	mg DM mg⁻¹ P (10^{-3})
		g 6 plants⁻¹			mg 6 plants⁻¹						
Blocks	4	74.52*	3.55*	108.77*	3.75	5.03	0.70*	6.59*	13.12	28.05	0.326*
Genotypes	23	15.09*	9.64*	42.49*	2.79	6.34*	1.93*	2.44	20.53	84.25*	0.241*
Hybrids (H)	15	7.83*	4.71*	19.28*	2.76	4.54	1.10*	2.28	14.91	90.33*	0.144
Males (M)	3	20.59*	9.05*	48.51*	1.10	7.25	2.30*	3.97	23.06	202.30*	0.218
Females (F)	3	2.26	11.58*	19.38*	7.04	6.63	0.63	2.19	23.09	87.10*	0.318*
F × M	9	5.43	0.98	9.50*	1.89	2.94	0.86*	1.75	9.47	54.09*	0.062
Parents (P)	7	13.46*	5.19*	26.98*	3.15	7.99*	2.28*	1.18	18.09	41.91*	0.146
M	3	27.68*	5.22*	45.50*	5.15	3.12	0.63	1.92	10.30	36.52	0.092
F	3	3.72	5.52*	16.38*	1.99	10.94*	0.42	0.38	22.45	0.58	0.227
F vs. M	1	0.05	4.08*	3.23	0.61	13.73*	12.81*	1.38	28.38	182.09*	0.062
H vs. P	1	135.44*	114.73*	499.48*	0.60	21.88*	11.85*	13.64*	121.92*	289.29*	2.365*
Error	92	3.61	0.89	5.85	3.77	2.85	0.25	2.01	17.73	13.90	0.092
Total	119										
CV. %		11.6	10.4	9.5	20.0	40.1	33.4	22.1	19.3	44.5	24.8

* Significant at $P = 0.05$.

ground to pass a 0.5 mm screen. Samples of plant parts were pressed into 13 mm (100 mg) pellets and analyzed for P using energy-dispersive x-ray fluorescence spectrometry[18].

The data were subjected to analysis of variance, and Duncan's multiple range test was used to separate mean differences[25]. Significance for heterosis was determined by a paired t-test comparison of a hybrid and its parental mean value using the data from each replication.

Results and discussion

Mean square data from line × line crosses such as those in this experiment imply additive gene action when the general combining ability (GCA) effects of females and male are significant and imply nonadditive gene action when the specific combining ability (SCA) effects of females × males are significant[16]. In an ideal diallel analysis, a large random set of females is crossed with a large random set of males to obtain genetic interpretation of data[17]. Because of these restrictions, the interpretations of GCA and SCA effects in Tables 1 and 2 apply only to the germplasm used in this study and should not be extrapolated to

Table 3. Mean leaf, root, and whole plant dry matter yields of sorghum male and female parental lines and their F_1 hybrids grown in nutrient solution with 75 μmol P L^{-1}

Female		Male				
		NB9040	SC120-15	Plainsman	SC33-9-8-E4	Mean
Leaf dry matter (g plant^{-1})						
		(0.99 b f)	(0.84 hi)	(0.70 j)	(0.77 ij)	
CK60	(0.86 hi)*	1.10 ab	1.02 a–e	0.83 hi	0.98 c–f	0.98 B
KS35	(0.84 hi)	1.12 a	1.08 a–d	0.86 g i	0.90 f–h	0.99 B
Wheatland	(0.97 d–f)	1.01 a–e	1.05 a–e	0.85 hi	0.98 d–f	0.97 B
Redlan	(0.96 e g)	1.09 a c	1.06 a e	1.00 b–f	0.98 d f	1.04 A
Mean		1.08 A	1.05 A	0.89 C	0.96 B	
Root dry matter (g plant^{-1})						
		(0.86 d–g)	(0.60 i)	(0.64 i)	(0.61 i)	
CK60	(0.61 i)	0.97 a–c	0.80 gh	0.79 gh	0.83 e g	0.85 B
KS35	(0.62 i)	0.94 b–d	0.90 c–f	0.79 gh	0.73 h	0.84 B
Wheatland	(0.84 e g)	0.99 a–c	1.03 a	0.81 f–h	0.91 b–e	0.94 A
Redlan	(0.64 i)	1.00 ab	0.90 c–f	0.91 b f	0.84 d g	0.91 A
Mean		0.97 A	0.91 B	0.82 C	0.83 C	
Whole plant dry matter (g plant^{-1})						
		(1.86 e–f)	(1.44 jk)	(1.34 k)	(1.39 k)	
CK60	(1.46 h–k)	2.07 ab:	*1.81* ef	*1.62* g–i	*1.82* ef	1.83 B
KS35	(1.45 i–k)	2.06 a c	*1.98* a–e	1.65 fg	*1.63* gh	1.83 B
Wheatland	(1.80 ef)	2.00 a–d	2.08 ab	1.67 fg	*1.89* c–e	1.91 A
Redlan	(1.60 g i)	2.09 a	*1.97* a e	*1.91* b e	*1.85* dc	1.95 A
Mean		2.06 A	1.96 B	1.71 D	1.79 C	

* Means followed by common letters are not significantly different at $P = 0.05$. Parental values are enclosed in parentheses. Small letters refer to differences among parents and/or hybrids. Capital letters refer to mean in each parental group separately.
: Underlined hybrid values indicate significant heterosis for this trait in relation to its parental means.

other genotypes. The combining ability effects of selected low P tolerant and P efficient male parents compared to low P intolerant or P inefficient male parents crossed to similar sets of female parents, however, furnish considerable plant breeding information, even with a low number of parents.

Because error mean squares were not homogeneous, the two experiments were not combined for analysis. For the most part, data from growth at the 75 μmol P L^{-1} produced more significant mean squares of each trait for females, males, and females × males than when grown under the 150 μmol P L^{-1} conditions. The latter level seemed to be too high for good expression of genetic differences for P uptake, distribution, and efficiency that existed in the germplasm. Significance for a trait at the 150 μmol L^{-1} level usually produced significance for the same trait at the 75 μmol P L^{-1}, whereas, the reverse was less likely to be true. For this reason, the discussion of results largely will involve Tables 1, 3, and 4 more than Tables 2, 5, and 6.

Table 4. Mean P content, upper/lower leaf P ratio, and dry matter produced per unit P of sorghum male and female parental lines and their F$_1$ hybrids grown in nutrient solution with 75 μmol P L^{-1}

Female		Male				
		NB9040	SC120-15[‡]	Plainsman	SC33-9-8-E4	Mean
P content (mg plant^{-1})						
		(1.90 b–e)	(2.42 a)	(2.13 a–c)	(1.86 b–e)	
CK60	(2.20 ab)[†]	1.95 b–e	1.92 b–e	2.00 a–e	1.91 b e	1.95 A
KS35	(1.94 b e)	1.64 de	1.98 a–e	2.03 a e	2.10 a–d	1.94 A
Wheatland	(2.15 ab)	1.53 e	1.98 a–e	1.90 b–e	2.02 a d	1.93 A
Redlan	(2.26 ab)	1.83 b e	2.18 ab	1.67 c e	2.03 a–d	1.92 A
Mean		1.74 B	2.02 A	1.90 AB	2.02 A	
Upper leaf/lower leaf P ratio						
		(16.6 b–d)	(7.8 e–g)	(5.3 g)	(9.6 c g)	
CK60	(10.3 c–g)	11.4 b–g	16.7 b d	16.6 b d	8.6 d–g	13.3 A
KS35	(12.1 b g)	14.6 b f	27.2 a	13.1 b–g	6.9 g	15.4 A
Wheatland	(17.7 bc)	17.4 bc	18.8 b	9.9 c g	15.9 b–e	15.5 A
Redlan	(10.1 c–g)	16.8 b–d	16.0 b–e	16.7 b–d	9.6 c–g	14.8 A
Mean		15.0 B	19.7 A	14.1 C	10.2 D	
Dry matter produced per unit P (mg mg^{-1})						
		(998 c–g)	(620 j)	(638 ij)	(754 g–i)	
CK60	(665 h–j)	*1070* b–d[‡]	965 c–g	816 e j	968 c–g	955 A
KS35	(758 f j)	*1277* ab	*1009* c e	862 d i	789 e–j	984 A
Wheatland	(858 d i)	*1329* a	*1062* b–d	895 d–h	942 c g	1057 A
Redlan	(748 g j)	*1171* a c	904 d g	*1169* a c	910 d–g	1038 A
Mean		1212 A	985 B	936 C	902 D	

[†] Means followed by common letters are not significantly different at P = 0.05. Parental values are enclosed in parentheses. Small letters refer to differences among parents and/or hybrids. Capital letters refer to mean in each parental group separately.
[‡] Underlined hybrid values indicate significant heterosis for this trait in relation to its parental means.

P deficiency

Under the 75 μmol P L^{-1} level, which was assumed to be the preferred condition to test for P efficiency, the GCA effects for males with one exception (P content of upper leaves) always exceeded the GCA effects of females and the SCA effects of males × females, and often with considerable magnitude (Table 1). The large number of significant mean square values for females and males × females, however, indicated that GCA in females and SCA in hybrid combinations were also of some importance. The mean data on parents and hybrids in Tables 3 and 4, and to a lesser extent in Tables 5 and 6, substantiated the general findings on combining ability.

At the 75 μmol P L^{-1} level (Table 1) the partitioned hybrid mean squares for males, females, and females × males were all significant for dry matter yields of plant parts, as were the parent mean squares for males and females. With 150 μmol P L^{-1} level, significance did not exist for female mean squares of either hybrids or parents for dry matter yields

Table 5. Mean leaf, root, and whole plant dry matter yields of sorghum male and female parental lines and their F$_1$ hybrids grown in nutrient solution with 150 μmol P L^{-1}

Female		Male				
		NB9040	SC120-15	Plainsman	SC33-9-8-E4	Mean
Leaf dry matter (g plant^{-1})						
		(2.80 a–f)	(2.78 a–g)	(2.00 h)	(2.31 gh)	
CK60	(2.35 f–h)†	2.99 a–c	3.18 a	2.68 b–g	2.98 a–c	2.93 A
KS35	(2.40 f–h)	2.93 a–d	2.87 a–e	2.52 c–g	2.98 a–e	2.82 A
Wheatland	(2.68 b–g)	3.18 a	2.72 a–g	2.71 a–g	2.88 a–e	2.87 A
Redlan	(2.51 d–g)	3.01 ab	2.52 c–g	2.64 b–g	3.04 ab	2.80 A
Mean		3.03 A	2.82 BC	2.63 C	2.94 AB	
Root dry matter (g plant^{-1})						
		(1.49 ef)	(1.38 fg)	(1.34 fg)	(1.09 h)	
CK60	(1.20 gh)	1.83 ab	1.68 b–e	1.48 ef	1.56 c–f	1.63 B
KS35	(1.08 h)	1.56 c–f	1.50 d–f	1.44 f	1.48 ef	1.50 C
Wheatland	(1.47 ef)	1.99 a	1.75 bc	1.73 b–d	1.68 b–e	1.79 A
Redlan	(1.12 h)	1.75 bc	1.41 fg	1.56 c–f	1.49 ef	1.55 BC
Mean		1.87 A	1.59 B	1.55 B	1.55 B	
Whole plant dry matter (g plant^{-1})						
		(4.30 b–e)	(4.16 c–f)	(3.34 h)	(3.40 gh)	
CK60	(3.56 gh)	4.82 ab‡	4.86 ab	4.15 c–f	4.41 b–e	4.56 AB
KS35	(3.49 gh)	4.50 b–e	4.37 b–e	3.96 d–g	4.46 b–e	4.32 B
Wheatland	(4.15 c–f)	5.17 a	4.47 b–e	4.44 b–e	4.57 b–d	4.66 A
Redlan	(3.62 f–h)	4.76 a–c	3.93 e–g	4.20 c–f	4.53 b–e	4.35 B
Mean		4.81 A	4.41 B	4.18 C	4.49 B	

† Means followed by common letters are not significantly different at $P = 0.05$. Parental values are enclosed in parentheses. Small letters refer to differences among parents and/or hybrids. Capital letters refer to mean in each parental group separately.
‡ Underlined hybrid values indicate significant heterosis for this trait in relation to its parental means.

of leaves (Table 2). At high P, significance for males *vs* females was found for dry matter yields of roots, whereas, under low P, no differences existed. For this trait, the parental mean values differed by 0.11 g plant^{-1} under the 150 μmol P L^{-1} level (Table 5), whereas, no mean difference existed under the 75 μmol P L^{-1} level (Table 3).

P content

When grown under 75 μmol P L^{-1}, the P content of lower leaves and roots of males of hybrids had significant mean square values whereas females of hybrids did not. At both P levels, males but not females showed significance for P content in the lower leaves of hybrids. Both parental groups alone were significant for this trait at the 75 μmol P L^{-1} level. A male × female interaction was significant at both P levels for P content of lower leaves only. Phosphorus content in the lower leaves was a better trait than P content of roots for measuring genotypic differences, and P content in upper leaves was of no practical use.

Table 6. Mean P content, upper/lower leaf P ratio, and dry matter produced per unit P of sorghum male and female parental lines and their F_1 hybrids grown in nutrient solution with 150 μmol P L^{-1}

Female		Male				
		NB9040	SC120-15	Plainsman	SC33-9-8-E4	Mean
P content (mg plant^{-1})						
		(3.55 ab)	(3.95 ab)	(3.93 ab)	(3.51 ab)	
CK60	(3.84 ab)†	3.66 ab	3.17 b	3.85 ab	3.19 b	3.47 A
KS35	(4.04 ab)	3.44 ab	3.84 ab	3.77 ab	3.40 b	3.61 A
Wheatland	(3.68 ab)	3.58 ab	3.05 b	3.22 b	3.31 b	3.29 A
Redlan	(4.49 a)	3.91 ab	3.61 ab	3.91 ab	3.37 b	3.70 A
Mean		3.65 A	3.42 A	.69 A	3.32 A	
Upper leaf/lower leaf P ratio						
		(12.1 b–d)	(8.2 a–h)	5.8 e–h)	(7.2 d–h)	
CK60	(4.6 gh)	8.2 d–h	10.4 c–f)	5.2 e–h	16.6 ab	10.1 B
KS35	(3.9 h)	9.5 d–h	10.6 c–e	4.8 f–h	10.7 c–e	8.9 C
Wheatland	(3.9 h)	14.3 bc	4.5 gh	9.6 c–g	19.3 a	11.9 A
Redlan	(3.9 h)	7.7 d–h	6.0 e–h	5.9 e–h	8.2 d–h	7.0 D
Mean		9.9 B	7.9 C	6.4 D	13.7 A	
Dry matter produced per unit P (mg mg^{-1})						
		(1235 a–g)	(1089 b–g)	(917 e–g)	(1002 c–g)	
CK60	(940 d–g)	1356 a–e	*1560* a‡	1083 b–g	*1388* a–d	1347 B
KS35	(892 fg)	1398 a–d	1188 a–g	1082 b–g	*1321* a–f	1247 C
Wheatland	(1290 a–f)	1565 a	1521 ab	1416 a–c	*1436* a–c	1484 A
Redlan	(806 g)	1243 a–g	1109 a–g	1080 b–g·	*1363* a–e	1199 D
Mean		1390 A	1344 A	1165 A	1377 A	

† Means followed by common letters are not significantly different at $P = 0.05$. Parental values are enclosed in parentheses. Small letters refer to differences among parents and/or hybrids. Capital letters refer to mean in each parental group separately.
‡ Underlined hybrid values indicate significant heterosis for this trait in relation to its parental means.

For upper/lower leaf P content ratios when grown at $75\,\mu\text{mol}\,\text{P}\,\text{L}^{-1}$ (Table 1), male mean squares (either hybrids or parents) were significant, whereas females were not. At $150\,\mu\text{mol}\,\text{P}\,\text{L}^{-1}$ (Table 2), females of hybrids were significant. Male × female interactions were significant at both P levels. P content of upper leaves was not an effective trait to measure P efficiency and tended to invalidate the use of P content of upper/lower leaf as a component for P nutrition.

Dry matter produced per unit P was another trait where parental performance was different under the two P levels. Males of hybrids were significant at the $75\,\mu\text{mol}\,\text{P}\,\text{L}^{-1}$ level and females were not (Table 1), whereas, at the $150\,\mu\text{mol}\,\text{P}\,\text{L}^{-1}$ level (Table 2) females were significant and males were not. These results indicated that a parent × treatment interaction existed, and the male parents to perform best under P deficiency stress conditions may not necessarily be the best when P is not as stressed. For example, NB9040 and SC120-15 produced more dry matter per unit P at $75\,\mu\text{mol}\,\text{P}\,\text{L}^{-1}$ than Plainsman and SC33-9-8-E4, but at $150\,\mu\text{mol}\,\text{P}\,\text{L}^{-1}$ all the males were statistically equal although Plainsman tended to be the poorest. SC33-9-8-E4, classed as a P inefficient male, seemed to be particularly responsive to the higher P level.

Performance of females for dry matter per unit P also varied. At the $150\,\mu\text{mol}\,\text{P}\,\text{L}^{-1}$ level (Table 6), the range in hybrid performance was not quite as wide as it was at the $75\,\mu\text{mol}\,\text{P}\,\text{L}^{-1}$ level (Table 4), but hybrids based on females diverged more at the former than at the latter level. The results illustrated that female parents under non-stressed P deficiency conditions may not perform the same as under P deficiency stress conditions.

Strain differences

Overall, the data from these experiments indicated that NB9040 and SC120-15 were more P efficient males than Plainsman and SC-33-9-8-E4, and that Wheatland and Redlan were somewhat more P efficient females than CK60 and KS35 (Tables 4 and 6). The four hybrids involving the efficient parents usually produced the most favorable values (higher dry matter yields, P contents, upper/lower leaf P ratios, and dry matter produced per unit P) in Tables 3 and 4 as well, but sometimes the values were not significantly different from hybrids involving one of these P efficient parental groups with the corresponding less efficient opposite parental group. These differences were indications of SCA.

The experiments demonstrated that performance of parental lines *per se* seemed to be a fairly good indicator of parental performance in hybrids, espeeially at the $75\,\mu\text{mol}\,\text{P}\,\text{L}^{-1}$ level (Tables 1, 3, and 4). The hybrid performances also tended to be more reliable for the male parents than for the female parents in these experiments. This was likely because of the wider range of P nutritional traits exhibited by the males than by the females.

Heterosis, defined in relation to the parental average, was observed for most hybrid crosses and their respective parents. Under the low P treatment, heterosis for whole plant dry matter yields was significant for the F_1 hybrids, except for Wheatland crossed with Plainsman (Table 3.)

Other studies in our laboratory on the heterosis of root growth in sorghum hybrid crosses grown in nutrient solution (A M C Furlani and R B Clark, unpublished data) showed that Wheatland and its hybrids performed better and had advantages over the best parents for several root characteristics (length of seminal roots, growth rate of adventitious roots, root volume, and lateral root branches) when compared to hybrids crossed with CK60. In the studies reported here, Wheatland and Redlan hybrids exhibited an overall greater heterosis for root dry matter yields than CK60 and KS35 hybrids (Table 3). Heterosis for P efficiency values was significant for all hybrids involving the P efficient male parents NB9040 and SC120-15 (Table 4).

Conclusions

Improving sorghum hybrids for ability to grow and perform well under low levels of P appears feasible. The favorable traits of male parents that had greater ability to grow with low P were transferred to their corresponding hybrids and indicated that these traits were heritable and at least partially dominant. This did not mean, however, that only male parents carried desirable genes since the male genotypes used in these studies were specifically selected because of their P efficiency and inefficiency traits[12]. Some of the female parents also showed favorable efficiency characteristics for P. Inheritance appeared to be more complex than a single gene. Additive genes appeared to be partially responsible for the variability of P efficiency in sorghum. Heritability needs to be further investigated, however, and a large random sample of sorghum genotypes must be tested to determine the heritability and predicted gains for these traits.

Improving sorghum for performance under low P conditions appears to be a practical approach to help solve some of the problems on many infertile soils throughout the world. These initial results with sorghum showed that genotype differed in their ability to grow and perform with low levels of P, the traits appeared to be heritable, and plants with appropriate characteristics such as high dry matter production and large root systems might be used as criteria for selection. The results, however, need to be verified with field responses.

References

1. Baker D E, Jarrell A E, Marshall L E and Thomas W I 1970 Phosphorus uptake from soils by corn hybrids selected for high and low phosphorus accumulation. Agron. J. 62, 103–106.
2. Barber S A 1980 Soil-plant interactions in the phosphorus nutrition of plants. pp 591–615. *In* Role of Phosphorus in Agriculture. Eds. F E Khasawneh, E C Sample and E J Kamprath. Am. Soc. Agron., Madison, WI.
3. Bernardo L M, Clark R B and Maranville J W 1984 Nitrate ammonium ratio effects on nutrient solution pH, dry matter yield, and nitrogen uptake of sorghum. J. Plant Nutr. 7, 1384–1400.
4. Brown J C, Clark R B and Jones W E 1977 Efficient and inefficient use of phosphorus by sorghum. Soil Sci. Soc. Am. J. 41, 747–750.
5. Brown J C and Jones W E 1977 Fitting plants nutritionally to soils. III. Sorghum. Agron. J. 69, 410–414.
6. Clark R B 1982 Nutrient solution growth of sorghum and corn in mineral nutrition studies. J. Plant Nutr. 5, 1039–1057.
7. Clark R B and Brown J C 1974 Differential mineral uptake by maize inbreds. Commun. Soil Sci. Plant Anal. 5, 213–277.
8. Clark R B and Brown J C 1974 Differential phosphorus uptake by phosphorus-stressed corn inbreds. Crop Sci. 14, 505–508.
9. Clark R B, Maranville J W and Gorz H J 1978 Phosphorus efficiency of sorghum grown with limited phosphorus. pp 93–99. *In* Plant Nutrition 1978. Eds. A R Ferguson, R L Bieleski and I B Ferguson Proc. 8th Int. Colloq. Plant Anal. Fert. Prob., Auckland, New Zealand.
10. Clark R B, Maranville J W and Ross W M 1977 Differential phosphorus efficiency in sorghum. pp 1–2. *In* Proc. 10th Grain Sorghum Util. Conf., Wichita, KS.
11. Epstein E and Jefferies R L 1964 The genetic basis of selective ion transport in plants. Annu. Rev. Plant Physiol. 15, 169–184.
12. Furlani A M C, Clark R B, Maranville J W and Ross W M 1984 Sorghum genotype differences in phosphorus uptake rate and distribution in plant parts. J. Plant Nutr. 7, 1113–1126.
13. Furlani A M C, Clark R B, Sullivan C Y and Maranville J W 1986 Induction of leaf 'red-speckling' by phosphorus on sorghum grown under controlled conditions. Crop Sci. 26, 551–557.
14. Gabelman W H 1976 Genetic potentials in nitrogen, phosphorus, and potassium efficiencies. pp 205–212. *In* Plant Adaptation to Mineral Stress in Problem Soils. Ed. M J Wright. Cornell Univ. Agric. Exp. Stn., Ithaca, NY.
15. Gerloff G C 1976 Plant efficiencies in the use of nitrogen, phosphorus, and potassium. pp 161–173. *In* Plant Adaptation to Mineral Stress in Problem Soils. Ed. M J Wright. Cornell Univ. Agric. Exp. Stn., Ithaca, NY.
16. Griffing B 1956 A generalized treatment of the use of diallel crosses in quantitative inheritance. Heridity 10, 31–50.
17. Griffing B 1956 Concept of general and specific combining ability in relation to diallel crossing systems. Aust. J. Biol. Sci. 9, 463–493.
18. Knudsen D, Clark R B, Denning J L and Pier P A 1981 Plant analysis of trace elements by X-ray. J. Plant Nutr. 3, 61–75.
19. Lindgren D T, Gabelman W H and Gerloff G C 1977 Variability of phosphorus uptake and translocation in *Phaseolus vulgaris* L. under phosphorus stress. J. Am. Soc. Hort. Sci. 102, 674–67.
20. Makmur A, Gerloff G C and Gabelman W H 1978 Physiology and inheritance of efficiency in potassium utilization in tomatoes grown under potassium stress. J. Am. Soc. Hort. Sci. 103, 545–549.
21. O'Sullivan J, Gabelman W H and Gerloff G C 1974 Variations in efficiency of nitrogen utilization in tomatoes (*Lycopersicon esculentum* Mill.) grown under nitrogen stress. J. Am. Soc. Hort. Sci. 99, 543–547.
22. Rice R M 1974 Physiology and inheritance of differential growth response under high phosphorus levels among different lines of beans, *Phaseolus vulgaris* L. Ph.D. Thesis, Univ. of Wisconsin, Madison. Diss. Abstr. 35, 5220-B (1975).

23 Shea P F, Gabelman W H and Gerloff G C 1967 The inheritance of efficiency in potassium utilization in snap beans. (*Phaseolus vulgaris* L.). Proc. Am. Soc. Hort. Sci. 91, 286-293.
24 Snaydon R W and Bradshaw A D 1962 Differences between natural populations of *Trifolium repens* L. in response to mineral nutrients. I. Phosphate. J. Exp. Bot. 13, 422-434.
25 Steele R G D and Torrie J H 1980 Principles and procedures of statistics. McGraw-Hill Book Co., New York.
26 Whiteaker G, Gerloff G C, Gabelman W H and Lindgren D 1976 Intraspecific differences in growth of beans at stress levels of phosphorus. J. Am. Soc. Hort. Sci. 101, 472-475.

External phosphorus requirements of five tropical grain legumes grown in flowing-solution culture

A.J. FIST, F.W. SMITH* and D.G. EDWARDS
Department of Agriculture, University of Queensland, St. Lucia, Queensland, Australia 4067
CSIRO Division of Tropical Crops and Pastures, St. Lucia, Queensland, Australia 4067

Key words Critical external phosphorus concentration Flowing-solution culture Tropical grain legumes

Summary Five tropical grain legume species were grown for periods from 4 to 20 days in flowing-solution culture at 7 maintained phosphorus (P) concentrations, ranging from $0.25 \mu M$ to $16 \mu M$. Critical external P requirements were $0.8 \mu M$ for cowpea cv. Vita 4 and soybean cv. Fitzroy, $1.0 \mu M$ for pigeon pea cv. Royes, $2.0 \mu M$ for mungbean cv. Regur and $3.0 \mu M$ for guar cv. Brooks. Plant responses to P deficiency included reduced growth rate, increased root percentage, and increased P uptake potential. The long-term P uptake rates of guar plants were lower than those of the other species at each external P concentration. Guar plants had a low P uptake potential as indicated by short-term ^{32}P-labelled uptake rate studies from $15 \mu M$ P solutions. Cowpea by contrast had high short-term uptake rates indicating a high P uptake potential.

Introduction

The external phosphorus (P) requirements of plants are difficult to determine using still-solution culture or sand culture methods because depletion of P occurs from the rooting medium adjacent to the plant roots. Solution P concentrations much higher than those found in soil solutions are normally used to maintain supply of P as depletion through plant uptake occurs. Investigations of the capacity of plants to absorb P require that a known concentration of P be maintained continuously at the plant root surface. This requirement is met by using flowing-solution culture, in which the plant roots are bathed in a solution which is replenished continuously with P at a rate which is equal to the rate of P uptake[9]. In this way, P concentrations similar to those found in the soil solution can be accurately maintained, and realistic estimates of the external P requirements of plants can be made.

This paper describes growth responses of 5 tropical grain legumes to maintained external P concentrations ranging from $0.25 \mu M$ to $16 \mu M$.

Materials and methods

A glasshouse trial was conducted at the University of Queensland, Brisbane, in which 5 tropical grain legume species were grown at 7 continuously maintained P concentrations. The species were soybean (*Glycine max* (L.) Merrill) cv. Fitzroy, cowpea (*Vigna unguiculata* (L.) Walp. subsp. *unguiculata*) cv. Vita 4, pigeon pea (*Cajanus cajan* (L.) Millsp.) cv. Royes, mungbean (*Vigna mungo* (L.) Hepper) cv. Regur, and guar (*Cyamopsis tetragonoloba* (L.) Taub.) cv. Brooks.

H.W. Gabelman and B.C. Loughman (Eds.), Genetic aspects of plant mineral nutrition.
ISBN-13: 978-94-010-8102-3
© *1987 Martinus Nijhoff Publishers, Dordrecht/Boston/Lancaster.*

Plant culture

Seeds were sterilized with sodium hypochlorite solution and germinated on paper soaked with 0.2 mM $CaSO_4$. When the radicles were 1 to 2 cm long, the seedlings were transferred to flow-culture units. The principles of operation and design of flow-culture units have been described elsewhere[2,9]. A basal nutrient solution of the following composition was used:(μM) N (as NO_3^-) 500, Ca 1000, S 976, K 250, Mg 100, Fe 20, Na 20, Cl 10, B 3, Zn 0.5, Mn 0.15, Cu 0.1, Mo 0.02. Solution pH was maintained at 5.5 $\pm$ 0.1 with automatic addition of 0.15 M HNO_3, and solution temperature was maintained at 27 $\pm$ 1 C. The mean minimum and maximum daily air temperatures experienced during the experiment (26 Jan–15 Feb 1984) were 22.6 C and 33.9 C. Solutions were continuously aerated with filtered air bubbled into each culture vessel.

The 7 P treatments investigated were 0.25, 0.5, 1.0, 2.0, 4.0, 8.0, and 16.0 μM. Phosphorus was supplied as KH_2PO_4 and the concentration was measured daily[11]. Peristaltic pumps were used to add the amount of P expected to be taken up by plants each day and thus maintain solutions at the desired concentrations. Where P concentrations were found to be less than desired, an immediate adjustment was made. Where excesses of P were found, a deduction was made from the amount to be added over the next 24 hours.

Harvest and post-harvest procedures

Harvests of 3 replicates were taken at 4-day intervals over a period of 20 days. The first harvest was a thinning on day 4 when the number of plants per culture was reduced to 3 for soybean and cowpea, 4 for mungbean and pigeon pea and 5 for guar. For all harvests, roots were excised from shoots and then blotted dry. Roots and shoots were weighed, and, after drying at 70 C for 48 hours, dry matter was determined. Plant material was dry-ashed at 480 C for 18 hours, and the ash dissolved in 0.1 M HCl. Tissue P was determined by a molybdenum blue method[16].

Short-term ^{32}P labelled P uptake studies

On day 12, uptake rates of ^{32}P labelled P were determined on cultures grown at 0.25 and 8.0 μM. Plants were initially transferred to 1.0 l solutions of complete nutrient solution containing 15 μM P for 2 minutes. They were then placed in a similar 1.0 l solution containing phosphorus labelled with ^{32}P. After 20 minutes, plants were returned to unlabelled solutions for 2 minutes to enable exchange of ^{32}P from the free space of the root. Environmental conditions during the uptake periods were the same as those for the plant culture: solution temperatures were 27 $\pm$ 1 C and the solution pH was 5.5. Post-harvest procedures were the same as in other harvests. Uptake of labelled P was determined from radioassay of 0.5 ml aliquots of uptake solution and ashed plant samples dissolved in 0.1 M HCl. Solutions were counted in Opti-fluor scintillant in a Packard Tri-Carb 4430 liquid scintillation spectrometer.

Results

Effect of external P concentration on growth

Figure 1 shows shoot and root dry weights 16 or 20 days after planting. The results are shown for 20 days for guar because treatment differences were slower to develop in this species than in the other species. External P requirements are the external P concentrations at which the greatest change of slope in the response curve occurred. These were derived from shoot dry weight data in the harvests taken after 16 and 20 days. Cowpea and soybean had the lowest critical external P requirement, about 0.8 μM. Pigeon pea required a P concentration of 1.0 μM while mungbean and guar were less efficient requiring 2.0 and 3.0 μM respectively for near maximum growth.

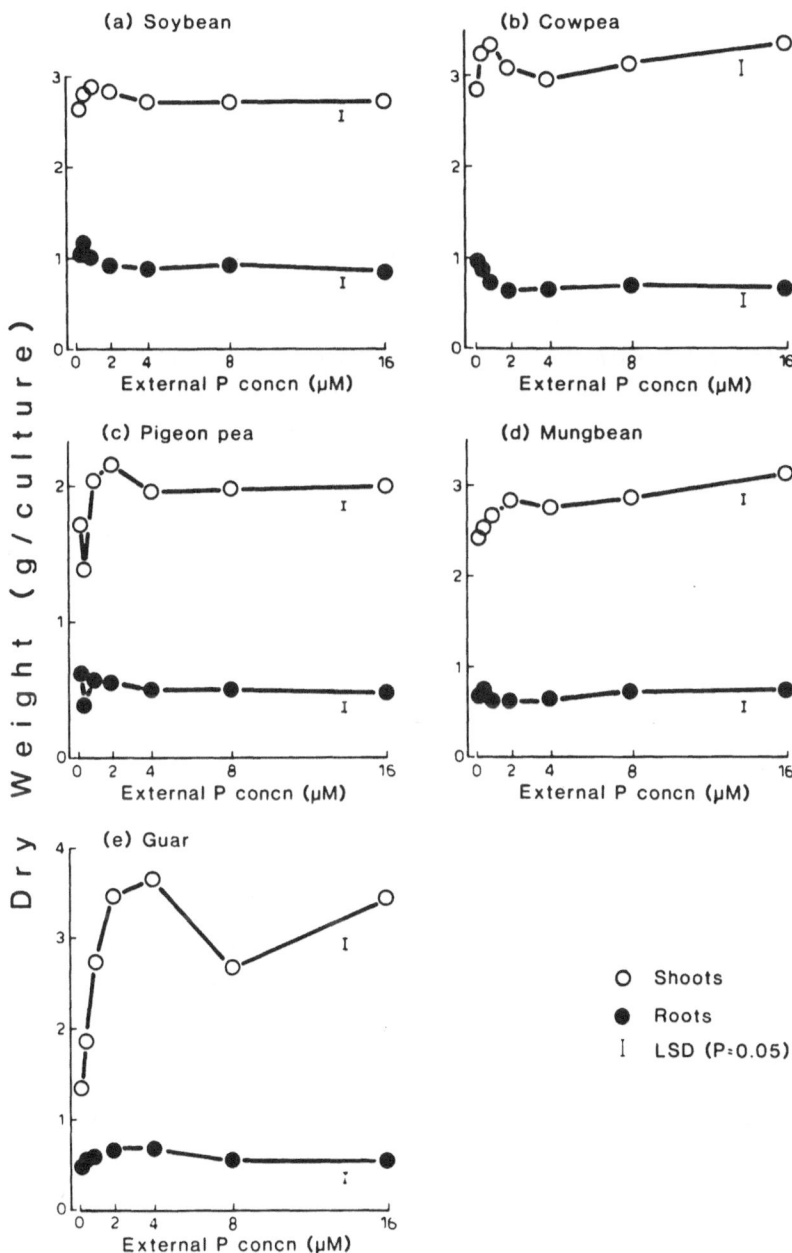

Fig. 1. Growth response of tropical grain legumes to external P concentration. Dry weight per culture of shoots and roots after 16 days for soybean, cowpea, pigeon pea and mungbean, and after 20 days for guar.

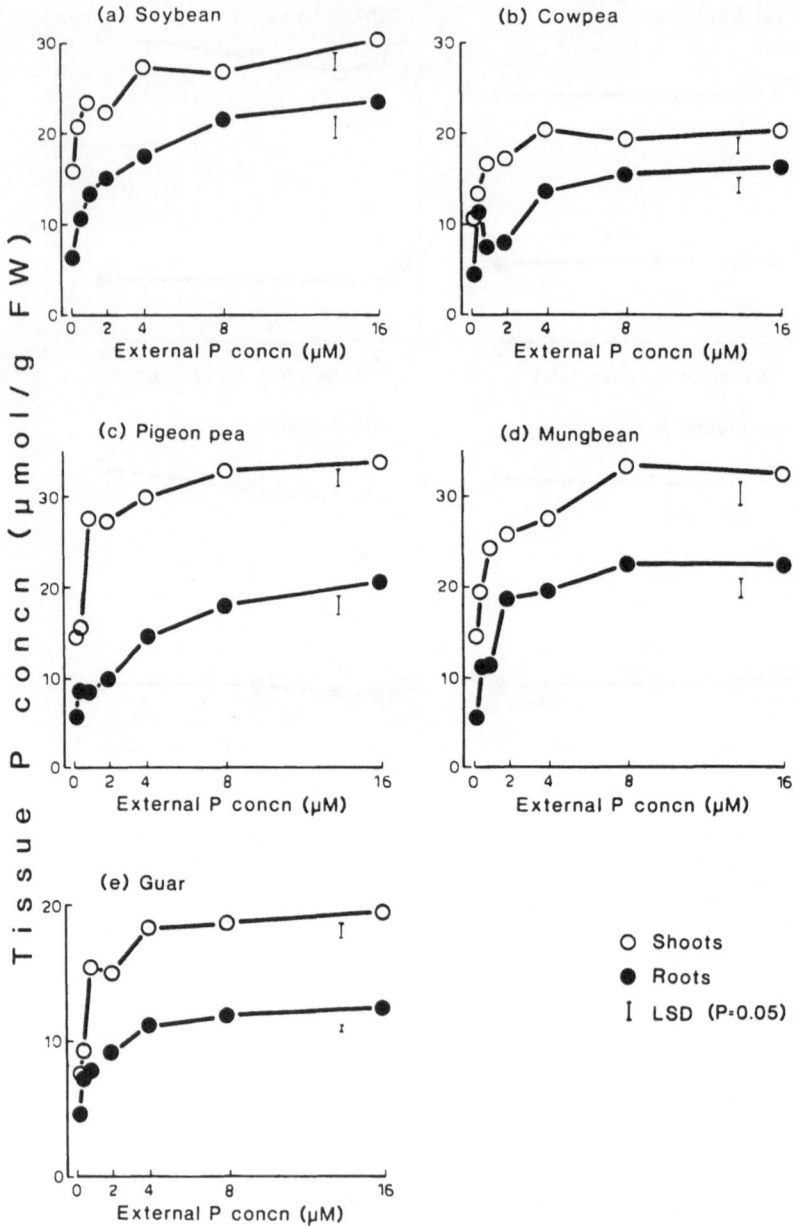

Fig. 2. P concentrations in plant tissue 16 days after planting.

Root tissue formed a large proportion of the plant weight at low than at high external P concentrations (Table 1). On a dry weight basis after 16 days the proportion of root varied from a low of 0.16 for cowpea grown at $16 \mu M$ P, to a high of 0.29 for guar at $0.25 \mu M$ P and 0.30 for soybean plants grown at $0.5 \mu M$ P. The absolute root weights increased with P deficiency in all species, except guar, at most harvests.

Table 1. Root weight ratios[a] 16 days after planting

External P concentration (μM)	Species				
	Soybean	Cowpea	Pigeon pea	Mungbean	Guar
0.25	0.27	0.26	0.27	0.22	0.29
0.5	0.30	0.21	0.23	0.23	0.24
1.0	0.26	0.18	0.22	0.19	0.20
2.0	0.24	0.17	0.21	0.18	0.18
4.0	0.24	0.18	0.20	0.19	0.19
8.0	0.25	0.18	0.20	0.20	0.20
16.0	0.24	0.16	0.19	0.19	0.18
LSD (P = 0.05)	0.03	0.02	0.02	0.02	0.02

[a] Calculated using the formula Wr/(Ws + Wr) where Ws and Wr are the dry weights of the shoots and roots respectively.

Effect of external P concentration on tissue P concentration

Figure 2 shows the tissue P concentrations present after 16 days. There was an increase in both root and shoot tissue P concentrations with increasing external P concentration, but the rate of increase declined markedly once adequate external P concentrations were reached. Guar and cowpea showed less luxury uptake than the other species, maximum shoot P concentrations rising to only 19.4 and 20.5 μmoles g^{-1} fresh weight respectively.

Relative growth rate

Relative growth rates (RGR) of each species between 8 and 12 days after planting, the period of greatest growth are presented in Table 2. Similar patterns were exhibited during other harvest intervals. Pigeon pea had the lowest RGR of the species evaluated, 0.21 day^{-1}. Cowpea had the highest RGR of 0.43 day^{-1} in the 16 μM treatment. A separate analysis of the RGR of roots and shoots showed that, in plants grown

Table 2. Mean relative growth rates[a] with standard errors of whole plants between 8 and 12 days after planting (day^{-1})

External P concn (μM)	Species				
	Soybean	Cowpea	Pigeon pea	Mungbean	Guar
0.25	0.24 ± 0.02	0.30 ± 0.09	0.20 ± 0.03	0.23 ± 0.01	0.22 ± 0.02
0.5	0.26 ± 0.03	0.40 ± 0.04	0.20 ± 0.02	0.23 ± 0.03	0.24 ± 0.02
1.0	0.26 ± 0.001	0.34 ± 0.05	0.22 ± 0.01	0.30 ± 0.01	0.23 ± 0.02
2.0	0.25 ± 0.01	0.40 ± 0.06	0.21 ± 0.01	0.28 ± 0.01	0.23 ± 0.03
4.0	0.26 ± 0.02	0.37 ± 0.06	0.21 ± 0.02	0.30 ± 0.02	0.29 ± 0.03
8.0	0.24 ± 0.02	0.26 ± 0.01	0.21 ± 0.04	0.25 ± 0.01	0.22 ± 0.03
16.0	0.26 ± 0.02	0.44 ± 0.07	0.25 ± 0.02	0.31 ± 0.01	0.25 ± 0.01

[a] Dry weight basis

Table 3. Mean long-term P uptake rates between 8 and 12 days after planting. The values of a and c shown in the table are parameters of curves of the form $Y = a(1 - e^{-cX})$ fitted to the relationship between long-term P uptake rate and external P concentration by a computerised non-linear least squares method

External P concentration (μM)	Species				
	Soybean	Cowpea	Pigeon pea	Mungbean	Guar
	Uptake rates (μmol g^{-1} FW root h^{-1})				
0.25	0.29	0.31	0.21	0.32	0.10
0.5	0.45	0.53	0.23	0.38	0.16
1.0	0.53	0.58	0.39	0.65	0.22
2.0	0.62	0.74	0.43	0.65	0.29
4.0	0.70	0.79	0.50	0.71	0.42
8.0	0.73	0.54	0.72	0.79	0.36
16.0	0.72	0.89	0.59	0.89	0.48
Curve fitting parameters:					
a	0.69 ± 0.02	0.73 ± 0.06	0.60 ± 0.05	0.77 ± 0.04	0.42 ± 0.03
c	1.82 ± 0.26	2.16 ± 0.74	0.89 ± 0.26	1.59 ± 0.35	0.77 ± 0.18
R^2	0.93	0.69	0.83	0.87	0.91

at low P concentrations, the RGR of the roots was greater than that of the shoots (results not presented).

Long-term P uptake rates at each external P concentration

Mean P uptake rates (I_M) by plants between harvests were calculated using the formula[17]:

$$I_M = \frac{\ln R_2 - \ln R_1}{t_2 - t_1} \times \frac{M_2 - M_1}{R_2 - R_1}$$

This arbitrarily assumes a linear relationship between nutrient content (M) and root weight (R). Any errors introduced by this assumption would not appear to materially affect the relative results produced by this formula. The uptake rates between 8 and 12 days after planting are shown in Table 3. Similar patterns were observed in the other harvest intervals.

The mean long-term P uptake rates of guar were lower than those of other species at any particular P concentration in all harvest intervals. At low P concentrations the uptake rates of guar and pigeon pea were similar but maximum uptake rates (parameter a) of pigeon pea were similar to those of soybean, cowpea, and mungbean. Both pigeon pea and guar had low c parameters indicating a response in uptake rate over a wider range of P concentrations than in other species. The c parameters of the five curves were significantly different in the first harvest interval but not in later intervals where the differences were smaller. In all species the mean long-term P uptake rates were greatest between 8 and 12 days

Table 4. Mean short-term uptake rates of 15 μM ^{32}P labelled P by plants grown at two external P concentrations

External P concentration during growth (μM)	Species				
	Soybean	Cowpea	Pigeon pea	Mungbean	Guar
Roots	μmol g^{-1} FW root h^{-1}				
0.25	1.85	2.60	1.86	1.57	1.29
8.00	0.72	0.92	0.80	0.83	0.54
Shoots	μmol g^{-1} FW shoot h^{-1}				
0.25	1.54	1.88	1.67	1.23	1.34
8.00	0.54	0.51	0.55	0.58	0.43

LSD (P = 0.05) for uptake rate per gram of root: 0.23
for uptake rates per gram of shoot: 0.30

from planting. In guar the uptake rates declined only slightly in subsequent periods.

Short-term ^{32}P labelled P uptake rates

Short-term ^{32}P labelled P uptake rates are an indication of the potential of the plants to absorb P. This potential is not realized in solutions of low P concentrations because the rate of uptake is dependent upon the P concentration in solution. When labelled P is supplied to plants at a concentration high enough to almost saturate the P uptake system, indications of the plants' maximum P uptake rates are obtained.

In all species, there were large and significant differences between short-term P uptake rates of plants grown in the two phosphorus concentrations (Table 4). Uptake rates of plants grown at 0.25 μM P were from 1.9 to 2.8 times those of plants grown at 8.0 μM. Cowpea had the highest uptake rate for both treatments and also exhibited the largest difference in uptake rates between the 2 treatments. Guar had the lowest uptake rate in each treatment, but had a higher root weight ratio than other species (Table 1). This would compensate for the lower uptake rate per gram of root in guar. Thus, when uptake rates were expressed on a fresh weight of shoot basis, differences between the species were reduced.

Discussion

The relationship between growth of plants and nutrient concentrations can be expressed in several ways. The external nutrient requirement is the minimal nutrient concentration in solution which results in maximal growth[13]. Because of the hyperbolic shape of response curves, it is difficult to determine the exact nutrient concentration required for maximum growth. Instead, the critical nutrient concentration of a plant, with respect to growth, may be defined either as the nutrient concentration that is just deficient for maximum growth, or that which is just adequate for maximum growth, or as the concentration separating the zone of deficiency from the zone of adequacy[15]. Although the three definitions represent different viewpoints, in practice they are approximately the same concentrations. Experimentally, it is easier to determine the nutrient concentration that is just deficient (as indicated by the rapid change in slope of the response curve), than the nutrient concentration that is just sufficient (as indicated by a zero slope in the response curve). Many authors use the external nutrient concentration which provides 90 to 95 percent of maximum growth as the critical external nutrient concentration[6,10]. This can be misleading since the concentration reported can be well above, or below, the zone of transition. Therefore, in this study, critical external nutrient concentrations are reported as the external nutrient concentration separating the zone of deficiency from the zone of adequacy.

A variety of responses to P deficiency was produced by plants in this study. These results relate only to the cultivars grown and are not necessarily species differences. Plant yield response occurred up to external P concentrations of $3 \mu M$. Root weight ratios were larger in P-deficient plants; in soil-grown plants such a response would permit better acquisition of P by enabling the plant to exploit a larger volume of soil. An increase in P uptake potential in P-stressed plants was also shown.

As experimental techniques have improved in the last 30 years, the critical external P requirements reported for plant species have become lower[1] and nearly all species have been found to have external P requirements of the same order as those found in soil solutions. Bulk soil solution P concentrations are typically in the range of 0.6 to $3 \mu M$[3,14], but lower P concentrations would be found near plant root surfaces due to depletion and the establishment of a P concentration gradient.

Tropical pasture legumes, with the exception of greenleaf desmodium, were found to have external P requirements similar to those of the grain legumes investigated in this study[6]. Endeavour stylo required $0.24 \mu M$ P for maximum growth whilst siratro, centro, phasey bean, and tinaroo glycine all required about $3 \mu M$ for maximum growth. Greenleaf desmodium was found to require the very high P concentration of $264 \mu M$ for maximum growth. However, this result could not be confirmed in a recent experiment where the external P requirement was found to be of the same order as those of other tropical pasture legumes (F. W. Smith pers. comm.).

In this investigation wide differences in response to external P concentration were found between related plants. Cowpea and mungbean, both members of the genus *Vigna*, had quite different external P requirements. Guar was found to differ in several ways from other species investigated in this study. Its tissue P requirement was less than that of most species grown, yet it required a higher external P concentration for maximum growth. On being supplied with a surplus of P, tissue P concentrations in guar rose little above those of deficient plants, while those of other species rose to luxury concentrations.

Derivation of long-term P uptake rates showed that guar absorbed less P per gram of root than other species at all external P concentrations. Pigeon pea had low P uptake rates at low external P concentrations but had similar uptake rates to cowpea, mungbean and soybean at the higher P concentrations. The low external P requirement of pigeon pea is probably obtained at the cost of having a low growth rate.

Plants supplied with a large concentration of ^{32}P labelled P for a short interval provided an indication of the potential of the roots to absorb P in conditions where uptake rate was not limited by concentration. Short-term, ^{32}P labelled, P uptake rates were determined at an external P concentration of 15 μM, a concentration approximately three times the Michaelis constant (K_m) and approaching the saturation concentration of the high affinity P uptake systems[4,5]. P-stressed cowpea plants had the greatest short-term uptake rates, and guar, the species with the highest external P requirement, had the lowest short-term uptake rate. Pigeon pea had a high short-term uptake rate. This suggests, since long-term uptake rates were quite low at low P concentrations, that K_m of the P uptake system of pigeon pea roots is relatively high.

The short-term uptake rate studies showed that plants grown under P deficiency developed a higher capacity to absorb P than control plants. Plants maintain adequate internal P concentrations by regulation of root uptake potential. Lefebvre and Glass[12] found that short-term uptake rates increased up to 6-fold in P deficient barley plants compared with controls grown in 15 μM P solutions. The P deficient plants therefore had a greater capacity to absorb P. On returning plants to a solution adequate in P, short-term uptake rates were found to fall dramatically within 2 hours, indicating a decrease in the P uptake potential as the deficiency was alleviated. Other workers have reported similar results for barley and tomato[7], and potato[8].

Acknowledgements The authors are grateful for the technical assistance provided by G L Walters, P J Vanden Berg and Mrs H Damirchi.

References

1 Asher C J and Loneragan J F 1967 Response of plants to phosphate concentration in solution culture: 1. Growth and phosphorus content. Soil Sci. 103, 225 233.
2 Asher C J, Ozanne P G, and Loneragan J F 1965 A method for controlling the ionic environment of plant roots. Soil Sci. 100, 149 156

3 Barber S A, Walker J M and Vasey E H 1963 Mechanisms for the movement of plant nutrients from the soil and fertiliser to the plant root. J. Agric. Food Chem. 11, 204–207
4 Borkert C M and Barber S A 1983 Effect of supplying P to a portion of the soybean root system on root growth and P uptake kinetics. J. Plant Nutr. 6, 895–910
5 Carter O G and Lathwell D J 1967 Effects of temperature on orthophosphate absorption by excised corn roots. Plant Physiol. 42, 1407–1412
6 Chantkam S, Edwards D G and Asher C J 1983 Response of selected tropical pasture legumes grown in flowing nutrient culture to constant solution phosphorus concentrations: I. Growth and phosphorus concentration. Thai J. Agric. Sci. 16, 217–231
7 Clarkson D T and Scattergood C B 1982 Growth and phosphorus transport in barley and tomato plants during the development of, and recovery from, phosphate-stress. J. Exp. Bot. 33, 865–875
8 Cogliatti D H Clarkson D T 1983 Physiological changes in, and phosphate uptake by potato plants during development of, and recovery from phosphate deficiency. Physiol. Plant. 58, 287–294
9 Edwards D G and Asher C J 1974 The significance of solution flow rate in flowing culture experiments. Plant and Soil 41, 161–175
10 Jintakanon S, Edwards D G and Asher C J 1979 An anomalous, high external phosphorus requirement for young cassava plants in solution culture. *In* V. Int. Symp. Tropical Root Crops. Eds. E H Belen and M Villanueva. pp 507–518 PCARR Los Banos, Philippines
11 Jintakanon S, Kerven G L, Edwards D G and Asher C J 1975 Measurement of low phosphorus concentrations in nutrient solutions containing silicon. Analyst 100, 408–414
12 Lefebvre D D and Glass A D M 1982 Regulation of phosphate influx in barley roots: Effects of phosphate deprivation and reduction of influx with provision of orthophosphate. Physiol. Plant. 54, 199–206
13 Loneragan J F 1968 Nutrient requirements of plants. Nature 220, 1307–1308
14 Reisenauer H M 1966 Mineral nutrients in soil solution *In* Environmental Biology. Eds. P L Altman and D S Dittmer. pp 507–508. Fed. Am. Soc. Exp. Biol., Bethesda
15 Ulrich A 1952 Physiological bases for assessing the nutritional requirements of plants. Annu. Rev. Plant Physiol. 3, 207–228
16 Watanabe F S and Olsen S R 1965 Test of an ascorbic acid method for determining phosphorus in water and $NaHCO_3$ extracts from soil. Soil Sci. Soc. Am. Proc. 29, 677–678
17 Williams R F 1948 The effects of phosphorus supply on the rates of intake of phosphorus and nitrogen and upon certain aspects of phosphorus metabolism in gramineous plants. Aust. J. Sci. Res. B1, 333–361

Genetics and physiology of low-phosphorus tolerance in a family derived from two differentially adapted strains of tomato (*Lycopersicon esculentum* Mill.)

R.R. COLTMAN, W.H. GABELMAN, G.C. GERLOFF, AND S. BARTA
Department of Horticulture, University of Hawaii, Honolulu, HI 96822, USA

Key words Genetics Low-P tolerance Phosphorus Physiology Sand-alumina Tomato

Summary Genetic control of low-P tolerance in tomato was investigated with shoot dry weight (SDW) data from parental, F_1, backcross, and F_2 generations of a cross between two differentially tolerant strains in sand-alumina cultures. F_1 SDWs exceeded parental SDWs by 89% in low P cultures, but were 20% less than parental SDWs in high P cultures. Two exceptional progenies from the backcross to the tolerant parent produced more SDW with low P than any of the parental or F_1 plants growing with high P. These progenies exhibited high total P acquisition, high uptake of P per m of root, and efficient internal utilization of P. Nonadditive genetic variance was more important than additive genetic variance in the expression of low-P tolerance. However, the range of phenotypic variation in the backcross generations suggests recombination and complimentation among a relatively small number of genes. Prospects for breeding genotypes more tolerant to low P appear favorable, although such genotypes may be at a selective disadvantage in P-rich environments.

Introduction

Phosphorus (P) deficiency is one of the most widespread soil constraints in the tropics. Approximately 82% of the land area of the American tropics is deficient in P in its natural state[17]. Tropical soils also frequently have high capacities for P fixation. Amelioration of P deficiency by the application of massive doses of costly P fertilizer is not an option open to many of the predominantly subsistence farmers of the tropics.

Strategies to improve agricultural production on P deficient soils have become focused on making the most efficient use of available soil P so that crop production can be sustained with minimum P applications. A principal component of these strategies is the selection and development of species and varieties that can grow well at lower levels of available soil P. The identification of genetic variation in the ability of crops to grow acceptably on P-deficient soils is necessary if breeding for low-P tolerance is to proceed. Differential responses to soil P deficiency have been reported among strains corn, beans, rice, wheat and white clover[2,3,17].

An understanding of the relative magnitude of additive, dominance and epistatic genetic variances and their interaction with nonheritable agencies improves the ability of the plant breeder to make accurate

H.W. Gabelman and B.C. Loughman (Eds.), Genetic aspects of plant mineral nutrition.
ISBN-13: 978-94-010-8102-3
© *1987 Martinus Nijhoff Publishers, Dordrecht/Boston/Lancaster.*

decisions about the most effective breeding practices[1]. Genetic studies of the control of tolerance to P deficiency in the field appear to be limited to investigations with corn[13,16,18]. The genetic control of certain physiological parameters thought to be important in conditioning low-P tolerance has been investigated in beans[8,12,19] and tomatoes[10].

We recently reported variation between strains of tomato in the ability to grow under P deficiency imposed in sand-alumina culture[5]. The sand-alumina culture technique mimics conditions of low P availability in soil while avoiding some of the problems encountered in that complex medium[6]. We now report on the genetic control of low P tolerance in a family derived from two of the differentially tolerant tomato strains.

Materials and methods

Strains 55 (P.I. 102886) and 214 (P.I. 126409) were shown previously to differ in growth in sand-alumina cultures with low P levels, and were designated as tolerant (t) and intolerant (i) of low P stress, respectively. The strains grew similarly with high levels of P in culture. The inheritance of low-P tolerance was investigated in this study in parental (P_1 = strain 55 and P_2 = strain 214), F_1, reciprocal F_1 (F_1'), backcross (BCP_1 and BCP_2), reciprocal backcross (BCP_1' and BCP_2'), and F_2 populations (Table 1).

Ten individual plants of each parental, F_1 and F_1' population, 20 plants of each F_2 population, and 15 plants of each backcross population were grown at low P ($10 \mu M$), while 3 plants of each parent and 4 plants of each F_1 population were grown at high P (+P, about $1000 \mu M$) in a completely randomized design. As in previous work, the experiment was conducted in the University of Wisconsin Biotron in rooms maintained under 16 hr photoperiods with 27 C day and 20 C night temperatures, 70% relative humidity, and a photosynthetic photon flux density of about $500 \mu mol\, m^{-2} sec^{-1}$. Nutrient solution compositions, sand-alumina preparation, and transplanting procedures were as described previously[5,6]. Fourteen-day-old seedlings were raised in perlite mois-

Table 1. Pedigree, shoot dry weight (SDW), and coefficients of variation (CV) for generations evaluated with low P and high P in sand-alumina

Generation	Pedigree	Mean SDW (g)	CV	Pooled CV*
P_1	55	2.04	27%**	
P_2	214	0.83	16%**	
F_1	55 × 214	2.36	16%**	
F_1'	214 × 55	3.05	15%**	
F_2	$(22 × 214)^2$	1.93		29%
F_2'	$(214 × 55)^2$	1.73		
BCP_1	55 × (55 × 214)	2.35	30%	
BCP_1'	(55 × 214) × 55	3.37	31%	
BCP_2	214 × (214 × 55)	1.83		30%
BCP_2'	(214 × 55) × 55	1.57		
$P_1(+P)$	55	3.54	16%**	
$P_2(+P)$	214	3.73	17%**	
$F_1(+P)$	55 × 214	2.65		19%**
$F_1'(+P)$	214 × 55	3.18		

* Pooled CV for reciprocal populations not differing according to t-test at $P < .05$.
** Mean CV of nonsegregating generations = 18%.

tened with a high-P-containing nutrient solution. Shoots of plants were harvested and dried at 65°C after 21 days of growth in sand-alumina cultures. Shoot P contents were determined using a vanadomolybdate procedure. Total root lengths for each plant were determined using a line-intercept method.

Statistical and genetic analyses were performed on shoot dry weight (SDW) data. Reciprocal differences in F_1 and BCP_1 populations were detected with t-tests (Table 1). The identity of these populations was retained in subsequent analyses, whereas the data from reciprocal F_2, BCP_2, and F_1 (+P) populations were pooled and reciprocal designations in nomenclature dropped. The coefficients of variation (CV) of the nonsegregating P_1, P_2, F_1, F_1', $P_1(+P)$, $P_2(+P)$, and $F_1(+P)$ populations were relatively constant and averaged 18 percent, indicating that the environmental variance increased proportionally with the population means. Environmental variance components of segregating populations were estimated by assuming 18 percent CVs from environmental influences. Genetic variance was calculated as the difference between the average phenotypic variance of the segregating populations, and their estimated environmental variance. Broad sense heritablity was calculated as the ratio of genetic variance to total genetic plus environmental variance. Genetic variance in the F_2 was further separated into additive and dominance variance[20]. The single estimate of phenotypic variance in the BCP_1 population required in these calculations was obtained by averaging phenotypic variances of the reciprocal populations.

Results

Growth and distributions

Strains 55 and 214 produced similar SDWs with high P in culture, but with low P in culture the low-P-tolerant strain 55 produced 246 percent more SDW than the low-P-intolerant strain 214 (2.04 g vs 0.83 g, Table 2). Significant heterosis was apparent in the F_1 generations with low P: the SDW of F_1 and F_1' exceeded the midparent SDW (1.44 g) by 64 and 212 percent, respectively. With high P, the F_1 and F_1' SDWs were depressed 27 and 12 percent, respectively, from the midparent SDW (3.63 g).

As expected, the CVs for segregating populations generally were larger than for parental and F_1 populations (Table 1). The mean of the BCP_2 population digressed toward the P_2 mean from the F_1. The mean of the BCP_1 and BCP_1' populations remained equivalent to the low-P tolerant F_1' population. The BCP_1' population mean was only 7% less than the high midparent (3.37 g vs 3.63 g, Table 2). Two exceptional plants of BCP_1' produced more SDW at low P than any of the parental or F_1 plants growing with high P. Growth of F_1' plants with low P was as good as F_1 plants with the high level of P.

Genetic analyses

Genetic variance was larger than environmental variance in the segregating generations (Table 3), hence broadsense heritability estimates were moderate to high. Genetic variance in the backcross generations equalled or exceeded genetic variance in the F_2. Consequently, a negative estimate of additive variance was obtained, and estimated dominance variance was large. These estimates were made on the assumption of no epistasis.

Table 2. Distribution of shoot dry weights (SDWs) for the family strain 55 × strain 214 when grown with low P or high P in sand-alumina cultures

Generation	SDW mean (g)	Number of plants in class midpoint						Total plants
		1.0	2.0	3.0	4.0	5.0	6.0	
$P_1 = 55$	2.04	2	6	2				10
$P_2 = 214$	0.83	10						10
F_1	2.36	8	2					10
F_1'	3.05	1	8	1				10
F_2	1.83	12	25	3				40
BCP_1	2.35	1	8	5	1			15
BCP_1'	3.37		3	7	3	1	1	15
BCP_2	1.70	2	17	2				30
$P_1(+P)$	3.54			2	1			3
$P_2(+P)$	3.73			2	1			3
$F_1(+P)$	2.92		2	5	1			8

Phosphorus uptake, utilization and root extension

Differences in SDW production between the segregating populations (Table 2) appeared more related to differences in abilities to acquire P than differences in internal P utilization (Table 4). Low-P-grown progeny exhibited similar P utilization ratios (PUR) as the low-P midparent, so the considerably greater P uptake of F_1' and BCP_1' resulted in greater SDWs. Root extension ratios (RER) of progeny consistently tended toward the P_1 RER value. As expected, high-P receiving generations showed greater P uptake, lower SDW conversion efficiency (PUR), and less proportionally extensive root systems than plants grown at low P.

The two exceptionally high yielding plants in the BCP_1' (Table 2) acquired nearly as much P from the low-P medium as did the plants from the high-P medium (Table 4). In contrast to the high-P plants, however, these BCP_1' plants retained the high-P utilization efficiency characteristic of the other plants in the BCP_1'. The high P uptake capacities of the exceptional BCP_1' plants was not associated with proportionally greater root extension.

Table 3. Estimated shoot dry weight variances of segregating populations among progeny of tomato strains 55 and 214 together with additive and dominance genetic components. Plants were grown at low P

Generation	Estimated variances*					Broadsense heritability
	V_P	V_E	V_G	V_A	V_D	
BCP_1	0.481	0.169	0.312			65%
BCP_1'	1.064	0.348	0.717			67%
BCP_2	0.254	0.094	0.160			63%
F_2	0.278	0.109	0.169	−0.337	0.506	61%

* Subscripts: P = phenotypic, E = environmental, G = genetic, A = additive, D = dominance.

Table 4. Phosphorus uptake, P utilization ratios, and root extension ratios for progeny of tomato strains 55 and 214 evaluated in sand-alumina with low P and with high P (+P)

Generation	P uptake (mg/plant)	P utilization ratio (mg SDW/mg P)	Root extension ratio (m/g SDW)
P_1	5.97 ± 0.59	346 ± 13	35.2 ± 2.2
P_2	1.98 ± 0.10	418 ± 7	47.8 ± 2.7
midparent	*3.98*	*381*	*42.0*
F_1	5.90 ± 0.35	404 ± 9	36.4 ± 1.0
F_1'	8.22 ± 0.47	376 ± 15	31.5 ± 0.8
F_2	4.96 ± 0.21	370 ± 6	35.3 ± 0.8
BCP_1	6.60 ± 0.53	362 ± 9	36.2 ± 1.2
BCP_1'	9.38 ± 0.73	361 ± 12	34.9 ± 1.6
BCP_2	4.81 ± 0.41	369 ± 10	38.6 ± 1.3
$P_1(+P)$	20.84 ± 2.23	171 ± 2	22.8 ± 2.2
$P_2(+P)$	18.03 ± 0.62	206 ± 15	27.0 ± 2.4
midparent	*19.44*	*187*	*24.9*
$F_1(+P)$	15.05 ± 0.91	194 ± 10	23.5 ± 1.4
Exceptional plants			
BCP_1' 1	13.95	357	33.8
BCP_1' 2	15.63	370	29.7

Standard errors of the mean are given with population mean values.

Discussion

The maternal effects detected between F_1 versus F_1', and BCP_1 versus BCP_1' generations do not appear to be cytoplasmic in origin because, according to the pedigree (Table 1), BCP_1 and BCP_1' share the same cytoplasm. With the exception of the F_2, the data are consistent with strain 55 being a poorer seed parent than strain 214.

Significant heterosis was exhibited in the SDW production of F_1 generations growing at low P. In contrast, the F_1 generations produced less SDW than both parents when P was in abundant supply. This result supports Lafever's suggestion[11] that nutritionally stress tolerant strains can be at a selective disadvantage when nutritionally luxuriant conditions prevail.

Genetic analyses confirmed the importance of dominance genetic variance in the observed patterns of segregation (Table 3). The estimated 'dominance' variance also would include components of epistatic variance. Dominance and epistatic variance were found more important than additive variance for root development in beans and cucumbers under P stress[7,9], and in P utilization efficiency in bean[8]. Other workers also have reported epistasis to be important in potassium and nitrogen utilization in tomatoes[14,15]. The exploitation of nonadditive genetic variance in breeding strategies is somewhat more difficult that that of additive genetic variance. Nonetheless, the range of phenotypic variation in the backcross generations suggested significant effects of recombination and complimentation among a relatively small number of genes. Thus the outlook for breeding genotypes more tolerant to low P appears favorable.

Large differences between and within generations appeared to be related primarily to differences in abilities to acquire P. The exceptional BCP'_1 progeny had the remarkable capability of acquiring nearly as much P from the low P medium as the generations receiving high P (Table 4). Because the enhanced P acquisition of these progenies was not associated with more extensive rooting, high uptake rates per m of root must have been responsible for their high total P acquisition. The effectiveness of increased uptake per unit of root for improving P acquisition has been questioned[4].

The high SDW yields of the BCP'_1 population in general, and the exceptional BCP'_1 progeny in particular, were not obtained solely by virtue of high P acquisition. Internal P reserves of comparable magnitude to those in the exceptional BCP'_1 progeny were also present in $P_1(+P)$, $P_2(+P)$, and $F_1(+P)$ plants (Table 4). However, such large reserves of P were associated with greatly reduced P utilization efficiency in all $+P$ populations, including the F_1. The backcross to P_1 apparently produced gene recombination or complimentation arrangements such that high P utilization ratios were maintained even in the presence of high internal P reserves. Alternatively, high conversion efficiency may have been induced by the low external P supplies. Mechanisms of this kind should make it possible to substantially improve the use of soil P where P availability is limiting by increasing P uptake and utilization. Depending on the nature of the control of P utilization in the presence of high external P concentrations, it also may be possible to improve the utilization of plant P where soil P levels are high enough to support large internal P reserves.

Acknowledgements Journal Series 2995 of the Hawaii Institute of Tropical Agriculture and Human Resources. This work was partially supported by National Science Foundation Grants PCM-7909808 and DAR-791226. The authors wish to express their gratitude to Mark Weisensel and the staff of the University of Wisconsin Biotron for technical assistance.

References

1. Allard R W 1960 Principles of Plant Breeding. John Wiley & Sons, Inc., New York, 485 p.
2. Caradus J R 1982 Genetic differences in the length of root hairs in white clover and their effect on phosphorus uptake. *In* Plant Nutrition 1982. Vol. 1. Ed. A Scaife. CAB, Slough, UK, 84–88.
3. Centro Internacional de Agricultura Tropical 1981 CIAT Report 1981. Cali, Colombia.
4. Chapin F S III and R L Bieleski 1982 Mild phosphorus stress in barley and a related low phosphorus-adapted barley grass. Phosphorus fractions and phosphate absorption in relation to growth. Physiol. Plant. 54, 309–317.

5   Coltman R R, Gerloff G C and Gabelman W H 1985 Differential tolerance of tomato strains to maintained and deficient levels of phosphorus. J. Am. Soc. Hort. Sci. 110, 140 144.
6   Coltman R R, Gerloff, G C and Gabelman W H 1982 A sand culture system for simulating plant responses to phosphorus in soil. J. Am. Soc. Hort. Sci. 107, 938 942.
7   Fawole I, Gabelman W H and Gerloff G C 1982 Genetic control of root development in beans (*Phaseolus vulgaris* L.) grown under phosphorus stress. J. Am. Soc. Hort. Sci. 107, 98 100.
8   Fawole I, Gabelman W H, Gerloff G C and Nordheim E V 1982 Heritability of efficiency in phosphorus utilization in beans (*Phaseolus vulgaris* L.) grown under phosphorus stress. J. Am. Soc. Hort. Sci. 107, 94–97.
9   Ghaderi A and Lower R L 1979 Gene effects of some vegetative characters of cucumber. J. Am. Soc. Hort. Sci. 104, 141 144.
10 Hochmuth G J 1980 An intact-plant screenign technique and its use in studying variation and physiology of phosphorus uptake and translocation in strains of tomato. PhD Thesis Univ. of Wisconsin, Madison.
11 Lafever H N 1981 Genetic differences in plant response to soil nutrient stress. J. Plant Nutr. 4, 89–109.
12 Lindgren D T, Gabelman W H and Gerloff G C 1977 Variability of phosphorus uptake and translocation in *Phaseolus vulgaris* L. under phosphorus stress. J. Am. Soc. Hort. Sci. 102, 674–677.
13 Lyness A S 1936 Varietal differences in the phosphorus feeding capacity of plants. Plant Physiol. 11, 665 668.
14 Makmur A, Gerloff G C and Gabelman W H 1978 Physiology and inheritance of efficiency in potassium utilization in tomatoes grown under potassium stress. J. Am. Soc. Hort. Sci. 103, 545 549.
15 O'Sullivan J, Gabelman W H and Gerloff G C 1974 Variations in efficiency of nitrogen utilization in tomatoes (*Lycopersicon esculentum* Mill) grown under nitrogen stress. J. Am. Soc. Hort. Sci. 99, 543–547.
16 Pulam T 1978 Genetic and agronomic studies of efficiency in phosphate utilization by corn (*Zea mays* L.). PhD Thesis Univ. of Hawaii, Honolulu.
17 Sanchez P A and J G Salinas 1981 Low input technology for managing Oxisols and Ultisols in tropical America. *In* Advances in Agronomy. Vol. 34. Amer. Soc. Agron., Madison, Wisconsin.
18 Smith S N 1934 Response of inbred lines and crosses in maize to variations of nitrogen and phosphorus supplied as nutrients. J. Am. Soc. Agron. 26, 785 805.
19 Whiteaker G P, Gerloff G C, Gabelman W H and Lindgren D 1976 Intraspecific differences in growth of beans at stress levels of phosphorus. J. Am. Soc. Hort. Sci., 101, 472 475.
20 Warner J N 1952 A method of estimating heritability. Agron. J. 44, 427 430.

Genotypic differences in the chemical composition of maize plants grown on a calcareous chernozem

T. ZAHARIEVA
N. Poushkarov Institute of Soil Science and Yield Prediction, Sofia, Bulgaria

Key words Calcareous chernozem Iron efficiency Maize hybrids Nutrient contents Nutrient ratios Zinc efficiency

Summary A field experiment was carried out with 20 maize hybrids growing on a calcareous chernozem and their chemical composition was determined at the 8–10 leaf stage. The nutrient concentration varied from hybrid to hybrid but were within the sufficiency range. Of the elements studied only N has shown a correlation between the concentration and yield (r = 0.53). The hybrids are characterized by comparatively constant percentage ratios of macronutrients and micronutrients, while the macro and micronutrient totals as well as N/K, K/Ca, N/P, Fe/Zn and Fe/Mn ratios vary. No regular relationship has been observed between the efficiency of the hybrid, assessed on the basis of its ability to accumulate a nutrient, and the yield it produces.

Introduction

It has been proved that the hybrid is undoubtedly a major factor in intensive maize production. The conditions of soil and climate as well as economic and agrotechnical factors are considered to be important when establishing the varietal distribution of maize in Bulgaria.[6].

Calcareous soils represent a great part of the agriculture area in Bulgaria on which maize is planted. Our previous studies[5,8,10] have shown that high carbonate contents combined with intensive P-fertilizing induce Zn and Fe chlorosis in maize.

Recently there has been a growing interest in genetic specificity of plant mineral nutrition. Maize lines differ markedly in their uptake, translocation and utilization of mineral elements[2,9]. It has been assumed that a given maize line is more effective with respect to a certain nutrient, if it accumulates higher concentrations of this nutrient at a certain level of the nutrient in the soil[1]. Fitting the plant to the soil can be another step forward in achieving maximum crop production with greater efficiency.

The object of this study is to compare the chemical composition of 20 maize hybrids at some indicative phenophase (8–10 leaf stage of growth) by using field experiments on a calcareous chernozem and on the basis of the results recommend certain hybrids as most suitable for the conditions characteristic of the calcareous soils.

Materials and methods

A field experiment was carried out on a calcareous chernozem (Table 1). The following fertilizer rates were used: N–200 kg/ha, P_2O_5 — 120 kg/ha and K_2O — 100 kg/ha. P and K fertilizers were

H.W. Gabelman and B.C. Loughman (Eds.), Genetic aspects of plant mineral nutrition.
ISBN-13: 978-94-010-8102-3
© *1987 Martinus Nijhoff Publishers, Dordrecht/Boston/Lancaster.*

Table 1. Chemical analysis of the calcareous chernozemic soil

Factor	Value
$CaCO_3$ (%)	14.8
pH (1:2.5 water)	8.1
Organic matter (%)	3.13
P (mg kg^{-1})	34.4
K (mg kg^{-1})	251.5
Fe (mg kg^{-1})	2.8
Zn (mg kg^{-1})	1.24
Mn (mg kg^{-1})	20.0

Table 2. Chemical composition of leaves at 8-10 leaf stage and yield of maize hybrids

Hybrids	N	P	K	Ca	Mg	Fe	Mn	Zn	Cu	Yield
	%					mg/kg				t/ha
Early and mid-early										
Russe 91	4.38	0.36	3.33	0.57	0.39	265	139	37	22	16.75
Pioneer 39787	4.06	0.37	2.18	0.52	0.41	205	118	27	24	15.50
BC 383	4.12	0.35	2.63	0.73	0.45	220	160	34	28	14.75
Pioneer 3747	3.85	0.29	1.53	0.57	0.47	200	118	32	27	13.25
Knezha 430	3.92	0.37	2.15	0.49	0.40	180	117	38	23	10.75
Knezha 470	3.61	0.32	2.09	0.61	0.46	200	135	39	25	10.75
Mid-late and late										
NSSC 606	4.42	0.34	2.65	0.59	0.44	225	160	45	24	15.75
Pioneer 3186	4.18	0.35	2.63	0.71	0.45	240	131	28	25	15.50
BC 622	4.19	0.34	2.94	0.61	0.41	235	155	40	24	15.25
Knezha 611	4.06	0.35	2.46	0.54	0.41	190	120	36	29	13.75
H 708	4.12	0.26	3.05	0.56	0.34	195	124	32	21	13.50
Knezha (HP) 633	4.05	0.32	1.80	0.72	0.50	190	130	33	23	13.00
KWS 533	3.69	0.33	2.77	0.51	0.32	210	117	29	20	12.75
Knezha (HP) 556	3.97	0.35	2.24	0.60	0.38	200	114	37	22	12.50
Pioneer 3184	4.46	0.36	2.15	0.70	0.50	240	123	31	27	12.50
Px 715	3.96	0.39	3.19	0.60	0.39	280	180	28	29	12.50
Px 723	4.52	0.36	2.72	0.67	0.49	295	165	29	25	12.50
Russe 504	3.88	0.30	3.12	0.50	0.31	248	132	30	24	12.50
BC 778	4.23	0.38	2.23	0.67	0.57	225	160	32	24	12.00
Knezha 530	3.79	0.36	2.55	0.56	0.39	200	114	38	25	10.75

applied in August and the N fertilizer rate was split up, applying 100 kg/ha before surviving in April and the rest in June. Soil samples were taken in April and analyzed for available P, K, Fe, Zn and Mn. Available P and K were determined in an acetate-lactate extract[3], Fe, Zn and Mn — by an EDTA extraction at pH 8.6[7] (Table 1). Twenty maize hybrids belonging to different maturity groups were used in the experiment (Table 2). Leaf samples were taken at the 8-10 leaf stage of growth at the end of August washed with distilled water and dried at normal temperature. The concentrations of P, K, Ca, Mg, Fe, Zn, Mn and Cu in leaves were determined after dry ashing and dissolving the residue in 20% HCl. Potassium was determined by flame photometry, Ca, Mg, Fe, Zn, Cu and Mn by atomic absorption spectrophotometry, N by the Kjeldahl method and P by the vanadate–molybdate method.

Results and discussion

The data in Table 2 show that grain yields range between 10.75 and 16.75 t/ha. The highest yield, obtained from the mid-early hybrid Russe

91, is 60% higher than the lowest yield, obtained from the hybrid Knezha 470 which belongs to the same maturity group according to the FAO classification.

The nutrient concentrations vary from hybrid to hybrid but are within the sufficiency range (Table 2)[4]. No relationship has been found between nutrient concentrations at this stage and the yield, except for N, but even in this case the correlation coefficient was relatively low (r = 0.53, F = 7.13839***).

In Tables 3 and 4 the date are shown on the percentage ratios of macro and micronutrients, the totals of which are assumed to be 100. The totals of macro and micronutrients also vary from hybrid to hybrid. On the whole, the hybrids producing very high yields (over 15 t/ha), have a total concentration of macronutrients of 8.36 on the average, whereas with hybrids giving the lowest yields (Knezha 430, Knezha 470 and Knezha 530) it averages 7.36.

The highest total concentration of macronutrients in the Russe 91 hybrid is, undoubtedly, a factor contributing to this hybrid producing the maximum yield. This index, however, should not be overestimated, since with the hybrids producing medium yields (12–14 t/ha) the macronutrient totals are both above and below 8.

Table 3. Totals and ratios of macronuttrients in leaves of maize hybrids at 8–10 leaf stage

Hybrids	$\sum$ (g 100 g DW)	N	P	K	Ca	Mg	$\frac{N}{K}$	$\frac{K}{Ca}$	$\frac{N}{P}$
		% of total							
Early and mid-early									
Russe 91	9.03	49	4	37	6	4	1.3	6.2	12.2
Pioneer 3978	7.54	54	4	29	7	5	1.9	4.1	10.8
BC 383	8.28	50	4	32	9	5	1.6	3.6	12.5
Pioneer 3747	6.71	57	4	24	8	7	2.4	3.0	14.2
Knezha 430	733	53	5	29	7	6	1.8	4.1	10.6
Knezha 470	7.09	51	5	29	9	6	1.8	3.2	10.2
Mid-late and late									
NSSC 606	8.44	52	4	31	7	6	1.7	4.4	13.0
Pioneer 3186	8.32	51	4	32	9	6	1.6	3.6	12.7
BC 622	8.49	49	4	35	7	5	1.4	5.0	12.2
Knezha 611	7.82	52	4	31	7	6	1.7	4.4	13.0
H 708	8.33	49	3	37	7	4	1.3	5.3	16.3
Knezha (HP) 633	7.39	55	4	24	10	7	2.3	2.1	13.7
KWS 533	7.62	48	4	36	7	5	1.3	5.1	12.0
Knezha (HP) 556	7.54	53	5	30	8	4	1.8	3.7	10.6
Pioneer 3184	8.17	55	4	26	9	6	2.1	2.9	13.7
Px 715	8.53	46	4	37	7	6	1.2	5.3	11.5
Px 723	8.76	52	4	31	8	5	1.7	3.9	13.0
Russe 504	8.11	48	4	38	6	4	1.3	6.3	12.0
BC 778	8.02	53	5	28	8	6	1.9	3.5	10.6
Knezha 530	7.65	50	5	33	7	5	1.5	4.7	10.0

Table 4. Totals and ratios of micronutrients in leaves of maize hybrids at 8–10 leaf stage

Hybrids	$\sum$ mg kg DW	Fe	Mn	Zn	Cu	$\frac{Fe}{Mn}$	$\frac{Fe}{Zn}$
		% of total					
Early and mid-early							
Russe 91	463	57	30	8	5	1.9	7.1
Pioneer 3978	378	54	30	7	7	1.7	7.7
BC 383	441	50	36	8	6	1.4	6.2
Pioneer 3747	368	54	30	9	7	1.8	6.0
Knezha 430	358	50	33	11	6	1.5	4.5
Knezha 470	399	50	34	10	6	1.5	5.0
Mid-late and late							
NSSC 606	454	50	35	10	5	1.4	5.0
Pioneer 3186	424	57	30	7	6	1.8	8.1
BC 622	454	52	34	9	5	1.5	5.8
Knezha 611	375	50	32	10	8	1.5	5.1
H 708	372	52	33	9	6	1.6	5.8
Knezha (HP) 633	376	50	35	9	6	1.5	5.7
KWS 533	376	56	31	8	5	1.8	7.0
Knezha (HP) 556	373	54	30	10	6	1.7	5.4
Pioneer 3184	421	57	30	7	6	2.0	8.1
Px 715	517	54	35	5	6	1.5	10.8
Px 723	514	57	32	6	5	1.8	9.5
Russe 504	434	57	30	7	6	1.8	8.1
BC 778	441	51	36	7	6	1.4	7.3
Knezha 530	377	53	30	10	7	1.8	5.3

Unlike the totals of macro and micronutrients, the percentage ratios (Tables 3 and 4) of the nutrients vary within a much narrower ragne. For example, P (with only one exception) ranges between 4 and 5%, Fe between 50 and 57%. A relatively wider variation range has been detected only for Zn: its percentage in Knezha 430 is more than twice higher than in Px 715. The results show that the percentage ratios of the elements remain relatively constant and do not depend on the hybrid or the yield produced.

Greater differences between the hybrids have been found in the N/K, K/Ca, Fe/Zn and Fe/Mn ratios which are used in plant diagnostics because the elements included in the ratios vary within a wider range. Every hybrid is characterized by fixed ratios (Tables 3 and 4). Closer values have been found only for Russe 91 and Russe 504, in spite of the fact that they produce different yields. It is difficult to say whether a fixed combination of ratios is a necessary condition for obtaining a maximum yield. If Zn-efficient hybrids are examined on the basis of Zn contents in plants, the NSSC 606 and BC 622 hybrids fall into this category. They have produced relatively high yields (over 15 t/ha) in the experiment used. On the other hand, the hybrids Pi 3978 and Pi 3186 have lower Zn-efficiency, judging by Zn-concentration, but have also produced high yields of more than 15 t/ha under the conditions of the experiment. If assessed on the basis of their Fe-concentration, Pi 3186, Pi 3184 and BC 622 have equal Fe-efficiencies but eventually give different yields (Table 2). As for Ca-efficiency, the least Ca-efficient hybrid (an advantage in calcareous soils) Knezha 430 gives the lowest yield.

The above results indicate that chemical analyses at the 8–10 leaf stage distinguishes between Zn-efficient, Fe-efficient, *etc.*, hybrids. There is no regular relationship, however, between the efficiency of a hybrid regarding its ability to accumulate a given nutrient (*e.g.* Zn and Fe in a calcareous chernozem) and the yield it produces. Obviously, this relationship is much more intricate and invites further investigations.

Acknowledgements The author would like to acknowledge the assistance of Yanko Stoev (Agro-Industriual Complex of Novi Pazar) and Zlatka Marinova (Shoumen District Soil Test Service) in the agronomic part of this work. Thanks also go to Dr. B C Loughman for his critical review of the manuscript.

References

1. Clark R B and Brown J C 1975 Corn lines differ in mineral efficiency. Ohio Report. 60, 83–86
2. Clark R B 1983 Plant genotype differences to uptake, translocation accumulation and use of mineral elements required for plant growth. Plant and Soil 72, 175 196.
3. Ivanov P 1984 New acetate-lactate method for determining soil phosphorus and potassium available to plants. Pochvoznanie i agrokhimia. 19, 88–98.
4. Jones J B Jr 1967 Interpretation of plant analysis for several agronomic crops. *In* Soil Testing and Plant Analysis Part II, SSSA Spec. Publ. Ser., Madison, Wis.
5. Sotyanov D and Peneva N 1985 Zinc fertilizing. Pochvoznanie, agrokhimia i rastitelna zashtita. 20, 74–81.
6. Tomov N and Y Yordanov 1984 Corn in Bulgaria. Sofia.
7. Trierweiler J F and Lindsay W L 1969 EDTA ammonium carbonate soil test for Zn. Soil Sci. Soc. Am. Proc 33, 49–54.
8. Zaharieva T 1978 Fe-chlorosis in maize plants grown on a calcareous chernozem. Pochvoznanie i agrokhimia. 13, 56–63.
9. Zaharieva T 1982 Differential response of corn genotypes to applied Fe-EDDHA. J. Plant Nutr. 5, 897–904.
10. Zaharieva T 1984 Response of corn to application of Fe and Zn to calcareous soils. Proc. 9th World Federation Congr., Budapest, II, pp 95 98.

Genotypic differences in subcellular compartmentation of K^+: Implications for protein synthesis, growth and yield

ABDUL RAZAQUE MEMON* and ANTHONY D.M. GLASS
Department of Botany, University of British Columbia, Vancouver, B.C., Canada V6T 2B1

Key words Barley varieties Potassium *In vivo* protein synthesis Subcellular compartmentation

Summary The subcellular distribution of K^+ in roots of three barley varieties, which was previously estimated by compartmental analysis[16], clearly indicated that varietal differences in K^+ utilization are associated with differences in the allocation of this ion between cytoplasmic and vacuolar compartments. In all varieties vacuolar $[K^+]$ increased with increasing availability of K^+, whereas cytoplasmic $[K^+]$ changed only slightly. Nevertheless, Betzes showed deficiency symptoms when grown at 10 mmol m^{-3} $[K^+]_o$ even though it accumulated more K^+ than Compana. Betzes appeared to be unable to mobilize K^+ from the vacuole to the cytoplasm and demonstrated a significant increase in cytoplasmic $[K^+]$ when $[K^+]_o$ was raised to 100 mmol m^{-3}. By contrast Compana showed no change in cytoplasmic $[K^+]$ under these conditions. Rates of ^{3}H-leucine incorporation into proteins were severely reduced in the roots and shoots of Betzes at 10 mmol m^{-3} $[K^+]_o$. On the other hand at 100 mmol m^{-3} $[K^+]_o$, Betzes showed significantly higher (40 to 90%) rates of protein synthesis in roots compared to Compana. No significant differences were found in protein synthesis rates between the shoots of these varieties at this concentration of $[K^+]_o$. These results clearly indicate that Betzes requires higher cytoplasmic K^+ than Compana for optimal protein synthesis. We propose that K^+ utilization efficiency in these varieties is a function of subcellular distribution of K^+ and differential K^+ requirements for protein synthesis. The higher $[K^+]_{vac}$ in Betzes at 10 mmol m^{-3} $[K^+]_o$ despite the failure to maintain cytoplasmic $[K^+]$ might be due to characteristic properties of the tonoplast transport mechanism and/or to lack of available sugars to provide an alternative source of turgor.

Introduction

The efficiency with which mineral nutrients are used is an important factor in determining overall efficient nutrition, especially when the nutrient supply from the soil is suboptimal[7,8]. Evidence showing that efficiency of utilization varies with genotypes of crop plants has been reported in our laboratory[15,17] and Gerloff and Gabelman's group[8,9]. It is clear that efficiency of this kind is an inheritable trait, but the underlying mechanisms which determine it are unclear.

In an earlier study, directed toward resolving genotypic differences in K^+ utilization, we observed that, *in vitro*, K_m values for K^+-activation of pyruvate kinase from selected barley varieties were not significantly different. The average K_m value for all 12 varieties (3.9 mol m^{-3}) was so far below current estimates of cytoplasm $[K^+]$ (100–200 mol m^{-2}) that

---

* Present address: Department of Biology, Middle East Technical University, Ankara, 06531, Turkey.

differential requirements for activation of particular enzymes were considered to be unlikely causes of differences in utilization efficiency[15].

We considered, therefore, that differential patterns of K^+ allocation between subcellular compartments (vacuole and cytoplasm) in different barley genotypes might contribute to the observed variations in utilization. We estimated the subcellular distribution of K^+ in the roots of three barley varieties, grown at low and high external K^+ by compartmental analysis[16]. The three varieties which were selected on the basis of their different rates of K^+ utilization, showed distinct differences in the allocation of K^+ between cytoplasm and vacuole[16]. At 10 mmol m^{-3} $[K^+]_o$ Betzes (an inefficient responder) exhibited typical K^+ deficiency symptoms (yellowing of leaf tips) while Fergus (an efficient responder) and Compana (an efficient non-responder) did not, even though Betzes had the higher $[K^+]$ in shoots and roots. The inefficient utilization of K^+ in Betzes appears to be associated with a failure to mobilize vacuolar K^+ into the cytoplasmic compartment. Fergus and Betzes, which demonstrate pronounced growth responses to increased $[K^+]_o$ between 10 and 100 mmol m^{-3}, showed significant increases of cytoplasmic $[K^+]$ in this range of $[K^+]_o$. By contrast, cytoplasmic $[K^+]$ in Compana, a variety whose growth is not stimulated by increased $[K^+]_o$ (from 10 to 100 mmol m^{-3}), showed virtually no increase. On the basis of these results, we suggested that the efficiency of K^+ utilization and the growth response to $[K^+]_o$ in these varieties are functions of the subcellular distribution of this ion between cytoplasm and vacuole[16].

Although K^+ contributes to the osmotic potential of the cytoplasm[3,18], an important function in this compartment is thought to be associated with the activation of enzymes[1,6,15] and protein synthesis[2,10,11,13,18]. Wyn Jones et al.[18] and Leigh and Wyn Jones[12] have proposed that high cytoplasmic $[K^+]$ (100–200 mol m^{-3}) which characterizes most eukaryote cells is determined by the requirement for protein synthesis. In the view of these reports and considering the results of our compartmental analyses[16], we considered that the efficient performance of Compana at low available levels of K^+ might be associated with a lower requirement for protein synthesis than that of Betzes.

In this communication we report *in vivo* rates of protein synthesis in leaves and roots of these two barley varieties grown at 10 and 100 mmol m^{-3} $[K^+]_o$.

Materials and methods

Plant material

Two barley (*Hordeum vulgare* L.) varieties were selected according to their relative growth and utilization efficiencies[15,16]. Betzes (inefficient responder) and Compana (efficient non-responder) were germinated in sand for 3 days and then supported on plexiglass discs and treated for 2 weeks with 10 mmol m^{-3} and 100 mmol m^{-3} [K$^+$]$_o$, respectively, in 36 l hydroponic tanks. The nutrient solution used in these experiments was 0.01 strength modified Johnson's solution[4] and micronutrients (of 0.1 strength Johnson's) without K$^+$. Nutrient levels were maintained by continuous additions of nutrients using peristaltic pumps which fed into the hydroponic tanks (for detailed procedures of growth and maintenance of nutrient levels see [15,16,17]).

The plants were maintained in a growth room on a 16 h light — 8 h darkness cycle at 25 ± 2 C and 70% relative humidity. Light was provided at 90 W m^{-2} (plant level) by 'vita-lite' fluorescent tubes. All nutrient solutions were equilibrated to growth room temperature one day prior to experiments. All experimental manipulations were performed under the same conditions of light and temperature as described above.

^{3}H-leucine uptake and incorporation

The uptake and incorporation of leucine into whole plants and excised leaves of Betzes and Compana barley varieties, grown at 10 and 100 mmol m^{-3} K$^+$, were measured at growth room temperature for 2, 5 and 10 h.

Plants were first pretreated for 1 h in sterilized 0.01 strength Johnson's solution with 10 or 100 mmol m^{-3} K$^+$. Tetracycline (6.0 mg l^{-1}) was added in these solutions to control bacterial contamination. The discs, containing three plants each, were then transferred to uptake solution (containing 0.01 strength Johnson's solution and micronutrients of 0.1 strength Johnson's without K$^+$, 10 or 100 mmol m^{-3} K$^+$, and 200 mmol m^{-3} unlabelled ^{12}C-leucine) labelled with L-[4,5 — ^{3}H] leucine (0.5 μCi/ml; specific activity 64 Ci mmol^{-1}; Amersham U.K). During uptake periods of 2, 5 and 10 h, K$^+$ in uptake solutions was analysed every 30 minutes and K$^+$ levels of 10 or 100 mmol m^{-3} were maintained by adding appropriate amount of sterilized K$_2$SO$_4$ solution. After uptake, plants were desorbed in cold nutrient solution containing 1 mol m^{-3} unlabelled leucine for 5 minutes and shoots and roots were separated and fresh weights determined. Immediately after weighing, plant parts were frozen in liquid nitrogen and were stored at −86 C.

Protein extraction and separation

Soluble proteins were extracted according to the method of McNeil et al.[14] except that 1.0 mol m^{-2} phenylmethylsulphonil fluoride (Sigma Chemical Co.) was added to the extraction buffer. The insoluble material was removed by centrifugation at 35,000 g for 15 min and the soluble fraction was used for determining the total ^{3}H-leucine uptake and incorporation into proteins.

For determining total ^{3}H-leucine uptake, 0.2 ml of the soluble fraction were dispensed into glass scintillation vials. In the case of root extracts they were digested with 2 cm^3 of protosol (500 mol m^{-3} quaternary ammonium hydroxide, New England Nuclear) for 5 h at 55 C and were counted in 10 cm^3 of aqueous counting scintillant (ACS, Amersham) in a Searle Isocap 300 scintillation counter. In the case of leaves, samples were first digested overnight with 0.4 cm^3 of H$_2$O$_2$ (for removing chlorophyll) and then further digested with 3 cm^3 of protosol for 4 to 5 h at 55 C. These samples were counted in 10 cm^3 of ACS. For protein separation 0.5 cm^3 of the soluble fraction was taken in centrifuge tubes and 1.0 cm^3 of ice cold 10% TCA (W/V) and 2 mol m^3 unlabelled leucine was added. Precipitation of protein occurred overnight at 1 C. The precipitate was compacted by centrifuging at 2500 rpm for 10 minutes in the swing-out head of the Sorvall GLC-1 centrifuge and the pellet washed twice with 5 cm^3 of cold 5% trichloroacetic acid. The pellet was resuspended in 5 cm^3 of 5% TCA and heated at 90 C for 15 minutes. The insoluble material was centrifuged down and washed with a further 5 cm^3 of cold 5% TCA. The precipitate was extracted with 5 cm^3 of diethyl ether/ethanol (1:1, V:V) for 15 minutes at 30 C and washed with 10 cm^3 ethanol. The final pellet was dissolved in 2 cm^3 of protosol and transferred to glass scintillation vials. These samples were counted in 10 cm^3 of ACS solutions in a scintillation counter. Counting error in all samples was less than 1.5%.

Each experiment was replicated three times and each replication was duplicated. The results given are the means of these replications ± s.e., expressed on a fresh weight basis.

K$^+$ concentrations shown in Tables 1 and 3 were taken from ref.[16].

Table 1. Mean root and shoot fresh weight ± s.e. (g) and $[K^+]_i$ ± s.e. (μmoles g^{-1}.f.w)* of two barley varieties (Compana and Betzes) grown for two weeks at two different $[K^+]_o$ levels

Variety		$[K^+]_o$ (mmol m^{-3})			
		10		100	
		Fresh weight	$[K^+]_i$	Fresh weight	$[K^+]_i$
Compana	Root	1.70 ± 0.14	25.2 ± 1.4	1.47 ± 0.16	62.0 ± 3.3
	Shoot	5.21 ± 0.50	77.4 ± 2.1	4.09 ± 0.19	96.9 ± 1.3
Betzes	Root	1.12 ± 0.11	44.4 ± 1.4	1.75 ± 0.12	59.3 ± 3.9
	Shoot	2.86 ± 0.29	93.2 ± 4.3	3.77 ± 0.18	100.2 ± 4.8

* $[K^+]_i$ (see Memon et al.[16])

Results and discussion

The data in Table 1 present the mean root and shoot fresh weights and $[K^+]_i$ in two barley varieties (cv. Compana and Betzes) grown at 10 and 100 mmol m^{-3} $[K^+]_o$.

Compana (an efficient non-responder) failed to respond positively in growth to increasing $[K^+]_o$; rather roots and shoots growth decreased by 15–20% at high $[K^+]_o$. On the other hand, Betzes (an inefficient responder) showed substantial increases in roots (56%) and shoots (32%) growth at high $[K^+]_o$ (Table 1).

A comparison between these two varieties showed clearly that, at low $[K^+]_o$, Compana performed better than Betzes and its root and shoot growth were respectively about 52 and 82% higher than those of Betzes. These results confirm our previous observations[15]. Data for the root and shoot K^+ concentrations were taken from our previous experiments[16] which were conducted under identical conditions. It is apparent from the results that at low $[K^+]_o$ Compana has lower root and shoot $[K^+]$ than that of Betzes. By contrast, at high $[K^+]_o$ no significant differences were found in the root and shoot $[K^+]$ between these varieties (Table 1).

Results for in vivo ^{3}H-leucine incorporation in roots and shoots of Compana and Betzes, grown at 10 mmol m^{-3} $[K^+]_o$, are shown in Table 2. Generally shoots and roots of Compana grown at 10 mmol m^{-3} $[K^+]_o$ showed higher incorporation of ^{3}H-leucine in protein than the shoots and roots of Betzes at all three uptake times. Shoots of Compana showed around 60 to 80% higher rates of incorporation of ^{3}H-leucine into protein than did shoots of Betzes (Table 2). Our previous report[16] (see also Table 4) showed that cytoplasmic concentration of $[K^+]$ does not vary significantly between Compana and Betzes when grown at 10 mmol m^{-3} $[K^+]_o$. By contrast, rates of protein synthesis were significantly higher in Compana than Betzes (Table 2). It seems likely that the optimum cytoplasmic K^+ requirement for protein synthesis in Compana

Table 2. Incorporation of ^{3}H-leucine into proteins extracted from the shoots and roots of two barley varieties (Compana and Betzes) grown at 10 mmol m^{-3} [K$^+$]$_o$ (% of total uptake ± s.e.)

Uptake time (h)	Compana		Betzes	
	Shoots	Roots	Shoots	Roots
2	4.9 ± 0.3	1.2 ± 0.1	3.1 ± 0.3	1.0 ± 0.1
5	5.4 ± 0.8	1.1 ± 0.1	3.1 ± 0.3	0.7 ± 0.1
10	6.0 ± 0.1	2.9 ± 0.1	3.7 ± 0.4	2.6 ± 0.3

is lower than that in Betzes. A more definitive test of this hypothesis will require that *in vitro* measurements of protein synthesis be obtained. It is quite interesting that, in spite of a higher vacuolar K$^+$ content in Betzes, this variety was unable to maintain an appropriate cytoplasmic K$^+$ concentration for metabolic processes. In spite of higher root and shoot [K$^+$]$_i$ and higher influx, net flux and xylem flux than Fergus and Compana, Betzes showed severe symptoms of K$^+$ deficiency in its leaves[16]. It seems evident that the deficiency symptoms of this K$^+$ inefficient variety (Betzes) are not caused by its inability to absorb and translocate K$^+$. Epstein has reported similar results for a K-inefficient tomato mutant[5]. Despite the presence of 0.1 to 0.2 mol m^{-3} K$^+$ in the medium, the leaves of this mutant showed severe K$^+$ deficiency symptoms. The gross tissue K$^+$-concentration in poorly growing leaves of the mutant was as high or higher than that in the wild-type tomato and only very high inputs of exogenous K$^+$ (20 mol m^{-3}) kept it alive. Epstein suggested that the leaf cells of the mutant sequester K$^+$ with unusual avidity within the vacuoles, rendering the cytoplasm K$^+$-deficient except under conditions of an abnormally high K$^+$ supply. Leigh and Wyn Jones[12] have proposed that there is a finite lower limit for vacuolar [K$^+$] and when this limit is reached there is no further net movement of K$^+$ from vacuole to cytoplasm. This minimum [K$^+$] ([K$^+$]$_{vac}^{min}$) value seems to vary from variety to variety[16]. For example Compana with 21 mol m^{-3} in the vacuole (Table 4) can still maintain a cytoplasmic [K$^+$] which is adequate for protein synthesis. On the other hand, Betzes with 33 mol m^{-3} in the vacuole cannot maintain the required cytoplasmic level for metabolic processes. It is apparent from these results that [K$^+$]$_{vac}^{min}$ value and the

Table 3. Incorporation of ^{3}H-leucine into proteins extracted from the shoots and roots of two barley varieties (Compana and Betzes) grown at 100 mmol m$_{-3}$ [K$^+$]$_o$ (% of total uptake ± s.e.)

Uptake time (h)	Compana		Betzes	
	Shoots	Roots	Shoots	Roots
5, Exp. 1	27.3 ± 3.4	3.5 ± 0.3	26.2 ± 2.7	5.1 ± 0.1
Exp. 2	--	2.6 ± 0.2	--	4.3 ± 0.3
10, Exp. 1	38.5 ± 1.4	5.0 ± 0.3	44.3 ± 1.2	9.6 ± 0.2
Exp. 2	--	3.6 ± 0.3		6.7 ± 0.5

Table 4. Cytoplasmic and vacuolar concentrations of K^+ (mol m^{-3} ± s.e.) in the roots of two barley varieties (Compana and Betzes) grown at 10 and 100 mmol m^{-3} $[K^+]_o$.*

Varieties	$[K^+]_o$ (mmol m^{-3})			
	10		100	
	Cytoplasm	Vacuole	Cytoplasm	Vacuole
Compana	133 ± 8	21 ± 0.5	140 ± 10	61 ± 1.6
Betzes	127 ± 14	33 ± 1.7	187 ± 12	56 ± 5.5

*K^+ concentrations were estimated by the method of compartmental analysis which has been described in detail in Memon et al.[16]

optimal K^+ requirement for protein synthesis in cytoplasm is higher in Betzes than Compana.

Data in Table 3 show the ^{3}H-leucine incorporation rates into proteins in the shoots and roots of Compana and Betzes grown at 100 mmol m^{-3} $[K^+]_o$. It is noteworthy that the incorporation of ^{3}H-leucine into root proteins was 44% and 90% greater, at 5 and 10 h respectively, in Betzes than in Compana. On the other hand, there were no significant differences in rates of incorporation in the leaves of these two varieties (Table 3).

Interestingly, when $[K^+]_o$ was increased to 100 mmol m^{-3}, cytoplasmic K^+ in the roots of Betzes increased significantly compared to the value obtained at 10 mmol m^{-3} $[K^+]_o$. By contrast, Compana failed to show significant increase in cytoplasmic K^+ over the same range of $[K^+]_o^{16}$ (see also Table 4). This again indicates clearly that the K^+ requirement for optimal protein synthesis is higher in Betzes than in Compana. Although, at 100 mmol m^{-3} $[K^+]_o$, rates of protein synthesis in both varieties increased compared to those at 10 mmol m^{-3} $[K^+]_o$, the extent of this increase varied with variety. Rates of protein synthesis in the roots of Compana were about 1.2 to 2 times higher at 100 mmol m^{-3} $[K^+]_o$ than at 10 mmol m^{-3} $[K^+]_o$ whereas in Betzes there was about a 5 fold increase of protein synthesis at the higher level of $[K^+]_o$. In agreement with these values for rates of protein synthesis, our previous report[15] noted that the total protein concentration of the roots and shoots of BT334 (a variety belonging to the same group as Betzes) was much lower than that of Compana when grown at 10 mmol m^{-3} $[K^+]_o$, but increased when this variety was grown at 100 mmol m^{-3} $[K^+]_o$.

These results are consistent with the hypothesis that plants require a cytoplasmic solute environment that is comaptible with ribosomal stability and translation of m-RNA in plants for protein synthesis[2,10]. Thus deviations in cytoplasmic K^+ concentration due to external ionic perturbations may cause severe disturbances of the potential for protein synthesis, and for growth. If these perturbations are maintained for long periods then varieties like Betzes and BT334 will show deficiency symptoms and stunted growth.

The model proposed by Leigh and Wyn Jones[12] provides a general basis for the high cytoplasmic $[K^+]$ observed in eukaryotic cells. In addition, we suggest that $[K^+]_{vac}^{min}$ level is a function of genotype and the variations in $[K^+]_{vac}^{min}$ among genotypes may be an important component of the documented differential response to applied K^+. The high $[K^+]_{vac}^{min}$ values in inefficient varieties (at low $[K^+]_o$) might be due to characteristic properties of the tonoplast transport mechanism and/or the lack of available sugars to provide an alternative source for turgor.

References

1. Besford R T and Maw G A 1975 Some properties of pyruvate kinase extracted from *Lycopersicon esculentum*. Phytochemistry 14, 677–682.
2. Brady C J, Gibson T S, Barlow E W R, Speirs J and Wyn Jones R G 1984 Salt-tolerance in plants. I. Ions, compatible organic solutes and the stability of plant ribosomes. Plant Cell Environ. 7, 571 578.
3. Cram W J 1976 Negative feedback regulation of transport in cells: The maintenance of turgor, volume and nutrient supply. *In* Encyclopedia of Plant Physiology, N.S., Vol. 2A: Transport in plants II: Cells. Eds. U Lüttge and M G Pitman. Springer-Verlag, Berlin, Heidelberg, New York, 284 315.
4. Epstein E 1972 Mineral Nutrition of Plants: Principles and Perspectives. John Wiley and Sons, New York, N.Y.
5. Epstein E 1978 An inborn error of potassium metabolism in the tomato, *Lycopersicon esculentum*. Plant Physiol. 62, 582–585.
6. Evans H J and Wildes R A 1971 Potassium and its role in enzyme activation. *In* Potassium in Biochemistry and Physiology. International Potash Institute, 13 39.
7. Fawole I, Gabelman W H, Gerloff G C and Nordheim E V 1982 Heritability of efficiency in phosphorus utilization in beans (*Phaseolus vulgaris* L.). J. Am. Soc. Hortic. Sci. 107, 94 97.
8. Gabelman W H and Gerloff G C 1983 The search for and interpretation of genetic controls that enhance plant growth under deficiency levels of a macronutrient. Plant and Soil 72, 335 350.
9. Gerloff G C and Gabelman W H 1983 Genetic basis of inorganic plant nutrition. *In* Encyclopedia of Plant Physiology, N.S., Vol. 15B: Inorganic Plant Nutrition. Eds. A Läuchli and R L Bieleski. Springer-Verlag, Berlin, Heidelberg, New York, Tokyo, pp 453–480.
10. Gibson T S, Speirs J, and Brady C J 1984 Salt-tolerance in plants. II. *In vitro* translation of m-RNAs from salt-tolerant and salt-sensitive plants on wheat germ ribosomes. Responses to ions and compatible organic solutes. Plant Cell Environ. 7, 579–587.
11. Knypl J S and Chylinska, K M 1972 Comparison of the stimulatory effect of potassium on growth, chlorophyll and protein synthesis in the lettuce cotyledons with the effects produced by other univalent ions. Biochem. Physiol. Pflanz. 163, 52 63.
12. Leigh R A and Wyn Jones R G 1984 A hypothesis relating critical potassium concentrations for growth to the distribution and functions of this ion in the plant cell. New Phytol. 97, 1 13.
13. Lubin M and Ennis H L 1964 On the role of intracellular potassium in protein synthesis. Biochem. Biophys. Acta 80, 614 631.
14. McNeil P H, Foyer C H, Walker D A, Bird I F, Cornelius M J and Keyes A J 1981 Similarity of ribulose-1,5-bisphosphate carboxylases of isogenic diploid and tetraploid ryegrass (*Lolium perenne* L.) cultivars. Plant Physiol. 67, 530 534.
15. Memon A R, Siddiqi M Y and Glass A D M 1985a Efficiency of K utilization by barley varieties: activation of pyruvate kinase. J. Exp. Bot. 36, 79–90.
16. Memon A R, Saccomani M and Glass A D M 1985b Efficiency of potassium utilization by barley varieties: the role of subcellular compartmentation. J. Exp. Bot. 36, 1860–1876.
17. Siddiqi M Y and Glass A D M 1983 Studies of the growth and mineral nutrition of barley varieties. I. Effect of potassium supply on the uptake of potassium and growth. Can. J. Bot. 61, 671–678.
18. Wyn Jones R G, Brady C J and Speirs J 1979 Ionic osmotic relations in plant cells. *In* Recent Advances in the Biochemistry of Cereals. Eds. D L Laidman and R G Wyn Jones. Academic Press, London, New York, pp 63-103.

Root morphological effects on Mg uptake in five tall fescue lines

J.F. PEDERSEN, J.H. EDWARDS and H.A. TORBERT
Department of Agronomy and Soils, Alabama Agricultural Experiment Station, and U.S. Department of Agriculture, Agricultural Research Service, Auburn University, AL 36849, USA

Key words *Festuca arundinacea* Magnesium nutrition Nutrient concentration Root diameter Xylem diameter

Summary A greenhouse experiment was conducted with tall fescue (*Festuca arundinacea* Schreb.) lines to determine the influence of root diameter and Mg concentration in nutrient solution on Mg uptake into shoots and roots. Propagules of 4 clonal tall fescue lines differing in root morphology, and a selection of 'Kentucky 31' (Ky 31), were grown for 39 or 70 days in 12-liter tanks containing a complete nutrient solution and Mg concentrations of 3, 21, 42, 125, 250, and 500 μM as $MgSO_4$. Root diameters averaged 0.98 mm for line AU-7; 0.83 mm for line AU-264; 0.72 mm for line AU-718; 0.72 mm for line AU-5; and 0.69 mm for Ky 31. At 39 days, leaf Mg concentration increased from about 1200 $\mu g/g$ at the 3 μM concentration to about 2200 to 2400 $\mu g/g$ at the 125 μM Mg concentration. Consistently, the large diameter root (LDR) lines AU-7 and AU-264 contained less Mg than the small diameter root (SDR) lines AU-5 and AU-718 and the selection of Ky 31 at both 39 and 70 days. Root Mg concentration was 50% of leaf Mg concentration. An Eadie-Hofstee plot indicated that influx of Mg proceeds via dual uptake mechanisms. The influx mechanism of the tall fescue line AU-7 appears to be saturated at a lower Mg concentration than the other fescue lines. SDR lines AU-5 and AU-718 have a larger capacity to accumulate Mg from solution.

Introduction

Consideration of root/soil interactions is essential in adapting tall fescue (*Festuca arundinacea* Schreb.) to the soil and climatic conditions of the Coastal Plain region of the Southeastern United States. Most of these soils are highly susceptible to formation of compaction layers, or plowpans[5]. The limitation of root growth due to compaction layers has been cited as a causal factor of drought stress and resulting stand decline which is characteristic of tall fescue in that area[2].

Williams *et al.*[16] demonstrated that morphological differences in roots exist between tall fescue lines, and that these differences are associated with drought resistance due to differential penetration of plowpans. A line with large diameter roots (LDR) was able to penetrate the plowpan to access subsoil water and survive severe drought stress better than a small diameter root (SDR) line which is restricted to the top 25 cm of soil.

The effects of differences in root morphology among tall fescue lines are not limited to differential abilities to penetrate compaction layers. Elkins *et al.*[4] showed a differential effect of nematode infection on Mg

H.W. Gabelman and B.C. Loughman (Eds.), Genetic aspects of plant mineral nutrition.
ISBN-13: 978-94-010-8102-3
© *1987 Martinus Nijhoff Publishers, Dordrecht/Boston/Lancaster.*

uptake in SDR and LDR tall fescue. The ability of LDR fescue to penetrate deeply into the soil reduced susceptibility to injury from plant-parasitic nematodes because less of the root system grew in nematode-infested soil[4]. Torbert et al.[15] showed differences in soil solution concentrations of NO_3-N under LDR and SDR tall fescue. However, they concluded that monitoring soil solution concentrations of K, Mg, and Ca in the field was not an adequate indicator of differences in tall fescue root activity.

Much effort has been expended in selecting tall fescue lines that will accumulate adequate Mg to prevent hypomagnesemic tetany in ruminant animals[8,10,12,13,14]. Magnesium accumulation in tall fescue lines requires a large volume of roots without suberized endodermis[3]. Fescue root anatomy (diameter, xlyem diameter, number of xlyem elements, and surface area of xylem) may have an affect on Mg uptake. Selection for LDR lines may alter all of above, as well as altering the plant's ability to extract water and nutrients from deeper levels in the soil profile. These factors in turn may alter the hypomagnesemic tetany potential of this species because Mg movement to plant roots is primarily by mass flow[1,7,9,11]. Therefore, the following experiment was conducted to characterize tall fescue roots and to study root anatomy effects on Mg influx under controlled conditions.

Materials and methods

Five clonal lines of tall fescue were used in this study: from breeding lines AU-5, AU-7, AU-264, AU-718, and a selection from Kentucky 31 (Ky 31). The clonal material was preconditioned in nutrient solution to produce propagules with roots free of soil contamination. Uniform single shoot propagules were removed from the 'parent' clones, washed for 2 hours in distilled water, and transferred into 12-liter tanks in the greenhouse.

Nutrient concentrations were: $0.25\,mM$ KCl, $0.25\,mM$ KH_2PO_4, $0.25\,mM$ NH_4NO_3, $0.5\,mM$ $CaCl_2$, $180\,\mu M$ FeDTPA (diethylene triaminepentacetic acid), $46\,\mu M$ B, $9\,\mu M$ Mn, $0.8\,\mu M$ Zn, $0.3\,\mu M$ Cu, and $0.05\,\mu M$ Mo. Magnesium concentrations of 3, 21, 42, 125, 250, and $500\,\mu M$ were imposed on individual tanks by the addition of $MgSO_4$. The SO_4 concentration was adjusted with Na_2SO_4 to give a constant SO_4 concentration of $500\,\mu M$. Nutrient concentrations were monitored every 2 days by removing 50 ml of solution from each tank and determining the nutrient concentrations by standard methods. Nutrient concentrations were maintained by addition of nutrients as required and the concentrations did not vary more than 5% during the course of the experiment.

Solution pH was measured daily and maintained at 5.6 to 5.8 by adding HCl or NaOH. To minimize the fluctuation of the solution pH, the sodium salt of [2-(N-Morpholino) ethanesulfonic acid] (pH 6.15) was added to the nutrient solution for a final concentration of $1\,mM$. All tanks were vigorously aerated, and nutrient solutions were changed every 7 days. Temperature was maintained at 24 ± 5 C, and sunlight was supplemented with fluorescent light to produce a minimum of 250–$300\,\mu Em^{-2}s^{-1}$ at the canopy for a 16-hour day.

An experimental unit consisted of two propagules (paired) of each clonal line at each Mg level. Individual propagules were supported by foam rubber collars in No. 6 plastic stoppers in a 0.5-cm-thick black plexiglass tank cover. After growing for 39 days, one propagule was harvested. The other propagule was harvested at 70 days. Roots were washed in diluted Ca solution ($0.1\,mM$) for 15 minutes in an attempt to remove all Mg in the free space of root cells. The propagules were separated into shoots (leaf blades, leaf sheaths, and stems) and roots, freeze-dried, weighed, and ground to pass a 40-mesh screen. Root volume was determined at harvest using a water-displacement method. Concentrations of Mg in the tissue were determined by Inductively Coupled Argon Plasma (ICAP). Net influx rates (I_m) of Mg were calculated from the change in total Mg content and the change in fresh weight of tall fescue propagule roots using the following equation:

$$I_m = \frac{M_2 - M_1}{WR_2 - WR_1} \cdot \frac{\ln(WR_2/WR_1)}{t_2 - t_1}$$

where I_m is uptake rate per gram fresh weight of root, M is total elemental content in tall fescue propagule (leaves + roots), WR is fresh root weight, and t is time (days). I_m for the first growth period (subscripts 1 and 2) denotes initial and first harvest, and for the second growth period denotes initial and second harvest.

The kinetic constants K_m and V_{max} were calculated for each fescue line by plotting I_m/[Mg concentration]. These plots are called Eadie-Hofstee plots and they magnify departures from linearity which is not apparent in a double-reciprocal plot. The kinetic constant V_{max} is obtained by extrapolating to the Y axis (I_m) and the slope of the plot is $-K_m$.

The experimental design was a randomized complete block with three replications. Treatments were arranged in a factorial design (5 clones × 6 Mg concentrations). Data were analyzed using standard analysis of variance and regression analysis when differences due to Mg concentration were significant.

Roots for characterization were grown in the 12-liter tanks as previously described. Each tank contained 10 propagules of a single tall fescue clone. The experiment was replicated 10 times (50 tanks) with a completely randomized arrangement of the tanks in the greenhouse. The nutrient solution was the same as the standard solution, with $84\,\mu M$ of Mg as $MgSO_4$. Twenty randomly selected roots were taken from each tank of propagules (a total of 200 roots) for characterization of each of the five clones. Roots removed were maintained in a petri dish with moist filter paper and root diameters were measured within 4 hours of removal from the clone. A hand-sectioned sample of 0.01 cm thickness was taken 7.5 cm behind the root apex, and examined under a light microscope. A scaled eyepiece was used to measure root and xylem diameters.

Results and discussion

Root anatomy

The 5 tall fescue lines were clustered into 3 groups on the basis of root diameter and xylem organization (Table 1). Tall fescue lines AU-7 and AU-264 had large diameter roots (LDR) with root diameters averaging 0.98 and 0.83 mm. Two lines (5 and 718, SDR) had root diameters averaging 0.72 mm. The Ky 31 clone root diameters averaged 0.69 mm. The reproducibility of the root diameter measurements was excellent, with CV's averaging less than 20%. At 39 days, root volume of the SDR lines (AU-5 and AU-718) were approximately double the root volume of the LDR (AU-7 and AU-264) (Table 2). However, at 70 days only the SDR line (AU-5) root volume was double the LDR lines. The larger root volume of the AU-5 line was a result of a greater number of secondary roots/cm of primary root than the other SDR line and both LDR lines.

Xylem organization was different for LDR's, SDR's, and Ky 31 (Fig. 1, Table 1). The LDR xylem elements were approximately equal in diameter and were arranged in a polyarch pattern. The average xylem

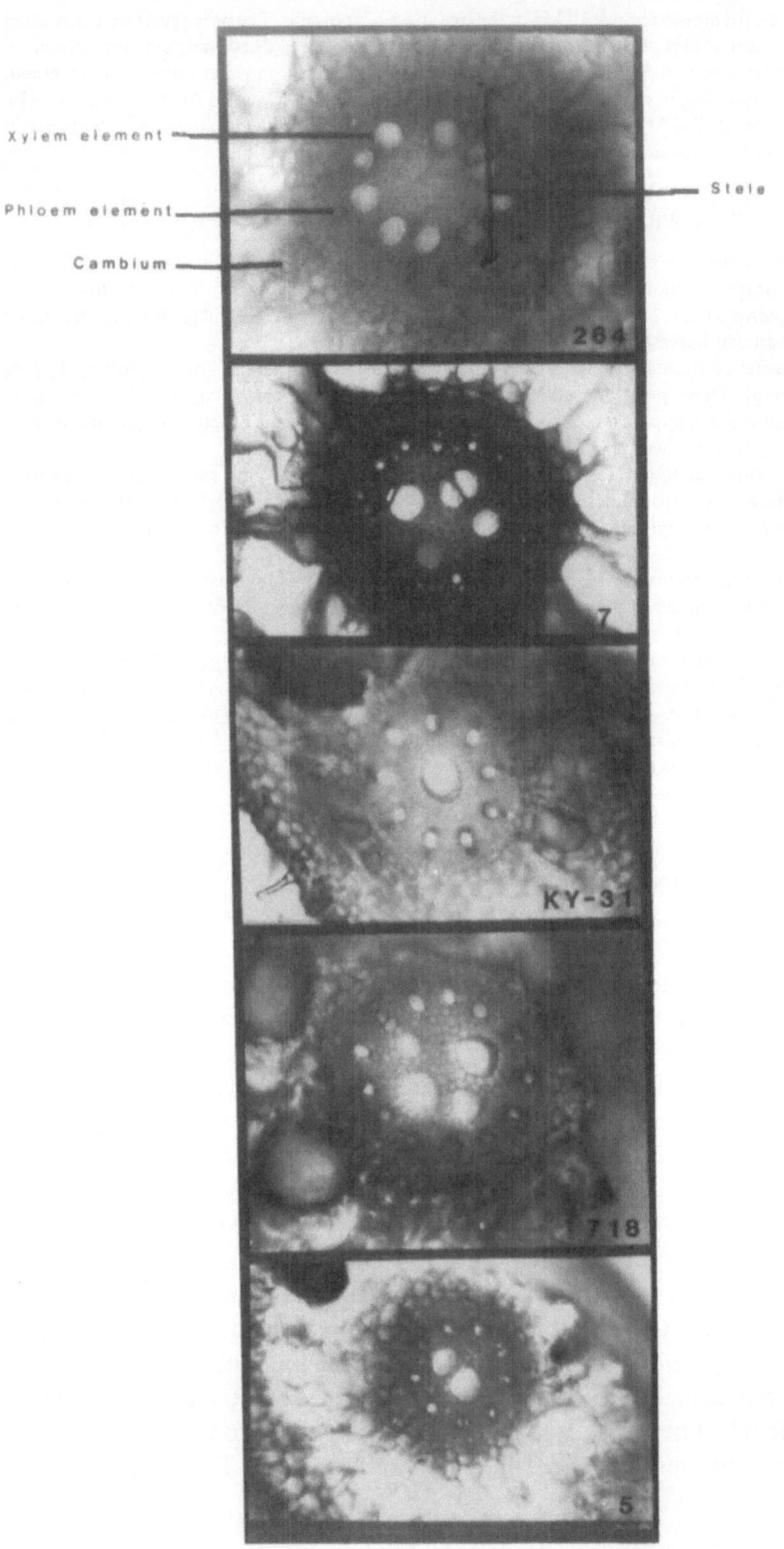

Table 1. Root morphology of tall fescue clones

Fescue lines	Root diameter (mm)		Xylem diameter (mm)
	Average	Range	Average
AU-7 (LDR)	0.98a'	1.04–0.92	0.18a
AU-264 (LDR)	0.83b	0.88–0.76	0.16a
AU-718 (SDR)	0.72c	0.76–0.66	0.14b
AU-5 (SDR)	0.72c	0.76–0.68	0.14b
Ky 31	0.69c	0.73–0.64	0.11c

' Fisher's protected LSD_{05} level.

diameters for LDR were 0.18 and 0.16 mm for lines AU-7 and AU-264, while the average xylem diameter for lines AU-5 and AU-718 was 0.14 mm and the single clone selection of Ky 31 was 0.11 mm. Xylem diameter of Ky 31 was smaller than either LDR or SDR lines and its organization was a monarch pattern of xylem arrangement and a relatively large xylem element in the center of the xylem.

Mg influx

Net influx rates of Mg were generally higher at all solution Mg concentrations in the first harvest period than in the second harvest period (Fig. 2). This supports the theory proposed by Elkins et al.[3] that the volume of root tissue without suberized endodermis is critical for Mg uptake, and that relative proportion of suberized endodermis increases with physiological age. Interactions between lines and harvest dates were significant at the 05 level of probability, and are shown graphically (Fig. 2). At harvest 2, both LDR lines exhibited lowest Mg uptake. However, at harvest 1, AU-264 was intermediate in uptake at low and medium solution Mg concentrations and highest in Mg influx rate at high solution Mg concentrations. Clearly, root morphology does affect Mg influx rate, but the effect is different as fescue roots mature.

Eadie-Hofstee constants

The K_m and V_{max} obtained from Eadie-Hofstee plots for AU-264 (Fig. 3) and a complete set of K_m and V_{max} for all five fescue lines are given in Table 3. At low Mg concentration for harvest 1, line AU-7 appears to have the lowest K_m for Mg influx. Although line AU-264 is classified as LDR, the K_m and V_{max} appear to parallel the SDR line AU-5. Lines AU-718 and Ky 31 appear to have similar mechanisms of uptake at low Mg concentrations in solution.

Fig. 1. Cross-section of mature roots of tall fescue lines AU-264, AU-7, Ky 31, AU-718 and AU-5 7.5 cm behind root apex.

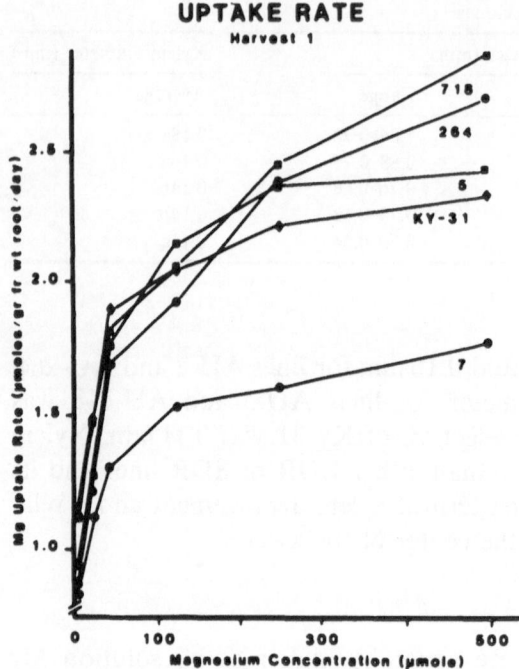

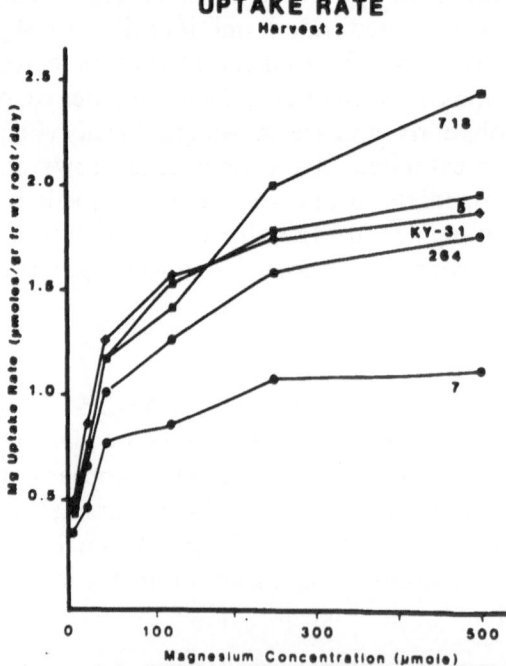

Fig. 2. Magnesium uptake for tall fesuce lines after 39 (harvest 1) or 70 (harvest 2) days of growth in nutrient solution with selected Mg concentration.

The low K_m of line AU-7 is also illustrated in the high solution Mg concentration range for harvest 1. Lines AU-5 and Ky 31 have similar K_m and V_{max}. Line AU-264 appears to have uptake characteristics similar to line AU-718.

For all but one fescue line, K_m increased while V_{max} decreased between harvest 1 and 2. The K_m of line AU-264 remained constant at high solution Mg between harvest 1 and 2. The low K_m for the low Mg concentration mechanism suggests that it is located at the plasmalemma, while the high concentration mechanism appears to be located at the tonoplast membrane[6].

Leaf Mg concentrations

Clonal Mg concentrations in leaf tissue at varying solution Mg concentrations generally follow the curves and ranking described for Mg influx (Fig. 4). Therefore, Mg uptake may be a valid indicator of leaf Mg concentration and tetany potential. Leaf Mg concentration was not strongly responsive to solution Mg concentration changes at intermediate levels, especially for line AU-7. The LDR lines were lower in leaf Mg concentration at all solution Mg concentrations at both harvest dates. This suggests that Mg tetany potential in fescue may be increased by selection for lines with LDR.

Root Mg concentration

Root Mg concentration was quite variable at low solution Mg concentrations at harvest 1 (Fig. 3). Smooth quadratic curves for the relationship of root Mg concentration and solution Mg concentration were obtained with the more mature roots at harvest 2. However, the relationship between root diameters and tissue Mg concentration was altered in root tissue compared to leaf tissue. Lines AU-7 and AU-718 were lowest in root Mg concentrations at harvest 2, while lines AU-5 and AU-264 were highest. Root Mg concentration was much lower than leaf Mg concentration. Neither the lowered Mg concentration of the roots nor the lack of a consistent relationship of this character with root morphology is of practical importance when considering Mg tetany potential in fescue, since ruminants consume only the above-ground portion of the plant.

In summary, differences in root morphology of tall fescue lines were related to differences in Mg tissue concentrations and Mg influx rate, with LDR plants accumulating less tissue Mg. Physiological age of the tall fescue root was shown to affect the K_m and V_{max} with younger plants being more effective in Mg uptake. The root volume for tall fescue lines AU-5 and AU-718 was approximately 50% greater than lines AU-7 and

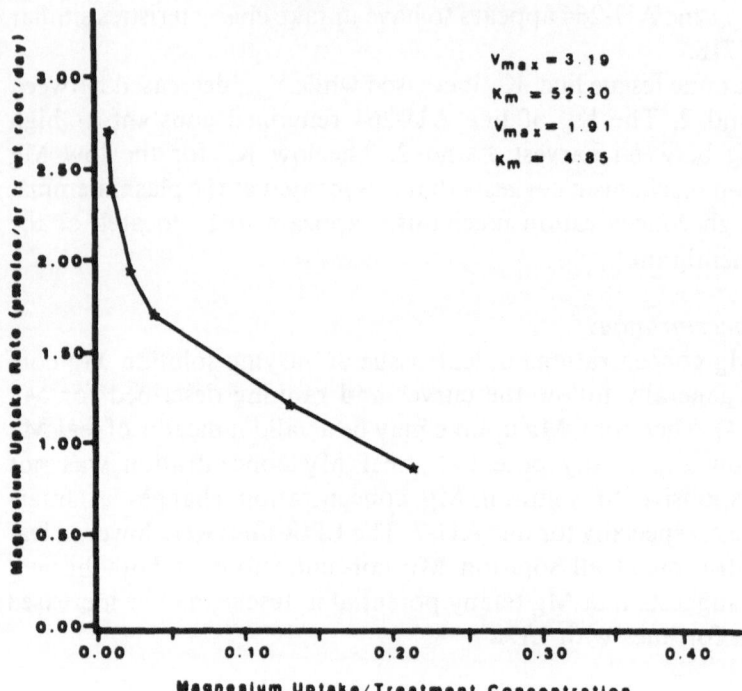

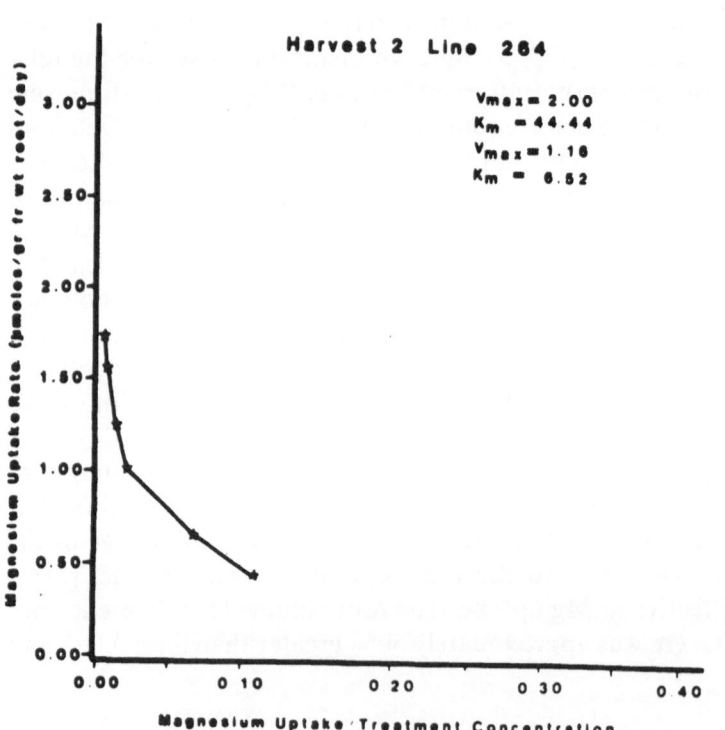

Table 2. Root volume (cm^3) of tall fescue lines after 39 or 70 days of growth in nutrient solution containing selected Mg concentrations

Fescue lines	Mg concentration (μM)						line mean
	3	21	42	125	250	500	
39 days							
AU-7 (LDR)	7.3	7.0	8.3	4.3	7.3	7.3	7.0
AU-264 (LDR)	8.0	7.3	5.6	7.3	5.3	6.0	6.6
AU-5 (SDR)	16.3	15.0	14.3	12.3	14.6	19.0	15.3
AU-718 (SDR)	18.6	16.3	12.7	10.0	10.6	10.0	13.0
Ky 31	11.7	14.3	10.3	8.6	10.7	10.0	10.9
Concentration means	12.4	12.1	10.3	8.5	9.7	10.4	
70 days							
AU-7 (LDR)	31.6	31.0	17.6	24.3	17.0	30.6	25.4
AU-264 (LDR)	36.3	29.0	21.3	31.0	28.3	22.6	28.1
AU-5 (SDR)	40.6	54.0	61.3	41.0	46.0	62.6	50.9
AU-718 (SDR)	49.0	43.3	31.3	32.6	33.3	45.6	39.2
Ky 31	42.3	33.0	47.3	31.6	34.3	29.6	36.4
Concentration means	40.00	38.06	35.8	32.1	31.8	38.3	

39 days FLSD$_{05}$ lines = 2.41
39 days FLSD$_{05}$ lines = 2.64
70 days FLSD$_{05}$ lines = 8.65
70 days FLSD$_{05}$ lines = NS

Table 3. K_m and V_{max} for Mg uptake for tall fescue lines grown in selected Mg concentrations for 39 or 70 days

Fescue lines	Low concentration		High concentration	
	K_m'	V_{max}'	K_m	V_{max}
39 days				
AU-7 (LDR)	2.86	1.46	15.25	1.86
AU-264 (LDR)	4.85	1.91	52.30	3.19
AU-5 (SDR)	4.64	1.97	20.38	2.65
AU-718 (SDR)	3.11	1.97	49.24	3.25
KY 31	3.15	2.06	13.77	2.41
70 days				
AU-7 (LDR)	6.85	0.89	34.21	1.30
AU-264 (LDR)	6.52	1.16	44.44	2.06
AU-5 (SDR)	9.22	1.41	39.46	2.21
AU-718 (SDR)	7.03	1.35	90.00	3.06
KY 31	8.15	1.50	26.62	2.05

' K_m — μMoles
' V_{max} — μMoles/g fr root wt/day

Fig. 3. A Eadie-Hofstee plot of tall fescue line 264 to obtain K_m and V_{max} for 39 or 70 day of growth in nutrient solution containing selected Mg Concentration.

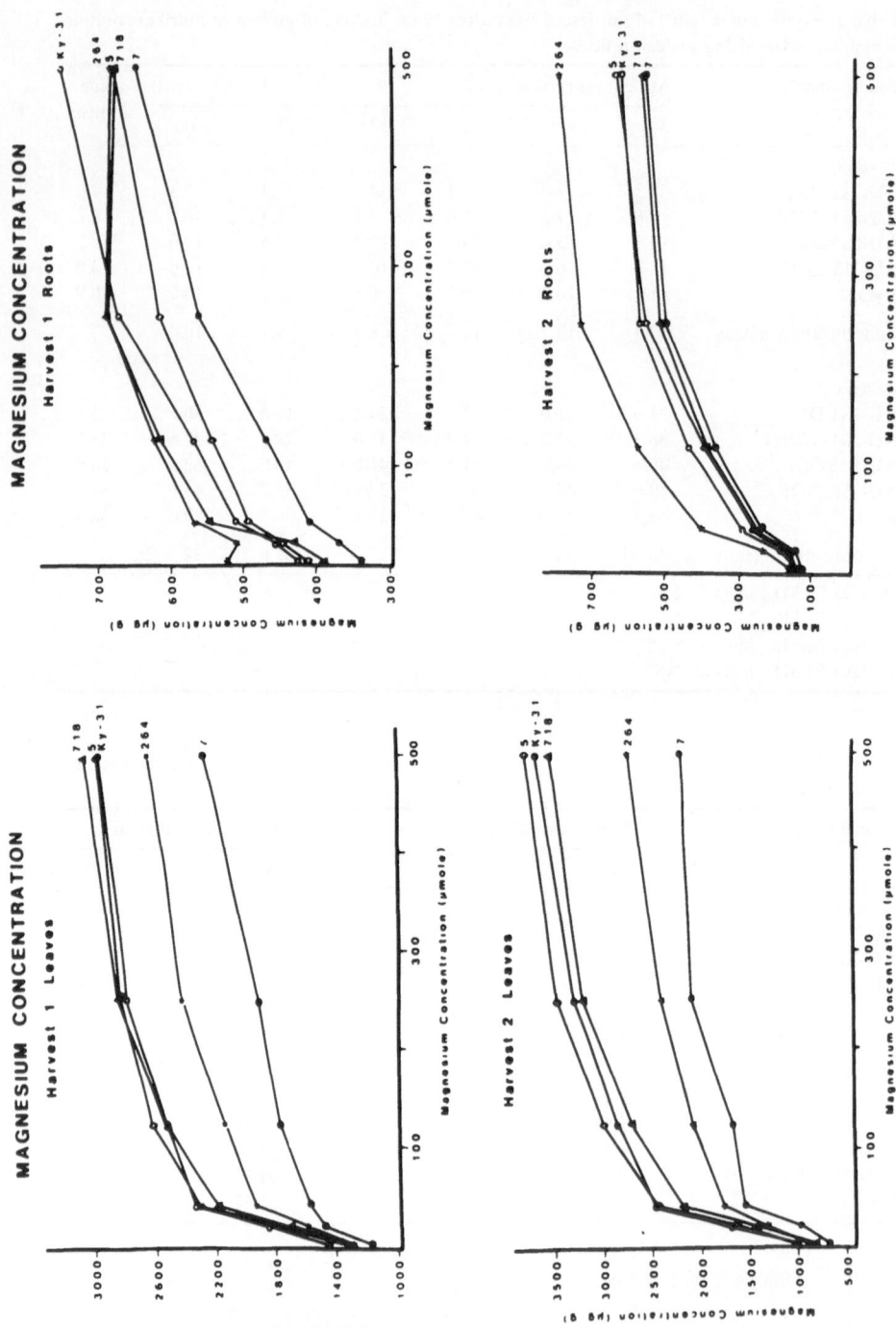

AU-264 when grown in nutrient solutions. If these relationships hold for a root system growing under confined conditions, such as caused by compaction layers in the field, Mg accumulation in LDR plants may be lower than in SDR plants. The possible benefits of LDR penetration into lower soil profiles and the resulting access to additional nutrients cannot be addressed by this study.

References

1. Barber S A 1974 Influence of plant root on ion movement in soil. pp 525–564 *In* The Plant Root and its Environment. Ed. E W Carsun. University Press of Virgina, Charlottesville, Va.
2. Edwards J H and Pedersen J F 1985 Why won't tall fescue persist in the Coastal Plain region of the Southeast? Proc. Amer. Forage and Grassland Cong., Hershey Park, Penn. pp 129–136.
3. Elkins C B, Haaland R L, and Hoveland C S 1983 Biological availability of magnesium to plants. pp 55–60 *In* Role of Magnesium in Animal Nutrition. Eds. J P Fontenot, G E Bunce, K E Weeb, Jr., and Vivien G Allen. Virginia Polytechnic Institute and State Univ., Blacksburg, Va.
4. Elkins C B, Haaland R L, Rodriguez-Kabana R, and Hoveland C S 1979 Plant-parasitic effects on waste use and nutrient uptake of small-and large-rooted tall fescue genotypes. Agron. J. 71, 497–500.
5. Elkins C B, Thurlow D L, and Hendrick J G 1983 Conservation tillage for long-term amelioration of plow pan soils. J. Soil Water Conserv. 38, 305–307.
6. Epstein E 1976 Kinetics of ion transport and the carrier concept. pp 70–94 *In* Encyclopedia of Plant Physiology, New series, Vol. 2, Transport in Plants II. Part B. Tissues and organs. Eds. U Luttge and M G Pitman. Springer-Verlag, Berlin.
7. Follett R F, Power J F, Grunes D L and Kleen C A 1977 Effect of N, K, and P fertilization, N source, and clipping on potential tetancy of bromegrass. Plant and Soil 48, 485–508.
8. Haaland R L, Elkins C B and Hoveland C S 1978 A method for detecting genetic variability for grass tetany potential in tall fescue. Crop Sci. 18, 339–340.
9. Hannaway D B, Leggett J E, Bush L P, and Shuler P E 1984 Magnesium (Mg) and Rubidium (Rb) absorption by tall fescue. J. Plant Nut. 7, 1127–1147.
10. Nguyen H T and Sleper D A 1981 Genetic variability of mineral concentration in *Festuca arundinacea* Schreb. Theor. Appl. Genetics 59, 57–63.
11. Oliver S and Barber S A 1966 An evaluation of the mechanism governing the supply of Ca Mg, K and Na to soybean roots (*Glycine max.*) Soil Sci Soc Am. Proc. 30, 82–86.
12. Sleper D A 1979 Plant breeding, selection, and species in relation to grass tetany. pp 63–77 *In* Grass Tetany. Eds. V V Rendig and D L Grunes. Am. Soc. Agron., Madison, Wis.
13. Sleper D A, Garner G B, Asay K H, Boland R, and Pickett E E 1977 Breeding for Mg, Ca, K, and P content in tall fescue. Crop Sci. 17, 433–438.
14. Sleper D A, Garner G B, Nelson C J, and Sebaugh J L 1980 Mineral concentration of tall fescue genotypes grown under controlled conditions. Agron. J. 72, 720–722.
15. Tobert H A, Edwards J H, and Pedersen J F 1985 Relative nutrient absorption by fescue lines. Proc. Amer. Forage and Grassland Cong., Hershey Park, Penn. pp 259–265.
16. Williams C B, Elkins C B, Haaland R L, Hoveland C S, and Rodriguez Kabana R 1981 Effects of root diameter, nematodes and soil compaction on forage yield of two tall fescue genotypes. Proc. Inter. Grassland Congr. June 15–24, Lexington, Ky. pp 121–124.

Influence of different *Triticum aestivum* L. genomes and chromosomes on the assimilation of the main nutrient elements

B. BOCHEV, E. NEIKOVA-BOCHEVA, N. MITREVA and G. GANEVA
Institute of Genetics, Bulgarian Academy of Sciences, Sofia Institute of Soil Science and Yield Programming 'N. Poushkarov' — Sofia, Bulgaria

Key words Aneuploidy Chromosomes Fertilizer application Genomes Nutrient elements Phosphates Plant organs Wheat

Summary A ditelocentric aneuploid series of cv. Chinese Spring (*T. aestivum* L.) was used for the pot experiment aiming to assess the effect of individual chromosomes and genomes on the uptake of the main nutrient elements (N, P, K, Ca, Mg, Mn, Na, Zn, Fe and Cu) by wheat plants. It was found that most of the chromosomes take part in the genetic control of nutrient element uptake, but their activity in regard to the various elements is not the same. Highest genetic activity was observed in the chromosomes of second and fourth homeologic groups and in the A-genome.

Individual chromosomes have an effect on the uptake of several nutrient elements: 3D–P, K, Mn and Zn; 5B–Ca, Mg and Mn; 1B–Zn and Mn; 2D inhibits the uptake of almost all nutrient elements except Na.

It is assumed that the effect of the 3D chromosome could be of practical interest for genotypes grown on carbonate soils, since this chromosome leads to the uptake of higher P, K, Zn and Mn quantities and inhibits Ca uptake. Chromosome 2A could be important for the genotypes growing on acid soils because it enhances the uptake of Ca and Mg by the grain and inhibits the uptake of Mn.

Introduction

Wheat aneuploid series and the use of new cytogenetic stocks derived from intercultivar, interspecific and intergeneric chromosome manipulations are an important prerequisite for such work. Combined research efforts of genetics, soil science and plant physiology are required to determine the genetic basis of wheat mineral nutrition.

Most available data demonstrate general genotypic effects on yield and quality[1,2,3], but research on the genetic control of assimilating individual nutrient elements is limited[4,5,6,7].

The present investigation was carried out to assess the influence of individual *T. aestivum* chromosomes on the content of the main nutrient elements (N, P, K, Ca, Mg, Mn, Na, Zn, Fe, Cu) in the vegetative and reproductive organs.

Material and methods

A pot experiment was carried out with ditelocentric lines of cv. Chinese Spring (Fig. 1), kindly supplied by Dr. E.R. Sears.

Plants were grown on leached chernozem soil at optimal phosphorus level (0.2 mg/l P) in 0.01 *M*

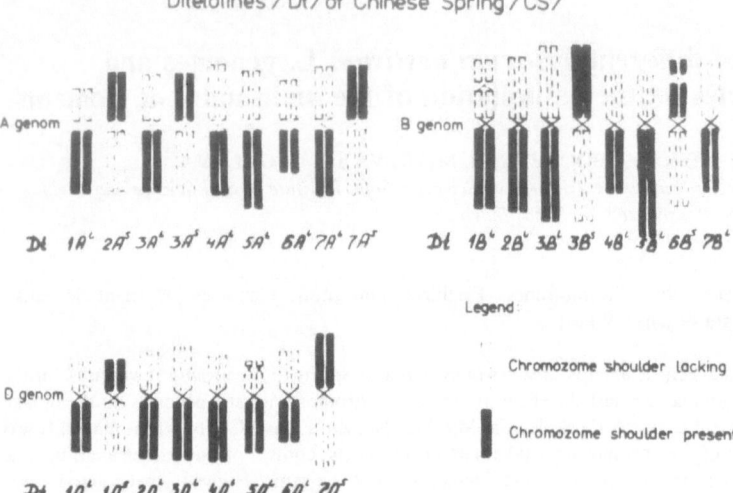

Fig. 1. Ditelolines of Chinese Spring. ::::: chromosome arm lacking; ▬▬▬ chromosome arm present.

$CaCl_2$ after the method of Neikova-Bocheva (8), 0.3 mg/kg N and K_2O. Phosphorus was applied as $Ca(H_2PO_4)_2$, nitrogen and potassium — as NH_4NO_3 and K_2SO_4, respectively. After harvesting the plants were assessed for total quantity of dry matter, nitrogen by wet ashing and Kjeldahl, phosphorus by dry ashing and colorimetrically after the vanadomolybdate method and K, Ca, Mg, Na, Zn, Mn, Cu and Fe — by atomic adsorption spectroscopy.

To compare results, individual nutrient elements assimilated by grain and straw were re-estimated in scale units, corresponding to one fifth of the total variation for a whole series of lines tested and the control. The effect of the individual chromosomes on the nutrient elements taken up by grain and straw was assessed in existing scale differences from the control (euploid cv. Chinese Spring).

Results

The investigations show that although the uptake of the individual elements in the main reproductive and vegetative plant organs is complex, individual chromosomes of wheat have a considerable effect on the uptake of nutrients.

A large number of chromosomes have an effect on nitrogen uptake by the grain and straw (Fig. 2). The absence of short arms in several chromosomes and of the long arm of 2A resulted in a pronounced increase of nitrogen percentage in the grain. It should be noted that nitrogen percentage involvement in the grain and straw varies in the different DT lines from 1.611% to 2.713% and from 0.288% to 0.549% respectively and in the control from 1.709% to 0.275%, respectively.

This indicates that genes with an inhibiting effect on that process are found in these arms of the respective chromosomes. The genetic factors in the indicated arms of 2A, 2B, 4D, 7A and 5A chromosomes confer the

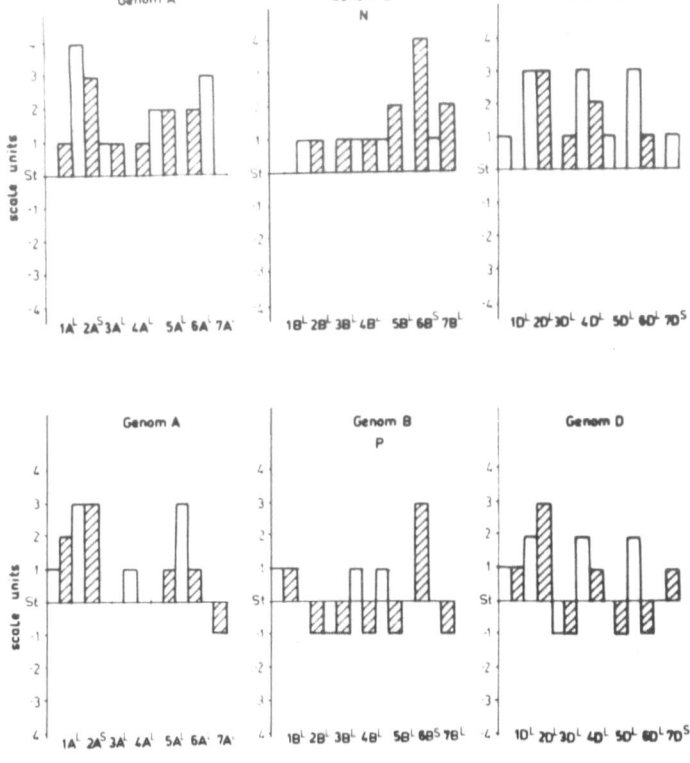

Fig. 2. Effect of different chromosomes of genom A, B and D on the uptake of N and P in *T. aestivum* L. ☐, grain; ▨, straw.

greatest inhibition of N uptake. Most of the chromosomes also have an influence on nitrogen assimilation in straw. The short arms of chromosomes 3B and 3D are critical for phosphorus entry into the grain and their absence leads to reduced P uptake (Fig. 2). Gene complexes in the long arm of 2A and in the short arms of chromosomes 6A, 2D, 4D, 6D reduced grain P uptake whereas the long arms of chromosomes 2A and 6B and the short arms of chromosomes 2D, 1A, 1B, 4D, 5A, 6A reduced straw P uptake. Overall P varied in the grain of the different lines from 0.318% to 0.54% and in the straw from 0.062% to 0.150%, (control, 0.390 and 0.062% respectively).

Genetic factors in the short arms of chromosomes 7A, 3A, 3D, 4B and 7B appeared critical for potassium content in the grain (Fig. 3). Gene complexes of the long arm of 2A and the short arm of 2D had an inhibiting effect on that element's content in the grain and the short arms of 2B, 3A and 3B chromosomes affected K uptake to the straw. K varied in the different DT lines from 0.30% to 0.61% for grain, from 1.09 to 2.45% for straw and in the control — 0.50% and 1.58% respectively.

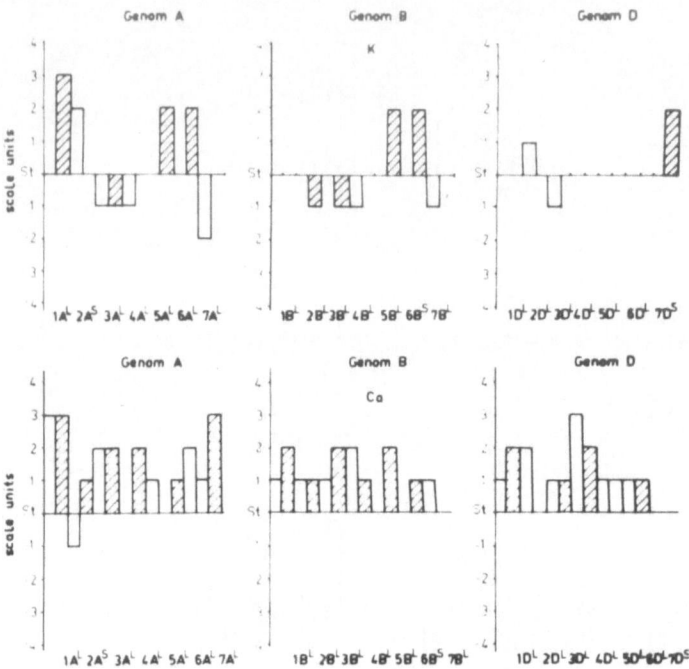

Fig. 3. Effect of different chromosomes of genom A, B and D on the uptake of K and Ca in *T. aestivum* L. ☐, grain; ▨, straw.

The long arm of chormosome 2A and the short arm of 5B (Fig. 3) were critical for calcium uptake to the grain. The gene complexes in the short arms of 1A, 4D, 2D, 4A and 4B chromosomes had an opposing effect. In respect to calcium in the straw the genes in the short arms of 1A, 7A, 1B, 1D, 3A, 3B, 4A, 4D, 5B and 6A chromosomes had the highest inhibiting effect. Ca percentage involvement in the different DT-lines varied from 0.021% to 0.040% for grain and from 0.37% to 0.76% in straw, while in the control it is 0.025% and 0.37% respectively.

The critical genes affecting Mg content in the grain were those of the long arm of 2A and the short arm of 5B (Fig. 4). Lines $1B^L$, $2B^L$, $3D^L$, $5A^L$, $5B^L$, $5D^L$, $7B^L$ and $7D^L$ were similar to the control. The remaining lines possessed a higher magnesium content in the grain. Most of the lines had higher magnesium content in the straw than the control. Mg percentage involvement in the different DT lines varied in grain from 0.100% to 0.205%, in straw from 0.135% to 0.230% and in the control 0.135% and 0.150% respectively.

Most of the lines do not differ from the control in sodium percentage involvement in the grain (Fig. 4). The short arm of 1D chromosome appeared critical in its effect on sodium uptake in straw.

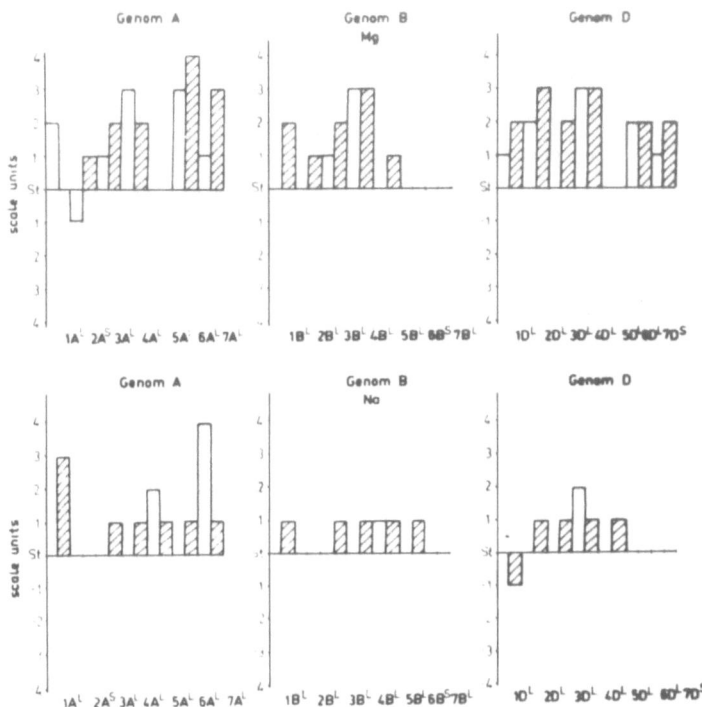

Fig. 4. Effect of different chromosomes of genom A, B and D on the uptake of Mg and Na in *T. aestivum* L. ▭, grain; ▨, straw.

The short arms of chromosomes 1B, 3D and 4A (Fig. 5) were important for zinc content in the grain. The most extreme inhibiting effects were in the short arms of chromosomes 6A and 7A and the long arm of 2A.

Zinc uptake in the straw was controlled mainly by genetic factors in the short arms of 4B, 2D and 7A chromosomes. Zn percentage in the grain of DT-lines varied from 21 to 47, in straw from 20 to 51 and of the control — 26 and 22 respectively.

The short arms of 3D and 6A and the long arm of 2A chromosomes had the greatest effect on manganese uptake by the grain. Manganese content in straw was controlled mainly by the gene complexes of the short arms of 4A, 5D, 3B, 4D, 6A and 7A chromosomes and in the long arm of chromosomes 7D. Mg percentage in the grain of the different DT lines varies from 36 to 74, in straw — from 126 to 216 and in the control is 52 and 150 respectively.

Data in Fig. 6 show that the greater part of ditelo-lines possess a higher content of iron in the grain than the control, suggesting that

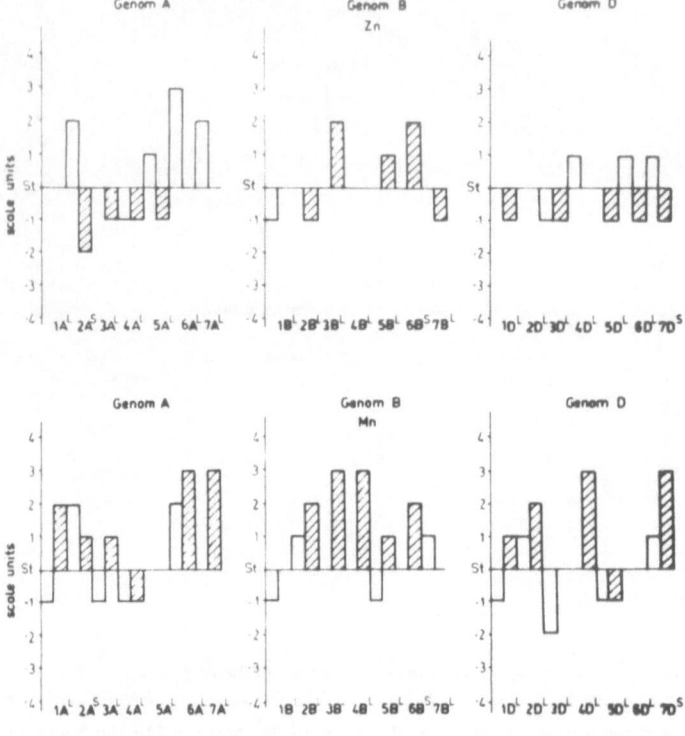

Fig. 5. Effect of different chromosomes of genom A, B and D on the uptake of Zn and Mn in *T. aestivum* L. ☐ , grain; ▨ , straw.

genetic factors with inhibiting effect are localized in the missing arms of the respective chromosomes. A relatively small part of the chromosomes has an opposite effect. Fe percentage in the grain of the different DT lines varies from 20 to 60, in straw — from 10 to 69 and in the control it is 30 and 10 respectively.

Discussion

Analysis of data shows the complicated mode of genetic control on the uptake of individual nutrient elements in the main reproductive and vegetative organs of wheat plants. Most of the chromosomes have an influence on the content of the individual nutrient elements in wheat grain and straw. The effect of different chromosomes is not the same. Probably individual chromosomes have a definite effect on the formation and activity of the various plant organs.

In most cases the effect of individual chromosomes on the content of a given nutrient element is not unidirectional. The absence of arms in a

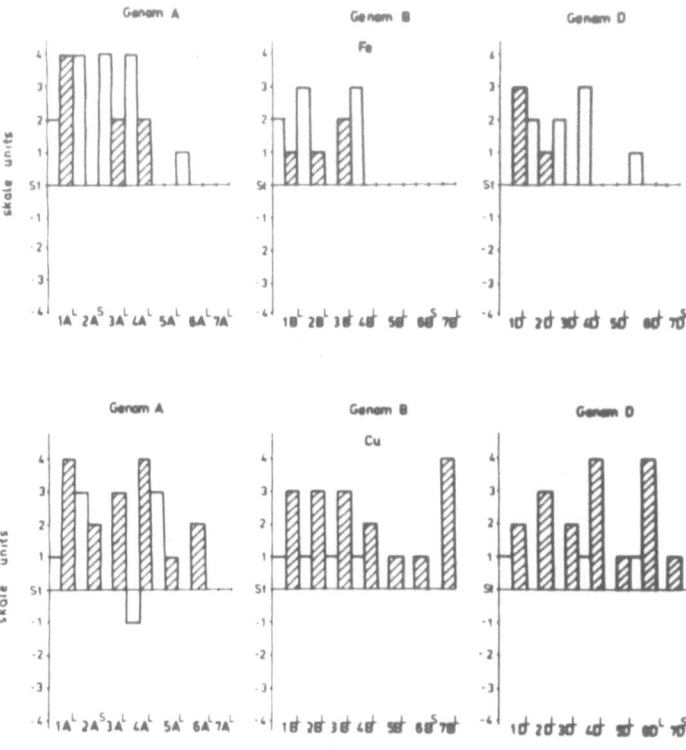

Fig. 6. Effect of different chromosomes of genom A, B and D on the uptake of Fe and Cu in *T. aestivum* L. ▢, grain; ▨, straw.

number of critical chromosomes causing considerable reduction of a definite nutrient element in grain and straw is insignificant, as compared with the number of chromosomes with inhibiting effect.

Chromosomes of different homeological groups of wheat do not participate to the same extent in the genetic control of the content of individual elements in grain and straw. The chromosomes of the second and fourth homeological groups have an influence on the genetic control of most nutrient elements in the grain (excepting magnesium and sodium). Along with these two groups, the chromosomes of other homeological group of importance to the uptake of the individual nutrient elements, are: nitrogen — the fifth group, phosphorus — sixth, potassium — third, calcium — almost all groups, magnesium — first and partially third and sixth, sodium — fifth, sixth and seventh, zinc — two chromosomes of the seventh group, manganese — two chromosomes of third, of sixth and of seventh group, copper — first and iron — first and two chromosomes of sixth group. A similar phenomenon was observed for the participation of chromosomes from the various groups in the control of individual nutrient elements in straw, but in this case more chromosomes were involved. The fact that separate chromosomes take part simultaneously in the genetic control of the content of several nutrient elements is of importance. Chromosome 3D, for instance, is involved simultaneously in the control of phosphorus, potassium, zinc and manganese, 5B — of calcium, magnesium and manganese, 4A — of potassium, manganese and copper, 2A — of calcium and magnesium, 3A — of potassium and manganese, 1B — of zinc and manganese.

The influence of specific homological chromosomes inhibiting various nutrient elements uptake is considerable. Chromosome 2D, for instance, had an inhibiting effect on the uptake of almost all nutrient elements (excepting sodium) in the grain. Other chromosomes had a unidirectional inhibiting effect on the content of most nutrient elements. It is of interest to note that individual chromosomes have an opposite effect on certain nutrient elements, known as antagonists. For example, chromosome 3D contributed to the uptake of more phosphorus, potassium, zinc and manganese in the grain, but inhibited the uptake of calcium and copper. Chromosomes 5B and 2A had opposing effects on phosphorus and calcium uptake. The effect of chromosome 3D in wheat genotypes grown on carbonate soils would be of particular importance, because it would stimulate the uptake of phosphorus, potassium, zinc and manganese and could inhibit the uptake of calcium. The influence of chromosome 2A would be of interest for acid soils, because it facilitates the uptake of more calcium and magnesium in the grain and inhibits manganese uptake.

The different genomes manifest unequal genetic activity in respect to the assimilating ability of the individual nutrient elements. The effect of separate genomes is more pronounced in most cases in straw than in the grain. Highest positive effect in the grain was produced by the A genome.

Results presented indicate a genetic basis of wheat mineral nutrition. This is important for developing desired genotypes utilizing nutrient elements in soils with normal and with disturbed nutrient regime. The complexity of the problem suggests the need to unite the efforts of pedologists, agrochemists, physiologists, geneticists and plant breeders.

References

1 Klimashevskiy E L 1974 Fisiologiya genotipicheskoy spetsifiki kornevogo pitaniya rasteniy. Fiziologiya rastenii v pomosht selektsii. Isdat. 'Nauka', Moskva, 226–241.
2 Libbert E 1976 Fiziologiya rastenii, German translation, izdat. 'Mir' M.
3 Strelnikova M M Vliyanie oudobrenii na hlebopekarnie kachestva zerna pshenitsi. Agrohimiya, No. 3.
4 Gerloff G C 1976 Plant efficiencies in the use of nitrogen, phosphorus and potassium. p. 161. Plant adaptation to Mineral stress in Problem soils. Proc. of a workshop held at the National Agricultural Library Beltsville Maryland November 22–23.

5   Clark R B 1976 Plant efficiencies in the use of calcium, magnesium and molybdenum p. 175. Plant adaptation to mineral stress in problem soils. Proc. of a workshop held at the National Agr. Library Beltsville Maryland Nov. 22-23.
6   Lafever H N, Campbell L G 1978 Inheritance of aluminium tolerance in wheat. Can. J. Gen. Cytol., 20, *3*, 355-364.
7   Loneragan J P 1976 Plant efficiencies in the use of B, Co, Cu, Mn and Zn. p. 193. Plant adaptation to mineral stress in problem soils. Proc. of a workshop held at the National Agricult. Library Beltsville Maryland November 22-23.
8   Neikova-Bocheva E, Sadovski A N, Dabcheva P L, Novkova E K, Nachkova T V and Turpanova H G 1980 A method of determining P fertilizer rates at different soils and different P levels. Compt. Rend. Acad. Bulg. Sci., 33, No. 277-279.

The accumulation and distribution of sodium in tomato strains differing in potassium efficiency when grown under low-K stress

SCOTT S. FIGDORE, W.H. GABELMAN, and G.C. GERLOFF
Departments of Horticulture and Botany, University of Wisconsin, Madison, WI 53706, USA

Key words   *Lycopersicon esculentum*   Na substitution   Tomato

Summary   Plants of five tomato strains were grown under low-K stress at three Na levels. These plants were harvested at three time intervals, and Na accumulation and distribution were measured in their tissues. Strain differences were observed for the ability to substitute Na for K under low-K stress. In two strains with high Na-substitution capacity, efficiency in substitution was associated with the accumulation of more Na and the maintenance of higher Na concentrations in shoot tissues than in other strains. In a third strain which also had a relatively high Na-substitution capacity at the highest solution Na level, an unusual efficiency in Na substitution was indicated, because the strain neither accumulated Na nor maintained high tissue Na levels.

Introduction

The identification of germplasm better adapted to low-K conditions should be useful in helping understand the physiological roles of K as well as in providing material which plant breeders can use in areas where soils are low in K and where K fertilizers may be too costly or unavailable. In a previous study in this project, strains of tomato (*Lycopersicon esculentum* Mill.) were identified that differed in dry matter accumulation when grown under low-K stress[5]. Partial substitution of Na for K contributed to these strain differences. Although the specific plant functions in which Na replaces K under low-K stress are unknown, the possibility that Na replaces K in non-specific functions in plant cell vacuoles has been suggested[2,6]. Thus, the extent of Na substitution for K would depend on Na uptake by plant roots and subsequent translocation to the shoots.

Sodium accumulation, in cells adjacent to the xylem vessels of roots and lower stems, may limit Na translocation to the shoots. This has been observed in maize (*Zea mays* L.)[4,8] and beans (*Phaseolus vulgaris* L.)[3], and has been implicated in tomatoes[1,7]. A salt-tolerant accession of *Lycopersicon cheesmanii* (Hook) C.H. Mull. accumulated more Na in shoot tissues than the salt-sensitive *L. esculentum* Mill. cv. Walter when grown under high-salt conditions[7]. The salt-tolerant accession also had greater Na-substitution capacity when K was in limited supply. In another study, when the tomato cultivar Amberly Cross was grown

H.W. Gabelman and B.C. Loughman (Eds.), Genetic aspects of plant mineral nutrition.
ISBN-13: 978-94-010-8102-3
© *1987 Martinus Nijhoff Publishers, Dordrecht/Boston/Lancaster.*

Table 1. Description of the five strains of tomato (*Lycopersicon esculentum* Mill.) studied

Accession number[a]	Identification	Country of origin
576[b]	PI 126452	Peru
571[c]	Oshogbo	Nigeria
349	PI 273033	El Salvador
203	PI 124163	Guatemala
546[d]	PI 289286	Hungary

[a] University of Wisconsin Horticulture Department seed collection numbers
[b] Line 98 in Makmur *et al*[5].
[c] Line 42 in Makmur *et al*[5].
[d] Line 94 in Makmur *et al*[5].

under low-K conditions with Na available, most of the Na translocated to the leaves had accumulated in the petioles, rather than in the leaf laminae[1].

Our study was initiated to measure Na accumulation and distribution in plant tissues of tomato strains which differ in dry matter accumulation when grown under low-K stress and to assess the nature of Na substitution in these strains.

Materials and methods

Seeds of five inbred tomato strains (Table 1), established as differing in K efficiency (*i.e.* dry matter produced per unit of potassium supplied in culture), were germinated in a modified half-concentration Hoagland solution containing 2.6 mM K. After fourteen days, the seedlings were transplanted individually into pots containing 1.8 liters of a nutrient solution containing the following salts: 7.0 mM Ca(No$_3$)$_2 \cdot$4H$_2$O, 2.0 mM MgSO$_4 \cdot$7H$_2$O, 1.0 mM NH$_4$H$_2$PO$_4$, 40 μM FeSO$_4 \cdot$7H$_2$O, 25 μM H$_3$BO$_3$, 5 μM MnSO$_4 \cdot$H$_2$O, 2 μM ZnSO$_4 \cdot$7H$_2$O, 0.5 μM CuSO$_4 \cdot$5H$_2$O, and 0.015 μM (NH$_4$)$_6$Mo$_7$O$_{24} \cdot$4H$_2$O. Free-acid EDTA (H$_4$-EDTA), instead of Na$_2$H$_2$EDTA, was used to complex the iron in solution. All pots were initially provided with 5.0 mg of K as KCl (71 μM), a low-K stress level. Plants of each strain were grown at three Na levels, by addition of either 0 mg Na/pot (0.014 mM Na, from uncontrolled sources), 10 mg Na/pot (0.25 mM Na), or 160 mg Na/pot (3.9 mM Na) using a NaCl stock.

The experiment was carried out in a growth room under fluorescent lights (150–180 μE $\cdot$ m$^{-2} \cdot$ s^{-1}), with a 16-h light cycle. Temperatures ranged from 30–33°C during the light and from 21–24°C during the dark. Distilled water was added to the pots, during the course of the experiment, to maintain a solution volume of 1.8 liters.

Plants of each strain at each Na level were harvested at 16, 23, and 30 days after transplanting. The entire 5.0 mg K was taken up by all plants, and severe K deficiency symptoms were observed on all plants, after thirty days. Plants were partitioned into roots, hypocotyls, lower stems (the first three internodes above the cotyledons), upper stems (the remaining stem sections), lower-leaf laminae and petioles (the lowest three leaves), and upper-leaves (the remaining expanded leaves). Dry weights for each partition were measured. Sodium and potassium were extracted from tissues with 0.2 N HCl, and analyzed by atomic absorption spectrophotometry. The potassium efficiency ratio (KER) was calculated as the total mg of plant dry weight divided by the total mg K present in a plant.

The experiment was set up as a randomized complete block design with four replications. A factorial treatment design was used, with five strains, three Na levels, and three harvests. Separate analyses of variance for each Na level were calculated for most variables studied. Strains, Na levels, and harvests were analyzed as fixed effects, and blocks were analyzed as random effects.

Table 2. Mean plant dry weight of five tomato strains grown under low-K stress at three Na levels (30 days after transplanting)

Strain	Plant dry weight (g)[y]		
	0 mg Na/pot	10 mg Na/pot	160 mg Na/pot
576	1.09	1.48a	1.58a
571	0.89	1.20b	1.37b
349	0.94	1.06c	1.23bc
203	0.83	0.94d	1.15c
546	0.82	0.88d	0.92d
$LSD_{0.05}$	[z]	0.10	0.14

[y] Mean of four replications.
[z] LSD not appropriate here, since the line × harvest interaction was not significant at the 5% level.

Results

The dry weights and KER values for the tomato strains differed when the plants were grown under low-K stress without added Na, these differences were maximal in the last harvest, as reported in Tables 2 and 3. Some strains, *e.g.* 576, appear to possess K efficiency that is unrelated to Na use. The KER values also differed among strains grown under low-K stress, with either 10 or 160 mg Na/pot present. The mean KER for strains 571 and 576 increased by 35–37%, with the addition of 10 mg Na/pot over the treatment without added Na. The KER of strains 203, 571, and 576 increased by 40–51% with the addition of 160 mg Na/pot over the treatment without added Na. Thus, these strains appeared to have a higher Na-substitution capacity than strains 349 and 546, for which the mean KER increased only by 16–22% with the addition of 160 mg Na/pot.

Total plant Na accumulation differed in tomato strains grown under low-K stress at both 10 and 160 mg Na/pot, the largest differences being observed in the last harvest (Table 4). The mean Na accumulation per

Table 3. Mean potassium efficiency ratio (KER = mg plant dry weight/mg K in plant) of five tomato strains grown under low-K stress at three Na levels (30 days after transplanting)

Strain	0 mg Na/pot	10 mg Na/pot		160 mg Na/pot	
	KER[z]	KER	% increase over no Na level	KER	% increase over no Na level
576	200a	274a	37%	293a	47%
571	164bc	221b	35%	248b	51%
349	177b	192c	9%	216c	22%
203	151c	175d	16%	213c	40%
546	152c	166d	9%	176d	16%
$LSD_{0.05}$	16	16		23	

[z] Mean of four replications.

Table 4. Mean total plant Na accumulation of five tomato strains grown under low-K stress with either 10 or 160 mg Na/pot (30 days after transplanting)

Strain	Total plant Na (mg)[′]	
	10 mg Na/pot	160 mg Na/pot
576	7.76a	19.0a
571	5.31b	15.2b
349	2.43c	7.2c
203	1.72d	6.4c
546	1.76d	6.3c
$LSD_{0.05}$	0.61	2.9

[′] Mean of four replications

plant was two to three times higher in strains 571 and 576, which had high Na-substitution capacity, than in the other strains, for both the 10 and 160 mg Na/pot treatments.

Higher Na concentrations were generally found in the hypocotyl, lower stem, upper stem, and upper leaf partitions of strains 571 and 576, relative to the other strains, at all harvests, and for both the 10 and 160 mg Na/pot treatments. The analyses for upper-leaf Na concentrations (Figures 1 and 2) are representative of this trend.

Over all harvests, the Na concentrations in the lower-leaf laminae and petioles of strains 571 and 576 also were higher than comparable values for the other strains, at the 160 mg Na/pot level (Fig. 3). However, at the 10 mg Na/pot level, the Na concentration in the lower-leaf laminae and petioles were higher only in strain 576 at all harvests (Fig. 4).

The Na concentrations in the roots differed among tomato strains grown under low-K stress with either 10 mg Na/pot or 160 mg Na/pot (Fig. 5). However, strains 349 and 546, which had low Na-substitution capacity, did not appear to maintain a higher Na concentrations (*i.e.* sequester Na) in the roots relative to the other strains, which had higher Na-substitution capacity.

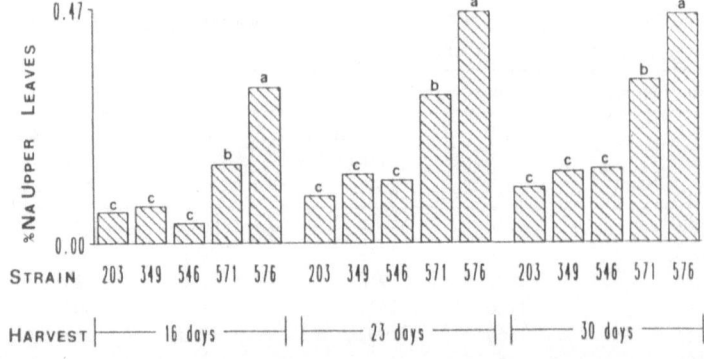

Fig. 1. Sodium concentration in the upper leaves of five tomato strains grown under low-K stress with 10 mg Na/pot at 16, 23, and 30 days after transplanting (mean of four replications).

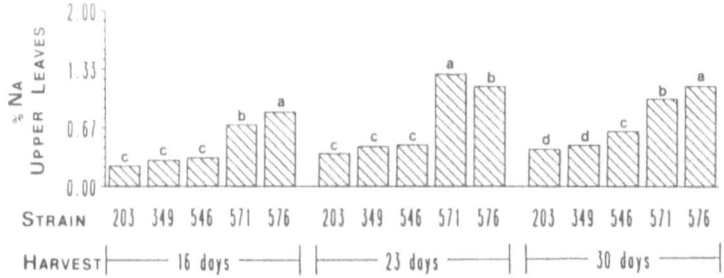

Fig. 2. Sodium concentration in the upper leaves of five tomato strains grown under low-K stress with 160 mg Na/pot at 16, 23, and 30 days after transplanting (mean of four replications).

The proportion of the total plant Na maintained in the roots (mg Na in the root/mg Na in the plant) was smaller in strains 571 and 576 than in the other strains at all harvests when grown under low-K stress with 10 mg Na/pot (Fig. 6). The proportion of total plant Na in the roots was also lower in strains 571 and 576, as compared to the other strains, in the 160 mg Na/pot treatment, except for strain 546 at the second and third harvests (Fig. 7).

The mean KER for strains 203 and 546 did not differ at the third harvest, when grown under low-K stress without added Na (Table 3). However, with the addition of 160 mg Na/pot, the mean KER for strain 203 was significantly higher than the KER for strain 546 by the third harvest. The higher KER in strain 203 at the high Na level, *i.e.* higher Na-substitution capacity, occurred without a difference in Na accumulation per plant between the two strains (Table 4). Thus, strain 203 appears to use the limited Na it accumulates more efficiently, in dry matter production, than does strain 546 when given 160 mg Na/pot.

The primary difference in Na distribution in the shoots of strains 203 and 546, for the high Na treatment at the third harvest, was in the lower-leaf laminae (Fig. 3a) and the upper leaves (Fig. 2). Surprisingly, the Na concentrations in these tissues were higher in strain 546 than in strain 203.

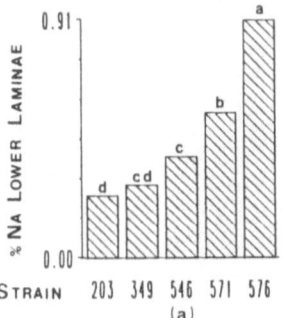 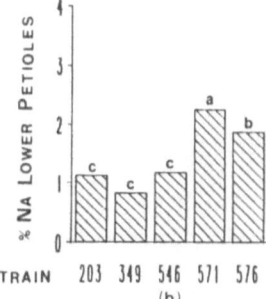

Fig. 3. Sodium concentrations in (a) the lower-leaf laminae and (b) the lower-leaf petioles of five tomato strains grown under low-K stress with 160 mg Na/pot at 30 days after transplanting (mean of four replications).

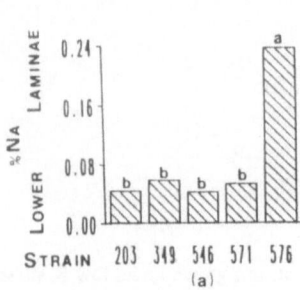
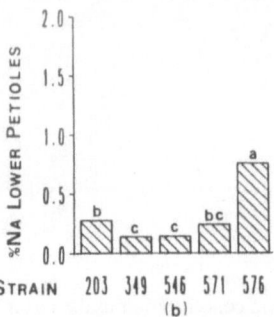

Fig. 4. Sodium concentrations in (a) the lower-leaf laminae and (b) the lower-leaf petioles of five tomato strains grown under low-K stress with 10 mg Na/pot at 30 days after transplanting (mean of four replications).

Discussion

Differences in dry matter accumulation per unit K, that are unrelated to the use of Na, appear to exist among the five tomato strains grown under low-K stress. However, these differences are relatively small in comparison to the differences in dry matter accumulation per unit K among the strains at either 10 or 160 mg Na/pot. Thus, a greater ability to partially substitute Na for K under low-K stress appears to be the primary factor in the observed strain differences. It should be noted that strain differences observed in this study are due solely to efficient K utilization. No differences in K acquisition were detected, as the entire 5.0 mg K available was taken up by all plants.

Strains 571 and 576, with high Na-substitution capacity, accumulated more Na and maintained higher Na concentrations in most shoot tissues

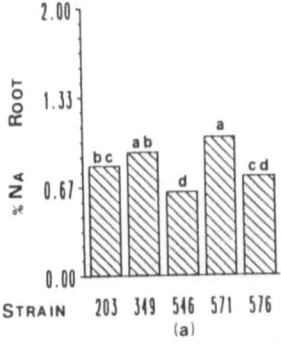
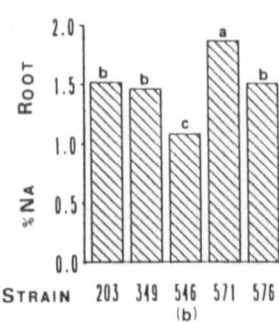

Fig. 5. Sodium concentrations in the roots of five tomato strains grown under low-K stress with either (a) 10 mg Na/pot or (b) 160 mg Na/pot at 30 days after transplanting (mean of four replications).

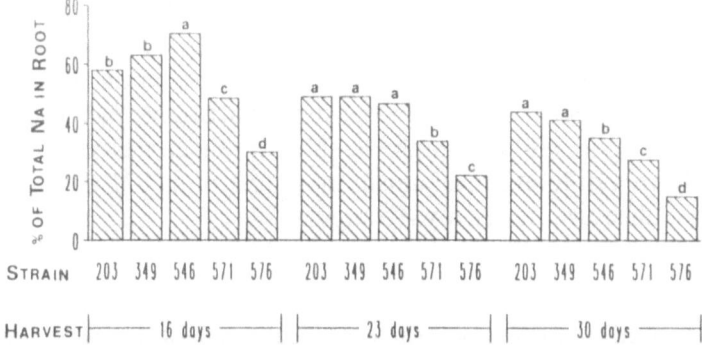

Fig. 6. Proportion of total plant Na in the roots [(mg Na in root/mg Na in plant) × 100] of five tomato strains grown under low-K stress with 10 mg Na/pot at 16, 23, and 30 days after transplanting (mean of four replications).

in both the 10 and 160 mg Na/pot treatments. These strains also had the largest increase in KER for both the 10 and 160 mg Na/pot levels, relative to the 0 mg Na/pot level. Maintenance of high Na levels in the shoot tissue thus appears to allow for greater Na-substitution capacity in these strains when grown under low-K stress. These results are similar to those obtained in comparative studies between a salt-tolerant wild tomato accession and a commercial cultivar for Na accumulation and Na substitution[7].

Sodium concentrations in the roots gave no indication that Na sequestration occurred in the root tissues of strains with poor Na-substitution capacity grown under low-K stress in the presence of added Na. However, the proportion of total plant Na in the roots was generally smaller in strains 571 and 576 (high Na-substitution capacity) than in the other strains. This seems to indicate that a larger proportion of the total plant Na was translocated to the shoots of strains 571 and 576, where it substitutes for K under low-K stress.

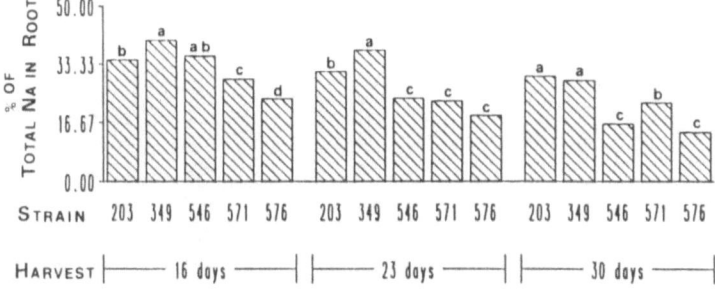

Fig. 7. Proportion of total plant Na in the roots [(mg Na in root/mg Na in plant) × 100] of five tomato strains grown under low-K stress with 160 mg Na/pot at 16, 23, and 30 days after transplanting (mean of four replications).

Strains 203 and 546 accumulate similar amounts of Na per plant, when grown under low-K stress. However, strain 203 appeared to substitute Na for K more efficiently than did strain 546. Also, strain 203 did not maintain high Na concentrations in the shoot partitions, as did the other strains (*i.e.* 571 and 576) with high Na-substitution capacity. Thus, strain 203 may have a different mechanism than do strains 571 and 576 for effective Na utilization under low-K stress.

References

1. Besford R T 1978 Effect of replacing nutrient potassium by sodium on uptake and distribution of sodium in tomato plants. Plant and Soil 50, 399–409.
2. Flowers T J and Läuchli A 1983 Sodium versus potassium: substitution and compartmentation. *In* Encyclopedia of Plant Physiology, New Series, Vol. 15B: Inorganic Plant Nutrition. Eds. A Läuchli and R L Bieleski. Springer-Verlag, Berlin-Heidelberg. pp 651–681.
3. Jacoby B 1964 Function of bean roots and stems in sodium retention. Plant Physiol. 39, 445–449.
4. Johanson J G and Cheeseman J H 1983 Uptake and distribution of sodium and potassium by corn seedlings. Plant Physiol. 73, 153–158.
5. Makmur A, Gerloff G C and Gabelman W H 1978 Physiology and inheritance of efficiency in potassium utilization in tomatoes (*Lycopersicon esculentum* Mill.) grown under potassium stress. J. Am. Soc. Hort. Sci. 103, 545–549.
6. Marschner H 1971 Why can sodium replace potassium in plants? Colloq. Int. Potash Inst. 8, 50–63.
7. Rush D W and Epstein E 1981 Comparative studies on the sodium, potassium, and chloride relations of a wild halophytic and a domestic salt-sensitive tomato species. Plant Physiol. 68, 1308–1313.
8. Shone M G T, Clarkson D T and Sanderson J 1969 The absorption and translocation of sodium by maize seedlings. Planta 86, 301–314.

Vegetative adaptation to N stress regimes in two barley cultivars with different N requirement

HARALD PERBY and PAUL JENSÉN
Department of Plant Physiology, P.O. Box 7007, S-220 07 Lund, Sweden

Key words Barley Cultivar Main stem N requirement N stress Partitioning Tiller Vegetative adaptation

Summary Plants of two barley cultivars, differing in requirement for N, were grown in water culture at combinations of stressing and non-stressing rates of N supply. At all N regimes, Kajsa, the cultivar with the lowest N requirement, produced more dry matter and used N more efficiently than Hellas. For both cultivars the morphological development was strongly influenced by the N supply, mainly seen as effects on tillering. Under N stress, tillering was more inhibited in Kajsa than Hellas. At higher rates of N supply Kajsa formed fewer, but on average large tillers than Hellas.

Introduction

Variation in mineral nutrition of barley cultivars have mainly been related to genetic variation in basic processes involved in mineral physiology[9], while the influence of morphological differences has received less attention. The structure and the size of the roots vary between cultivars[7,9]. Tillering is important for the vegetative development of cereal shoots. The rate of this process is influenced by the supply of N and the choice of cultivar[1,2,11]. Cereals adapt to high or low minerals supply by regulating tillering. Plants with few tillers and a low growth rate survive under mineral stress[3], whereas plants with many tillers and a high growth rate grow vigorously on large mineral resources.

Vegetative development, dry matter production and N accumulation have been compared for two cultivars of barley. Developmental changes were correlated with adaptation to low-N treatments.

Methods

The experiment was performed with two cultivars of barley (Hordeum vulgare L.). Hellas, a fairly late cultivar, has been grown in south Sweden and Kajsa, a very early cultivar, in north Sweden. In the field, Hellas requires more N than Kajsa (G. Persson, Svalöf AB, Sweden, personal communication).

Seeds of both cultivars were imbibed at 10 C in distilled water for one day and germinated on moist filter papers in petri dishes for two and a half days at 20 C. The seedlings were transferred to black plastic discs, 35 mm in diameter. Two discs with three plants each were placed together in a 2 l black-painted beaker containing 1800 ml nutrient medium.

H.W. Gabelman and B.C. Loughman (Eds.), Genetic aspects of plant mineral nutrition.
ISBN-13: 978-94-010-8102-3
© *1987 Martinus Nijhoff Publishers, Dordrecht/Boston/Lancaster.*

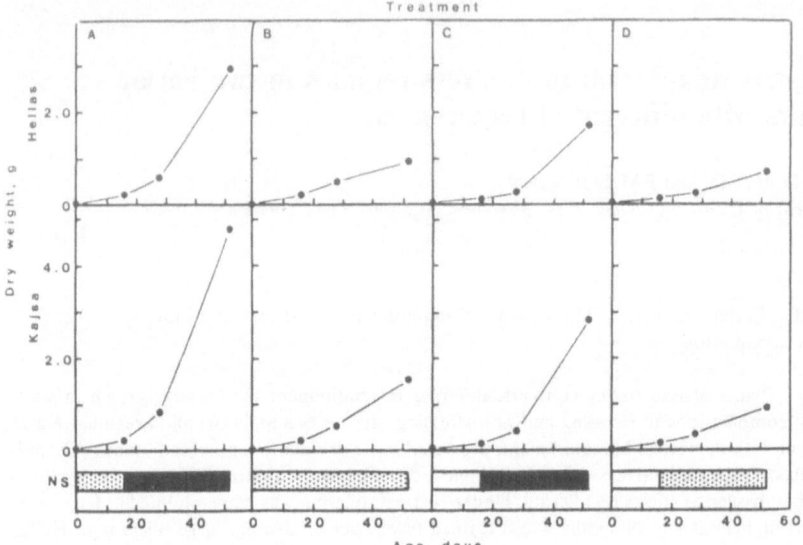

Fig. 1. Dry weight production per plant for two barley cultivars. Plants were cultivated with various NO_3 supplies (NS). Standard medium = black area, 10% medium = dotted area and 2.5% medium = white area. SE was usually less than 8% of mean.

The standard medium was composed of 1.6 mM KNO_3, 0.8 mM $Ca(NO_3)_2$, 0.8 mM $MgSO_4$, 0.8 mM KH_2PO_4, 0.4 mM Na_2HPO_4 and 0.16 mM Fe-EDTA. The concentrations of micronutrients (except Fe) were 80 percent of those recommended[8] and the initial pH was about 6.1. Plants were grown at two other NO_3^- concentrations: 10% and 2.5% of the NO_3^- concentration in the standard medium. The remainder of the NO_3^- was then replaced by SO_4^{2-}, thereby maintaining the same K^+ and Ca^{++} concentrations. The nutrient solutions were continuously aerated and replaced every fourth day. Water levels in the beakers were adjusted daily with distilled water. The design of the experiment is shown in Fig. 1.

The plants were kept in a greenhouse where sunlight was complemented with mercury lamps (Philips 400 W, total irradiance 80–90 Wm^{-2}) on a 16 h day. The temperature was 16 ± 2 C and relative humidity ca 65 percent.

The appearance of leaves on the main stem and the start of tillering were recorded. Six groups of each cultivar and treatment were harvested after 16, 28 and 52 days. 16-day-old plants were divided into shoots and roots. On older plants tillers were counted, thereafter roots, main stems, tillers and, if present, ears were separated[1]. Dry leaves on the oldest plants were not separable and were included in the main stem fraction. The introduced error is negligable. The plant parts were dried at 60 C for two days before determination of dry weights. Total contents of N were determined by an automatic Kjeldahl technique modified for efficient reduction of NO_3^-.

Results

Vegetative development

Plants of both cultivars given the highest NO_3^- supply developed eight leaves on the main stem (10% — S; Table 1). At other combinations of NO_3^- supply before and after day 16 only seven leaves developed on the main stem. A low N supply during any part of the cultivation period

Table 1. Influence of NO_3^- supply on leaf emergence on main stem, first tiller and total number of vital tillers. Code left of hyphen indicates NO_3^- supply day 0 to 16 and right of hyphen supply day 16 to 52 (S = standard medium). H = Hellas and K = Kajsa. Times for emergence of leaves and tillers are given as the median for 54 (day 0 to 16), 36 (day 16 to 28) and 18 (day 28 to 52) plants per treatment and cultivar. Stars indicate flag leaf. Total number of vital tillers are given as means per plant ± SE. Leaves number 1 and 2 always emerged on day 0 and day 4, respectively

NO_3^- supply	Cultivar	Leaf number:						First tiller, day	Number of vital tillers,	
		3 day	4	5	6	7	8		day 28	day 52
10%–S	H	8	14	18.5	24	32	38*	11	4.5 ± 0.3	10.1 ± 0.6
	K	8	14	18.5	24	31	36*	11	3.2 ± 0.2	6.4 ± 0.2
10%–10%	H	8	14	19.5	27	37*		11	2.8 ± 0.4	3.3 ± 0.3
	K	8	14	19.5	26.5	32*		11	1.8 ± 0.2	0.4 ± 0.2
2.5%–S	H	11.5	18	23	30.5	35*		21	1.9 ± 0.3	5.9 ± 0.5
	K	11	18	22	28	34*		20	2.2 ± 0.3	5.2 ± 0.4
2.5%–10%	H	11.5	18	23	31	36.5*	–	27	1.8 ± 0.2	2.1 ± 0.2
	K	11	18	23	29	35*		28	1.4 ± 0.2	1.0 ± 0.1

delayed expansion of leaves, Hellas generally being somewhat behind Kajsa. Similarly, tillering was delayed among plants supplied with 2.5% medium during the first 16 days. The number of vital tillers on 28- and 52-day-old plants depended on the N supply before and after day 16. These differences were significant (0.05 > P). Kajsa usually had fewer tillers than Hellas.

Dry weight and content of N

When treated equally from day 16, dry weight and nitrogen accumulation in 28- and 52-day-old plants were affected by the NO_3^- supply before day 16 (Figs 1 and 2). Similarly, the NO_3^- regime after day 16 greatly influenced growth and N accumulation. Kajsa generally grew better and accumulated in three cases of four more N than Hellas. For 52-day-old plants, cultivar differences in dry weight were always significant (Table

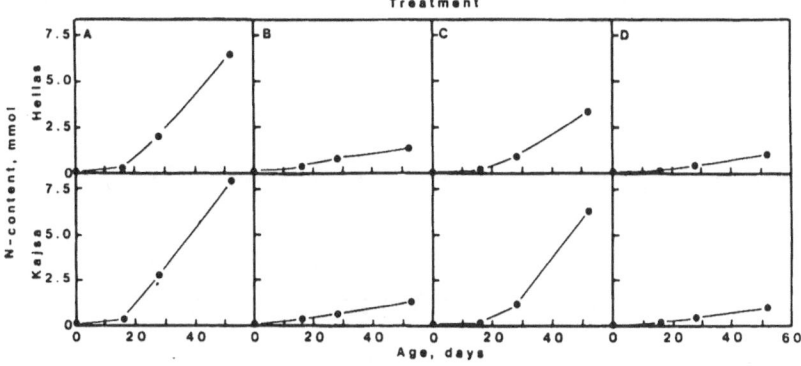

Fig. 2. Content of nitrogen in two barley cultivars. Treatments as in Fig. 1. SE was usually less than 5% of mean.

Table 2. Dry weights and N contents of 52-day-old plants. Levels of significance for cultivar differences are indicated by stars. * $0.05 > P > 0.01$ and ** $P < 0.01$. Otherwise as for Table 1.

NO_3^- supply	Cultivar	Dry weight (g)	N-content (mmol)
10% S	H	2.93	6.65
	K	4.80**	8.07
10% 10%	H	0.94	1.27
	K	1.52**	1.31
2.5% S	H	1.70	5.16
	K	2.82**	6.36*
2.5% 10%	H	0.68	1.07
	K	0.93*	1.04

2). N contents differed significantly between Hellas and Kajsa only for 2.5%-S treatment.

The main stems and roots of 52-day-old plants were between 10 and 55 percent smaller on plants that were supplied 2.5% medium initially or supplied 10% medium after day 16 (Figs 1 and 3) than on those receiving 10% medium before day 16 and standard medium thereafter. The percentages of the dry matter and total N content (Fig. 4) found in roots were about the same in the two cultivars, but varied according to the NO_3^- supply. Dry matter and N accumulation in tillers on 52-day-old plants decreased by 90 to 99 percent if grown at 10% medium from day 16.

Levels of N

Levels of N (Fig. 5) increased between day 16 and 28 in plants transferred from 10% and 2.5% media to the standard medium. For plants continuously supplied 10% medium, N levels in roots and main stems decreased as the plants aged. In plants supplied 2.5% medium initially and 10% thereafter, the N level in the main stems increased

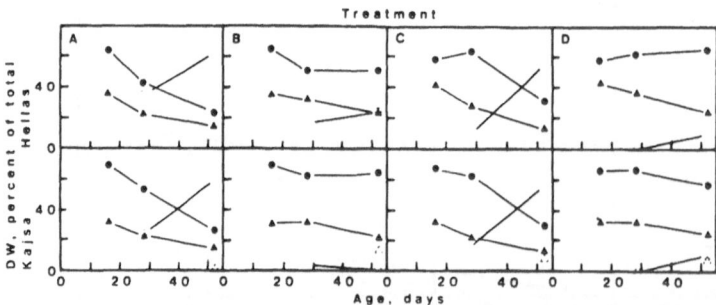

Fig. 3. Partitioning of dry matter between main stem, tillers, ears and root in two barley cultivars. Values are given as percent of total dry weight. Filled circles indicate main stem, unfilled circles tillers, filled triangles root and unfilled triangles ears.

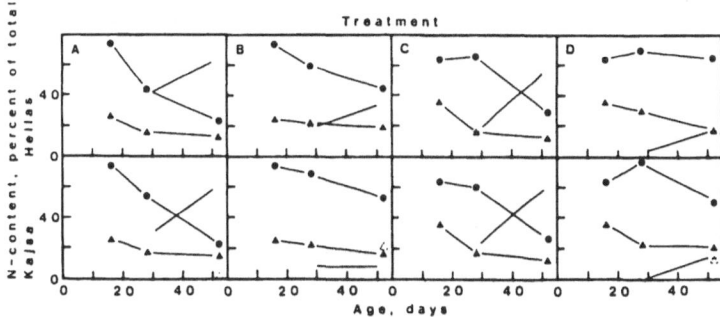

Fig. 4. Partitioning of N between main stem, tillers, ears and root in two barley cultivars. Otherwise as for Fig. 3.

between day 16 and 28, while it was almost constant in roots. In tillers, levels of N were usually higher than in main stems and roots. With one exception, the N levels in tillers decreased between day 28 and 52. While N levels in 16- and 28-day-old plants of Kajsa and Hellas were almost the same, N levels in 52-day-old plants were usually higher in Hellas.

Discussion

Determination of NO_3^- concentrations in used nutrient solutions with an ion selective electrode[10] showed that the standard medium supplied NO_3^- to the plants at a non-stressing rate (data not shown). Seedlings almost completely deplete a 10% medium of NO_3^-, and early growth decreases to some extent (Fig. 1). If older plants are kept on 10% medium, N stress becomes severe due to an increasing demand and the 2.5% medium induces severe N stress.

The growth rate of barley seedlings is lower when application of NO_3^- is delayed beyond unfolding of the first leaf[4,5,6]. N stress before or after day 16 decreases dry weight production and N accumulation (Figs. 1 and 2). The effect of the early NO_3^- supply is still apparent on day 52. Although the main stems are smaller on N stressed plants, the effect on tillers is more important (Figs. 1 and 3). Under N-stress, the plants strongly give priority to the main stem, in contrast to the situation at higher rates of NO_3^- supply (Figs. 1–4). N levels are generally highest in tillers and lowest in roots (Fig. 5), a result of the more juvenile condition of the tillers[11].

Kajsa produces more dry matter than Hellas at all rates of NO_3^- supply (Fig. 1 and Table 2). In three cases out of four, Kajsa accumulates more N (Fig. 2 and Table 2). Among the oldest plants, N levels are usually lowest in Kajsa, indicating a more efficient use of the element. In

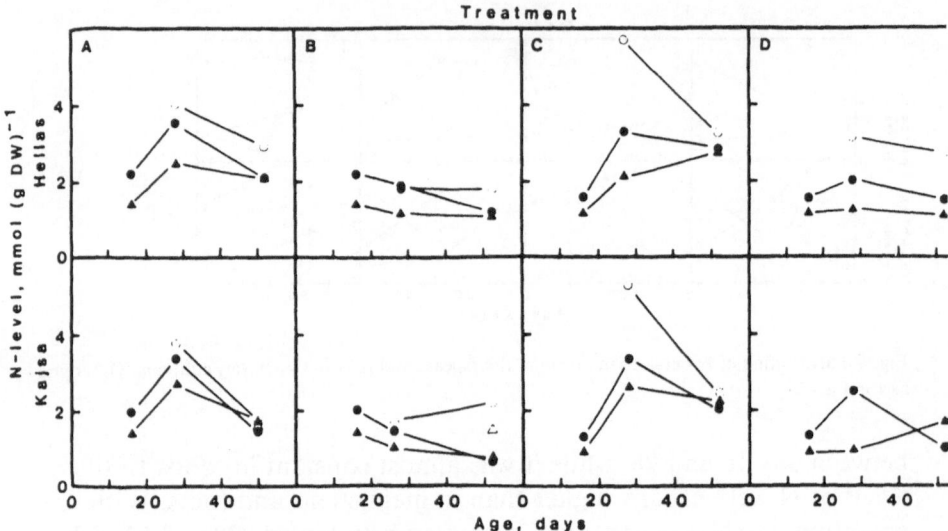

Fig. 5. N levels in main stems, tillers, ears and roots for two barley cultivars. Symbols as in Fig. 3. Treatments as in Fig. 1. SE was usually less than 8% of mean.

52-day-old plants supplied standard medium from day 16, dry matter and N is partitioned in a similar manner in both cultivars (Figs. 3 and 4). The inhibition of tiller growth, induced by severe N stress after day 16, is most prominent for Kajsa. Kajsa produces fewer, but on average larger tillers than Hellas (Figs. 1 and 3, Table 1). Similar cultivar differences in tillering have been described[2,11].

The nitrogen supply affects the leaf emergence rate and the day for the start of tillering (Table 1). If there are any cultivar differences in this respect, the very early Kajsa is always ahead of the farily late Hellas.

In summary, the cultivar differences are best visualized by the oldest plants[10]. The strategies for adaptation to N stress differ between cultivars. Kajsa, known to be more adapted to low N supply in the field, produces more dry matter, and uses it more efficiently than Hellas (Figs. 1, 2 and 5). Differences in N accumulation is not significant in three cases of four (Table 2). Exposure to N stress inhibits expansion of tillers in Kajsa more than Hellas (Fig. 3 and Table 1). If NO_3^- is supplied at a non-growth-limiting rate, tillers contribute to the biomass to the same extent in both cultivars. It can be concluded that genetic differences in morphology contribute to the differences between these two barley cultivars in nitrogen nutrition.

Acknowledgements We would like to thank Mr Fredrik Owman for skillful technical assistance and Prof. Terence Murphy for critical reading of the manuscript. The work was supported by grants from Ö.E. and Edla Johanssons Foundation, The Swedish Society for Natural Conservation and Hierta-Retzius Foundation.

References

1. Briggs D E 1978 Barley. Chapman and Hall Ltd, London. pp 14–19. ISBN 0-412-11870-X.
2. Cannell R Q 1969 The tillering pattern in barley varieties. 1. Production, survival and contribution to yield by component tillers. J. Agric. Sci., Camb. 72, 405–422.
3. Chapin F S 1983 Adaptation of selected trees and grasses to low availability of phosphorus. Plant and Soil 72, 283–287.
4. Dale J E 1972 Growth and photosynthesis in the first leaf of barley. The effect of time of application of nitrogen. Ann. Bot. 36, 967–979.
5. Dale J E 1976 Nitrate reduction in the first leaf and roots of barley seedlings grown in sand and in culture solution. Ann. Bot. 40, 1177–1184.
6. Dale J E, Felippe G M, Marriott C 1974 An analysis of the response of young barley seedlings to time of application of nitrogen. Ann. Bot. 38, 575–588.
7. Hackett C 1968 A study of the root system of barley. I. Effects of nutrition on two varieties. New Phytol. 67, 287–300.
8. Hewitt E J and Smith T A 1975 Plant Mineral Nutrition. The English Universities Press Ltd, London. p. 32. ISBN 0-340-05086-1.
9. Perby H and Jensen P 1983 Varietal differences in uptake and utilization of nitrogen and other macroelements in seedlings of barley, *Hordeum vulgare*. Physiol. Plant. 58, 223–230.
10. Perby H and Jensén P 1984 Net uptake and partitioning of nitrogen and potassium in cultivars of barley during ageing. Physiol. Plant. 61, 559–565.
11. Thorne G N 1962 Survival of tillers and distribution of dry matter between ear and shoot of barley varieties. Ann. Bot. 26, 37–54.

Genetic differences among wild oat lines in potassium uptake and growth in relation to potassium supply

M.Y. SIDDIQI, A.D.M. GLASS, A.I. HSIAO* and A.N. MINJAS**
Department of Botany, University of British Columbia, Vancouver, B.C., Canada V6T 2B1

Key words Genetic variation Growth Potassium uptake Wild oats

Summary Short-term (10 min) K^+ (^{86}Rb) influxes (ϕK_{oc}^+), rates of net K^+ uptake (ϕK_{net}^+) and growth in relation to K^+ supply were studied in genetically pure lines of wild oat. The wild oat lines employed in this study showed substantial differences in these traits. ϕK_{oc}^+ was higher in the AN lines (AN5 1, AN 474) than CS 40 and SH lines (SH 319, SH 430) in plants grown under both low and high K^+ conditions. Kinetic constants V_{max} and K_m for ϕK_{oc}^+, of selected lines and the rates of change of these constants with root K^+ concentration ([K^+]) showed that ϕK_{oc}^+ in AN 51 was consistently higher than in CS 40 and SH 319. ϕK_{net}^+ was also generally higher in AN 51. ϕK_{net}^+ (at different growth stages), unlike ϕK_{oc}^+, failed to correlate with root [K^+]; ϕK_{net}^+ values were low during the first two weeks despite lower root [K^+]. CS 40 showed the highest utilization efficiency, produced the largest amount of biomass, absorbed most K^+ and flowered earliest (by day 30). AN 51 had not produced flowers at day 42.

Introduction

Even under conditions in which soil tests indicate relatively high K^+ availabilities, plant growth and crop yield demonstrate pronounced positive responses to applied potassium[23]. Under such conditions (*i.e.*, when K^+ availability imposes a limit upon plant growth) it might be anticipated that competition for the available K^+, arising from the interactions of two or more genotypes growing in close proximity, might have a strong influence on growth. Indeed, previous experiments[22] conducted in pot trials at both low and high levels of K^+ availability have demonstrated that growth of barley (*Hordeum vulgare* L.) and of wild oat (*Avena fatua* L.) may be severely reduced by the competition for K^+. These studies indicated that the extent of reduction of growth of barley and wild oat was extremely sensitive to the identity of the barley variety.

It is recognized that plants require a certain minimum tissue concentration of ions, [ion]$_{crit}$, to achieve and maintain their potential growth rates[1,19,26] and that [ion]$_{crit}$ may differ from species to species within a certain range[1]. The initial rates of uptake of ions (when the seedling is in a so-called 'low-salt' state) must clearly be in excess of rates required to compensate for dilution due to growth so that internal

---

* Agriculture Canada, Research Station, Box 440, Regina, Saskatchewan, Canada S4P 3A2.
** Department of Crop Science, Sokoine, University of Agriculture, Morogoro, Tanzania.

H.W. Gabelman and B.C. Loughman (Eds.), Genetic aspects of plant mineral nutrition.
ISBN-13: 978-94-010-8102-3
© 1987 Martinus Nijhoff Publishers, Dordrecht/Boston/Lancaster.

concentration of ion, $[ion]_i$, can increase toward $[ion]_{crit}$. The rapidity with which $[ion]_i$ will approach $[ion]_{crit}$ will obviously be the balance between rates of uptake and growth. However, as $[ion]_i$ increases, the rate of uptake decreases exponentially as a result of negative feedback processes (for references see 3, 5). Plants also differ in the rates of decline of flux per unit of increase in $[ion]_i$[20] (henceforth referred to as sensitivity to negative feedback). These feedback effects serve to maintain $[ion]_i$ near $[ion]_{crit}$ under a wide range of external concentrations[5,21]. At steady-state therefore, uptake just compensates for any dilution due to growth. Thus, initial rates of uptake[15] and the sensitivity of the uptake system to negative feedback from $[ion]_i$, should be critical in determining the rapidity with which $[ion]_{crit}$ (thus, potential growth rate) is achieved. Indeed two plants (with similar potential growth rates) may have similar steady-state relative growth rates (RGR), $[ion]_i$ and rates of uptake and yet possess quite different biomass if one plant had achieved its $[ion]_{crit}$ (thus, potential growth rate) earlier than the other since growth is an exponential function of time (cf. 19).

In a situation where two (or more) genotypes compete, even a small advantage of growth at the initial stage of competition (as a result of higher uptake rate) may, in time, translate into a large effect because of the subsequent effects in competition for light in the shoot zone. Martin and Snaydon[13] have isolated the effects of competition in rhizosphere from the effects of competition in shoot zone alone; the former (root competition) had a much greater effect on the relative performance of barley and field beans. We have demonstrated substantial differences among barley cultivars in their competitiveness against wild oat (genetically pure line CS 40) in relation to K^+ supply which were related to the differences in K^+ uptake characteristics of the barley cultivars[22].

It is known that there is a high degree of genetic variation in wild oats and several genetically pure lines have been isolated[12,14,16]. However, studies on genetic differences among wild oat lines have centred on seed dormancy and very little (if anything) is known regarding genetic variation in their nutrient uptake and growth characteristics. With these considerations in mind, we have studied short-term K^+ influxes and their regulation, long-term K^+ accumulation and growth in relation to K^+ supply in selected genetically pure lines of wild oat.

Materials and methods

Growth of plants

The wild oat lines were genetically uniform populations[7,14,16] grown under closely regulated conditions in plots at the Agriculture Canada Regina Station. Seeds were dehulled and germinated in moist sand in the dark for 3–4 days. The seedlings were then transferred to plexiglas hydroponic tanks containing appropriate nutrient solutions in a growth room[19]. The growth room was maintained on a 16 h day:8 h night cycle at 25 ± 2 C temperature and 70% relative humidity. The light regime was provided at 5 mW cm^{-2} (plant level) by 'Vita-lite' fluorescent tubes having spectral composition similar to that of sunlight. Nutrient concentrations in growth tanks were maintained by continuous additions from two separate stock solutions, by means of peristaltic pumps: one stock solution contained K_2SO_4 and the other was modified Johnsons Solution (without K^+), except when otherwise described. Solution $[K^+]$ and $[NO_3^-]$ were measured in the tanks daily and the pump speed and/or concentrations of the stock solutions adjusted to maintain these ions at prescribed levels. All other nutrients were therefore supplied at the rate required to maintain $[NO_3^-]$.

Experiment 1. K^+ influx into intact roots of five wild oat lines

The plants were grown in modified 0.01 Johnson's nutrient solution with $[K^+]$ at 6 or 60 μM[19] as described above. Short term (10 min) K^+ influxes (ϕK_{oc}^+) into intact roots of 10-day old seedlings were determined from a solution containing 0.5 mM $CaSO_4$ and 0.05 mM K_2SO_4 following the procedure of Siddiqi and Glass[20]. The roots were prewashed for 5 min in non-radioactive solution (identical to uptake solution in all other respects) maintained at 25 ± 1 C and aerated continuously. The roots were then transferred to ^{86}Rb-labelled uptake solution for 10 min at 25 ± 1 C and aerated continuously. The uptake period was followed by a 5 min desorption in identical, but non-radioactive, solution maintained at 1 C. The volumes of prewash, uptake and desorption solutions were 1 l for approximately 2 g fresh weight of roots so that there were no appreciable changes in solution concentrations during the 5-10 min treatment periods. After determining their fresh weights, the roots were ashed at 500 C for 24 h. The ash was dissolved in 10 ml of distilled water and the radioactivity obtained by Cerenkov counting in a Searle Isocap 300 scintillation spectrometer. Root $[K^+]$ of these samples were determined using a flame photometer (Instrumentation Laboratory, Model 443).

Experiment 2. Influence of root $[K^+]$ on the kinetic constants of selected wild oat lines

Plants were grown in modified 0.01 Johnson's nutrient solution with $[K^+]$ at 6 μM as described above. On day 11 (from sowing), K^+ influxes were determined from modified Johnson's solution containing 0.01, 0.02, 0.04, 0.12 or 0.2 mM K^+ (in the form of sulfate) as outlined above. These were referred to as t_o samples. Immediately after t_o samples were removed, 3 mM K_2SO_4 was added to the growth solution. At intervals after the solution $[K^+]$ was increased to 6 mM, K^+ influxes from the aforementioned solutions were measured at 3 h (t_1) and 5 h (t_2). The radioactivities and root $[K^+]$ of these samples were determined. The kinetic 'constants', V_{max} and K_m, were determined at each time interval by Hofstee plots (Influx (V) vs V/$[K^+]_o$). The theoretical maximum V_{max} (Max V_{max}), slope relating V_{max} to root $[K^+]$ (b), theoretical minimum K_m (Min K_m), and slope relating K_m to root $[K^+]$ (b') were estimated by exponential regressions of V_{max} and K_m values, respectively, with root $[K^+]$[18].

Experiment 3. Growth of three wild oat lines in response to K^+ supply

Plants were grown in modified 0.01 Johnson's nutrient solution with $[K^+]$ at 5, 10, 20 or 100 μM as described above. Replicated samples (3) of each line (each replicate consisting of three plants) from each treatment were harvested at 0, 2, 4 and 6 weeks from the time of transferring the seedlings to the hydroponic tanks. After obtaining their fresh weights, roots and shoots were placed in an oven at 80 C for 72 h and dry weights determined. The roots and shoots were then ashed and their K^+ contents measured as described above. From the data obtained, utilization efficiency [biomass (g)^2/mmol K^+ as defined by Siddiqi and Glass[17]] and net K^+ uptake rates (ϕK_{net}^+) as defined by Williams[24] were calculated.

All experiments were repeated. In each experiment, treatments were replicated 3 times. In Experiments 1 and 2, each replicate consisted of 8-10 plants whereas in Experiment 3, it consisted of 3 plants.

Table 1. Root [K$^+$] (μmol g^{-1} fwt) and K$^+$ influx (ϕK$_{oc}^+$) (μmol g^{-1} h^{-1}) of 10-day old seedlings of 5 genetically pure lines of wild oats grown at 2 constant [K$^+$]$_o$ (see text). The data are means $\pm$ S.E. of 3 replicated samples, each sample consisting of 10 plants

Wild oat	[K$^+$] in the growth solution			
	6 μM		60 μM	
	Root [K$^+$]	ϕK$_{oc}^+$	Root [K$^+$]	ϕK$^+$
AN 51	29.3 $\pm$ 3.5	8.29 $\pm$ 0.33	59.4 $\pm$ 2.4	1.51 $\pm$ 0.05
AN 474	28.7 $\pm$ 2.5	8.95 $\pm$ 0.41	57.3 $\pm$ 2.2	1.79 $\pm$ 0.06
CS 40	29.6 $\pm$ 2.3	5.88 $\pm$ 0.44	53.0 $\pm$ 2.6	1.23 $\pm$ 0.02
SH 319	27.6 $\pm$ 3.5	6.26 $\pm$ 0.18	52.1 $\pm$ 0.6	0.68 $\pm$ 0.04
SH 430	25.8 $\pm$ 1.2	6.54 $\pm$ 0.15	52.0 $\pm$ 3.0	0.90 $\pm$ 0.04

Results

Short-term influxes of K^+ (ϕK_{oc}^+ and their regulation (Experiments 1 and 2)

Values for K$^+$ influx differed substantially among genetically pure lines of wild oat when grown under conditions of low and high [K$^+$]$_o$ (Table 1). Both AN lines, AN 51 and AN 474, showed a higher ϕK$_{oc}^+$ than the other lines. Root [K$^+$] was similar among these lines at both low and high [K$^+$]$_o$ (Table 1). In a detailed study involving AN 51, CS 40 and SH 319, AN 51 showed higher V$_{max}$ than the others at all times (except CS 40 at t$_1$) despite the fact that the former had a higher root [K$^+$] (Table 2). At t$_o$, K$_m$ for ϕK$_{oc}^+$ was lowest in AN 51 but at t$_1$ and t$_2$ (corresponding to higher [K$^+$]$_i$) it was lowest in SH 319 (Table 2). V$_{max}$ and K$_m$ showed significant exponential relationships (negative and positive respectively)

Table 2. V$_{max}$ and K$_m$ for K$^+$ influx at varying [K$^+$]$_i$ for three genetically pure lines of wild oat, AN 51, CS 40 and SH 319. (r^2 values from Hofstee plots are also given). The values are means $\pm$ S.E. of 3 replicated samples, each sample consisted of 8–10 plants

Time	Root [K$^+$] (μmol g^{-1})	V$_{max}$ for K$^+$ influx (μmol g^{-1} h^{-1})	K$_m$ for K$^+$ influx (mM)	r^2
AN 51				
t$_0$	35.8 $\pm$ 2.3	7.851	0.0119	0.98
t$_1$	48.8 $\pm$ 2.4	2.935	0.0176	0.77
t$_2$	55.1 $\pm$ 1.1	2.143	0.0437	0.98
CS 40				
t$_0$	29.9 $\pm$ 2.8	7.424	0.0148	0.90
t$_1$	42.7 $\pm$ 3.1	2.974	0.0179	0.82
t$_2$	47.9 $\pm$ 0.9	1.363	0.0303	0.73
SH 319				
t$_0$	30.9 $\pm$ 2.7	6.350	0.0134	0.87
t$_1$	39.4 $\pm$ 2.1	2.443	0.0159	0.96
t$_2$	49.1 $\pm$ 1.5	1.107	0.0220	0.97

Table 3. Theoretical maximum V_{max} (Max V_{max}), slope relating V_{max} to root [K^+] (b), theoretical minimum K_m (Min K_m) and slope relating K_m to root K^+ (b') obtained from exponential regressions[18] (see text). r^2 values for these regressions are also given

Wild oat	Max V_{max}	b	r^2	Min K_m	b'	r^2
AN 51	89.25	−0.069	0.99	0.0012	0.0618	0.83
CS 40	115.00	−0.090	0.96	0.0049	0.0351	0.77
SH 319	116.01	−0.096	0.99	0.0056	0.0274	0.98

with root [K^+] in all cases (Table 3). Max V_{max} of AN 51 was somewhat lower than in the other lines but the rate of decline of V_{max} with increasing [K^+]$_i$ (b) was substantially lower than CS 40 and SH 319 which had similar Max V_{max} and b values (Table 3). AN 51 also had the lowest Min K_m but its rate of increase with increasing [K_+]$_i$ (b') was also the highest; SH 319 had the lowest b' value (Table 3).

Growth and potassium uptake of three wild oat lines (Experiment 3)

In all the wild oat lines, particularly after 6 weeks growth, there was a significant increase in tillering (Table 4) and biomass as [K^+]$_o$ was increased from $5\,\mu M$ to $10\,\mu M$ (Fig. 1). There was little difference in biomass production between $10\,\mu M$ and $20\,\mu M$ treatments, particularly at the final (6 weeks) harvest. CS 40 produced the highest and AN 51 the lowest amount of biomass. In all cases root weight ratios (root weight:whole plant weight) at final harvest were less at $5\,\mu M$ than at higher [K^+]$_o$ treatments (Fig. 2). At 5–20 μM [K_+]$_o$, root weight ratios

Table 4. Mean numbers of shoots and inflorescences, dry weights and time of emergence of inflorescences of wild oat lines. The values are means ± S.E. of 3 replicated samples, each sample consisting of 3 plants

[K^+]$_o$ μM	No. of shoots	No. of Infl. plant^{-1}	Dry wt. of Infl. (g plant^{-1})	Time of emergence (days)
CS 40				
5	3.2 ± 0.2	1 ± 0	0.20 ± 0.03	30. fully emerged at 6 weeks
10	4.9 ± 0.1	1.33 ± 0	0.38 ± 0.01	35. fully emerged at 6 weeks
20	5.0 ± 0	0.89 ± 0.11	0.14 ± 0.01	38. fully emerged at 6 weeks
100	5.3 ± 2	0.67 ± 0.19	0.13 ± 0.05	Enclosed in sheath at 6 weeks
SH 319				
5	2.8 ± 0.1	0.78 ± 0.11	0.13 ± 0.01	38. partially emerged at 6 weeks
10	3.3 ± 0.1	0.67 ± 0	0.10 ± 0.01	Enclosed in sheath at 6 weeks
20	3.4 ± 0.1	0.78 ± 0.11	0.11 ± 0.01	Enclosed in sheath at 6 weeks
100	5.0 ± 0.4	0	0	Did not produce
AN 51				
5	1.9 ± 0.1	Did not produce inflorescences in any treatment.		
10	3.6 ± 0.1			
20	4.0 ± 0			
100	4.4 ± 0.1			

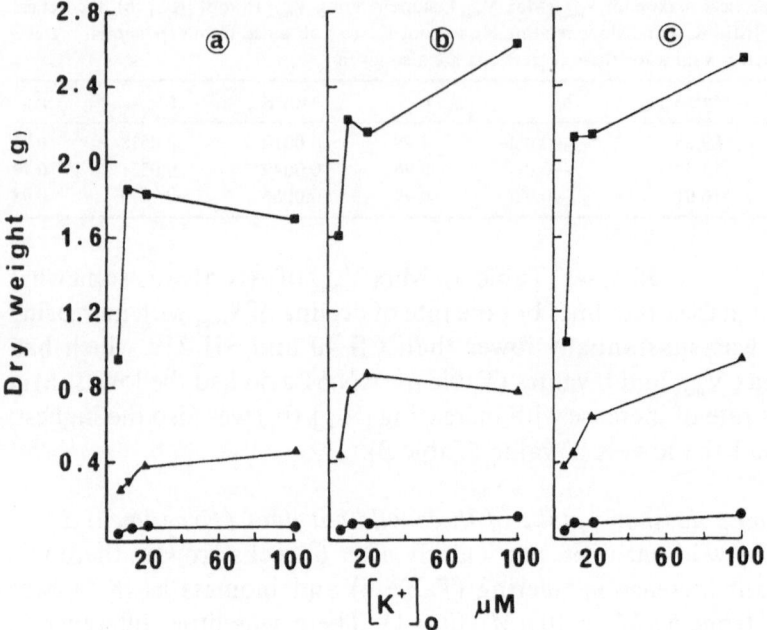

Fig. 1. Mean dry weight plant^{-1} (g) of wild oat lines: AN 51 (a), CS 40 (b) and SH 319 (c) in relation to K$^+$ supply ($[K^+]_o$) after 2 (●), 4 (▲), and 6 (■) weeks growth. (Standard errors within 15%).

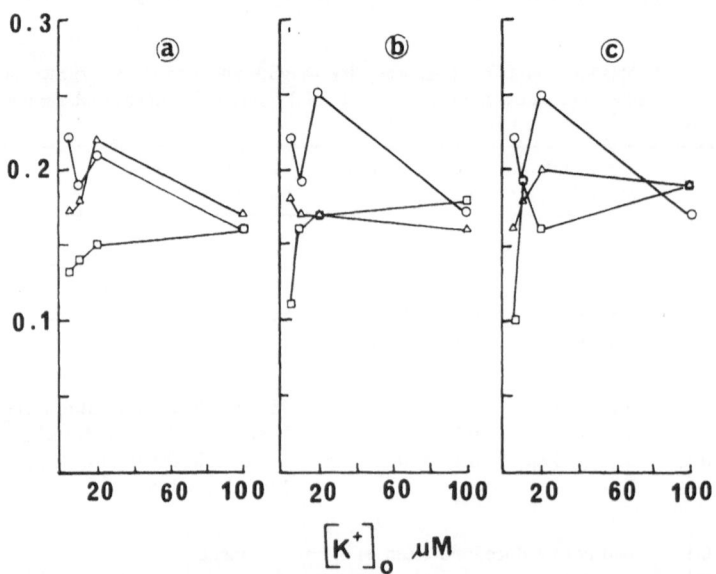

Fig. 2. Mean root weight ratio, RWR (root weight/total plant weight) of wild oat lines: AN 51 (a), CS 40 (b) and SH 319 (c) in relation to K$^+$ supply after 2 (○), 4 (△), and 6 (□) weeks growth.

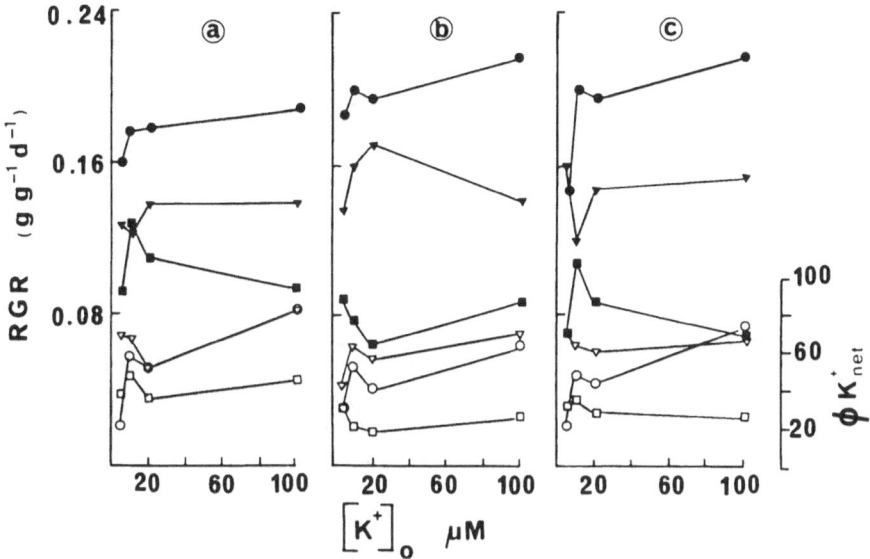

Fig. 3. Mean relative growth rates (RGR) (closed symbols) and mean net K^+ uptake rates, ϕK^+_{net}, (μmol g^{-1} d^{-1}) (open symbols) of wild oat lines: AN 51 (a), CS 40 (b) and SH 319 (c) in relation to K^+ supply over growth periods 0–2 weeks (●, ○), 2–4 weeks (▲, △), and 4–6 weeks (■, □).

were high at two weeks and then decreased to steady values; at 100 μM they remained constant from the initial time.

RGR decreased with time; in CS 40 and SH 319 the decline was more pronounced during the last two weeks (Fig. 3). In all cases average RGR over 0–2 weeks increased with increasing $[K^+]_o$ (from 5 to 10 and 20 to 100 μM). During this period RGR of these lines were in the order: at 5 μM, CS 40 > AN 51 > SH 319 and at 10–100 μM CS 40 ~ SH 319 > AN 51. Average net fluxes of K^+ (ϕK^+_{net}) increased with increasing $[K^+]_o$ (from 5 to 10 μM and 20 to 100 μM) particularly during the initial period (Fig. 3). At 5–20 μM $[K^+]_o$ ϕK^+_{net} increased from initial to intermediate period (2–4 weeks), the increase being much more pronounced at 5 μM than at higher concentrations. During the final period, it declined dramatically, more so in CS 40 and SH 319 than AN 51. At 10–100 μM, AN 51 had higher ϕK^+_{net} than the other lines at all times examined.

Both root and shoot $[K^+]$ increased from 0 to 2 and 2 to 4 weeks much more so in AN 51 and SH 319 (Fig. 4). From 4 to 6 weeks shoot $[K^+]$ generally declined. Up to 4 weeks, AN 51 and SH 319 had higher tissue $[K^+]$ than CS 40.

Utilization efficiency, UE (g^2 mmol^{-1} K^+) was highest in CS 40 and lowest in AN 51 (Fig. 5). In all the cases, UE increased with time. At the final harvest with respect to $[K^+]_o$, UE increased from 5 to 10 μM. While

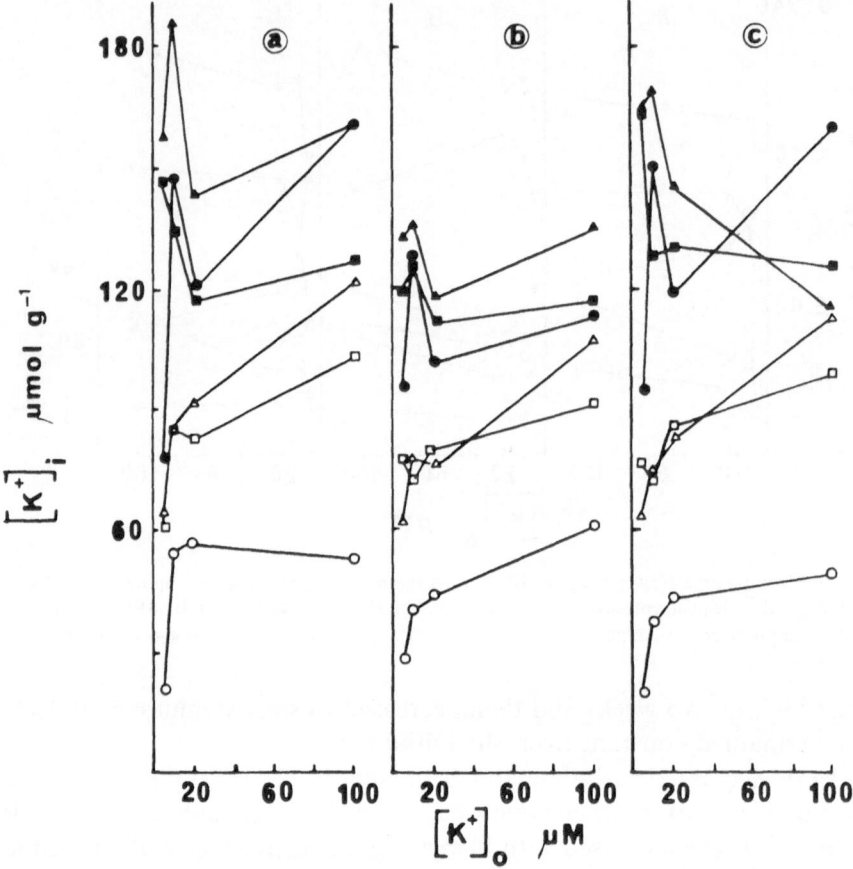

Fig. 4. Mean shoot (closed symbols) and root (open symbols) potassium concentration ($[K^+]_i$, μmol g^{-1} fresh weight) of wild oat lines: AN 51 (a), CS 40 (b), and SH 319 (c). Symbols as in Fig. 3. (Standard errors within 10%).

SC 40 and SH 319 showed a further increase from 20 to 100 μM, UE of AN 51 was essentially constant from 10 to 100 μM.

At final harvest (6 weeks), CS 40 and SH 319 (except at 100 μM) but not AN 51 had produced inflorescences. In CS 40 which flowered earlier than SH 319, mean number of inflorescences and their dry weights increased from 5 μM to 10 μM, beyond which they decreased to values even lower than at 5 μM (Table 4). SH 319 showed little differences at 5–20 μM and failed to produce any infloresence at 100 μM. In both cases, inflorescences emerged earlier at lower $[K^+]_o$.

Discussion

There were substantial differences among wild oat lines in short-term fluxes, ϕK^+_{oc} (Tables 1,2) which appear to be genetically based. Similar

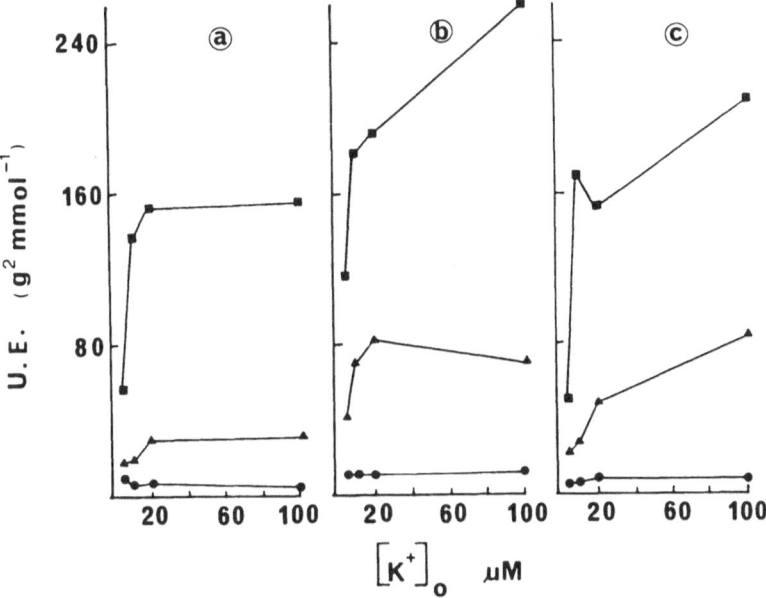

Fig. 5. Utilization efficiency (UE) of wild oat lines: AN 51 (a), CS 40 (b) and SH 319 (c). Symbols as in Fig. 1.

cultivar differences in K^+ uptake rates have been observed in crop species, e.g., barley[4,9,20] and wheat[25]. In the few wild oat lines we have studied, ϕK^+_{oc} correlate well with their degree of seed dormancy, i.e., the more dormant lines (AN 51, AN 474) had higher fluxes than the less dormant lines (SH 319, SH 430), with CS 40 being intermediate in both characteristics. During the initial period (0–2 weeks), ϕK^+_{net} was also higher (except at $5\,\mu M$) in AN 51 than in CS 40 and SH 319 (Fig. 3). At present the implication of this correlation is not known. In all cases ϕK^+_{oc} was negatively correlated with root $[K^+]$[4,9,20]. Moreover, these wild oat lines showed differences in the extent of reduction of ϕK^+_{oc} by root $[K^+]$[20]; the rate of reduction was less in AN 51 than others implying that the former would have higher ϕK^+_{oc} than the latter at all root $[K^+]$ possibly encountered by the plants. Surprisingly, however, ϕK^+_{net} estimated over periods of 2 weeks (Fig. 3) showed no correlation with root $[K^+]$ (Fig. 4). ϕK^+_{net}, particularly at $5-20\,\mu M\,[K^+]_o$, increased from the initial (0–2 weeks) to intermediate (2–4 weeks) period despite a substantial increase in root $[K^+]$ during the latter period. It appears that in wild oats, the developmental stage of the plant has an overriding effect on the rate of K^+ uptake. This effect may be brought about by changes in root morphology and/or physiology associated with the different developmental stages. Wild oat seedlings produce few seminal roots; prolific crown root growth occurs later in the development[2,11] which might

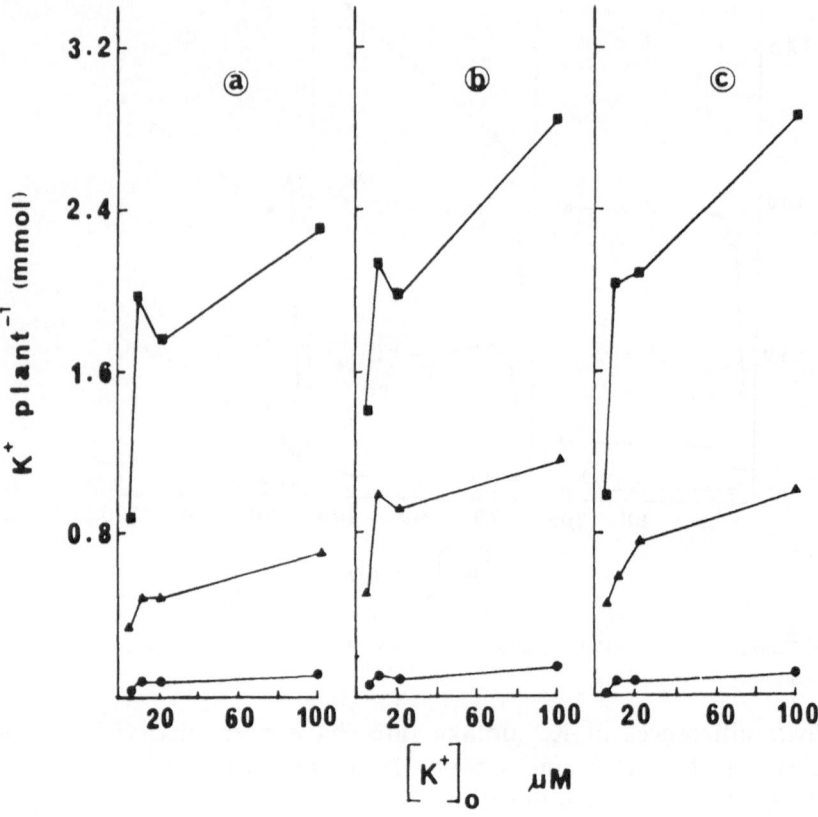

Fig. 6. Mean K^+ content plant^{-1} (mmoles) of wild oat lines: AN 51 (a), CS 40 (b) and SH 319 (c). Symbols as in Fig. 1.

substantially change the surface area/fresh weight relationships. Similar changes in ϕK^+_{net} have been demonstrated in barley, radish and ryegrass[26]. These effects, however, were directed towards the attainment and maintenance of tissue $[K^+]$. This should have serious implications in competition: low potential ϕK^+_{net} during the early seedling growth suggest that wild oats would be relatively less effective in competition during this period[2,11]. Indeed we have demonstrated that efficient K^+ absorbing barley cultivars (e.g., Fergus) could outcompete wild oats when planted together at the same seed rate[22]. During the final growth period (4–6 weeks), both RGR and ϕK^+_{net} declined despite little change in root $[K^+]$[8]. It is noteworthy that this decline in RGR and ϕK^+_{net} during weeks 4–6 was greater in the early flowering lines, CS 40 and SH 319 (which had produced seeds during this period), than in AN 51 which had not produced flowers by 6 weeks (Table 4). In annuals, the production of flowers and seeds mark the beginning of the termination of the life cycle.

It is, therefore, not surprising that the physiological activity of roots should decrease concomitant with reduced demand.

The wild oat lines studied also showed substantial differences in biomass production and RGR particularly after two weeks growth. Growth of these lines was not related to their K^+ fluxes. This is because their potential growth rates (Fig. 1) and utilization efficiencies (UE) (Fig. 5) were substantially different[19,20]: a higher root growth rate may more than offset the lower fluxes. For example, despite its lower ϕK_{oc}^+ and ϕK_{net}^+, CS 40 generally removed equivalent or even greater amounts of K^+ from the medium than AN 51 (Fig. 6). That AN 51, with its higher fluxes and lower growth rate, should have higher tissue $[K^+]$ than the others was also generally borne out by the data (Fig. 4).

The growth responses to increasing $[K^+]_o$ differed among these wild oat lines and these differences were time dependent (Fig. 1). All lines responded to increasing $[K^+]_o$ from 5 to $10\,\mu M$ at all stages. AN 51 did not respond to further increases of $[K^+]_o$. CS 4–0 (only at 4–6 weeks) and SH 319 (both at 2–4 weeks and 4–6 weeks) showed increases of growth from 20 to $100\,\mu M\,[K^+]_o$. CS 40 generally had a higher biomass than the other two lines. Interestingly, the time of flowering was negatively related to $[K^+]_o$, occurring earlier at the lower $[K^+]_o$ (Table 4). It is known that weeds and plants native to stressed environment flower early and allocate relatively more energy towards seed production[6,10] under limited supply of nutrients (and water). The numbers and dry weights of the inflorescences also generally showed this trend with the notable exception of CS 40 at $5\,\mu M\,[K^+]_o$. However, the differences in dry weights may also be a reflection of the differences in the degree of seed maturation (seed filling).

In conclusion, our results show substantial differences among genetically pure lines of wild oats in K^+ uptake rates, growth, capacity to remove K^+ and time of flowering. Based upon these characteristics, CS 40 appears to be potentially the most effective competitor among these lines. It is interesting to note that, compared to AN and SH lines, CS 40 encounters more severe competition from other weeds in its habitat[7]. It is particularly noteworthy that the developmental stage of plants had an overriding effect on K^+ uptake which might have serious implications for the ultimate outcome of competition with other plants. It is suggested that wild oats can be suppressed by cereals at an early stage but not later due to the prolific growth of the crown root[2]. Our results suggest that some physiological changes may also accompany these morphological changes.

Acknowledgement. This work is part of a project being carried out under contract from Agriculture Canada (DSS Contract NO. 0 1SG.0 1706-2-WO7).

References

1. Asher C J and Ozanne P G 1967 Growth and potassium content of plants in solution cultures maintained at constant potassium concentration. Soil Sci. 103, 155 161.
2. Chancellor R J and Peters N C B 1976 Competition between wild oats and crops. *In* Wild Oats in World Agriculture. Ed. D P Jones, Agriculture Research Council, London, pp 99–112.
3. Cram W J 1976 Negative feedback regulation of transport in cells. The maintenance of turgor, volume and nutrient supply. *In* Transport in Plants. II. Part A, Cells. Encyclopedia of Plant Physiology, New Series, Vol. 2. Eds. U Lüttge and M G Pitman. Springer-Verlag, Berlin, pp 284–316.
4. Glass A D M and Perley J E 1980 Varietal differences in potassium uptake by barley. Plant Physiol. 65, 160–164.
5. Glass A D M and Siddiqi M Y 1984 The control of nutrient uptake rates in relation to the inorganic composition of plants. Adv. Plant Nutr. 1, 103–147.
6. Grime J P 1974 Vegetation classification by reference to strategies. Nature London 250, 26–31.
7. Jana S and Naylor J M 1980 Dormancy studies in seed of *Avena fatua*. 11. Heritability for seed dormancy. Can. J. Bot. 58, 91–93.
8. Jensen P 1980 Control of K^+ (Rb^+) influx and transport with changed internal K^+ concentration and age in two varieties of barley. Physiol. Plant. 49, 291–295.
9. Jensen P and Pettersson S 1980 Varietal variation in uptake and utilization of potassium (rubidium) in high-salt seedlings of barley. Physiol. Plant. 48, 411–415.
10. Kimaro S J M 1977 A comparative study of the growth rates of *Eleusine indica* (L.) Gaertn. and *Eleusine coracana* (L.) Gaertn. M.Sc. Thesis, University of Dar es Salaam.
11. Koch W and Rademacher B 1966 Competition between crop plants and weeds. I. Absolute and relative development of cereals and some weed species. Weed Res. 6, 243–253.
12. Lee G A, Coleman-Harrell M E and Mundt G A 1980 Wild Oat. Competition and crop loss. University of Idaho, College of Agriculture, Current Information Series No. 541, 3 pp.
13. Martin M P L D and Snaydon R W 1982 Root and shoot interactions between barley and field beans when intercropped. J. Appl. Ecol. 19, 263–272.
14. Naylor J M and Fedec P 1978 Dormancy studies in seed of *Avena fatua*. 8. Genetic diversity affecting response to temperature. Can. J. Bot. 56, 2224–2229.
15. Nielsen N E 1979 Plant factors determining the efficiency of nutrient uptake from soils. Acta Agric. Scand. 29, 81–84.
16. Sawhney R and Naylor J M 1979 Dormancy studies in seed of *Avena fatua*. 9. Demonstration of genetic variability affecting the response to temperature during seed development. Can. J. Bot. 57, 59–63.
17. Siddiqi M Y and Glass A D M 1981 Utilization index: a modified approach to the estimation and comparison of nutrient utilization efficiency in plants. J. Plant Nutr. 4, 289–302.
18. Siddiqi M Y and Glass A D M 1982 Simultaneous consideration of tissue and substrate potassium concentration in K^+ uptake kinetics: a model. Plant Physiol. 69, 283–285.
19. Siddiqi M Y and Glass A D M 1983 Studies of the growth and mineral nutrition of barley varieties. I. Effect of potassium supply on the uptake of potassium and growth. Can. J. Bot. 61, 671–678.
20. Siddiqi M Y and Glass A D M 1983 Studies of the growth and mineral nutrition of barley varieties. II. Potassium uptake and its regulation. Can. J. Bot. 61, 1551–1558.
21. Siddiqi M Y and Glass A D M 1986 A model for the regulation of K^+ influx, and tissue potassium concentrations by negative feedback effects upon plasmalemma influx. Plant Physiol. 81, 1–7.
22. Siddiqi M Y, Glass A D M, Hsiao A I and Minjas A N 1985 Wild oat/barley interactions: varietal differences in competitiveness in relation to K^+ supply. Ann. Bot. 55, 17.
23. Walker D R 1979 Potassium fertilization of rapeseed and barley in central Alberta. Proceedings of the Workshop, Nov. 27–28, 1979, Saskatoon, Sask., Potash and Phosphate Inst. of Canada, pp 29–34.
24. Williams R F 1948 The effects of phosphorus supply on the rates of intake of phosphorus and nitrogen and upon certain aspects of phosphorus metabolism in gramineous plants. Aust. J. Sci. Res. Ser. B. 1, 333–361.

25 Woodend J J, Glass A D M and Person C O 1987 Genetic variation in the uptake and utilization of K^+ in wheat. Plant and Soil 99, ...
26 Woodhouse P J, Wild A and Clement C R 1978 Rate of uptake of potassium by three crop species in relation to growth. J. Exp. Bot. 29, 885–894.

Genetic variation in the uptake and utilization of potassium in wheat (*Triticum aestivum* L.) varieties grown under potassium stress

J.J. WOODEND, A.D.M. GLASS and C.O. PERSON
Department of Botany, University of British Columbia, Vancouver, B.C., Canada V6T 2B1

Key words Average net flux Efficiency ratio Potassium Short-term net flux Utilization efficiency Wheat

Summary Three-week-old plants of twenty four wheat varieties were evaluated for potassium uptake and utilization under conditions of potassium stress. Both short-term and average net fluxes over a 16-day period were determined. Utilization was expressed as shoot fresh weight, efficiency ratio (g mmol^{-1}) and utilization efficiency (g^2 mmol^{-1}). To compare varieties developed during different periods in the history of wheat breeding, the varieties were assigned to five groups on the basis of height and origin, viz. tall Indian and Mexican types, semidwarfs, double dwarfs and triple dwarfs. Significant differences in short-term net flux, root and shoot weights, efficiency ratio and utilization efficiency were observed. Differences in average net fluxes were also considerable but unrelated to short-term net fluxes. Although short-term net fluxes were negatively correlated with root potassium concentration, some of the differences were not related to root potassium concentration thus indicating that they are genetically determined. Both short-term and averaged net fluxes were not correlated with root weight. Variability for efficiency ratio was slight but much more evident for shoot weight and utilization efficiency. Despite substantial variation within groups, considerable differences between groups were observed. The tall varieties, particularly those of Indian origin, and the triple dwarfs, were superior to the semidwarfs and double dwarfs for both short-term net fluxes and efficiency ratio. The tall Indian varieties exhibited the highest shoot weight and utilization efficiency. The poorest performers for both uptake and utilization were the double dwarfs. Intermediate in performance between the double dwarfs and the tall Indian varietes were the semidwarfs.

Introduction

Remarkable progress has been made in the improvement of wheat yields over the last century. Perhaps the most notable success was the development of short-statured varieties which, under optimal conditions of fertility and water supply, far outyield the tall traditional types which were widely grown in some countries prior to the introduction of the dwarf types[1,2,3]. However, without fertilizer their yield advantage is at best marginal[2].

A number of studies have been undertaken to compare the performance of the dwarf types with that of the tall traditional types. Most of these have been agronomic studies with the main emphasis being on yield response to increasing soil fertility. There is clearly a paucity of information on comparative physiological studies particularly with respect to mineral nutrition.

H.W. Gabelman and B.C. Loughman (Eds.), Genetic aspects of plant mineral nutrition.
ISBN-13: 978-94-010-8102-3
© *1987 Martinus Nijhoff Publishers, Dordrecht/Boston/Lancaster.*

Table 1. Varietal grouping on the basis of height and origin

Group	Height (cm)	Varieties
TI: Tall, India	>91	NP 52, NP 718, Pusa 4, Pb 8A, Pb 9D, C306, K-13
TM: Tall, Mexico	>91	Yaqui 50, Yaqui 54, Chapingo 53, Nainari 60
SD: Semidwarf	71-90	Pitic 62, Penjamo 62, Sonora 64, Lerma Rojo 64, Kalyansona, Siete Cerros, Pavon 76, Jupateco 73
DD: Double dwarf	61-70	Yecora 70, Tesia 79, Arjun
TD: Triple dwarf	<60	Moti, UP301

Considerable intraspecific variation in mineral nutrition has been observed in a number of species[5,7,8,13,18]. The extent of this variation raises the possibility of developing strains exhibiting improved nutrient uptake and utilization; strains which might be expected to perform better under conditions of nutrient limitation.

The aim of this study was to evaluate a representative sample of tall and dwarf wheat varieties for potassium uptake and utilization under conditions of potassium stress. Utilization was determined on the basis of vegetative growth. We also undertook this investigation to identify genotypes suitable for inclusion in genetic studies on potassium uptake and utilization.

Materials and methods

Limited seed quantities of a number of wheat varieties were obtained from the International Maize and Wheat Improvement Centre (CIMMYT) in Mexico and the Agriculture Research Institute in New Delhi, India. The varieties consisted of tall traditional types of Indian and Mexican origin as well as various dwarfs, most of which carry 'Norin-10' dwarfing genes. The twenty four varieties selected for the study were assigned to five groups on the basis of height and origin (Table 1). The groups consisted of tall Indian (TI) and Mexican (TM) varieties, semidwarfs (SD), double dwarfs (DD) and triple dwarfs (TD). This group according to morphological type corresponds with breeding efforts designed to produce varieties suited to increasing fertility levels.

Seeds were surface-sterilized in 1% sodium hypochlorite, thoroughly rinsed and germinated overnight on moist filter paper in sterile petri dishes. Germinated seeds were placed on the nylon mesh of plexiglass discs (5-6 seeds per disc) embedded in trays of moist sterile sand. Three days later the discs were removed from the trays and washed free of sand in running water. They were then placed in 72 l hydroponic tanks equipped with circulation pumps (Model 1C-2, Brinkman Instruments) and overflow drains. After 3-4 days the plants were thinned to three even-sized seedlings per disc. Three replications were used in all experiments.

A maintained potassium concentration of $10 \mu M$ was used as the limiting level. Previous observations indicated that plants grown at this level developed typical potassium deficiency symptoms by two weeks. The initial concentrations (micromolar) of the other nutrients were as follows: N, 160; P, 20; Ca, 65; Mg, 10; B, 0.25; Mn, 0.02; Cu, 0.005; Fe, 0.4; Cl, 0.5. A solution containing all nutrients in proportion to their initial concentrations was continuously pumped into the tanks using peristaltic pumps (Fluid Metering, Inc.). The potassium level was monitored daily with a flame photometer (Instrumentation Laboratory 443) and the pumping rate adjusted, if necessary, to maintain the concentration at approximately $10 \mu M$. During the course of the experiments the potassium concentration ranged from 7–12 μM. Hydroponic solutions were aerated with a single airstone placed close to the circulation pump. The plants were grown in a growth room on a 16h/8h day/night cycle and 25/18 C day/night temperature regime. Lighting was provided by 'Vita-lite' fluorescent tubes at a photosynthetic flux density of $180 \mu E\,m^{-2}s^{-1}$ (tank level).

After three weeks short-term net potassium fluxes were determined by depletion from a solution containing 200 μM KNO_3 plus 0.5 mM $CaSO_4$ at a temperature of 21 C. Uptake experiments were done starting five hours after the beginning of the light cycle. Plants were initially placed in a large volume of aerated uptake solution for 10 minutes to allow for equilibration of the free space. They were then gently placed in 190 ml of aerated uptake solution. Depletion was determined by drawing 1 ml samples after 30 and 60 minutes. Roots were then excised, spun in a basket centrifuge for 20 seconds and weighed. Both roots and shoots were ashed in a muffle furnace at 450 C for 24 and 48 hours respectively and their potassium contents determined by flame photometry. Potassium utilization was expressed as shoot fresh weight, efficiency ratio (KER = g shoot fresh weight per mmol K^+)[8] and utilization efficiency (KUE = g^2 shoot fresh weight per mmol K^+)[15].

Average net fluxes over a 16-day period (5–21 days after germination) were also determined. To do so, average root and shoot fresh weights and potassium contents per plant at 5 and 21 days after germination were determined and average fluxes on an hourly basis then calculated according to the method of Williams[19]. Because average values were used for each harvest, statistical analysis for average net flux was not done.

Results

K^+ uptake

Although short-term net fluxes were measured over two time periods (0–30, and 0–60 minutes), only the rates for the 0–30 minute period are presented (Table 2). Significant differences among varieties were evident in this data and uptake rates for the 1 h period were similar to, or only slightly lower than values reported for the 30 minute period. Variation in short-term net uptake exceeded 300%. The highest net flux was observed in NP 52, a tall Indian variety, and the lowest net flux recorded was for the semidwarf Penjamo 62 (Table 2). Average net fluxes also showed considerable variation although this was not as marked as for short-term net flux (Table 2). Average net fluxes in the tall varieties ranged from 1.5 to 1.9 $\mu mol\,g^{-1}\,h^{-1}$. Average net fluxes were considerably lower than short-term net fluxes, more so for some varieties than others (Table 2), and the correlation between short-term net flux and average net flux was rather poor (r = 0.25, not significant at 5% level). The correlation between both fluxes and root weight per plant was also poor (0.04 for short-term net flux and -0.12 for average net flux).

Differences in root potassium concentration were of the magnitude of 160% (Table 2). The reported values are after adjustment for potassium taken up during the uptake experiments. Genotypic differences in root weight per plant were as much as two fold (Table 2).

Despite substantial variation within the groups of varieties, significant differences between the groups were observed (Tables 4 and 5). Overall the tall varieties exhibited high net and average fluxes although the triple dwarfs (TD) did surpass the tall Mexican varieties in net flux. The double

Table 2. Short-term net potassium fluxes, average net fluxes, root potassium concentrations ($\pm$ SE) and root fresh weights per plant of the 24 wheat varieties listed according to decreasing short-term net flux

Variety	Height group	Net K$^+$ flux (μmol g^{-1} h^{-1})		Root [K$^+$] (μmol g^{-1})	Root weight (g plant^{-1})
		Short-term	Average		
NP 52	TI	7.3^a	1.9	38.7 $\pm$ 1.36	0.73^{hi}
Pitic 62	SD	5.8^b	1.5	33.2 $\pm$ 4.47	1.07^{bcd}
Yaqui 54	TM	5.3^{bc}	1.6	43.4 $\pm$ 1.15	1.19^{ab}
Moti	TD	5.2^{bc}	1.7	47.2 $\pm$ 3.18	0.72^{hi}
NP 718	TI	5.2^{bc}	1.8	44.0 $\pm$ 2.31	0.61^i
Sonora 64	SD	5.2^{bc}	1.4	40.1 $\pm$ 2.36	0.77^{hi}
C306	TI	4.8^{bcd}	1.6	41.7 $\pm$ 2.70	1.09^{bcd}
Pusa 4	TI	4.8^{bcd}	1.5	38.2 $\pm$ 1.48	1.04^{b-e}
K-13	TI	4.8^{bcd}	1.5	39.1 $\pm$ 2.00	1.10^{bcd}
Nainari 60	TM	4.7^{bcd}	1.9	45.2 $\pm$ 1.28	0.85^{e-h}
UP 301	TD	4.6^{bcd}	1.0	40.4 $\pm$ 1.87	0.90^{a-h}
Pb 8A	TI	4.4^{b-e}	1.6	43.7 $\pm$ 0.87	1.00^{b-f}
Pb 9D	TI	4.4^{b-e}	1.5	35.0 $\pm$ 2.37	1.16^{abc}
Kalyansona	SD	4.2^{c-f}	1.3	47.5 $\pm$ 1.84	0.80^{f-i}
Pavon 76	SD	3.9^{c-g}	1.4	44.3 $\pm$ 5.90	0.96^{c-g}
Jupateco 73	SD	3.6^{d-h}	1.4	48.9 $\pm$ 2.21	0.78^{ghi}
Chapingo 53	TM	3.6^{d-h}	1.6	44.1 $\pm$ 1.15	1.32^a
Yaqui 50	TM	3.6^{d-h}	1.8	53.4 $\pm$ 2.14	0.70^{hi}
Arjun	DD	3.1^{e-h}	1.3	45.6 $\pm$ 1.88	0.78^{ghi}
Siete Cerros	SD	3.0^{e-h}	1.7	46.4 $\pm$ 0.95	0.78^{ghi}
Yecora 70	DD	3.0^{e-h}	1.3	40.5 $\pm$ 0.97	0.75^{ghi}
Tesia 79	DD	2.7^{gh}	1.4	45.6 $\pm$ 1.52	0.62^i
Lerma Rojo 64	SD	2.6^{gh}	1.4	44.9 $\pm$ 1.75	1.15^{abc}
Penjamo 62	SD	2.4^h	1.9	53.7 $\pm$ 2.49	0.91^{d-h}

Values followed by a different superscript are significantly different ($P \leqslant 0.05$) according to Duncan's multiple range test.

dwarfs (DD) showed the lowest short-term and average net fluxes while the semidwarfs (SD) were intermediate between the tall Mexican varieties and the double dwarfs. Exceptions to the overall trend are evident in Table 2. For example the semidwarfs Pitic 62 and Sonora 64 exhibited net fluxes comparable to most of the tall varieties. The tall varieties had the highest root weight per plant but were not significantly different from the semidwarfs. The double dwarfs which exhibited the lowest short-term net flux, also had the lowest root weight per plant. However, they were not significantly different from the triple dwarfs for root weight (Table 5).

K$^+$ utilization

Genotypic differences in shoot weight per plant, potassium efficiency ratio and utilization efficiency were also significant. The greatest amount of variation was recorded for utilization efficiency and the lowest for

Table 3. Shoot fresh weights per plant, potassium efficiency ratios and utilization efficiencies of the 24 wheat varieties

Variety	Height group	Shoot weight (g plant^{-1})	Efficiency ratio (g mmol^{-1})	Utilization efficiency (g^2 mmol^{-1})
NP 52	TI	1.78^a	9.5^{ah}	16.9^a
Pitic 62	SD	1.70^{abc}	9.7^{ah}	16.4^{ab}
Yaqui 54	TM	1.73^{ah}	8.0^{fg}	13.8^{bcd}
Moti	TD	1.43^{def}	9.3^{abc}	13.3^{cd}
NP 718	TI	1.35^{efg}	8.6^{b-f}	11.8^{d-g}
Sonora 64	SD	1.30^{e-h}	7.8^{fg}	10.1^{e-i}
C306	TI	1.90^a	9.2^{a-d}	17.3^a
Pusa 4	TI	1.74^{ab}	8.7^{b-f}	15.1^{abc}
K-13	TI	1.56^{b-e}	8.2^{c-f}	12.4^{c-f}
Nainari 60	TM	1.67^{a-d}	7.7^{f-g}	12.9^{c-e}
UP 301	TD	1.09^h	8.5^{c-f}	9.2^{g-i}
Pb 8A	TI	1.71^{ab}	8.6^{b-f}	14.6^{a-d}
Pb 9D	TI	1.79^a	9.2^{a-d}	16.5^{ab}
Kalyansona	SD	1.20^{fgh}	8.0^{fg}	9.7^{f-i}
Pavon 76	SD	1.45^{c-f}	9.1^{a-e}	13.3^{cd}
Jupateco 73	SD	1.15^{gh}	7.8^{fg}	9.0^{ghi}
Chapingo 53	TM	1.64^{a-d}	7.1^{gh}	11.7^{d-h}
Yaqui 50	TM	1.22^{fgh}	7.2^{gh}	8.8^{hi}
Arjun	DD	1.07^h	8.0^{fgh}	8.6^i
Siete Cerros	SD	1.23^{fgh}	7.8^{fg}	9.6^{f-i}
Yecora 70	DD	1.11^{gh}	8.3^{def}	9.2^{ghi}
Tesia 79	DD	1.05^h	7.2^{gh}	7.6^i
Lerma Rojo 64	SD	1.51^{b-e}	8.5^{cde}	12.9^{cde}
Penjamo 62	SD	1.36^{efg}	6.6^e	9.1^{ghi}

Values followed by a different superscript are significantly different ($P \leq 0.05$) according to Duncan's multiple range test.

efficiency ratio (Table 3). Variation in shoot weight was intermediate between that observed for efficiency ratio and utilization efficiency. The tall varieties had the highest shoot weight per plant which was significantly greater than that of all the dwarf types (Table 5). The double dwarfs had the lowest shoot weight per plant although it was not significantly different from that of the triple dwarfs (Table 5). The tall Mexican varieties had the lowest efficiency ratio which was only slightly exceeded by that of the double dwarfs (Table 5). The triple dwarfs and

Table 4. Analyses of variance for short-term net flux, root and shoot fresh weight per plant, efficiency ratio (KER) and utilization efficiency (KUE)

Source	DF	Variance ratio				
		Net flux	Root weight	Shoot weight	KER	KUE
Groups	4	17.01**	11.99**	39.85**	17.48**	32.81**
Varieties within groups	19	5.41**	8.79**	5.65**	5.40**	6.78**
Error	48					

** Significant at 1% level

Table 5. Comparison of group means for short-term net flux, average net flux, root and shoot weight per plant, efficiency ratio (KER) and utilization efficiency (KUE)

Height group	Net K^+ flux (μmol g^{-1} h^{-1})		Root weight (g plant^{-1})	Shoot weight (g plant^{-1})	KER (g mmol^{-1})	KUE (g^2 mmol^{-1})
	Short-term	Average				
TI	5.1^a	1.65	0.96^{ab}	1.68^a	8.8^a	15.0^a
TM	4.3^{ab}	1.72	1.02^a	1.57^a	7.5^c	11.8^b
SD	3.8^{bc}	1.49	0.90^{ab}	1.36^b	8.2^b	11.3^b
DD	2.9^c	1.32	0.72^c	1.08^c	7.8^{bc}	8.5^c
TD	4.9^{ab}	1.35	0.81^{bc}	1.26^{bc}	8.9^a	11.3^b

Values within a column followed by the same superscript are not significantly different ($P \leq 0.05$) according to Scheffé's test.

tall Indian varieties were similar and showed the highest efficiency ratios. Intermediate between the tall Mexican and Indian varieties were the semidwarfs.

Variation in utilization efficiency exceeded 225% with the highest value being recorded for the tall Indian variety C306 (Table 3). The lowest utilization efficiency was observed in the double dwarf Tesia 79 and was only slightly larger than its efficiency ratio. Overall the tall Indian varieties had the highest, and the double dwarfs the lowest, utilization efficiencies (Table 5). The tall Mexican varieties, semidwarfs and triple dwarfs were similar and intermediate between the tall Indian varieties and double dwarfs.

Discussion

Our study has demonstrated that there is considerable variation for potassium uptake rate and utilization among varieties of wheat. Furthermore, because the material used in this study is representative of wheat varieties developed over the last eighty years, it enabled us to examine the changes which have occurred over this period with regard to potassium uptake and utilization. The tall, traditional varieties were widely grown under conditions of low to moderate fertility[12] prior to the introduction of the dwarf varieties. The semidwarfs such as Sonora 64 and Lerma Rojo 64, in conjunction with improved soil fertility, ushered in what is popularly known as the Green Revolution[3]. Subsequently, in response to the need for varieties better able to withstand higher fertilizer levels, double dwarfs were produced. More recently triple dwarfs have been developed for culture under even higher fertility levels. It is interesting to note that these varieties have not been widely grown because of fertilizer shortages and their lack of adequate disease resistance[14]. We should point out that our system of grouping these varieties does not necessarily mean that the shorter varieties are more recent nor that they are more productive. For example the double dwarf Yecora 70 was released before a number of the semidwarfs and the semidwarf Pavon 76 is a better performer than some of the double dwarfs[1].

The variability observed in short-term net fluxes is considerable but should be viewed in conjunction with the results obtained for root potassium concentration (Table 2). It is well established that potassium uptake is influenced by root potassium concentration[9]. Consequently, varietal differences in uptake may reflect differences in root potassium concentration rather than actual genetic differences. Our results indicated that there was a significance negative correlation ($r = -0.57$, significant at 5% level) between short-term net flux and root potassium concentration. Therefore, some of the observed differences in short-term net fluxes may be due to differences in root potassium concentrations. However, this is clearly not so where higher uptake rates are associated with higher root potassium concentration or where large differences in uptake are associated with small differences in root potassium concentration. These differences must therefore be heritable and could reside in the the uptake mechanism per se and/or in the extent of the absorptive surface. Similarly, Glass and Perley[10], and Siddiqi and Glass[17] concluded that some of the variation in potassium influx observed in barley is likely to be heritable because the differences were not related to internal root potassium status.

In addition there were substantial differences in average net fluxes over a 16-day period. However, average net fluxes were poorly correlated with short-term net fluxes. Comparison of short-term net fluxes with average net fluxes indicates that for most of the varieties, average net fluxes were considerably lower than short-term net fluxes (Table 2). This is to be expected since average fluxes are acclimated fluxes from a solution of low potassium concentration whereas short-term net fluxes were determined from solutions of much higher concentration. Petterssson and Jensén[13] also reported very little variation in average net flux but considerable differences in influx (^{86}Rb) in barley seedlings grown for seven days. On the other hand Siddiqi and Glass[16] reported substantial differences in average potassium flux in barley varieties grown for a longer period of time.

Barring some exceptions we have observed that the tall varieties were superior in both net and average fluxes although they were not much better than the triple dwarfs with regard to short-term net flux. Thus although there appears to be a decreasing trend in both short-term and average net fluxes, the performance of the triple dwarfs is not consistent with this trend. In contrast, Cacco et al.[4] did a comparative study of 13 maize hybrids developed over a 45-year period and reported increased sulphate and nitrate uptake with improvements in yield potential.

To enable better evaluation of genotypic differences in potassium acquisition, both uptake rates and the size of root systems were examined. Considerable variation in root weight per plant was observed but both short-term and average net fluxes were not correlated with root weight. However, the correlation between root weight per plant and flux should be examined with caution becasue differences in shoot growth or demand were not taken into consideration. Also of importance are the compensatory changes in root and shoot growth which occur in response to nutrient limitation.

With regard to potassium utilization we chose to use three indices of efficiency: shoot fresh weight, efficiency ratio (KER), and utilization efficiency (KUE). Utilization efficiency is the product of the efficiency ratio and biomass produced per plant and has been recommended as a more appropriate measure of nutrient utilization[15]. Studies in barley have shown that variation in efficiency ratio tends to be slight whereas intervarietal differences tend to be much more pronounced when expressed as utilization efficiency[15]. We made similar observations in this study. Furthermore, comparison of the group means for shoot weight per plant, efficiency ratio (KER) and utilization efficiency (KUE) indicates that the main reason for the differences in utilization efficiency was the variation in shoot weight per plant rather than efficiency ratio (Table 5). Thus the tall Indian varieties which generally grew more vigorously than the other varieties, exhibited the highest utilization efficiency although they were no better than the triple dwarfs in efficiency ratio (Table 5). However, when grown to maturity under field conditions, differences in total biomass production between the tall and dwarf types have been reported to be negligible[12].

It has been suggested that selection under nutrient-rich conditions may have resulted in the diminution of some characteristics responsible for efficient nutrient utilization[6]. In view of the dramatic changes in fertilizer application that have been associated with the development of the varieties used in this study, it is tempting to relate our findings to the likely effect of selection under different nutrient regimes. Any attempt to do so should keep in mind the genetic diversity of the material, the major selection pressures applied, *i.e.* selection for improved yield under high nitrogen levels, and the number of genotypes examined. For example, some of the morphological types, particularly the triple dwarfs were not adequately represented in our study. Therefore, although we have demonstrated that there are significant differences among groups, we are hesitant to associate this variation with selective forces that operated under different nutrient regimes. A valid test of the effect of selection under different nutrient conditions would require that a variable population be initially subjected to selection under different nutrient regimes.

Thereafter the selected lines would be evaluated for characteristics considered essential to efficient nutrient acquisition and utilization.

Acknowledgements We wish to thank Drs. V P Kulshrestha and P Brajcich for providing the seed, and Dr. M Y Siddiqi for critically reading the manuscript. The award of a Graduate Fellowship to J Woodend is greatly appreciated.

References

1. Annual Report 1979 International Maize and Wheat Improvement Centre, Mexico.
2. Arnon I 1978 Fertilizer use as a lead practice in modernizing agriculture. *In* Proceedings of 11th Congress Intl. Potash Inst., Berne, Switzerland, pp 453–478.
3. Borlaug N E 1968 Wheat breeding and its impact on world food supply. *In* Third Intl. Wheat Genetics Symposium. Eds. F W Finlay and K W Shephard. Canberra, Australia, pp 1–36.
4. Cacco G M, Saccomani M and Ferrari G 1983 Changes in the uptake and assimilation efficiency for sulfate and nitrate in maize hybrids selected during the period 1930 through 1975. Physiol. Plant. 58, 171–174.
5. Clark R B 1983 Plant genotype differences in the uptake, translocation, and use of mineral elements required for plant growth. Plant and Soil 72, 175–196.
6. Clarkson D T and Hanson J B 1980 The mineral nutrition of higher plants. Annu. Rev. Plant Physiol. 31, 239–298.
7. Epstein E 1972 Mineral Nutrition of Plants: Principles and Perspectives. John Wiley and Sons, New York.
8. Gerloff G C 1976 Plant efficiencies in the use of nitrogen, phosphorus, and potassium. *In* Plant Adaptation to Mineral Stress in Problem Soils. Ed. M J Wright. Cornell Agri. Expt. Stn., Ithaca, NY.
9. Glass A D M 1978 The regulation of potassium influx into intact roots of barley by internal potassium levels. Can. J. Bot. 56, 1759–1764.
10. Glass A D M and Perley J E 1980 Varietal differences in potassium uptake by barley. Plant Physiol. 65, 160–164.
11. Kulshrestha V P and Tsunoda S 1981 The role of 'Norin 10' dwarfing genes in photosynthetic and respiratory activity of wheat leaves. Theor. Appl. Genet. 60, 81–84.
12. Kulshrestha V P and Jain H K 1982 Eighty years of wheat breeding in India: Past selection pressures and future prospects. Z. Pflanzenzüchtg. 89, 19–30.
13. Pettersson S and Jensén P 1983 Variation among species and varieties in uptake and utilization of potassium. Plant and Soil 72, 231–237.
14. Rao M V 1975 Wheat agronomy in India. *In* Bread: Social, Nutritional and Agricultural Aspects of Wheaten Bread. Ed. A Spicer. Applied Science. Publishers Ltd., London, pp 303–316.
15. Siddiqi M Y and Glass A D M 1981 Utilization index: a modified approach to the estimation and comparison of nutrient utilization efficiency in plants. J. Plant. Nutr. 4, 289–302.
16. Siddiqi M Y and Glass A D M 1983 Studies on the growth and mineral nutrition of barley varieties I. Effect of potassium supply on the uptake of potassium and growth. Can. J. Bot. 61, 671–678.
17. Siddiqi M Y and Glass A D M 1983 Studies on the growth and mineral nutrition of barley varieties. II. Potassium uptake and its regulation. Can. J. Bot. 61, 1551–1558.
18. Vose P B 1963 Varietal differences in plant nutrition. Herbage Abstr. 33, 1–13.
19. Williams R F 1948 The effects of phosphorus supply on the rates of intake of phosphorus and nitrogen and upon certain aspects of phosphorus metabolism in gramineaceous plants. Aust. J. Sci. Res. Ser. B. 1, 333–361.

Inheritance of response of sunflower inbreds to a low calcium/magnesium ratio

E. ALCANTARA and M.D. DE LA GUARDIA
Cátedra de Fisiolología Vegetal, Escuela Técnica Superior de Ingenieros Agrónomos, Universidad de Córdoba, Córdoba, Spain

Key words Calcium efficiency *Helianthus annuus* Inheritance Magnesium efficiency Sunflower

Summary The inheritance of response to a low Ca/Mg ratio in nutrient solution has been studied in sunflower. The parental lines RHA 274(P_1) and CMS HA 290(P_2) were chosen for their opposite response. Afterwards, F_1, F_2, BC_1 ($P_1 \times F_1$) and BC_2 ($P_2 \times F_1$) generations were obtained and genetically investigated with a low Ca test.

The Ca stress was brought about by decreasing the Ca concentration and by increasing the Mg concentration in the nutrient solution. Severity of foliar Ca deficiency symptoms in seedlings grown under Ca stress was used as an index of tolerance.

Results show that epistasis was not important, and a simple additive-dominance model was considered adequate to explain differences in tolerance to Ca stress, with additive gene effects highly significant. The high broad-sense heritability and the simplicity of the test suggest that screening for tolerant genotypes should be possible.

Introduction

Differences in the response to Ca stress or in Ca efficiency have been described in several species, such as *Zea mays*[4] and *Solanum lycopersicum*[7,9]. Sometimes the response to Ca stress is inversely related with the response to Mg level. In maize, high levels of Mg interact with Ca producing more detrimental effects in Ca inefficient than in Ca efficient genotypes[4].

In sunflower (*Helianthus annuus* L.) genetic differences between inbred lines, hybrids and cultivars as to their response to grow at low Ca level were found[6]. Later two inbred lines (one Ca efficient and the other Ca inefficient) were grown in a wide range of Ca and Mg concentrations in nutrient solution and were studied for their response in dry matter accumulation and for Ca and Mg content. The Ca efficient line had maximum yield with a lower level of Ca than the Ca inefficient one, while the opposite occurred in relation to the level of Mg[1]. It was concluded that Ca stress in sunflower can be produced by a high Mg concentration or by a low Ca/Mg ratio[1].

Severity of foliar deficiency symptoms in seedlings grown in low Ca nutrient solution has been used in this study as an index of the response. Visual symptoms have been used widely as an index to investigate the inheritance of some characters[5,11]. The objective of this research was to

H.W. Gabelman and B.C. Loughman (Eds.), Genetic aspects of plant mineral nutrition.
ISBN-13: 978-94-010-8102-3
© *1987 Martinus Nijhoff Publishers, Dordrecht/Boston/Lancaster.*

study the inheritance of the response to a low Ca/Mg ratio in sunflower when Ca efficient and Ca inefficient lines are crossed.

Material and methods

The original materials were the American lines CMS HA 290 and RHA 274 developed cooperatively by North Dakota and Texas Agricultural Experiment Stations and the Agricultural Research Service, USDA. Both lines were chosen because they had similar response to a low Ca/Mg ratio as the two lines used in an earlier report[1]. RHA 274(P_1) is Ca inefficient and CMS HA 290(P_2) is Ca efficient. Afterwards, F_1, F_2, BC_1 ($P_1 \times F_1$) and BC_2 ($P_2 \times F_1$) generations were obtained in a greenhouse. These generations were genetically investigated in 2 separate experiments.

The experiments were conducted in a growth chamber at 22 C during the day/18 C during the night with a 13 h photo-period and a photosynthetic irradiance of 60 Wm^{-2} (cool white VHO fluorescent tubes, Sylvania). Seed treated with sodium hypochlorite (0.5% v/v) were germinated for 5 days at 25 C in Perlite irrigated with 5 mM $CaCl_2$. Seedlings were transferred to 800 ml glass flask wrapped in aluminium foil (4 seedlings/flask). The flasks contained a standard nutrient solution of the following composition: 5 mM $Ca(NO_3)_2$; 5 mM KNO_3; 2 mM $MgSO_4$; 1 mM KH_2PO_4; 25 mM H_3BO_3; 2 μM $MnSO_4$; 2 μM $ZnSO_4$; 0.5 μM $CuSO_4$; 0.4 μM $(NH_4)_6Mo_7O_{24}$; 20 μM Fe-EDDHA. The pH was adjusted to 5.5 with KOH. The nutrient solution was aerated daily. After 10 days the Ca test was applied in the following way. The solution was changed to a lower Ca/Mg ratio. The composition of this new solution was: 0.5 mM $Ca(NO_3)_2$; 5 mM KNO_3; 2 mM $MgSO_4$; 3 mM $Mg(NO_3)_2$; 1 mM KH_2PO_4, with the other elements and pH being identical to the standard solution. The plants were grown in this solution for 13 days and then classified with an index of visual symptoms of calcium deficiency on a 0-5 scale:

0 — No symptoms.
1 — Light symptoms in the third pair of leaves.
2 — Intermediate symptoms in the third pair of leaves.
3 — Light symptoms in the second pair and intermediate symptoms in the third pair of leaves.
4 — Intermediate symptoms in both second and third pair of leaves.
5 — Intermediate symptoms in the second pair and heavy symptoms in the third pair of leaves.

The number of plants per generation (N) was different in the two experiments and from one generation to another, as it appears in Table 1. In the second experiment some plants were discharged from the analysis because they presented abnormal growth not related with the test. To avoid differences in the environment the position of the flasks in the growth chamber was changed every day.

Means, variances and standard deviations were calculated from individual plant data. Statistical and genetic analyses were carried out on the basis of the means and variances of the P_1, P_2, F_1, F_2, BC_1 and BC_2 generations. These data were used for tests of non allelic interaction[14] and for estimates of m, d and h parameters[12]. These estimates were done by weighted least squares, taking as weights the reciprocal of the variance of each mean. Broad-sense heritability was estimated from phenotypic and genetic variances according to the method described by Allard[2].

Results

The frequency distribution of plants for the classes of visual symptoms of calcium deficiency, the means and the standard deviations of each population are summarized in Table 1. Results from both experiments were similar so we will refer to them in general.

Table 1. Number of plants per class across classes of visual symptoms of calcium deficiency in two experiments, means and standard deviations (SD).

Generation	Number of plants per class						N	Mean	SD
	0	1	2	3	4	5			
Experiment I									
P_1					3	5	8	4.63	0.52
P_2	5	3					8	0.38	0.52
F_1			5	6	1		12	2.67	0.65
BC_1				2	7	3	12	4.08	0.67
BC_2	4	3	4	1			12	1.17	1.03
F_2	1	4	12	18	10	3	48	2.85	1.11
Experiment II									
P_1				1	18	13	32	4.38	0.55
P_2	8						8	0.00	0.00
F_1			3	8			11	2.73	0.47
BC_1				8	4		12	3.33	0.49
BC_2	3	6	2				11	0.91	0.70
F_2	3	10	14	19	1		47	2.09	1.02

The means of the F_1 and F_2 populations were intermediate between the two parents. The backcrosses tended toward their respective parents. The data suggest mainly additive inheritance.

The tests of non-allelic interaction[14] gave no significant deviations from zero (Table 2) suggesting that epistasis was not involved in the inheritance of Ca efficiency in the sunflower material studied. Consequently, the additive-dominance model was considered adequate for further analysis.

Estimates of the magnitude of the parameters m (mean), d (cumulative additive effect) and h (cumulative dominance effect), and their standard deviations from generation means[12] are summarised in Table 3. The estimates of d were significantly different from zero whereas h did not differ significantly from zero. The model is adequate as shown by the values of χ^2 and P (Table 3). It was concluded that additive gene action had a preponderant importance in the inheritance of Ca efficiency in the sunflower material studied.

Broad sense heritability estimates were computed as the ratio of total

Table 2. Test of non-allelic interaction

Experiment	Parameter	Mean	SD
	A	0.86	1.58
I	B	−0.71	2.22
	C	1.05	4.74
	A	−0.45	1.22
II	B	−0.91	1.48
	C	−1.48	4.22

Table 3. Estimates (±SD) of gene effects in a 3 parameters model

Experiment	m	d	h	χ^2	P
I	2.54 ± 0.35	2.20 ± 0.35	0.24 ± 0.71	0.480(3)*	0.95-0.90
II	1.70 ± 0.75	2.58 ± 0.74	0.96 ± 1.00	0.443(2)	0.90-0.75

( )* Degrees of freedom

genetic variance to phenotypic variance[2]. The estimates listed in Table 4 show high values for the heritability of the character Ca efficiency.

Discussion

The technique applied in these experiments to evaluate the response to Ca stress has an advantage as it uses seedlings and is non destructive. Once the plant has been classified according to the scale for visual symptoms, it can be transferred to a medium with enough Ca where the plant recovers from the deficiency and completes its life cycle by producing seeds.

It seemed better to impose the Ca stress when the seedlings had grown a few days than when they had just begun to grow. So we grew the seedlings in a standard solution for the first 10 days, after which the Ca stress was imposed. Previously we had shown that Ca stress in sunflower could be produced by high Mg concentration[1]. For this reason in the present study the Ca stress was brought about not only by decreasing the Ca concentration (from 5 to 0.5 mM) but by increasing the Mg concentration (from 2 to 5 mM).

With this technique, we found that response to Ca stress in the genetic material studied seems to be polygenically inherited, with epistatic gene effects being insignificant. The generation means analysis using the additive-dominance model indicated additive gene action to be of major importance (Table 3).

Giordano et al.[9] found that efficiency in Ca utilization by tomato plants appeared to be controlled by simple additive-dominance systems, with additive effects being highly significant. Preponderant additive gene effects have also been reported in the inheritance of characters like accumulation of mineral elements in tall fescue[13]. However, characters like efficiency in P utilization in beans are more complexly inherited[8].

The high broad sense heritability values in Table 4 and the simplicity of the test based on visual symptoms of Ca deficiency permit screening sunflower genotypes tolerant to Ca stress so as to produce cultivars that

Table 4. Broad sense heritability estimates (%)

Experiment I	74.0
Experiment II	82.7

are probably better adapted to certain soils and environmental conditions.

It is probable that a relationship exists between seedlings response in Ca test and the performance of mature plants in the field with a low Ca/Mg ratio. In a previous work with young plants, we found that the response to Ca and Mg levels in the nutrient solution of a Ca-efficient line of *Helianthus annuus* was similar to the response described by Madhok and Walker[10] for *H. bolanderi* exilis, which is endemic of serpentinic soils[1]. In maize, differences in Ca and Mg accumulation by mature field grown plants could be predicted from the behaviour of young plants grown under greenhouse conditions[3].

Any breeding program with the purpose of screening sunflower genotypes tolerant to Ca stress should consider the great importance of additive gene effects in the inheritance of tolerance.

Acknowledgements The cooperation of Dr. A Porras (Facultad de Ciencias de Córdoba), the Department of Genetics (Escuela T.S. de Ingenieros Agrónomos, Córdoba) and Prof. E Fereres is gratefully acknowledged. This work was financially supported by Comisión Asesora de Investigación Científica y Técnica. Project 4697/79.

References

1 Alcántara E and De la Guardia MD 1983 Genotypic differences in calcium and magnesium nutrition in sunflower. *In* Genetics Aspects of Plant Nutrition. Eds M R Sarić and B C Loughman. pp 87–91. Martinus Nijhoff/Dr W Junk, Publ., The Hague, Boston, Lancaster.
2 Allard R W 1960 Principles of Plant Breeding. pp 89–98. John Wiley and Sons, New York.
3 Baker D E, Thomas W I and Gorsline G W 1964 Differential accumulations of strontium, calcium and other elements by corn (*Zea mays* L.) under greenhouse and field conditions. Agron. J. 56, 352–355.
4 Clark R B 1978 Differential response of corn inbreds to calcium. Commun. Soil Sci. Plant Anal 9, 729–744.
5 Coyne D P, Korban S S, Knudsen D and Clark R B 1982 Inheritance of iron deficiency in crosses of dry beans (*Phaseolus vulgaris* L.) J. Plant Nutr. 5, 575–585.
6 De la Guardia M D, Saiz de Omenaca J A, Pérez Torres E B and Montes Agustí F 1980 Diferentes respuestas de las líneas, híbridos y variedades de girasol a la nutrición con bajo nivel de calcio. IX Conferencia Internacional de Girasol, vol. 1, 239–248. Servicio de Publicaciones Agrarias, Madrid.
7 English J E and Maynard D N 1981 Calcium efficiency among tomato strains. J. Am. Soc. Hort. Sci. 106, 552–557.
8 Fawole I, Gabelman W H, Gerloff G C and Nordheim E V 1982 Heritability of efficiency in phosphorus utilization in beans (*Phaseolus vulgaris* L.) grown under phosphorus stress. J. Am. Soc. Hort. Sci. 107, 94–97.
9 Giordano L B, Gabelman W H and Gerloff G C 1982 Inheritance of differences in calcium utilization by tomatoes under low-calcium stress. J. Am. Soc. Hort. Sci. 107, 664–669.
10 Madhok O P and Walker R B 1969 Magnesium nutrition of two species of sunflower. Plant Physiol. 44, 1016–1022.
11 Mahadevappa M, Ikehashi M and Aurin P 1981 Screening rice genotypes for tolerance to alkalinity and zinc deficiency. Euphytica 30, 253–257.
12 Mather K and Jinks J L 1971 Biometrical Genetics. pp 65–82 Chapman and Hall, London.
13 Sleper D A, Garner G B, Asay K H, Boland R and Pickett E E 1977 Breeding for Mg, Ca, K and P content in tall fescue. Crop Sci 17, 433–438.
14 Veen J H van der 1959 Tests of non-allelic interaction and linkage for quantitative characters in generations derived from two diploid pure lines. Genetica 30, 201–232.

Genetic differences in the ear-leaf nutrient content of inbred lines of corn (*Zea mays* L.)

V. KOVACEVIC, L. J. RADIC and N. VEKIC
Agricultural Institute, Osijek, Yugoslavia

Key words Calcium Corn Copper Ear-leaf General combining ability Inbred line Magnesium Manganese Nitrogen Zinc

Summary Thirty corn inbred lines were grown under identical field conditions. The ear-leaf N, Ca, Mg, Mn, Zn and Cu contents were widely different, depending on genotype. For example, when the highest value was designated as 100 the lowest was 48 (N), 39 (Cu), 35 (Ca), 14 (Mg), 12 (Zn) and 9 (Mn). Comparison of mean grain yields of single-cross hybrids by ear-leaf chemical composition, showed significant correlations only in the case of nitrogen ($r = 0.44^+$). A significant correlation was found between ear-leaf Ca and Mg ($r = 0.67^{++}$).

Introduction

Corn inbred lines differ in their ability to take up and use mineral nutrients[2,3,4,5,8,9,10,11,12 etc]. These differences are important in programs designed to develop plants with increased utilization of particular nutrients. In addition, nutrient status could possibly be useful for estimation of genetic similarity.

The purpose of this study was to examine any connection between ear-leaf mineral composition and performance of tested inbred lines, as well as coincidence with their combining ability towards grain yields of their single cross hybrids.

Materials and methods

Thirty inbred corn (*Zea mays* L.) lines of different origin (FAO group 500) were grown for one year on the selection field of the Agricultural Institute Osijek. Each inbred line was grown in a 6.3 m row (20 plant/row and two plants in each sowing site). General fertilizers were broadcast during autumn and before sowing (141 kg N:146 kg P_2O_5:180 kg K_2O/ha).

Collection and analysis of samples

The ear-leaf was taken from 20 plants when the corn was in the tasseling stage (early August). The leaves were washed with deionized water, dried at 60 C over night and prepared for chemical analysis by grinding. Each sample was analysed in duplicate.

N was determined by the micro Kjeldahl procedure and for Ca, Mg, and micronutrients the samples were dry-ashed, digested by nitric acid and analysed by atomic absorption. The individual results of two replicated analyses deviated from the mean value by less than ±2.5%.

Soil pH was determined electrochemically[13] and humus content colorimetrically[6]. Plant available nutrients were extracted with 1 M NaCl (calcium, 0.5 M NaCl (magnesium), 1 M KCl (zinc), 1 M HCl (copper) and 1 M NH_4OAc (Manganese) and analysed by atomic absorption.

H.W. Gabelman and B.C. Loughman (Eds.), Genetic aspects of plant mineral nutrition.
ISBN-13: 978-94-010-8102-3
© *1987 Martinus Nijhoff Publishers, Dordrecht/Boston/Lancaster.*

Estimation of general combining ability
The general combining ability of inbred corn lines was estimated by means of mean grain yield of their singe-cross hybrids sown on the selection field for creation of commercial corn hybrids. These selection materials were constituted from more genotypes than those included in the ear-leaf analysis. Depending on inbred lines, 10 to 168 single-cross hybrids (mean value was 55) were used for estimation of combining ability.

Results and discussion

Chemical characteristics of soil

The corn inbred lines were grown on an eutric cambisol originating from a calcareous loess substrate. The soil was inadequate in humus and plant-available manganese and zinc, while levels of exchangeable calcium, magnesium and copper were high (Table 1).

Table 1. Chemical properties of soil

Soil pH		Humus content (%)	Amount of available nutrients (μg/g of dry soil)				
H_2O	KCl		Ca	Mg	Mn	Zn	Cu
7.80	6.60	1.80	674	32	3.0	0.87	4.40

Ear-leaf composition

The ear-leaf composition of inbred corn lines was widely different. (Table 2.) For example, when the highest value was designated as 100, the lowest corresponding value was 48 (nitrogen), 39 (copper), 35 (calcium), 14 (magnesium), 12 (zinc) and 9 (manganese).

Genetic differences in ear-leaf composition among corn genotypes were found by Barber and Olson[1] but their range of variation was narrow compared to our results.

Our material included some American inbred lines that were tested by Clark[2]. In both studies the C123 line showed low Ca and Mg efficiency, while the A632 line exhibited low Zn efficiency. The general combining ability of inbred lines, as a function of the mean grain yield of their single-cross hybrids, also ranged widely (Table 2). However, only in the case of nitrogen did we find a significant connection with mean grain yield ($r = 0.44^+$). When the ear-leaf mineral composition was compared, the only significant correlation was found between Ca and Mg ($r = 0.67^{++}$).

By testing eight inbred corn lines and their diallel progeny it was found that ear-leaf P, K, Ca and Mg were inherited on their single-cross hybrids, but only when the inbred lines were the female parent[9,10].

Table 2. The ear-leaf composition of corn inbred lines and mean grain yield of their single-cross hybrids

Inbred line	The ear-leaf composition on dry weight basis						Grain yield of single-cross hybrids (t/ha)
	%			µg/g			
	N	Ca	Mg	Mn	Zn	Cu	
Os5	2.35	7.12	0.41	89.9	5.1	5.3	9.24
Os17tc	2.44	1.61	0.27	103.1	16.5	9.3	9.16
Os28A	2.39	0.92	0.14	169.2	15.6	5.0	10.28
Os44	2.66	1.19	0.18	34.0	18.9	4.7	9.49
Os51	2.59	1.35	0.35	69.8	12.3	5.8	11.41
Os56	3.11	0.99	0.32	77.3	17.1	7.1	10.19
Os57	2.35	1.68	0.48	56.3	10.4	5.2	10.28
Os59	3.49	1.58	0.36	103.3	18.3	5.1	10.84
Os73	2.95	1.68	0.62	14.5	12.7	8.6	8.80
Os74	2.39	1.23	0.42	44.6	16.3	5.7	9.63
Os78/1	3.07	1.58	0.56	79.9	23.4	7.0	11.13
Os79	2.21	1.69	0.16	55.2	5.7	5.1	9.13
Os83	2.37	1.69	0.35	103.5	12.5	7.9	9.54
Os86	2.13	0.73	0.09	52.5	4.0	6.6	9.20
Os87	3.05	2.06	0.51	163.5	13.5	6.8	8.98
Os88	2.39	1.67	0.63	73.2	14.6	7.0	8.93
Os89	2.80	2.11	0.64	155.1	16.0	6.0	10.19
Os94	2.07	1.33	0.41	128.4	7.5	3.6	8.82
Os95	2.32	1.04	0.17	105.4	11.0	6.6	9.29
Os97	2.27	1.33	0.32	51.3	32.2	6.4	7.45
A239	2.23	1.33	0.46	95.5	11.2	8.8	10.43
A257	2.32	1.60	0.63	73.6	25.9	8.8	8.27
A295	1.95	1.12	0.40	40.6	24.4	5.3	8.28
A619	2.38	1.31	0.18	30.4	14.8	6.1	10.02
A632Ht	2.53	1.23	0.36	42.5	12.9	6.5	9.83
Oh43	1.96	1.12	0.33	92.0	14.0	7.0	10.03
Oh43lg_2	2.16	1.01	0.26	39.5	12.1	7.4	10.34
Oh43o$_2$	1.68	1.12	0.22	63.1	15.2	6.7	8.34
Cl14	2.74	1.15	0.39	48.3	16.6	4.7	10.14
Cl23	2.45	1.04	0.24	136.4	12.7	4.4	10.55
Mean	2.46	1.37	0.36	80.0	14.8	6.3	9.61

References

1. Barber S A and Olson R A 1968 Fertilizer use on corn. Changing patterns in fertilizer use. Soil Science Society of America, Madison, Wisconsin.
2. Clark R B 1974 Efficiency of corn inbreds and sorghum lines for nutrient elements. Proceeding of the Twenty-eight Corn and Sorghum Research Conference. Seed Trade Association. Publ. No 28, Washington D.C. 144–160.
3. Clark R B 1975 Differential magnesium efficiency in corn inbreds 1. Dry matter yields and mineral element composition. Soil Sci. Soc. Am. Proc. 39, 488–491.
4. Clark R B 1978 Differential response of corn inbreds to calcium. Comm. Soil. Sci. Plant Anal. 9, 729–744.
5. Callaher R N 1969 The utilization of potassium by corn inbreds and hybrids. M.S. Thesis. University of Tennessie, Knoxville.

6   Janeković Dj, Resulović H and Šestić S 1958 Metoda za odredivanje humusa u tlu, Zemljište i Biljka, Beograd.
7   Judel G K 1969 Mineralstoffanalyse von biologischen Material mit Hilfe der Atomabsorptionspektroskopie. Zeitschrift Pflanzenernaehr. Bodenkd. 124, 1.
8   Kovačević V 1980 Proučavanje specifičnosti samo-oplodnih linija kukuruz u odnosu na mineralnu ishranu (doktorska disertacija), Zbornik radova Poljoprivrednog instituta Osijek X (2).
9   Kovačević V 1983 The ear-leaf percentage of nitrogen, phosphorus and potassium in maize (*Zea mays* L.) inbred lines and their diallel progeny. Genetic Aspects of Plant Nutrition, Martinus Nijhoff/Dr W Junk Publ. The Hague pp 471–475.
10 Kovačević V 1984 The ear-leaf percentage of calcium and magnesium in maize inbred lines and their diallel progeny. Theor. Appl. Genet. 68, 521–523.
11 Sarić M 1981 Genetic specifity in relation to plant mineral nutrition. J. Plant Nutr. 3, 743–766.
12 Sarić M and Kovačević V 1978 Physiologic-genetical aspects of content of mineral nutrition elements in maize (*Zea mays* L.). Plant Nutrition 1978 DSIR Inf. Ser. 134. Wellington New Zealand. pp 439–448.
13 Škorić A 1965 Pedološki praktikum, Sveučilište u Zagrebu.

Section 4B

Genotypic response to nutrient deficiency: Trace elements

Transfer to wheat of the copper efficiency factor carried on rye chromosome arm 5RL

ROBIN D. GRAHAM, JULIE S. ASCHER, P.A.E. ELLIS and K.W. SHEPHERD
Department of Agronomy, Waite Agricultural Research Institute, University of Adelaide, Glen Osmond, S.A., 5064, Australia

Key words   Copper deficiency   Copper uptake   Root growth   Wheat-rye chromosome translocation

Summary   Rye carries a gene(s) on the long arm of chromosome 5 which confers the ability to tolerate soils too copper-deficient for wheat. Because many South Australian soils are low in copper, copper deficiency in wheat is common. To overcome this problem, wheats were bred having the rye chromosome arm (5RL) attached to a wheat chromosome. The presence of the rye 5RL chromosome segment in four different wheat cultivars increased grain yield on copper-deficient soils by more than 100% on average. Effects in vegetative yields were also significant at stem extension. Copper concentrations were on average little higher in plant tissues of 5R lines than in the controls but copper uptake was greater, in proportion to yield. Possible mechanisms of the copper efficiency factor are discussed.

Introduction

Copper deficient soils in southern Australia usually have sufficient total copper for thousands of wheat crops[1] but the availability to the common cultivars of the crop is inadequate for normal growth and development, especially pollen production and seed-set. Moreover, although copper fertilizers are effective and have good residual value[2], many low copper areas in Australia remain untreated. Additionally, crops on such soils are more susceptible to disease[3]. There is potential then, for improving the tolerance of wheat to copper deficiency by breeding[4].

The copper efficiency of rye is well established and is expressed in the rye-wheat hybrid, triticale[5]. Earlier studies[1,2,6] have shown that gene(s) controlling copper efficiency are carried on the long arm of chromosome 5R of rye. Chinese Spring wheat carrying the 5RL chromosome segment as a translocation was available[7,8] but because this cultivar was a poor agronomic type, a series of backcrosses was made to locally adapted wheats to produce material more suitable for testing the value of this rye chromosome segment on copper deficient soils. This paper describes the field results over two seasons and some attempts to elucidate the mechanism of action of the factor carried on the 5RL chromosome segment.

In this paper we define copper efficiency as the ability of a genotype to yield well in copper deficient soil, and also, taking account of differing yield potentials, as the relative yield of paired $-$Cu and $+$Cu plots, *i.e.*, $(-Cu/+Cu) \times 100$.

H.W. Gabelman and B.C. Loughman (Eds.), Genetic aspects of plant mineral nutrition.
ISBN-13: 978-94-010-8102-3
© *1987 Martinus Nijhoff Publishers, Dordrecht/Boston/Lancaster.*

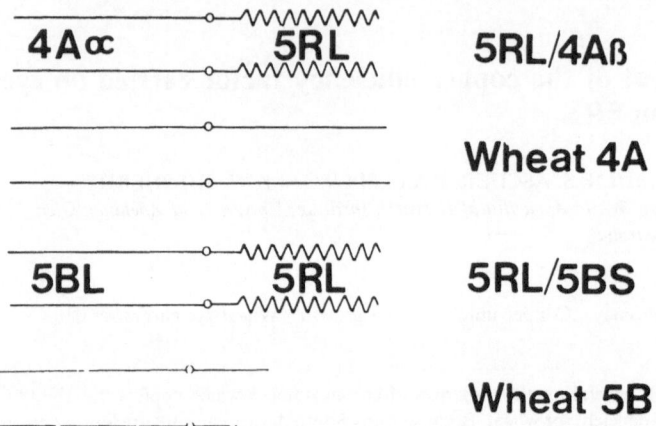

Fig. 1. Structure of the two translocations used in this study. The 5RL chromosome segment (from the long arm of chromosome 5R of rye) is joined to the β arm of wheat chromosome 4A after some of the latter was lost, and the second translocation is to the short arm of the 5B chromosome of wheat. For comparison the normal 4A and 5B chromosomes of wheat are shown.

Materials and methods

Genotypes

The 5RL chromosome arm carries, besides the copper efficiency character, a morphological marker Hp (hairy peduncle). This was used to identify 5RL-carrying progeny in the back-crossing program. Two translocation lines were available, 5RL/4A$^\gamma$, and 5RL/5BS^8 both in Chinese Spring wheat background (see Figure 1). These were crossed to a number of locally adapted Australian wheats (Warigal, Oxley, Timgalen, Champlein-Pitic (CP)), and hairy peduncle types retained through three backcrosses to the Australian parent in each case. After two generations of selfing, seed of four lines of hairy peduncle types which bred true were bulked for field testing. These were tested in the field with a range of Australian wheats including the recurrent parent cultivars.

Field experiments

The 1983 trial was conducted on a lateritic ironstone sandy loam at Ungarra on Eyre Peninsula, South Australia and in 1984, the trial was located on a loamy terra rossa at Keppoch, in the SE district of South Australia. Thirty genotypes were sown at each site in five replicates. Each entry was assigned at random to each block and sown in paired plots each four rows × 5 mm long. One plot of each pair received a foliar application of copper at tillering; the other did not. The use of paired −Cu and +Cu subplots for each genotype gave the most accurate assessment of relative yield since soil differences between the two were minimized by their proximity. At the same time as the copper treatment was applied, all plots received a basal foliar application of other essential trace elements. Superphosphate (150 kg/ha) was drilled with the seed, and at Ungarra, 80 units of nitrogen fertilizer was broadcast at tillering. The nitrogen status at Keppoch was high. During growth visual scores of symptom expression and responsiveness to copper spray was recorded. All plots were quadrat sampled at early tillering, stem extension and maturity (Ungarra). At maturity, grain yield was determined using a small plot harvester. Plant material and grain were subsampled and digested in nitric-perchloric acid for determination of copper concentration by atomic absorption spectrophotometry.

Results and discussion

The performance of the 5RL lines in comparison with their recurrent

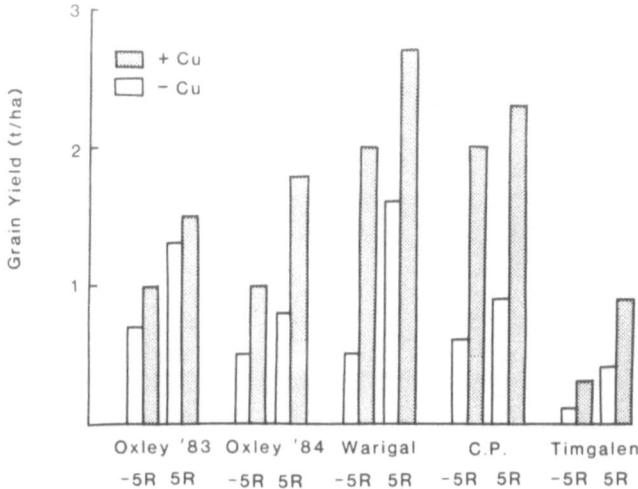

Fig. 2. Grain yields of 5RL translocation lines in comparison to their recurrent parent lines with and without supplemental copper fertilizer at Ungarra (1983) and Keppoch (1984) in South Australia.

parents is summarized in Figure 2. In both $+\text{Cu}$ and $-\text{Cu}$ treatments, the 5RL lines outyielded the parent cultivars. In $-\text{Cu}$ plots, grain yield was increased by 1.5–4 fold over the parental lines but the 5RL lines did not reach their potential: that is, there was a further increase in yield of 5RL lines with the application of foliar copper. That the 5RL lines outyielded the parent lines in $+\text{Cu}$ treatments was unexpected, contrasting with their approximately equal yields measured at other copper-adequate sites. This indicates that the copper spray was either inadequate or too late to fully overcome copper deficiency in both the 5RL and parental lines, neither therefore reaching their yield potential. The fact that on non-limiting sites, parent and 5RL lines had similar yields suggests that the presence of the 5RL genes from rye and the absence of some of the $4A\beta$ chromosome of wheat are neither advantageous nor deleterious except the former where the soil is copper-limiting.

Without added copper, Oxley 5RL/4A produced the top yield of 1.3 t/ha at Ungarra in 1983 and this yield was matched by Kite, a wheat not having a 5RL translocation (Table 1). However, in 1984 at Keppoch, Kite produced only 0.5 t/ha whereas Oxley 5RL/4A yielded 1.0 t/ha grain and other 5RL line, Warigal 5RL/4A, produced the best yield of the trial at 1.6 t/ha (Table 2). These yields compare favourably with the South Australian average yield of wheat of 1.2 t/ha, despite the severe copper deficiency which caused total grain failure in some cultivars. Of the 30 entries at each site, the four highest yields (without copper spray) at keppoch and two of the best three at Ungarra were 5RL/4A translocation lines. In fact all the 5RL/4A lines entered in both trials were found

Table 1. Vegetative performance and grain yields of wheat and wheat-rye translocation lines on a copper deficient soil at Ungarra, South Australia in 1983

Genotype		Vegetative yield 7 weeks t/ha	Vegetative yield 17 weeks			Visual score*		Grain yield at maturity		
			t/ha		Efficiency $\frac{-Cu}{+Cu} \times 100$	Vigour 17 weeks	Response	t/ha		Efficiency $\frac{-Cu}{+Cu} \times 100$
			−Cu	+Cu				−Cu	+Cu	
Dundee		0.32	2.7	5.1	53	2.9	1.0	0.6	1.1	62
Kite		0.36	3.3	4.1	80	3.5	1.0	1.3	1.5	87
Gatcher		0.37	3.0	4.6	65	2.3	1.2	0.6	1.1	54
Oxley		0.41	2.8	4.5	63	2.7	1.0	0.7	1.0	69
Oxley 5RL/4A		0.34	3.4	4.1	83	3.7	0.6	1.3	1.5	84
CP 5RL/4A		0.35	3.6	4.5	80	4.0	0.4	1.1	1.5	78
LSD ($P < 0.05$)	Genotype	NS	0.62					0.35		
	Cu		0.22					0.10		
	Genotype × Cu		0.78					NS		

NS = Not significant

Visual Score* vigour on a scale 1 → 6. 1 = pale, floppy appearance, 6 = maximum (−Cu plots). response to Cu scale 0 → 2. 0 = no response measured on height and vigour basis.

Table 2. Vegetative performance and grain yields of wheat and wheat-rye translocation lines on a copper deficient soil at Keppoch, South Australia in 1984

Genotype	Vegetative yield 8 weeks t/ha	Vegetative yield 17 weeks t/ha −Cu	+Cu	Efficiency $\frac{-Cu}{+Cu} \times 100$	Visual score* Vigour 17 weeks	Response	Grain yield at maturity t/ha −Cu	+Cu	Efficiency $\frac{-Cu}{+Cu} \times 100$
Aroona	0.44	2.0	3.9	53	3.0	1.6	0.3	1.4	24
Kite	0.38	1.6	3.1	51	2.6	1.4	0.5	1.1	41
Spear	0.48	2.4	4.0	59	4.1	1.4	0.7	2.1	32
Bindawarra	0.36	2.4	4.6	53	4.4	1.2	0.8	1.9	39
Warigal	0.45	2.2	4.3	50	3.4	1.6	0.5	2.0	24
Warigal 5RL/4A	0.37	2.7	3.9	68	5.5	0.2	1.6	2.7	61
Warigal 5RL/5BS	0.31	2.7	3.9	68	4.1	0.2	1.0	1.2	84
Oxley	0.30	1.7	3.7	46	2.9	1.6	0.5	1.0	53
Oxley 5RL/4A	0.39	3.6	3.0	117	4.0	1.0	0.8	1.8	42
CP	0.53	2.4	4.6	52	3.6	1.6	0.6	2.0	30
CP 5RL/4A	0.58	3.3	4.5	75	3.8	1.0	0.9	2.3	39
Timgalen	0.42	1.7	4.0	43	2.5	1.8	0.1	0.3	38
Timgalen 5RL/5BS	0.39	3.5	4.7	74	3.6	1.0	0.4	0.9	45
LSD ($P < 0.05$) Genotype	0.04	0.22					0.25		
Cu		0.08					0.09		
Genotype × Cu		0.28					0.33		

* Visual score: Vigour — Cu plots. 1 = pale, wilting appearance; 6 = green, erect.
Response of plants to foliar Cu spray: 0 = no response; 2 = response in height and erectness.

Table 3. Cu concentration and uptake of wheats and wheat-rye translocation lines grown on a copper deficient soil at Keppoch, South Australia in 1984

Genotype	Cu concentration (μg/g)		Cu uptake (g/ha)	
	$-$Cu	$+$Cu	$-$Cu	$+$Cu
Aroona	0.66	0.93	1.30	3.50
Kite	0.83	1.01	1.34	2.95
Spear	0.79	0.96	1.88	3.79
Bindawarra	0.88	0.85	2.08	3.90
Warigal	0.69	0.87	1.51	3.72
Warigal 5RL/4A	0.77	1.01	1.98	3.94
Warigal 5RL/5BS	0.91	1.07	2.43	4.37
Oxley	0.74	0.82	1.22	3.14
Oxley 5RL/4A	0.74	0.96	2.75	2.96
CP	0.74	0.97	1.75	4.21
CP 5RL/4A	0.83	1.13	2.73	5.06
Timgalen	0.71	0.92	1.22	3.74
Timgalen 5RL/5BS	0.91	1.16	2.11	5.23
LSD ($P < 0.05$) Genotype	NS		0.92	
Cu	0.04		0.24	
Genotype $\times$ Cu	NS		NS	
LSD ($P < 0.05$) 5RL effect*	0.07		0.23	

NS = not significant
* Main effect of 5RL, excluding first 4 genotypes from the analysis

in the highest yielding bracket. Of the copper-sprayed plots, Champlein-Pitic 5RL/4A and Oxley 5RL/4A in 1983 and Warigal 5RL/4A in 1984 produced the best overall yields in their respective trials.

The 5RL/5BS lines were poorer by comparison with the 5RL/4A lines but still remarkably more copper efficient than the recurrent parents, Warigal and Timgalen (Table 1 and Table 2). This suggests that although the 5RL/5BS translocation does effectively confer copper efficiency, yields are low because of other deleterious effects. These conclusions are also supported by the greater concentrations and uptakes of copper by Timgalen 5RL/5BS compared to the parent lines (Table 3).

The vegetative harvests (Table 1 and Table 2) show that although no differences due to the presence of 5RL were measurable at early tillering, the patterns seen in the grain yields were well established by stem extension (17 weeks). Vegetative yields were doubled in some cases by the presence of 5RL, an increase of up to 2 t/ha. Table 3 shows the concentrations of copper in the tops at stem extension in the Keppoch experiment (1984). All values are low but generally slightly higher in 5RL/4A and 5RL/5BS lines than in the corresponding recurrent parents. A clear increase due to the copper spray is also apparent. Under copper limiting conditions such as these, copper concentrations are not increased much by copper efficient cultivars; rather any increment in absorbed copper is converted into yield at or near the critical concentration of copper for growth. For this reason we advocate the use of copper uptake (concentration × yield) as a better index of copper efficiency than either yield or concentration alone. Paired − Cu and + Cu subplots determine this copper efficiency index with high biometrical precision in this design. The copper uptakes (Table 3) of the 5RL lines under copper deficiency are 50–100% greater than those of their recurrent parents. These increases in the limiting factor for growth are ultimately converted to even greater advantage in grain yield because of the critical role of copper in pollen viability[9].

A pot experiment showed that at 7 weeks, four 5RL lines (those used at Keppoch 1984) averaged 40% more root weight than the recurrent parents; however, at that time shoot yield was only 10% greater (not significant) and copper concentrations in tops less than 1% greater in 5RL lines than in parents. Whether greater root growth is the cause or the result of copper efficiency in 5RL lines cannot be established from such an experiment. In earlier studies in non-limiting solution cultures[6] we showed that rye and triticale had longer, finer, more branched roots than wheat in the seedling stage, but at that stage no differences in root geometry could be measured between a 5RL/4A line and its parent, Chinese Spring wheat. Rye in those studies clearly had higher rates of absorption of copper per unit of root but differences between a variety of 5R lines and wheat were inconclusive. It must be recognised that a very small increase in rate of absorption per unit of root may be sufficient to compound into large differences in yield over the length of a growing season in a nutrient-limiting environment.

Higher uptake of copper by efficient lines has been the consistent pattern we have seen in all studies[2,5,10]. Like Snowball and Robson[11] we have no evidence to support the hypothesis that copper efficiency in cereals can be due to a lower internal requirement for the element.

Although even the 5RL lines in our field trials did not produce maximum grain yield without supplemental copper, these two sites were chosen for the severity of the copper deficiency in order to magnify genotypic differences. We have reason to believe that the level of copper efficiency conferred by the 5RL/4A translocation would be sufficient to overcome limitation due to copper deficiency in most farmers' fields, certainly those where the deficiency is subclinical. Moreover, our results suggest that genetic copper efficiency can complement fertilizer copper in augmenting yield in severe cases of deficiency.

Acknowledgements The authors acknowledge the provision of land at Ungarra by Mr. Dene Roediger and at Keppoch by Mr. Ian Ward, and technical assistance of Miss Jane Harbard. Thanks are due to staff of Adelaide and Wallaroo Fertilizers, Ltd. for assistance and special fertilizer mixtures.

References

1 Graham R D 1978 Nutrient efficiency objectives in cereal breeding. *In* Plant Nutrition 1978. Proc. 8th Coll. Plant Anal. Fert. Probl., Auckland, N.Z., pp 165–170.
2 Gartrell J W 1981 Distribution and correction of copper deficiency in crops and pastures. *In* Copper in Soils and Plants. Eds. J F Loneragan, A D Robson and R D Graham. Academic Press, Sydney.
3 Graham R D 1983 Effects of nutrient stress on susceptibility of plants to disease with particular reference to the trace elements. Adv. Bot. Res. 10, 221–276.
4 Graham R D 1984 Breeding for nutritional characteristics in cereals. Adv. Plant Nutr. 1, 57–102.
5 Graham R D and Pearce D T 1979 The sensitivity of hexaploid and octoploid triticales and their parent species to copper deficiency. Aust. J. Agric. Res. 30, 791–799.
6 Graham R D, Anderson G D and Ascher J S 1981 Absorption of copper by wheat, rye and some hybrid genotypes. J. Plant Nutr. 3, 679–686.
7 Driscoll C J and Sears E R 1965 Mapping of a wheat-rye translocation. Genetics 51, 439–443.
8 Sears E R 1967 Induced transfer of hairy neck from rye to wheat. Z. Pflanzenzuchtung 57, 4 25.
9 Graham R D 1975 Male sterility in wheat plants deficient in copper. Nature 271, 542–543.
10 Harry S P and Graham R D 1981 Tolerance of triticale, wheat and rye to copper deficiency and to low and high pH. J. Plant Nutr. 3, 710–730.
11 Snowball K and Robson A D 1984 Comparison of the internal and external requirements of wheat, oats and barley for copper. Aust. J. Agric. Res. 35, 359 65.

Physiology of genotypic differences in zinc and copper uptake in rice and tomato

JOHN E. BOWEN
Botany Department, Beaumont Agricultural Research Center, University of Hawaii, 461 W. Lanikaula Street Hilo, HI 96720-4094, USA

Key words Copper Genotypes Rice Tomato Zinc

Summary Excised roots of rice (*Oryzae sativa* L.) cv IR26 absorbed both Zn^{2+} and Cu^{2+} from 0.01 mM to 0.50 mM external solutions at rates twice those of cv M101 over a 30-min period. However, the latter have a two-fold greater affinity ($1/Km$) for Zn^{2+} and Cu^{2+} than do those of the former. Zn^{2+} and Cu^{2+} mutually and competitively inhibited uptake of each other, indicating that both micronutrient cations are absorbed through the same uptake mechanism or carrier sites. Further, these differences in uptake rates are restricted to roots but they cannot be explained by variations in root surface areas.
   Excised roots of tomato (*Lycopersicon esculentum* L.) cv Kewalo absorbed Zn^{2+} and Cu^{2+} much more rapidly than did cv Sel 7625-2. Uptake of each cation was competitively and reciprocally inhibited by the other, so Zn^{2+} and Cu^{2+} are seemingly accumulated through the same uptake system in tomato also. Tomato cultivars Kewalo and Sel 7625-2 did not differ with regard to affinities of the root apices for Zn^{2+} and Cu^{2+}, however. Vmax values for Zn^{2+} and Cu^{2+} uptake by roots of cv Kewalo were three-fold greater than those for cv Sel 7625-2.

Introduction

The plant nutrition literature abounds with reports of differences in rate of uptake and accumulation of specific essential nutrients by cultivars of economically-important crops[2,3,6,8,10,12,13,17,18]. Differences in Zn^{2+} uptake rates, for example, have been studied to some extent in cultivars of such crops as rice[3,12,13], sugarcane[2,17] and barley[8].

Excised roots of rice cv IR26 absorbed Zn^{2+} at a rate approximately twice as fast as that of cv M101[3]. However, cv M101 roots had a two-fold greater affinity ($1/Km$) for Zn^{2+} than did those of cv IR26[3]. Furthermore, the increased Zn^{2+} uptake rate in cv M101 was restricted to the roots. There were no differences in Zn^{2+} uptake rates by leaf blade tissue from these two cvs[3]. Also, the rate difference between cvs M101 and IR26 could not be explained by equivalent differences in surface area of the roots.

Differential Cu^{2+} uptake among cultivars of tropical and sub-tropical crops has received little attention[10] although Cu^{2+} and Zn^{2+} are absorbed through the same mechanism or carrier sites in sugarcane[1,2], for example.

Journal Series 2991 of the Hawaii Institute of Tropical Agriculture and Human Resources. Supported by USDA/CSRS Grants Program in Tropical and Subtropical Agriculture (83-CSRS-2-2245).

H.W. Gabelman and B.C. Loughman (Eds.), Genetic aspects of plant mineral nutrition.
ISBN-13: 978-94-010-8102-3
© 1987 Martinus Nijhoff Publishers, Dordrecht/Boston/Lancaster.

The purpose of the present study was to extend earlier work done in this laboratory[3], as well as that of other workers[9,12,13], in a continuing effort to gain further knowledge of the physiological mechanisms of genotypic differences in Zn^{2+} and Cu^{2+} uptake in rice and tomato cultivars.

Materials and methods

Plant material

Two rice (*Oryzae sativa* L.) cultivars were used in these experiments: cv IR26, considered to have a high Zn^{2+} requirement; and cv M101, a cultivar that requires much less Zn^{2+} for optimal growth (G. S. Khush, personal communication, and[3]). Rice cv IR26 is considerably more susceptible to Zn^{2+} deficiency under field and greenhouse conditions than is cv M101, as measured by dry matter accumulation and visual manifestation of Zn^{2+} deficiency symptoms[3].

Apical 1-cm sections of roots were excised from 10-day-old hydroponically-cultured seedlings and prepared for experimental use as described previously[3,4].

Leaf blade sections (300 μ × 14 mm) were cut from the center third of elongating blades of 8-week-old rice plants grown in vermiculite in the greenhouse in the same manner as was done earlier with sugarcane[1].

Apical 1-cm portions of tomato (*Lycopersicon esculentum* L.) root tips were excised from 5- to 7-day-old seedlings, also hydroponically cultured in the greenhouse. Two cultivars were used: cv Kewalo and Sel 7625-2. The former required three-fold less Zn^{2+} in greenhouse growth trials than did the latter, as measured by dry matter accumulation. Also, Zn^{2+} deficiency symptoms appeared on Sel 7625-2 plants much more readily than on those of Kewalo (Bowen, unpublished data).

Disks of tomato leaf blade tissue 1 cm in diameter were punched from recently matured leaves of 6-week-old plants grown in a mixture of 60 percent volcanic cinder and 40 percent Hawaiian tree fern (hapuu).

Excised root apices and leaf blade sections of both rice and tomato were immediately placed in constantly-aerated 0.5 mM $CaSO_4$ solution and held there for a maximum of 1 h before the start of an experiment.

Experimental methods

Plant tissue [approximately 1 g fr wt in cheesecloth 'teabags'[1,2,3,4] was placed for up to 2 h in continuously-aerated solutions containing 0.5 mM $CaSO_4$ and the following cations either singly or in various combinations: Zn^{2+} as $ZnCl_2$, Cu^{2+} as $CuSO_4$, and Mn^{2+} as $MnCl_2$. Concentrations of the latter three cations ranged from 0.01 to 0.50 mM. This range of concentrations has been widely used by many researchers in studies of micronutrient cation uptake[1,2,3,15]. It was previously reported[1] that Zn^{2+} and Cu^{2+} levels below 0.50 mM do not affect photosynthetic O_2 evolution or respiratory CO_2 evolution in sugarcane leaf tissue over a 2-h period. This was confirmed manometrically[1] for the rice and tomato tissues used in the present experiments.

The solutions for the Zn^{2+} uptake experiments also contained $^{65}Zn^{2+}$ (0.75 μc $^{65}Zn^{2+}$/μmole non-radioactive Zn^{2+}; sp. act. of $^{65}Zn^{2+}$ was 178 μc/mole) in addition to 0.5 mM $CaSO_4$ and non-radioactive Zn^{2+}. Metabolic inhibitors were added as indicated.

The initial pH of all solutions was 5.7 and this varied by a maximum of ± 0.2 during the uptake period. The temperature was 26° ± 1°.

Uptake experiments were terminated by rinsing the tissue three times for 1 min each in flowing distilled water, and then desorbing reversibly-adsorbed cations for 30 min in aerated 0.5 mM $CaSO_4$ at 4°[1,3,15]. At the end of the desorption period, the plant tissue was rinsed in distilled water again and then carefully blotted dry. This procedure has been shown to remove adsorbed cations quite effectively[1,15]. The Zn^{2+}, Cu^{2+} and Mn^{2+} uptake rates reported thus pertain only to metabolically-mediated active uptake across the plasmalemma[1,3,4,15].

The plant tissue was dried overnight at 75°, and then ashed at 400° for 12 h. The ash was re-dissolved in 0.1 N HCl, and analyzed for Zn^{2+}, Cu^{2+} and/or Mn^{2+}.

Zn^{2+} absorption was measured by assaying a 1-ml aliquot of the re-dissolved ash preparation for radioactivity with a scintillation probe equipped with a NaI (Tl) crystal (3 cm diameter × 2.5 cm thick) connected to a scaler.

The Cu^{2+} and Mn^{2+} concentrations in the tissues were measured by atomic absorption spectrophotometric analysis of the re-dissolved ash preparations.

Each experiment contained a minimum of four replicates and was run at least three times.

Root diameters, lengths and surface areas were estimated with the method of Dittmer[7] and also by direct measurement from photographs of excised root apices.

Kinetic analyses

The methods of Lineweaver and Burk[14] and Hofstee[11] were used to estimate Km and Vmax for the uptake processes and to establish the specific mechanisms of inhibition where applicable. Affinity constants of tissues for specific cations were calculated as reciprocals of the Km values[2].

Statistical analyses

Computerized linear and curvilinear regression techniques were used to plot the curves on the graphs. Student's 't-test' was used for evaluating the statistical significance of differences in Km, Vmax and the affinity constants between treatments, tissues and cultivars[16].

Seed sources

Seeds of rice cvs IR26 and M101 were provided by the International Rice Research Institute and by Dr. J Neil Rutger of the USDA-ARS. Tomato seeds (cvs Kewalo and Sel 7625-2) were obtained from Dr. Ken Takeda of the Horticulture Department, University of Hawaii.

Results and discussion

Cu^{2+} uptake by excised rice roots as a function of external Cu^{2+} concentration

Cu^{2+} uptake by rice root apices from 0.10 mM Cu^{2+} solutions, like the uptake of Zn^{2+} (ref.[3]), was linear with time for at least 2 h (data not presented). When measured as a function of the external Cu^{2+} concentration over the range of 0.025 mM to 0.50 mM Cu^{2+}, the Cu^{2+} uptake mechanism was saturated at 0.10 mM to 0.20 mM Cu^{2+} (Fig. 1). The Zn^{2+} uptake mechanism was likewise saturated at 0.10 mM to 0.20 mM external Zn^{2+} (ref[3]). Cultivar IR26 absorbed Cu^{2+} twice as fast as did the root apices of cv M101 throughout this concentration range, however (Fig. 1). These aspects of Cu^{2+} uptake closely parallel those of Zn^{2+} absorption reported earlier[3].

Kinetics of Cu^{2+} uptake by rice cvs IR26 and M101

Kinetic constants were calculated for the uptake of Cu^{2+} by root apices of each rice cultivar from the data in Figure 1 (Fig. 2 for cv M101). Vmax was 1.55 μmoles Cu^{2+} absorbed/g fr wt/30 min for cv M101 and 3.07 μmoles/g fr wt/30 min for cv IR26. The Km values were 0.007 and 0.016 μmoles for cvs M101 and IR26, respectively.

Based upon the reciprocals of these Km values, it is estimated that the affinity of M101 root apices for Cu^{2+} is 2.3-fold greater than that of IR26

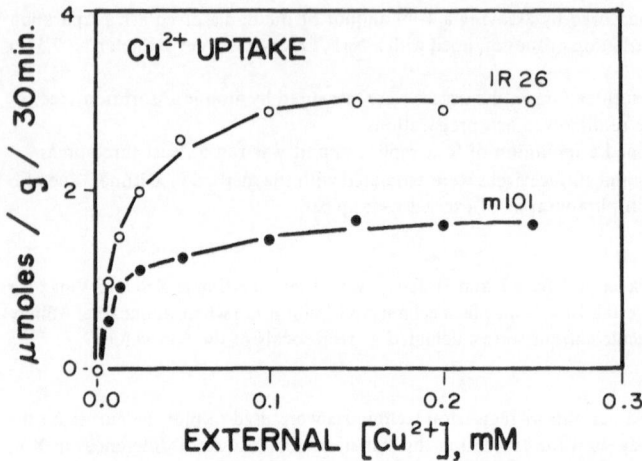

Fig. 1. Uptake of Cu^{2+} by excised root apices of rice cvs M101 and IR26 as a function of the external Cu^{2+} concentration. Initial pH 5.7 ± 0.2; temperature 26° ± 1; Ca^{2+} concentration, 0.50 mM; uptake period, 30 min.

roots. Roots of cv M101 also have a much greater affinity for Zn^{2+} than do those of IR26[3].

The total root surface area per gram fr wt of roots did not differ significantly between the two cultivars (12.3 cm^2 for cv M101 and 14.8 cm^2 for cv IR26), so differences in Cu^{2+} uptake rates could not be attributed to differences in root surface areas.

There was no apparent difference in the capacity of leaf blade tissue from rice cvs M101 and IR26 to absorb Cu^{2+}. Uptake rates from a

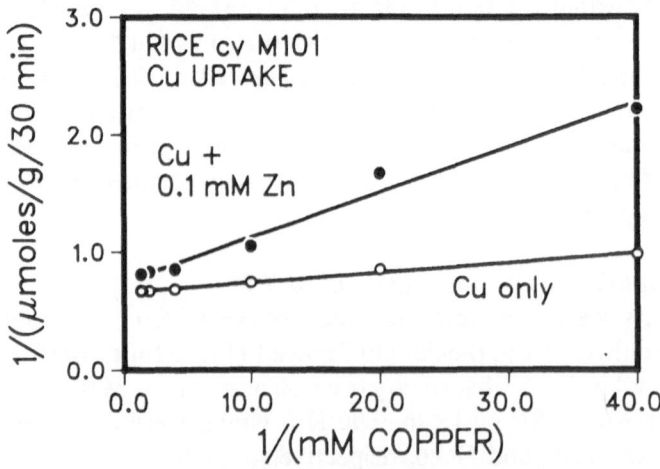

Fig. 2. Double reciprocal plots for the absorption of Zn^{2+} by rice cv M101 in the absence and presence of 0.10 mM Cu^{2+}. All other conditions as in Fig. 1.

Table 1. Mutual effects of Zn^{2+}, Cu^{2+} and Mn^{2+} in their absorption by excised root apices of rice cvs M101 and IR26. The concentration of each ion was 0.10 mM; Ca^{2+}, 0.50 mM; initial pH, 5.7; and temperature, 26 ± 1

Ions present	Uptake (μmoles/g fr wt/30 min)		
	Zn^{2+}	Cu^{2+}	Mn^{2+}
cv M101			
Zn^{2+}	1.39	-	-
Cu^{2+}	-	1.46	-
Mn^{2+}	-	-	2.02
$Cu^{2+} + Zn^{2+}$	1.11	1.01	-
$Cu^{2+} + Mn^{2+}$	-	1.49	1.98
$Zn^{2+} + Mn^{2+}$	1.38	-	2.07
$Zn^{2+} + Cu^{2+} + Mn^{2+}$	1.04	1.00	1.97
cv IR26			
Zn^{2+}	2.74	-	-
Cu^{2+}	-	2.87	-
Mn^{2+}	-	-	2.13
$Cu^{2+} + Zn^{2+}$	2.04	2.01	-
$Cu^{2+} + Mn^{2+}$	-	2.77	2.00
$Zn^{2+} + Mn^{2+}$	2.78	-	1.99
$Zn^{2+} + Cu^{2+} + Mn^{2+}$	2.02	1.97	2.07

0.10 mM Cu^{2+} solution averaged 0.38 and 0.34 μmoles Cu^{2+} absorbed/ g fr wt/30 min for cvs IR26 and M101, respectively.

Interactions in the uptake of Zn^{2+}, Cu^{2+} and Mn^{2+} by root apices of rice cvs IR26 and M101

The possible occurence of interactions and mutual competitions in uptake of Zn^{2+}, Cu^{2+} and Mn^{2+} in rice root apices was investigated in a series of factorially-designed experiments. Each solution contained 0.5 mM Ca^{2+} and one or more of the above three cations at 0.10 mM.

Mn^{2+} had no significant effect on uptake of Zn^{2+} or Cu^{2+} by root apices of either cv IR26 or M101 (Table 1). Uptake of Zn^{2+} and Cu^{2+} did manifest mutual interferences in each cultivar, however (Table 1). Further experiments showed that Zn^{2+} and Cu^{2+} competitively inhibited the absorption of each other in cv M101 (Fig. 2) as well as in cv IR26 (data not presented).

Mn^{2+} uptake rates did not differ in excised root apices from rice cvs IR26 and M101, nor did the presence of Mn^{2+} in the Zn^{2+} and Cu^{2+} uptake solutions interfere with absorption of the latter two cations (Table 1).

Effect of metabolic inhibitors on uptake of Zn^{2+} and Cu^{2+} by excised roots of rice cvs M101 and IR26

Uptake of both Zn^{2+} and Cu^{2+} was strongly reduced by 0.001 mM

Table 2. Effects of metabolic inhibitors on the uptake of Zn^{2+} and Cu^{2+} (μmoles absorbed/g fr wt/ 30 min) by excised root apices of rice cvs M101 and IR26. Cation concentration, 0.10 M; Ca^{2+}, 0.50 mM; initial pH 5.7; temperature, 26 $\pm$ 1

Additive	$Zinc^{2+}$		$Copper^{2+}$	
	M101	IR26	M101	IR26
Control	1.43	2.87	1.49	2.73
$10^{-6} M$ DNP	0.57	1.13	0.69	1.29
$10^{-5} M$ DNP	0.23	0.99	0.18	1.08
$10^{-4} M$ NaN$_3$	0.46	0.96	0.58	1.34
$10^{-4} M$ CN	0.78	1.67	0.81	1.63
$10^{-3} M$ Arsenate	0.61	0.97	0.54	1.08

2,4-dinitrophenol (DNP), 0.1 mM Na azide, 0.1 mM cyanide and 1 mM arsenate (Table 2). The decrease in uptake attributable to each inhibitor, measured as a percentage of the uninhibited control, was similar in both cv M101 and cv IR26 (Table 2).

Uptake of Zn^{2+} and Cu^{2+} in each cultivar is sensitive to metabolic inhibitors so the process is apparently metabolically-mediated. Rice cultivars M101 and IR26 both actively absorb Zn^{2+} and Cu^{2+} through one system and accumulate Mn^{2+} through another. The differences in uptake rates between the two cultivars are restricted to roots; uptake by leaf blade tissue was equal in both cultivars.

The major difference in Zn^{2+} and Cu^{2+} uptake between rice cvs M101 and IR26 lies in the relative affinities (1/Km) of the root apices for these cations. Roots of cv M101 have a two-fold greater affinity for Zn^{2+} and Cu^{2+} than do those of cv IR26. Therefore, rice cv M101 would be expected to grow and yield better under conditions of marginal Zn^{2+} or Cu^{2+} deficiency than would cv IR26. This has been shown to be the case in greenhouse experiments (data not presented).

Zn^{2+} and Cu^{2+} uptake by tomato cv Kewalo and cv Sel 7625-2

Uptake of Zn^{2+} by excised roots of tomato cvs Kewalo and Sel 7625-2 as a function of the external Zn^{2+} concentration is shown in Figure 3. Uptake by cv Kewalo was 2- to 4-fold greater than that by cv Sel 7625-2 roots at each concentration used in these experiments (Fig. 3).

It is also apparent that Zn^{2+} uptake in roots of each tomato cultivar was reduced by the presence of 0.10 mM Cu^{2+} in the external solution when the external Zn^{2+} concentration is less than 0.10 mM (Fig. 3).

Kinetic constants for Zn^{2+} absorption were calculated from the data depicted in Figure 3 (Fig. 4). Km values were 0.050 mM and 0.057 mM for cvs Kewalo and Sel 7625-2, respectively. The Vmax values were 3.61 μmoles Zn^{2+} absorbed/g fr wt/30 min for cv Kewalo, and 1.17 μmoles/g fr wt/30 min for cv Sel 7625-2.

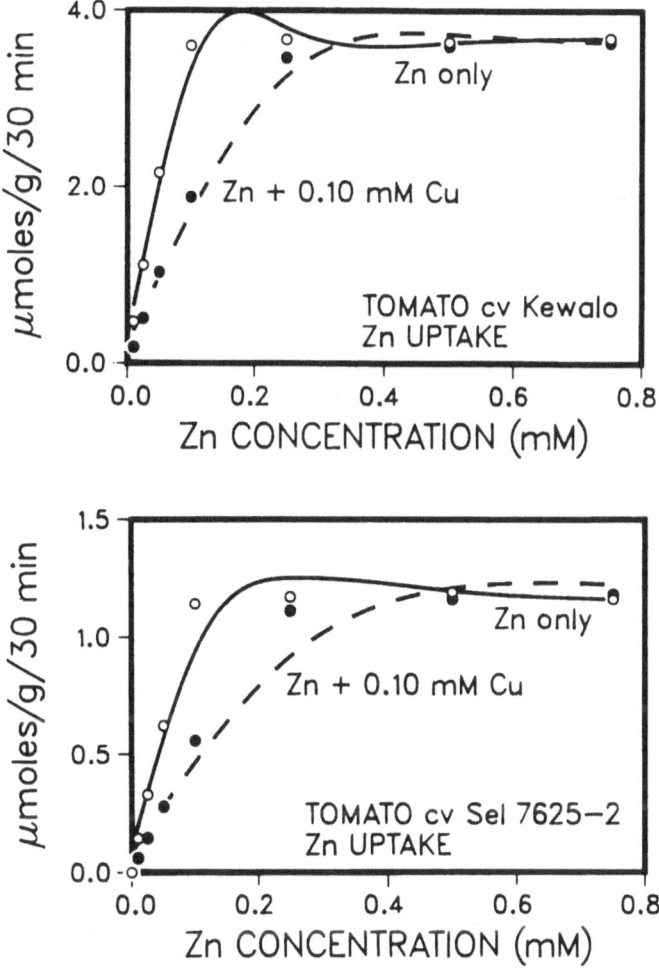

Fig. 3. Uptake of Zn^{2+} by excised root apices of tomato cvs Kewalo and Sel 7625-2 as a function of the external Zn^{2+} concentration, and the effect of 0.10 mM Cu^{2+} thereon. All conditions as in Fig. 1.

By interpreting the reciprocals of the Km values as affinity constants, K_{aff} values of 20.0 and 17.5 are obtained for cvs Kewalo and Sel 7625-2, respectively. Thus, unlike the situation with rice cvs M101 and IR26, tomato cvs Kewalo and Sel 7625-2 manifested no statistically significant differences insofar as affinity of root apices for Zn^{2+} is concerned. The Vmax for Zn^{2+} uptake is three-fold greater in cv Kewalo than in cv Sel 7625-2, however. Cv Kewalo thus has a significantly (1 percent confidence level) greater capacity for Zn^{2+} absorption than does Sel 7625-2 but this difference in absorptive capacities cannot be explained by a difference in affinity of the roots for the ion.

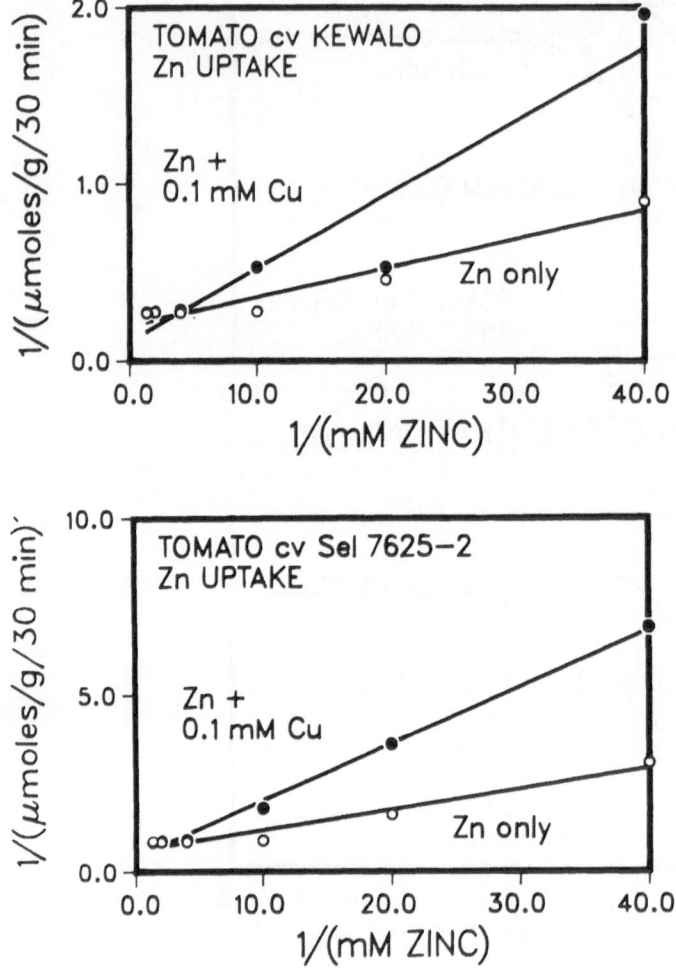

Fig. 4. Double reciprocal plots for the uptake of Zn^{2+} by tomato cvs Kewalo and Sel 7625-2 in the absence and presence of $0.10\,mM$ Cu^{2+}. All other conditions as in Fig. 1.

As was the case in the two rice cultivars, the Zn^{2+}-uptake rate differential was limited to roots in tomato also. Km values for Zn^{2+} uptake by tomato leaf blade tissue were $0.019\,mM$ and $0.017\,mM$ for cvs Kewalo and Sel 7625-2, respectively. Vmax for cv Kewalo was $1.73\,\mu moles\,Zn^{2+}$ absorbed/g fr wt/30 min and, for cv Sel 7625-2, $1.68\,\mu moles/g\,fr\,wt/30\,min$/These differences were not statistically significant.

Neither could the difference in Zn^{2+} uptake rates between tomato cvs Kewalo and Sel 7625-2 be explained by the statistically non-significant difference in root surface area ($18.6\,cm^2/g\,fr\,wt$ for cv Kewalo, and $17.0\,cm^2/g\,fr\,wt$ for cv Sel 7625-2). The Vmax values on a surface area

Table 3. Mutual effects of Zn^{2+}, Cu^{2+} and Mn^{2+} in their absorption by excised root apices of tomato cvs Kewalo and Sel 7625-2 (μmoles uptake/g fr wt/30 min). The concentration of each ion was 0.10 mM; Ca^{2+}, 0.50 mM; initial pH 5.7; temperature, 26 ± 1

Ions present	Zn^{2+}	Cu^{2+}	Mn^{2+}
cv Kewalo			
Zn^{2+}	3.54	–	–
Cu^{2+}	–	3.49	–
Mn^{2+}	–	–	2.80
$Cu^{2+} + Zn^{2+}$	1.91	2.02	–
$Cu^{2+} + Mn^{2+}$	..	3.37	2.67
$Zn^{2+} + Mn^{2+}$	3.42	–	2.62
$Zn^{2+} + Cu^{2+} + Mn^{2+}$	1.99	1.91	2.61
cv Sel 7625-2			
Zn^{2+}	1.19	–	–
Cu^{2+}	–	1.26	–
Mn^{2+}	–	–	2.64
$Cu^{2+} + Zn^{2+}$	0.68	0.65	–
$Cu^{2+} + Mn^{2+}$	–	1.17	2.60
$Zn^{2+} + Mn^{2+}$	1.10	–	2.61
$Zn^{2+} + Cu^{2+} + Mn^{2+}$	0.61	0.54	2.67

basis were, for cv Kewalo, 0.194 μmoles Zn^{2+} absorbed/cm^2/30 min and, for cv Sel 7625-2, 0.069 μmoles/cm^2/30 min.

The uptake of Cu^{2+}, a cation that appeared to interfere with uptake of Zn^{2+} by tomato root apices (Fig. 3), was similar in every respect to Zn^{2+} absorption. Cu^{2+} uptake was approximately three-fold greater in roots of cv Kewalo as compared to that by cv Sel 7625-2 roots throughout the external concentration range of 0.01 mM to 0.50 mM (data not presented for brevity). Cu^{2+} uptake was also reduced by addition of 0.10 mM Zn^{2+} to the external solution when the Cu^{2+} concentration was equimolar or less.

The kinetic constants for Cu^{2+} uptake were very similar to those for Zn^{2+} absorption. Km_{Cu} for cv Kewalo was 0.062 mM and, for cv Sel 7625-2, 0.054 mM. $Vmax_{Cu}$ was 3.45 and 1.24 μmoles/g fr wt/30 min for cvs Kewalo and Sel 7625-2, respectively. There was thus no apparent difference between the two cultivars insofar as affinity of the roots for Cu^{2+} was concerned (16.1 for cv Kewalo vs 18.5 for cv Sel 7625-2). The sole kinetic difference between the two tomato cultivars in regard to their capacity to absorb Cu^{2+} was in the relative rates at which each absorbed the ion; *i.e.*, Vmax.

Mutual effects of Cu^{2+}, Zn^{2+} and Mn^{2+} in their absorption by root apices of tomato cvs Kewalo and Sel 7625-2

Data in Figures 3 and 4 indicated that Cu^{2+} and Zn^{2+} mutually interfered with uptake of the other. This, and the potential effect of Mn^{2+}

on Zn^{2+} and Cu^{2+} uptake, was studied further in experiments utilizing external solutions containing 0.50 mM Ca^{2+} and one or more of the three cations (Zn^{2+}, Cu^{2+} or Mn^{2+}) at 0.10 mM.

Zn^{2+} and Cu^{2+} were, in fact, mutually inhibitory insofar as their uptake by roots of each tomato cultivar was concerned (Table 3). Further, the inhibition of Zn^{2+} uptake by Cu^{2+} was apparently competitive, as evidenced by the double reciprocal plot of the data from Figure 3 (Fig. 4). Cu^{2+} absorption was likewise competitively inhibited by Zn^{2+} (data not presented).

Mn^{2+} had no effect on uptake of either Zn^{2+} or Cu^{2+} by root apices of either tomato cultivar (Table 3).

It was thus concluded that Zn^{2+} and Cu^{2+} are absorbed through common carrier sites in both tomato cultivars; *i.e.*, Kewalo and Sel 7625-2, regardless of the relative capacities of the cultivars to absorb these micronutrients. Mn^{2+}, on the other hand, is absorbed through a separate and independent system or mechanism.

$Zinc^{2+}$ and Cu^{2+} uptake are closely allied. That is, excised roots of cv Kewalo absorb both Zn^{2+} and Cu^{2+} much faster than do those of cv Sel 7625-2.

The physiological mechanisms of these differences in absorption rates differ in rice and tomato, however. The greater Zn^{2+} and Cu^{2+} uptake rates in rice are apparently effected through a greater affinity of the carrier sites in root apices for these ions in cv M101 as compared to cv IR26. But, there are no differences between the two tomato cultivars insofar as affinity of roots for Zn^{2+} and Cu^{2+} is concerned. In tomato, rapid uptake of Zn^{2+} and Cu^{2+} appears to be due to a greater capacity to absorb these ions; *i.e.*, a higher concentration of carrier molecules in the roots, perhaps.

Differences in Zn^{2+} and Cu^{2+} absorption rates are limited to roots of both rice and tomato, though. Leaf blade tissues manifested no evidence of these differences between cultivars.

References

1. Bowen J E 1969 Absorption of copper, zinc and manganese by sugarcane leaf tissue. Plant Physiol. 44, 255–261.
2. Bowen J E 1973 Kinetics of zinc absorption by excised roots to two sugarcane clones. Plant and Soil 39, 125–129.
3. Bowen J E 1986 Kinetics of zinc uptake by two rice cultivars. Plant and Soil 94, 99–107.
4. Bowen J E and Nissen P 1976 Boron uptake by excised barley roots. I. Uptake into the free space. Plant Physiol. 57, 353–357.
5. Brar M S and Sekhon G S 1976 Interaction of zinc with other micronutrients cations. I. Effect of copper on zinc^{65} absorption by wheat seedlings and its translocation within the plant. Plant and Soil 45, 137–143.

6 Brown J C, Ambler J E, Chaney R I and Foy C D 1972 Differential responses of plant genotypes to micronutrients. *In* Micronutrients in Agriculture. Eds. J J Mortvedt, P M Giordano and W L Lindsay. Soil Sci. Soc. Am., Madison, Wisc. pp 389 418.
7 Dittmer H J 1937 A quantitative study of the roots and root hairs of winter rye plants. Am. J. Bot. 24, 417 420.
8 Fleming A L and Foy C D 1982 Differential response of barley varieties to iron stress. J. Plant Nutr. 5, 457 468.
9 Giordano P M, Noggle J C and Mortvedt J J 1974 Zinc uptake by rice as affected by metabolic inhibitors and competing cations. Plant and Soil 41. 637–646.
10 Graham R D 1981 Absorption of copper by plant roots. *In* Copper in Soils and Plants. Eds. J F Loneragan, A D Robson and R D Graham. Academic Press, Sydney, Australia. pp 141–160.
11 Hofstee B F J 1952 On the evaluation of the constants V_m and K_m in enzyme reactions. Science 116, 329 331.
12 International Rice Research Institute 1977 Annual Report for 1976. pp 97 104.
13 International Rice Research Institute 1978 Annual Report for 1977. pp 124–125.
14 Lineweaver H and Burk D 1936 The determination of enzyme dissociation constants. J. Am. Chem. Soc. 56, 658 666.
15 Schmid W E, Haag H P and Epstein E 1965 Absorption of zinc by excised barley roots. Physiol. Plant, 18, 860–869.
16 Snedecor G W 1948 Statistical Methods Applied to Experiments in Agriculture and Biology. Iowa State College Press, Ames, Iowa. 485 p.
17 Takahashi D T 1968 Zinc deficiency — varietal differences. Hawaiian Sugar Planters' Assoc. Expt. Sta. Ann. Rpt. 1968, 17 18.
18 Wallace A 1981 Some physiological aspects of iron deficiency in plants. J. Plant Nutr. 3, 637–642.

Variability and correlation of iron-deficiency symptoms in a sorghum population evaluated in the field and growth chamber

E.P. WILLIAMS, R.B. CLARK, W.M. ROSS,
Department of Agronomy and U.S. Department of Agriculture, Agricultural Research Service, University of Nebraska, Lincoln, NE 68583, USA

G.M. HERRON and M.D. WITT
Kansas Agricultural Experiment Station, Garden City, KS 67846, USA

Key words Chlorosis Genetic and phenotypic correlations Heritability Iron-deficiency ratings Population improvement *Sorghum bicolor* (L.) Moench

Summary One hundred S_1 families from a random-mating sorghum [*Sorghum bicolor* (L.) Moench] population (NP21R) were rated visually for differences in iron-deficiency chlorosis in the field and in a growth chamber. Heritabilities on a family-mean basis were estimated for the first (0.63 ± 0.15), maximum (0.79 ± 0.15), and average chlorosis rating (0.84 ± 0.15) recorded over S_1 families in the field. Heritability on an individual plant basis was estimated for the average chlorosis ratings in the growth chamber (0.65 ± 0.16). Phenotypic and genotypic correlations were large and negative between field chlorosis ratings and yield or yield-related traits, which indicated that the severity of chlorosis was an important factor in determining S_1 family agronomic performance. Correlations between chlorosis traits over all S_1 families in the growth chamber and field generally were small and nonsignificant.

Introduction

Iron-deficiency chlorosis (commonly known as lime-induced chlorosis or iron chlorosis) is a mineral element deficiency disorder often observed in plants grown on calcareous soils. Sorghum grows well in the relatively arid environments commonly associated with calcareous soils, but it is particularly susceptible to chlorosis. Soil or foliar iron amendments often are applied to correct chlorosis so that the crop will produce economic yields, but these correction methods are not permanent solutions for correcting the disorder.

Many crops have been evaluated for tolerance to iron chlorosis in field and greenhouse studies. Few studies have been concerned with genetic improvement of plants for tolerance to these conditions. Through an extensive improvement program, soybean [*Glycine max* (L.) Merr.] genotypes have been developed which tolerate known iron deficiency (chlorosis) conditions induced by soils in Iowa[1,2,3,9,10,13].

Controlled environment and laboratory experiments can reduce the time and labor needed too evaluate genetic sources of resistance to mineral nutrient disorders. Because of possible differences in tolerance mechanisms, it is necessary to compare results from field and laboratory experiments before practical applications are made of laboratory techniques.

H.W. Gabelman and B.C. Loughman (Eds.), Genetic aspects of plant mineral nutrition.
ISBN-13: 978-94-010-8102-3
© *1987 Martinus Nijhoff Publishers, Dordrecht/Boston/Lancaster.*

The objectives of this study were to (1) determine the genetic variation available for developing sorghum genotypes tolerant to iron chlorosis, (2) determine the theoretical gains from selection possible through population improvement, (3) correlate and determine the relationships between chlorosis ratings and agronomic traits in the field, and (4) correlate field and growth chamber responses to iron chlorosis.

Materials and methods

Experimental design

The 100 S_1 families chosen for this study were a random sample from the third cycle of selection for increased 100-seed weight from the sorghum population NP21R. The population was formed from a composite of 100 F_1 hybrids[14] and used the ms_c gene for recombination.

The experimental design used in both the field and growth chamber experiments was a blocks-within-replications design[6]. The S_1 families were randomly grouped into five blocks of 20 families each. The same 20 families were randomized and remained together in the allocation of blocks within replications and repetitions over years. Two replications of the design were used in both the growth chamber and field experiment.

Field experiment

The field experiment was conducted on a Manter sandy loam (coarse-loamy, mixed, mesic Aridic Argiustoll) which had pH of 8.4, 1.2% organic matter, 836 meq Ca, 650 mg K, 24.5 mg P, 10.5 mg Mn, 7.4 mg Fe, and 9.9 mg Zn kg^{-1} soil.

Field plantings were made 21 June 1983 and 23 May 1984 at the Kansas Agricultural Experiment Station, Garden City, KS. Prior to planting, 135 kg N ha^{-1} was applied each year as anhydrous ammonia. Weeds were controlled with standard recommended rates of post-planting applications of propachlor (2-chloro-N-isopropylacetanilide)s in 1983 and propachlor and atrazine [2-chlor-4-(ethylamino)-6-(isopropylamino)-s-triazine] in 1984. The experimental area was irrigated once 44 days after planting in 1983 and once 48 days after planting in 1984. Field evaluations and yield data were collected on 4.4-m rows planted 0.76 m apart with an average spacing of 15 cm between plants. Plant densities were approximately 87 700 plants ha^{-1}.

Growth chamber experiment

Seeds of S_1 families treated with captan, N-[(trichloromethyl)thio]-4-cyclohexene-1,2-dicarboximide, were germinated between rolled paper towels kept moist with aerated, distilled water in a growth chamber under the same light, temperature, and humidity conditions as described for the experiments. Four-day-old seedlings were transferred to plastic plates supporting 126 plants and grown for 5 days in 6 L of pretreatment nutrient solutions with 49 μmol Fe L^{-1}. These seedlings were transferred to plastic plates (45 cm × 78 cm) supporting 80 plants and grown another 5 days in 40 L of pretreatment solution. The 10 plates, containing 80 14-day-old seedlings, were transferred briefly to distilled water to rinse the roots and then placed into plastic-lined container holding 40 L of treatment solutions. Plants were grown in treatment solutions for 12 days. Four S_1 plants spaced 5.7 cm apart represented experimental units. Experimental units were spaced 7.6 cm apart, resulting in a plant density of 236 seedlings m^2. Plants were supported with loosely wrapped sponge rubber in 1.9-cm diameter holes.

The composition of nutrient solutions (μmol element L^{-1}) was 22 900 NO_3-N, 7540 Ca, 7240 K, 2780 NH_4-N, 1940 Cl, 1820 S, 1550 Mg, 65 P, 50 B, 18 Mn, 4.6 Zn, 1.6 Mo, 1.2 Cu, and 49 Fe as FeHEDTA (ferric hydroxyethylethylenediaminetriacetate)for the pretreatment nutrient solution. The same nutrient solution composition was used for the treatment solution except FeHEDTA was omitted, and NO_3-N was the only source of N[4,15].

Growth chamber conditions were 12 h light at 450 μE m^{-2} s^{-1} (110 cm below the lamp), 32 ± 1 C, and 30 ± 3% relative humidity. Darkness was for 12 h at 24 ± 1 C and 40 ± 3% relative humidity. Light was provided by high-pressure sodium and metal halide lamps.

Iron-deficiency chlorosis ratings

Chlorosis ratings were made in 0.5 increments on a scale of 1.0 (no chlorosis) to 5.0 (severe chlorosis)[15]. In the field, chlorosis symptoms were distinguishable and leaf areas were sufficiently large to determine seedling differences at 21 days after planting. Chlorosis ratings were made every other day for 36 days (19 ratings) until 57 days after planting. Visual ratings were made on recently emerged leaves of plants in each experimental unit.

In the growth chamber, chlorosis appeared on leaves 3 days after plants were placed in treatment solutions. The severity of chlorosis was rated on the most recently fully expanded leaf of individual plants every day from day 3 to day 12.

Three chlorosis traits were used to evaluate S_1 families in the field: (1) the first chlorosis rating recorded 21 days after planting, (2) the maximum chlorosis rating recorded during the evaluation period, and (3) the average chlorosis rating for each S_1 family over the evaluation period. Each of the daily chlorosis ratings (CR1 through CR10) and the average chlorosis rating for each S_1 family over the 10-day evaluation period were used to evaluate S_1 families in the growth chamber.

Statistical analysis

The chlorosis and agronomic traits in the field were analyzed statistically to estimate phenotypic and genotypic variance on a S_1 family-mean basis. Gross-product analyses were used to estimate genetic and phenotypic correlations among traits in the field experiment.

The 19 sequential field and 10 sequential growth chamber chlorosis ratings were analyzed as a split-plot in time for the growth chamber and each year of the field experiment so that LSD values could be computed to compare chlorosis ratings. All effects within the experimental model were considered to be random except for the chlorosis ratings.

Results and discussion

Field experiment

Figure 1 shows the seasonal trend for the population mean chlorosis ratings in 1983 and 1984. The chlorosis symptoms were more severe in 1983 than in 1984. The highest mean population chlorosis rating was 3.5 in 1983 and 3.2 in 1984. The differences in chlorosis symptoms may be attributed to changes in the environment between years. In 1983 a drought occurred at Garden city, KS, and the mean growing season temperature was 25.6°C, which was 1.1°C higher than in 1984. Prior to the irrigation 44 days after planting, no significant precipitation had been recorded. As a result of the irrigation, the population mean chlorosis rating decreased from 3.5 to 3.0 within 2 days. The population mean chlorosis rating continued to decrease in 1983 until the last mean chlorosis rating value recorded was 2.6. In 1984, environmental conditions were not as severe, and significant precipitation was recorded intermittently during the growing season.

The phenotypic and genotypic correlations between the first, average, and maximum chlorosis ratings and their correlations with agronomic traits are given in Table 1. Since the first chlorosis and the maximum chlorosis ratings are components of the average chlorosis rating, they correlated better with the average chlorosis rating than with each other. Genotypically, the maximum chlorosis ratings correlated perfectly

Table 1. Genotypic (upper value) and phenotypic (lower value) correlations between field chlorosis ratings and agronomic traits over 2 years

Rating	Rating or trait									
	Max.	First	Flowering (days)	Plant height (cm)	Grain yield (kg ha^{-1})	Grain protein (%)	Grain yield plant^{-1} (g)	Grain yield head^{-1} (g)	Seeds plant^{-1} (no.)	Seeds head^{-1} (no.)	100-seed weight (g)
Average	1.00	0.86	0.39	−0.59	−0.98	0.07	−0.89	−1.19	−0.73	−0.66	−0.05
	0.96**	0.81**	0.32**	−0.51**	−0.60**	0.15	−0.52**	−0.51**	−0.49**	−0.44**	−0.01
Maximum		0.83	0.40	−0.62	−0.98	0.10	−0.90	−1.27	−0.72	−0.68	−0.08
		0.72**	0.32**	−0.53**	−0.56**	0.15	−0.47**	−0.48**	−0.44**	−0.40**	−0.01
First			0.34	−0.46	−0.68	−0.03	−0.68	−0.79	−0.48	−0.36	−0.20
			0.26**	−0.32**	−0.37**	0.00	−0.33**	−0.31**	−0.31**	−0.27**	−0.07

** Significant at the 0.01 probability level.

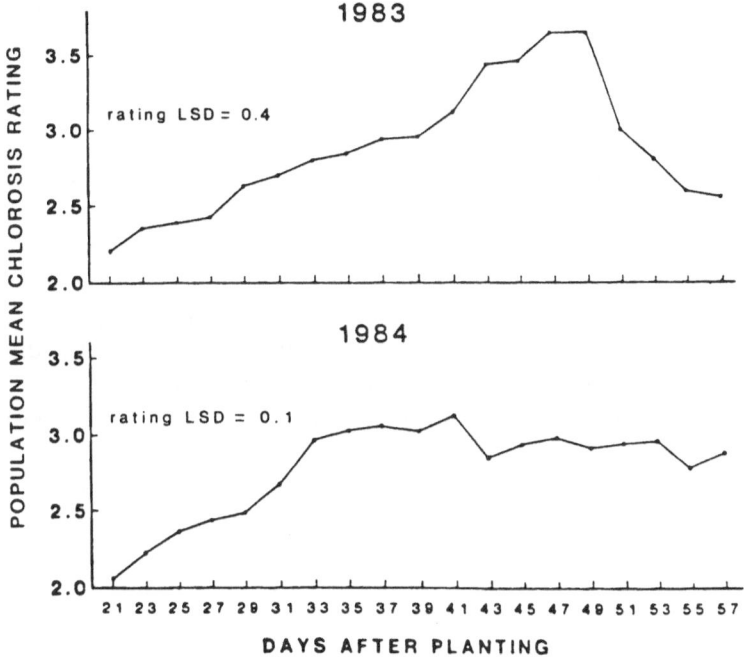

Fig. 1. Field chlorosis ratings.

($r = 1.00$) with the average chlorosis ratings. Trial selections verified that the same two S_1 families most tolerant to iron chlorosis in each block were identified using either rating. Using the 2-year average of the average chlorosis rating as a standard, the maximum chlorosis rating gave the best predictions of superior individuals when selections were made on a yearly basis.

Table 2 shows parameters for the field chlorosis traits. Heritability tended to be highest for the average and maximum chlorosis ratings. Over the 2 years, the only trait with a higher, though notsignificantly different, heritability than the average chlorosis rating was flowering.

Table 2. Parameters, including standard errors (SE), for field chlorosis ratings over 2 years

Parameter	Rating		
	Average	Maximum	First
Mean ± SE[†]	2.8 ± 0.1	3.4 ± 0.2	2.1 ± 0.2
High	3.6	4.4	3.0
Low	1.9	2.3	1.5
Genetic variance ± SE[†]	0.13 ± 0.02	0.17 ± 0.03	0.08 ± 0.02
Phenotypic variance ± SE[†]	0.15 ± 0.02	0.20 ± 0.03	0.10 ± 0.01
Heritability ± SE[†]	0.86 ± 0.15	0.85 ± 0.15	0.73 ± 0.16
CV (%)[‡]	7.6	8.3	15.6

[†] Estimated from S_1 family means.
[‡] Coefficient of variation.

Even though the average chlorosis ratings had a higher heritability value (Table 2), the maximum chlorosis rating would be more practical for field evaluations since only a few observations during a period of severe chlorosis would be necessary. The effectiveness of the maximum chlorosis rating, however, would depend upon the ability of the investigator to record a maximum chlorosis rating for each S_1 family at a given time. For example, in 1983 it was possible to represent the maximum chlorosis rating for the greatest number of S_1 families with a single evaluation at CR15; 88% of the S_1 family plots had maximum readings at CR15. If the evaluation period was extended to include CR14, CR13, and CR12, the number of S_1 family plots represented by a maximum chlorosis ratings increased to 94, 98, and 100%, respectively. Thus, it was possible to observe the maximum chlorosis rating for any S_1 family plot over 6 days in 1983.

The representation of maximum chlorosis ratings was not accomplished over such a narrow period of time in 1984. CR11 represented a maximum chlorosis rating for the greatest number of S_1 family plots (85%). Moreover, 100% of the S_1 family plots had maximum chlorosis ratings represented only if the evaluation period was extended from CR3 through CR19 in 1984 (26 days).

Genotypic correlations between the chlorosis traits and plant height, grain yield, seeds head^{-1}, seeds plant^{-1}, grain yield head^{-1}, and grain yield plant^{-1} were relatively high in magnitude and negative. These correlations indicated that the severity of chlorosis was a major factor limiting sorghum performance. The absolute values of genotypic correlations between the average and maximum chlorosis ratings and grain yield head^{-1} were greater than 1.0. Absolute values greater than 1.0 often represent poor estimates and were probably due to high error variances for heads plant^{-1}, high error variances for grain yield, and low yield caused by the drought and early frost in 1983 (data not shown).

Growth chamber experiment

Figure 2 shows the population mean chlorosis ratings for daily observations of plants grown in the growth chamber. A continual and consistent increase in chlorosis symptoms was noted 7 days after plants were placed in treatment solutions and this continued throughout the experiment. The experiment was terminated when the majority of the plants showed severe chlorosis symptoms (rating of 5.0) which were observed after 12 days in treatment solutions.

Table 3 gives parameters for each daily chlorosis rating (CR1 through CR10) during the course of the experiment and the average chlorosis rating. Though none of the heritabilities were significantly different, the

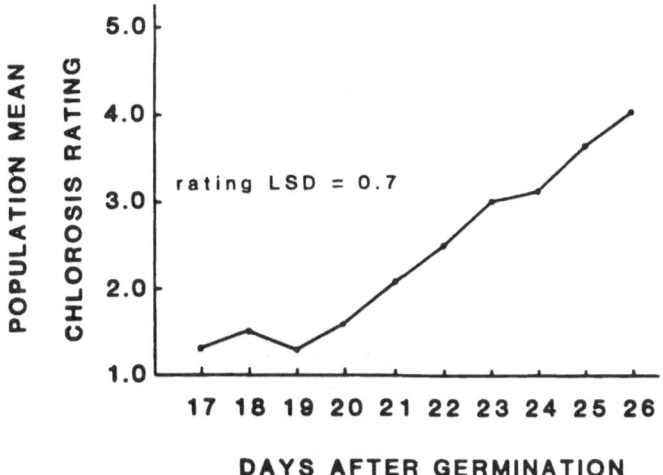

Fig. 2. Growth chamber chlorosis ratings.

highest values were for the average, CR9, CR7, and CR6, respectively. Genetic variances increased markedly at CR6 and thereafter.

Because of the nature of chlorosis development under growth chamber conditions, an optimum time for observing variation among S_1 families for chlorosis would be expected. Initially, all plants started with the lowest rating (1.0), and eventually all plants showed severe chlorosis (5.0). Since the scale values were restricted to a minimum of 1.0 and a maximum of 5.0, the population showed a skewed distribution for

Table 3. Parameters, including standard errors (SE), for growth chamber chlorosis ratings

Day of rating	Chlorosis rating parameter						
	Mean ± SE	High	Low	Genetic variance[†] ± SE	Phenotypic variance[†] ± SE	Heritability[‡] ± SE	CV[§]
							%
1	1.3 ± 0.1	2.1	1.0	0.89 ± 0.46	2.52 ± 0.39	0.35 ± 0.18	22.7
2	1.5 ± 0.1	2.5	1.1	1.90 ± 0.61	3.60 ± 0.55	0.53 ± 0.17	20.5
3	1.3 ± 0.1	1.9	1.0	0.90 ± 0.42	2.34 ± 0.36	0.38 ± 0.18	21.0
4	1.6 ± 0.2	2.5	1.2	1.58 ± 0.66	3.73 ± 0.57	0.43 ± 0.18	19.4
5	2.1 ± 0.2	3.0	1.5	1.72 ± 0.95	5.15 ± 0.79	0.33 ± 0.18	17.7
6	2.5 ± 0.2	3.5	1.3	5.54 ± 1.53	9.26 ± 1.42	0.60 ± 0.17	16.3
7	3.0 ± 0.2	4.0	1.7	7.00 ± 1.89	11.49 ± 1.76	0.61 ± 0.16	14.0
8	3.2 ± 0.2	4.3	1.9	5.58 ± 1.70	10.09 ± 1.55	0.55 ± 0.17	13.4
9	3.7 ± 0.2	4.6	2.3	6.47 ± 1.72	10.47 ± 1.61	0.62 ± 0.16	11.0
10	4.1 ± 0.2	4.9	2.5	4.69 ± 1.56	9.13 ± 1.77	0.51 ± 0.17	9.6
Average	2.4 ± 0.1	3.1	1.8	2.62 ± 0.65	4.02 ± 0.62	0.65 ± 0.16	10.1

[†] Estimated from S_1 family means. Values presented are actual × 10^2.
[‡] Estimated from S_1 family means.
[§] Coefficient of variation.

chlorosis symptoms when ratings were first and last recorded. The skewed distribution became normally as more S_1 families developed chlorosis symptoms. The distribution became skewed again as the majority of the plants developed severe chlorosis. In this experiment, the genetic variance and distribution for chlorosis attained a maximum at CR7.

Theoretical gains from selection

Table 4 shows the predicted gains from selection (or the theoretical decrease in the population mean chlorosis rating) over four cycles of recurrent selection for various field and growth chamber chlorosis traits. The calculations were based on a 10% selection intensity in a sample of 100 S_1 families. The values for gains from selection (Δg) indicated that the maximum chlorosis rating for the field population theoretically could be reduced below 1.0 with four cycles of selection. Since population improvement on an S_1 family basis requires three growing seasons per cycle of selection, 12 growing seasons would be required to reduce the population mean chlorosis rating below 1.0. Because the rating scale is restricted by a minimum value of 1.0, it would be impossible to differentiate superior individuals beyond a certain point in population improvement. Progress would require developing techniques capable of further differentiating superior individuals. Such a problem has been observed in developing soybeans tolerant to iron chlorosis[1,9,13]. As a result, field techniques were developed to further differentiate tolerant genotypes[9]. Graded nutrient solution stress treatments have also been used to differentiate sorghum genotypes for tolerance to iron chlorosis[8].

The theoretical estimates for Δg assume that S_1 family variance largely

Table 4. Predicted gains from selection (Δg) and the population mean chlorosis ratings in subsequent cycles (C)

Site and rating	Δg[†]	Cycle				
		C_0	C_1	C_2	C_3	C_4
		Population mean chlorosis rating				
Growth chamber						
CR7	0.36	3.0	2.6	2.2	1.9	1.5
CR9	0.35	3.7	3.4	3.0	2.7	2.3
Average	0.23	2.4	2.2	2.0	1.8	1.5
Field						
Average	0.58	2.8	2.2	1.7	1.1	0.5[§]
Maximum	0.66	3.4	2.8	2.1	1.5	0.8[§]
First	0.40	2.1	1.7	1.3	0.9[§]	0.5[§]

[†] Gains from selection calculated on a 10% selection intensity from a sample of 100 families.
[‡] CR7 and CR9 = the 7th and 9th chlorosis rating, respectively.
[§] The minimum rating possible in the experiment was 1.0.

represents additive genetic variance in a self-pollinating species like sorghum. Because of other types of gene action, the S_1 family genetic variance is a biased estimate of additive genetic variance[5,7]. The best estimates of additive genetic variance are calculated from realized gains from selection, but S_1 family estimates are still useful indicators of the relative variances of traits within a population and can be used to develop recurrent selection programs.

The theoretical decrease in the population mean chlorosis rating was estimated at nearly 0.7 units per cycle of selection for the maximum chlorosis rating. The observed decrease in the mean chlorosis rating per cycle of selection in soybeans, based on the same selection criteria, was 0.2 units[13]. It appears that greater progress from selection may be possible in sorghum unless heritability estimates are greatly biased. Estimates of heritability on a S_1 family basis typically are greater than those predicted on an individual-plant basis[11]. Realized gains from selection often can be expected to be about half of the estimated gains from selection among S_1 families (W.M. Ross, personal communication). Even if realized gains were half the estimated gains from selection, progress for developing tolerance to iron chlorosis in sorghum would be greater compared to those realized in soybeans.

Correlations between field and growth chamber chlorosis ratings

The S_1 family mean correlations over the popultion as a whole between growth chamber and the three field chlorosis traits were low. Only the correlation between the average chlorosis rating in the field and CR5 in the growth chamber (r = 0.23) was significant.

Correlation coefficients increased as the unit of experimental design over which they were calculated decreased in size (data not shown). However, no apparent characteristics identified specific replications or blocks in the growth chamber experiment which had higher correlations between field and growth chamber chlorosis ratings.

The highest correlations and greatest number of significant correlations over blocks of 20 S_1 families were either between the first growth chamber chlorosis ratings and the first field chlorosis rating or between later growth chamber chlorosis ratings and the average or maximum field chlorosis ratings. This was probably the result of the first chlorosis rating in the field and growth chamber being recorded for plants of similar maturity. Since these experiments were conducted independently, different mechanisms for resistance to iron chlorosis in seedlings and in more mature plants may have been involved.

Phenotypic correlations between field chlorosis ratings in 1983 and 1984 (Table 5) were large and significant. The consistency between years in the field experiment and lack of correlation between field and growth

Table 5. Phenotypic correlations between 1983 and 1984 field chlorosis ratings

1984 rating	1983 rating		
	Average	Maximum	First
Average	0.74**	0.68**	0.54**
Maximum	0.73**	0.67**	0.51**
First	0.58**	0.55**	0.52**

** Significant at the 0.01 probability level.

chamber chlorosis traits indicated discrepancies between the two techniques. Reasons for the low correlation may be inherent with the stage of plant development and the variation present when evaluations were made for chlorosis symptoms in the growth chamber. Variability between blocks was evident in the growth chamber experiment. Some researchers have dealt with the large experimental error typically associated with iron chlorosis symptoms by using large numbers of replications[9,12]. Experimental error should be reduced either through additional replications or through improved techniques before laboratory results can be widely applied to field research.

References

1. Cianzio S R de and Fehr W R 1980 Genetic control of iron deficiency chlorosis in soybeans. Iowa State J. Res. 54, 367-375.
2. Cianzio S R de and Fehr W R 1982 Variation in the inheritance of resistance to iron deficiency chlorosis in soybeans. Crop Sci. 22, 433-434.
3. Cianzio S R de, Fehr W R and Anderson I C 1979 Genotypic evaluation for iron deficiency chlorosis in soybeans by visual scores and chlorophyll concentration. Crop Sci. 19, 644-646.
4. Clark R B, Yusuf Y, Ross W M and Maranville J W 1982 Screening for sorghum differences to iron deficiency. J. Plant Nutr. 5, 587-604.
5. Cockerham C C 1983 Covariances of relatives from self-fertilization. Crop Sci. 23, 1177-1180.
6. Eckebil J P, Ross W M, Gardner C O and Maranville J W 1977 Heritability estimates, genetic correlations, and predicted gains from S_1 progeny tests in three sorghum random-mating populations. Crop Sci. 17, 373-377.
7. Empig L T, Gardner C O and Compton W A 1972 Theoretical gains for different population improvement procedures. Neb. Agric. Exp. Sta. MP26 (rev.).
8. Esty J C, Onken A B, Hossner L R and Matheson R 1980 Iron use efficiency in grain sorghum hybrids and parental lines. Agron. J. 72, 589-592.
9. Fehr W R, Froehlich D M and Ertl D S 1985 Iron-deficiency chlorosis of soybean cultivars injured by plant cutoff and defoliation. Crop Sci. 25, 21-23.
10. Froehlich D M and Fehr W R 1981 Agronomic performance of soybeans with differing levels of iron deficiency chlorosis on calcareous soil. Crop Sci. 21, 438-441.
11. Kofoid K D 1979 Estimates of genetic parameters for agronomic, nutritional, and production traits in sorghum using S_1 family testing. Ph.D. Dissertation. Univ. of Nebraska, Lincoln. (Diss. Abstr. 40, 525B-526B).
12. McKenzie D B, Hossner L R, and Newton R J 1984 Sorghum cultivar evaluation for iron chlorosis resistance by visual scores. J. Plant Nutr. 7, 677-685.

13 Prohaska K R and Fehr W R 1981 Recurrent selection for resistance to iron deficiency chlorosis in soybeans. Crop Sci. 21, 524–526.
14 Ross W M, Gorz H J, Haskins F A, Hookstra G H, Rutto J K and Ritter R 1983 Combining ability effects for forage residue traits in grain sorghum hybrids. Crop Sci. 23, 97-101.
15 Williams E P, Clark R B, Yusuf Y, Ross W M and Maranville J W 1982 Variability of sorghum genotypes to tolerate iron deficiency. J. Plant Nutr. 5, 553-567.

Sorghum genotype differences in uptake and use efficiency of mineral elements

N. SEETHARAMA,
Sorghum Program, ICRISAT (International Crops Research Institute for the Semi-arid Tropics) Patancheru P.O., Andhra Pradesh 502 324, India

R.B. CLARK and J W MARRANVILLE
U.S. Department of Agriculture, Agricultural Research Service, and Department of Agronomy, University of Nebraska, Lincoln, NE 68583, USA

Key words Al Ca Cl Cu Fe Harvest index K Mg Mineral translocation index Mn N P S Si Zn *Sorghum bicolor*

Summary Twelve sorghum [*Sorghum bicolor* (L.) Moench] genotypes consisting of hybrids, varieties from tropical and temperate regions, and land races of India and West Africa were evaluated for mineral element uptake from the soil, efficiency of dry matter production per unit of element absorbed, and efficiency of mineral element transfer from vegetative parts to the grain. The genotypes were grown on an Alfisol in India during a normal rainy season under five levels of applied N (0 to 100 kg ha^{-1}) and P (0 to 26 kg ha^{-1}). At physiological maturity, grain and aboveground vegetative parts were analyzed separately for N, P, K, S, Mg, Si, Cl, Ca, Al, Mn, Fe, Cu, and Zn.

Genotype × fertility interactions were low indicating that genotypes could be evaluated for uptake and efficiency at various levels of soil fertility. Differences among genotypes to each element were noted for each of above mentioned traits. No genotype showed high efficiency for dry matter production per unit element absorbed for each element, but the West African cultivar Naga White showed highest efficiency relative to five mineral elements. Genotypes varied extensively in their ability to translocate different mineral elements from the vegetative parts to the grain. Except for Naga White, the degree of translocation of mineral elements and harvest index were highly correlated. Genotypes excelling in individual mineral element uptake or use efficiency were generally the land races with lower grain yields and lower translocation efficiencies. Thus, a future challenge is to combine desirable mineral element traits into specific few sorghum lines for improvement purposes.

Introduction

In recent years, crop improvement research specifically directed towards increasing the efficiency of mineral nutrition of plants has received increased attention[2,3,5]. However, in most of these studies only seedlings or young plants grown in either nutrient solution or in plots were used. Nutritional differences in genotypes have rarely been related to final economic yield. Efficient production of cereals may be visualized as the product of (a) uptake and accumulation of mineral elements by the plant, (b) production of dry matter per unit element assimilated, and (c) translocation of elements from vegetative parts to the grain. Genotype differences to N for these above traits have been reported in sorghum [*Sorghum bicolor* L.) Moench][1]. However, information is lacking on

H.W. Gabelman and B.C. Loughman (Eds.), Genetic aspects of plant mineral nutrition.
ISBN-13: 978-94-010-8102-3
© *1987 Martinus Nijhoff Publishers, Dordrecht/Boston/Lancaster.*

genotype differences to each mineral element and for genotypes grown under various levels of applied fertilizer. In this study differences among a set of 12 genotypes grown under five levels of N and P fertility were compared for the three above mentioned traits. Thirteen elements were used to evaluate the genotypic responses.

Materials and methods

Treatments and collection of data

Twelve sorghum genotypes were selected for this study. They consisted of three hybrids and two varieties released for commercial cultivation in India (CSH1, CSH5, CSH6, CSV3 and CSV5), inbred lines selected for wide adaptability from a random-mating population (RS1 × VGC, FLR53, and FLR101), tropical-temperate conversion lines (SC108-3 and SC108-8-4), and land races of sorghum from India (Patancheru Local) and West Africa (Naga White). Seed of these genotypes were provided by ICRISAT (International Crops Research Institute for the Semi-Arid Tropics) Center, Patancheru, India. The experiment was located on an Alfisol at ICRISAT Center. Plants were grown during the normal rainy season. A split-plot design with four replications was used. Subplots consisted of genotypes planted in four rows of 5 m length. The main plots consisted of five levels of applied fertilizer: F_1 = No fertilizer applied, F_2 = 20 kg N and 9 kg P ha^{-1}, F_3 = 20 kg N and 26 kg P ha^{-1}, F_4 = 60 kg N and 26 kg P ha^{-1}, and F_5 = 100 kg N and 26 kg P ha^{-1}.

At physiological maturity, 10 randomly selected plants (aboveground parts only) from each subplot were combined as a single bulk sample. Grain and non-grain parts (chaff, leaves, and culms) were separated and dried in a forced-air oven at 80 C for a minimum of 48 hours and weighed. The samples were ground to pass a 1.0 mm screen. Nitrogen concentrations were determined by micro-Kjeldahl. The concentrations of P, K, S, Mg, Si, Cl, Ca, Al, Mn, Fe, Cu, and Zn were determined by energy dispersive x-ray fluorescence spectrometry[4].

Characteristics used to compare genotypes

The traits used to characterize and compare genotypes were (i) total biomass production (grain plus non-grain dry weights), (ii) harvest index (grain dry weight divided by total plant dry weight × 100), (iii) mineral element contents of grain and non-grain parts (plant part dry weight multiplied by its element concentration), (iv) total plant content of mineral element (total mineral element content in grain plus non-grain plant parts), (v) mineral element use efficiency (biomass divided by the total mineral element content in the plant), and (vi) mineral element translocation index (mineral element content in the grain divided by the total mineral element content in the plant × 100).

Results and discussion

The analysis of variance for mineral element concentrations in the plants at maturity did not show significant effects from fertilizer treatment except for N, Mg, Cu, and Zn. Consistent and substantial increases in mineral element concentrations were noted only for N; increases of 5, 22, and 45% over the no fertilizer treatment were observed with the addition of 20, 60, and 100 kg N ha^{-1}, respectively.

Except for Cu, fertilizer x genotype interactions were not significant. This indicated that genotypes could be evaluated at various levels of applied fertilizer. Hence, mean performance of genotypes across five fertility levels were considered.

Table 1. Mean, range, and LSD values for mineral element uptake, use efficiency, and translocation index

Mineral element	Uptake (mg plant^{-1})			Use efficiency (g dry wt g^{-1} element)			Translocation index (%)		
	Mean	Range	LSD$_{0.05}$	Mean	Range	LSD$_{0.05}$	Mean	Range	LSD$_{0.05}$
N	443	866–261	88	136	167–110	17	49.9	65.6–9.4	5.1
P	64.2	113.6–43.5	19.9	1002	1787–697	309	73.4	93.1–15.3	8.3
K	614	1544–288	171	101	136–86	11	14.4	19.8–1.7	2.5
S	53.4	134.9–26.9	14.5	1174	1458–875	154	29.1	50.4–4.0	5.3
Mg	130	291–68	32	471	580–398	47	24.2	41.0–3.4	4.1
Si	2139	5610–1000	605	30.5	36.5–24.0	4.6	0.47	1.33–0.05	0.04
Cl	181	438–81	53	359	430–301	65	5.02	9.95–0.99	2.3
Ca	183	510–97	55	347	418–278	50	2.83	4.74–2.47	1.3
Al	49.6	84.1–34.2	21.9	3370	2400–100	106	8.4	14.4–2.1	7.3
Mn	3.49	7.01–1.92	1.12	19500	23300–15800	4100	9.4	16.7–1.6	2.9
Fe	14.1	23.0–8.9	6.5	5230	8580–3090	2050	6.4	12.3–2.2	3.8
Cu	0.44	0.96–0.28	0.11	139000	181000–112000	18000	19.0	27.7–2.6	4.3
Zn	1.45	3.35–0.76	0.40	42900	58400–37100	5000	27.2	43.5–4.7	5.4

Genotype differences in mineral uptake

Table 1 shows the mean and the range in the uptake (content) of 13 mineral elements by sorghum plants at maturity. Genotype differences were significant in each case. High variations among genotypes were noted in the contents of Si, Cl, K, and Ca. The uptake of Al, Fe, P, and N showed the least amount of variation. Even though genotype differences in mineral element uptake were noted, most of the variation for each element could be accounted for by plant size. Thus, Patancheru Local with its high biomass contained the highest amount of each mineral element studied, and FLR101 (low biomass) contained the lowest amount of each element (Table 2). However, genotypes with similar biomass also differed in mineral element contents. For example, Naga White had a greater biomass production than CSV5, but the latter contained higher contents of each mineral element, except Ca, Al and Fe. Superiority of hybrids over varieties or inbred lines was not noted for uptake of the mineral elements.

Genotype differences in mineral element use efficiency

Mineral element use efficiency was highest for Cu, and lowest for K. The variations among genotypes for utilization efficiency of individual mineral elements was generally higher for those elements which had low variability in uptake (Al, Fe, P and S), and lower for those elements which had high variability. Unlike uptake, no consistent relationships were noted among the rankings for use efficiency of individual mineral elements and biomass of genotypes (Tables 2 and 3).

Table 2. Rankings of sorghum genotypes for mineral element uptake and biomass production

Genotype	Mineral element uptake (Ranking[†])												Biomass production	
	N	P	K	S	Mg	Si	Cl	Ca	Al	Mn	Fe	Cu	Zn	g plant[-1]	ranking[†]
CSH1	5	6	7	7	7	9	4.5	9	11	9	9	5	8	46.0	8
CSH6	6	4	6	5	8.5	10	8	8	8	8	7	6	6	47.5	6
CSH5	3	2	3	4	3	4	3	3	4	3	4	3	3	68.8	3
RS1 × VGC	10	7	9	9	8.5	7	7	7	6	11	12	11	7	41.6	9
FLR53	8	9	5	6	6	6	10	6	7	4	6	10	5	47.3	7
FLR101	12	12	12	12	12	12	12	12	12	12	11	12	12	29.9	12
SC108-3	11	10	11	10.5	10	8	9	10	5	10	5	9	10	36.8	11
SC108-4-8	9	11	10	10.5	11	11	11	11	9	7	10	7	11	39.2	10
CSV3	2	5	2	2	2	2	2	2	2	2	2	2	2	94.7	2
CSV5	4	3	4	3	4	3	4.5	5	10	5	8	4	4	54.2	5
Patancheru Local	1	1	1	1	1	1	1	1	1	1	1	1	1	130.5	1
Naga White	7	8	8	8	5	5	6	4	3	6	3	8	9	57.4	4

[†] Rankings are 1 = highest and 12 = lowest for element uptake and biomass production.

Table 3. Rankings of sorghum genotypes for mineral element use efficiency

Genotype	Mineral element use efficiency (Ranking[*])												
	N	P	K	S	Mg	Si	Cl	Ca	Al	Mn	Fe	Cu	Zn
CSH1	12	9	6	8	7	2	12	2	2	2	6	10	9
CSH6	11	12	3	6	2	1	4	1	7	5	8	8	8
CSH5	4	4	8	2	4	7	7	7	5	8	5	2	4.5
RS1 × VGC	6	7	9	4	10	6	11	5	11	1	3	6	12
FLR53	5	5	7	5	5	8	1	6	6	12	9	3.5	10
FLR101	10	8	5	10	9	5	3	8	9	11	11	10	6
SC108-3	9	10	4	9	8	10	8	10	12	9	12	10	4.5
SC108-4-8	8	6	2	7	3	3	6	4	10	10	7	12	3
CSV3	3	1	10	3	12	12	10	11	3	6	2	3.5	2
CSV5	7	11	12	12	11	9	5	9	4	7	4	7	11
Patancheru Local	1	2	11	11	6	11	9	12	1	3	1	5	7
Naga White	2	3	1	1	1	4	2	3	8	4	10	1	1

[*] Rankings are 1 = highest and 12 = lowest for use efficiency.

Individual genotypes did not show high use efficiency relative to each mineral element (Table 3). The West African land race Naga White had the highest use efficiency values for five (K, S, Mg, Cu, and Zn) out of 13 elements. Patancheru Local had the highest use efficiency value for N, Al, and Fe, but the lowest for Ca; it also ranked next to the lowest for K, S, and Si. The improved cultivars were not necessarily inferior to the land races for each of the mineral elements studied. The opposite rankings between the Indian land race Patancheru Local and the hybrids CSH1 and CSH6 were noted for use efficiency. That is, Patancheru Local had low use efficiency for Cu and Si, but CSH6 had the highest use efficiency for these mineral elements. The opposite was noted for N and P use efficiency, and CSH6 had the lowest use efficiency value for N. SC108-3 and FLR101 generally showed low use efficiency for most mineral elements.

Genotype differences in mineral element translocation index

The translocation of mineral elements to the grain from the vegetative plant parts was highest for P (73%) followed by N (50%), and lowest for Si (0.5%) and Ca (2.8%) (Table 1). Variations in the translocation of individual mineral elements among genotypes were high for Si, S, and Mg, and low for Ca, Fe, and P. The rankings of genotypes for translocation index values of the elements were generally highly correlated with their ranking for harvest index. Thus, CSH6 with the highest harvest index ranked first or second for translocation index values of the mineral elements, and Patanchery Local (harvest index = 3.5%) had the lowest translocation index values of the elements, except for Al (Table 4). Naga

Table 4. Rankings of sorghum genotypes for mineral element translocation index and harvest index

Genotype	Translocation index (Ranking[†])												Harvest index	
	N	P	K	S	Mg	Si	Cl	Ca	Al	Mn	Fe	Cu	Zn	(%)	Ranking[†]
CSH1	2	4	2	2	2	2	5	2	1	3	3	3	2	38.1	2
CSH6	1	1	1	1	1	1	2	1	2	1	2	1	1	41.7	1
CSH5	5	8	8	7	8	8	3	7	6	8	4	7	8	28.6	5
RS1 × VGC	6	6	6	4	5	4	6	5	4	2	1	4	5.5	31.3	3
FLR53	9	9	9	9	9	9	8	9	10	9	10	8	9	21.0	9
FLR101	7	7	4	3	4	3	7	4	7	4	6	6	4	30.6	4
SC108-3	8	5	7	6	7	7	9	8	8	6	7	10	7	27.1	7
SC108-4-8	4	2	5	5	3	6	4	6	9	7	5	5	5.5	28.3	6
CSV3	11	10	11	11	11	11	11	10	12	11	11	11	11	6.9	11
CSV5	10	11	10	10	10	10	10	12	11	10	8	9	10	13.8	10
Patancheru Local	12	12	12	12	12	12	12	11	3	12	12	12	12	3.5	12
Naga White	3	3	3	8	6	5	1	3	5	5	9	2	3	26.9	8

[†] Rankings are 1 = highest and 12 = lowest for translocation and harvest indices.

White was an exception to this; it ranked eighth out of 12 for harvest index, had higher translocation index values, and ranked first in translocation index for Cl, second for Cu, and third for N, P, K, and Zn.

Conclusions

Genotype differences were noted for mineral element uptake, use, and translocation in plants. The genotypes tested represented broad genetic backgrounds. The land races were generally superior to the more improved cultivars for uptake and use efficiency of the mineral elements, but the improved genotypes had higher harvest and translocation indices and higher grain yields. Thus, increases in grain yield and higher amounts of nutrients in the grain have generally come from genotype ability to distribute mineral elements in the grain rather than from increased element uptake or use efficiency. A challenge for the future will be to combine favorable traits into a relatively few agronomically desirable sorghum genotypes. However, this may be impractical if attempts are made to improve the efficiency of each mineral element in the same genotype. Improvement of genotypes for those mineral elements which are severely limiting in specific areas and which are not readily supplied by fertilization should be emphasized. Cost benefits for applied elements should also be considered. In selecting parents and testing prospective genotype products, it is important to monitor each of the mineral elements to ensure that efficiency traits for particular mineral elements are not lost at the expense of others. Fortunately, techniques to analyze several mineral elements simultaneously are readily available and accomplishment of these goals appears practical.

References

1. Alagarswamy A and Seetharama N 1983 Biomass and harvest index as indicators of nitrogen uptake and translocation to the grain in sorghum genotypes. pp 423 427. *In* Genetic Aspects of Plant Nutrition. Eds. M R Saric and B C Loughman. Martinus Nijhoff/Dr. W. Junk Publ., The Hague, The Netherlands.
2. Devine T E 1982 Genetic fitting of crops to problem soils. pp 143-173. *In* Breeding Plants for less Favorable Environments. Eds. M B Christiansen and C F Lewis. John Wiley & Sons, New York.
3. Gabelman W H (Ed.) 1986 Proceedings of Second International Symposium on Genetic Aspects of Plant Mineral Nutrition. Martinus Nijhoff/Dr. W. Junk Publ., The Hague, The Netherlands.
4. Knudsen D, Clark R B, Denning J L and Pier P A 1981 Plant analysis of trace elements by x-ray. J. Plant Nutr. 3, 61 75.
5. Saric M R and Loughman B C (Eds.) 1983 Genetic Aspects of Plant Nutrition. Martinus Nijhoff/Dr. W. Junk Publ., The Hague, The Netherlands.

Resistance to the European corn borer, *Ostrinia nubilalis* (Hubner), in maize, *Zea mays* L., as affected by soil silica, plant silica, structural carbohydrates, and lignin

JAMES G. COORS
Department of Agronomy, University of Wisconsin, Madison, WI 53706, USA

Key words European corn borer Lignin Neutral detergent fiber *Ostrinia nubilalis* (Hubner) Silica Structural carbohydrates *Zea mays* L.

Summary The study examined the effectiveness of silica content of whorl and leaf sheath tissues in maize, *Zea mays* L., as a soil-moderated defense against the European corn borer, *Ostrinia nubilalis* (Huber). Results were based on data from two test sites in Wisconsin at which 15 maize genotypes were grown with artificial insect infestation. At Madison [Dresden silt loam (fine-loamy over sandy, mixed mesic Mollic Hapludalf), SiO_2 = 20 ppm], approximately 49 percent of the total variation in first generation damage ratings could be accounted for by silica and lignin content in whorl tissue. No significant relationships between whorl composition and first generation insect damage were observed at Hancock [Plainfield sand (mixed, mesic, Typic Udipsamment), SiO_2 = 5 ppm]. Silica and neutral detergent fiber in leaf sheath tissue could account for 71 percent of the variation measured for second generation insect damage at Madison, while at Hancock, plant silica content was not a significant factor. Instead, neutral detergent fiber and lignin accounted for an equivalent 71 percent of variation in insect damage. The most resistant genotypes at either location depended upon elevated levels of total structural carbohydrates, lignin, and silica to provide adequate defense against the second generation of the European corn borer.

Introduction

The European corn borer, *Ostrinia nubilalis* (Hubner), is one of the most destructive pests of corn, *Zea mays* L., in the United States. Resistance to leaf feeding of first generation corn borers has been primarily associated with high levels of a cyclic hydroxamate, 2,4-dihydroxy-7-methoxy-1,4-benzoaxin-3-one (DIMBOA) in the leaves throughout the whorl stage of development[16]. However, several resistant maize lines have levels of DIMBOA in the whorl as low as that found in susceptible lines[30], and recently it has been demonstrated that lignin and silica content jointly contribute to first generation resistance[27].

Second generation corn borers initially feed on the leaf sheath and collar tissue, then bore into the stalks often causing stalk breakage[10]. DIMBOA content typically decreases with age of plant and provides little protection from the second generation. The silica content of some maize germplasm increases with age to the point where it may become the dominant factor promoting resistance at later stages of plant development[27].

Similar positive correlations between plant silica content and insect and disease resistance occur in several species of the Gramineae. Silica content of rice has been associated with resistance to the Asiatic rice stem borer, *Chilo suppressalis* (Walker)[4,29], the yellow rice borer, *Scirpophaga incertulas* (Walker)[24], the leaf-roller, *Cnaphalocrocis medinalis* (Guenee)[12], the field slug, *Agriolimax reticulatus* (Muller)[33], and the blast fungus, *Pyricularia oryzae* (Cav.)[13,32]. Direct soil application of a siliceous blast furnace slag to rice paddy fields can decrease stem borer damage[29]. In wheat, silica content of stem tissue may be related to stem strength, resistance to the cereal leaf beetle, *Oulema melanope* (L.)[35], and resistance to the Hessian fly, *Mayetiola destructor* (Say)[20,22,25]. In oats and barley, silica has been associated with resistance to the hessian fly and *Erisiphe graminis* (D. C.)[7,8,34]. The chinch bug, *Blissus leucopterus* (Say) is affected by plant silica content in sorghum[17], and some sugarcane cultivars have high enough levels of silica to deter *Melanopsis glomerata* (G.)[1]. Recently, there have been reports of strong correlations in *Lolium* spp. between high silica and resistance to the frit fly, *Oscinella frit* (L.) and related species[33].

Plant silica content is usually lower in dicotyledonous species than in grasses. Nevertheless, there have been reports relating plant silica content to pest resistance in dicots. High silica content in cucumber leaves may reduce the incidence of powdery mildew[34], and aphids and plant lice may be adversely affected by high silica in *Lithospermum arvense*[20].

Silica is present in the soil solution in the form of monosilic acid, $Si(OH)_4$. It is this form that is taken up by the roots and moved throughout the above ground portions of the plant where it is deposited in the form of opal, $SiO_2 nH_2 O$[15,19]. In oats, silica accumulation may be a more active process conditioning an inducible defense against herbivory[21].

In the Gramineae, silica is deposited primarily in the epidermal layers of the leaf sheath and blade, and in the xylem vessels and mestome sheath of the vascular system[3,6,15,17,18]. It is the silica deposits in the epidermal regions which are thought to condition resistance to disease and insect pests. Marked mandible wearing has been observed in *C. suppressalis* after feeding on high silica rice[4,12,29]. Fungal pathogens may not be able to penetrate thickened and silicified cell walls[15].

Soil type and environmental conditions markedly influence plant composition[11]. In particular, plant silica content depends to a large extent on soil mineral content, temperature, and amount of solar radiation[15,17,18]. Research was undertaken to investigate the effect of growing environments on plant silica content and cell wall composition of several genotypes of maize, and to relate these plant constituents to first and second generation European corn borer damage.

Materials and methods

In the summer of 1984, 14 inbred lines of maize and one composite population, ECBR (developed at Cornell University from selections from crosses of adapted inbred lines with San Juan and Antiqua populations), were planted in a randomized block design with two replicates at each of four locations in Wisconsin. The plots consisted of three 24-plant rows with 22.9 cm between plants and 76.2 cm between rows. The locations were Madison [Dresden silt loam (fine-loamy over sandy, mixed mesic Mollic Hapludalf)], Hancock [Plainfield sand (mixed, mesic, Typic Udipsamment)], Lancaster [Fayette silt loam (fine silty, mixed, mesic, Typic Hapludalf)], and Marshfield [Withee silt loam (fine-loamy, mixed, frigid, Aeric Glossoboralf)]. Plots were fertilized according to soil test recommendations.

At the Hancock and Madison sites, when plants were at the mid-whorl stage of development (about 50–70 cm in height), the first row of each plot was infested two times, two days apart by dropping two European corn borer (ECB) egg masses into the whorl of each plant, simulating a first generation infestation. On the first day of infestation, samples of whorl tissue were collected from 10 plants from the third row in each plot. Approximately 20 days after infestation, leaf feeding damage was evaluated for infested rows using the 1–9 scale described by Guthrie[9]. Ratings of 1 and 9 represented the most and least resistant classes, respectively.

When plants reached mid-silk stage, 10 plants in the second row were infested with ECB egg masses simulating a second generation infestation. Two egg masses were pinned to each of three leaves: the leaf just above the top ear shoot, the leaf below the top ear shoot, and the second leaf below the top ear shoot. Infestation was repeated after a two day interval. On the first day of infestation, samples of leaf sheath (plus collar) and leaf blade tissue were collected from five plants in the third row of each plot. Each maize genotype had a different flowering data and, consequently, a different infestation and sampling data. After approximately 50 days, 10 infested plants were cut at soil level, their stalks split, and the number of tunnels determined. Because of height differences among genotypes, the total number of tunnels was adjusted according to the height of the sampled plants. Second generation infestation is, therefore, expressed as the number of tunnels per meter of stalk tissue.

Plant samples were washed to remove any soil residue, dried at 55°C, and ground through a 1.0-mm screen using a Christy and Norris hammer mill. Acid detergent fiber (ADF), cellulose, and lignin content were measured on a dry matter basis using the sequential detergent and $KMnO_4$ lignin determinations of Robertson and Van Soest[26]. Biogenic silica was estimated by wetting the residual ash with 48 percent HBr for one hour, washing with acetone, and ashing briefly at 550°C. A separate measurement for neutral detergent fiber (NDF) was obtained for leaf blade and leaf sheath tissues from the two infested sites, Madison and Hancock. Consequently, hemicellulose content could also be estimated at these sites by subtracting ADF from NDF content. Available soil silica at all four sites was determined by the method of Lanning et al.[18].

Analysis of variance and multiple regression procedures were used to analyze the data. Tissues and locations were considered to be fixed variables, while genotypes and blocks (within locations) were considered to be random variables. F tests of significance were calculated based upon the expected mean squares using the above assumptions. For genotype means at Madison and Hancock, stepwise multiple regression was used to examine the association between first and second generation ECB damage and plant composition of the whorl and leaf sheath tissue, respectively. Components were entered in the overall regression model if the probability of obtaining a larger F value was less than 0.15.

Results

Tissues, locations, and genotypes differed significantly for all parameters measured. Furthermore, nearly all two and three factor interactions were significant showing the complexity of plant composition as expressed in different plant parts and in different environments. The

Table 1. First generation ECB damage, cell wall composition, and silica content of whorl tissue at Madison and Hancock

Entry	Location	Damage rating	ADF (%)	Cellulose (%)	Lignin (%)	Silica (%)
Oh43	Madison	1.9	30.5	28.0	2.5	0.7
	Hancock	1.6	30.2	27.6	2.6	0.4
W64A	Madison	4.1	30.3	27.6	2.7	0.4
	Hancock	4.3	29.2	26.8	2.4	0.2
Oh51A	Madison	2.6	30.8	28.1	2.8	0.6
	Hancock	2.1	28.9	26.5	2.4	0.3
W182BN	Madison	3.9	28.2	25.9	2.3	0.2
	Hancock	2.2	31.6	28.1	3.5	0.1
W540	Madison	3.3	30.2	28.0	2.2	0.5
	Hancock	3.1	30.2	28.1	2.1	0.3
W845	Madison	1.6	30.2	27.2	3.0	0.5
	Hancock	1.9	30.2	27.6	2.7	0.2
A619	Madison	1.3	30.3	27.7	2.6	0.7
	Hancock	1.5	26.4	24.0	2.4	0.2
W548	Madison	2.8	29.5	27.4	2.2	0.4
	Hancock	2.1	30.6	28.1	2.5	0.1
W552C2	Madison	4.4	30.1	27.8	2.3	0.6
	Hancock	2.5	30.5	28.0	2.5	0.2
W590	Madison	4.1	30.7	28.8	1.9	0.4
	Hancock	3.9	31.0	28.8	2.1	0.1
W117	Madison	4.0	27.2	25.2	2.0	0.3
	Hancock	2.0	28.2	25.6	2.6	0.2
B86	Madison	2.2	31.0	28.3	2.6	0.6
	Hancock	1.7	29.9	27.4	2.5	0.2
W61BR3	Madison	2.1	30.6	28.1	2.5	0.4
	Hancock	1.9	29.6	27.3	2.3	0.2
W22G	Madison	2.1	32.7	29.9	2.8	0.5
	Hancock	3.0	30.4	28.0	2.4	0.2
ECBR	Madison	1.6	32.9	30.8	2.1	0.8
	Hancock	1.7	32.7	30.4	2.3	0.3
Mean		2.6	30.2	27.7	2.5	0.4
LSD (0.05)		0.7	1.6	1.6	0.4	0.5

nature of these interactions will be dealt with in a separate publication. The results specifically pertaining to the two sites with extreme values for soil silica content, Madison (SiO_2 = 20 ppm) and Hancock, (SiO_2 = 5 ppm) will be examined in this report.

First and second generation ECB damage ratings were not significantly correlated at either location (r = 0.28 at Madison and r = 0.21 at Hancock). Therefore, the associations between plant constituents and insect resistance were analyzed for each generation individually.

The contrast between the Madison and Hancock sites for first generation ECB damage and whorl tissue composition is provided in Table 1. There were no consistent differences for either insect damage or plant composition, with the exception of plant silica content. The average

Table 2. Multiple regression analysis for first generation ECB damage versus silica and lignin content of whorl tissue at Madison

Source	df	M.S.	F	PROB F	R^2
Model	2	4.03	5.77	0.02	
Silica	1	5.10	7.32	0.03	0.31
Lignin/Silica	1	2.95	4.23	0.06	0.49
Error	12	0.70			

silica content of plants at Madison and Hancock were 0.5 and 0.2 percent, respectively, which corresponds to the different soil silica values at the two locations.

Within locations, genotypes differed significantly for all measures, but there did not appear to be a strong association of any plant component with insect damage. The multiple regression analysis at Madison did suggest that plant silica was the most important single component, followed by lignin (Table 2). Lignin and silica together accounted for 49 percent of the total variation in damage ratings observed at Madison, and the best regression equation developed from Madison data produced negative regression coefficients for both lignin and silica (Table 5). In contrast, at Hancock, neither silica nor any other plant component was related to insect damage.

Second generation ECB damage at both Madison and Hancock was strongly associated with plant cell wall composition and silica content. Leaf sheath composition, in general, was more highly correlated with insect damage than was leaf blade composition, as might be expected from the feeding behavior of second generation larvae. The relationships between insect damage and plant composition of the leaf sheath were typical of those also seen for the leaf blade, and the following discussion will deal only with leaf sheath composition.

Second generation insect damage was consistently higher at Hancock (Table 3). In general, values for NDF, hemicellulose, and lignin were higher, and values for ADF, cellulose, and silica were lower at Hancock.

The contrasting situations at Madison and Hancock are graphically illustrated in Figures 1 and 2. At Madison, plant silica alone accounted for 60 percent of the variation seen in the number of insect tunnels. Total structural carbohydrate content as measured by NDF was of secondary importance, but nonetheless accounted for an additional 11 percent, for a total of 71 percent of ECB variation (Table 4). Lignin was not correlated with insect damage. At Hancock, NDF was the most important component, and silica was not significantly correlated to insect damage. Together, NDF and lignin accounted for 71 percent of the variation in second generation damage. Silica, NDF, and lignin had negative regression coefficients at those locations where they were significant (Table 5).

Table 3. Second generation ECB damage, cell wall composition, and silica content of leaf sheath tissue at Madison and Hancock

Entry	Location	No. tunnels	NDF (%)	ADF (%)	Cellulose (%)	Hemi-cellulose (%)	Lignin (%)	Silica (%)
Oh43	Madison	5.7	65.7	37.0	33.9	28.6	3.1	1.9
	Hancock	6.5	66.3	34.0	30.6	32.3	3.4	1.0
W64A	Madison	4.0	65.0	35.6	32.5	29.3	3.1	1.4
	Hancock	5.3	66.6	34.2	30.4	32.4	3.8	0.6
Oh51A	Madison	5.1	62.0	35.1	31.8	26.9	3.3	1.7
	Hancock	6.5	63.2	33.3	29.3	30.0	4.0	0.8
W182BN	Madison	11.7	62.5	35.1	31.5	27.3	3.7	0.7
	Hancock	9.1	62.3	34.2	29.8	28.2	4.3	0.6
W540	Madison	4.5	65.8	35.4	32.1	30.4	3.2	1.6
	Hancock	6.5	65.6	33.4	30.4	32.2	3.1	0.6
W845	Madison	3.5	61.6	34.2	31.4	27.4	2.8	2.2
	Hancock	6.0	64.4	31.8	28.8	32.6	3.1	0.8
A619	Madison	5.9	62.8	35.2	31.9	27.6	3.2	2.1
	Hancock	7.0	66.1	34.2	31.4	31.9	2.8	0.4
W548	Madison	4.2	66.1	36.8	33.8	29.4	2.9	1.4
	Hancock	5.6	70.4	37.4	34.7	33.1	2.6	0.2
W552C2	Madison	3.2	61.6	35.6	31.5	26.0	4.2	1.9
	Hancock	5.6	64.1	35.0	30.4	29.1	4.6	1.0
W590	Madison	5.6	68.2	38.1	35.0	30.2	3.1	1.5
	Hancock	9.1	69.0	35.9	32.5	33.1	3.4	0.7
W117	Madison	8.6	58.7	32.2	29.5	26.6	2.7	1.5
	Hancock	12.0	58.6	30.7	27.5	27.9	3.2	0.8
B86	Madison	1.4	70.2	39.4	34.5	30.8	4.8	2.5
	Hancock	1.9	72.0	39.3	33.5	32.7	5.8	0.8
W61BR3	Madison	6.1	65.5	38.2	34.0	27.3	4.2	1.3
	Hancock	7.3	66.0	35.1	31.4	30.9	3.7	0.6
W22G	Madison	7.0	61.7	35.6	32.4	26.1	3.2	1.7
	Hancock	8.8	61.7	34.0	29.6	27.7	4.4	0.7
ECBR	Madison	2.9	62.9	39.0	35.4	23.8	3.6	2.4
	Hancock	2.2	70.9	39.3	34.5	31.6	4.8	1.1
Mean		6.0	64.9	35.5	31.8	29.4	3.6	1.2
LSD (0.05)		2.5	2.6	1.6	1.6	1.9	0.4	0.5

The greater amount of structural carbohydrates and lignin seen in plant tissue at Hancock possibly prevented further ECB damage which might have resulted from low plant and soil silica. This was shown most clearly by the genotypes with the greatest resistance at both locations, B86 and ECBR. Neither genotype had a silica content significantly higher than the majority of other genotypes at Hancock, but both had significantly higher values of NDF than all but two other entries. The lignin content of the most resistant genotype, B86, was significantly higher than any other genotype. On the other extreme, the most susceptible lines at Hancock, W117 and W182BN, had low NDF and/or low lignin content.

In comparing plant responses at the two locations, the resistant to

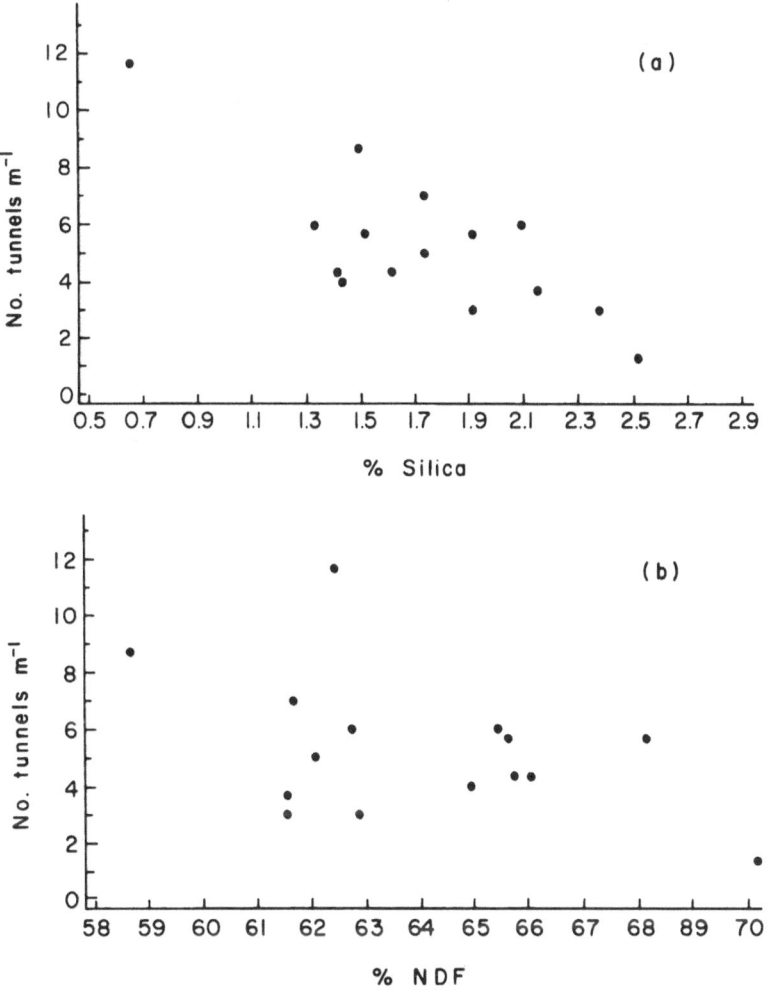

Fig. 1. Genotype means at Madison for number of second generation ECB tunnels versus a) leaf sheath silica content (r = −0.77**), and b) leaf sheath neutral detergent fiber (r = −0.43).

moderately resistant genotypes showed greater increase in NDF and lignin at Hancock than susceptible lines. ECBR was particularly notable in this regard. W117 and W182BN appeared to lack this capacity. It was evident that silica, NDF and lignin are complimentary in either environment. For any particular genotype, one component may have been near the upper extreme (e.g. NDF for W548 at Hancock), but the lack of ECB resistance could be attributed to a deficiency in some other constituent (e.g. lignin and possibly silica for W548 at Hancock). The most resistant genotype at both locations, B86, had elevated values for all three components; silica, NDF, and lignin.

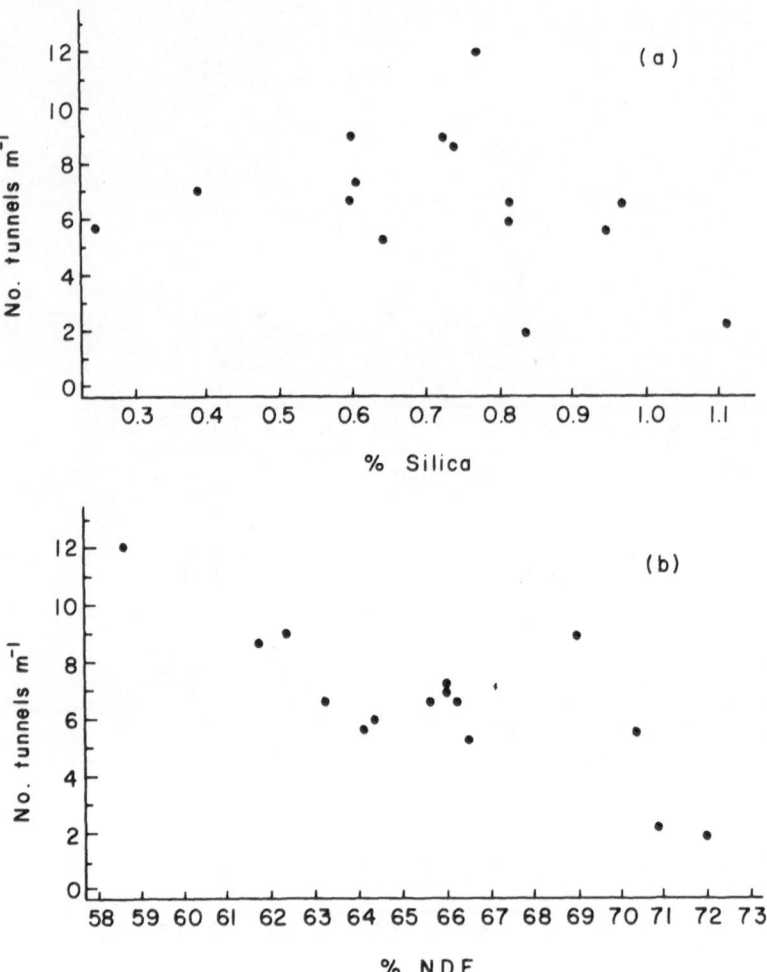

Fig. 2. Genotype means at Hancock for number of second generation ECB tunnels versus a) leaf sheath silica content (r = −0.28), and b) leaf sheath neutral detergent fiber (r = −0.78**).

Discussion

The results are in substantial agreement with those of Rojanaridpiche et al.[27], who were the first investigators to suggest a role for silica and lignin in ECB resistance. Although DIMBOA was not measured in the current study, two of the lines most resistant to first generation ECB, A619, and Oh43, were shown previously to have high DIMBOA levels, and the least resistant line, W64A, had low DIMBOA levels[27]. It is expected that DIMBOA would account for much of the unexplained variation among genotypes for first, but not second generation damage. Data regarding second generation ECB damage in the present study are

Table 4. Multiple regression analysis for second generation ECB damage versus silica, lignin, and NDF content of leaf sheath tissue at Madison and Hancock

Source	df	M.S.	F	PROB > F	R^2
Madison					
Model	2	31.12	14.50	0.00	
Silica	1	52.69	24.54	0.00	0.60
NDF/Silica	1	9.55	4.45	0.06	0.71
Error	12	2.15			
Hancock					
Model	2	33.11	14.87	0.00	
NDF	1	56.47	25.36	0.00	0.61
Lignin/NDF	1	9.76	4.38	0.06	0.71
Error	12	2.23			

more conclusive than those previously reported in that plant silica content has been shown to vary with soil silica content, and that, in the absence of silica, both total structural carbohydrate content, as measured by NDF, and lignin exert strong influence on second generation ECB resistance.

Maturity effects are as yet undetermined, and until appropriate data are available, the true extent of genotypic variation for cell wall composition and silica content is not completely known. In many grass species, NDF, lignin, and silica steadily increase throughout the growing season[11,31]. In this regard, it is noted that the two most resistant genotypes with the highest levels of these constituents, B86 and ECBR, were also the latest maturing genotypes under test. These genotypes were derived from germplasm selected for second generation ECB resistance. For the maize breeder, it remains to be seen whether it is possible to accelerate silica uptake, as well as cell wall formation and lignification in earlier maturing genotypes.

In certain grass species which are subject to heavy grazing, increased silica content can be an induced response to herbivory[21]. Grass leaf

Table 5. Regression coefficients at Madison and Hancock relating plant composition of the whorl and leaf sheath tissue to first and second generation ECB damage.

Location	b ± standard error			
	Intercept	NDF	Lignin	Silica
First generation damage — whorl:				
Madison	8.1	–	–1.4 ± 0.7	–3.4 ± 1.4
Hancock	–	–	–	–
Second generation damage — leaf sheath:				
Madison	30.0	–0.3 ± 0.1	–	–3.9 ± 0.8
Hancock	43.4	–0.5 ± 0.1	–1.0 ± 0.5	–

silicification can reach such extreme levels that grazing ungulates experience marked abrasion of tooth enamel[2], silica urolithiasis[36], and esophogeal cancer[28]. No conclusions are yet possible in maize with respect to induction as a response to insect pressure, but it is likely that resistance attributable to either cell wall composition or plant silica content is not related to direct exposure to predators, but rather a defense mechanism resulting from coevolutionary adaptation to insect pests. Silica, in particular, would appear to be a very effective mechanism for reducing insect damage. Should its uptake and translocation in the corn plant be primarily passive, comparatively little energy need be devoted to uptake and deposition in target tissues. If the mechanism providing resistance is due to mandible abrasion and/or destruction of gut tissue, silica-based resistance may extend to a broad array of boring and chewing insects. Gross morphological changes in insects are probably necessary to overcome such a feeding deterrent, and resistance may have a more stable adaptive value than one conferred by toxic metabolites. The formation of complex carbohydrates may well require more energy and lead to a depletion of carbon reserves otherwise dedicated to agronomically desirable attributes such as seed yield. Nonetheless, NDF and lignin are effective digestibility reducing components which may be useful for pest resistance in what Feeny[5] would call 'apparent' species such as cultivated maize. The combination of feeding deterrents, digestibility reducing substances, and toxic metabolites should provide a highly effective barrier to insect predation.

Acknowledgements The author is grateful to Anita Brown for providing assistance with insect infestation, and to Teri White and Tanis Cuff for data collection and laboratory analyses. Special appreciation is given to Mario Mardones for all aspects of field preparation, planting, and cultural management. Financial support was from the University of Wisconsin-Madison Graduate School, Project Number 150396.

References

1　Agarwal R A 1969 Morphological characteristics of sugarcane and insect resistance. Ent. Exp. Appl. 12, 767–776.
2　Baker G, Jones L H P and Wardrop I D 1959 Cause of wear in sheep's teeth. Nature (London) 164, 1583 1584.
3　Blackman E and Parry D W 1968 Opaline silica deposition in rye (*Secale cereale* L.). Ann. Bot. (London) 32, 199 206.
4　Djamin A and Pathak M D 1967 Role of silica in resistance to Asiatic rice borer, *Chilo suppressalis* (Walker), in rice varieties J. Econ. Entomol. 60, 347 351.
5　Feeny P P 1976 Plant apparency and chemical defense. Rec. Adv. Phytochem. 10, 1–40.
6　Geis J E 1978 Biogenic opal in three species of gramineae. Ann. Bot. (London) 42, 1119 1129.
7　Germar B 1934 Uber einige Wirkungen der Kieselsaure in Getreidepflanzen,insbesondere auf dèren Resistenz gegenuber mehltau. Z. Pflanzenernaehr. Dung. Bodenkd. 35, 102 155.
8　Grosse-Brauckmann E 1957 Uber den Einfluss der Kieselsaure auf den Mehltaubefall von Getri ede bei unterschiedlicher Stickstoffdungung. Phytopathol. Z. 30, 112 116.

9. Guthrie W D, Dickie F F and Neiswander C R 1960 Leaf and sheath feeding resistance to the European corn borer in eight inbred lines of dent corn. Ohio Agric. Exp. Sta. Res. Bull. 860.
10. Guthrie W D, Huggans J L and Chatterji S M 1970 Sheath and collar feeding resistance to the second-brood European corn borer in six inbred lines of dent corn. Iowa State J. Sci. 44, 297–311.
11. Handreck K A and Jones L H P 1968 Studies of silica in the oat plant IV. Silica content of plant parts in relation to stage of growth, supply of silica, and transpiration. Plant and Soil 29, 449–450.
12. Hanifa A M Subramaniam T R and Ponnaiya B W X 1974 Role of silica in resistance to the leaf roller, *Cnaphalocrocis medinalis* Guenee, in rice. Indian J. Exp. Biol. 12, 463–465.
13. Izawa G and Kume I 1959 Influence of silica on the growth of rice plant in relation to the nitrogen levels. Hyogo Noka Daigaku Kenkyu Hokoku Nogu Kagaku 4, 13–17.
14. Jones L H P and Handreck K A 1965 Studies of silica in the oat plant III. Uptake of silica from soils by the plant. Plant and Soil 23, 79–96.
15. Jones L H P and Handreck K A 1967 Silica in soils, plants, and animals. Adv. Agron. 19, 107–149.
16. Klun J A, Tipton C L and Brindley T A 1967 2,4-Dihydroxy-7-methoxy-1, 4-benzoxazin-3-one (DIMBOA), an active agent in the resistance of maize to the European corn borer. J. Econ. Entomol. 60, 1529–1533.
17. Lanning F C and Kinko Y 1961 Absorption and deposition of silica by four varieties of sorghum. J. Agric. Food. Chem. 9, 463–465.
18. Lanning F C, Hopkins T L, and Loera J C 1980 Silica and ash content and depositional patterns in tissues of mature *Zea mays* L. plants. Ann. Bot. (London) 45, 549–554.
19. Lewis J and Reiman B E F 1969 Silicon and plant growth. Annu. Rev. Plant. Physiol. 20, 289–304.
20. McCollough J W 1923 The resistance of wheat to the hessian fly — A progress report. J. Econ. Entomol. 16, 293–298.
21. McNaughton S J, Tarrants J L, McNaughton M M and Davis R H 1985 Silica as a defense against herbivory and a growth promotor in African grasses. Ecology 66, 528–535.
22. Miller B S, Robinson R J, Johnson J A, Jones E T and Ponnaiya B W X 1960 Studies on the relation between silica in wheat plants and resistance to hessian fly attack. J. Econ. Entomol. 53, 995–999.
23. Moore D 1984 The role of silica in protecting Italian ryegrass (*Lolium multiflorum*) from attack by dipterous stem-boring larvae (*Oscinella frit* and other related species). Ann. Appl. Biol. 104, 161–166.
24. Panda N, Pradham B, Samalo A P and Prakasa Rao P S 1975 Note on the relationship of some biochemical factors with resistance in rice varieties to yellow rice-borer. Indian J. Agric. Sci. 45, 499–501.
25. Refai F Y, Jones E T and Miller B S 1955 Some biochemical factors involved in the resistance of the wheat plant to attack by hessian fly. Cereal Chem. 32, 437–454.
26. Robertson J B and Van Soest P J 1980 Detergent system of analysis and its application to human foods. p. 123–158. *In* The Analysis of Dietary Fiber in Food. Eds. W F James and O Theander. Marcel Dekker, Inc., New York.
27. Rojanaaridpiched C, Gracen V E, Everett H L, Coors J G, Pugh B F and Bouthyette P 1984 Multiple factor resistance in maize to European corn borer. Maydica 24, 305–315.
28. Sangster A G, Hodson M J and Parry D W 1983 Silicon deposition and anatomical studies in the inflorescence bracts of four *Phalaris* species with their possible relevance to carcinogenesis. New Phytol. 93, 105–122.
29. Sasamoto K 1958 Studies on the relation between the silica content in the rice plant and insect pests VII: Effects of water glass and slag on the damage of rice plant caused by the rice stem borer. Jpn. J. Appl. Entom. and Zool. 3, 153–156.
30. Scriber J M, Tingey W M, Gracen V E and Sullivan S E 1975 Leaf feeding resistance to the European corn borer in genotypes of tropical (low-DIMBOA) and U.S. inbred (high-DIMBOA) maize. J. Econ. Entomol. 68, 823–826.
31. Soest P J van 1982 Nutritional Ecology of the Ruminant. O & B Books, Inc. Corvalis. 374 p.

32   Volk R J, Kahn R P and Weintraub R L 1958 Silicon content of the rice plant as a factor influencing its resistance to infection by the blast fungus, *Piricularia oryzae*. Phytopathology 48, 179–184.
33   Wadham M D and Parry D W 1981 The silicon content of *Oryza sativa* L. and its effect on the grazing behavior of *Agriolimax reticulatus* Muller. Ann. Bot. (London) 48, 399–402.
34   Wagner F 1940 Die Bedeutung der Kieselsaure fur das Wachstum einiger Kulturpflanzen, ihren Nahrstoffhaushalt und ihre Anfalligkeit gegen echte Mehltaupilze. Phytopathol. Z. 12, 427–479.
35   Wellso S G 1973 Cereal leaf beetle: Larval feeding orientation, development, and survival on four small grain cultivars in the laboratory. Ann. Entomol. Soc. Am. 66, 1201–1208.
36   Whiting F, Connell F R and Forman S A 1958 Silica urolithiasis in beef cattle. Can. J. Comp. Med. Vet. Sci. 22, 332–337.

Differential response of sunflower genotypes to iron deficiency

E. ALCANTARA and M.D. DE LA GUARDIA
Cátedra de Fisiología Vegetal, Escuela Técnica Superior de Ingenieros Agrónomos, Universidad de Córdoba, Córdoba, Spain

Key words Iron efficiency   Plant breeding   Sunflower

Summary   In recent years chlorotic plants have appeared in the hybrid sunflower (*Helianthus annuus* L.) crop in several Spanish growing areas. This prompted us to study the physiology and genetic variability of some of the hybrids and inbred parental lines in relation to their efficiency in iron nutrition.

Seedlings were grown in glass flasks hydroponically or in pots with Perlite and irrigated with nutrient solution. Iron was supplied as $FeCl_3$ or FeEDDHA at low concentration (2×10^{-6} or $10^{-5} M$). Plants were grown in a growth chamber or in a greenhouse and between 3 and 5 weeks they were evaluated by visual assessment of chlorosis. Although the results were variable between treatments they showed the existence of differential response to Fe deficiencies among inbred lines and among individuals of some inbred lines. Response of individual plants grown under iron stress showed differences among inbred lines in the decrease in pH of the external solution and in the capacity for reducing iron.

Introduction

Sunflower was developed as a crop for edible oil in Russia where a large number of local cultivars existed. In the 1920's Pustovoit developed a highly successful method for improving sunflower cultivars based on recurrent selection, progeny evaluation and subsequent cross pollination among progenies with superior characteristics[1]. The cultivars obtained were populations in the genetic sense, with wide genetic variability.

Until the mid-seventies the cultivars were used in most of the world sunflower area. Since then hybrid cultivars, obtained by crossing two parental lines (one cytoplasmic male sterile and other a fertility restorer) have been displacing the old cultivars in most countries. Parental lines are inbred lines obtained through several generations of self-pollination. When hybrids or parental lines are grown in other areas they can show different responses for stresses to which they have not been selected, such as iron deficiency in calcareous soils. Chlorotic plants appearing in Spain recover after application of FeEDDHA either to the leaves or the soil.

In other species breeding lines or mutants exist with different response to Fe-stress. In soybean T203 and in tomato T3238 fer a recessive gene controls the iron deficiency response[8,9]. In sorghum[10], oats[5] and barley[2] differential response has been reported among genotypes although the inheritance has not been stablished.

Sunflower has usually been classified as a Fe-efficient species and rapid responses to Fe-stress include increased reducing capacity of the roots,

H.W. Gabelman and B.C. Loughman (Eds.), Genetic aspects of plant mineral nutrition.
ISBN-13: 978-94-010-8102-3
© *1987 Martinus Nijhoff Publishers, Dordrecht/Boston/Lancaster.*

proton extrusion causing a fall in the pH of the medium, differentiation of peripheral cells into transfer cells and development of abundant root hairs in a swollen zone near the root apex[4,6].

However, differences have been reported in susceptibility to Fe-chlorosis between sunflower cultivars Hesa and HebH in West Germany[7] and in the response to Fe-stress among cultivars PKV-7237, Morden and EC-68414 grown in India[3].

Several inbred parental lines and one hybrid have been studied with several Fe-stress conditions and the first results are presented in this paper.

Materials and methods

Several American parental lines (females: 89, 290, 300, 303; males: 271, 273, 274, 276, 297, 298, 299), a Spanish parental line (female 478) and the American hybrid 89 × 273 were studied in different assays. Seeds treated with sodium hypochlorite (0.5%, v/v) were germinated for 2 or 5 days at 25 C in Perlite irrigated with 5 mM $CaCl_2$.

In one assay seeds germinated for 2 days were transferred to 500 ml plastic pots (3 plants/pot) filled with Perlite and were irrigated with nutrient solution. In the other assays, 5 days old seedlings were transferred to 650 ml flasks wrapped in aluminium foil (80 ml in the assay for determination of reducing capacity).

The standard nutrient solution used for all the assays had the following composition (after Römheld and Marschner[6]): 2 mM $Ca(NO_3)_2$, 0.75 mM K_2SO_4, 0.5 mM KH_2PO_4, 0.65 mM $MgSO_4$, 10 μM H_3BO_3, 1 μM $MnSO_4$, 0.05 μM $(NH_4)_6Mo_7O_{24}$, 0.5 μM $CuSO_4$ and 0.5 μM $ZnSO_4$

In the assay for the determination of reducing capacity 2 × 10^{-5} M FeEDDHA or no iron was added and the pH was adjusted to 6.0 with NaOH. In the other assays $FeCl_3$ or FeEDDHA was added at the concentrations indicated and the pH was adjusted to 5.5. Nutrient solutions were aerated daily and changed every 3 or 5 days.

Plants used for determinations of pH and reducing capacity were grown in a growth chamber at 22 C day/18 C night, with a 13 h photoperiod and an irradiance of 40 Wm^{-2} (fluorescent tubes Sylvania: 1/2 Cool white and 1/2 Grolux). In the other assays plants were grown in a greenhouse at 23 ± 5 C.

The reducing capacity of the roots was determined in the dark with intact plants using (3 × 10^{-4} M) Ferrozine (3-(2-pyridyl)-5,6-bis(4-phenylsulfonic acid)-1,2,4,-triazine sodium salt) or BPDS (4,7-di(4-phenylsulfonate)-1,10 phenanthroline, bathophenanthroline disulphonic acid) reagents and Fe-III-EDTA (10^{-4} M) as the source of Fe-III, and measuring the absorbance at 562 or 535 nm respectively after 7.5 h.

Results

Two experiments were carried out with different genotypes growing with Fe as $FeCl_3$ to determine whether they had the capacity to reduce and absorb Fe-III. The results showed variability of response among lines and among individual plants (Table 1). Line 89 was chlorosis susceptible since most of plants showed severe chlorosis in both experiments. Line 274 was Fe-efficient although it appeared heterogeneous in the Fe-stress response in both experiments. Lines 290 and 271 appeared chlorosis resistant, and line 299 was intermediate. Hybrid 89 × 271 was Fe-efficient although with heterogeneous response.

Table 1. Frequency distribution of chlorosis score of plants (0 = no chlorosis; 5 = severe chlorosis plus necrosis) of parental lines or hybrids in two separate experiments. Plants were grown for 35 days in plastic pots with Perlite and irrigated with a standard nutrient solution with Fe as $FeCl_3$ ($4 \times 10^{-6} M$, Exp. 1 or $2 \times 10^{-6} M$, Exp. 2)

Parental lines or hybrid	Number of plants with score						Mean score
	0	1	2	3	4	5	
Experiment 1							
89						5	5.0
274	5	1	1	1	1		1.1
290		9					1.0
Experiment 2							
89				5	7	2	3.8
274	8	2	2	3			1.0
271	13	2					0.1
299	2	1	1	11			2.4
89 × 271	9	2	1	2	1		0.9

When plants were grown in culture solution containing FeEDDHA for 15 days and thereafter in solution containing $FeCl_3$ chlorosis appeared quickly in some lines. In some of them the degree of chlorosis increased with time in $FeCl_3$ (lines 300 and 89) and in other lines regreening was observed (lines 298 and 271) after 4 or 6 days of chlorosis. Figure 1 shows the change with time of the chlorosis of four lines and Table 2 shows the results of all lines at day 10. Lines 89, 271 and 274 gave

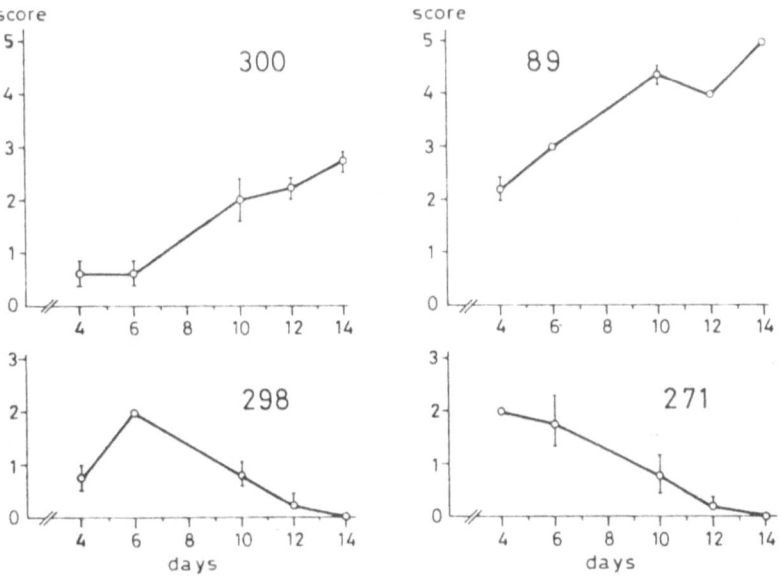

Fig. 1. Time course of chlorosis score of parental lines 300, 89, 298 and 271. At day 0, 14 days old plants grown in standard nutrient solution with $2 \times 10^{-5} M$ FeEDDHA, were transferred to a standard nutrient solution with $10^{-5} M$ $FeCl_3$. Each point represents the mean of 4 replicates $\pm$ SE.

Table 2. Frequency distribution of chlorosis score of plants (0 = no chlorosis; 5 = severe chlorosis plus necrosis) of parental lines. Plants were grown individually in glass flasks for 14 days with standard nutrient solution plus FeEDDHA ($2 \times 10^{-5} M$). On the 15th day the solution was changed to the standard plus $FeCl_3$ ($10^{-5} M$) and 10 days later the chlorosis score was assessed

Parental line	Number of plants with score						Mean score
	0	1	2	3	4	5	
89					3	2	4.4
300		1	2	1			2.0
303	3	1			1		1.0
478	1	4	2	1			1.4
271	2	2	1				0.8
273	5	2	1				0.5
274	2	2	1				0.8
276	2	2		2	2		2.0
297		4	1				1.2
298	1	3					0.8
299				2	2	1	3.8

chlorosis scores similar to the ones obtained in earlier experiments presented in Table 1. Line 89 was susceptible and lines 271 and 274 were resistant, although some of the plants were less Fe-efficient. Again line 299 appeared intermediate and line 276 had a high proportion of susceptible plants. The other lines were resistant although some of them were not uniform in their response.

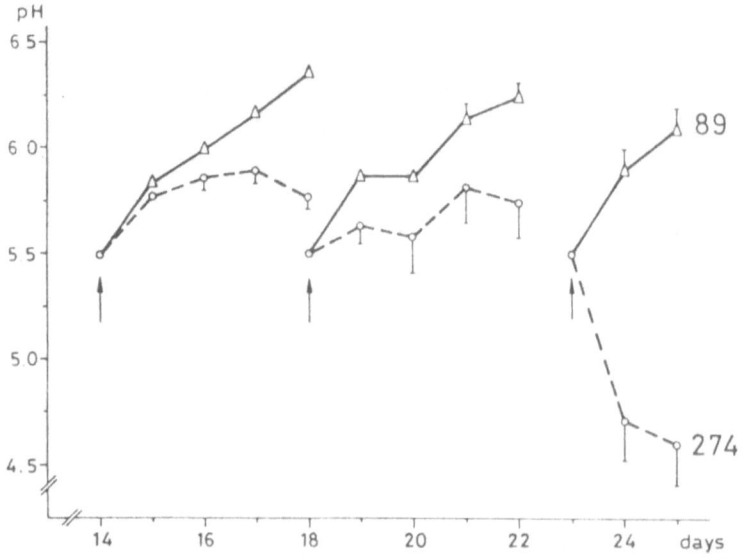

Fig. 2. Effect of Fe deficiency on the pH of the nutrient solution. Until day 14 plants were in standard nutrient solution plus $2 \times 10^{-5} M$ FeEDDHA. Arrows indicate days (14, 18 and 23) at which nutrient solution was replaced. Between day 14 to 23 plants were deprived of iron and on day 23 $10^{-6} M$ FeEDDHA was added. Each point represents the mean of 4 replicates ± SE.

Table 3. Effect of iron pretreatment on the reducing capacity of roots of two parental lines measured at 7.5 h, using Ferrozine or BPDS as reagents

Pretreatment	Reducing capacity (μmole Fe^{II}/g FW roots)	
	Ferrozine	BPDS
89 + Fe	0.22	0.18
− Fe	0.14	0.15
478 + Fe	0.25	0.34
− Fe	2.39	2.95

The change of pH with time in the nutrient solution was followed in two lines with opposite susceptibility to chlorosis (Fig. 2). Roots of line 274 (Fe-efficient) decreased the pH, releasing H-ions to the medium, while roots of line 89 (Fe-inefficient) did not have this capacity.

Line 478 (resistant) and line 89 (susceptible) were grown with or without iron and tested for their ferric reducing capacity in two different assays (Table 3). When plants were supplied with adequate iron, the roots showed little reducing capacity, but when they were grown without iron line 478 showed a ten fold increase.

Discussion

Sunflower has been classified as a chlorosis resistant species but our results show that differences exist in the response to Fe-stress among breeding lines and among plants of the same inbred line.

There is variability in the response of one particular line to Fe-stress in the different tests or between different replicates of the same test, (especially in the chlorosis score test) probably due to the influence of environmental conditions (light, temperature). However line 89 behaved like a susceptible line in all the tests: 1) a high chlorosis score 2) no decrease in the pH of the nutrient solution 3) very low reducing capacity in the roots. This is an important point because line 89 is a female line widely used to produce commercial hybrids. The other lines studied did not show such clear effects. Lines 478 and 274 appear to be chlorosis resistant, although they show some heterogeneity of response in the chlorosis test. The response of lines 271 and 298 in the chlorosis test (Fig. 1), which recover from chlorosis 4 to 6 days after resupplying iron is in agreement with the results of Kannan[3] with three sunflower cultivars.

Expansion in the cultivated area of sunflower hybrids on a world wide scale has produced a great interchange of parental lines and hybrids between countries. Parental lines with a narrow genetic base, selected in a particular environment, are crossed and their hybrids are cultivated in very diverse areas in which nutrient stress such as Fe-stress can appear.

Variability in the response to Fe-stress in sunflower has been demonstrated and although more studies are needed to determine the best method for selecting chlorosis resistant genotypes and to relate the response with the response in the field, it now seems feasible to select for several generations and to obtain parental lines and hybrids with uniform resistance to chlorosis.

Acknowledgements The work was supported by Comisión Asesora de Investigación Científica y Técnica, project 2011-83.

References

1. Fick G N 1978 Breeding and Genetics. *In* Sunflower Science Technology. Agronomy 19. Ed. J F Carter. Am. Soc. of Agronomy. Madison Wisconsin, pp 279-338.
2. Fleming A L and Foy C D 1982 Differential response of barley varieties to Fe stress. J. Plant Nutr. 5, 457–468.
3. Kannan S 1984 Studies on Fe deficiency stress response in three cultivars of sunflower (*Helianthus annuus* L.). J. Plant Nutr. 7, 1203–1212.
4. Kramer D, Römheld V, Landsberg E and Marschner H 1980 Induction of transfer-cell formation by iron deficiency in the root epidermis of *Helianthus annuus* L. Planta 147, 335-339.
5. McDaniel M E and Brown J C 1982 Differential iron chlorosis of oat cultivars, a review. J. Plant Nutr. 5, 545–552.
6. Römheld V and Marschner H 1979 Fine regulation of iron uptake by the Fe-efficient plant *Helianthus annuus*. *In* The Soil-Root Interface. Eds J L Harley and R. Scott Russel. Academic Press, London pp 405-417.
7. Scherer H W and Höfner W 1980 Einfluß von Fe- und Mn-Mangel auf den Kationen- und Anionengehalt von Mais und Sonnen-blumen. Z. Pflanzenernaehr. Bodenkd. 143, 26–37.
8. Wann E V and Hills W A 1973 The genetics of boron and iron transport in the tomato. J. Heredity 64, 370–371.
9. Weiss M G 1943 Inheritance and physiology of efficiency in iron utilization in soybeans. Genetics 28, 253–268.
10. Williams E P, Clark R B, Yusuf Y, Ross W M and Maranville J W 1982 Variability of sorghum genotypes to tolerate iron deficiency. J. Plant Nutr. 5, 553-567.

Soybean genetic differences in response to Fe and Mn: Activity of metalloenzymes

E.O. LEIDI, M. GOMEZ and M.D. DE LA GUARDIA*
*U.E.I. Fisiología Vegetal, Estación Experimental del Zaidín (CSIC), Granada, Spain *Cátedra de Fisiología Vegetal, E.T.S. Ingenieros Agrónomos, Universidad de Córdoba, Spain*

Key words Catalase Genetic differences Iron Manganese Peroxidases Soybean Superoxide dismutases

Summary Metalloenzymes as biochemical markers of nutritional status have been studied for 30 years since Brown and Hendricks[5] proposed the use of enzyme activities as indicators for the diagnosis of micronutrient deficiencies. Enzyme activities can be used in other lines of research, such as the characterization of particular reactions of genotypes to nutritive imbalances.

Catalase, peroxidase and superoxide dismutase activities were assayed in soybean cultivars growing in nutrient solutions with different Fe/Mn ratios. Considerable differences were observed between cultivars. Relationships between growth and enzyme activity were also studied.

Although more information is required, our results provide an introduction to the knowledge of the mechanisms developed by genotypes in response to Fe-Mn imbalances.

Introduction

Information about the specific reactions of plant genotypes to nutritional stress is the first approach to improve the efficiency of plant production under limiting conditions[21].

Iron and manganese nutritional disorders are rather common in acid or alkaline soils. Early reports have indicated the existence of a certain Fe-Mn balance for optimum growth[22,23] but the real importance of the Fe-Mn relationship remains unclear[18], and the situation is complicated by the different responses of genotypes to Fe and Mn nutrition. Several authors have pointed out differences in terms of efficiency in iron absorption[6,24] or in Mn use[20], as well as in tolerance to excess of Fe and Mn nutrient supplies[3,7,8,17,20].

The use of metalloenzymes as biochemical markers of plant nutrient status has the advantage of the high sensitivity and specificity of enzymatic analysis[2,9]. The activities of the metalloenzymes catalase, peroxidase and superoxide dismutase were found to be related to Fe and Mn supply in soybean[19] and in other species[9], and thus could be used as indicators for the appraisal of Fe and Mn nutrient status.

In the present work, we analyze some metalloenzyme activities from soybean cultivars showing diverse responses to Fe and Mn nutrition in an attempt to characterize the different behaviours of the genotypes on a biochemical basis.

H.W. Gabelman and B.C. Loughman (Eds.), Genetic aspects of plant mineral nutrition.
ISBN-13: 978-94-010-8102-3
© 1987 Martinus Nijhoff Publishers, Dordrecht/Boston/Lancaster.

Materials and methods

Plant material

Soybean (*Glycine max* (L.) Merr.) cultivars with different reponse to Fe and Mn (Bragg, Lee, Forrest, T-203) were grown in aerated nutrient solutions with the following composition in meq/l: NO_3^-, 10; $PO_4H_2^-$, 0.5; SO_4^{2-}, 4; K^+, 5; Ca^{2+}, 9; Mg^{2+}, 3; and in ppm: B, 0.5; Cu, 0.05; Zn, 0.05; Mo, 0.03. The initial pH was 5.5. Nine treatments were used varying Fe (Fe^0: 0.25; Fe_1: 1.0; Fe_2: 2.5 ppm) (as Fe-EDDHA) and Mn (Mn_0: 0.0; Mn_1: 0.1; Mn_2: 5.0 ppm). Plants were grown in a growth chamber, CONVIRON mod.PGW-36, under 14 h light and 10 h dark periods, at 30°C and 20°C, respectively. Photosynthetic active radiation was approximately $390\,\mu E.\,s^{-1}.\,m^{-2}$.

Enzyme assays

The first trifoliate leaf, at the second trifoliate stage (14 days old), was sampled for analysis. Catalase was determined polarographically by the oxygen electrode method[10], peroxidase was estimated using o-dianisidine as H-donor[11] and superoxide dismutase by a photochemical method[16]. Mn-superoxide dismutase isozyme (Mn-SOD) was differentiated from total superoxide dismutase (Total-SOD) by performing the activity assay in the presence and in the absence of 1 mM NaCN.

Other determinations

Dry weight of the plant material was determined after 48 h in a forced draft oven at 70°C. Tissue Fe and Mn were determined by atomic absorption spectrophotometry.

Experimental design

A factorial design (3 Mn × 3 Fe levels) with three replications was used.

Results and discussion

Symptoms of Fe deficiency appeared on leaves of cultivar T-203 grown in solutions with low (0.25 ppm) or medium (1.0 ppm) Fe concentrations. In Bragg, Lee and Forrest, symptoms resembling Fe chlorosis and Mn toxicity (crinckle leaf) were observed on leaves of plants grown in solutions of 5.0 ppm Mn with 0.25 ppm Fe. Dry matter values of the first trifoliate leaves are shown in Table 1.

The catalase activity of the different soybean cultivars is shown in Fig. 1. Increases in Fe supply produced different effects with varying Mn concentrations. At low Mn supply, there was a general trend to an increase in catalase activity when Fe concentration was raised. However, at medium or high Mn concentrations, the response of catalase activity to increasing Fe supplies depended on the cultivars, remaining positive only in Forest and T-203. At each Fe concentration, the increase in Mn supply determined a different behaviour for catalase activity in the cultivars. Only at the highest Fe level, the increase in Mn caused a general tendency (except in Forrest) to depress catalase activity. Previously, Agarwala *et al.*[1] had observed higher catalase activity at medium Mn supply than at lower or higher concentrations in the entire range of Fe supply. Other authors have shown a relation between catalase activity and Fe levels suggesting the possibility of using that

Table 1. Dry weights of first trifoliate leaves of soybean cultivars grown with different Fe and Mn levels. Values followed by the same letter within columns are not significantly different at the 5% level of probability according to Duncan's Multiple Range Test

Nutrient level (ppm)		Leaf growth (mg. leaf^{-1})			
Mn	Fe	Bragg	Lee	Forrest	T-203
0.0	0.25	95.5 de	78.4 b	59.0 b	118.5 d
0.0	1.0	112.4 bcd	75.9 b	91.7 a	115.6 d
0.0	2.5	117.9 bc	86.1 b	102.0 a	135.2 cd
0.1	0.25	141.8 a	111.6 a	109.9 a	109.1 d
0.1	1.0	122.2 abc	118.9 a	109.7 a	173.2 ab
0.1	2.5	130.5 ab	116.8 a	111.5 a	184.1 a
5.0	0.25	45.8 f	67.4 b	36.9 b	122.5 d
5.0	1.0	106.9 cd	81.9 b	99.5 a	116.4 d
5.0	2.5	86.4 e	79.8 b	102.9 a	150.2 bc
Significance of treatments differences ($P < \ldots$)					
Mn		0.001	0.001	0.001	0.001
Fe		0.001	n.s.	0.001	0.001
Mn × Fe		0.001	n.s.	0.001	0.001

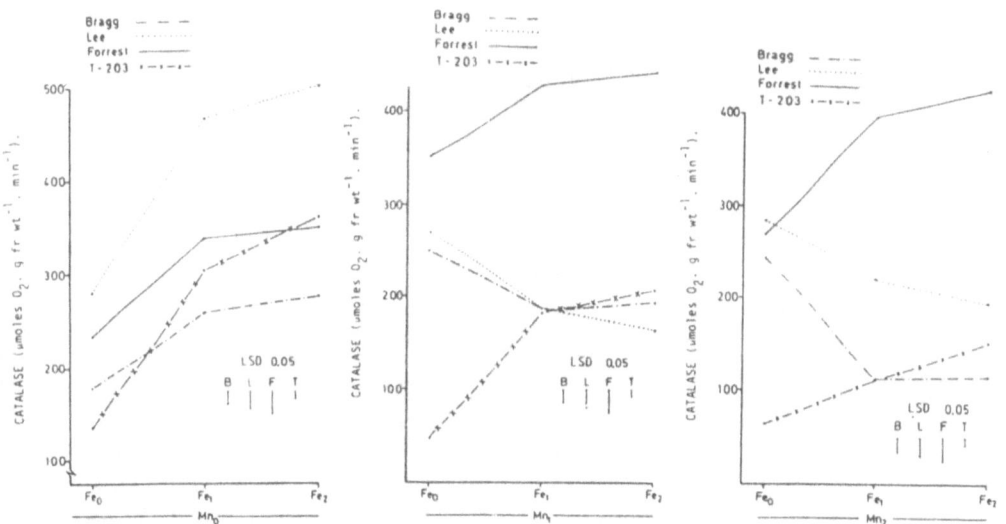

Fig. 1. Leaf catalase activity of soybean cultivars grown under different Fe and Mn levels.

enzyme as a functional indicator of the Fe involvement in plant metabolism processes[9,11]. We could not observe a relationship between the leaf Fe concentration (Table 2) and the catalase activity whereas Mn concentration (Table 3) was correlated to the enzyme in Bragg (correlation

Table 2. Iron concentration in first trifoliate leaves of soybean cultivars grown with different Fe and Mn levels. Values followed by the same letter within columns are not significantly different at the 5% level of probability according to Duncan's Multiple Range Test

Nutrient level (ppm)		Iron concentration (μg.g dry wt.$^{-1}$).			
Mn	Fe	Bragg	Lee	Forrest	T-203
0.0	0.25	84.9 cd	74.5 e	95.1 cde	51.1 b
0.0	1.0	125.4 ab	142.7 a	96.7 cde	57.2 ab
0.0	2.5	114.3 ab	101.5 bcde	102.1 cd	80.1 a
0.1	0.25	89.8 cd	77.1 de	117.6 bc	81.9 a
0.1	1.0	129.9 a	107.2 bc	169.6 a	121.1 c
0.1	2.5	181.5 e	130.4 ab	135.4 ab	83.3 a
5.0	0.25	77.8 d	106.4 bc	133.3 ab	62.3 ab
5.0	1.0	99.8 bc	82.5 cde	78.1 e	57.9 ab
5.0	2.5	99.9 bc	84.2 cde	87.4 de	61.7 ab
Significance of treatments differences ($P < \ldots$)					
Mn		0.001	n.s.	0.001	0.001
Fe		0.001	0.01	n.s.	n.s.
Mn × Fe		0.05	0.001	0.001	n.s.

Table 3. Manganese concentration in first trifoliate leaves of soybean cultivars grown with different Fe and Mn levels. Values followed by the same letter within columns are not significantly different at the 5% level of probability according to Duncan's Multiple Range Test

Nutrient level (ppm)		Manganese concentration (μg.g dry wt.$^{-1}$).			
Mn	Fe	Bragg	Lee	Forrest	T-203
0.0	0.25	15.7 b	18.5 a	14.8 a	12.3 c
0.0	1.0	14.3 b	10.9 b	14.0 a	9.5 d
0.0	2.5	12.3 c	14.3 c	16.2 a	12.6 c
0.1	0.25	56.8 a	69.2 d	58.8 bc	61.7 b
0.1	1.0	52.0 a	52.6 e	64.2 b	51.1 b
0.1	2.5	52.9 a	38.1 f	54.6 c	33.4 e
5.0	0.25	846.8 d	1179.4 g	70.6 d	285.6 a
5.0	1.0	624.1 e	868.9 h	665.1 d	472.7 f
5.0	2.5	324.8 f	407.2 i	389.8 e	291.1 a
Significance of treatments differences ($P \times \ldots$)					
Mn		0.001	0.001	0.001	0.001
Fe		0.001	0.001	0.001	0.01
Mn × Fe		0.001	0.001	0.001	0.001

coefficient, r = −0.497**), Lee (r = −0.551**), T-203 (r = −0.657***) and Forrest (parabolically, y = 322.092 + 0.849.x − 0.001.x^2, Fc = 12.3). The expression of catalase activity depended on various factors (Fe and Mn supply; cultivar assayed) but apparently there was a more direct relation of the enzyme activity with Fe supply in the Fe-inefficient cultivars (Forrest, T-203) (Fig. 1).

Manganese stress (deficiency or toxicity) increased peroxidase activity (Fig. 2) although individual responses of cultivars also were observed. Peroxidase activity was higher in the Mn-intolerant cultivars (Bragg,

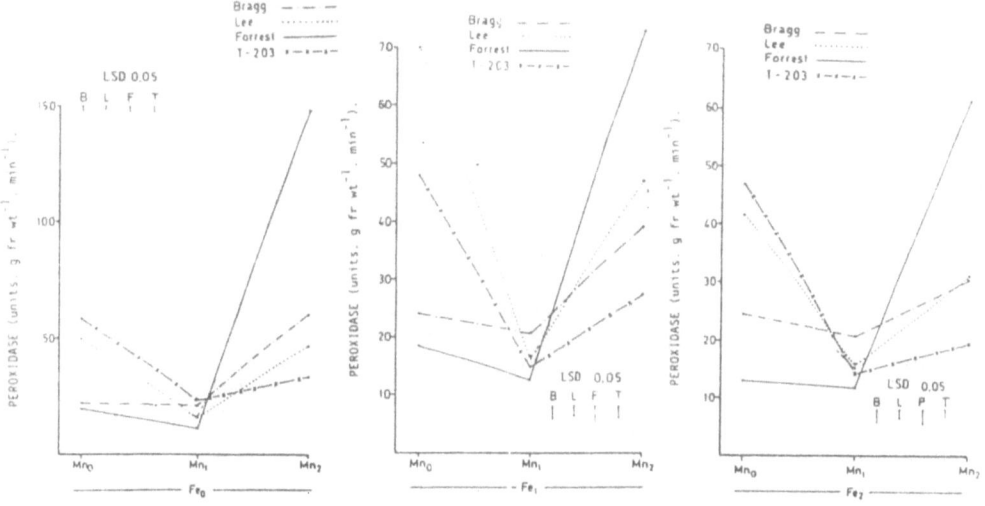

Fig. 2. Leaf peroxidase activity of soybean cultivars grown under different Fe and Mn levels.

Forrest) under Mn-toxicity, while the same was detected for the Mn-tolerant cultivars (Lee, T-203) under Mn-deficiency (Mn_0). The increase in Fe supply at each Mn level failed to produce increases in the peroxidase activity as it was shown in other species[2,9]. Foliar Fe concentration (Table 2) was negatively correlated to peroxidase activity in Bragg ($r = -0.446*$) and T-203 ($r = -0.503**$) cultivars. Leaf Mn concentration (Table 3) was positively related to the enzyme in Bragg ($r = 0.952***$) and Forrest ($r = 0.920***$), but negatively in T-203 ($r = -0.566**$) cultivars.

An interesting point was the negative correlation between growth (Table 1) and peroxidase activity in the cultivars assayed (Bragg, $r = -0.810***$; Lee, $r = -0.713***$; Forrest, $r = -0.643***$; T-203, $r = -0.581***$) in agreement with those reports describing the involvement of peroxidases in plant growth processes[15].

Treatments affected superoxide dismutase activities (Total-SOD and isozyme Mn-SOD) in a different manner (Tables 4 and 5). Total-SOD activity was always affected by Fe levels and in a variable way by Mn or Fe-Mn interactions, but on the contrary, Mn-SOD activity was only modified by Mn levels. Foliar Fe concentration was negatively correlated to Total-SOD activity in Bragg ($r = -0.513**$) while Mn concentration was positively correlated to Mn-SOD activity in Bragg ($r = 0.635***$), Lee ($r = 0.721***$), Forrest ($r = 0.705***$) and T-203 ($r = 0.625***$). Previous works have indicated a dependence of Mn-SOD activity on the Mn nutrient supplies[13] and the induction of a new Mn-SOD isozyme by high nutrient levels of Mn[12].

Table 4. Total-SOD activity in leaves of soybean cultivars grown with different Fe and Mn levels. Values followed by the same letter within columns are not significantly different at the 5% level of probability according to Duncan's Multiple Range Test

Nutrient level (ppm)		Total-SOD activity (units.ml^{-1}).			
Mn	Fe	Bragg	Lee	Forrest	T-203
0.0	0.25	39.05 ab	39.69 ab	39.57 ab	37.73 ab
0.0	1.0	37.67 b	41.56 ab	40.04 ab	40.83 ab
0.0	2.5	37.83 b	38.21 b	35.93 a	34.76 ac
0.1	0.25	44.90 c	44.38 c	40.82 ab	41.62 ab
0.1	1.0	38.03 b	38.52 b	39.13 ab	38.66 ab
0.1	2.5	38.89 ab	38.79 b	37.78 ab	30.50 c
5.0	0.25	50.46	42.94 ac	42.55 b	42.63 b
5.0	1.0	43.15 ac	44.79 c	42.43 b	43.35 b
5.0	2.5	37.88 b	38.17 b	38.51 ab	42.30 b
Significance of treatments differences ($P < \ldots$)					
Mn		0.001	n.s.	n.s.	0.05
Fe		0.001	0.01	0.05	0.05
Mn × Fe		0.01	0.05	n.s.	n.s.

Table 5. Mn-SOD activity in leaves of soybean cultivars grown with different Fe and Mn levels. Values followed by the same letter within columns are not significantly different at the 5% level of probability according to Duncan's Multiple Range Test

Nutrient level (ppm)		Mn-SOD activity (units.ml^{-1})			
Mn	Fe	Bragg	Lee	Forrest	T-203
0.0	0.25	20.94 a	25.58 ab	22.65 a	22.58 ab
0.0	1.0	18.49 a	23.04 a	22.36 a	20.39 a
0.0	2.5	21.79 a	27.83 abc	22.23 a	22.96 ab
0.1	0.25	37.64 b	36.19 cd	33.73 bc	26.25 ab
0.1	1.0	26.42 ac	29.43 abc	31.68 c	30.33 ab
0.1	2.5	30.93 bc	29.29 abc	29.07 ac	22.76 ab
5.0	0.25	34.16 bc	42.28 d	37.36 bc	29.41 ab
5.0	1.0	32.38 bc	36.19 cd	30.24 ac	35.05 b
5.0	2.5	33.95 bc	33.22 bcd	40.90 b	33.64 b
Significance of treatments differences ($P < \ldots$)					
Mn		0.001	0.001	0.001	0.01
Fe		n.s.	n.s.	n.s.	n.s.
Mn × Fe		n.s.	n.s.	n.s.	n.s.

The correlation studies showed positive coefficients between Total-SOD and peroxidase activities in Bragg (r = 0.705***) and Forrest (r = 0.422*). A negative correlation was found between catalase and Mn-SOD activity in Lee (r = −0.403*), or catalase and Total-SOD activity in T-203 (r = −0.406*). Only in Lee cultivars's could a positive relation between catalase and peroxidase activities be registered (r = 0.625***).

The enzymes studied, all involved in oxygen metabolism, were affected in different ways by treatment in the cultivars. The conditions of deficiency and toxicity of each metal could generate stress situations by inducing the production of oxygen free radicals[14]. Active oxygen species such as metal-oxygen complexes could also be formed in certain biochemical systems from iron, copper and manganese[25]. Brown and Devine[4] indicated that the form of Mn (Mn^{4+} or Mn^{2+}) inside the plant might be a controlling factor of Mn tolerance. However, considering our results, other factors could be involved.

Whether the behaviour of soybean genotypes under metallic stresses could be related to the particular induction of protective enzymes (genetically coded) or to differential metabolic production of toxic radicals are problems that remain to be clarified in the future.

Acknowledgements The authors thank JA Jacobs and DR Erickson (INTSOY, University of Illinois at Urbana-Champaign) for the cultivars supplied and to Dr LA Del Río (Estación Experimental del Zaidín, Granada) for his comments and revision of the manuscript. E.O.L. acknowledges a scholarship from the Instituto de Cooperación Iberoamericana (Spain).

References

1 Agarwala S C, Sharma C P and Kumar A 1964 Interrelationship of iron and manganese supply in growth, chlorophyll, and iron porphyrin enzymes in barley plants. Plant Physiol. 39, 603–609.
2 Bar-Akiva A 1971 Functional aspects of mineral nutrients in use for the evaluation of plant nutrient requirement. In Recent Advances in Plant Nutrition vol. 1. Ed. R Samish. Gordon and Breach Science Publishers, New York, pp 115-142.
3 Bataglia O C and Mascarenhas H A A 1981 Toxidade de ferro em soja. Bragantia 40, 199–203.
4 Brown J C and Devine T E 1980 Inheritance of tolerance or resistance to Mn toxicity in soybeans. Agron. J. 72, 898–904.
5 Brown J C and Hendricks S B 1952 Enzymatic activities as indicators of copper and iron deficiencies in plants. Plant Physiol. 27, 651- 660.
6 Brown J C, Holmes R S and Tiffin L O 1958 Iron chlorosis in soybean as related to the genotype of the rootstalk. Soil Sci. 86, 75–92.
7 Brown J C and Jones W E 1977 Manganese and iron toxicities dependent on soybean variety. Commun. Soil Sci. Plant Anal. 8, 1–15.
8 Carter O G, Rose I A and Reading P F 1975 Variation in susceptibility to manganese toxicity in thirty soybean genotypes. Crop Sci. 15, 730–732.
9 Del Río L A 1983 Metalloenzymes as biological markers for the appraisal of micronutrient imbalances in higher plants. Life Chem. Rep. 2, 1–34.
10 Del Río L A, Gómez M, Leal A and López Gorge J 1977 A more sensitive modification of the catalase assay with the Clark oxygen electrode. Application to the kinetic study of the pea leaf enzyme. Anal. Biochem. 80, 409–415.
11 Del Río L A, Gómez M, Yáñez J, Leal A and López Gorge J 1978 Iron deficiency in pea plants. Effect on catalase, peroxidase, chlorophyll and proteins of leaves. Plant and Soil 49, 343–353.
12 Del Río L A, Sandalio L M, Yáñez J amd Gómez M 1985 Induction of a manganese containing superoxide dismutase in leaves of Pisum sativum L. by high nutrient levels of zinc and manganese. J. Inorg. Biochem. 24, 25–34.
13 Del Río L A, Sevilla F, Gómez M, Yáñez J and López J 1978 Superoxide dismutase: An enzyme system for the study of micronutrient interactions in plants. Planta 140, 221–225.

14 Elstner E F 1982 Oxygen activation and oxygen toxicity. Annu. Rev. Plant Physiol. 33, 73–96.
15 Gaspar T, Penel C, Thorpe T and Greppin H 1982 Peroxidases 1970–1980. A survey of their biochemical and physiological roles in higher plants. University of Genève, Switzerland.
16 Giannopolitis C N and Ries S K 1977 Superoxide dismutases. I. Occurrence in higher plants. Plant Physiol. 59, 309–314.
17 Heenan D P and Carter O G 1976 Tolerance of soybean cultivars to manganese toxicity. Crop Sci. 16, 389–391.
18 Kohno Y and Foy C D 1983 Manganese toxicity in bush bean as affected by concentration of manganese and iron in the nutrient solution. J. Plant Nutr. 6, 363–386.
19 Leidi E O and Gómez M 1983 Studies on the manganese and iron nutrition of soybean: Biochemical and physiological approximations. *In*. Mineral Nutrition of Plants vol. 4. Eds. T. Kudrev, V G Georgieva, S Pandev and S Kamenova-Youchimenko. Publishing House Central Cooperative Union, Sofia, pp 39–42.
20 Ohki K, Wilson D O and Anderson O E 1980 Manganese deficiency and toxicity sensitivities of soybean cultivars. Agron. J. 72, 713–716.
21 Sarić M R 1983 Theoretical and practical approaches to the genetic specificity of mineral nutrition of plants. Plant and Soil 72, 137–150.
22 Somers I I and Shive J W 1942 The iron-manganese relation in plant metabolism. Plant Physiol. 17, 582–602.
23 Twyman E S 1951 The iron and manganese requirements of plants. New Phytol. 50, 210–226.
24 Weiss M G 1943 Inheritance and physiology of efficiency in iron utilization in soybean. Genetics 28, 253–268.
25 Youngman R J 1984 Oxygen activation: is the hydroxyl radical always biologically relevant? Trends Biochem. Sci. 9, 280–283.

Ultrastructure of mesophyll cells grown on different levels of selenium of two pea genotypes

S.M. FATALIEVA
Institute of Botany, Azerbaijan Academy of Sciences, Baku, USSR

Key words Chloroplasts Grana Lamellae Maize Membrane Pea Selenium Stroma Thylakoid

Summary It is established that changes in Se doses result in slight changes in ultrastructure of mesophyll cells. An attempt is made to correlate the changes in ultrastructure with water deficiency.

Introduction

Selenium is an essential microelement for animals, therefore in most of the investigations of Se content the aerial parts of the fodder plants were studied[1,2]. The plants display varying sensitivity to the presence of Se in the growth medium and this depends on the species[3,4,4] and is genetically determined. The significance of Se for plants has not been studied in depth, and little is known about its influence on the photosynthetic apparatus.

The aim of the present work is to reveal the influence of different Se concentrations on the ultrastructure of plants of differing sensitivity to the element.

Materials and methods

Seedlings of two pea genotypes (*Pisum sativum* L) -var. 'Kurdamirsky uluchenny', var. 'Aznichi' and of one maize (*Zea mays*) var. Zakatalskaya uluchenny' were used in the experiments. Seeds were moistened and grown on solution containing varying levels of Se. Seedlings were grown on 1/5 strength Knop's nutrient mixture. The following concentrations were used:
  I. − Se (control);
  II. +0.5 mg Se/l (promoting dose);
  III. +5 and 10 mg Se/l for pea and maize respectively (inhibitory doses).
   The tissues were fixed in 3% glutaraldehyde in phosphate buffer (pH 7.4). After OsO_4 fixation the material was embedded into a mixture of EPON + araldite. Ultrathin sections were obtained by means of the LKB ultramicrotome. The sections were contrasted with 2% uranyl acetate solution and lead hydroxide according to Reynolds and examined in the electron microscope (JEM-100B).

Results and discussion

The plants grown in variant II were stronger than the controls except the pea var. Aznihi. Leaves of plants grown in variant III were smaller than control leaves but they became rough in maize. The growth of all

H.W. Gabelman and B.C. Loughman (Eds.), Genetic aspects of plant mineral nutrition.
ISBN-13: 978-94-010-8102-3
© *1987 Martinus Nijhoff Publishers, Dordrecht/Boston/Lancaster.*

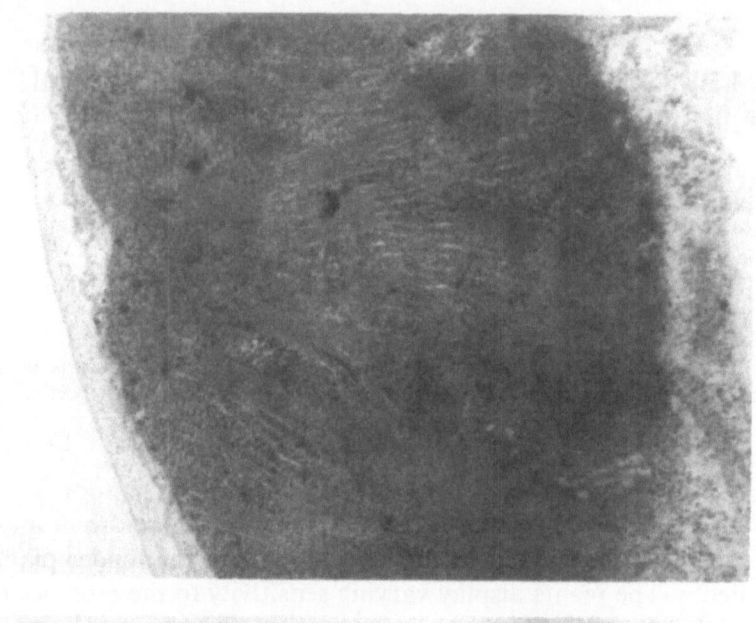

Fig. 1. Fragments of mesophyll cells I variant. a — Kurdamirsky uluchenny (×70,000) b — Zakatalskaya uluchenny (×20,000).

three species in variant III was depressed — in maize and pea var. Kurdamirsky uluchenny to approximately 50% and more than 50% in Aznihi.

Mesophyll chloroplasts of control plants (−Se) are associated with the perimeter of the cell and their quantity in the cell varies. Maize chloroplasts are of the lens type with a stretch form, while in two pea genotypes they are oval. The stroma consists of a small granular matrix; a well-developed membrane system and the number of thylakoids fluctuate between 15 and 20 (Fig. 1a, b). Chloroplasts of variant II differ

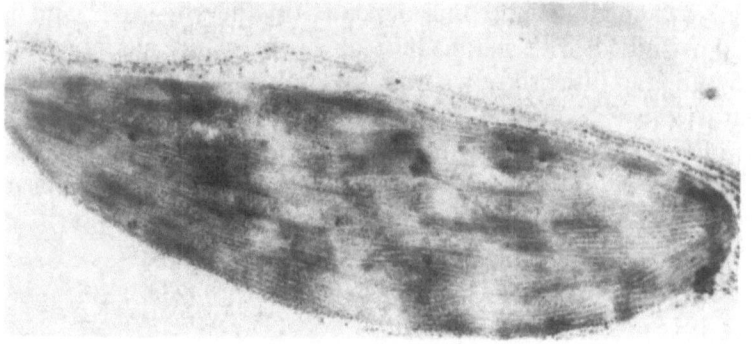

Fig. 2. Fragments of mesophyll cells II variant. a — Kurdamirsky uluchenny (×60,000) b — Aznihi (×70,000) c — Zakatalskayauluchenny (×20,000).

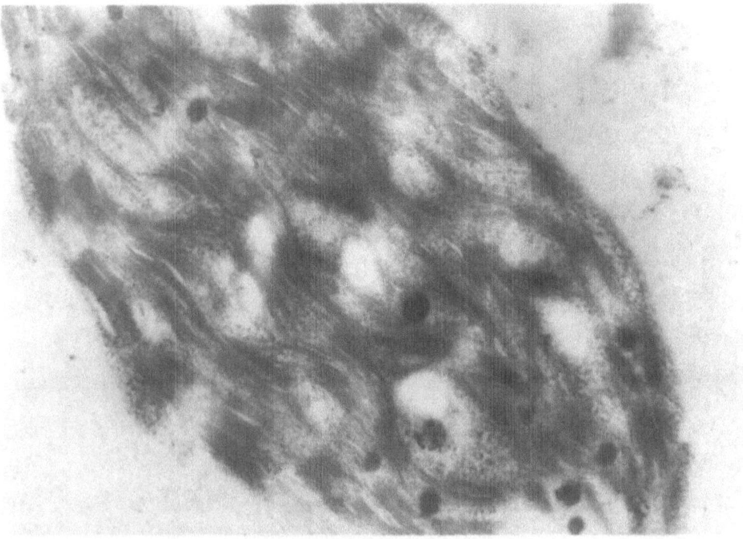

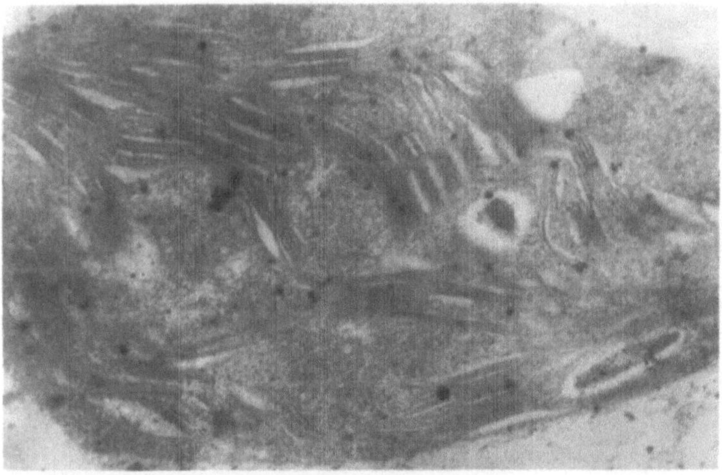

EFFECT OF SE ON THE ULTRASTRUCTURE OF PEA

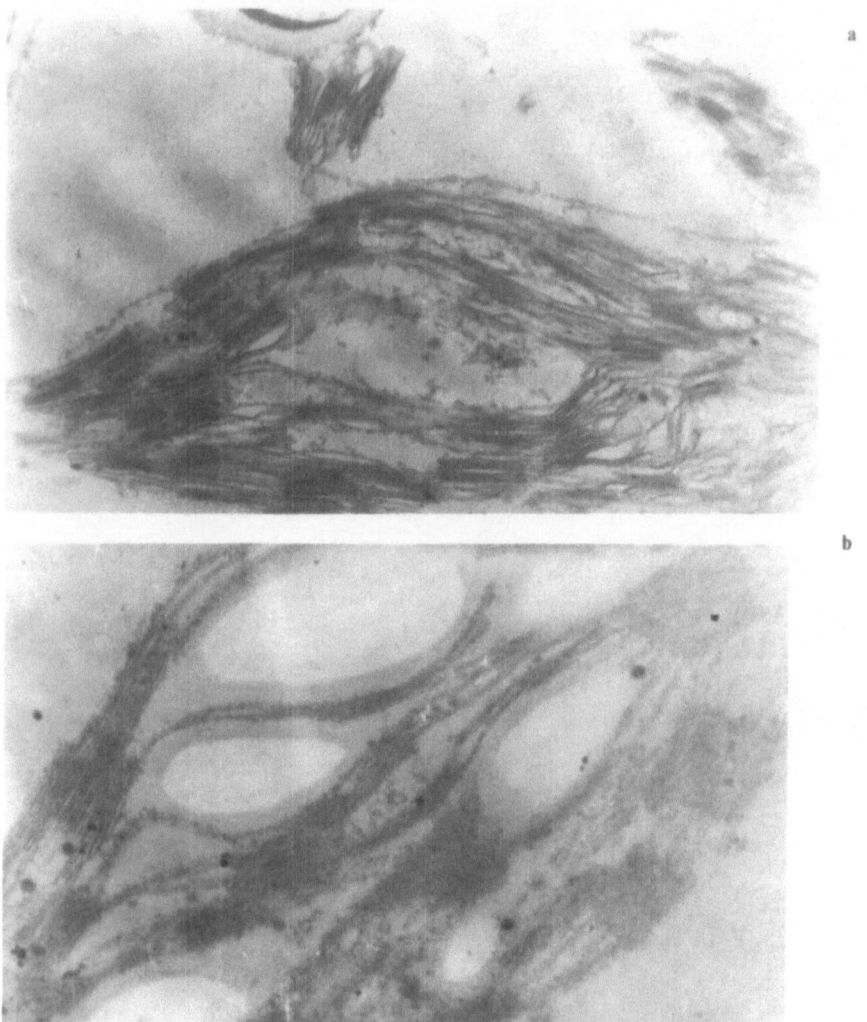

Fig. 3. Fragments of mesophyll cells III variant. **a** — Kurdamirsky uluchenny (× 60,000) **b** — Aznihi (× 70,000) **c** — Zakatalskay uluchenny (× 60,000).

from the control and between species. In the relatively resistant pea species swelling of chloroplasts and a slight change of stroma correlation and membrane structure was observed. Grana size and quantity of discs were decreased but many chloroplasts remain unchanged (Fig. 2a, b, c). In sensitive species the changes are more intensified: the number of starch grain and osmiophilic globules is increased and the grana have increased interthylakoid spaces. Maize chloroplasts in variant II differ slightly from controls (Fig. 2c).

Figure 3 shows the decreased number of chloroplasts in variant III. Chloroplasts have increased interthylakoid spaces anddegradation of

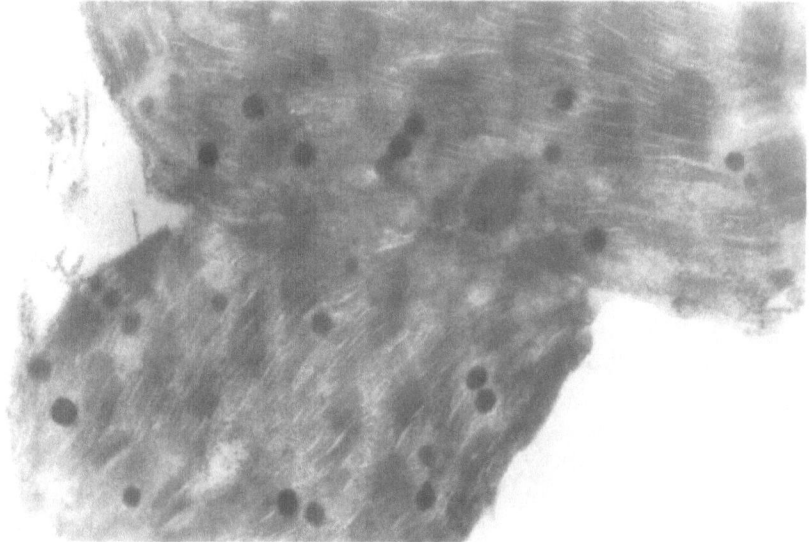

chloroplasts with a simplified lamellae system is observed together with an increased quantity and size of starch grains.

The data point to the influence of Se on the structure of the photosynthetic apparatus even when growth is promoted (Variant II). Similar changes in chloroplasts can be observed with different environmental factors including water stress. In our early work we reported that, with the increase of Se in the growth medium, the water content of the plants decreased[6]. On reviewing the literature it may be stated that grasses grown in well-watered regions contain less Se than in dry regions[7,8]. It has been established also that Se accumulating plants are on the whole xerophytes such as Astragalus[9]. We supposed that one of the factors in the resistance to the presence of Se is a balanced action of water relations. The increase in the quantity and size of the starch grains as growth decreases points to a reduction in flow of assimilates that are not used for growth.

References

1 Dubikovski G P and V P Lebedev 1974 Se v pochvah i lugovyh rasteniyaeh BSSR. Agrochimya 5, pp 113 119.
2 Ermakov V V Kovalsky V V 1974 Biologicheskoe znachenie sekena. Izd. Nauka M.
3 Shrift A 1969 Aspects of selenium metabolism in higher plants. Annu. Rev. Plant physiology, 20, 475 494.
4 Fatalieva S M 1978 Deistvie selenita natriya na rost i energoobmen kornei gorocha i kukuruzy. Zh. Sel'sko hozyaistvennaya biologia, 12.
5 Brown T A and A Shrift 1982 Selenium- toxicity and tolerance in higher plants. Biol. Rev. 57, 59 84.

6 Dekov P, L Pe trova T Kudrev 1984 Ultrastructural changes in chloroplasts of maize plants grown under potassium deficiency and water stress conditions. Mineral nutrition of plants, vol 4, Sofia, pp 99–102.
7 Fatalieva S M, Gadjeva E A and Zeinalova G R 1981 Deistvie selenita natriya na rost i vodoobmen. Selen v biologii, 3, s. 207–211, Izd. 'ELM', Baku.
8 Koval'skiy V V 1968 Geochimicheskaya ekologiya organizmov pri povyshennom soderzanii Se. Izd. 'Nauka'.
9 Trelease S F and Trelease H M 1939 Physiological determination within Astragalus with reference to selenium. Am. J. Bot. 26, 530–535.

Section 5

Genetic variation in microorganism host interactions in mineral nutrition

Host plant control of symbiotic N_2 fixation in grain legumes

F.A. BLISS
Department of Horticulture, University of Wisconsin, Madison, WI 53706, USA

Key words Grain legumes Legume-rhizobial symbiosis Nitrogen fixation

Summary A successful legume-rhizobial symbiosis implies that a beneficial plant response occurs in the form of increased N_2 fixation. The criteria used to measure this response should include: total N_2 fixed per plant, percentage of total plant N from fixation, and plant biomass and/or seed yield resulting from fixed N_2. Estimation of these criteria is done most effectively using direct measurements, but because of high cost and the technical requirements of most direct methods, indirect measures of fixation variables have been used. The addition of highly effective rhizobial strains in differentiating among plant genotypes that vary in their ability to fix N_2, and the availability of improved plant genotypes provide greater opportunity to realize the full value of improved rhizobia in the field. Several breeding strategies for selecting plants able to fix more N_2 have been used, and depending on the legume species, the rhizobial population and edaphic factors common to the production system, improved plants have resulted from selection.

Introduction

Most plant-microorganism interactions are either of limited importance or deleterious to the plant, while only a few are beneficial and symbiotic. Among the latter, the ecological uniqueness of root nodules associated with the Leguminosae was noted as early as the 16th century, and the soil improving properties of legumes have been known for many years[2].

Legumes are utilized widely in crop rotations and rhizobia are cultured on a commercial scale for inoculant production. Because of host-microsymbiont specificity the proper rhizobial strains either must be present in the native population or applied. The knowledge that rhizobia differ in their host specificity, infectivity, competitiveness with other rhizobia and soil organisms, and effectiveness with different host genotypes has contributed to the development of improved strains for commercial use. To date most attempts to increase the beneficial symbiotic plant response have emphasized rhizobial improvement and the use of better agronomic practices.

Accurate estimates of the amount of N_2 fixed under field conditions are available for relatively few crops[26,40]. Observations and indirect estimates suggest that increased N_2 fixation resulting from inoculation often is variable, and that although cost of inoculant is minimal, inoculation may not be profitable. The performance of 'super' rhizobial strains has been promising in the laboratory but often disappointing under field conditions[32]. While poor performance has been attributed to lack of

H.W. Gabelman and B.C. Loughman (Eds.), Genetic aspects of plant mineral nutrition.
ISBN-13: 978-94-010-8102-3
© *1987 Martinus Nijhoff Publishers, Dordrecht/Boston/Lancaster.*

competitive ability against indigenous rhizobial populations[36], use of an inefficient host cultivar could attribute to a limited plant response.

The plant member of the legume-rhizobial symbiosis must be efficient also if maximum plant response is to be achieved. A better knowledge of the host's role in N_2 fixation will allow more effective genetic manipulation of plant traits affecting symbiosis, using either conventional breeding methods or recombinant DNA techniques. The best delivery system for utilizing N_2 fixation in widely differing field cropping systems is through high yielding legume cultivars that can fix large amounts of N_2 with either indigenous rhizobia or improved inoculant.

Plant symbiotic response

Unambiguous measurement of plant symbiotic response is necessary to establish the value of each host genotype, rhizobial strain and host-strain combination. The criteria used should include: 1) total N_2 fixed per plant (or per unit area if plant population is specified), 2) percentage of total plant N derived from fixation, and 3) plant biomass and/or seed yield resulting from fixed N_2. These estimates provide a measure of the amount of N_2 fixed and the N utilization efficiency.

Indirect methods for estimating fixation variables, or the potential for N_2 fixation rather than the actual amount of N_2 fixed, have been used because of simplicity and low cost[24,26,39,44]. Such methods include: 1) plant growth and total plant mass; 2) visual evaluation of the nodulated root system; 3) determination of nodule dry mass and/or nodule number; 4) acetylene reduction (nitrogenase) activity (ARA) by the nodulated root system or by a given nodule mass and 5) ureide concentration of the xylem sap. Although used widely, these methods have several disadvantages. All are plant destructive, which precludes retention of superior individual plants for further study or for breeding purposes except by cuttings. Each method provides only a single point estimate of a variable that is descriptive of fixation potential rather than actual N_2 fixed. Unless numerous measurements are made throughout the plant's growth cycle, a poor estimate of total N_2 fixation is likely to be obtained. The single point estimates are affected greatly by physical and environmental factors that often occur sporadically and contribute to poor estimates of total N_2 fixed. The correlations between indirect and direct methods for measuring N_2 fixation have been found to vary substantially.

Indirect methods can be valuable for determining the relative fixation potential of different plant genotypes, particularly in field breeding studies requiring rapid screening of many lines. Usually nodule dry mass, acetylene reduction (nitrogenase) activity (ARA), and visual evaluation of the nodulated root have been found to be correlated positively with each other, but may or may not be correlated with plant dry mass, nodule number and ureide in the xylem sap[43]. While a high ureide concentration in the xylem sap usually indicates that plants are fixing N_2 actively[39,49], ureide assays have not been a useful criterion for accurately measuring different levels of N_2 fixation among genotypes. Although used widely as an estimate of N_2 fixation, the acetylene reduction assay of disturbed nodule systems has been found to produce misleading results, in part because of a rapid decline in nitrogenase activity caused by the addition of acetylene. Minchin et al.[33] suggest that results from experiments in which the conventional acetylene reduction assay was used should be viewed with caution.

The use of direct methods for estimating N_2 fixation[39] provides the most accurate determinations of plant response for the three criteria mentioned previously. Direct methods include: 1) growth of plants on N-free media, 2) estimation of ^{15}N levels in fixing and non-fixing plants and, 3) comparison of the amount of total N accumulated in fixing plants with that in comparable non-fixing plants. Estimates of N_2 fixed by plants grown on N-free media can be obtained only in non-field environments, and then the applicability of those results to field situations still must be determined. However, some soils contain such low levels of available N, that they approach an N-free medium. Available N may be reduced to low levels by adding organic amendments, e.g. sawdust, sugar etc.

Determination of whether plant N came from soil N, fertilizer N or from N_2 fixation can be made by comparing the level of ^{15}N in actively fixing plants with that in non-fixing plants. ^{15}N-enriched or ^{15}N-depleted fertilizer is applied to experimental plants and non-fixing reference plants at levels suitable to ensure that fixation is not inhibited. An important consideration when using ^{15}N isotope methods is the choice of an appropriate non-fixing reference crop, since N uptake by the reference plants affects the estimates of N_2 fixed by the fixing legume. Cereal plants such as barley or wheat have been used, but a non-nodulating genotype of the test legume (e.g. non-nodulating soybean) is preferable when one is available[9]. The isotope methods appear to be the most suitable for estimating actual N_2 fixed, but limitations result from the need for a mass spectrophotometer, the cost of ^{15}N-labelled fertilizers and the time required for analyses of plant samples. In breeding programs where numerous experimental plants must be analyzed routinely, the resulting costs may be prohibitive.

The estimation of N_2 fixed by the difference method also relies on the use of an appropriate non-fixing reference crop. Although it is not possible to distinguish between plant N derived from soil and from fertilizer using the difference method, it may be useful in programs having limited resources. The primary requirement is the ability to determine total plant N, which can be done in most laboratories using one of the standard chemical methods, *e.g.*, Kjeldahl analysis.

Currently, there are few published estimates of the amounts of N_2-fixed under field conditions[26,40]. The magnitude of symbiotic plant response differs among lines, suggesting that the N_2 fixed is a genetically-controlled trait. In most experiments the only plants studied have been commercial cultivars, and the extent of possible improvement from hybridization and selection has been reported in only a few cases[5,31,42].

Plant-rhizobial associations

A successful plant-rhizobial symbiosis implies a beneficial plant response in the form of increased N_2 fixation, which in turn may produce other beneficial effects such as greater plant growth and/or more seed. Since the plant response results from attributes of both the plant and microsymbiont, either alone or in combination, a clear understanding of those attributes is useful when assessing genotypic variation in the host plant and for determining how the variation might affect the plant response.

Corresponding attributes of rhizobia and plant hosts are shown in Table 1. Various combinations may produce no plant response, a limited

Table 1. Symbiotic plant responses resulting from different interactions between rhizobia and legume plants

Rhizobia	Plant	Symbiotic plant response
Noninvasive	Resistant	None
	Susceptible	None
Invasive	Resistant	None
	Susceptible	Lim./Bene.
Specific	Specific	Beneficial
	Promiscuous	Lim./Bene.
Versatile	Specific	Lim./Bene.
	Promiscuous	Beneficial
Competitive	Receptive	Beneficial
	Nonreceptive	Limited
Noncompetitive	Receptive	Limited
	Nonreceptive	Limited
Ineffective	Inefficient	Lim./None
	Efficient	Lim./None
Effective	Inefficient	Limited
	Efficient	Beneficial

response, either a limited or beneficial response, or a beneficial response. The extent to which response is limited and the magnitude of a beneficial response can vary depending on different host-symbiont combinations and external factors. A primary goal of host improvement through breeding is to increase the beneficial response of the host without producing a negative effect on another plant trait that may offset the realized gain.

Considering the results expected from various host-rhizobial combinations, determination of plant genotypic variability must be made in association with suitable rhizobial strains. For example, to distinguish between resistant and susceptible plants requires the use of rhizobia that are invasive, and the distinction between inefficient and efficient plant genotypes requires testing with effective rhizobial strains. Consideration of host attributes should aid also in screening among rhizobial strains. Differentiation between effective and ineffective strains and determination of the level of effectiveness can be accomplished best using an efficient plant genotype.

Host plant resistance and susceptibility

A legume plant is susceptible, when upon being exposed to an invasive rhizobium, the plant becomes infected and nodulation occurs. Distinction between resistant and susceptible host genotypes can be made only when exposed to an invasive strain, since no plant response will result from a non-invasive strain regardless of the plant genotype.

Simply inherited Mendelian genes which preclude nodulation by either one or many strains of rhizobia have been identified in several grain legumes including pea, soybean, chickpea, and peanut[7,20,35,52]. Since those genes prevent nodulation by otherwise invasive rhizobia, they can be designated as resistance genes in the host plant.

Questions arise concerning the degree of susceptibility, occurrence of

Table 2. Susceptibility of three common bean host plant genotypes to *Rhizobium phaseoli* based on nodule and shoot characteristics. Values are means for three rhizobial strains, CE-3, Kim 5 and CIAT 899. Plants grown in growth room were harvested at 28 days. Madison, Wisconsin, 1985

Host plant	Nodule		Shoot		
	Number (total·plant^{-1})	Dry wt. (mg·plant^{-1})	Dry wt. ($-$mg·plant^{-1})	Total N	N conc. (%)
Puebla 152	150	146	640	22.8	3.30
Sanilac	110	121	790	23.1	2.82
57-1	51	53	360	12.9	3.43
lsd$_{0.05}$	35	42	130	7.7	0.62

Table 3. Nodule and plant growth characteristics of three common bean host plant genotypes infected with three strains of *R. phaseoli* (CE-3, Kim 5 and CIAT 899), indicating different measures of susceptibility. Plants produced in growth room were harvested at 28 days. Madison, Wisconsin, 1985

Host plant	Nodule No.			Nodule dry wt. (mg · plant^{-1})		
	CE-3	Kim 5	CIAT 899	CE-3	Kim 5	CIAT 899
Puebla 152	131	173	147	154	103	180
Sanilac	123	107	98	118	65	179
57-1	74	46	33	42	50	66
lsd$_{0.05}$		60			72	

Host plant	Shoot dry wt. (mg · plant^{-1})			Shoot N (mg · plant^{-1})			Shoot N conc (%)		
	CE-3	Kim 5	CIAT 899	CE-3	Kim 5	CIAT 899	CE-3	Kim 5	CIAT 899
Puebla 512	550	420	940	18.4	9.5	40.4	3.35	2.26	4.30
Sanilac	490	300	1570	15.6	6.9	46.8	3.19	2.29	2.98
57-1	260	250	580	10.0	8.1	20.5	3.85	3.25	3.53
lsd$_{0.05}$		225			13.3			1.07	

genotypic differences for susceptibility and whether increased susceptibility will produce a quantitatively greater N_2 fixation response. Several criteria can be used to measure susceptibility. Based on mean response to three rhizobial strains, nodule number and nodule mass per plant allow discrimination among the three plant genotypes, with Puebla 152 being the most susceptible. However, for total shoot dry wt. or shoot N, Sanilac produced the greatest mean response (Table 2). Percentage N provided an entirely different ranking. The conclusions drawn concerning susceptibility are affected also by the strain of *R. phaseoli* used (Table 3). There does not appear to be a single 'best' criterion for measuring degree of susceptibility[38].

It is not clear whether selection for increased susceptibility to rhizobia will result in greater N_2 fixed per plant, % of total N from fixation and dry mass or grain yield resulting from fixation. In alfalfa selection in the glasshouse for criteria indicative of susceptibility has produced mixed results in field trials[3].

Plants that form ineffective nodules may be judged to be susceptible also, but a limited or nil response results because the nodules fail to develop fully. Nodule number as a selection criterion in this case would be of little practical value.

Host plant specificity and promiscuity

Some plant genotypes show resistance to most rhizobial strains, but are susceptible to one or more specific strains, and thus can be classified as strain specific. Soybean plants that are genotypically $rj_1 rj_1$, are resistant to most indigenous rhizobia, but some strains of *Bradyrhizobium japonicum* have been found that are invasive to $rj_1 rj_1$, genotypes and produce symbiotically active nodules[8]. There are other mutant genes (Rj_2, Rj_3, Rj_4) in soybean that produce ineffective nodules with indigenous strains but effective nodules with some *B. japonicum* strains, giving rise to specificity for ineffectiveness[6,50,51]. Such is likely to be the case with many single gene mutants when challenged with a wide array of rhizobial strains.

Single gene host specificity has been reported also in pea[10,20]. In addition, Lie[28] reported that three genes which control nodulation specificity in primitive pea lines from the Middle East, also confer resistance to rhizobial strains that nodulate cultivated peas from Europe. Lie[27] has speculated that centers of diversity for wild pea forms may be important centers of diversity for rhizobia as well.

In addition to genetic control of host specificity when challenged with single rhizobial isolates, there is host control of strain preference among heterogenous populations, *e.g.*, in forage legumes[15,16]. In pea genotypes of Middle Eastern origin, nodulation by an invasive rhizobium strain was found to be suppressed by the presence of a non-nodulating strain[53]. Although preferential nodulation may be influenced by the host, nodulation may not always be by the most efficient strains.

While some legumes show specific susceptibility to certain rhizobia, others are susceptible to a wide range of types. Those legumes have been termed promiscuous. Most soybean cultivars developed for production in the USA fail to nodulate effectively in tropical soils without being inoculated with competitive strains of *B. japonicum*[25]. However, soybean cultivars of Asian and African origin nodulate profusely by many rhizobia of the cowpea miscellaneous group found in tropical soils[41]. In addition those 'promiscuous' lines can be nodulated effectively by introduced strains of *B. japonicum*. Promiscuous nodulation is controlled genetically and there are now efforts underway to combine that trait into agronomically superior cultivars[25].

Host plant receptivity and non receptivity

Nodulation, nodule development, activity and senescence vary during the course of plant growth. Nodule distribution is usually nonuniform along individual roots and the pattern of nodulation varies within the root profile. Bhuvaneswari *et al.*[4] found that in soybean the infectible root cells are located above the zone of root elongation and below the area of the smallest root hairs at time of inoculation. The infectibility of any given cell seemed to last only a few hours.

The relative importance of early-formed nodules in the crown area and on primary roots compared to those that develop later on lateral roots is not known. There is evidence that rhizobia added in the seed furrow or placed on the seed at planting move only short distances in the soil[29]. Therefore it is likely that most of the lateral root nodules result from native rhizobia. If the nodules on the lateral roots are important suppliers of fixed N_2 to the plant, particularly during seed filling, and if strains added as inoculum do not contribute substantially to lateral root nodulation, this may explain in part the poor performance of 'super' rhizobial strains under field conditions.

The extent of plant control of nodulation pattern is not known, in part because the growth medium surrounding the root can produce large variation. Nutman[37] found that in clover, selection for increased nodule number also resulted in plants with smaller nodules, more secondary roots and a greater proportion of nodules on secondary roots. In common bean, genotypes with higher ARA values at advanced growth stages also had greater nodulation scores on lateral roots (Author's unpublished results).

The N nutrition of the legume seedling is important for early growth. N deficiency is often observed after the seed reserves have been depleted and prior to the export of fixed N from the functioning nodules. Sprent and Thomas[45] related the different expressions of seedling N deficiency to type and placement of cotyledons, patterns of nodule development, and synchrony of nodule and leaf development. Timely application of starter fertilizer often is used to alleviate this stress. It is not known whether seedlings produced from seeds that contain more N reserves (*e.g.*, high protein phenotyes) would show less N stress prior to active transport of fixed N.

Although early nodulation has been noted on some plants[13], there appears to be limited genetic variation for this trait. In the few cases studied little change in amount of N_2 fixed has been attributed to early nodulation[37]. The duration of maximum nodule activity is important provided there is a substantial active nodule mass. Graham[11] suggested that a few days delay in the onset of flowering in common bean can produce sizeable increases in the amount of N_2 fixed. Usually it has been concluded from estimates of nitrogenase activity based on ARA, that fixation in grain legumes declines rapidly after the onset of flowering (R1 growth stage) and the beginning of pod fill (R3 stage). However, the timing of maximum nodule activity and the duration of activity vary considerably. From estimates of N_2 fixation using ^{15}N-depleted $(NH_4)_2SO_4$, it appeared that in the common bean breeding line U.W. 24-17, almost 3/4 of the total N_2 fixation occurred after the R_3 stage, while in the parent lines Puebla 152 and Sanilac a smaller proportion of N_2 was fixed after the R3 stage[46].

Edaphic factors such as temperature extremes, moisture stress excessive or inadequate element availability (*e.g.*, P, Fe and Al), high levels of exogenous N, and non-optimum soil pH markedly affect symbiosis and decrease N_2 fixation. Graham[11] summarized the effects of these factors on N_2 fixation in common bean and suggested that genotypic variability in the host and selection for increased rhizobial strain tolerance can be used to improve the potential for increased fixation. Soybean genotypes resistant to iron chlorosis have been identified and used in breeding programs. In several cases legume genotypes able to grow and fix N_2 at low P levels and high Al levels have been identified[11].

The presence of large concentrations of mineral N is usually accompanied by reduced N_2 fixation. Several mechanisms for the inhibitory effects of NO_3^- have been proposed, with NO_3^- being shown to inhibit separately root hair infection, nodule growth and development, and nitrogenase activity[19]. When more shoot growth is produced with greater amounts of NO_3^-, there may be increased competition between the growing shoot and the functioning nodule for photosynthate.

There are both interspecific and intraspecific differences for ability to tolerate NO_3^- without substantial reduction in fixation. Harper and Gibson[19] suggested that attempts to overcome the inhibitory effect of NO_3^- on nodulation should concentrate on limiting the uptake and/or metabolism of NO_3^-.

Induced mutants of pea (*Pisum sativum* L.) with modified symbiotic properties have been characterized for response to NO_3^-. The mutant E_1 showed greatly reduced NO_3^- reductase activity compared to the standard cultivar, but despite these differences, both had reduced nodule mass and number in the presence of 15 mM KNO_3. When NO_3^- was added to nodulated plants grown previously on N-free medium, N_2 fixation of the standard cultivar was reduced by 47% and in the E_1 mutant by only 19%[21]. Jacobsen[21] concluded that either NO_3^- reduction or the reduced products, rather than NO_3^- per se caused the inhibition. A second mutant, nod_3, nodulated vigorously when grown on NO_3^--containing as well as NO_3^- — free medium in contrast to the standard cultivar in which nodulation was reduced in the presence of 15 mM KNO_3^{22}. The mutant plants also showed a modified root branching pattern. The nod_3 mutant, which is controlled by a single recessive allele, has been suggested for use in breeding programs, since mutant plants would have the potential for high nodulation and fixation both on soils where N is limiting and where N is relatively abundant[21].

Host plant efficiency and inefficiency

The efficiency of fixation depends greatly on the root and shoot characteristics of the host plant. The supply of carbohydrates available for nodule growth and function relative to that used for plant growth and seed development is an important determinant of the total amount of N_2 fixed. Time of flowering affects the amount of N_2 fixed, with the peak level of nitrogenase activity usually occurring during the early part of the reproductive stages when pods are still small. As seed development continues, acetylene reduction activity usually drops rather sharply. In most cases, plants that flower and mature early fix less N_2. Hardy et al.[17] showed that early flowering soybeans usually fixed less N_2 than later cultivars; and in common bean fixation almost doubled when flowering was delayed a few days by manipulating the photoperiod[12].

As nitrogenase activity declines during seed development, leaf senescence also begins, particularly in early maturing genotypes[18,47]. Presumably the declining nodule activity is due in part at least to declining leaf photosynthesis. In soybeans, some plant genotypes maintain green, photosynthetically-active leaves after the seeds are mature[1]. However, in field studies, genotypes showing delayed leaf senescence (DLS) fixed amounts of N_2 similar to normal counterparts, but yielded somewhat less[40]. Leaf area duration is important, where in peanut 70–75% of the variation in nodulation and fixation was attributed to differences in duration[54].

Studies using different techniques such as leaf and pod removal, grafting, modification of light and CO_2 enhancement suggest that photosynthate supply to the nodules greatly affects N_2 fixation in legumes[34]. Total size of the plant canopy is often related to the plant growth habit which is usually correlated with total fixation potential. In common bean, indeterminate, climbing plant habit is related to increased fixation potential, while plants with a determinate bush phenotype usually fix less N_2[11]. However, because plant growth habit is an important agronomic trait, it is not feasible to introduce a climbing bean into a cropping system where bush beans are grown and vice versa. It is important to consider whether the efficiency of each plant type can be increased to produce greater N_2 fixation. Since length of growing season is often limited by environmental factors ranging from temperature extremes to lack of moisture, breeders must attempt to improve each plant type required for different agroecological systems.

Not only total photosynthate but also the form in which it is available for partitioning between shoot, root and nodule are important efficiency traits of the host plant. Carbohydrates accumulate in the roots of some inefficient plants, and apparently are unavailable to the functioning nodule. Climbing beans appear to mobilize a greater proportion of nonstructural carbohydrate to the nodules than do cultivars of other growth habits[11]. The importance of root size varies also since cultivars with different root mass accumulate similar amounts of total carbohydrates in the roots and apportion it differently to the nodules[14].

The transport of N in various forms from the root to the shoot appears to be related to N_2 fixation as well as overall N uptake and utilization[48]. Although there were no apparent differences in N composition of the xylem sap between high and low fixing common bean lines, high fixing lines generally showed a greater rate of total N translocation regardless of whether the N source was fertilizer or N_2 fixation[49].

Host plant selection for increased symbiotic response

There is substantial genetic variability for ability to fix N_2 in the germplasm of most grain legumes. However, since many of the estimates of genetic variability have relied on indirect measures of N_2 fixation potential, they should be viewed with caution when speculating on the amount of improvement that is possible through breeding. Because of the complexity of the trait and the difficulty and expense of estimating N_2 fixation by direct methods, there have been few selection experiments for increased N_2 fixation conducted within genetically-variable breeding populations. In soybean populations, total N and fixed N in the seeds were shown to be heritable traits and there was a strong positive correlation between seed dry weight and fixed N content of the seed[42].

In common bean, inbred backcross lines from populations resulting from crosses between parents with different fixation potential were selected for increased N_2 fixation potential using indirect methods of estimation[30]. The BC_3S_2 lines selected for high ARA values, superior seed yield, suitable plant and seed type and moderate earliness were superior to the recurrent parent when grown on soils with low N. Plant response criteria of the selected breeding lines were determined in later experiments using ^{15}N-depleted $(NH_4)_2SO_4$. With a non-nodulating soybean used as a reference crop, selected breeding lines exceeded the recurrent parent Sanilac, for total N_2 fixed per plant, % of total N derived from fixation and total seed yield (Table 4). These data show that selection for improved host plants having increased symbiotic response was effective.

Genetic control of host genotype specificity[8] and promiscuity[25] have been proposed for use in breeding programs to improve N_2 fixation of soybeans. In the former approach strain-specific cultivars that exclude nodulation by 'undesirable' indigenous rhizobia and favor nodulation by

Table 4. Total N fixed, %N derived from atmosphere (%NDFA) and seed yield of selected inbred backcross lines and parents. ^{15}N-depleted $(NH_4)_2SO_4$ applied to field-grown plants, Wisconsin, 1984

Parent or line	50% Bloom (R3 Stage)		Maturity (R9 Stage)		
	Total N fixed (mg·plant^{-1})	% N DFA	Total N fixed (mg·plant^{-1})	% N DFA	Seed yield (g·plant^{-1})
Puebla 152	300	42	852	57	38
Porr. Sint.	175	33	558	48	28
21–16	187	33	779	50	35
21–38	178	33	779	56	32
21–43	213	36	741	54	33
21-58	243	38	712	54	30
Sanilac	4	2	76	12	18
24-17	46	13	583	48	31
24-21	98	25	216	25	19
24–48	54	14	211	22	23
24–55	91	24	192	22	23
24 65	44	12	279	31	25
Non-Nod. S.B.	0	0	0	0	13
Duncan Critical Value 0.05	96	18	180	9	8

Source: DuBois et al., personal communication, 1984.

inoculated strains would be selected. In the latter approach, agronomically improved promiscuous cultivars would be developed to favor the indigenous strains and reduce the need for inoculant where it is difficult to obtain.

Although the practical consequences (*e.g.*, increased N content, plant growth, yield) of most legume-rhizobial symbiotic relationships are measured by the beneficial response of the host, the improvement of the plant genotype to increase the magnitude of N_2 fixation has been largely ignored. The limited studies to date have shown that there is considerable genetic variability for amount of N_2 fixed, as well as for plant traits that either limit or enhance the potential for fixation. When appropriate methods for measuring fixation or fixation potential have been combined with effective breeding methods, selection has produced experimental lines able to fix more N_2. In addition, use of these improved lines allows better discrimination between rhizobial strains able to fix different amounts of N_2.

Acknowledgements The author wishes to acknowledge the contributions of J DuBois and R H Burris, Dept. of Biochemistry and Dina St Clair, J C Rosas and K A Kmiecik, Dept. of Horticulture, University of Wisconsin. Support for this research was received from the USAID/CSRS Special Grant No.59-2551-1-5-006-0, the Bean/Cowpea CRSP funded through Grant No. AID/DSAN-XII-G-0261 and the College of Agricultural and Life Sciences, the University of Wisconsin-Madison.

References

1. Abu-Shakra S S, Phillips D A and Huffaker R C 1978 Nitrogen fixation and delayed leaf senescence in soybeans. Science 199, 973–975.
2. Allen O N and Allen E K 1981 The Leguminosae, University of Wisconsin Press, Madison, 812 p.
3. Barnes D K, Heichel G H, Vance C P and Ellis W R 1984 A multiple-trait breeding program for improving the symbiosis for N_2-fixation between *Medicago sativa* L. and *Rhizobium meliloti*. Plant and Soil 82, 303–314.
4. Bhuvaneswari T V, Turgeon B G and Bauer W D 1980 Early events in the infection of soybean (*Glycine max* L. Merr) by *Rhizobium japonicum*. Plant Physiol. 66, 1027–1081.
5. Bliss F A 1984 Breeding for enhanced dinitrogen fixation potential of common bean. *In* Nitrogen Fixation and CO_2 Metabolism. Eds. P W Ludden and J E Burris. Elsevier, New York pp 303–310.
6. Caldwell B E 1966 Inheritance of a strain specific ineffective nodulation on soybeans. Crop Sci. 6, 427–428.
7. Davis T M, Foster K W and Phillips D A 1985 Nodulation mutants in chickpea. Crop Sci. 25, 345–347.
8. Devine T E and Weber D F 1977 Genetic specificity of nodulation. Euphytica 26, 527–535.
9. Fried M, Danso S K A and Zapata F 1983 The methodology of measurement of N_2 fixation by nonlegumes as inferred from field experiments with legumes. Can. J. Microbiol. 29, 1053–1062.
10. Gelin O and Blixt S 1964 Root nodulation in peas. Agri. Hortique Genetica 22, 149–159.
11. Graham P H 1981 Some problems of nodulation and symbiotic nitrogen fixation in *Phaseolus vulgaris* L: A review. Field Crops Res. 4, 93–112.

12. Graham P H 1982 Plant factors affecting symbiotic nitrogen fixation in legumes. *In* Biological Nitrogen Fixation Technology for Tropical Agriculture. Eds. P H Graham and S C Harris. CIAT, Cali, Colombia pp 27–37.
13. Graham P H 1983 Plant factors affecting nodulation and symbiotic nitrogen fixation in legumes. *In* Biological Nitrogen Fixation and its Ecological Basis. Ed. M Alexander. Plenum, New York pp 1–40.
14. Graham P H and Halliday J 1977 Nodulation and nitrogen fixation the genus *Phaseolus*. *In* Exploiting the Legume-Rhizobium Symbiosis in Tropical Agriculture. Eds. J M Vincent *et al.* Univ. Hawaii Coll. Tropic. Agr. Misc. Publ. 145, 313–334.
15. Hardarson G and Jones D G 1979 The inheritance of preference for strains of *Rhizobium trifolii* by white clover (*Trifolium repens*). Ann. Appl. Biol. 92, 329–333.
16. Hardarson G, Heichel G H, Barnes D K and Vance C P 1982 Rhizobial strain preference of alfalfa populations selected for characteristics associated with N_2 fixation. Crop Sci. 22, 55–58.
17. Hardy R W D, Burns R C and Holsten R D 1973 Applications of the acetylene-ethylene assay for measurement of nitrogen fixation. Soil Biol. Biochem. 5, 47–51.
18. Harper J E 1974 Soil and symbiotic nitrogen requirements for optimum soybean production. Crop Sci. 14, 255–260.
19. Harper J E and Gibson A H 1984 Differential nodulation tolerance to nitrate among legume species. Crop Sci. 24, 797–801.
20. Holl, F B 1975 Host plant control of the inheritance of dinitrogen fixation in the Pisum-Rhizobium symbiosis. Euphytica 24, 767–770.
21. Jacobsen E 1984 Modification of symbiotic interaction of pea (*Pisum sativum* L.) and *Rhizobium leguminosarum* by induced mutations. Pland and Soil 82, 427–438.
22. Jacobsen E and Feenstra W J 1984 A new pea mutant with efficient nodulation in the presence of nitrate. Plant Sci. Letters 33, 337–344.
23. Kneen B Y E and LaRue T A 1984 EMS derived mutant of *Pisum sativum* resistant to nodulation. *In* Advances in Nitrogen Fixation Research. Eds. C. Vleeger and W E Newton. Martinus Nijhoff/Dr. W. Junk Publishers, Dordrecht.
24. Knowles R 1981 The measurement of nitrogen fixation. *In* Current Perspective in Nitrogen Fixation. Proc. 4th Int. Symp. Nitrogen Fixation. Eds. A H Gibson and W E Newton. Canberra. pp 327–333.
25. Kueneman E A, Root W R, Dashiell K E and Hohenberg J 1984 Breeding soybeans for the tropics capable of nodulating effectively with indigenous *Rhizobium* spp. Plant and Soil 82, 387–396.
26. LaRue T A and Patterson T G 1981 How much nitrogen do legumes fix? Adv. Agron. 34, 15–38.
27. Lie T A 1981 Gene centres, a source for genetic variants in symbiotic nitrogen fixation: host-induced infectivity in *Pisum sativum* L. ecotype *fulvum*. Plant and Soil 61, 125–134.
28. Lie T A 1984 Host genes in *Pisum sativum* L. conferring resistance to European *Rhizobium leguminosarium* strains. Plant and Soil 82, 415–425.
29. Madsen E L and Alexander M 1982 Transport of *Rhizobium* and *Pseudomonas* through soil. Soil Sci. Soc. Am. J. 46, 557–560.
30. McFerson J R 1983 Genetic and breeding studies of dinitrogen fixation in common bean, *Phaseolus vulgaris* L., Ph.D. Thesis, University of Wisconsin, Madison, 147 p.
31. McFerson J R, Bliss F A and Rosas J C 1982 Selection for enhanced nitrogen fixation in common beans, *Phaseolus vulgaris*. *In* Biological Nitrogen Fixation Technology for Tropical Agriculture. Eds. P H Graham and S C Harris. CIAT, Cali, Colombia. pp 39–44.
32. Maier R G and Brill W J 1978 Mutant strains of *Rhizobium japonicum* with increased ability to fix nitrogen for soybean. Science 201, 448–450.
33. Minchin F R, Sheehy J E and Witty J F 1985 Factors limiting N_2 fixation by the legume-*Rhizobium* symbiosis. *In* Nitrogen fixation Research Progress. Eds. H J Evans, P J Bottomley, and W E Newton. Martinus Nijhoff Publishers. Dordrecht pp 285–291.
34. Minchin F R, Summerfield R J, Hadley P, Roberts E H, and Rawsthorne S 1981 Carbon and nitrogen nutrition of nodulated roots of grain legumes. Plant Cell Envir. 4, 5–26.

35 Nambiar P T, Nigam S N, Dart P J and Gibbons R W 1983 Absence of root hairs in non-nodulating groundnut. *Arachis hypogea* L. J. Exp. Bot. 34, 484 488.
36 Noel K D and Brill W J 1980 Diversity and dynamics of indigenous *Rhizobium japonicum* populations. Appl. Environ. Microbiol. 40, 931 938.
37 Nutman P S 1967 Varietal differences in the nodulation of subterranean clover. Aust. J. Agric. Res. 18, 381–425.
38 Pacovsky R S, Bayne H G and Bethlenfalvay G J 1984 Symbiotic interactions between strains of *Rhizobium phaseoli* L. Crop. Sci. 24, 101 105.
39 Patterson T G and LaRue T A 1983 N_2-fixation (C_2H_2) and ureide content of soybeans: Ureides as an index of fixation. Crop Sci. 23, 825 831.
40 Phillips D A 1980 Efficiency of symbiotic nitrogen fixation in legumes. Ann. Rev. Plant Physiol. 31, 29 49.
41 Rao V R, Thollapilly G and Ayanaba A 1982 Studies on the persistence of introduced strains of *Rhizobium japonicum* in soil during fallow and the effect on soybean growth and yield. *In* Biological Nitrogen Fixation Technology for Tropical Agriculture. Eds. P H Graham and S C Harris. CIAT, Cali. pp 309 315.
42 Ronis D H, Sammons D J, Kenworthy W J and Meisinger J J 1985 Heritability of total and fixed N content of the seed in two soybean populations. Crop Sci. 25, 1 4.
43 Rosas J C 1983 Partitioning of dry matter, nitrogen fixation and seed yield of common bean (*Phaseolus vulgaris* L.) influenced by plant genotype and nitrogen fertilization. Ph.D. Thesis, University of Wisconsin, Madison, 127 p.
44 Rosas J C and Bliss F A 1986 Host plant traits associated with estimates of nodulation and nitrogen fixation in common bean. HortScience 21(2): (*In press*).
45 Sprent J I and Thomas R J 1984 Nitrogen nutrition of seedling grain legumes: some taxonomic, morphological and physiological constraints. Plant Cell Envir. 7, 637–645.
46 St. Clair D A, Rosas J C, Bliss F A, DuBois J D and Burris R H 1985 Evaluation of N_2 fixation and N-partitioning in common bean using ^{15}N-depleted $(NH_4)_2SO_4$ and acetylene reduction assay. HortScience 20, 586.
47 Thibodeau P S and Jaworski E G 1975 Patterns of nitrogen utilization in the soybean. Planta 127, 133 147.
48 Thomas R J and Schrader L E 1981 Ureide metabolism in higher plants. Phytochemistry 20, 361–371.
49 Thomas R J, McFerson J R, Schrader L E and Bliss F A 1984 Composition of bleeding sap nitrogen from lines of field-grown *Phaseolus vulgaris* L. Plant and Soil 79, 77 78.
50 Vest G 1970 Rj_3 — a gene conditioning ineffective nodulation in soybean. Crop Sci. 10, 34–35.
51 Vest G and Caldwell B F 1972 Rj_4 — a gene conditioning ineffective nodulation in slybean. Crop Sci. 12, 692 693.
52 Williams L F and Lynch D L 1954 Inheritance of a non-nodulating character in the soybean. Agron. J. 46, 28–29.
53 Winarno R and Lie T A 1979 Competition between Rhizobium strains in nodule formation: interaction between nodulating and non-nodulating strains. Plant and soil 51, 135 142.
54 Wynne J C, Ball S T, Elkan G H, Islieb T G and Schneeweis T J 1982 Host and host plant factors affecting nitrogen fixation in peanut. *In* Biological Nitrogen Fixation Technology for Tropical Agriculture. Eds. P H Graham and S C Harris. CIAT, Cali, Colombia, pp 67–75.

Specific relations between some strains of diazotrophs and corn hybrids

M.R. SARIĆ, ZORA SARIĆ* and M. GOVEDARICA*
Institute of Biology, Faculty of Natural Sciences and Institute of Field and Vegetable Crops, Faculty of Agriculture Novi Sad, Yugoslavia*

Key words Associations Corn hybrids Diazotroph strains Genotype N-concentration

Summary We tested the effect of a large number of diazotrophic strains from the genera Azotobacter, Klebsiella, Escherichia, Derxia, Beijerinckia, and Azospirillum, on nitrogen concentration in four corn hybrids. The range obtained, from highly significant positive to highly significant negative effects, indicates that the strains studied differed in their characters, including the capacity to form useful associations with certain hybrids. The forming of positive associations depended on the specificity of diazotrophic strains and corn hybrids.

Introduction

Many papers cover the morphology, physiology, and ecology of diazotrophs[4,7,10,15,22]. Recently attention has been focused on the relationship between strains of nitrogen-fixing microorganisms and plant species[2,6,12,14,16,25].

We continued our investigations of genetic specificity of plant nutrition[12,20], with plant cultivars, *i.e.*, genotypes, and microorganism strains[21]. These investigations dealt specifically with the relationships between Azotobacter strains on one side and corn and wheat genotypes on the other. Later on, however, the investigations were extended to include not only Azotobacter but other diazotrophs as well. The main objective was to investigate the effect of numerous strains of different genera on genotypes of cultivated plants in order to find the best strain/genotype combinations with respect to the efficiency of atmospheric nitrogen fixation.

Materials and methods

The investigations included 21 Azotobacter strains, 18 Klebsiella strains, 12 Escherichia strains, 7 Derxia strains, 4 Azospirillum strains, and 4 Beijerinckia strains isolated from specific zones of corn rhizosphere. The seed of four corn hybrids (NSSC-606, NSSC-78, NSSC-530, and NSSC-425) was inoculated simultaneously with all diazotrophic strains at the time of sowing. To each pot was added 30 ml of inoculum derived from corresponding liquid media: Fiodorov's nutritive medium for Azotobacter strains, a combined lactose and yeast extract medium for Escherichia strains, a combined sodim citrate and yeast extract medium for Klebsiella strains, Jenssen's medium (1960) for Derxia strains, Becking's medium (1961) for Beijerinckia strains, and Fiodorov's medium with malate for Azospirillum strains.

The inoculation was prepared at 30°C, with a vibrator and all strains were in the initial, stationary phase. The generative periods of Azotobacter, Escherichia, and Klebsiella, in the respec-

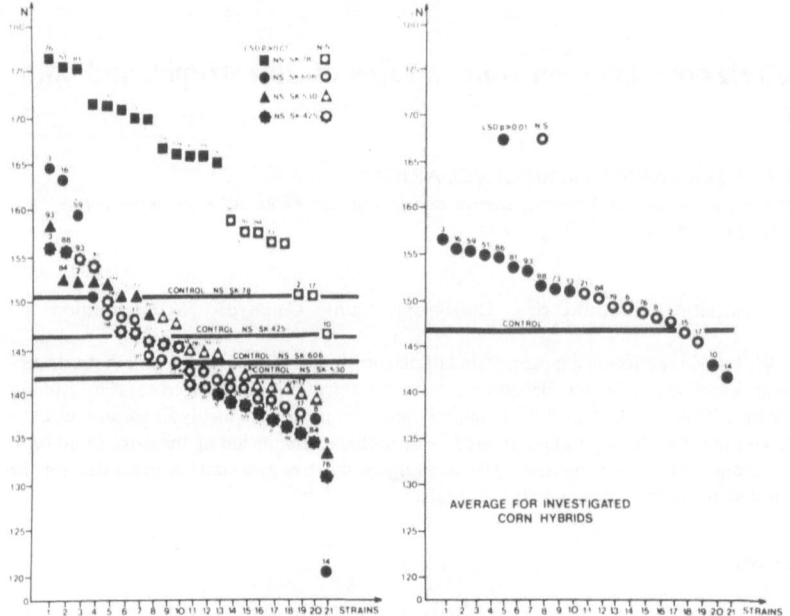

Fig. 1. The influence of different Azotobacter strains on the concentration of nitrogen in above-ground parts of corn hybrids.

tive media, were shorter than those of the other three genera. The number of cells in the inoculum ranged between 10^8 and 10^9/ml. The trials were carried out in a greenhouse, in sand culture, adding only distilled water, without nutrients, in the first 30 days of plant growth. Each treatment included five replicates, 10 plants per replicate. Although we analysed several parameters, this paper deals only with nitrogen concentration in the roots and above-ground parts of the experimental plants. Nitrogen was determined after Kjeldahl.

Results and discussion

Azotobacter

The 21 Azotobacter strains displayed large variability regarding their effect on nitrogen concentration in the above-ground parts and roots of the corn hybrids. Eighteen strains caused significant increases in nitrogen concentration in the above-ground parts of at least one hybrid, one strain caused a statistically significant reduction in one hybrid, while the effects of two strains varied from significant to statistically insignificant (Fig. 1). However, out of the 18 strains which brought highly significant increases, eight did not cause a highly significant reduction in a single hybrid.

The largest number of Azotobacter strains (13) caused highly significant increases in nitrogen concentration in the bybrid NSSC-78; eight strains caused highly significant increases in NSSC-530, four strains in

RELATIONS BETWEEN DIAZOTROPHS AND CORN   497

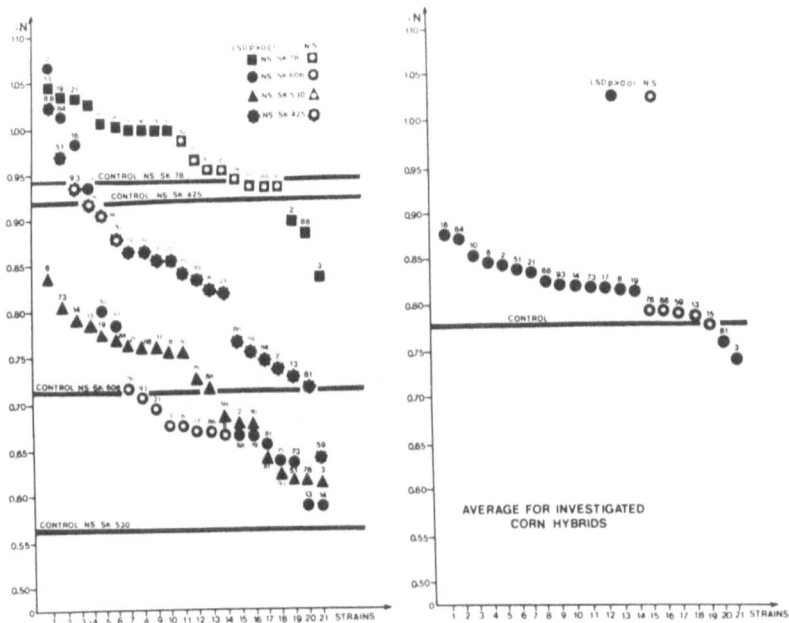

Fig. 2. The influence of different Azotobacter strains on the concentration of nitrogen in roots of corn hybrids.

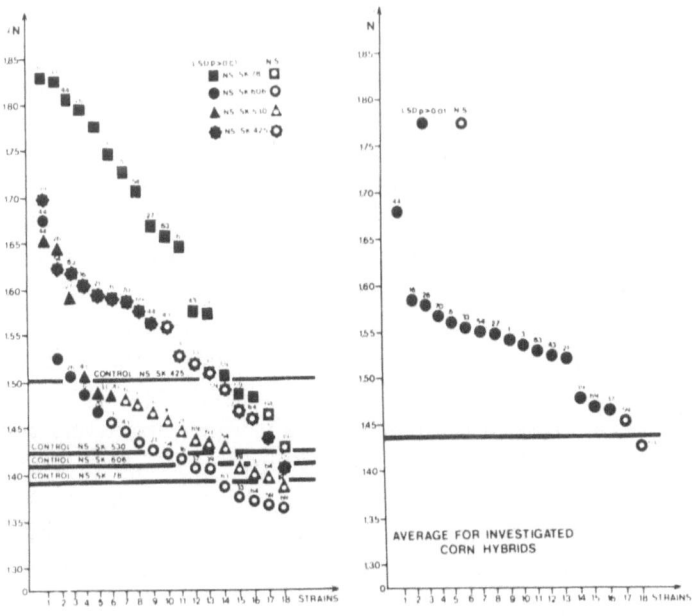

Fig. 3. The influence of different Klebsiella strains on the concentration of nitrogen in above-ground parts of corn hybrids.

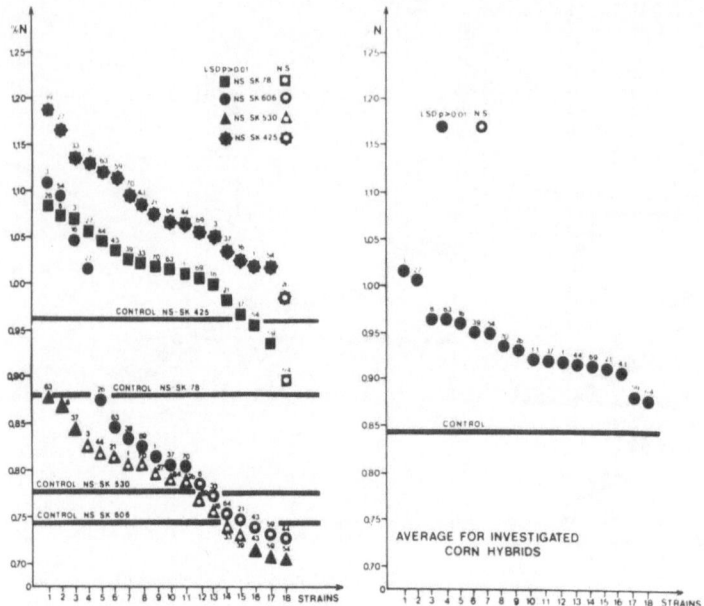

Fig. 4. The influence of different Klebsiella strains on the concentration of nitrogen in roots of corn hybrids.

NSSC-606, and two strains in NSSC-425. Conversely, the largest number of strains (9) reduced the concentration of nitrogen in NSSC-425.

Some strains affected several hybrids positively. The strain 59b brought highly significant increases in three hybrids, the strains, 3, 51, 73, 81, 86, 88, and 93 caused positive effects in two hybrids; the strain 15 did not bring a significant positive effect to a single hybrid. Some strains, *e.g.* 8 and 14, had highly significant negative effects on NSSC-606 and NSSC-530 but also highly significant positive effects on NSSC-78. Regarding their effects on nitrogen concentration in the roots, all the strains caused a highly significant increase in at least one hybrid (Fig. 2). The results display the scope of reaction of the corn hybrids to the strains of Azotobacter. All the strains caused highly significant increases in nitrogen concentration in the roots of NSSC-530, 10 strains in the roots of NSSC-78, six strains in the roots of NSSC-606, and only two strains in the roots of NSSC-425. Only three strains did not affect negatively the concentration of nitrogen in roots of a single hybrid.

It was evident that specific relationships exist between certain Azotobacter strains and certain corn hybrids. Those specific relationships were reflected on the concentration of nitrogen not only in the above-ground plant parts but in the roots as well.

Strains 10 and 51 induced highly significant positive effects in three hybrids; a number of strains caused positive effects in two hybrids. No single strain caused negative effects in all hybrids but certain strains caused negative effects in two hybrids at the most. The strain 2 was specifically positive for NSSC-606 and specifically negative for NSSC-78 and NSSC-425.

Klebsiella

All 18 Klebsiella strains had highly significant positive effects on nitrogen concentration in the above-ground parts and roots of at least one of the tested hybrids.

Regarding the concentration of nitrogen in the above-ground plant parts, 16 strains had highly significant positive effects on NSSC-78, nine strains on NSSC-425, six strains on NSSC-530, and only five strains on NSSC-606. Negative effects were found with only two strains (1, 26), on NSSC-425 (Fig. 3).

Strains 44 and 70 incited highly significant positive effects in all four hybrids, the strains 6 and 26 in three hybrids, and the strains 1, 16, 21, 27, 33, 37, 43, 54, 63, and 69 in two hybrids. The strains 3, 39, and 59 incited highly significant positive effects in only one hybrid, the strain 64 in neither one of the tested hybrids. Similarly to the behaviour of some Azotobacter strains, certain Klebsiella strains (1, 26) caused both highly significant positive effects (in NSSC-606 and NSSC-78) and highly significant negative effects (in NSSC-425).

Regarding the concentration of nitrogen in the roots, all Klebsiella strains had highly significant positive effects on at least one hybrid (Fig. 4). Seventeen strains induced highly significant positive effects in NSSC-78 and NSSC-425, 11 strains in NSSC-606, and three strains in NSSC-530. The strains 63 and 37 caused highly significant positive increases in nitrogen concentration in the roots of all four hybrids, the strains 1, 3, 6, 16, 27, 39, 54, 69, and 70 caused the same effects in three hybrids, the strains 21, 26, 33, 43, 44, and 59 in two hybrids. There were no negative effects of the Klebsiella strains on nitrogen concentration in the roots.

The results show that the Klebsiella strains affected specifically the concentration of nitrogen in the above-ground parts and roots of the hybrids tested. The Klebsiella strains deserve more attention since they were more universal than the Azotobacter strains.

Escherichia

The strains of this genus differed in their effect on the concentration of nitrogen in the above-ground parts and roots of the corn hybrids. Eight strains caused highly significant increases in nitrogen concentration in the above-ground parts of NSSC-606 and NSSC-425, seven strains caused highly significant increase in NSSC-530, three strains in

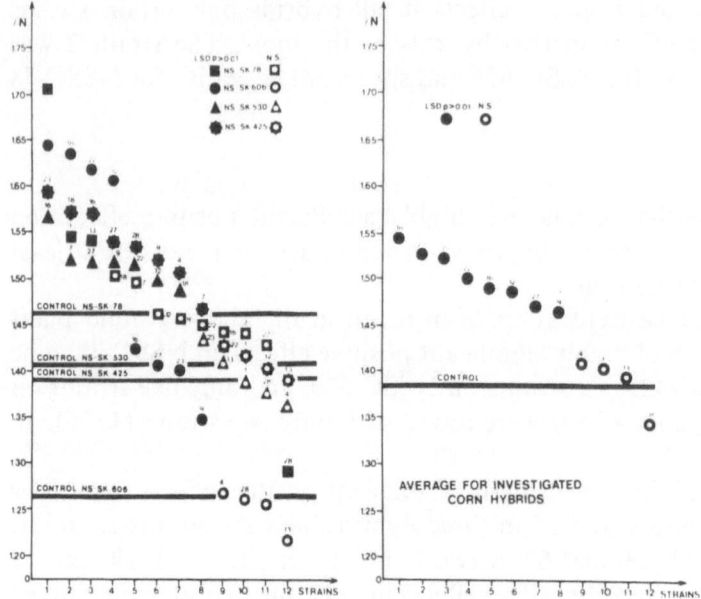

Fig. 5. The influence of different Escherichia strains on the concentration of nitrogen in above-ground parts of corn hybrids.

NSSC-78 (Fig. 5). A negative effect was induced by one strain in NSSC-78.

None of the strains affected all four hybrids positively. However, strains 7, 9, 16, 18, 23, and 27 induced highly significant positive effects in three hybrids, strains 22 and 32 in two hybrids. The strain 28 caused highly significant positive and negative effects in NSSC-425 and NSSC-78, respectively.

The strains differed regarding their effect on the concentration of nitrogen in the roots. Eleven strains caused highly significant increases in NSSC-78, ten in NSSC-606, six in NSSC-425, and two in NSSC-530 (Fig. 6). Three strains caused highly significant decreases in NSSC-530, one strain in NSSC-425. Neither one of the strains caused highly significant increases in all four hybrids; however, six strains (1, 4, 18, 22, 27, and 32) caused highly significant increases in three hybrids and the other six strains (7, 9, 13, 16, 23, and 28) in only two hybrids.

Strains 1 and 4 caused highly significant increases in NSSC-606, NSSC-78, and NSSC-425, but the contrary effect in NSSC-530. Strain 18 incited highly significant increases in NSSC-530, NSSC-606, and NSSC-78, but a decrease in NSSC-425.

In spite of the observed specificity regarding the effect on nitrogen concentration in the tested hybrids, virtually all of the strains affected positively at least two hybrids, either their above-ground parts or roots.

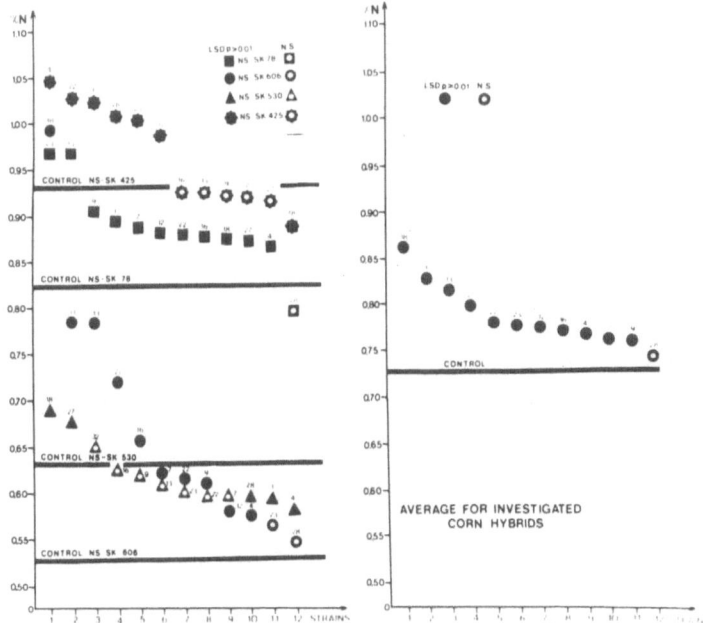

Fig. 6. The influence of different Escherichia strains on the concentration of nitrogen in roots of corn hybrids.

The only exception was strain 28 which was highly effective only with NSSC-425. Similarly to the previously discussed genera, the Escherichia strains differed in their effect from highly positive to highly negative.

Derxia

All of the tested strains caused highly significant increases in the concentration of nitrogen in the above-ground parts or roots of at least one hybrid. With respect to the concentration of nitrogen in the above-ground plant parts, four strains incited highly significant increases in NSSC-606 and one in each, NSSC-530 and NSSC-425 (Fig. 7). There were no positive effects with NSSC-78. Negative effects were found with all hybrids, especially with NSSC-78 which was negatively affected by four strains.

Neither one of the tested strains affected positively four or three hybrids. Strain 7 had highly significant positive effects on nitrogen concentration in the above-ground parts of two hybrids. The strain 14 had highly significant negative effects on three hybrids, the strains 1, 5, and 8 on two hybrids. The strains 5 and 8 incited highly significant positive effects in NSSC-606 and highly significant negative effects in NSSC-425 and NSSC-78.

Regarding the concentration of nitrogen in the roots, highly significant increases were incited in NSSC-425 by four strains, in NSSC-606 by

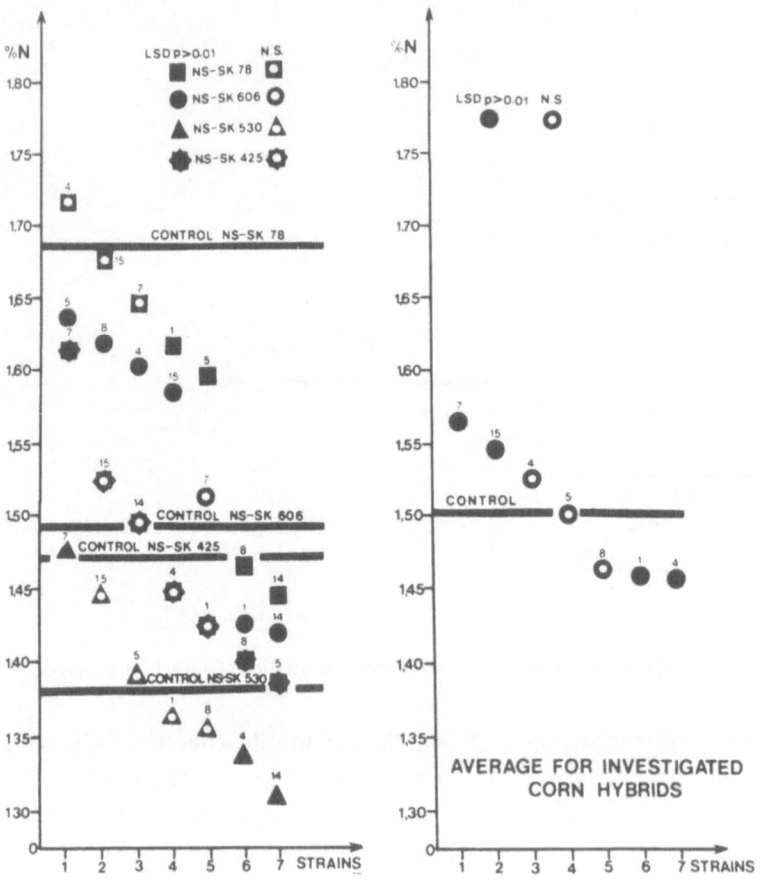

Fig. 7. The influence of different Derxia strains on the concentration of nitrogen in above-ground parts of corn hybrids.

three strains, and in each, NSSC-78 and NSSC-530, by one strain (Fig. 8). Negative effects were incited in NSSC-78 by three strains and in NSSC-530 by one strain.

Similarly to the situation with the above-ground parts, neither one of the examined strains increased nitrogen concentration in the roots of all four hybrids. The strains, 1, 14, and 15 brought increases in two hybrids. On the other hand, the strains 7 and 14 caused highly significant increases in NSSC-606 but they also brought decreases in NSSC-78.

The results show that the Derxia strains were as diverse as the previously discussed genera regarding the relationship strain — hybrid.

Beijerinckia

Four Beijerinckia strains caused highly significant increases in root nitrogen in NSSC-425 and NSSC-530, three strains in NSSC-78, and one

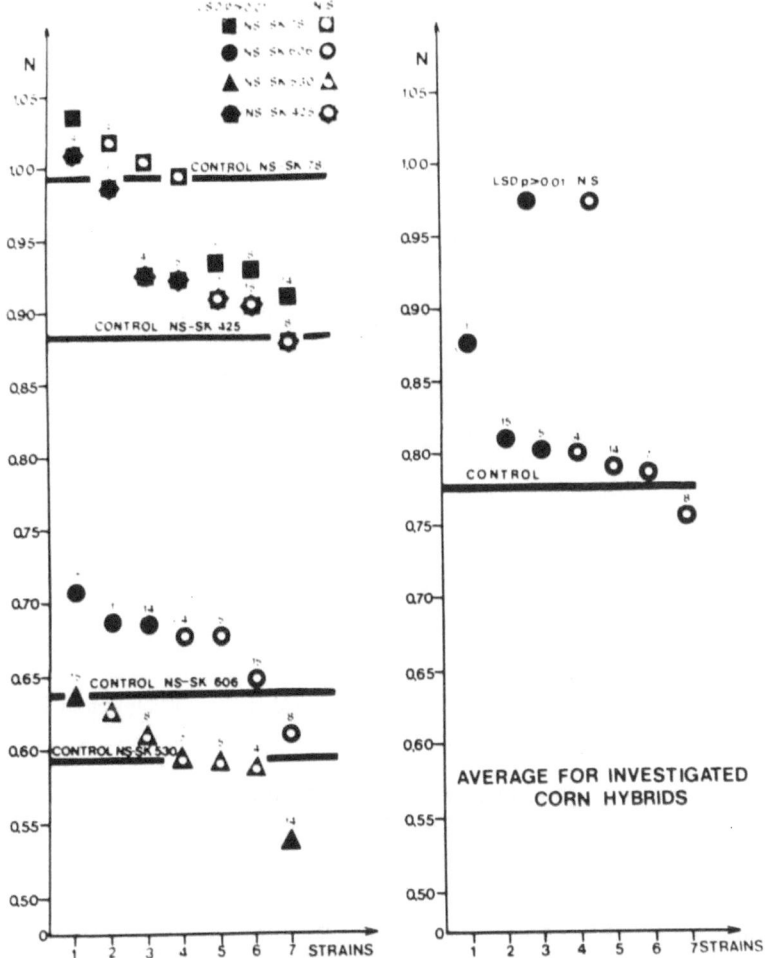

Fig. 8. The influence of different Derxia strains on the concentration of nitrogen in roots of corn hybrids.

strain in NSSC-606 (Fig. 9). Negative effects were not found. The strain 4 caused highly significant positive effects in all four hybrids, the strains 5 and 8 in three hybrids, and the strain 2 in two hybrids.

In the roots, two strains caused large positive effects in NSSC-530 and one strain in each, NSSC-425 and NSSC-606 (Fig. 10). On the other hand, highly significant negative effects were incited in NSSC-606, NSSC-78, and NSSC-425 by one strain. The strain 8 had positive effects on all hybrids. The strain 4 incited a highly significant increase in NSSC-530 and a highly significant decrease in NSSC-78.

Azospirillum

Regarding the concentration of nitrogen in the above-ground plant

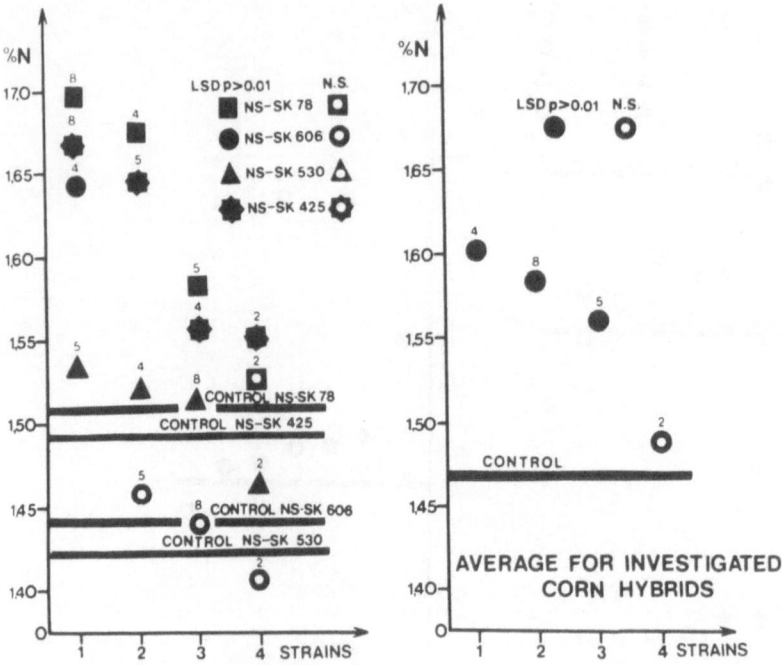

Fig. 9. The influence of different Beijerinckia strains on the concentration of nitrogen in above-ground parts of corn hybrids.

parts, two strains brought about highly significant increases in NSSC-425, and four strains in NSSC-530 (Fig. 11). There were no highly significant increases in NSSC-606 and NSSC-78. The strain 3 had highly significant effects on NSSC-530 and NSSC-425.

Three strains brought highly significant positive effects on root nitrogen in NSSC-530 and NSSC-606, one strain in NSSC-425 (Fig. 12). There were no highly significant increases in NSSC-78. Also, neither one of the tested Azospirillum strains had a highly significant negative effect on the hybrids. The strain 3 incited highly significant increases in three hybrids, the strains 2 and 8 in two hybrids.

The diazotrophs studied exhibited positive or negative effects on nitrogen concentration in the above-ground parts and roots of corn hybrids. These results indicate that large differences exist among diazotrophs which inhabit the rhizosphere and soil. It is the target of selection programs to isolate strains with a positive effect which could be used in plant production.

Specific relations between the diazotrophic strains and certain corn hybrids were clearly distinguished, expressed as positive or negative effects on nitrogen concentration, not only in a single hybrid but also in individual plant organs.

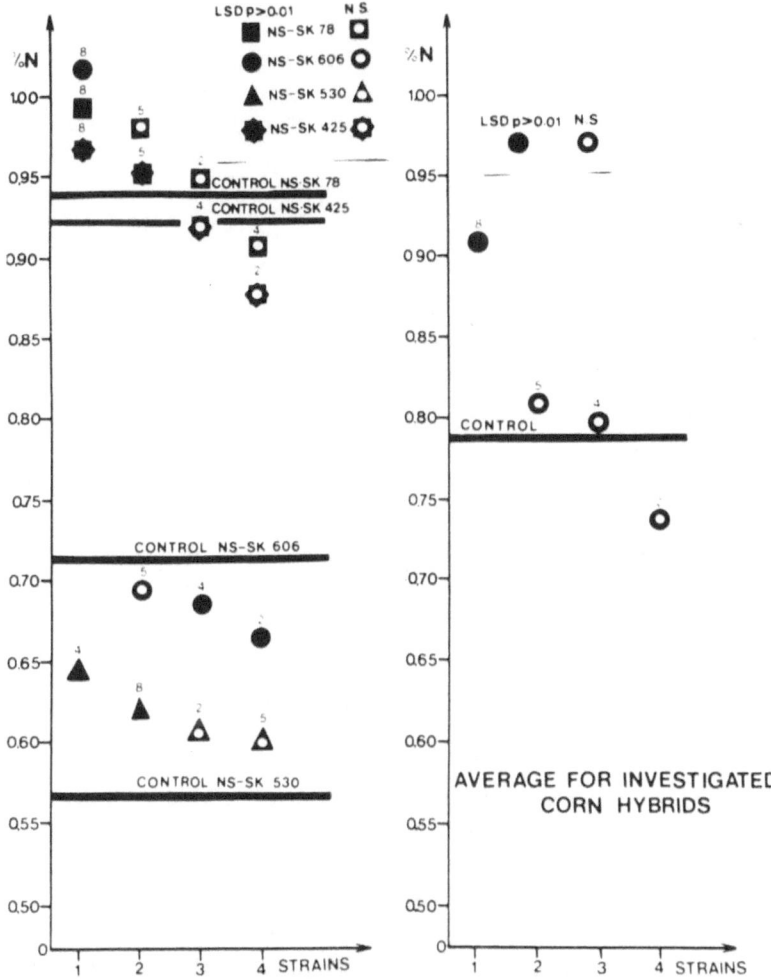

Fig. 10. The influence of different Beijerinckia strains on the concentration of nitrogen in roots of corn hybrids.

The capacity of certain combinations to produce positive associations depended on the variability of diazotrophic strains and the characters of corn genotypes. Some genotypes were capable of forming positive associations with a large number of diazotrophic strains, *e.g.*, NSSC-78 and several Klebsiella strains.

The genera of diazotrophs studied featured various degrees of adaptability to corn hybrids. For example, Klebsiella, Escherichia, and Derxia strains were each associated with a larger number of hybrids than Azotobacter strains. The last strains had a narrow specificity, *i.e.*, they were most frequently associated with only one hybrid.

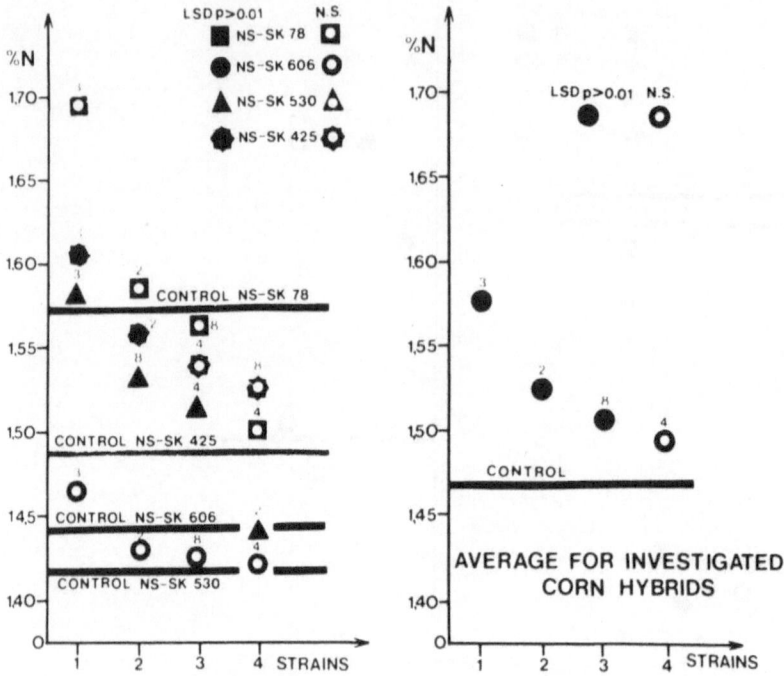

Fig. 11. The influence of different Azospirillum strains on the concentration of nitrogen in aboveground parts of corn hybrids.

The results show clearly that diazotrophs are capable of inducing positive effects on nitrogen concentration in corn hybrids. Therefore, they could be used to enhance plant growth. To obtain positive effects, however, it is necessary to find compatible partners, a certain genotype and a certain diazotrophic strain, to form associations.

Over the last decade major attention was paid to the acetylene — ethylene method of detecting nitrogenase activity[3,24,26] in a few strains from the Azospirillum genus. Positive results were obtained with these strains in Brasil[25] and Israel[12], in field conditions. An opinion was voiced that Azospirillum strains are most suitable for use in agriculture[27]. However, negative or nonsignificant effects were also obtained with Azospirillum strains[1,6,8,18]. These results indicate that apparently Azospirillum strains might be used advantageously in warmer climates.

Our results show that positive effects may be expected from different diazotrophic genera (Tables 1 and 2). Our attention should not be limited to Azospirillum alone. However, it is a painstaking endeavour to select appropriate diazotrophs for each plant genotype of a certain environment. Some researchers studying the use of N-fixing associations in agriculture believe Azotobacter strains to be less convenient for use than the strains of other diazotrophs[5]. On the other hand, our results

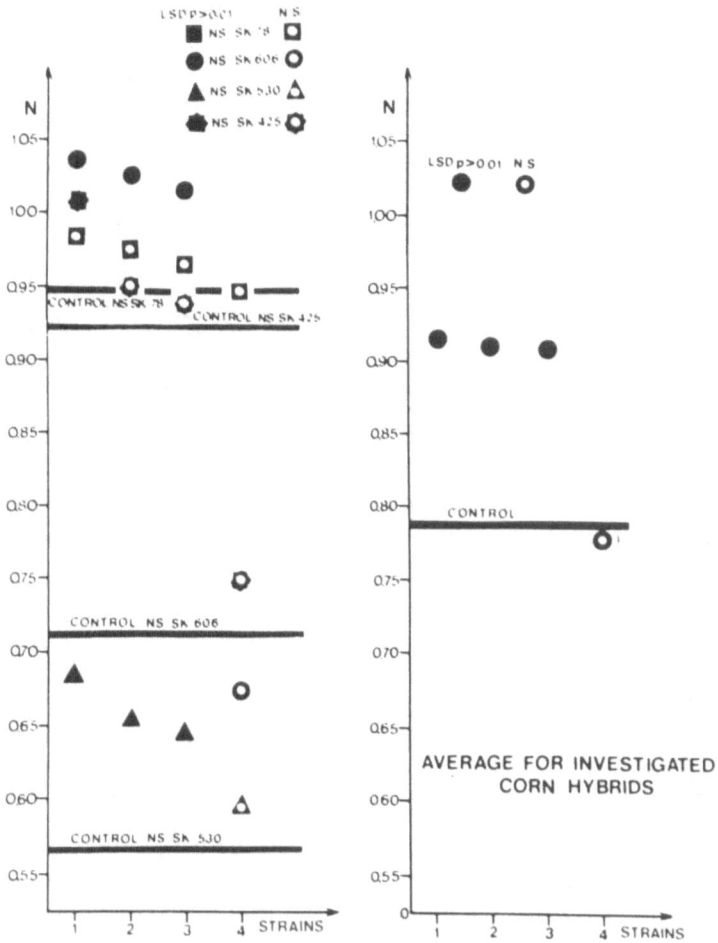

Fig. 12. The influence of different Azospirillum strains on the concentration of nitrogen in roots of corn hybrids.

Table 1. Percentage of strains of tested genera diazotrophs with highly significant positive effects on nitrogen concentration in the above-ground part of the hybrids

Genera	Corn hybrids								Total	
	606		78		530		425		Number	(%)
	Number	(%)	Number	(%)	Number	(%)	Number	(%)		
Azotobacter	4	14.8	13	48.2	8	29.6	2	7.4	27	23.9
Klebsiella	5	13.9	16	44.4	6	16.7	9	25.0	36	31.9
Escherichia	8	30.8	3	11.5	7	26.9	8	30.8	26	23.0
Derxia	4	66.7	0	0	1	16.6	1	16.6	6	5.3
Beijerinchia	1	8.3	3	25.0	4	33.3	4	33.3	12	10.6
Azospirillum	0	0	0	25.0	4	50.0	2	25.0	6	5.3
Total	22	19.5	35	31.0	30	26.5	26	23.0	113	100.0

Table 2. Percentage of strains of tested genera diazotrophs with highly significant positive effects on nitrogen concentration in the root of the hybrids

Genera	Corn hybrids								Total	
	606		78		530		425		Number	(%)
	Number	(%)	Number	(%)	Number	(%)	Number	(%)		
Azotobacter	7	17.5	10.0	25.0	21	52.5	2	5.0	40	29.0
Klebsiella	11	22.9	17	35.4	3	6.3	17	35.4	48	34.8
Escherichia	10	34.5	11	37.9	2	6.9	6	20.7	29	21.0
Derxia	3	33.3	1	11.1	1	11.1	4	44.4	9	6.5
Beijerinchia	1	20.0	1	20.0	2	40.0	1	20.0	5	3.6
Azospirillum	3	42.9	0	0	3	42.9	1	14.2	7	5.1
Total	35	25.4	40	29.0	32	23.2	31	22.5	138	100.0

show that Azotobacter strains are as diverse as the strains of other genera. The association between Paspalum notatum and Azotobacter[9] warns that Azotobacter should not be excluded from selection, especially not because in regions with the continental climate its rate of incidence varies from one agricultural crop to another[22].

Problems involved in the selection of diazotrophs are more serious than those involved in the selection of any plant species. It is expected that diazotrophs not only fix nitrogen but also to positively affect the yield, *i.e.*, the ultimate goal of agricultural production. The basic problem encountered in the selection of diazotrophs is the capacity of efficient strains to compete against the other members of microflora inhabiting the rhizosphere of a certain plant genotype.

It may be assumed that strains from prolific genera in a particular rhizosphere have a genetic advantage, *i.e.*, an acquired capacity to develop faster in that rhizosphere. Further along that line of thinking, it would be simpler to isolate efficient strains adapted to the genotype in question, from the prolific genera. However, we submit that the selection of diazotrophs should not be limited to acetylene reduction or % Na but it should include a larger number of indicators[21]. Differences in nitrogen concentration in the above-ground parts and roots of a certain hybrid imply differences in the mechanism of diazotrophic action, uptake of fixed N by roots, and nitrogen transport to the above-ground plant parts.

We may hope that further studies will supply more information on the specificity and differences in the mechanism of action of active associations between N_2-fixing strains and plant genotypes.

References

1 Albrecht S L, Okon Y and Burris R H 1977 Effects of light and temperature on the association between *Zea mays* and *Spirillum lipoferum*. Plant Physiol. 60, 528–531.

2. Baltenspreger A A, Schank S C, Smith R L, Litell R C, Bouton J H and Dudeck A E 1978 Effect of inoculation with Azospirillum and Azotobacter on turf-type Bermuda genotypes. Crop Sci. 18, 1043–1045.
3. Barber L E, Tjepkema J D, Russell S A and Evans H J 1976 Acetylene reduction (nitrogen fixation) associated with corn inoculated with Spirillum. Appl. Environ. Microbial. 82, 108–113.
4. Boddey R M and Dobereiner J 1982 Association of Azospirillum and other diazotrophs with tropical Gramineae in non-symbiotic nitrogen fixation and organic matter in the tropics. 12th Intl. Congress of Soil Sci, New Delhi, India, 28–48.
5. Brown M E 1982 *In* Bacteria and Plants Rhodes-Roberts and F A Skinner. Eds. M E Academic Press, London, 25–41.
6. Burris R H, Okon Y and Albrecht S L 1978 Environmental role of nitrogen-fixing blue-green algae and asymbiotic bacteria. Granhall U, Bull. Ecol., Stockholm, 26, 353–363.
7. Dobereiner J, Marriel I E and Nery M 1976 Ecological distribution of *Spirillum lipoferum* Beijerinck. Can. J. Microbiol. 22, 1464–1473.
8. Dobereiner J 1981 Emerging technology based on BNF by associative N_2 fixing organisms. Presented at Intl. Workshop on BIological Nitrogen Fixation Technology for Tropical Agriculture, Cali, Colombia.
9. Dobereiner J 1966 *Azotobacter paspali* sp. n., uma bacteria fixadora de nitrogenio na rizosfera de Paspalum. Pesq. agropec. bras. 1, 357–365.
10. Dommergues Y, Balandreau J, Rinando G and Weinhard P 1973 Non-symbiotic nitrogen fixation in the rhizospheres of rice, maize and different tropical grosses. Soil Biol. Biochem. 5, 83–89.
11. De-Polli H, Boyer C H and Neyra C A 1982 Nitrogenase activity associated with roots and stems of field-grown corn (*Zea mays* L.) Plants. Plant Physiol. 70, 1609–1613.
12. Ela W S, Anderson M A amd Brill W J 1982 Screening and selection of maize to enhance associative bacterial nitrogen fixation. Plant Physiol. 70, 1564–1567.
13. Kapulnik Y, Sarig S, Nur I, Okon Y, Kigel J and Henis Y 1981 Yield increases in summer cereal crops of Israel in fields inoculated with Azospirillum. Expl. Agric. 17, 179–187.
14. Mishustin E N and Yemtsev Y T 1982 Anaerobic nitrogen fixation and plant nutrition in non-symbiotic nitrogen fixation and organic matter in the tropics. 12th Inter. Congress of Soil Sci. New Delhi, India, pp 48–54.
15. Mulder E G 1975 Physiology and ecology of free-living, nitrogen-fixing bacteria, IBP, Volume 6, Cambridge University, Printed in Great Britain. pp 3–28.
16. Nur I, Okon Y and Henis 1980 An increase in nitrogen content of *Seracia italica* and *Zea mays* inoculated with Azospirillum. Can. J. Microbiol. 26, 482.
17. Okon Y 1984 Response of cereal and forage grasses to inoculation with N_2-fixing bacteria. *In* Advances in Nitrogen Fixation Research. Eds. C Veeger and W E Newton, Martinus Nijhoff, Dordrecht. pp 303–309.
18. Owens L 1977 Genetic Engineering for Nitrogen Fixation. Ed. A. Hollaender Plenum Press, New York. pp 473–482.
19. Sarić M R 1981 Genetic specificity in relation to plant mineral nutrition J. Plant Nutr. 3, 743–766.
20. Sarić M R and Loughman B C (Eds) 1983 Genetic Aspects of Plant Nutrition. Martinus Nijhoff, Dordrecht. pp 495.
21. Sarić Z, Sarić M R, Govedarica M, Krstić B, Barašević B and Stanković Ž 1984 Efficiency of nitrogen fixation by Azotobacter depending on maize form and mineral nitrogen fertilization. Third Intern. Symp. on Nitrogen Fixation with Non-Legumes, Helsinki.
22. Sarić Z 1978 The influence of mineral fertilizers on the population of Azotobacter and Oligonitrophilic bacteria in chernozem. Mikrobiologia, 15, 2, 153–166.
23. Tilak K V B R, Singh C S, Roy N K and Subba Rao N S 1982 *Azospirillum brasilense* and *Azotobacter chroococcum*: effect on yield of maize (*Zea mays*) and sorghum (*Sorghum bicolor*). Soil Biol. Biochem. 14, 417–418.
24. Tjepkema J and Van Berkum P 1977 Acetylene reduction by soil cores of maize and sorghum in Brazil. Appl. Environ. Microbiol. 33, 626–629.

25 Von Bülow J F W and Döbereiner J 1975 Potential for nitrogen fixation in maize genotypes in Brazil. Proc. Nat. Acad. Sci. USA. 72, 2389–2393.
26 Wood L V, Klucas R V and Shearman R C 1981 Nitrogen fixation (acetylene reduction) by *Klebsiella pneumoniae* in association with 'Park' Kentucky blue-grass (*Poa pratensis* L). Can. J. Microbiol. 27, 52–56.

Nitrogen fixation in soybean as influenced by cultivar and Rhizobium strain

S.K.A. DANSO,
Joint FAO/IAEA Division, Wagramerstrasse 5, P.O.B. 100, A-1400 Vienna, Austria

C. HERA
Research Institute for Cereals and Industrial Crops, Fundulea, Romania

and C. DOUKA
Nuclear Research Centre 'Democritos', Aghia Paraskevi, Attiki, Greece

Key words *Bradyrhizobium japonicum*   Dry matter yield ^{15}N-substratum labeling technique   Soybean varieties

Summary   The ^{15}N-substratum labeling technique and other indirect methods were used to compare nitrogen (N_2) fixation in soybean varieties grown in the field in Greece and Romania.

Significant variation in the amount (Ndfa) and proportion of N derived from fixation (% Ndfa) was found in different varieties. With 20 kg N/ha applied to soil, N_2 fixed ranged from 22 to 236 kg N/ha in Greece and from 17 to 132 kg N/ha in Romania. In general, varieties or treatments with higher dry matter yield supported greater fixation. Also, varieties with high Ndfa had high % Ndfa and *vice versa*. Breeding N_2-fixing legumes for high yields at low soil N levels therefore appears to be a reasonable strategy for enhancing N_2 fixation.

Heavy applications of inorganic N fertilizer severely depressed N_2 fixation in two out of the three varieties used in Romania. One variety, F 74-412, however, derived slightly higher amounts of N_2 from fixation at 100 kg N/ha rate than when fertilized with 20 kg N/ha. In Greece, Chippewa, Williams and Amsoy-71 inoculated with a Nitragin inoculant fixed similar amounts of N_2 at both 20 and 100 kg N/ha fertilizer rates. However, when Chippewa and Williams were inoculated with another, locally-isolated Rhizobium strain, N_2 fixation was substantially depressed at the higher N rate.

Introduction

Application of inorganic N fertilizer has been identified as one of the most important non-biological contributors to increased cereal yields after World War II[12]. However, the energy crisis of the seventies stimulated great interest in the exploitation of alternative or supplemental sources of N (such as biological nitrogen fixation) for high crop yields.

Although the genes that code for nitrogenase, the enzyme complex that converts N_2 into ammonia, are restricted to procaryotic organisms, the energy needed for the process and C skeletons required for amino acid synthesis are supplied by the host plant[12,20]. Thus, several bacterial and plant genes are involved in biological nitrogen fixation[14]. Enhanced nitrogen fixation would therefore require selecting for, or improving upon, the most important genetic traits in plants that influence the process.

H.W. Gabelman and B.C. Loughman (Eds.), Genetic aspects of plant mineral nutrition.
ISBN-13: 978-94-010-8102-3
© *1987 Martinus Nijhoff Publishers, Dordrecht/Boston/Lancaster.*

This paper reports the results of experiments using the N-15 labeling technique[7] to measure differences in N_2 fixation in soybean cultivars inoculated with different Rhizobium strains, and examines under field conditions the interaction between N_2 fixation and application of fertilizer N.

Materials and methods

The experiments were conducted in 1981 in Greece (Experiment I) and Romania (Experiment II) as part of an FAO/IAEA Co-ordinated Research Programme on N_2 fixation in grain legumes. The soil in Greece was a clay loam on a drained-out lake with a pH (in water) of 7.6 for the top soil, and that in Romania was a medium leached Chernozem on loess parent material, pH (in water), 7.3.

Three N_2-fixing soybean varieties were used in each experiment. The Greece study involved Chippewa, Williams and Amsoy-71, with a non-nodulating Chippewa isoline as reference crop, and those examined in Romania were Hodgson, F74-4124 and Flora with a non-nodulating soybean isoline Clay as reference crop.

In Experiment I, two *Bradyrhizobium japonicum* inoculants were used, one obtained from Nitragin Co., Milwaukee, U.S.A., and the other was a local isolate designated as Strain D. The two inoculants used in Experiment II were both local *B. japonicum* isolates designated as SO_{30} and SO_{618}. In all cases, Rhizobium strains were inoculated onto surface-sterilized seeds as peat-based water slurry just before sowing.

The ^{15}N substratum labeling equation of Fried and Middelboe[7] was adopted to estimate N_2 fixed under two N fertilizer rates, 20 and 100 kg N/ha in the form of 5 and 1% N-15 atom excess ammonium sulphate, respectively, in Experiment I. In Experiment II, similar ^{15}N enrichments were used but the fertilizer was double-labeled ammonium nitrate ($^{15}NH_4\ ^{15}NO_3$). No other fertilizer was applied in Experiment I, but the soil in Experiment II received an additional 90 kg P_2O_5/ha.

The experimental design in each study was a randomized block with N level as main plot treatment and the varieties and Rhizobium inoculants as sub-plots replicated six times. Crops were sown in Greece on 23 April and on 16 May in Romania. At various times during growth of the crops, plants were harvested and nodule number, mass and interior color were assessed and N_2 fixation was determined on plants harvested at physiological maturity. The aerial parts were dried in an oven at 70°C and % N was determined on a Kjeldahl digest[2]. Nitrogen isotope ratio analysis of samples in both Experiments was performed on a mass spectrometer[5].

Results

Experiment I

All plants were well nodulated except for Chippewa inoculated with Strain D which formed very few nodules with high dry weights (Table 1). The interior of most of the nodules examined was pink suggesting active N_2 fixation. The ranking for number of nodules formed by varieties was, Amsoy-71 > Williams > Chippewa. Nitragin inoculant induced production of more nodules than Strain D. Nodule development was, in general, higher in soil fertilized at the 20 kg N/ha rate than at the higher N rate especially when inoculated with Nitragin. Nodule numbers and weights increased during the period from 30/6/81 to 3/8/81.

The dry matter yield of Chippewa at maturity was substantially less

Table 1. Number and dry weight of nodules formed by Chippewa, Williams and Amsoy-71 inoculated with Nitragin or Strain D in soil fertilized with 20 and 100 kg N/ha

Variety	Rhizobium inoculant	Number of nodules/plant		Nodule dry weight (g/plant)	
		30/6/81	3/8/81	30/6/81	3/8/81
20 kg N/ha					
Chippewa	Nitragin	27	57	0.159	0.611
Williams	Nitragin	31	87	0.088	0.448
Amsoy-71	Nitragin	33	110	0.199	0.647
Chippewa	Strain D	2	8	0.025	0.125
Williams	Strain D	6	37	0.046	0.239
100 kg N/ha					
Chippewa	Nitragin	26	42	0.140	0.346
Williams	Nitragin	24	60	0.091	0.340
Amsoy-71	Nitragin	21	70	0.096	0.453
Chippewa	Strain D	1	10	0.006	0.089
Williams	Strain D	2	38	0.017	0.210

Table 2. Total dry matter yield (kg/ha) of Chippewa, Williams and Amsoy-71 inoculated with two different strains of Rhizobium

Variety	Nitrogen rate (kg N/ha)	
	20	100
Nitragin		
Chippewa	4702 a*	4771 a
Williams	7740 b	8098 b
Amsoy-71	7658 b	8090 b
Strain D		
Chippewa	4846 a	3894 a
Williams	7521 b	7060 b
Non-Nod. Chippewa	3522 a	4214 a

* All values followed by the same letter do not differ significantly as determined by Duncan's Multiple Range Test ($P < 0.05$).

than that of Williams or Amsoy-71, which gave reasonably similiar yields, especially in the Nitragin-inoculated treatments (Table 2). The type of Rhizobium strain used did not markedly affect dry matter yield of Chippewa or Williams, and the dry matter yield of each variety at the 100 kg N/ha rate was essentially similar to that at the lower N rate.

The N yield data in Table 3 show that like dry matter yields, Amsoy-71 and Williams contained almost equal amounts of total N, and significantly greater than that in Chippewa. In addition, Chippewa obtained a greater proportion of its total N from soil and fertilizer than Amsoy-71 and Williams (Table 4). The total N yield of each N_2-fixing variety was

Table 3. Total nitrogen content (kg N/ha) of Chippewa, Williams and Amsoy-71 inoculated with Nitragin or Strain D

Variety	Nitrogen rate (kg N/ha)	
	20	100
Nitragin		
Chippewa	217.9 cd*	228.0 d
Williams	373.7 fg	383.7 gh
Amsoy-71	363.9 fg	395.9 h
Strain D		
Chippewa	193.2 bc	167.6 ab
Williams	347.4 ef	320.7 e
Non-Nod. Chippewa	145.7 a	176.2 b

* All values followed by the same letter do not differ significantly as determined by Duncan's Multiple Range Test ($P < 0.05$).

Table 4. Nitrogen derived from fertilizer (Ndff) and soil (Ndfs) in kg N/ha or percent in Chippewa, Williams and Amsoy-71 inoculated with two strains of Rhizobium

Variety	Nitrogen rate (kgN/ha)			
	20		100	
	Ndff	Ndfs	Ndff	Ndfs
Nitragin				
Chippewa	3.9 a (1.8)t	149.5 a (69.8)yz*	21.8 b (9.7)xy	149.1 a (65.6)xy
Williams	4.2 a (1.2)s	160.8 a (44.8)vw	24.1 b (6.3)vw	159.4 a (42.7)uvw
Amsoy-71	3.3 a (0.9)r	125.0 a (34.6)uv	18.6 b (4.8)v	126.5 a (32.8)u
Strain D				
Chippewa	4.3 a (2.3)tu	167.0 a (87.3)z	20.4 b (12.2)yz	142.1 a (84.4)yz
Williams	3.8 a (1.1)rs	147.0 a (42.5)uvw	23.7 b (7.7)wx	159.8 a (53.6)wx

* Values in brackets indicate percentages derived from fertilizer or soil, and all values of either Ndff or Ndfs followed by the same letter do not differ significantly as determined by Duncan's Multiple Range Test ($P < 0.05$).

not increased much by raising the N fertilizer rate from 20 to 100 kg N/ha and total N yield in Chippewa and Williams was higher when Nitragin instead of Strain D was inoculated (Table 3).

With the exception of Chippewa at 100 kg N/ha, substantial N_2 fixation was detected in all varieties and treatments confirming that the nodules were active in fixation (Table 5). Amounts or proportion of N_2 fixed, however, varied greatly between treatments and ranged from almost undetectable to about 251 kg N/ha. A variety effect on fixation was apparent, with Amsoy-71 deriving the most from fixation, followed reasonably closely by Williams, and least was Chippewa in which fixation was substantially lower than in Williams or Amsoy in all treatments

Table 5. Nitrogen derived from fixation (kg N/ha or percent) in Chippewa, Williams and Amsoy-71 inoculated with Nitragin and strain D

Variety	Nitrogen rate (kg N/ha)	
	20	100
Nitragin		
Chippewa	64.5 b (28.4)xyz*	57.1 b (24.7)xy
Williams	208.6 d (54.1)yz	200.2 d (51.1)yz
Amsoy-71	235.6 d (64.5)z	250.8 d (62.3)yz
Strain D		
Chippewa	21.8 ab (10.4)wxy	5.1 a (0.0)w
Williams	196.9 d (56.3)yz	137.2 c (38.7)yz

* Numbers in brackets represent % N derived from fixation, and all values of either Ndfa or % Ndfa followed by the same letter do not differ significantly as determined by Duncan's Multiple Range Test ($P < 0.05$).

(Table 5). A Rhizobium strain effect was also observed. While Ndfa or % Ndfa at the 20 kg N/ha rate was similar in Williams inoculated with either Nitragin (208.6 kg N/ha) or Strain D (196.9 kg N/ha), Nitragin inoculation resulted in about a three-fold greater fixation in Chippewa (64.5 kg N/ha) than Strain D (21.8 kg N/ha). At the higher fertilizer rate, however, N_2 fixation was appreciably lower in both Chippewa and Williams inoculated with Strain D than Nitragin.

Nitrogen fixation in each of the three fixing varieties inoculated with Nitragin inoculant was similar at both N rates, but when strain D was used, N_2 fixation declined at the higher N fertilizer rate, although this did not reach a significant level with Chippewa (Table 5), suggesting a Rhizobium Strain × N fertilizer interaction.

Experiment II

Nodules were formed in all treatments (Table 6), and the interior of most of these were pink, suggesting that N_2 fixation took place. Nodule numbers in each treatment increased between the V_3 to the R_2 stage[4].

Except for the treatment with Strain SO_{618} at the 100 kg N/ha rate, in which dry matter yields of all nodulated varieties were essentially similar, the dry weight of Flora exceeded those of F 74-412 and Hodgson (Table 7). In general, F 74-412 also yielded more than Hodgson, which produced the lowest yield. Nitrogen fertilizer rate did not appreciably affect the dry matter yields of F74-412 and Flora, but Hodgson produced a significantly increased dry matter yield at the higher N fertilizer rate (Table 7). Hodgson also demonstrated a Rhizobium strain effect and yielded better when inoculated with strain SO_{30}.

Table 6. Number and dry weight of nodules formed by Hodgson, F74-412 and Flora inoculated with SO_{30} or SO_{618} in soil fertilized with 20 and 100 kg N/ha

Variety	Rhizobium strain	Nodule No./plant		Nodule wt (g/plant)	
		Growth stage			
		V_3	R_2	V_3	R_2
20 kg N/ha					
Hodgson	SO_{30}	18.8	30.0	0.11	0.15
F 74-412	SO_{30}	18.8	65.5	0.12	0.37
Flora	SO_{30}	15.3	28.6	0.12	0.14
Hodgson	SO_{618}	20.7	32.6	0.10	0.20
F 74-412	SO_{618}	23.8	64.1	0.16	0.39
Flora	SO_{618}	47.9	32.6	0.09	0.20
100 kg N/ha					
Hodgson	SO_{30}	12.1	25.7	0.05	0.15
F 74-412	SO_{30}	17.0	31.5	0.07	0.21
Flora	SO_{30}	9.1	23.2	0.05	0.14
Hodgson	SO_{618}	16.8	24.7	0.05	0.13
F 74-412	SO_{618}	17.2	32.2	0.09	0.20
Flora	SO_{618}	12.6	19.1	0.06	0.14

Table 7. Total dry matter yields (kg/ha) of Hodgson, F 74-412, and Flora and non-nodulating isoline Clay in soil fertilized with 20 and 100 kg N/ha

Variety	Nitrogen rate (kg N/ha)	
	20	100
SO_{30}		
Hodgson	7185 a*	8387 b
F 74-412	9613 cde	9253 bc
Flora	10680 f	10340 ef
SO_{618}		
Hodgson	8223 b	9850 def
F 74-412	9055 bcd	9917 def
Flora	9512 cde	9940 cde
Non-nod. Clay	8715 ab	8201 ab

* All values, followed by the same letter do not differ significantly as determined by Duncan's Multiple Range Test ($P < 0.05$).

The total N yield results presented in Table 8 show a wide variation between the various treatment combinations of the nodulated legumes (205 to 328 kg N/ha). On the average, Flora contained the highest N content, followed by F74-412 and Hodgson the least. Apart from Hodgson inoculated with Strain SO_{618}, the nitrogen yields of individual varieties were not significantly different at both N fertilizer rates. An examination of each individual variety inoculated with the different strains

Table 8. Total nitrogen content (kg N/ha) of Hodgson, F 74-412, Flora and non-nodulating isoline Clay in soil fertilized with 20 and 100 kg N/ha

Variety	Nitrogen rate (kg N/ha)	
	20	100
SO_{30}		
Hodgson	205.1 a*	226.2 ab
F 74-412	264.1 cd	258.8 bcd
Flora	326.0 e	327.9 e
SO_{618}		
Hodgson	220.8 a	283.8 d
F 74-412	263.0 bcd	293.9 de
Flora	281.1 d	282.2 d
Non-nod. Clay	226.3 abc	221.0 a

* All values followed by the same letter do not differ significantly as determined by Duncan's Multiple Range Test ($P < 0.05$).

Table 9. Nitrogen derived from fertilizer (Ndff) and soil (Ndfs) in kg N/ha in Hodgson, F 74-412 and Flora inoculated with strain SO_{30} or SO_{618} in soil fertilized with 20 or 100 kg N/ha

Variety	Nitrogen rate (kg/ha)			
	20		100	
	Ndff	Ndfs	Ndff	Ndfs
SO_{30}				
Hodgson	3.4 a (1.6)*st	130.3 a (62.9)st	35.2 c (15.6)u	171.4 abc (60.3)st
F 74-412	5.0 b (1.9)st	197.5 bc (74.0)u	31.2 c (12.0)v	152.3 ab (58.7)s
Flora	4.7 b (1.4)s	189.2 bc (57.3)st	43.4 cd (13.3)uv	209.6 bc (63.7)st
SO_{618}				
Hodgson	3.5 ab (2.0)t	164.2 abc (75.1)u	44.6 d (15.8)u	222.1 bc (70.8)tu
F 74-412	4.3 b (2.0)t	231.6 c (88.1)u	40.8 cd (14.0)uv	197.8 abc (70.8)tu
Flora	6.0 ab (2.3)t	173.4 abc (61.4)st	43.2 cd (15.4)uv	208.2 bc (73.6)tu

* Numbers enclosed in brackets represent percentages derived from fertilizer or soil, and all values of either Ndff or Ndfs (at both 20 and 100 kg N/ha) followed by similar letters do not differ significantly as determined by Duncan's Multiple Range Test ($P < 0.05$).

revealed no strain effect except for Flora which produced higher N yields at each of the N fertilizer rates when the inoculant was strain SO_{30} instead of SO_{618}. Soil N was a major contributor to the total N in all treatments (Table 9).

There was a large variation in N_2 fixation in various treatments (Table 10). The range in actual amounts fixed was from 17 to 132.1 kg N/ha, and the proportion fixed varied from 6 to 44.6%. On average, the highest N_2 fixation was achieved in Flora, with a mean of about 100 kg N/ha for all 4 treatments, or approximately 30% of the plant's N, and corresponds with the high ranking of Flora for dry matter (Table 7) and total N yield (Table 8). However, although F74-412 in general contained more N and

Table 10. Nitrogen fixed (kg/ha) in Hodgson, F 74-412 and Flora inoculated with strain SO_{30} or SO_{618} in soil fertilized with 20 or 100 kg N/ha

Variety	Nitrogen rate (kg N/ha)	
	20	100
SO_{30}		
Hodgson	71.4 bdc (35.5)wx*	19.6 a (7.4)v
F 74-412	61.6 abc (24.1)vwx	75.3 bcd (29.3)wx
Flora	132.1 d (41.2)x	35.7 ab (23.0)vwx
SO_{618}		
Hodgson	52.4 abc (22.9)vwx	17.0 a (5.9)v
F 74-412	25.4 ab (9.6)v	54.0 abc (16.2)vw
Flora	103.1 cd (37.0)wx	30.9 ab (11.0)vw

* Numbers in brackets represent percentage nitrogen derived from fixation, and all values of either Ndfa or % Ndfa followed by the same letter do not differ significantly as determined by Duncan's Multiple Range Test ($P < 0.05$).

produced more dry matter than Hodgson, the N_2 fixation trend was reversed at the lower N rate (Table 10).

A Rhizobium strain effect was obvious, with SO_{30} fixing slightly more N_2 in each of the three varieties at each N fertilizer rate (Table 10). Although, in general, N_2 fixation was severely reduced or inhibited at the higher N rate, it is significant that F74-412 in which N_2 fixation was lowest at the 20 kg N/ha rate irrespective of rhizobial strain, supported more fixation than any other variety at the 100 kg N/ha rate. The amounts fixed in F74-412 at the higher N rate were also a little more than the respective amounts fixed by either of the two Rhizobium strains at the 20 kg N/ha level (Table 10).

Discussion

Nodule number and mass have often been employed for indirect assessment of N_2 fixation[21,22]. In Experiment I in which there was profuse nodulation (Table 1), N_2 fixation estimates were also high, as much as 235 kg N/ha in one treatment (Table 5). In addition, based on nodule numbers, the N_2 fixation potential of varieties and strains at the 20 kg N/ha rate could be ranked as Amsoy-71 > Williams > Chippewa, and with Nitragin > Strain D, and this is in agreement with the estimates of N_2 fixed using ^{15}N. Dry weights of nodules on the other hand, did not show a good relationship with actual estimates of N_2 fixed. In Experiment II, there was no clear agreement between ranking made on actual amounts fixed (Table 10) and nodule numbers or dry weight (Table 6). For example, the F74-412 inoculated with SO_{30} combination had the highest nodule number at the R_2 stage (Table 6) and yet fixed the lowest amount of N_2 (Table 10), while Flora with less than half as much nodules derived about twice the amount of N_2 fixed in F74-412. These results emphasize the need for caution in using nodulation parameters as criteria in the final screening of legumes for N_2 fixing potentials. They are certainly useful in the initial screening of large collections of legumes and rhizobia, and especially for discarding obvious nonfixers.

The results presented in Tables 7 and 8 reveal a good link between dry matter and total N yield which is not surprising. In Experiment I, ranking varieties inoculated with Nitragin inoculant treatments according to dry matter yield (Table 2) was similar to ranking by actual N_2 fixed and proportion of N derived from fixation. Again, in Experiment, II Flora which had the highest dry matter (Table 7) and total N yields (Table 8) also gave the highest estimate of N_2 fixed as well as percentage N derived from fixation when SO_{30} was inoculated on plants grown with 20 kg N/ha fertilizer (Table 10). However, although F74-412 recorded substantially higher dry matter (Table 7) and total N yields (Table 8) than Hodgson, the latter fixed more N_2, thus showing the opposite trend. In addition, although the dry matter and especially N yields of F74-412 and Flora inoculated with SO_{618} were statistically similar in soil fertilized with 20 kg N/ha, the actual amount of N_2 fixed by Flora was about 300% that fixed by F74-412. Thus, both dry matter and N yield would be valuable tools in early stages of screening legumes for N_2 fixation abilities, but may not be reliable in more advanced or final stages.

The advantage in using ^{15}N is in its ability to distinguish between sources of N, and thus makes it possible to distinguish the amount of fixed N_2 in the plant from soil and fertilizer-derived N[6]. Thus, the ^{15}N methodology was capable of distinguishing differences in N_2 fixed by Flora and F74-412 (Table 10) even though they had similar total N contents (Table 8). In addition, by expressing N_2 fixation as the proportion of a plant's N that is derived from fixation, a more valid comparison of the N_2 fixing potentials of varieties or species with very different genetic backgrounds for dry matter and N yield can be made.

Several reports have shown that N_2 fixed by a Rhizobium strain is strongly influenced by the host plant[8,11,17,19], and that nitrogen fixation supporting traits[16] often vary among different hosts. Although only 3 varieties were used in each study, the results show a large spread in N_2 fixed between the three varieties in each study under identical soil N and Rhizobium treatment. In Experiment I, Amsoy-71 supported the highest N_2 fixation, and derived 236 kg N/ha or 64% of its total N from fixation in soil fertilized with 20 kg N/ha; Chippewa, on the other hand, fixed only 64 kg N/ha or 28% of total N under identical conditions (Table 5). In Experiment II similar large differences were found in the abilities of different soybean varieties to influence how much N_2 was fixed as well as the proportion of the plant's N that was derived from N_2 fixation (Table 10), stressing the importance of selecting for legume varieties with capability for high N_2 fixation.

Results obtained in these studies, in general, suggest that high yields may be an essential trait for enhanced N_2 fixation. In Experiment I, a reasonably good relationship was found between dry matter yields at both N rates (Table 2) and N_2 fixed in soybean at 20 kg N/ha (Table 5). Also in Experiment II, Flora on the average gave the highest dry matter yield (Table 7) as well as highest N_2 fixed in soil with 20 kg N/ha added (Table 10). The only notable exception in this relationship was the lower N_2 fixed by F74-412 despite its higher dry matter yield than Hodgson. This, however, does not invalidate the hypothesis that although Hodgson may genetically be capable of supporting high N_2 fixation, further enhancement of this capability could be realized by increasing the present yield potential. It is likely that by increasing yield further, not only would actual N_2 fixed be greater, but the proportion of N derived from fixation may also be higher, since photosynthates may be less limiting, and more carbohydrates could be made available to the nodules for N_2 fixation than under low yields or limited supply of photosynthates. Thus, by increasing soybean yields through CO_2 enrichment, not only was total N_2 fixed increased, but % Ndfa was also increased from 25% to 84%[13]. In addition, results from Experiments I (Table 5) and II (Table 10) show that plants that fixed high amounts of N_2 in general, also obtained higher proportions of their N through fixation than those that fixed little N_2. In both experiments, the Rhizobium strain used influenced the amount of N_2 fixed in any given variety. Nitragin was overall a superior inoculant to strain D at both N rates in Experiment I (Table 5), and strain SO_{30} was slightly superior to SO_{618} in Experiment II (Table 10), indicating the importance of Rhizobium strain selection to optimise N_2 fixation[11].

Nitrogen fixation in Hodgson and Flora was severely inhibited by the application of 100 kg N/ha to soil (Table 10), in agreement with reports that high soil N suppresses N_2 fixation[9,10,18]. However, F74-412 fixed more N_2 at the higher N rate than at 20 kg N/ha rate, in support of reports indicating that some varieties are capable of fixing high amounts of N_2 in nitrogen-rich soils[9,10,18]. Such varieties are necessary for optimising N_2 fixed in fertile soils or in intercropped systems that require the use of fertilizer N for high yields of the associated non-fixing crops. Further evidence for substantial N_2 fixation in fertilized or N-rich soils is shown by results obtained in Experiment I, where N_2 fixation in each of the three varieties inoculated with Nitragin was identical at the two N rates

(Table 5). However, when inoculated with Strain D, N_2 fixed at 100 kg N/ha was lower than at 20 kg N/ha, and points to the importance of selecting not only varieties for high N_2 fixation in N-rich soils but also attaching great importance to the strain of Rhizobium used.

In conclusion, several reports have demonstrated that a great genetic diversity in legume and Rhizobium strains exists in nature and represents large untapped biological resources that should be exploited to improve significantly upon the present levels of nitrogen fixed in several ecosystems. The problem can, however, be tackled best through an inter-disciplinary approach and may need closer collaboration between plant breeders, microbiologists, plant physiologists and agronomists, *etc.*, since N_2 fixation can only be maximized when highly effective Rhizobium strains symbiose with the best yielding lines. Breeding legumes for high yields while retaining or even enhancing the ability to nodulate and fix N_2 is essential.

References

1 De Jong T M, Brewin N J, Johnston A W B and Phillips D A 1982 Improvement of symbiotic properties in *Rhizobium leguminosarum* by plasmid transfer. J. Gen. Microbiol. 128, 1829–1838.
2 Eastin E F 1978 Total nitrogen determination for plant material. Anal. Biochem. 85, 591–594.
3 Eriksen F I and Whitney A S 1984 Effects of solar radiation regimes on growth and N_2 fixation of soybean, cowpea, and bushbean. Agron. J. 76, 529–535.
4 Fehr W R C, Caviness C E, Burmood D T and Pennington J S 1971 Stage of development descriptions for soybean, *Glycine max.* (L) Merill. Crop Sci. 11, 929–931.
5 Fiedler R and Proksch G 1975 The determination of ^{15}N by emission and mass spectrometry in biochemical analysis. A review. Anal. Chem. Acta. 78, 1–62.
6 Fried M, Danso S K A and Zapata F 1983 The methodology of measurement of N_2 fixation by non-legumes as inferred from field experiments with legumes. Can. J. Microbiol. 29, 1053–1062.
7 Fried M and Middelboe 1977 Measurement of amount of nitrogen fixed by a legume crop. Plant and Soil 47, 713–715.
8 Graham P H and Rosas J C 1977 Growth and development of intermediate bush and climbing cultivars of *Phaseolus vulgaris.* inoculated with *Rhizobium*. J. Agric. Sci. Camb. 88, 503–508.
9 Hardarson G, Zapata F and Danso S K A 1984 Effect of plant genotype and nitrogen fertilizer on symbiotic nitrogen fixation by soybean cultivars. Plant and Soil 82, 397–405.
10 Hardarson G, Zapata F and Danso S K A 1984 Symbiotic nitrogen fixation by soybean varieties as affected by nitrogen fertilizer application. *In* Advances in nitrogen fixation Research. Eds. C Veeger and W E Newton, Martinus Nijhoff/Dr. W. Junk Publishers, Hague, The Netherlands p 34.
11 Hardarson G, Zapata F and Danso S K A 1984 Field evaluation of symbiotic nitrogen fixation by rhizobial strains using ^{15}N methodology. Plant and Soil 82, 369–375.
12 Hardy R W F and Havelka U D 1975 Nitrogen fixation research: A key to world food? Science 188, 633–643.
13 Hardy R W F and Havelka U D 1975 Photosynthate as a major factor limiting N_2 fixation by field-grown legumes with emphasis on soybeans. *In* Symbiotic nitrogen fixation in plants. Ed. P.S. Nutman. Cambridge University Press. London pp 421–439.
14 Holl F B and LaRue T A 1975 Genetics of legume plant hosts. Proc. Ist. Intern. Symp. N_2 fixation, Washington State University, Pullman pp 391–399.

15 Nutman P S 1984 Improving nitrogen fixation in legumes by plant breeding: the relevance of host selection experiments in red clover (*Trifolium pratense* L.) and subterranean clover (*T. subterranean* L.) Plant and Soil 82, 285–301.

16 Rennie R J 1981 Potential use of induced mutations to improve symbiosis of crop plants with N_2-fixing bacteria. *In* Proc. Intern. Symp. Induced Mutations: A Tool in Plant Research IAEA STI/PUB/591 Vienna pp 293–321.

17 Rennie R J and Kemp G A 1983 N_2-fixation in field beans quantified by ^{15}N isotope dilution. I. Effect of strains of *Rhizobium phaseoli*. Agron. J. 75, 640–644.

18 Rennie R J and Kemp G A 1983 N_2-fixation in field beans quantified by ^{15}N isotope dilution. II. Effect of cultivars of beans. Agron. J. 1975, 645–649.

19 Ruschel A P, Vose P B, Matsui E, Victoria R L and Tsai Saito S M 1982 Field evaluation of N_2-fixation and N-utilisation by *Phaseolus* bean varieties determined by ^{15}N isotope dilution. Plant and Soil 65, 397–407.

20 Vincent J M 1984 Potential for enhancing biological nitrogen fixation. *In* Crop Breeding A Contemporary Basis. Eds P B Vose and S G Blixt. Permagon Press. pp 185–215.

21 Weber C R 1966 Nodulating and non-nodulating soybean isolines: II. Response to applied nitrogen and modified soil conditions. Agron. J. 58, 46–49.

22 Westermann D T and Kolar J J 1978 Symbiotic $N_2(C_2H_2)$ fixation by bean. Crop Sci. 18, 986–990.

Intraspecific variability for VA mycorrhizal symbiosis in cowpea (*Vigna unguiculata* [L.] Walp.)

S. RAJAPAKSE and J.C. MILLER, JR.
Department of Horticultural Sciences, Texas A & M University, College Station, TX 77843, USA

Key words Cowpea Cultivar variability Nitrogen fixation Nitrogenase activity Phosphorus uptake Rhizobium Southern pea Vesicular-arbuscular mycorrhizae Mycorrhizal development pattern

Summary The effects of inoculation with *Glomus mosseae* or *G. fasciculatum* and Rhizobium on plant growth and N_2 fixation parameters of two cowpea (*Vigna unguiculata* (L.) Walp.) cultivars, Brown Crowder (high N_2 fixing) and Bush Purple Hull (low N_2 fixing), were studied in pot experiments. Inoculation with VAM fungi significantly increased the percentage of colonized roots, plant height, and percent shoot nitrogen (N) and phosphorus (P), while root length decreased. Influence of VAM fungi on shoot dry weight and nitrogenase activity was dependent on species of VAM fungi as well as host cultivar. Root weight, nodule number and weight, root N and P and seed yield were not affected by mycorrhizal inoculation. A significant difference between the two cultivars for percent root colonization at four weeks after inoculation was also found. The developmental pattern for hyphae, vesicles and arbuscules was not significantly different between cowpea cultivars or VAM species.

Introduction

The benefits of mycorrhizal symbiosis in agronomic and horticultural plants have been well documented[7,25]. Response to inoculation with vesicular-arbuscular mycorrhizal (VAM) fungi by the host plant differs markedly, depending on species and cultivar[1,7,19]. Legumes and citrus are generally more responsive to inoculation with VAM fungi than are grass species[16,19], and cultivar differences in response have been reported for corn (*Zea mays* L.)[7], wheat (*Triticum vulgare*)[1], soybean [*Glycine max* (L.) Merr.][21], clover (*Trifolium subterraneum* L.)[4], citrus (*citrus* species)[13], and cowpea[17].

Ollivier *et al.*[17] studied the mycorrhizae formed by 3 VAM fungi on 2 cowpea cultivars. Their study was directed mainly towards the enzymatic activity in the extracts of mycorrhizal roots. However, they failed to look at nitrogen (N_2) fixation, nodulation and root growth. Islam and Ayanaba[9] used 2 cowpea cultivars in 2 separate experiments, so that direct comparisons cannot be made between the 2 cultivars. Studies with a single cowpea cultivar have also been reported[2,8,10,11].

The species of VAM endophyte as well as the variety of host plant influences host response[4,17,21]; for example, *Gigaspora gigantea* performs better than *Glomus mosseae* in association with soybeans (cv. Bossier)[21], and similar interactions have been reported for white clover[4]. Mosse[15]

reported that progress of root colonization and net effects on host plants differ from one host-endophyte combination to another. Therefore, beneficial effects of the VAM symbiosis can be maximized by selecting appropriate VAM endophyte-host genotype combinations.

The extent to which mycorrhizae can improve growth depends on the level of root colonization[15]. The intensity of mycorrhizal colonization positively influences the development of nodules in legumes and favors an effective symbiosis[4,22]. Therefore, quantitative estimation of root colonization by VAM fungi is an important aspect of research on endomycorrhizae. The level of root colonization by mycorrhizal fungi is subject to change with growth of the host plant. Hence, estimating mycorrhizal colonization at successive stages of growth of the host plant should help in the understanding of possible variation in the development of root colonization by different VAM fungi on different host cultivars.

Development of VA mycorrhiza with time has been studied in several crop plants, including groundnut (*Arachis hypogaea* L.), pigeon pea (*Cajanus cajan* L.) and certain vegetables[20]. Time course studies of root colonization in onion (*Allium cepa* L.) and subterranean clover (*Trifolium subterraneum* L.) with 3 species of VAM fungi showed a marked variation in ontogeny of colonization between host-endophyte pairs[3]. As there are differences between host-endohyte pairs for colonization, cultivars of one host species may also exhibit differential colonization with time.

The research reported herein was undertaken to study the time course development of root colonization by 2 VAM fungi on high and low N_2 fixing cowpea cultivars and to study the effects on plant growth, nodulation, N_2 fixation and P uptake at a given time using two different symbionts.

Materials and methods

Time course study of root colonization

This experiment was conducted in the greenhouse using 'Brown Crowder' and 'Bush Purple Hull' cowpeas. These cultivars were selected because of their differential growth habit and ability to fix N_2. 'Bush Purple Hull' is a low N_2 fixing, determinate cultivar, whereas 'Brown Crowder' is a high N_2 fixing, indeterminate cultivar[27]. Seeds were planted in seedling trays containing a medium of 1:1 (v v) peat and perlite steamed for 3 hours. This medium contained < 6 ppm N and < 11 ppm P with a pH of 6.7. Daylength averaged 16 h and the daytime temperature during the light period ranged from 23–28 C. One week after planting, at the time of transplanting, seedlings were inoculated with either *G. mosseae* (Nicolson and Gerdemann) Gerdemann and Trappe or *G. fasciculatum* (Thaxt. *sensu* Gerdemann) Gerdemann and Trappe. Spore count per 40 g of *G. mosseae* and *G. fasciculatum* inoculum was 40 and 50, respectively. Inocula were prepared using Sudan grass (*Sorghum vulgare* Pers.) as the host plant. All plants were inoculated with rhizobial strain 32-H-1. Approximately 40 g of mycorrhizal inoculum and 5 g of peat-based rhizobial inoculum were added to 1 liter plastic pots and mixed with steamed (80 C, 225 min) perlite medium before seedling transplant. Treatments were arranged in a randomized block design with 4 replications. A nutrient solution lacking N and P was added 3 times, and plants were watered as required. Plants were harvested at 1, 2, 4, 6, 8 and 10 weeks after transplanting. At each harvest, plants were cut at the soil line with pruning shears. The roots were then carefully removed from the soil, placed in plastic bags and frozen for later assessment of colonization by VAM fungi.

Root colonization by VAM fungi was determined as follows. Roots were choped into approximately 1 cm segments and random samples were selected. These root segments were washed with distilled water, cleared, and stained according to the technique of Phillips and Hayman[18]. From each of the 4 replicates, 50 stained root segments/plant were examined in an open petri dish under a dissecting microscope. Segments containing VAM fungi were considered mycorrhizal. The percent colonization or the incidence of colonization was defined as [(number of colonized root segments/ total number of root segments observed) × 100]. Ten colonized root samples were selected out of 50, mounted on glass slides and examined under a light microscope to assess the intensity of colonization. The amount of hyphae, arbuscules or vesicles present was rated on a scale of 0-5, as described by Yost and Fox[26].

Study of growth effects

The same 2 cultivars and VAM fungi were used in this experiment and seedlings were started as described for the time course study. Both cowpea cultivars were inoculated with *G. mosseae* or *G. fasciculatum* or not inoculated (mycorrhizal control).

The 2 × 3 factorial experiment in a randomized block design included 4 replications. In the mycorrhizal treatments, seedlings were inoculated when transplanted: 25 g of inoculum, including soil, colonized roots, and spores were added to the medium. The inocula had been prepared the previous season, with Sudan grass (*Sorghum vulgare* Pers.) serving as the host plant. *G. mosseae* inoculum had 76 spores/100 g of soil, while *G. fasciculatum* inoculum consisted of 81 spores/100 g of soil. Seedlings were inoculated with a single strain (32-H-1) of Rhizobium. All pots were placed in a greenhouse, and nutrient solution, free of N and P, was applied as needed. The daytime temperature varied from 25 to 30 C.

In this experiment, there were 2 plants per treatment in each replication. Planted in separate pots, the two plants were harvested at different times; one at 8 weeks after planting to measure nitrogenase activity, plant height, dry weight, nodulation, root length, and N and P content of shoots and the other at 10 weeks to measure seed weight.

Plant height was measured at 8 weeks, using one plant/treatment/replication. Plants were severed at the soil line. Nitrogenase activity was measured by the acetylene-ethylene assay according to the method described by Zary *et al.*[27]. Dry weight of shoots (70 C, 24 h), fresh weight of roots, number and dry weight of nodules were also recorded. For shoot N and P determination, dried and ground samples were digested with sulfuric acid[24]. Total N and P determination based on colorimetric methods was carried out using a Technicon Auto Analyser II[23].

Roots from each plant were cut into 1 cm segments and placed evenly on a gridded plexiglass sheet containing 100 1 cm squares. The average number of roots per square was determined and multiplied by the total number of squares as a measure of the length of each plant's root system. Percent root colonization by VAM fungi was determined as described in the previous experiment.

Data were subjected to analysis of variance, and the means of the treatments were separated using the Protected Fisher's LSD test. Linear and quadratic regression analysis was used to describe the relationship between harvest time and root colonization. Percentile data were analyzed after arcsin transformation. Nonparametric analysis of variance was performed on the number of hyphae, arbuscules and vesicles.

Results and discussion

Time course study of root colonization

A significant quadratic regression relationship between percent VAM colonization and harvest time was found in both cowpea cultivars (Fig.

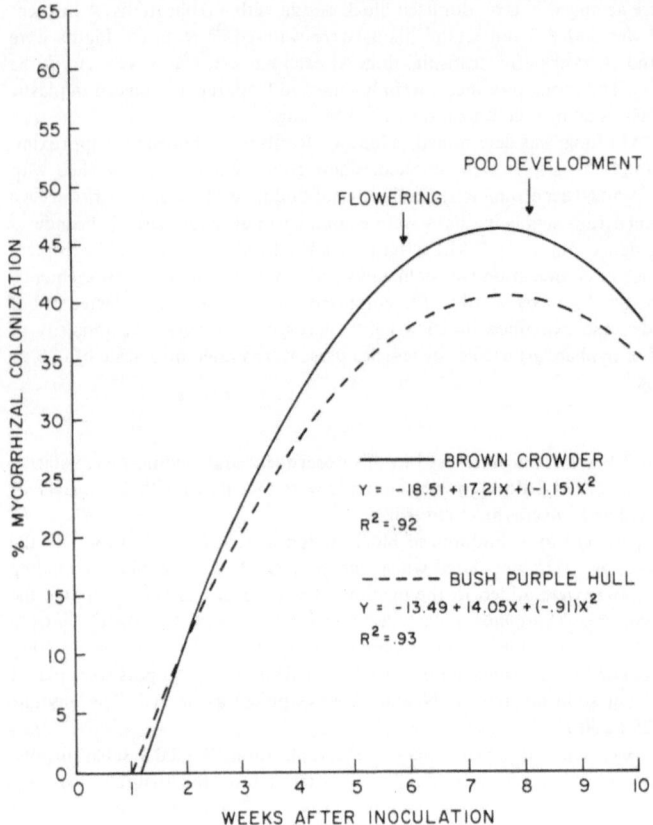

Fig. 1. Development of VAM colonization in cowpea with time. Data pooled from 2 VAM fungi.

1). Very few roots formed mycorrhizal associations during the first week after inoculation (lag phase). Entry of the fungi into the rootlets and initiation of the symbiosis appeared to take place during this period. From 1 to 6 weeks, *i.e.* seedling stage to flowering stage, percent root colonization increased rapidly, reaching a maximum at approximately 7 weeks. Both cowpea cultivars were colonized satisfactorily by both VAM fungi. By 8 weeks, at the beginning of pod development, root colonization started to decline. This decline in percent root colonization could be the result of reduced supply of organic carbon and other nutrients to the host, since most of the nutrients are diverted to the pods at this stage.

Three phases could be identified in the process of mycorrhizal development in cowpeas, *i.e.* 1) lag 2) rapid increase in colonization, 3) decline in colonization. Another type of 3 phase mycorrhizal development pattern was reported with field experiments on groundnuts[20]. In these crops a constant or plateau phase of colonization was observed.

Table 1. Summary of analysis of variance for development of mycorrhizal root colonization in cowpea with time

Source of variation	df	Mean squares			
		% VAM	Hyphae	Arbuscles	Vesicles
Rep	3	17.09	0.00	0.00	0.00
Genotype (G)	1	193.51*	11.34	18.38	2.04
VAM Fungi (M)	1	15.47	8.16	0.17	0.38
Time (T)	5	5293.91*	2330.47*	3219.75*	3878.28*
G × M	1	7.65	27.09	1.04	2.67
G × T	5	15.17*	540.12*	67.00	119.86*
M × T	5	52.54	85.55	76.33	27.40
G × M × T	5	17.52	44.25	150.21	46.11
Error	69	17.38	20.28	14.09	3.85

* Significant at 5% level.

Such a plateau was not observed in this cowpea pot experiment. In pot experiments, the plant roots and the mycelium of mycorrhizal fungi are confined. This restriction might have caused the observed reduction in colonization after the eighth week. Also, the difference in development patterns of root colonization between cowpea and groundnut may be due to the differences in time between occurrence of maximum root colonization and pod development. If pod development occurs soon after maximum root colonization, a truncated plateau can be observed as was in this cowpea experiment. If there is a delay between maximum root colonization and pod development, the plateau phase can be elongated as in groundnut.

No interaction was found between the VAM fungi and cowpea genotype for percent VAM colonization (Table 1). Cowpea cultivars showed a significant interaction with time, and percent VAM colonization was significantly different between the 2 cowpea cultivars (Table 1). 'Brown Crowder' had greater root colonization than 'Bush Purple Hull', but this difference was observed only between 6–8 weeks after inoculation (Fig. 1). Therefore, the best time to assess maximal mycorrhizal root colonization in cowpeas would be 6–8 weeks after inoculation (*i.e.* 7–9 weeks after seeding) or at the beginning of pod fill.

The developmental patterns for hyphae, arbuscules and vesicles differed. In some plants, hyphae and arbuscules were observed as early as one week after inoculation (Fig. 2). The amount of hyphae present in the roots increased up to 10 weeks. Arbuscules developed rapidly during the initial stages of colonization and were observed in approximately 75% of the root cortex at 4 weeks after inoculation (Fig. 2). Vesicles were not observed until the fourth week. Similar results on the appearance of vesicles were reported with groundnuts, with gradual development of

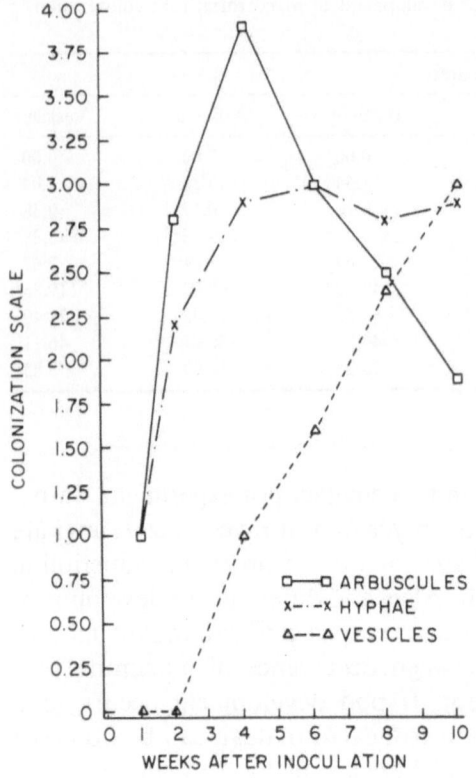

Fig. 2. Time course development of hyphae, arbuscules and vesicles in cowpea roots. Data pooled from 2 cowpea cultivars and 2 VAM fungi.

vesicles up to 12 weeks[20]. In our study, the number of vesicles present increased up to the end of the study at 10 weeks. While vesicles became more numerous as the plants matured, the number of arbuscules was greater during the early stages of root colonization. However, presence of hyphae, arbuscules and vesicles did not differ with cowpea cultivar or VAM fungus (Table 1).

The results of this investigation have shown that percent VAM colonization of cowpea roots varied between the 2 cultivars at a given stage of colonization, but not between the 2 VAM fungal species used. Maximum VAM colonization was observed 8 weeks after planting. Evidence is also provided that the degree of VAM colonization obtained during one particular time of plant growth may not reflect that of the whole growth period. Both percent root colonization and plant response may vary with time. Therefore, it is appropriate to study the variation in plant response to mycorrhizal colonization during different stages of plant growth.

Table 2. Significance levels for the effects of two VAM fungi on two cowpea cultivars

Dependent variable	Significance level		
	Cultivar	VAM fungus	Cultivar × VAM fungus
Percent colonization	NS	**	NS
Hyphae	NS	**	NS
Arbuscules	NS	**	NS
Vesicles	NS	**	NS
Root length	*	*	NS
Mycorrhizal root length	*	*	NS
Root weight	NS	NS	NS
Plant height	*	*	NS
Dry weight	*	*	*
Nodule number	*	NS	NS
Nodule weight	*	NS	NS
Nitrogenase activity	*	*	*
Seed weight	*	NS	NS
Shoot N%	NS	*	NS
Shoot P%	NS	*	NS
Root N%	NS	NS	NS
Root P%	NS	NS	NS

NS, *, ** Nonsignificant (NS) or significant at 5% (*), 1% (**) levels.

Study of growth effects

The results of the analysis of variance for this experiment are summarized in Table 2. Inoculation with mycorrhizal fungi resulted in root colonization in both 'Brown Crowder' and 'Bush Purple Hull' and noninoculated controls were not colonized by any contaminants. At harvest, cultivars did not differ in percentage of colonized roots and the number of hyphae, arbuscules and vesicles formed, although percent root colonization may have differed at other times during the growth period, as shown in the previous experiment. There was no significant difference between the 2 VAM fungi in percent root colonization, and the number of hyphae, arbuscules and vesicles formed. Contrasts between control vs *G. mosseae* or *G. fasciculatum* were significant at the 0.05 level while a contrast between *G. mosseae* and *G. fasciculatum* was not significant.

Despite the insignificant differences in percent root colonization, 'Brown Crowder' had significantly longer mycorrhizal roots (percent root colonization × root length) than did 'Bush Purple Hull' — because of its larger root system (Fig. 3). Daft and Nicolson[6] showed the importance of measuring the total root colonization rather than percent root colonization. They found that percentage of root colonization did not

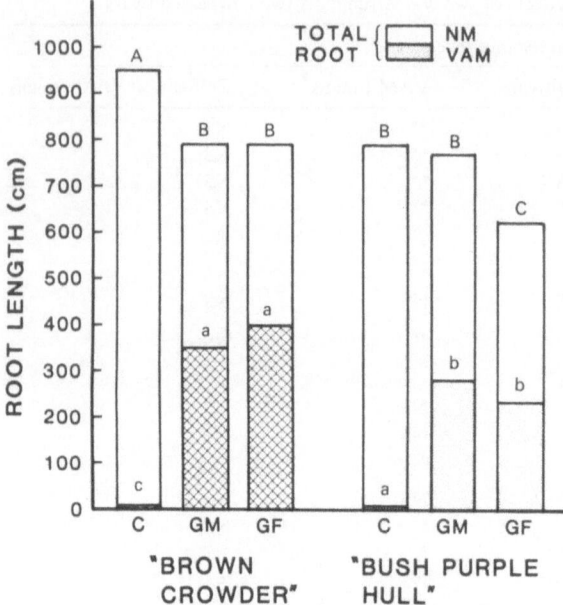

Fig. 3. Root length of cowpea with (VAM) and without (NM) mycorrhizal inoculation [Control (C), *G. mosseae* (GM), *G. fasciculatum* (GF)] (mycorrhizal root length = root length × percent of colonized root segments). Mean separation by Fisher's LSD test, 0.05 level.

correlate well with the shoot weight of maize and tomato plants, unless the size of the root system was considered.

Root length in plants inoculated with *G. fasciculatum* was significantly shorter than those of control plants of both cowpea cultivars (Fig. 3). One possible explanation is that hyphae act as extensions analogous to root hairs in exploring soil for water and certain nutrients, whereas the roots alone must meet the requirements of nonmycorrhizal plants. Another possibility is that mycorrhizal root systems decay faster than do nonmycorrhizal roots — a finding by Mosse[14] studying the root system of apple. However, inoculation with *G. mosseae* reduced root length only in 'Brown Crowder'. Root weight was not affected by mycorrhizal inoculation (Table 2).

Mycorrhizal plants were significantly taller than were controls in both cultivars (Fig. 4). The increase in plant height may be attributed to improved nutrient and water uptake by mycorrhizae. Similar increases in plant height of mycorrhizal cowpea and soybean have been reported[2,21].

Interactions between VAM inoculation and cowpea cultivar were significant for shoot dry weight (Fig. 5), with mycorrhizal inoculation increasing shoot weight of 'Brown Crowder' by 40% but having little effect on 'Bush Purple Hull'. The differential response of the two cowpea

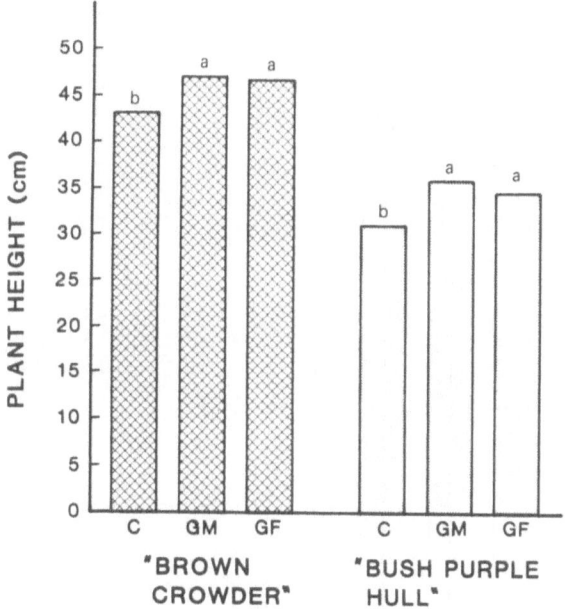

Fig. 4. Plant heights of mycorrhizal [*G. mosseae* (GM), *G. fasciculatum* (GF)] nonmycorrhizal (C) cowpea. Mean separation within a cultivar by Fisher's LSD test, 0.05 level.

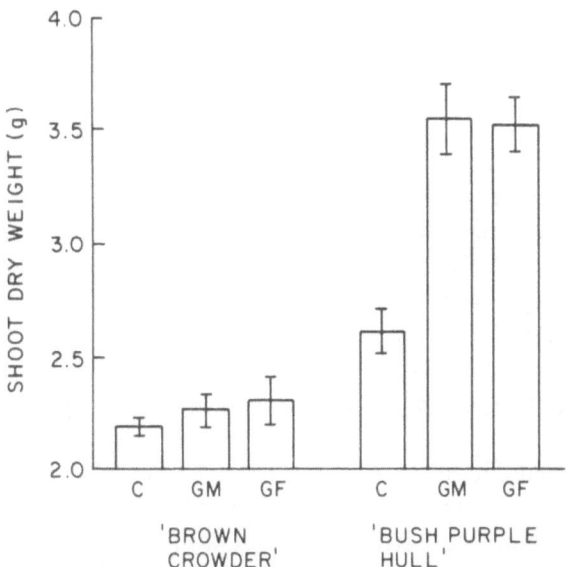

Fig. 5. Interaction between cowpea cultivar and VAM inoculation [Control (C), *G. mosseae* (GM), *G. fasciculatum* (GF)] for shoot dry weight. Each point represents mean of 4 replicates. Vertical bars represent standard error.

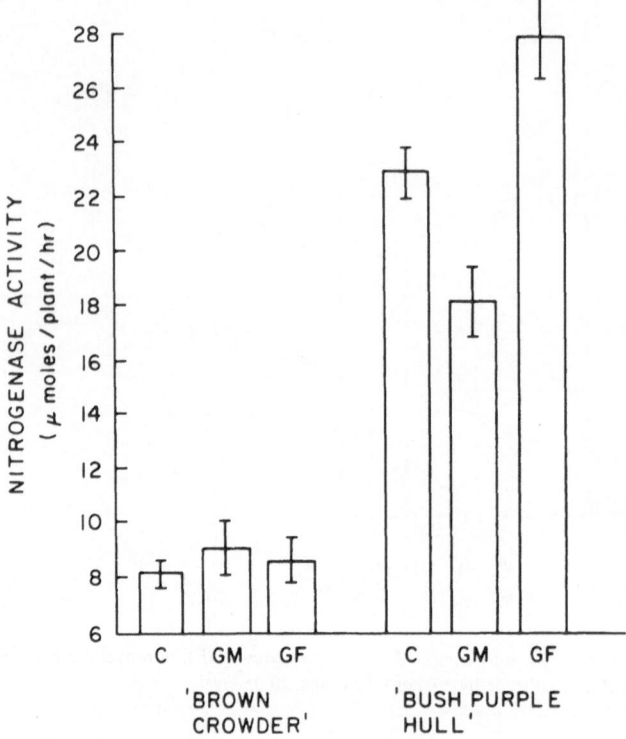

Fig. 6. Interaction between cowpea cultivar and VAM inoculation [Control (C), *G. mosseae* (GM), *G. fasciculatum* (GF)] for nitrogenase activity. Each point represents mean of 4 replicates. Vertical bars represent standard error.

varieties confirms the results reported by Ollivier et al.[17]. In that study, the species of VAM also influenced shoot growth.

In our study nodule number and weight were not affected by mycorrhizal inoculation; however, other studies have found that inoculating with VAM affects nodulation[9] and nodule mass[10]. Interaction between VAM inoculation and cowpea genotype was significant for nitrogenase activity (Fig. 6): *G. fasciculatum* significantly increased nitrogenase activity of 'Brown Crowder' compared with the control, whereas *G. mosseae* reduced the activity and neither one affected nitrogenase activity of 'Bush Purple Hull'.

Thus, our experiment provides clear evidence for differential response in cowpea cultivars to inoculation with mycorrhizal fungi. 'Brown Crowder' had enhanced nitrogenase activity when inoculated with *G. fasciculatum* and Rhizobium while nitrogenase activity of 'Bush Purple Hull' was not enhanced. Enhancement of N_2 fixation in plants inoculated with both Rhizobium and mycorrhizal fungi has also been reported for soybean[21], French bean[5] and clover[4].

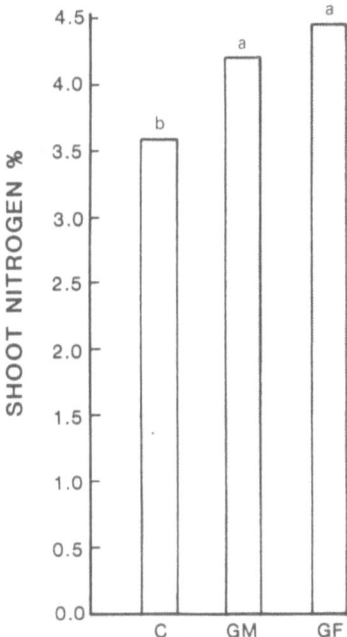

Fig. 7. The effect of VAM inoculation [Control (C), *G. mosseae* (GM), *G. fasciculatum* (GF)] on percent nitrogen in cowpea shoots (pooled data from two cultivars). Mean separation by Fisher's LSD test, 0.05 level.

Mycorrhizal inoculation increased both shoot N and P percentages significantly (Figs. 7 and 8). This influence was not dependent on species of VAM fungi or on cowpea cultivar. Increased nitrogenase activity of 'Brown Crowder' when inoculated with *G. fasciculatum* could explain the enhanced N content in the shoots of these plants, but both 'Bush Purple Hull' and 'Brown Crowder' inoculated with *G. mosseae*, and 'Bush Purple Hull' inoculated with *G. fasciculatum* also had increased percentages of N in their shoots without increased nitrogenase activity. This finding suggests that mycorrhizae may assist in absorption of N from the soil when soil N is limited. Also the increased plant height in mycorrhizal plants could be attributed to the improved N status in these plants. Despite increased P in all mycorrhizal plants, only the 'Brown Crowder' and *G. fasciculatum* combination increased nitrogenase activity significantly. Therefore, it appears that this host genotye and VAM endophyte combination is preferred to the other combinations used in this experiment. Root N and P percentages were not affected by inoculation with VAM fungi (Table 2). The above results also suggest that effect of VAM fungi on nitrogenase activity could not be attributed directly to P concentration of shoots or roots.

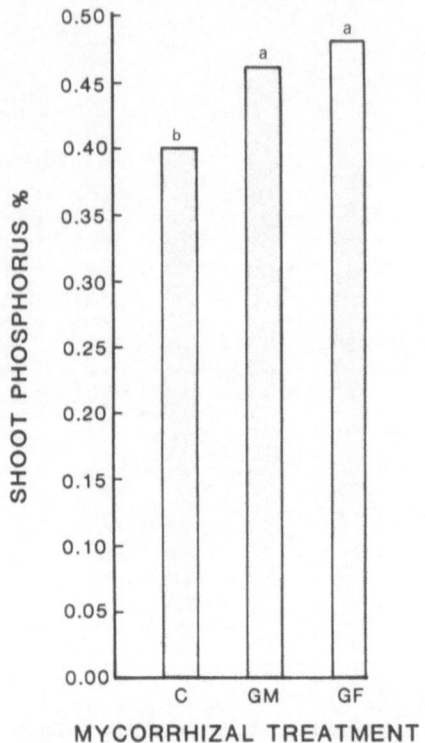

Fig. 8. The effect of VAM inoculation [Control (C), *G. mosseae* (GM), *G. fasciculatum* (GF)] on percent phosphorus in cowpea shoots (pooled data from two cultivars). Mean separation by Fisher's LSD test, 0.05 level.

The differential response of the two cultivars may be related to their growth habits. 'Brown Crowder' is an indeterminate cultivar which continues vegetative growth even after flowering and pod set; whereas, 'Bush Purple Hull' is a determinate cultivar which does not continue growth after pods are set. With determinate types, vegetative growth ceases at an earlier stage, so the benefits of VAM may be limited. Unlike P and N concentration in the plant, dry weight is more likely to be directly influenced by growth habit. Phosphorus and N concentrations were found to be influenced by VAM similarly in both cultivars, while dry weight was not. However, this should be studied further before attributing the effectiveness of VAM fungi to cultivar growth habit.

Despite reports[10] that inoculation increased seed yield of 'Pale Green' cowpeas, we found no significant increase in this variable supporting results of Kuyper and Lambeth[12] with 'Mississippi Silver'. These conflicting findings suggest that genotype of the host plant is an important factor in determining the benefits derived from mycorrhizal colonization, even though these experiments were conducted separately under different experimental conditions.

Both cowpea cultivars were capable of being colonized by *G. mosseae* and *G. fasciculatum*; however, 'Brown Crowder' responded better to inoculation with mycorrhizal fungi than did 'Bush Purple Hull'. Mycorrhizal colonization of cowpea can significantly influence shoot height and weight, root length, percentage of N and P in shoots, and nitrogenase activity.

Genetic variability in response to root colonization by VAM fungi was observed for shoot dry weight and nitrogenase activity. Interactions between cowpea cultivar and VAM endophyte were significant for these two traits. 'Brown Crowder', when inoculated with *G. fasciculatum*, performed especially well, exhibiting enhanced shoot weight and nitrogenase activity.

Acknowledgements These studies were supported in part by a grant under A.I.D. PASA AG/TAB (USDA-82-CSRS-5-0201). Texas Agricultural Experiment Station Technical Article 20757.

References

1. Azcon R and Ocampo J A 1980 Factors affecting the VA infection and mycorrhizal dependency of thirteen wheat cultivars. New Phytol. 87, 677–685.
2. Baradas S N and Halos P M 1980 Selection of mycorrhizal isolates for biological control of *Fusarium solani* f. sp. *Phaseoli* on *Vigna unguiculata*, pp 247–248. *In* Tropical Mycorrhizal Research. P Mikola. Ed. New York, Oxford Univ. Press.
3. Bevege D I and Bowen G D 1977 *Endogone* strain and host plant differences in development of vesicular-arbuscular mycorrhizas. pp 78–86. *In* Endomycorrhizas. F E Sanders, B Mosse and P B Tinker. Eds. Academic Press, London.
4. Crush, J R 1974 Plant growth responses to vesicular-arbuscular mycorrhiza. VII. Growth and nodulation of some herbage legumes. New Phytol. 73, 743–749.
5. Daft M J and El-Giahmi A A 1974 Effect of *Endogone* mycorrhiza on plant growth. VII. Influence of infection on the growth and nodulation in French beans (*Phaseolus vulgaris*). New Phytol. 73, 1139–1147.
6. Daft M J and Nicolson T H 1972 Effect of *Endogone* mycorrhiza on plant growth. IV. Quantitative relationships between the growth of the host and the development of endophyte in tomato and maize. New Phytol. 71, 287–295.
7. Hall I R 1975 Effect of vesicular-arbuscular mycorrhizae on two varieties of maize and one of sweet corn. New Zealand J. Agric. Res. 21, 517–519.
8. Islam R 1977 Effect of several *Endogone* spore types on the yield of cowpea (*Vigna unguiculata* (L.) Walp). Internal Report, Internat. Inst. Trop. Agric., Ibadan, Nigeria.
9. Islam R and Ayanaba A 1981 Effect of seed inoculation and preinfecting cowpea (*Vigna unguiculata*) with *Glomus mosseae* on growth and seed weight of the plants under field conditions. Plant and Soil 61, 341–350.
10. Islam R and Ayanaba A 1981 Growth and yield responses of cowpea and maize to inoculation with *Glomus mosseae* in sterilized soil under field conditions. Plant and Soil 63, 505–509.
11. Islam R Ayanaba A and Sanders F E 1980 Response of cowpea [*Vigna unguiculata* (L.) Walp] to inoculation with vesicular-arbuscular mycorrhizal fungi and to rock phosphate fertilization in some sterilized Nigerian soils. Plant and Soil 54, 107–117.
12. Kuyper B J and Lambeth V N 1986 Evaluation of four *Glomus* species with cowpea. J. Amer. Soc. Hort. Sci. (in press).

13 Menge J A Johnson E L V and Platt R G 1978 Mycorrhizal dependency of several citrus cultivars under three nutrient regimes. New Phytol. 81, 553–559.
14 Mosse B 1956 Studies on the endotrophic mycorrhiza of some fruit plants. Ph.D. Thesis, Univ. of London, London, England. 36 p.
15 Mosse B 1972 Effects of different *Endogone* strains on the growth of *Paspalum notatum*. Nature 239, 221–223.
16 Mosse B 1981 Vesicular-arbuscular mycorrhiza research for tropical agriculture. Hawaii Institute of Trop. Agric. and Human Res., Univ. of Hawaii, Honolulu, Res. Bull. 194.
17 Ollivier B Bertheau Y Diem H G and Gianinazzi-Pearson V 1983 Influence de la variete de *Vigna unguiculata* dans l'expression de trois associations endomycorrhiziennes a vesicules et arbuscules. Can. J. Bot. 61, 354–358.
18 Phillips J M and Hayman D S 1970 Improved procedures for clearing roots and staining parasitic and vesicular-arbuscular mycorrhizal fungi for rapid assessment of infection. Trans. British Mycol. Soc. 55, 158–161.
19 Powell C L and Sithamparanathan J 1977 Mycorrhizae in hill country soils. IV. Infection rate in grass and legume species by indigenous mycorrhizal fungi under field conditions. New Zealand J. Agric. Res. 20, 489–502.
20 Rao A A and Parwathi K 1982 Development of vesicular-arbuscular mycorrhiza in groundnut and other hosts. Plant and Soil 66, 133–137.
21 Skipper H D and Smith G W 1979 Influence of soil pH on the soybean-endomycorrhiza symbiosis. Plant and Soil 53, 559–563.
22 Smith S E and Daft M J 1977 Interactions between growth, phosphate content and nitrogen fixation in mycorrhizal and non-mycorrhizal *Medicago sativa*. Aust. J. Plant Physiol. 4, 401–413.
23 Technicon 1975 Industrial Method No. 369–75 A/A. Technicon Industrial systems, Tarrytown, New York.
24 Technicon 1977 Industrial Method No. 334–74 W/B. Technicon Industrial systems, Tarrytown, New York.
25 Timmer L W and Leyden R F 1978 Stunting of citrus seedlings in fumigated soils in Texas and its correction by phosphorus fertilization and inoculation with mycorrhizal fungi. J. Am. Soc. Hort. Sci. 103, 533–537.
26 Yost R S and Fox R L 1981 Influence of mycorrhizae on the mineral contents of cowpea and soybean grown in an oxisol. Agron. J. 74, 475–480.
27 Zary K W Miller J C Jr Weaver R W and Barnes L W 1978 Intraspecific variability for N_2 fixation in southernpea [*Vigna unguiculata* (L.) Walp]. J. Am. Soc. Hort. Sci. 103, 806–808.

Section 6

Germplasm resources and modifications

Sources of germplasm for research on mineral nutrition

W.H. GABELMAN
Department of Horticulture, University of Wisconsin, Madison, WI 53706, USA

Key words IBPGR   Landraces   Mineral nutrition

Summary   Phenotypic and genotypic variations have become useful tools in research on mineral nutrition of plants. Landraces, particularly from areas which practice low-input farming, are valuable experimental materials. Likewise, weedy exotic relatives of economic crops are useful, having evolved in low-nutrient environments. Collection, curation and documentation of germplasm have been made by many countries of the world. Publications of the International Board for Plant Genetic Resources, FAO, Rome, Italy are excellent sources of information on germplasm collections. A list of these collections is included here for the convenience of mineral nutrition research scientists.

Germplasm for mineral nutrition research

Worldwide collection, curation and maintenance of germplasm resources are carried out quite effectively even though most countries operate independently. Information on these collections is published but the distribution of this material is limited and many biologists are unaware of these listings. Germplasm generated in plant breeding and plant genetic programs is potentially important also, but seldom is it well documented. Information on materials from plant breeding programs can be obtained best through personal contact with the breeders. Breeding materials will reflect the environment in which the breeder practices selection. If materials are selected on soils high in fertility the derived materials will reflect genetic variation controlling plant response to high levels of ions in the rhizosphere, either avoiding or tolerating toxic concentrations of the ion(s) at the cellular level. If materials are selected on soils low in fertility the genetic system will reflect differences in acquisition, transport and use. A third source of unique germplasm can be found in cell and tissue cultures maintained by research workers in biotechnology.

Landraces developed over centuries of natural selection are excellent sources of genes conferring adaptation of plants to soils with adverse levels of pH, toxic elements and available essential nutrients. Countries, where the cost of commercial fertilizers and lime is prohibitive, have accepted low yields as a necessity. These same countries seldom have a well developed seed industry. Seed often is maintained by the farmer. Under these conditions the more productive plants tend to produce the

H.W. Gabelman and B.C. Loughman (Eds.), Genetic aspects of plant mineral nutrition.
ISBN-13: 978-94-010-8102-3
© *1987 Martinus Nijhoff Publishers, Dordrecht/Boston/Lancaster.*

most seed. Plants maintained under these conditions are continually selected for adaptation to either low-mineral nutrient environment or to conditions of ion toxicity. Landraces, in species that have a significant level of self pollination, are more apt to exhibit recombination of recessive genes which confer adaptation than do landraces in species that are largely cross pollinated. Our best sources of useful landraces have been the lesser developed agricultural systems of the world. National and international collections from these countries are very important. The challenges are to collect, identify, isolate and utilize these materials effectively.

1. Germplasm Collections

Many large, active collections of germplasm exist. Countries have formalized procedures for obtaining, cataloging, curating, preserving, evaluating and enhancing germplasm. The collections are important for the preservation of genetic diversity. Early efforts tried to prevent the importation of economic pests and to systematically record the accessions. The genetic diversity of these collections is a multifaceted resource. These sources of germplasm should be useful in identifying intraspecific differences in acquisition, transport and utilization of essential elements and mechanisms which exclude or tolerate toxic elements.

Germplasm collections are composed primarily of 4 types of materials; namely, exotic weedy materials, landraces of economic importance in developing agricultures, obsolete cultivars, and recent germplasm releases from breeding programs. Each type of material is useful but varies in potential importance.

The wild weedy species survive in a rigorous environment without assistance in the form of applied fertilizer, pesticides, irrigation and other cultural amenities. Exotic materials represent extremes in genetic variability, but are often difficult sources from which to transmit genetic information into an economic crop relative. Non-synchronous flowering, related to differences in photoperiod response, in a common nuisance. Evolutionary change, either genetic or chromosomal, which contributes to sterility is much more serious. Sexual barriers have been the greatest deterrent to the effective use of wild weedy germplasm in plant improvement. Plant breeders have used exotic materials most effectively for traits inherited qualitatively. Single genes can be transferred into adapted cultivars rapidly and predictably by systematically backcrossing the F_1 hybrid of the exotic and cultivated parents to the cultivated (recurrent) parent.

Germplasm released from major breeding programs tends to be a rather poor source of genetic variability related to mineral nutrition.

Seldom is breeding material evaluated on 'toxic' soils, although cultivars tolerant to heavy metals and salinity are being developed in some countries e.g. Brazil, Indonesia, Australia and the U.S. Similarly, few systematic breeding efforts are made to improve crop cultivars under conditions of marginal deficiency for one or more of the essential elements. Hence, cultivars and other recently released breeding materials would be poor genetic reservoirs of genes contributing to efficiency in acquisition and/or use at low nutrient levels These same materials, however, may be excellent sources of genes controlling the degree of enhanced growth resulting from increasing levels of the major elements — N, P, K, Ca, Mg — , reflecting the nutritional regimes employed by the breeder.

Breeders developing F_1 hybrids in cross-pollinating species develop inbred parents from which the hybrids are ultimately synthesized. The inbreeding process permits genetic recombination, fixing of homozygous recessives and permitting the expression of the recessive alleles. Inbreeding fixes homozygosity regardless of selection pressure for traits. Thus inbreds, derived in cross-pollinated species, may be valuable sources of genes controlling nutrient acquisition and use, even though selection pressure has not been applied for traits associated with mineral nutrition.

Obsolete cultivars and landraces are the most useful reservoirs of genes in germplasm collections contributing to plant productivity under stress. Cultivars developed by systematic breeding before 1930, were created before agriculture developed the concept that 'fertilizer is cheap insurance' for crop productivity. Like landraces in developing countries, these early cultivars were developed by selection pressure, preserving phenotypes (and genetic diversity) adapted to environments marginal for soil nutrients and often on soils containing toxic levels of one or more heavy metals.

Botanists, working in national germplasm laboratories, systematically collect in areas of greatest natural genetic diversity for one or more species. At the same time these collectors obtain seed samples of nearly all plants sold in the open markets. These markets are the principle source of landraces in the U.S. collections. Landraces are often mixtures of genetic materials. In highly outbreeding species each sample represents much genetic diversity in which dominant alleles mask the expression of their recessive counterparts. In highly inbreeding species each landrace may represent a heterogeneous collection of highly homozygous individuals. The inbreeding species, because of their homozygosity, express recessive as well as dominant genes. Heterozygosity does not hide potential genetic expression in the inbreeding species. Hence, inbreeding species are much easier sources from which to isolate genetic diversity to environmental variables.

In 1984 in the report of the 'First Decade of Service, 1974–1984,' issued by the Internatioonal Board for Plant Genetic Resources (IBPGR), Rome, Italy, 113 functional centers conserving crop germplasm were listed (Table 1). These centers represent 53 different national collections and 12 international organizations[5]. Additional genebanks are listed in Table 2. Since many biologists are not well informed about these collections these tables are included here to aid in the use of this germplasm for research on genetic aspects of mineral nutrition.

The Internatinal Board for Plant Genetic Resources (IBPGR) is an autonomous international scientific organization under the aegis of the Consultative Group on International Agricultural Research (CGIAR) of FAO, Rome, Italy. The basic function of the IBPGR is to promote and coordinate an international network of genetic resource centers to further the collection, conservation, documentation, evaluation and use of plant germplasm and thereby contribute to raising the standard of living and welfare of people throughout the world. Much of the data on the germplasm collections is now stored in computers which facilitates information retrieval on each germplasm accession. Most germplasm collections exist as seeds. IBPGR has also funded and published a comprehensive bibliography of crop genetic resources[6].

The amount of material in each germplasm collection far exceeds the capacity of any mineral nutrition laboratory to evaluate. The U.S. Plant Introduction System, one of the larger collections of germplasm in the world, has more than 500,000 recorded accessions. Of these about 300,000 accessions are active. If these large numbers of accessions are to be evaluated for genetic variation controlling mineral nutrition biologists must develop precise, rapid screening procedures that can be used early in the life of the plant.

Not all germplasm collections are readily available to all biologists. Policies based on political judgements, have created barriers to free exchange. However, germplasm is often traded between countries having needs that cannot be met by collection or by other means. Table 1 has been included in this manuscript in order that research workers can peruse the list to identify collaborative sources worldwide.

2. Use of germplasm collections for mineral nutrition research

Research on phenotypic responses to low-element levels of P, K and Ca in our program has necessitated the isolation of phenotypic variants at the onset of the research. These procedures have been described by Gerloff[4]. 'Efficient' strains of tomatoes and beans may produce twice as much dry matter as 'inefficient' strains when grown at a stress (low element) level of an essential nutrient but would produce similar

Table 1. List of centers conserving crop germplasm in medium- and long-term storage units (1984)

Country/Acronym	Center	Major Crops
Afghanistan	AFG01 Plant Genetic Resources Unit, Darulaman	Significant collections of local germplasm of barley, chickpea faba bean, lentil, pea and wheat, and introduced germplasm of maize, mung bean, *Phaseolus* and rice
Argentina	ARG05 Estación Experimental Regional Agropecuaria, Pergamino	Large collection of germplasm of barley, cotton, flax, forages, groundnut, maize, oat, *Phaseolus*, sorghum, soyabean, tomato and wheat
Australia	AUS01 Commonwealth Scientific and Industrial Research Organization, Canberra	Collection of introduced germplasm of forage grasses and legumes, soyabean, sunflower, rice, and indigenous wild *Glycine* species
	AUS02 Western Australian Department of Agriculture, South Perth	Large collection of introduced forage grasses and legumes
	AUS03 Australian Wheat Collection Department of Agriculture, Tamworth	Large collection of *Aegilops* and wheat and smaller collections of barley, rye and Triticale
	AUS04 Queensland Department of Primary Industries, Brisbane	Collection of cluster bean and safflower
	AUS06 South Australian Department of Agriculture, Adelaide	Large collection of wild species of forage legumes
	AUS08 CSIRO Division of Tropical Crops and Pastures, St Lucia	Large collection of tropical grasses and legumes
Austria	AUT01 Landwirtschaftlich-Chemische Bundesversuchsanstalt, Linz	Collection of local landraces and advanced cultivars of barley, oat and wheat and local landraces of *Phaseolus*
AVRDC	Asian Vegetable Research and Development Center, Taiwan, China	Large collection of Asiatic landraces, advanced cultivars and breeding lines of *Amaranthus*, black gram, chinese cabbage, other brassicas, mung bean, soyabean, sweet potato and tomato

Table 1. Continued

Country/Acronym	Center	Major Crops
Bangladesh	BGD01 Bangladesh Jute Research Institute, Dacca	Collection of landraces, wild species, advanced cultivars and mutants of species of *Corchorus* and *Hibiscus*
Belgium	BEL03 Faculté des sciences Agronomiques de l'Etat à Gembloux	Large number of wild species of *Phaseolus* and *Vigna* and related genera
Brazil	BRA03 National Genetic Resources Center, Brasilia	Local collections of brassicas, groundnut and related wild species, maize, *Phaseolus*, soyabean and tobacco and introduced germplasm of cowpea, rice, sesame and wheat. This collection is being expanded with material from active genebanks
Bulgaria	BGR01 Institute of Introduction and Plant Resources, Sadovo	Collection of barley, brassicas, *Capsicum*, chickpea, cowpea, cucurbis, faba bean, forage grasses and legumes, groundnut, lentil, maize, oat, onion, pea *Phaseolus*, potato, soyabean, sunflower, rice, rye, tomato, Triticale and wheat
Canada	CAN01 Plant Gene Resources of Canada, Ottawa	Large collections of introduced germplasm of barley, brassicas, maize, millets, oat, soyabean, tomato and wheat
Canada	CAN04 Agriculture Canada Research Station, Saskatoon	Diverse collection of introduced germplasm of landraces, breeding lines, advanced cultivars and related wild species of brassicas
	CAN05 Université Laval, Quebec	Collection of introduced germplasm of barley, faba bean, oat and wheat
CATIE	Centro Agronomico Tropical de Investigacion y Enseñanza, Turrialba.	Large collections of indigenous germplasm of landraces, wild species and advanced cultivars of *Capsicum*, cucurbits, *Phaseolus*, *Sechium edule*, *Solanun* and tomato
Chile	CHL03 Banco de Genes, Instituto de Producción y Sanidad Vegetal, Universidad Austral de Chile, Valdivia	Collection of landraces and wild species of potatoes

Table 1. Continued

Country/Acronym	Center	Major Crops
	CHL05 La Platina Experiment Station INIA, Santiago	Large collection of chickpea, lentil, *Phaseolus*, sunflower and wheat
China	CHN01 Chinese Academy of Agricultural Sciences, Beijing	Collection of indigenous brassicas and related wild species, maize, sorghum, rice and wheat Large collection of groundnut and related wild species
	CHN04 Beijing Vegetable Research Centre, Beijing	Range of local vegetable landraces
	TWN05 Taiwan Seed Service, Taichung, Taiwan	Collection of landraces, breeding lines and advanced cultivars of brassicas and advanced cultivars of *Capsicum*
CIAT	Centro Internacional de Agricultura Tropical, Cali, Colombia	Large collection of forage grasses and legumes Large collection of New World land races, advanced cultivars and related wild species of *Phaseolus coccineus, P. lunatus, P. vulgaris* and smaller collectins of cowpea and winged bean
CIMMYT	Centro Internacional de Mejoramiento de Maiz y Trigo, El Batan, Mexico	Large collection of indigenous land races of maize from Central and South America Large collection of cultivars and breeding material of wheat
CIP	Centro Internacional de la Papa, Lima, Peru	Large collection of indigenous Central and South American land races and advanced cultivars of potato and related wild species
Colombia	COL02 Instituto Colombiana Agropecuaria, Bogota	Large collection of landraces of *Allium*, barley, *Capsicum*, cucurbits, lupins, maize, *Phaseolus*, rice, sorghum, tuber-bearing solanums, tomato and wheat
Cyprus	CYP01 Agricultural Research Institute, Nicosia	Local collections of cereals and legumes

Table 1. Continued

Country/Acronym	Center	Major Crops
Czechoslovakia	CSK01 Research Institute of Plant Production, Bratislavska	Collections of barley, clover, lentil, lucerne, *Phaseolus*, soyabean, Triticale and wheat
	CSK02 Maize Research Institute, Trstinska	Collection of maize germplasm
Ecuador	ECU03 Instituto Nacional de Investigaciones Agropecuarias, Estación Experimental Santa Catalina	Collection of indigenous landraces of *Amaranthus*, *Lupinus* and quinoa
Ethiopia	ETH01 Plant Genetic Resources Center, Addis Ababa	Large collection of indigenous germplasm of barley, brassicas, castor oil, chickpea, cucurbits, faba bean, flax, lentil, millets, niger-seed, pea, safflower, sesame, sorghum, wheat, and introduced *Capsicum*, maize and *Phaseolus*
France	FRA03 Office de la Recherche Scientifique et Technique Outre-Mer, Bondy	African collections of millets, sorghum and rice
	FRA10 INRA station d'Amélioration des Plantes, Le Rheu	Large collection of landraces and breeding lines of cauliflower
	FRA11 INRA Station d'Amélioration des Plantes Maraichères, Montfavet to Avignon	Large collection of introduced landraces and related wild species of *Capsicum*, eggplant and melon and advanced cultivars and breeding lines of tomato
	FRA13 C.N.R.S. Laboratoire de Génétique et Physiologie du Développement des Plantes, Gif sur Yvette	Large collection of millets, rice and sorghum
German Democratic Republic	DDR01 Zentralinstitut für Genetik und Kulturpflanzenforschung, Gatersleben	World collections of *Aegilops*, barley, *Capsicum*, cucurbits, faba bean, flax, forage grasses and legumes, lentil, lupin, maize, medicinal plants, millets, oat, pea, poppy, Phaseolus, rye, sorghum, soyabean, spices, tobacco, tomato, vegetables and wheat

Table 1. Continued

Country/Acronym	Center	Major Crops
German Federal Republic	DEU01 Institut für Pflanzenbau und Pflanzenzüchtung, Braunschweig	Large collections of landraces and advanced cultivars of barley, beet, brassicas, clover, faba bean, lupin, oat, pea, poppy, rye, sorghum, tomato and wheat
GNPG	German-Netherlands Potato Genebank, Braunschweig	Large collection of introduced germplasm of potato and related wild species
Ghana	GHA07 Crops Research Institute, Bunso	Collection of local landraces of bambarra groundnut, cowpea, lima bean, okra and other local legumes and vegetables
Greece	CRC05 Greek Genebank, Thessaloniki	Locally collected germplasm of *Aegilos*, barley, beet, brassicas, chickpea, clover, cotton, faba bean, lentil, pea, *Phaseolus* and wheat
Hungary	HUN03 Research Centre for Agrobotany, Institute for Plant Production and Qualification, Tapioszele	Collection of landraces and advanced cultivars of barley, beet, brassicas, *Capsicum*, chickpea, clover, cucurbits, faba bean, forage legumes, groundnut, lentil, lupin, millets, oat, pea, *Phaseolus*, poppy, sorghum, soyabean, wheat and related wild species and locally collected material of flax, lettuce, maize, onion, sunflower and tomato
ICARDA	International Center for Agricultural Research in Dry Areas, Aleppo, Syria	Large collection of germplasm of *Aegilops*, barley, chickpea, faba bean, forage grasses and legumes, lentil, oat, pea, Triticale and wheat
ICRISAT	International Crops Research Institute for the Semi-arid Tropics, Hyderabad, India	Large collection of local and introduced germplasm of chickpea, groundnut, millets, pigeon pea and sorghum
IITA	International Institute of Tropical Agriculture, Ibadan, Nigeria	Large collection of African germplasm of african yam bean, bambarra groundnut, cowpea and related wild species, kersting's groundnut, lablab bean and rice Large collection of cowpea and related wild species, rice and soyabean

Table 1. Continued

Country/Acronym	Center	Major Crops
India	IND08 Central Rice Research Institute, Cuttack	Large collection of rice and related wild species
Indonesia	IDN02 National Biological Institute, Bogor	Seeds of indigenous tropical legumes
	IDN09 Research Institute for Food Crops, Bogor	Large collection of local landraces and advanced cultivars of rice
	IDN13 Research Institute for Food Crops, Sukamandi	Collection of local landraces and advanced cultivars of rice and introduced germplasm of mung bean
Iraq	IRQ01 Plant Genetic Resources Unit, Agricultural Research Centre, Baghdad	Significant collections of local germplasm of *Aegilops*, barley, chickpea, lentil, okra and wheat
IRRI	International Rice Research Institute, Los Banos, Philippines	Large collection of Asiatic and African landraces, advanced cultivars, breeding lines and wild species of rice
Israel	ISR02 Agricultural Research Organization, Bet Dagan	Collections of barley, brassicas, chickpea, cotton, forage legumes, lentil, maize, melon, onion, pea, *Phaseolus*, rice, rye, watermelon and wheat
Italy	ITA04 Instituto del Germoplasma, CNR, Bari	Large collection of landraces and cultivars of barley, faba bean, forage grasses and legumes maize, oat, pea, tomato and wheat
	ITA06 Instituto di Miglioramento Genetico and Produzione delle Sementi, Torino	Collection of landraces and advanced cultivars of *Capsicum*, eggplant and *Phaseolus*
Ivory Coast	CIV01 Office de la Recherche Scientifique et Technique d'Outre-Mer, Abidjan	Large collection of indigenous germplasm of cultivated and related wild species of okra and rice
Japan	JPN01 Plant Germplasm Institute, Kyoto	Large collection of introduced material of *Aegilops*, barley, *Capsicum*, *Phaseolus* and wheat

Table 1. Continued

Country/Acronym	Center	Major Crops
	JPN03 National Institute of Agrobiological Resources, Yatabe	Large collection of landraces, advanced cultivars and breeding lines of barley, beet, brassicas, cucurbits, eggplant, flax, forages, maize, millets, mung bean, oat, onion, pea, rice, sorghum, soyabean, sweet potato, tomato and wheat
	JPN05 Vegetable and Ornamental Crops Research Station, Mie	Large collection of landraces of brassicas and landraces, breeding lines and wild species of tomato
	JPN07 Tohoku University, Sendai	Large collection of brassicas and related wild species
	JPN08 Faculty of Agriculture, Kobe University, Kobe	Large collection of introduced germplasm of landraces of potato and related wild species
	JPN09 Institute for Agricultural and Biological Sciences, Okayama University, Kurashiki	Large collection of germplasm of cultivated barley and related wild species
	JPN11 Hokkaido National Agricultural Experiment Station, Shimamatsu, Hokkaido	Large collection of introduced germplasm of landraces and advanced cultivars of potato and smaller holdings of onion
	JPN12 Laboratory of Plant Breeding, Mie University	Collection of advanced cultivars and landraces of sweet potato
Kenya	KEN03 National Agricultural Research Station, Kitale	Large collection of maize, oat and tropical forage species
Korea, Republic of	KOR02 Germplasm Management Office, Office of Rural Development, Suweon	Large collection of indigenous land- races of rice and wild species of soyabean and locally collected germplasm of barley, maize, mung bean, red bean, sorghum and wheat
	KOR03 Agricultural Experiment Station, Suweon	Large collection of landraces and introduced cultivars of rice
Malawi	MWI02 Chitedze Agricultural Research Station, Lilongwe	Collection of landraces and wild species of cowpea, groundnut and other minor legumes, maize, millets, rice and sorghum

Table 1. Continued

Country/Acronym	Center	Major Crops
Malaysia	MYS05 Rice Research Centre, Bumbong Lima	Large collection of local landraces of rice
Mexico	MEX01 Instituto Nacional de Investigaciones Agricolas, Mexico City	Large collection of local germplasm of cucurbits, maize, *Phaseolus* and tropical legumes and introduced germplasm of chickea, lentil, medicago, millets, safflower, sesame, sorghum, soyabean and temperate grasses
Netherlands	NLD01 Institute of Horticultural Plant Breeding, Wageningen	Large collection of landraces, advanced cultivars and related wild species of brassicas, *Capsicum*, eggplant, lettuce, melon, pea and *Phaseolus*
	NLD07 Bejo-Zaden, Noordscharwoude	Collection of landraces and advanced cultivars of brassicas and onion
	NLD03 Foundation for Agricultural Plant Breeding, Wageningen	Large collections of barley, brassicas, faba bean, maize and wheat
NGB	Nordic Gene Bank, Lund, Sweden	Collection of local and introduced barley, beet, brassicas, clover, flax, forage grasses, oat, onion, pea, rye and wheat
Nigeria	NGA04 National Horticultural Research Institute, Ibadan	Collection of landraces, advanced cultivars and breeding lines of *Amaranthus, Capsicum*, cucurbits, okra, onion and tomato
Pakistan	PAK01 Pakistan Agricultural Research Council, Islamabad	Collection of germplasm of landraces of barley, brassicas, chickpea, forages, maize, okra, onion, pea, rice and wheat
Paraguay	PRY01 Instituto Agronomico Nacional, Cacupe	Collections of native landraces and introduced forage grasses and legumes and wheat
Peru	PER02 Programa de Investigaciónes de Maiz Universidad Nacional Agraria, La Molina, Lima	Large collection of indigenous germplasm of maize

Table 1. Continued

Country/Acronym	Center	Major Crops
	PER13 Programa de Investigación en Hortalizas, Universidad Nacional Agraria, La Molina, Lima	Large collection of indigenous germplasm of *Capsicum*, cucurbits and tomato
Philippines	PHL05 Institute of Plant Breeding, University of the Philippines, Los Banos	Large collection of brassicas, *Capsicum*, cowpea, cucurbits, eggplant, groundnut, mung bean, okra, pigeon pea, soyabean, tomato and winged bean
Poland	POL02 Institute for Potato Research, Roszalin	Collection of wild species of potato
Poland	POL03 Plant Breeding and Acclimatization Institute, Radzikow	Large collection of barley, faba bean, forages, oat, pea, rye, sorghum, soyabean and wheat
Portugal	PRT01 Maize Breeding Centre, Braga	Collection of Mediterranean maize
	PRT06 Departamento de Genética, Estaçao Agronómica Nacional, Oeiras	Collection of local germplasm of lupin, maize, *Phaseolus*, rye and wheat
Solomon Islands	SLB01 Dodo Creek Research Station, Ministry of Agriculture and Lands, Honiara	Collection of landraces of sweet potato
South Africa	ZAF01 Division of Plant and Seed Control, Pretoria	Large collection of introduced germplasm of advanced cultivars and landraces of barley, brassicas, cotton, cowpea, cucurbits, forage legumes and grasses, groundnut, lupin, oat, onion, *Phaseolus*, rye, sesame, sorghum, soyabean, sunflower, tomato and wheat
Spain	ESP04 Banco de Germoplasma INIA, Finca El Encin, Madrid	Collection of landraces, advanced cultivars and wild species of barley, chickpea, faba bean, forage legumes, lentil, lupin, melon and related species, oat, pea, *Phaseolus*, rye and wheat

Table 1. Continued

Country/Acronym	Center	Major Crops
	ESP05 Escuela T. S. de Ingenieros Agrónomos, Universidad Politécnica, Madrid	Large collection of brassicas
	ESP06 Centro Regional de Investigacion y Desarrollo Agrario del Ebro, Zaragoza	Large collection of *Capsicum*
Switzerland	CHE01 Station Fédérale de Recherches Agronomiques de Changins, Nyon	Large collection of forage grasses and wheat
Syria	SYR03 Agricultural Research Centre, Douma	Significant collection of local landraces of *Aegilops*, barley, chickpea, faba bean, forages, lentil, oat, safflower and wheat
Thailand	THA06 Thailand Institute of Scientific and Technological Research, Bangkok	Collection of Asiatic maize and winged bean
	THA07 Rice Division, Department of Agriculture, Bangkok	Large collection of local land-races, wild types and advanced cultivars of rice
Turkey	TUR01 Aegean Regional Agricultural Research Institute, Izmir	Large collection of local land-races of barley, beet, brassicas, *Capsicum*, chickpea, cucurbits, forage legumes, lentil, maize, oat, okra, onion, *Phaseolus*, poppy, rye, sesame, spinach, sunflower, tobacco, wheat and related wild species
Uganda	UGA01 Uganda Agriculture and Forestry Research Organisation, Soroti	Large collection of millets and sorghum
Union of Soviet Socialist Republic	SUN01 N.I. Vavilov Institute of Plant Industry, Leningrad	World collections of landraces of all major crop species. Includes the original collections of Vavilov
United Kingdom	GBR04 Royal Botanic Gardens, Kew	Collection of forage grasses and legumes

Table 1. Continued

Country/Acronym	Center	Major Crops
	GBR05 Plant Breeding Institute, Cambridge	Large collection of landraces and wild species of barley, maize, oat, rye and wheat
	GBR06 National Vegetable Research Station, Wellesbourne	Large collection of landraces and advanced cultivars of beet, brassicas, carrot, faba bean, lettuce and radish
	GBR10 Scottish Crop Research Institute, Pentlandfield	Collection of forage crucifers
	GBR11 John Innes Institute, Norwich	Large collection of landraces, advanced culivars and wild species of pea
	GBR14 Department of Applied Biology, University of Cambridge	Large collection of introduced germplasm of landraces and wild species of *Phaseolus*
	GBR16 Welsh Plant Breeding Station, Aberystwyth	Large collection of forage grasses
United States of America	USA03 North Eastern Region Plant Introduction Station, Geneva, New York	Large collection of germplasm of landraces, advanced cultivars and wild species of brassicas, celery, forage grasses and legumes, onion, pea and pumpkin
	USA06 National Seed Storage Laboratory, Fort Collins, Colorado	World collections of germplasm of barley, beet, brassicas, *Capsicum*, castor, cotton, cowpea, cucurbits, flax, forage grasses and legumes, groundnut, lentil, lettuce, maize, millets, oat, okra, onion, pea, *Phaseolus*, cultivated and wild potatoes, rice, rye, safflower, sesame, sorghum, soyabean, tobacco, tomato and wheat
	USA19 Southern Region Plant Introduction Station, Experiment, Georgia	Large collection of germplasm of landraces, advanced cultivars and wild species of *Capsicum*, annual clovers, cucurbits, egg- plant and other related *Solanum* species, groundnut, *Leucaena* millets, mung bean, okra and pigeon pea

Table 1. Continued

Country/Acronym	Center	Major Crops
	USA21 Department of Horticulture, Purdue University, Indiana	Large collection of landraces, advanced cultivars and breeding lines of lima bean
	USA24 USDA Vegetable Production Research Unit, Salinas, California	Collection of advanced cultivars and landraces of brassicas, chicory and lettuce
	USA23 North Central Region Plant Introduction Station, Ames, Iowa	Large collection of advanced cultivars and landraces of *Amaranthus*, brassicas, beet, carrot, cucumber, *Cucurbita*, *Lathyrus*, maize, *Medicago*, millets, radish, spinach, sunflower, sweet clover and tomato
	USA25 Western Region Plant Introduction Station, Pullman, Washington	Large collection of *Allium*, brassicas, chickpea, faba bean, forage grasses, lentils, lettuce lupin, *Phaseolus*, safflower and tef
	USA32 USDA Small Grains Collection, Plant Genetics and Germplasm Institute, Beltsville	Large collection of barley, oat, rice, rye, Triticale, wild wheat and related genera
	USA33 Southern Soybean Collection USDA-ARS Delta Branch Experiment Station, Stoneville	Large collection of soyabean
	USA36 USDA Soybean Laboratory, University of Illinois, Urbana	World collection of soyabean
WARDA	LBR04 West African Rice Development Association, Monrovia, Liberia	Large collection of African rice germplasm and introduced material from Asia

amounts of dry matter when grown in a standard Hoagland's solution culture. The primary source of our germplasm has been the U.S. Plant Germplasm Collection. Less than 200 strains of tomato and beans have given the program a wide range of phenotypes for each element. With about 10,000 bean and 5000 tomato accessions in the U.S. system the potential phenotypic variability should exceed greatly the maximum variances isolated thus far. The potential value of systematic evaluation of major economic crops, especially of inbreeding species, is difficult to envision. The biological, political and economic return of such a program would be most significant.

Germplasm collected throughout the world is not always in a readily usable form. Materials from the tropics are often adapted to short photoperiods and difficult to use in northern and southern latitudes unless converted to a more usable form. The inbred-backcross method can be used rather predictably to make these conversions[1].

3. Induce changes in germplasm

Induced mutagenesis has been envisioned as a reservoir of genetic change. Cell cultures of Nicotiana growing in the presence of Al have

Table 2. Institutes where genebanks are under construction or being upgraded (1984)

Country/Acronym	Institute
Austria	Central Agricultural Research Institute, Vienna
Bangladesh	Bangladesh Agricultural Research Institute, Dacca
Bolivia	Instituto Boliviano de Technologia Agropecuaria, La Paz
China	Chinese Academy of Agricultural Sciences, Beijing
CIMMYT	Centro Internacional de Mejoramiento de Maiz y Trigo, Mexico
Cuba	Academia de Ciencias de Cuba, Habana
Czechoslovakia	Research Institute of Plant Production, Praha
Ecuador	Instituto Nacional de Investigaciones Agropecuarias, Quito
Egypt	Agricultural Research Centre, Giza
Ghana	Crops Research Institute, Bunso Agricultural Experiment Station, Bunso
ICARDA	International Centre for Agricultural Research for Dry Areas, Aleppo
ILCA	International Livestock Centre for Africa, Addis Ababa, Ethiopia
India	National Bureau of Plant Genetic Resources, New Delhi
Iran	Seed and Plant Improvement Institute, Karadj
Ivory Coast	Ministère de Recherche Scientifique, Abidjan
Kenya	Kenya Agricultural Research Institute, Nairobi
Malaysia	Malaysian Agricultural Research and Development Institute, Kuala Lumpur
Mozambique	Universidade Eduardo Mondlane, Maputo
Netherlands	National Genebank, Wageningen
Niger	University of Niamey
Nigeria	National Horticultural Research Institute, Ibadan
Peru	Instituto Nacional de Investigación y Promocion Agropecuaria, Lima
Poland	Plant Breeding and Acclimatisation Institute, Radzikow
Sudan	Agricultural Research Centre, Wad Medani
Togo	Direction de la Recherche Agronomique, Lomé
Tunisia	Ministère de l'Enseignement Supérieur et de la Recherche Scientifique, Tunis
Upper Volta	Institut Voltãique de Recherche Agronomique et Zootechnique, Ouagadougou
Zimbabwe	Ministry of Agriculture Crop Breeding Institute, Harare

given rise to single gene mutants tolerant to Al-toxic conditions[3]. This type of assay, when successful, permits the screening of millions of cells in a limited laboratory space, but these isolates are not apt to exhibit phenotypic variability related to acquisition and transport. Higher plants have evolved as multicellular organisms dependent on specialized tissues and cells for specific functions. Responses of single cells in or on a synthetic medium may not reflect an adaptive value that will be important in a whole plant growing in a soil environment.

4. Recombinant phenotypes

Phenotypic variability within family structures used to estimate inheritance of differential responses may provide our most useful extremes in response. Research usually compares nonsegregating and segregating progenies in a common environment. Within the segregating progenies (backcross and F_2) a number of segregants are usually detected that exceed the phenotypic value of either parent. These segregants are valuable if heritability is high. In the report on P acquisition in tomato some BCP_1 plants yielded more than 2x the dry matter of either parent[2]. Genetic recombination permits the recovery of useful segregants beyond the phenotypic extremes in the parents.

In a survey of early work on P utilization by beans grown under low P a series of families composed of parents that were either (1) both inefficient, (2) both efficient, or (3) one efficient and one inefficient were studied[7]. Efficient parents produced about 2x more dry matter than the inefficient parents. Ten to 30% of the F_2 and backcross segregants were superior to the better parent used to create the family structure (Table 3). Since only 2 efficient parents (1 Family) were used in these studies comparisons between families should be viewed with caution. Several aspects are worth noting, however. First, the two efficient parents did not give a very high proportion of desired recombinants, about 10%; second, F2 progeny never produced as many desired recombinants as the BCP_1 and BCP_2 generations; and third, families of both inefficient x inefficient

Table 3. Proportion of segregating bean progenies grown under low-P stress that exceeded mean plant top dry weight of better parent

Parent phenotypes		No. of Families	Percent exceeding P_2 parent[ʹ]		
P_1	P_2		BCP_1	BCP_2	F_2
			%	%	%
I	E	5	17.8	24.0	12.2
I	I	6	28.6	28.5	11.4
E	E	1	–	10.0	8.7

[ʹ] Calculations based on (mean value for better parent + CV) × 110%

parents and inefficient x efficient parents provided a significant number of desired segregants. Since all comparisons were made against the performance of the better parent, the BCP_2 of I x E (inefficient x efficient) would seem to be a much more useful source of variation than the segregation progeny from E x E. The same efficient parent was the P_2 individual in these studies. Many of the best segregants exceeded the better parent by at least 50%. Natural occurrence of germplasm useful for research on mineral nutrition is plentiful. Progress has been made on experimental procedures for detecting useful phenotypic variation. Genetic analysis of these responses is rather easy also. Synthesis, via recombination and selection, should continue to provide materials which respond uniquely to mineral nutrient stress. These unique biotypes will be most useful experimental materials for study of mechanism by which plants acquire, transport and use essential elements.

References

1. Bliss F A 1982 The inbred backcross line method for improving quantitative traits of self-pollinated crops. HortScience. 17, 503.
2. Coltman R R, Gabelman W H, Gerloff G C and Barta S 1986 Genetic and physiology of low-phosphorus tolerance in a family derived from two differently adapted strains of tomato (*Lycopersicon esculentum* M.11) Plant and Soil (*In press*).
3. Conner A J and Meredith C P 1985 Large scale selection of aluminum-resistant mutants from plant cell culture: expression and inheritance in seedling. Theor. Appl. Genet. 71, 159–165.
4. Gerloff G C 1986 Intact-plant screening for tolerance of nutrient-deficiency stress. Plant and Soil 99, 3–16.
5. Hanson J, Williams J T and Freund R 1984 Institutes Conserving Crop Germplasm: The IBPGR Global Network of Genebanks, International Board of Plant Genetic Resources FAO, Rome, 25 p.
6. Hawkes J G, Williams J T and Croston R P 1983 A bibliography of Crop Genetic Resources. International Board of Plant Genetic resources, FAO, Rome, 442 p.
7. Whiteaker G, Gerloff G C, Gabelman W H and Lindgren D 1976 Intraspecific differences in growth of beans at stress level of phosphorus J. Am. Soc. Hort. Sci. 101, 472–475.

Incorporation of phosphorus efficiency from exotic germplasm into agriculturally adapted germplasm of common bean (*Phaseolus vulgaris* L.)

TERRY M. SCHETTINI, W.H. GABELMAN and G.C. GERLOFF
Departments of Horticulture and Botany, University of Wisconsin, Madison, WI 53706, USA

Key words Bean Breeding Inbred backcross Stress

Summary The growth of inbred backcross (IB) lines derived from the cross of the phosphorus (P)-efficient donor parent (PI 206002) and the recurrent parent (cultivar Sanilac) was measured in low-P nutrient solution culture and in a field nursery on a soil moderately deficient in P. Several IB lines that resembled 'Sanilac' in general morphology were identified as P-efficient in the nutrient solution culture (10 to 25% more shoot dry weight accumulation than 'Sanilac'). In general, these lines accumulated 30 to 50% more shoot dry weight and more P in the shoot tissue at first flower than 'Sanilac' in the low-P field plot but did not differ from 'Sanilac' in a field plot amended with P fertilizer. Some IB lines with seed yields higher than 'Sanilac' may have both the vegetative P efficiency of PI 206002 and the ability to convert this growth into seed production.

Transfer of a quantitative trait such as P efficiency using the inbred backcross breeding method and preliminary evaluation of the IB lines in nutrient solution culture before field testing were shown to be useful techniques for developing common bean germplasm tolerant to soils low in available P. These genetically related lines should also be useful for physiological studies of P nutrition in common bean.

Introduction

There has been an increased interest in various problems of crop production related to soil mineral stresses[12,15]. Selecting plants of common bean tolerant to soil mineral stresses, such as low phosphorus (P) availability, is one approach to improving productivity on soils low in available P and where fertilization is not practicable[5]. Cultivars of common bean, which were ranked for shoot and root growth rate and for nutrient absorption and translocation in a flowing nutrient solution culture system, gave the same ranks when evaluated for seed yield on soil low in available P[11]. We have selected previously bean accessions for efficient utilization of P and for large root development in a low-P nutrient solution culture[14]. These traits later were shown to be quantitatively inherited[6,7].

The objectives of this study were (1) to determine if the traits associated with P efficiency in common bean could be transferred effectively into the genetic background of an agriculturally useful parent (cv. Sanilac), using the inbred backcross line method;[1] (2) to use the inbred backcross lines to dermine if P efficiency measured in nutrient solution culture was correlated with field response; and (3) to identify genetically

H.W. Gabelman and B.C. Loughman (Eds.), Genetic aspects of plant mineral nutrition.
ISBN-13: 978-94-010-8102-3
© *1987 Martinus Nijhoff Publishers, Dordrecht/Boston/Lancaster.*

related lines of common bean differing in P efficiency which could be used for physiological studies of P utilization.

Materials

A pure line derived from the accession PI 206002 was characterized as P-efficient (high total plant P efficiency ratio and large root mass) in low-P nutrient solution culture[14]. It was used as the donor parent in an inbred backcross program with the cultivar Sanilac as the recurrent parent, to develop BC_2S_3 and BC_2S_4 populations for evaluation both in the growth room and in the field.

Methods

Growth room experiments

Seeds were germinated in polyethylene pouches lined with germination paper[10] which had been soaked with 15 ml of a modified Hoagland's nutrient solution (minus P) containing the following nutrient concentrations as mg/l: N, 210; K, 234; Ca, 200; Mg, 48; S, 64; Fe as Fe-EDTA, 4.0; Mn, 0.27; B, 0.27; Zn, 0.13; Cu, 0.03; Mo, 0.01. When the unifoliolate leaves were expanded (day 7 to 10) seedlings were transplanted into aerated, single-plant pot cultures containing 900 ml of the nutrient solution (minus P). Inorganic P (as KH_2PO_4) was added to each pot solution to complement the estimated P content of each seed, for a total P supply of 3 mg P per plant.

Environmental conditions during the three experiments were: light, 16 hour light period of 200 to 250 μmoles/m^2/s of photosynthetically active radiation from cool-white, fluorescent lamps; and temperature, 30–34 C during the light period and 23–27 C during the dark period.

Plants were harvested 26 to 28 days after imbibition, when 'Sanilac' began to flower. The number of nodes on the main stem, plant height and stage of plant development were recorded for each plant. The shoot and root of each plant were separated, dried in a forced-air oven for 3 days at 65 C, weighed, and total plant dry weights (TDW) were calculated from the shoot dry weight (SDW) and root dry weight (RDW) data.

Field experiments

The field experiments were on a Plano silt loam having an initial Bray 1 soil P test[2] of 29 kgP/ha. In both years the entire nursery was fertilized with 22 kg N/ha and 225 kg K/ha and half of the nursery with 64 kg P/ha, according to soil test recommendations.

Rows were 90 cm apart and plants were 10 cm apart in the row. The tops of five plants per row were cut off at the unifoliolate node at the R1 growth stage (when 50% of plants in the row had at least one open flower)[9] and were bulked. Ten plants per row were harvested in bulk when the seeds were mature.

The following measurements were made in the field plots: number of days from planting to the R1 growth stage (days to R1), dry weight of five tops at R1 (top DW), weight of 50 mature seeds (seed size) and total seed yield of ten plants.

The concentration of P in oven-dried tops at R1, harvested from the plots without added P, was determined by the vanado-molybdate method.[8] The P content of five tops at R1 (top P content) and the P efficiency ratio of the tops at R1 (top PER = g top DW/g top P content) were calculated from the data.

Statistical analysis

Parents The data were combined over environments (experiments in the growth room, years in the field) for the analysis of variance. The environment-by-line interactions were not significant ($P > 0.05$).

Inbred backcross (IB) population. In the experimental analysis a protected LSD value ($P = 0.05$) was calculated to compare means only when the variation due to line effects was

Table 1. Performance of 'Sanilac' and PI 206002 grown at 3 mg P per plant in three growth room experiments

Line	Dry weight, g per plant (mean ± SE)		
	Shoot	Root	Total plant
'Sanilac'	1.27 ± 0.02	0.73 ± 0.08	2.00 ± 0.09
PI 206002	1.40 ± 0.03	0.97 ± 0.09	2.37 ± 0.10
F test	**	*	**

*, ** Significant at 0.05 and 0.01 level, respectively.

significant ($P \leqslant 0.05$).[3] A two-tailed LSD value was calculated to compare the mean of the IB population with the mean of 'Sanilac' and a one-tailed LSD value was calculated to determine which IB lines had means greater than the mean of 'Sanilac.'

Results

Parents

Plants of PI 206002 had greater shoot, root and total plant dry weight than 'Sanilac' in all three growth room experiments (Table 1). In both field plots (with and without added P), PI 206002 flowered 2 to 3 days later than 'Sanilac', produced more shoot biomass and had larger seeds than 'Sanilac', but did not differ from 'Sanilac' for total seed yield. In the field plot without added P, PI 206002 accumulated more P in the shoot tissue than 'Sanilac', but the parents did not differ in top PER (Table 2).

Table 2. Performance (mean ± SE) of 'Sanilac' and PI 206002 grown in the field plots with and without added phosphorus (P) in 1982 and 1983

Treatment	Without added P 24 kg Bray 1 soil P/ha			With added P 72 kg Bray 1 soil P/ha		
	'Sanilac'	PI 206002	F test	'Sanilac'	PI 206002	F test
Days to R1'	42.1 ± 0.6	45.3 ± 0.7	***	42.1 ± 0.5	44.2 ± 0.7	***
Top dry wt. (g/5 plants)	33.8 ± 5.0	59.0 ± 1.6	**	51.3 ± 8.9	71.4 ± 8.8	NS
Top P content (mg/5 plants)	103 ± 13	191 ± 9	**	–		
Top PER[s] (Dry Wt./unit P)	319 ± 12	311 ± 9	NS			
Seed size (g/50 seed)	9.83 ± 0.23	13.03 ± 0.29	**	9.84 ± 0.16	13.70 ± 0.16	***
Seed yield (g/10 plants)	159 ± 10	136 ± 20	NS	197 ± 15	189 ± 9	NS

NS,**,*** Not significant or significant at 0.01 and 0.001 level, respectively.
' Number of days from planting to the R1 growth stage.
[s] Phosphorus efficiency ratio.

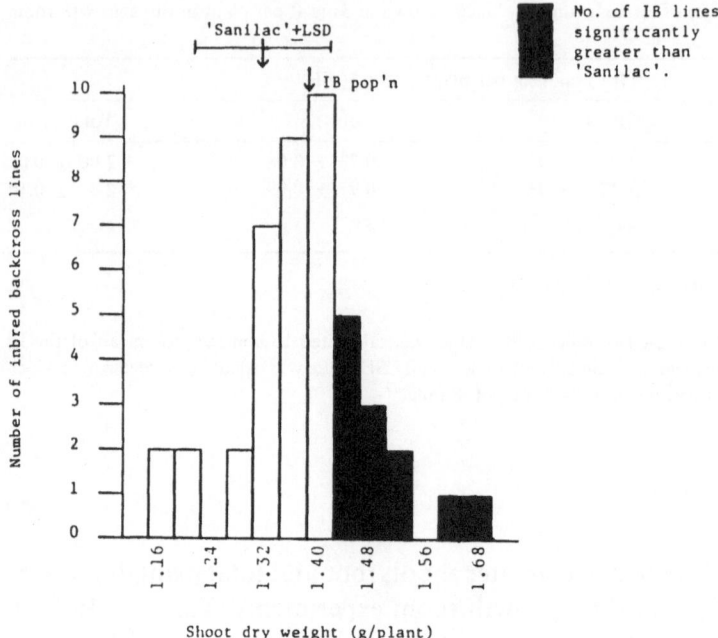

Fig. 1. Distribution of Sanilac inbred backcross (IB) population for mean shoot dry weight, grown at 3 mg P per plant in one growth room experiment.

IB population

The mean of the IB population did not differ from the mean of 'Sanilac' for shoot, root and total dry weight in the growth room (Figures 1 to 3). Similarly, the mean of the IB population did not differ from the mean of 'Sanilac' for top DW, top P content, seed yield (Figures 4 to 6), top PER and seed size in the field plots without added P, and for top DW and seed yield in the field plots with added P (Figures 7, 8). The mean flowering date of the IB population was later than 'Sanilac' only in 1982. The mean seed size of the IB population was larger than 'Sanilac' only in the 1983 field plot with added P[13].

While most of the IB lines resembled the recurrent parent in morphology (as expected), some of these lines differed from 'Sanilac' for each response measured in the growth room and field experiments. For example, the mean separation tests indicated that the SDW of 12 IB lines, the RDW of 6 IB lines and the TDW of 11 IB lines were greater than 'Sanilac' in the third growth room experiment (Figures 1 to 3).

In the field, about 35% of the IB lines flowered later than 'Sanilac'. In the field plots without added P there were differences among IB lines for top PER. In addition, a small number of IB lines had greater top DW than 'Sanilac' (Figure 4) and accumulated more P in their tops at the R1

PHOSPHORUS EFFICIENCY IN COMMON BEAN

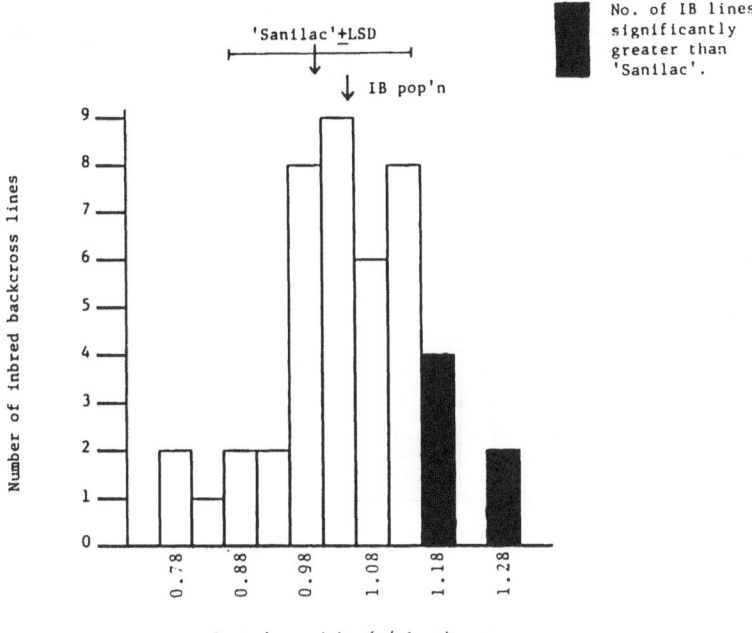

Fig. 2. Distribution of Sanilac inbred backcross (IB) population for mean root dry weight, grown at 3 mg P per plant in one growth room experiment.

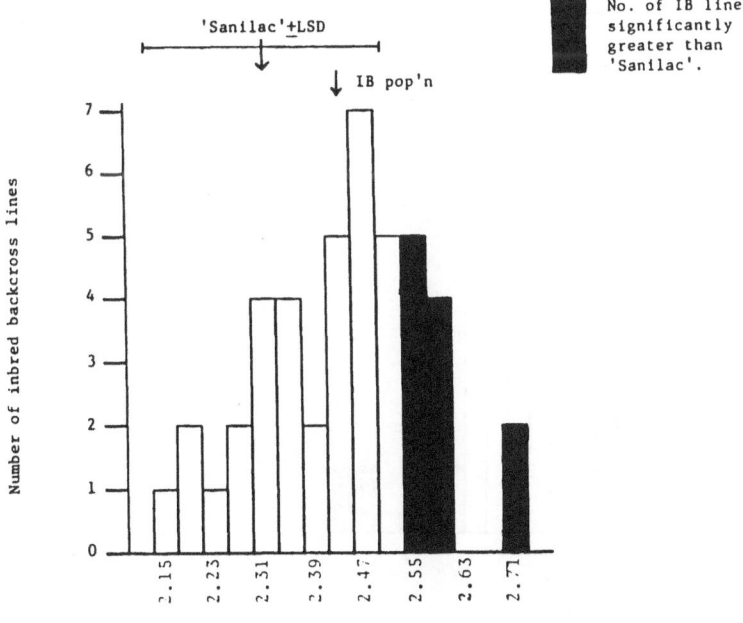

Fig. 3. Distribution of Sanilac inbred backcross (IB) population for mean total plant dry weight, grown at 3 mg per plant in one growth room experiment.

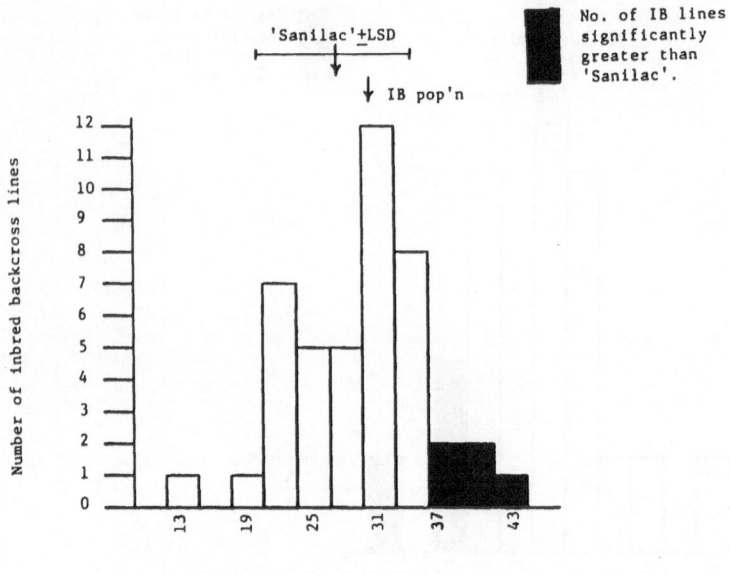

Fig. 4. Distribution of Sanilac inbred backcross (IB) population for mean top dry weight at the R1 growth stage, grown in the field plot without added P (21 kg P/ha, Bray 1) in 1982.

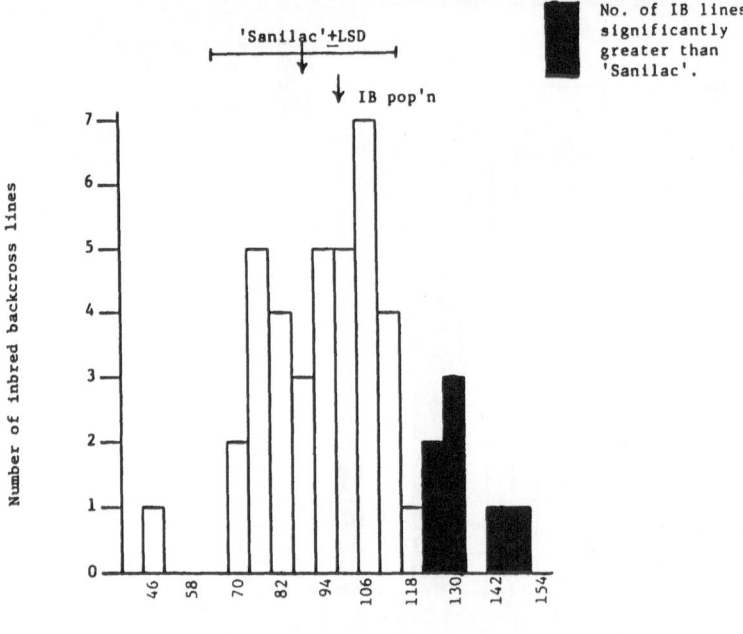

Fig. 5. Distribution of Sanilac inbred backcross (IB) population for mean top P content at the R1 growth stage, grown in the field plot without added P (21 kg P/ha, Bray 1) in 1982.

growth stage (Figure 5). About 10% of the IB lines had greater seed yield than 'Sanilac' under the low-P soil conditions. Approximately 30% of the IB lines had larger seeds than 'Sanilac'; however, the seed size of IB lines resembled more closely the recurrent parent versus PI 206002.

In the field plots with added P, certain IB lines continued to accumulate more top dry weight at the R1 growth stage (Figure 7) and to produce greater seed yield (Figure 8) than 'Sanilac'. The addition of P fertilizer brought the performance of 'Sanilac' closer to the best IB lines without necessarily enhancing the response of the latter. It appears that the best IB lines performed near their optimum at the low level of soil P whereas 'Sanilac' required the addition of P to respond in a similar fashion.

Discussion

Parents

PI 206002 is a snap bean cultivar from Sweden previously identified as P efficient as indicated by vegetative growth in low-P nutrient solution culture[14]. In the present study plants of PI 206002 produced larger shoots and roots than 'Sanilac', had larger leaves, were shorter and flowered later in low-P nutrient solution culture experiments.

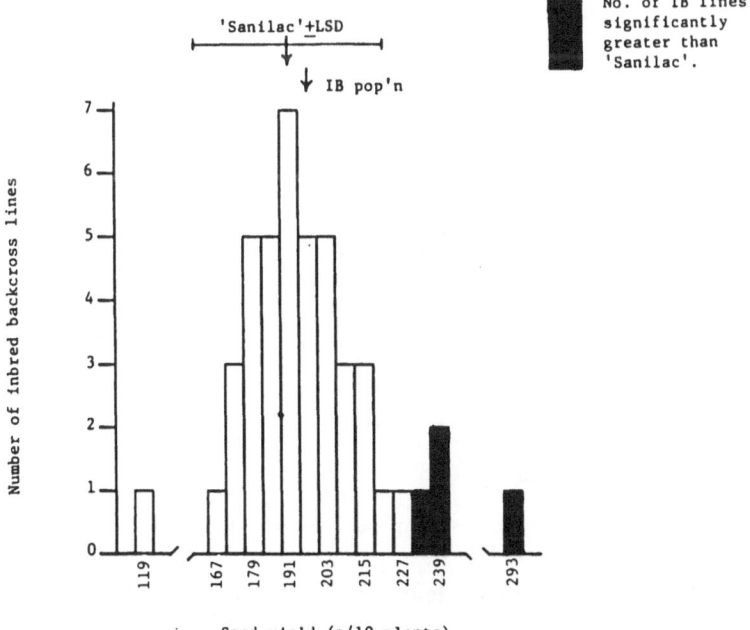

Fig. 6. Distribution of Sanilac inbred backcross (IB) population for mean seed yield, grown in the field plot without added P (21 kg P/ha, Bray 1) in 1982.

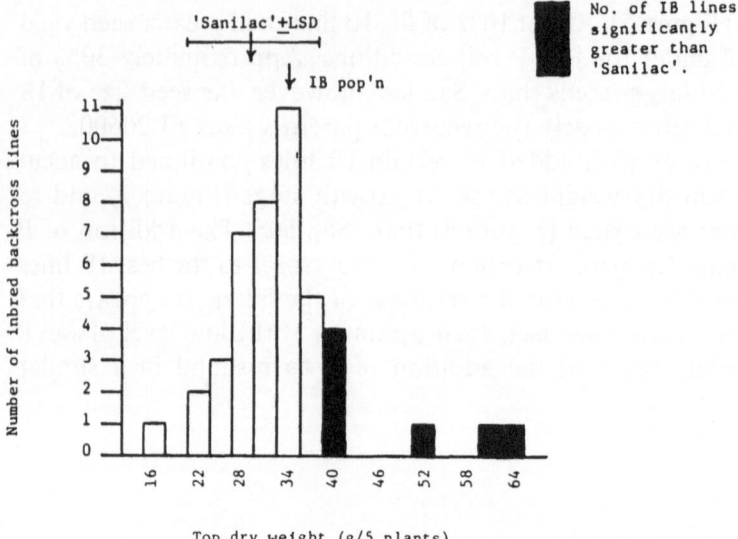

Fig. 7. Distribution of Sanilac inbred backcross (IB) population for mean top dry weight at the R1 growth stage, grown in the field plot with added P (44 kg P/ha, Bray 1) in 1982.

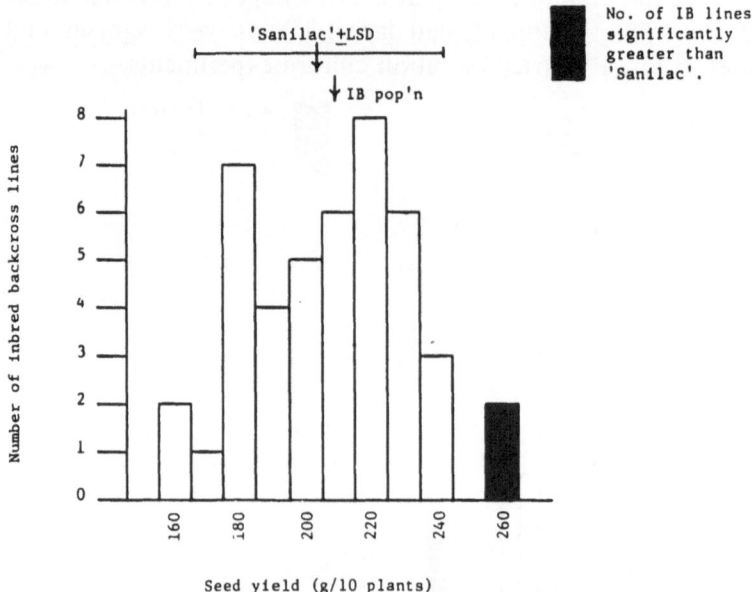

Fig. 8. Distribution of Sanilac inbred backcross (IB) population for mean seed yield, grown in the field plot with added P (99 kg P/ha, Bray 1) in 1983.

In the field, PI 206002 had larger leaves than 'Sanilac', and had a Type II plant habit[4], rather than the Type I habit of 'Sanilac'. PI 206002 accumulated more shoot biomass by the R1 growth stage than 'Sanilac' in plots both with and without added P. This difference was due in part to the two to three day delay in flowering of PI 206002. Since the two parents did not differ in top PER, their shoot dry weight differences can be attributed to differences in shoot P accumulation. The seeds of the pure line derivative of PI 206002 used in this study were large and elliptical with a white seed coat while seeds of 'Sanilac' were smaller, round and white. 'Sanilac' is an early maturing white navy bean cultivar of good commercial seed quality but with lower seed yields than newer cultivars. PI 206002 is a snap bean line with low seed yield in this area.

IB population

Transfer of P efficiency into 'Sanilac' background. Most of the IB lines were similar morphologically to 'Sanilac' in the growth room and in the field. Some of the IB lines grown in field plots without added P produced more shoot biomass and accumulated more P in the shoot at R1 than 'Sanilac'. Only a portion of these differences was associated with the later vegetative harvest dates of the improved lines. In the field plot with added P, the performance of 'Sanilac' improved to the level of the IB lines. These growth room and field studies support the conclusion that traits related to P efficiency have been incorporated into a 'Sanilac' genetic background.

Relationship between growth room and field response. Spearman rank correlation coefficients between field and growth room parameters were low (r less than 0.4). This may be due to the relative contribution to the plant's growth of P acquisition versus P utilization in the field and nutrient solution culture, respectively. However, testing the IB lines in a soil more severely deficient in available P than used in this study might increase the correlation observed between the severely P-stressed growth room environment and the field environment. Still, the superior performance of selected IB lines in the growth room and in the field supports the feasibility of discarding the worst lines in growth room studies before investing in field evaluation.

Potential of P-efficient IB-lines. A few IB lines performed consistently well in the various experiments. The mean SDW of three lines (S11-11, S11-24 and S11-35) was always greater than 'Sanilac' in the growth room experiments and line S11-26 always placed in the top-third of the population. In the field plots without added P, these four IB lines produced a large shoot biomass and accumulated a high shoot P content at the R1 growth stage. Lines S11-11 and S11-26 also had high seed yield in the low-P field plots. Line S11-26 was indistinguishable from, and line S11-24 closely resembled, 'Sanilac' for morphological and phenological traits, yet both produced high top DW and high seed yield when grown under moderately low P field conditions.

The data presented here indicate that quantitative traits can be transferred into an agriculturally useful genetic background using the inbred backcross line method. In general, lines which performed well in low-P nutrient solution culture also performed well in a field test in soil moderately deficient in P.

References

1 Bliss F A 1981 Utilization of vegetable germplasm. HortScience. 16, 129-132.
2 Bray R H and Kurtz L T 1945 Determination of total, organic and available forms of phosphorus in soils. Soil. Sci. 59, 39–45.
3 Carmer S G and Swanson M R 1973 An evaluation of ten pairwise multiple comparison procedures by Monte Carlo methods. J. Am. Stat. Assoc. 68, 66–74.
4 CIAT 1980 Appendix A: Description of growth habits of *Phaseolus vulgaris* L. used in this Annual Report. Bean Program 1979 Annual Report CIAT, Cali, Colombia. p 106.
5 CIAT 1982 Bean program-Annual report. CIAT, Cali, Colombia.
6 Fawole I, Gabelman W H, Gerloff G C and Nordheim E V 1982 Heritability of efficiency in phosphorus utilization in beans (*Phaseolus vulgaris* L.) grown under phosphorus stress. J. Am. Soc. Hort. Sci. 107, 94-97.
7 Fawole I, Gabelman W H and Gerloff G C 1982 Genetic control of root development in beans (*Phaseolus vulgaris* L.) grown under phosphorus stress. J. Am. Soc. Hort. Sci. 107, 98-100.
8 Jackson M L 1958 Soil Chemical Analysis. Prentice-Hall, Inc. pp. 151-153.
9 Lebaron M J 1974 Developmental stages of the common bean plant. Current Information Series #228. April 1974. Univ. of Idaho.
10 Northrup King Company, Attention Horticultural Supplies, 1500 Jackson Street NE, Minneapolis MN 55413. Plant Growth Pouch, 100 per packet, item #82700.
11 Salinas J G 1978 Differential response of some cereal and bean cultivars to aluminum and phosphorus stress in an Oxisol in Central Brazil. PhD Thesis, N. Carolina State University at Raleigh.
12 Saric M R and Loughman B C (Eds.) 1983 Genetic Aspects of Plant Nutrition. *Plant and Soil* 72, 137-396.
13 Schettini T M 1985 Field and growth room studies of phosphorus efficiency in two inbred backcross populations of common bean (*Phaseolus vulgaris* L.) PhD Thesis, U of W-Madison.
14 Whiteaker G, Gerloff G C, Gabelman W H and Lindgren D 1976 Intraspecific differences in growth of beans at stress levels of phosphorus. J. Am. Soc. Hort. Sci. 101, 472–475.
15 Wright M J (Ed) 1976 Plant Adaptation to Mineral Stress in Problem Soils. Cornell University Ag Expt Sta, Ithaca, NY.

Genetic variability for mineral element concentrations in smooth bromegrass related to dairy cattle nutritional requirements

M.D. CASLER and J.M REICH
Department of Agronomy, University of Wisconsin-Madison, Madison, WI 53706 USA and Cal/West Seeds Inc., Woodland, CA 95616, USA

Key words: Alfalfa/grass mixture *Bromus inermis* Leyss. Genetic gain *Medicago sativa* L. Selection Variance components

Summary Breeding for increased mineral element concentrations may be a method of improving the nutritional value of forage grasses. Genetic variability among 9 polycross families of smooth bromegrass (*Bromus inermis* Leyss.) was investigated for the concentration of 11 mineral elements and 3 elemental ratios. Estimates were obtained for smooth bromebrass forage at 7 spring growth stages, in pure stands and in binary mixtures with alfalfa (*Medicago sativa* L.). Family differences were generally significant for Ca and Ca/P, but generally were not for K, B, Mn, and Fe. Expected selection responses indicated selection in pure stands of smooth bromegrass could increase Cu and Ca P, and decrease N/S and K/(Ca + Mg) to adequate levels for a moderately-producing dairy cow. Other elements did not have high enough mean values or sufficient genetic variability to warrant selection. An 80% alfalfa/20% smooth bromegrass mixture, deficient in P, Zn, and Cu, but excessive in the N/S ratio, could not be improved enough to make breeding for these minerals a viable objective for either component of the mixture. The alfalfa/grass mixture was usually adequate for all other elements.

Introduction

The level of mineral elements in a feed is an important factor influencing performance of ruminants. Generations of natural and artificial selection have produced forage species that, when fed alone, cannot often sustain a healthy, growing, high-producing ruminant. For example, low Mg content of many grasses, particularly in combination with a high K/(Ca + Mg) ratio, induces hypomagnesaemia in grazing cattle[8,11]. Estimates of genetic variability and heritability in tall fescue (*Festuca arundinacea* Schreb.) have indicated that selection for increased Mg or reduced K concentration, or reduced K/(Ca + Mg) would be successful[13]. Selection for Mg content of Italian ryegrass (*Lolium multiflorum* Lam.) was successful in improving intake, apparent availability, and retention of Mg by grazing sheep[10].

Concentration of N, P, Cu, and Zn in timothy (*Phleum pratense* L.) were below those recommended for a moderately-producing dairy cow[1]. Selection for increased concentration in the forage was expected to provide adequate levels of N and Zn for a timothy/alfalfa (*Medicago sativa* L.) mixture with at least 50% alfalfa. However, acceptable P and Cu levels could not reasonably be reached by selection in timothy, alfalfa, or their mixture. In smooth bromegrass (*Bromus inermis* Leyss.), the concentration of cell wall constituents is higher when the grass is grown in association with alfalfa than in monoculture[12]. Therefore, greater genetic variability would be required to reduce cell wall constituents to a specified level in grass tissue obtained from a mixture compared to a monoculture. However, if that constituent were present in lower concentration in the alfalfa component, as is the case with some cell wall constituents, less genetic variability would be required in the grass species to develop a total mixture meeting animal requirements[6]. The purpose of this research was to predict selection responses for mineral element concentrations in a smooth bromegrass pure stand and an 80% alfalfa/20% smooth bromegrass mixture, which is considered the target for alfalfa/smooth bromegrass mixtures in the North Central U.S. Emphasis was placed on the potential for development of a crop able to meet minimum dairy cow requirements at various stages of growth in the spring.

Materials and methods

Nine polycross families of the B8HD experimental synthetic[4] were established in 0.9 × 3.0-m 5-row plots, both with and without an alfalfa (cv. 'Saranac') overseeding, at Hancock (mixed, mesic, Typic Udipsamment soil) and Lancaster (fine silty, mixed, mesic, Typic Hapludalf soil), Wisconsin in late August 1981. The experimental design was 4 randomized complete blocks in a split-plot arrangement with alfalfa/no alfalfa (stand type) whole-plots and polycross family sub-plots.

The smooth bromegrass and alfalfa were seeded at a rate of 15 kg ha^{-1} for each species at both locations. About 5 weeks after planting, all plots were topdressed with 29.5 kg P ha^{-1} and 166 kg K ha^{-1}, respectively. Smooth bromegrass plots without alfalfa overseeding were also topdressed with 50 kg N ha^{-1}. Plots at Hancock were irrigated with 2.5 cm of water weekly or as needed.

Random 0.1-m^2 sections of smooth bromegrass were sampled from the center three rows of each plot every 7 days from about the 15-cm height stage to approximately 1 week past anthesis, giving the following 7 growth stages: early vegetative (17 to 22 cm), late vegetative (35 to 40 cm), early jointing (50 to 55 cm), late jointing (65 to 70 cm), heading (65 to 75% headed), anthesis, and 1 week post-anthesis. Sampling began on 8 and 13 May 1982 and 11 and 14 May 1983 at Lancaster and Hancock, respectively. Sampling was conducted to avoid repeated harvests from the same area of a plot. Forage harvested from alfalfa/smooth bromegrass plots was hand separated to retain only the grass component. A composite alfalfa sample was also harvested from each location on each sampling date. All samples were harvested at 5.0-cm stubble height and dried in perforated paper bags at 55°C.

Following the 7th sampling date in 1982, all plots at each location were clipped to a 10.0-cm stubble height. For the remainder of 1982 the plots were managed as recommended for a 3-cut alfalfa/smooth bromegrass hay crop.

In both 1982 and 1983 the experimental area at both locations was fertilized with 88.5 kg P ha^{-1} and 498 kg K ha^{-1} split equally in 3 applications: the last week of April, following the 7th growth stage (first cutting) and following the 3rd cutting. Smooth bromegrass pure stand plots were also fertilized with 200 kg N ha^{-1} split equally in two applications: the last week of April and following the first cutting. All fertilization rates were based on estimated nutrient availability in the soil and nutrient removal throughout the season.

Forage samples were ground through a 40-mesh (1.0-mm) screen and stored in glass bottles. Concentrations of P, K, Ca, Mg, S, B, Cu, Fe, Mn, and Zn were determined on 0.5-g samples by inductively-coupled plasma emission spectroscopy. The concentration of N was determined on 0.25-g samples by the micro-kjeldahl procedure. Analyses were based on pairwise-composited replicates for each sampling date, reducing the effective number of replicates from 4 to 2.

Analyses of variance were computed separately for each stand type at each growth stage. All sources of variation were considered to have random effects. Each of the 14 stand type-growth stage combinations was analyzed as 2 randomized complete blocks in a split-plot arrangement with polycross family whole-plots and years within families as sub-plots. Expected mean squares for these analyses were given by Casler[2].

Multivariate analysis of variance was also computed for each growth stage-stand type combination using all mineral element concentrations plus the ratios Ca/P, N/S and K/(Ca + Mg). The Hotelling-Lawley trace was tested using the F-distribution as a measure of the overall significance of family and family × environment interaction effects[9]. Expected genetic gain from one cycle of phenotypic selection was computed as

$$\Delta G_x = kH\hat{\sigma}_p$$

where k is 1.76, the standardized selection differential for a selection intensity of σ_p 10%; H is narrow sense heritability for family means over 2 years, 2 locations, and 2 replicates; and $\hat{\sigma}_p$ is the phenotypic standard deviation. Genetic gain was expressed as a percentage of the difference between mean performance and the requirements of a moderately-producing dairy cow[11] using

$$\Delta G_b = 100(\Delta G_x)/(Y - B) \quad \text{and}$$

$$\Delta G_a = 100(0.2)(\Delta G_x)/[Y - (0.8A + 0.2B)]$$

where ΔG_b is the gain for a pure smooth bromegrass stand, ΔG_a is the gain for an 80% alfalfa/20% smooth bromegrass stand (based on total forage dry matter) due to selection based on the smooth bromegrass component, ΔG_x was computed from pure stand data for ΔG_b and mixed stand data for ΔG_a. A is the alfalfa mean, B is the smooth bromegrass mean for pure or mixed stands, and Y is the dairy cow requirement. The percentage contribution to total forage dry matter of alfalfa and smooth bromegrass was visually estimated to approximate the 80%/20% target in this study. Therefore, the 80%/20% mixture values were computed assuming that these data were representative of these species in such a mixture.

Results and discussion

Family differences, for the array of eleven mineral element concentrations and the 3 mineral ratios, were significant at the 0.05 probability level for 4 of 7 pure stand growth stages and 6 of 7 mixed stand growth stages (Table 1). Family × environment interaction tended to be unimportant since only 3 or 14 stand type-growth stage combinations had significant family × location or family × year interaction and only 1 had significant family × location × year interaction. Examination of analyses of variance for individual mineral element concentrations and ratios showed that no single mineral element appeared to contribute to the presence or absence of family × environment interaction. Each element showed some interaction, but usually in only a small number of stand type-growth stage combinations, and with no consistent trend. Total acid-insoluble ash (silica + minerals) also did not show family × environment interaction for these families[12]. These results are similar to results from alfalfa, which showed, for 11 minerals, some genoty-

Table 1. Effect of stand type and growth stage on probabilities of a greater F-value for the Hotelling-Lawley trace for 4 sources of variation of mineral element concentrations of smooth bromegrass

Stand Type	Growth stage	Families	Families × locations	Families × years	Families × locations × years
Pure	Early vegetative	< 0.001	0.900	0.398	0.987
	Late vegetative	< 0.001	0.066	0.024	0.059
	Early jointing	0.106	0.549	0.445	0.565
	Late jointing	0.001	0.022	0.345	0.137
	Heading	0.830	0.753	0.862	0.959
	Anthesis	0.002	0.162	0.461	0.231
	Post-anthesis	0.160	0.713	0.842	0.838
Mixed	Early vegetative	0.417	0.553	0.976	0.998
	Late vegetative	< 0.001	0.040	0.005	0.074
	Early jointing	0.042	0.901	0.981	0.872
	Late jointing	0.004	0.184	0.218	0.333
	Heading	< 0.001	0.002	0.001	0.006
	Anthesis	0.042	0.905	0.193	0.342
	Post-anthesis	0.018	0.373	0.371	0.255

pe × year interaction only for Ca, P, Ca/P, and B[7]. However in maize (*Zea mays* L.), all elements except Cu (8 of 9) showed some single cross × location interaction[5].

There were also no consistent stand type or growth stage tendencies for any of these 4 effects (Table 1). Selection for mineral element concentrations for most growth stages in pure or mixed stands appears to be feasible. The general absence of family × environment interaction indicated that progeny testing at 1 location for 1 year should be effective in maximizing expected selection responses per year.

Calcium concentration and Ca/P showed the greatest frequency of significant family differences (Table 2). For P, Mg, Zn, Cu, and K/(Ca + Mg), family differences were significant in at least 5 stand type-growth stage combinations, indicating that some selection should be successful for these elements. Overall, consistent trends due to stand type or growth stage could not be discerned. However, some elements (*e.g.* N and S) showed significant family differences in only one stand type. For N, this result was similar to previous work on smooth bromegrass, where phenotypic variances were over 3 × greater for N concentration and over 2 × greater for fiber percentage in mixed stands with alfalfa compared to pure stands[3]. At least for N, the competitive effects of alfalfa appear to allow greater genetic expression of smooth bromegrass genotypes, resulting in larger genetic differences in mixed stands.

Family differences for K, B, Mn, and Fe were so seldom significant that selection for these elements probably would be unsuccessful (Table 2).

Table 2. Estimated family variance components for concentration of mineral elements of smooth bromegrass forage, as affected by stand type and growth stage

Stand type	Growth stage	Character[z]												
		N	P ($\times 10$)	K	Ca ($\times 10$)	Mg ($\times 10$)	S ($\times 10$)	Zn	B	Mn	Fe	Cu	Ca/P ($\times 10^2$)	N/S	K/(Ca + Mg) ($\times 10$)[y]
Pure	Early vegetative	0.01	1.50**	0.32	2.28**	0.30*	1.02*	0.56	0.20	10.0	–	0.26	8.8*	0.22*	1.15
	Late vegetative	.	0.41	0.14	1.73*	–	0.03	0.24	–	4.0	8.61	0.26*	13.1	0.10	2.96
	Early jointing	0.87	0.54	0.09	1.77**	0.18	0.14	.	0.06	1.0	0.49	0.15*	18.0*	0.03	0.50
	Late jointing	0.52	–	0.02	2.01*	0.06	.	0.17	0.64	2.8	–	–	12.6	0.14	1.90*
	Heading	0.04	0.08	0.42	3.33**	0.37*	0.36*	0.67*	–	1.6	1.35	0.03	34.9**	0.06	4.26**
	Anthesis	0.40	0.42	0.26	2.74*	–	0.41*	0.40*	–	1.3	0.40	0.18	47.3*	–	4.16**
	Post-anthesis	0.22	–	–	3.31*	–	0.02	.	0.42	7.1	.	0.07	53.8*	0.09	1.44
Mixed	Early vegetative	0.35	1.38*	.	1.98*	0.47*	0.89	0.56*	0.55	1.6	–	0.10	11.8*	0.10	0.40
	Late vegetative	1.49**	0.46	0.10	1.44**	0.41**	0.50	0.44*	0.47	0.8	.	0.17*	7.1*	0.21*	5.02**
	Early jointing	1.46*	1.37*	0.92*	1.75**	0.22*	0.70	1.10*	0.03	10.5	9.65**	0.32**	6.5	0.10	2.68
	Late jointing	0.10	2.09**	–	1.85**	0.04	0.28	0.44	0.01	11.8	–	0.18**	12.9**	0.08	3.93**
	Heading	0.20	.	0.33	0.96	0.04	0.10	.	0.68	9.3	–	–	4.4	0.24*	1.75
	Anthesis	0.36	0.95*	0.20	1.34	0.00	0.29	0.70	.	7.5	0.49	0.10	20.2*	.	2.21
	Post anthesis	.	0.26	–	1.44	–	–	0.32	0.47	10.5	–	0.07	16.3	–	0.77

[z] Units are g kg^{-1} for N, P, K, Ca, Mg, and S and mg kg^{-1} for Zn, B, Mn, Fe, and Cu.
[y] Ratio expressed on a milli equivalent basis.
* Mean square for families significant at 5% level.
** Mean square for families significant at 1% level.
. Negative family variance component.

Table 3. Percentage of the differential between mean performance of pure smooth bromegrass or alfalfa/smooth bromegrass mixtures and moderately-producing dairy cow requirements expected to be realized by 1 cycle of phenotypic selection for specified mineral element concentration in smooth bromegrass

Stand type	Growth stage	Character (%)												
		N	P	K	Ca	Mg	S	Zn	Mn	Fe	Cu	Ca/P	N/S	K/(Ca + Mg)
Pure grass	Early vegetative	+	+	+	10	37	+	7	+	+	130	14	31	16
	Late vegetative	+	+	+	6	–	+	3	+	+	73	14	14	22
	Early jointing	+	111	+	8	–	+	–	+	+	37	24	5	5
	Late jointing	+	–	+	7	9	–	–	+	–	–	18	29	13
	Heading	4	2	+	11	3	28	2	+	–	6	75	28	28
	Anthesis	14	6	+	8	13	19	5	+	11	16	+	+	48
	Post-anthesis	6	–	+	9	–	1	3	+	3	7	+	+	29
Mixed	Early vegetative	+	+	+	+	+	+	4	+	+	51	+	4	+
	Late vegetative	+	+	+	+	+	+	2	+	+	16	+	7	+
	Early jointing	+	+	+	+	+	+	2	+	+	12	+	3	+
	Late jointing	+	27	+	+	+	+	1	+	+	5	+	3	+
	Heading	+	–	+	+	+	+	–	+	+	–	+	10	+
	Anthesis	+	4	+	+	+	+	1	+	+	4	+	–	+
	Post-anthesis	+	1	+	+	+	+	0	+	+	2	+	–	+

+ Mean performance within limits of moderately-producing dairy cow requirements[11].
– Negative family variance component indicated selection gain was not possible.

Estimated selection responses for 1 selection cycle, expressed as a percentage of the required response, are given in Table 3. Pure smooth bromegrass had adequate K and Mn at each growth stage; N, P, S and Fe at early growth stages; and Ca/P and N/S at later growth stages for a moderately-producing dairy cow. For most elements which were deficient, selection would not be anticipated to provide adequate levels within a reasonably few (3 to 6) cycles. Most elements could be increased only by 2 to 15% of the required amount per cycle. However, a relatively few cycles of selection would be expected to adequately adjust deficiencies in Cu and Ca/P, and excess levels of K/(Ca + Mg). This would be particularly important for K/(Ca + Mg), since the smooth bromegrass in this study was never below the recommended maximum for grazing of 2.2^8, ranging from 2.72 to 3.50. Although N/S was seldom significantly different among families, the relative consistency of expected gains and the small gains required (observed means of 12.0 to 15.1 compared to a recommended maximum of 12.0) indicated the possibility of adequate genetic progress.

For an 80% alfalfa/20% smooth bromegrass mixture, all elements were present in sufficient quantities, except P at later growth stages and Zn and Cu at all growth stages (Table 3). The ratio of N to S was excessive at all growth stages (Table 3). Copper at the early vegetative growth stage and P at late jointing were the only elements that could be adjusted to adequate levels by selection within smooth bromegrass in a reasonable time frame. Increasing dietary P, Zn, and Cu and decreasing the N/S ratio of the feed (probably by increasing S, since N was adequate) would seem to require fertilization or mineral supplementation. Selection for increased P, Cu, or Zn in the alfalfa component of the mixture would not be expected to reach required levels in a reasonable time frame[6], while data on S is unavailable.

Genetic correlation coefficients among mineral element concentrations and ratios generally appear to provide favorable expected correlated selection responses (Table 4). Selection for higher Cu, higher Ca/P, lower N/S, or lower K/(Ca + Mg) would be expected to have similar results for most elements. With few exceptions, these responses included increases in Ca, Mg, S, Zn, Cu, and Ca/P and decreases in N/S and K/(Ca + Mg). Each of these responses would be favorable to the development of a more nutritious smooth bromegrass forage. Selection for increased Cu concentration may be the most beneficial, because it had the strongest favorable genetic correlations with other deficient elements and it was also strongly correlated with N. These results were similar for smooth bromegrass in pure stands and mixtures with alfalfa. The relatively high frequency of mean correlations exceeding twice their standard error indicated reasonable consistency among growth stages for most of these relationships.

Table 4. Genetic correlation coefficients, averaged over growth stages, among mineral elements and ratios for smooth bromegrass in pure stands (above diagonal) and mixed stands with alfalfa (below diagonal)

Element or ratio	N	P	K	Ca	Mg	S	Zn	Mn	Fe	Cu	Ca/P	N/S	K/(Ca + Mg)
N		0.79**	1.32	0.38	1.04*	0.57	−0.24	−0.45	1.95**	1.75**	−0.24	0.29	0.02
P	0.74**		0.00	0.06	−0.56	0.16	0.71*	1.28**	0.47	0.37	−0.28	0.35	0.00
K	0.57*	−0.54		−0.50*	−0.50	−0.29	−0.80*	−0.14	−0.40	0.06	0.30	0.38	0.51*
Ca	0.41	0.21	−1.06		0.40	0.74**	0.58**	0.88**	0.26	0.51**	0.99**	−0.40	−0.32**
Mg	0.42	0.94	−3.16	0.97**		1.00**	1.16**	0.91*	1.02**	1.04**	0.39	−1.89**	−1.17**
S	0.54**	0.68**	−0.19	1.08**	1.30**		0.60**	0.47	0.39	1.02**	0.76**	−0.61	−0.68**
Zn	0.94**	0.65**	0.63**	1.00**	2.82	1.33**		−0.08	1.20**	0.63**	0.27	−0.22	−0.99**
Mn	0.69	−0.27	0.70	0.27	0.86**	1.19**	0.71*		−1.12	0.71	0.55*	−0.28	−0.90**
Fe	1.55**	1.13*	−0.53	2.49	14.93	1.69	1.31**	2.64		1.24**	0.18	3.19**	−0.08
Cu	0.98**	0.60**	0.25	0.45**	2.17	1.10**	1.13**	1.04**	1.49*		0.28	0.54	−0.52**
Ca/P	0.16	−0.30*	−1.08**	0.96**	0.75*	1.05**	0.63	0.46*	1.53	0.14		−0.54*	−1.00**
N/S	0.37	−0.18	0.88*	−0.80**	−1.22**	−0.54*	−0.37	−0.42	0.51	−0.38	−0.84**		0.40
K/(Ca + Mg)	0.02	−0.10	1.10**	−0.29**	−1.83**	−0.97**	−0.89**	−0.34	−1.74	−0.34**	−1.09**	1.09**	

*, **. Mean genetic correlation coefficient greater than its among-growth stages standard error by 2 or 3 times, respectively, as an approximate indication of significance.

In summary, smooth bromegrass pasture or hay fell short of a moderately-producing dairy cow's mineral requirements. Breeding could probably solve the Cu and Ca/P deficiencies plus the N/S and K/(Ca + Mg) excesses. This would be particularly important in reducing the potential for hypomagnesaemic tetany on smooth bromegrass pastures due to a high K/(Ca + Mg) ratio. Breeding smooth bromegrass to improve the mineral balance of an 80% alfalfa/20% smooth bromegrass mixture would not be an economically viable objective.

References

1. Berg C C and Hill R R Jr 1983 Quantitative inheritance and correlations among forage yield and quality components in timothy. Crop Sci. 23, 380–384.
2. Casler M D 1982 Genotype × environment interaction bias to parent-offspring regression heritability estimates. Crop Sci. 22, 540–542.
3. Churchill B R 1947 Productiveness of bromegrass strains from different regions when grown in pure stands and in mixture with alfalfa in Michigan. J. Am. Soc. Argon. 39, 750–761.
4. Ehlke N J Casler M D and Drolsom P N 1983 Selection for improved digestibility in smooth bromegrass. Proc. Amer. Forage and Grassl. Council, Eau Claire, WI, pp 54–58.
5. Gorsline G W Thomas W I and Baker D E 1964 Inheritance of P, K, Mg, Cu, B, Zn, Mn, Al, and Fe concentrations by corn (*Zea mays* L.) leaves and grain. Crop Sci. 4, 207–210.
6. Hill R R Jr and Guss S B 1976 Genetic variability for mineral concentration in plants related to mineral requirements of cattle. Crop Sci. 16, 680–685.
7. Hill R R Jr and Jung G A 1975 Genetic variability for chemical composition of alfalfa. I. Mineral elements. Crop Sci. 15, 652–657.
8. Kemp A and 't Hart M L 1957 Grass tetany in grazing milk cows. Neth. J. Agric. Sci. 5, 4–17.
9. Morrison D F 1976 Multivariate Statistical Methods. McGraw-Hill Book Co. New York, NY.
10. Moseley G and Griffiths D W 1984 The mineral metabolism of sheep fed high- and low-magnesium selections of Italian ryegrass. Grass Forage Sci. 39, 195–199.
11. National Academy of Sciences — National Research Council 1978 Nutrient requirements of dairy cattle. Pub. 3, 5th Ed. Washington, DC.
12. Reich J M and Casler M D 1985 Effect of maturity and alfalfa competition on expected selection response for smooth bromegrass forage quality traits. Crop. Sci. 25, 635–640.
13. Sleper D A Garner G B Asay K H Boland R and Pickett E E 1977 Breeding for Mg, Ca, K, and P content in tall fescue. Crop Sci. 17, 433–438.

Molecular cloning of the plant plasma membrane H^+-ATPase

T.K. SUROWY and M.R. SUSSMAN
Department of Horticulture, University of Wisconsin, Madison, WI 53706, USA

Key words Antibody H^+-ATPase Membrane Recombinant DNA Transport

Summary Mineral transport across the plasma membrane of plant cells is controlled by an electrochemical gradient of protons. This gradient is generated by an ATP-consuming enzyme in the membrane known as a proton pump, or H^+-ATPase. The protein has a catalytic subunit of $Mr = 100,000$ and is a prominent band when plasma membrane proteins are analyzed by sodium dodecyl sulfate-polyacrylamide gel electrophoresis.

We generated specific rabbit polyclonal antibody against the $Mr = 100,000$ H^+-ATPase and used the antibody to screen λgt11 expression vector libraries of plant DNA. Several phage clones producing immunoreactive protein, and presumably containing DNA sequences for the ATPase structural gene, were isolated and purified from a carrot cDNA library and a Arabidopsis genomic DNA library. These studies represent our first efforts at cloning the structural gene for a plant plasma membrane transport protein. Applicability of the technique to other transport protein genes and the potential for use of recombinant DNA technology in plant mineral transport research are discussed.

Introduction

In the past decade there have been major advances in our understanding of how biological membranes transport minerals and other solutes. These advances were made possible by theoretical insights concerning how ion gradients participate in energy coupling (Mitchell's 'chemiosmotic hypothesis') and by new biochemical techniques that facilitate the separation and purification of hydrophobic, membrane-bound proteins. Furthermore, recent advances in the isolation and sequencing of recombinant DNA have led to improvements in the speed and ease of determining the structure of transport proteins.

Most membrane transport proteins are co-transport proteins, otherwise known as 'leaks'[31]. These utilize the energy gradient established by a 'pump' (or primary active transport protein) to catalyze transport of specific minerals, ions and other solutes against a concentration gradient[31]. In the plant plasma membrane the pump is an ATPase which sets up an electrochemical gradient of *protons*[28]. The proton electrochemical gradient in plants and fungi performs the same function that a sodium gradient performs in all animal cells and some bacteria (Fig. 1). The proton motive force, as defined by Mitchell[18], is dependent on both the electric potential and the pH gradient across the membrane. The electric potential across the plasma membrane of plant and fungal cells is the highest of any cell in nature and values of $-300\,mV$ (interior negative)

H.W. Gabelman and B.C. Loughman (Eds.), Genetic aspects of plant mineral nutrition.
ISBN-13: 978-94-010-8102-3
© *1987 Martinus Nijhoff Publishers, Dordrecht/Boston/Lancaster.*

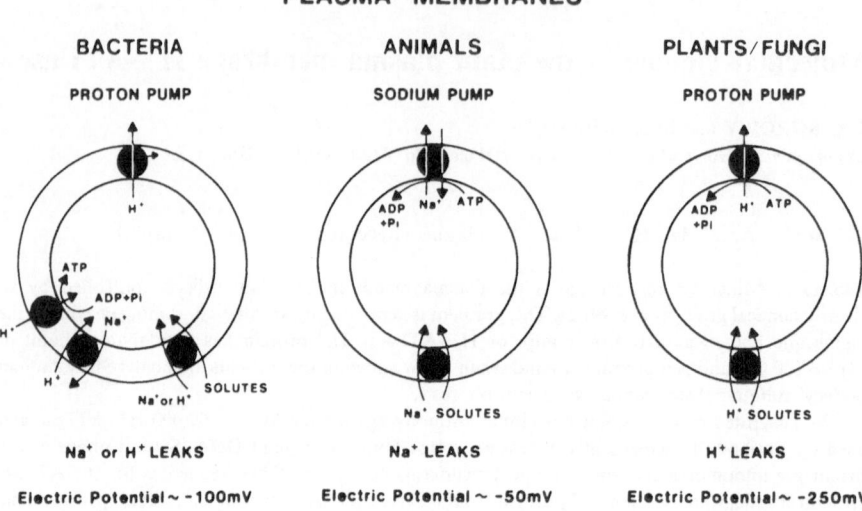

Fig. 1. Schematic representation of pump and leak proteins in the plasma membrane of bacteria (left), animal (middle), and plant/fungal (right) cells.

have been recorded (Fig. 1). An electric potential of 60 mV is approximately equivalent to a ten-fold chemical gradient[22]. The pH gradient is usually no more than 2 units, equivalent to a 100-fold chemical gradient. Together these could drive the inward concentration of a solute 10^7 times higher than the outside concentration if the electrochemical gradient were coupled to transport of that solute. An example would be the co-transport of glucose as a glucose · H^+ complex. Even charged ions such as phosphate can be coupled to the electric potential as long as a sufficient number (n) of protons are coupled to give the overall complex, $nH^+ \cdot X^-$, a net positive charge. There is accumulating evidence in the literature that co-transport systems in plants are dependent on an electrochemical proton gradient across the plasma membrane[22,31]. This has been shown to be set up by the plasma membrane H^+-ATPase[28]; no other primary active transport system has been found in plants. Given the great utility and importance of the protonmotive force, it is not surprising that the H^+-ATPase is a major consumer of cellular ATP. An estimate based on electrophysiological recordings indicates that in root hair cells of mustard, this enzyme alone consumes 25–33% of the cytoplasmic ATP[7].

For the membrane biochemist, an important practical difference between primary active transport (pump) proteins and co-transport (leak) proteins is their turnover number, the number of ions transported per protein molecule per unit time. Co-transport proteins can be extremely fast; some have turnover numbers as high as 10^6/second. In practice this means that co-transport proteins need be present only at very low concentration, as few as 10 molecules per cell. This makes it difficult to isolate enough of the transport protein for biochemical characterization. In contrast, pump proteins are fairly sluggish; the H^+-ATPase molecule has a turnover number of only several hundred/second[24]. As a consequence a greater number of pump protein molecules is needed per cell (relative to co-transport molecules) in order to generate the large energy gradient used by the co-transport proteins. It is therefore not surprising that the H^+-ATPase was the first plant plasma membrane transport protein to be isolated and characterized[21,32].

The plant plasma membrane H^+-ATPase is an outwardly-directed electrogenic proton pump, although the inward flux of potassium ions may play some role[28]. It has a catalytic subunit of Mr = $100,000^{21,32}$ and is phosphorylated during the reaction cycle[5,33]. As such it belongs to a group of Mr = 100,000 plasma membrane transport ATPases which includes the Na^+/K^+-ATPase of animal cells, the Ca^{2+}-ATPase of sarcoplasmic reticulum, the H^+/K^+-ATPase of gastric mucosa, the K^+-ATPase of bacterial cells and the H^+-ATPase of fungal cells. Methods used in the isolation of these other ATPases have provided a basis for isolation methods for the plant plasma membrane H^+-ATPase, although low protein levels and proteolysis problems still thwart the successful purification of biologically active ATPase with high specific activity.

An important goal in understanding further the role of the H^+-ATPase in plant plasma membrane mineral transport is determination of the structure of the protein. The plasma membrane ion-transporting ATPases are all large hydrophobic proteins which are difficult to sequence by standard techniques. The amino acid sequence of the Ca^+-ATPase of sarcoplasmic reticulum has been determined partially through conventional protein sequencing techniques[1] and we are using similar approaches for sequencing sections of the H^+-ATPase from plant plasma membranes. However, complete amino acid sequences, now known for the K^+-ATPase of *E. coli* and the Na^+/K^+- and Ca^{2+}-ATPases of animal cells, have only been obtained using DNA cloning techniques[11,15,23]. Comparison of the sequences of these ATPases revealed areas of sequence homology[11,15,23]. Recently, sequencing of short stretches of other plasma membrane ATPases through protein chemistry techniques has revealed conservation of sequences at the phosphorylation site[34] and an ATP-protectable fluorescein-5'-isothiocyanate binding site[6]. One of our main interests is in determining the sequence and proximities of the ATP binding site and the site in the ATPase molecule which binds DCCD and which is thought to be involved in ion translocation[2,19]. This information will help us learn how the protein uses energy released from the splitting of ATP to translocate protons and set up the electrochemical gradient used by co-transport membrane proteins. The recent determinations of amino acid sequences for the Na^+/K^+- and Ca^{2+}-ATPases have provided information on location and proximities of active site domains and have enabled predictions of secondary structure and of folding of the molecule in the membrane[15,23]. All these have led to further understanding of structure-function relationships for these transport proteins.

Since it has proved extremely difficult to obtain full-length sequences of plasma membrane ATPases through conventional protein sequencing techniques, we decided to pursue the DNA cloning approach for the plant plasma membrane H^+-ATPase and to back-up the protein sequence obtained from the DNA sequence with stretches of protein sequence obtained through conventional protein sequencing methods. One routine procedure has been to isolate genes from recombinant DNA libraries using oligonucleotide probes synthesized on the basis of a known short stretch of protein sequence[16]. Recently, however, methods have been developed in which genes can be isolated using antibody probes: recombinant DNA libraries are constructed in an expression vector system and clones that express the protein of interest are detected using the appropriate antibody. The best characterized and most successful of the expression vector systems is the λgtll system[35-37]. Here stretches of plant DNA (sheared genomic DNA or cDNA synthesized from mRNA) are ligated into a specific EcoRI restriction enzyme site in the λgtll vector. This site is located 53 base pairs upstream from the C-terminal end of the lac Z gene. Induction of expression of the lac Z gene (which codes for β-galactosidase) results in formation of large quantities of a fusion protein. The production of a specific fusion protein can be detected by using a specific antibody directed to the protein of interest. The recombinant DNA encoding this protein can then be isolated form the recombinant phage and sequenced. All that is needed to isolate a gene for a specific protein is enough of that protein to make a specific antibody. A few hundred micrograms is enough to make polyclonal rabbit antibody. For polyclonal or monoclonal antibodies made in mice, even less is needed (on the order of tens of micrograms) and monoclonal antibodies made using *in vitro* immunization techniques have been obtained using nanograms of antigen[14]. We describe here the production of polyclonal rabbit antibody directed to the Mr = 100,000 plant plasma membrane H^+-ATPase and its use in the immunoscreening of two λgtll recombinant DNA expression libraries, a genomic library from Arabidopsis and cDNA library from carrot somatic embryos.

Materials and methods

Generation of antibody

Plasma membranes from *Avena sativa* roots were isolated as described elsewhere[27,32]. Briefly, roots were homogenized in a buffer containing sucrose and the protease inhibitor PMSF. After an initial low-speed centrifugation at 8000 × g for 15 minutes to get rid of mitochondria and other large organelles, the microsomal fraction was pelleted by centrifugation at 48,000 × g for 1.5 hours. It was resuspended in a buffer containing glycerol and applied to a discontinuous sucrose gradient. After centrifugation at 200,000 × g for 1.5 hours, the plasma membrane fraction was isolated from the 33%/46% (w/w) sucrose interface. Loosely bound membrane proteins were removed by a wash in a low concentration of detergent (0.2% (w/v) deoxycholate) followed by centrifugation at 200,000 × g for 1 hour[27]. The membrane pellet was resuspended in SDS polyacrylamide gel electrophoresis sample buffer and the membrane proteins subjected to polyacrylamide gel electrophoresis in an 8% polyacrylamide gel, 1.5 mm thick[13]. After staining with Coomassie Brilliant Blue R-250 to localize the Mr = 100,000 plasma membrane H$^+$-ATPase, the stained Mr = 100,000 band was cut out, washed in water, homogenized in a solution containing 0.9% (w/v) NaCl and 1% (w/v) SDS and injected at multiple intradermal sites into a rabbit[27]. The Mr = 100,000 band was located by reference to molecular weight markers and evidence supporting its identity was obtained by demonstration that antibodies made to this band reacted specifically with purified plasma membrane H$^+$-ATPase[27].

Immunoblotting

Following SDS polyacrylamide gel electrophoresis, proteins were transferred electrophoretically onto nitrocellulose paper as described in detail elsewhere[27,30]. Immunological detection of transferred proteins was also as described previously[12,27]. Briefly, the nitrocellulose paper was incubated for 1 hour with rabbit antiserum (1:1000 dilution) directed to the Mr = 100,000 oat H$^+$-ATPase and then for 1 hour with alkaline phosphatase — conjugated goat antibody (1 μg/ml) directed to rabbit IgG. By adding the colormetric substrate mixture 5-bromo, 4-chloro, 3-indolyl phosphate/ nitrobluetetrazolium for 0.5–1 hour, immunoreactive protein was visualized as a purple band on the nitrocellulose paper.

Immunoscreening of λgtll recombinant DNA expression libraries

A λgtll cDNA library made from carrot somatic embryos was obtained from Dr. T. Thomas, Dept. of Biology, Texas A & M University. A λgtll genomic DNA library made from Arabidopsis plants was obtained from Drs. B. Kohorn and E. Tobin, University of California, Los Angeles. These libraries were screened for recombinants synthesizing hybrid proteins containing plant plasma membrane H$^+$-ATPase sequence by using the rabbit antiserum directed against the oat Mr = 100,000 ATPase and the immunoscreening procedure described below.

For immunoscreening, *E. coli* Y1090 (R-) bacteria were infected with λgtll recombinant phage and plated on LB agar[36,37] at a density of 50,000–100,000 phage per 85 mm dia. petri dish. After the plates were incubated at 42° for 3.5 hours, nitrocellulose filters, which had been soaked in 10 mM of the lac Z gene inducer isopropyl-β-D-thiogalactoside, were placed over the agar and incubation continued for a further 3.5 h at 37°. The filters were then incubated for 1 h with the rabbit polyclonal antibody directed to the oat Mr = 100,000 ATPase, used at a dilution of 1:300 relative to original antiserum concentration. (The antibody was passed through an affinity column prior to use to rid it of antibodies recognizing *E. coli* antigens). Nitrocellulose filters were then incubated with biotinylated goat anti-rabbit antibody obtained from Vector Labs (Burlingame, CA) for 30 minutes followed by a 30-minute incubation with Vectastain ABC reagent (biotin-avidin-conjugated horseradish peroxidase) from the same manufacturer. Visualization was with the substrate 4-chloro-1-napthol, incubation time 3–15 minutes[37]. Positive plaques were picked with a Pasteur pipet, replated at a 10-fold lower density of phage, re-screened as above and the procedure repeated until all plaques on the plate gave a positive immunochemical staining i.e. positive clones were purified to homogeneity.

Results and discussion

Characterization of the rabbit polyclonal antibody generated against the plant plasma membrane $Mr = 100,000$ H^+-ATPase

Specific activities of the H^+-ATPase in detergent-washed plasma membrane preparations of oat roots were ca. 2–4 µmoles/minute/mg protein[27]. This activity is resistant to 100 mM KNO_3, 5 mM NaN_3 and 0.1 mM ammonium molybdate, inhibitors of vacuolar, mitochondrial and nonspecific phosphatases, respectively[27]. Upon analysis by SDS polyacrylamide gel electrophoresis, a prominent Coomassie-stained band at $Mr = 100,000$, previously identified as the catalytic subunit of the ATPase, was observed (Figure 2A, arrow). Electroblotting of an equivalent plasma membrane preparation and detection of protein which reacted with the rabbit $Mr = 100,000$ polyclonal antiserum showed clearly that the antibody recognized only an $Mr = 100,000$ protein in the membrane preparation (Figure 2B, arrow). Specific antibody was obtained after only three injections of about 100 µg each of the

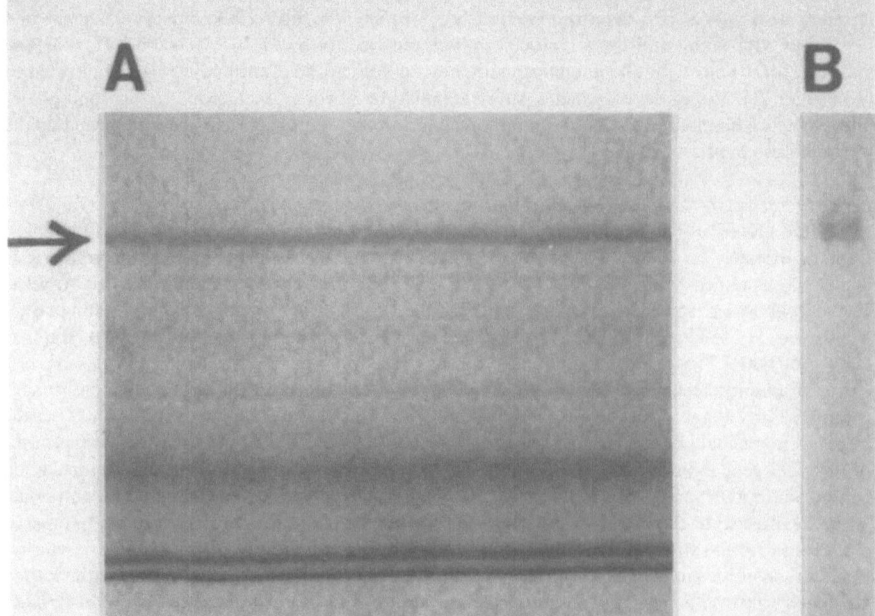

Fig. 2. Coomassie-stained (A) and immuno-stained (B) profile of plasma membrane polypeptides separated on the basis of molecular weight by sodium dodecyl sulfate-polyacrylamide gel electrophoresis. Arrow denotes position of the $Mr = 100,000$ H^+-ATPase polypeptide.

$Mr = 100,000$ band cut out of polyacrylamide gels. The antibody has been shown to bind specifically to purified biologically active H^+-ATPase from oat plasma membranes and to cross react with the $Mr = 100,000$ plasma membrane H^+-ATPase from other plant species[27]. The antibody is very sensitive and can detect 10–50 ng of the $Mr = 100,000$ ATPase in a gel lane.

Immunoscreening of λgt11 expression vector libraries of Arabidopsis genomic DNA and carrot somatic embryo cDNA

Figure 3 shows positive signals obtained from imunoscreening of the λgt11 expression vector library of Arabidopsis genomic DNA with the polyclonal $Mr = 100,000$ oat ATPase antibody. The nitrocellulose filter on the left shows the first screening where only 2–3 positive plaques can be seen amongst the 100,000 phage plated. The nitrocellulose filter on the right shows results obtained after two further screenings of the plug picked from a positive clone. Here about one fourth of the plaques on the plate were positive *i.e.* were synthesizing a fusion protein containing

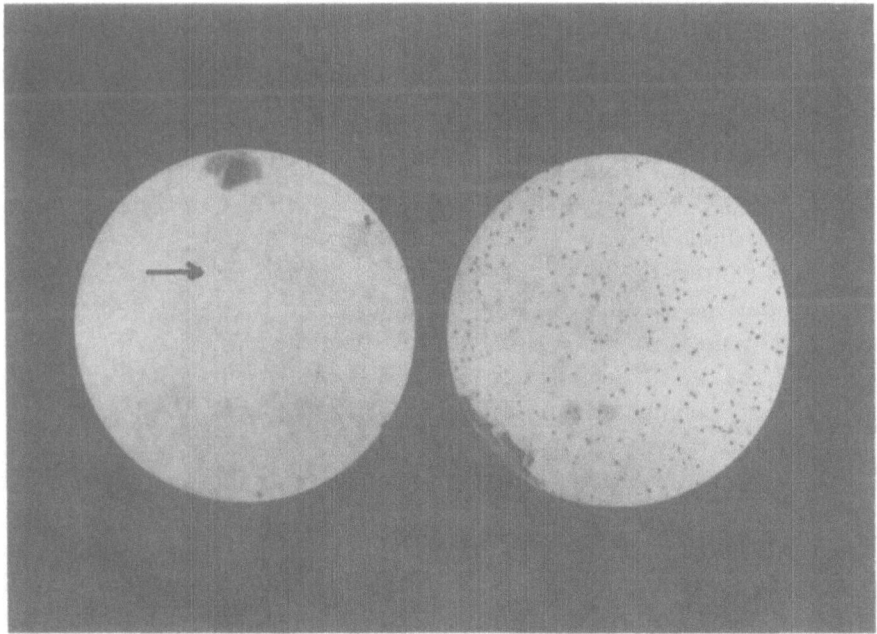

Fig. 3. Immunoscreening of an expression vector library of Arabidopsis genomic DNA using Rabbit polyclonal antibody generated against the plasma membrane H^+-ATPase. Both filters were blotted onto petri dishes containing recombinant phage plaques. Left filter shows results of the first screening. Arrow denotes position of immunopositive clone that was picked and used as source of phage for subsequent screenings. Right filter shows results after two subsequent screenings using this positive clone.

sequences recognized by the rabbit polyclonal antibody directed toward the Mr = 100,000 oat ATPase. After two more successive screenings the positive clone was purified to homogeneity. Similar results were obtained with the carrot cDNA library. Overall three positive clones were chosen from the Arabidopsis library for further characterization and two were chosen from the carrot library.

Further characterization of Mr = 100,000 ATPase clones obtained from immunoscreening of λgt11 expression vector libraries

Once positive clones have been isolated to homogeneity the next step is to isolate the DNA and remove and characterize the inserted DNA sequence that presumably contains at least part of the Mr = 100,000 ATPase gene. This insert can then be sequenced. If it is found to comprise only part of the sequence of the gene it will be used to screen other libraries for a full-length gene sequence by plaque hybridization[4].

We will also carry out further characterizations to determine whether positive clones we isolate do indeed carry sequences for the Mr = 100,000 ATPase gene of the plant plasma membranes. This will involve the following:

(i) The insert DNA which we have isolated will be used as a labeled probe in a Northern blot[16,29]. In this procedure, messenger RNA is electrophoresed on a gel and then transferred to nitrocellulose paper. The radiolabeled probe is hybridized to the mRNA under high stringency conditions. If indeed the insert DNA contains sequences for the Mr = 100,000 ATPase it should hybridize to mRNA of a size equal to or greater than 3 Kb.

(ii) The insert DNA will be used in hybrid-selected translation[9,16] to ensure that the specific mRNA to which it can hybridize does indeed encode a protein of Mr = 100,000. For this, the cloned insert DNA is immobilized on nitrocellulose paper and hybridized to mRNA in solution. After washing off non-bound mRNA, the hybridized mRNA is released and translated in a cell-free protein-synthesizing system. The size of *in vitro* translation products is determined after electrophoresis on an SDS polyacrylamide gel.

(iii) The positive clone will be used to make a lysogen which can synthesize large amounts of fusion protein[35]. This fusion protein will then be used in a competitive radioimmunoassay using the oat Mr = 100,000 ATPase antibody, and biologically active ATPase which has been purified by published procedures[21], and then radiolabeled. Displacement of label from antibody by increasing amounts of unlabeled fusion protein but not by β-galactosidase (lac Z gene product) will be taken as confirming evidence that the clone does contain plant plasma membrane ATPase sequences.

Clones which pass through the above tests and which are believed to be full length will be sequenced by standard techniques[17,20], the sequence translated into protein sequence and compared with the known sequences of the Mr = 100,000 subunit of the K^+-ATPase of *E. coli*[11] and the Na^+/K^+- and Ca^{2+}-ATPases of animal cells[15,23].

The techniques described above for isolating the structural gene for the plant plasma membrane H^+-ATPase can also be applied to the isolation of structural genes for other membrane proteins involved in mineral transport in plants. For some of these mineral transport proteins, expected to be present in very low amounts, a monoclonal antibody library made against cell surface proteins may be useful for screening expression vector libraries.

Uses of recombinant DNA technology in plant mineral transport studies

Information gained from the use of recombinant DNA technology in isolating the genes for plant mineral transport proteins and in determining their amino acid sequences will help determine the significance of the proteins and elucidate the control mechanisms governing them. Modification of plant transport characteristics through genetic engineering and breeding programs is one of the outstanding possibilities for the future.

(i) Analysis of transport protein structure. Membrane proteins, because of their low solubility in water, are notoriously difficult to analyze using standard protein chemistry techniques. As with the plasma membrane ATPase, it may be easier and faster to determine the amino acid sequence of other plasma membrane mineral transport proteins through DNA cloning and sequencing methods. With knowledge of the amino acid sequence, it will be possible to identify and locate active site(s), predict and test secondary structure and topography and overall gain a better understanding of how the protein functions as a specific transport carrier.

(ii) Measuring transcriptional activity of genes encoding transport proteins. It is likely that, like other enzymes, some plasma membrane transport proteins are inducible and/or repressible, *i.e.*, their gene activity is turned on or off depending on environmental conditions. A good example of this is the Kdp potassium transport system of *Escherichia coli*[11]. The Kdp operon contains three genes encoding the polypeptides of a potassium transport protein; one of these is the Mr = 100,000 K^+-ATPase polypeptide. Activity of this operon is stimulated during potassium starvation and osmotic imbalances[11].

Gene activity can be quantified using nucleic acid hybridization methods. For this, a specific radiolabeled nucleic acid probe is made on the basis of DNA sequence known for the gene. This is then used to measure changes in mRNA levels, for example, during various mineral feeding regimes. Changes in transcriptional activity of specific transport genes can then be correlated with adaptation of the plant to different mineral 'environments'.

(iii) Application to a breeding program. There are uses for recombinant DNA technology in genetic studies geared towards breeding plants with agriculturally useful changes in mineral transport activities.

An immediate application of recombinant DNA technology is the use of recombinant DNA "probes" to follow the inheritance of specific genes through the study of Restriction Fragment Length Polymorphism (RFLP). Enzymes exist, called restriction enzymes, which cleave DNA at specific sites, determined by the sequence of only a few base pairs at that site. Genetic changes in that site and therefore, in the patterns of restriction digestion, can be used to label the gene. This technique is gaining popularity amongst both breeders and molecular geneticists. It can be used to isolate and purify the gene(s) responsible for complex phenotypes[10] and to map and follow the inheritance of a gene locus even amongst individuals in a large population[3,26].

A more long-range application of recombinant DNA technology is in the genetic transformation of higher plants using cloned DNA. Using this, it may be possible to introduce new transport systems into a plant. An important consideration in the setting-up of transformation systems for membrane-bound transport proteins is that the proteins must be inserted correctly and functionally into the membrane. In eukaryotic cells, specific mechanisms, including proteolytic cleavage of leader sequences, transport through the endoplasmic reticulum and sometimes attachment of carbohydrate moieties, are involved. It is obvious from classical genetic studies in plants[8] and other eukaryotes[25], that there is a large potential for altering mineral nutrition needs through changes in mineral transport. The promise for using recombinant DNA technology in this area is attractive, but as yet, unrealized. Our work in progress involving cloning the first gene encoding a plant plasma membrane protein is an important first step in this direction.

Acknowledgements This work was supported by grants from the McKnight Foundation and the Department of Energy (DE-AC02-83ER13086). The technical assistance of Marie Martinell in the immunoscreening is gratefully acknowledged.

References

1   Allen G, Trinnaman B J and Green N M 1980 The primary structure of the calcium ion-transporting adenosine triphosphatase protein of rabbit skeletal sarcoplasmic reticulum. Biochem. J. 187, 591–616

2. Amzel L M and Pedersen P L 1983 Proton ATPases: structure and mechanism. Annu. Rev. Biochem. 52, 801–824
3. Beckmann J S and Soller M 1983 Restriction fragment length polymorphisms in genetic improvement: methodologies, mapping and costs. Theoret. Appl. Genet. 67, 34–43
4. Benton W D and Davis R W 1977 Screening λgt recombinant clones by hybridization to single plaques in situ. Science 196, 180–182
5. Briskin D P and Poole R J 1983 Plasma membrane ATPase of red beet forms a phosphorylated intermediate. Plant Physiol. 71, 507–512
6. Farley R A and Faller L D 1985 The amino acid sequence of an active site peptide from the H,K-ATPase of gastric mucosa. J. Biol. Chem. 260, 3899–3901
7. Felle H 1982 Effects of fusicoccin upon membrane potential. Resistance and current voltage characteristics in root hairs of *Sinapis alba*. Plant Sci. Lett. 25, 219–225
8. Gabelman W H and Gerloff G C 1983 The search for and interpretation of genetic controls 5–350
9. Goldberg M L, Lefton R P Stark G R and Williams J G 1979 Isolation of specific RNAs using DNA covalently linked to diazobenzyloxymethyl-cellulose paper. Methods Enzymol. 68, 206–220
10. Gusella J F, Wexler N S, Conneally P M, Naylor S L, Anderson M A, Tanzi R E, Watkins P C, Ottina K, Wallace M R, Sakaguchi A Y, Young A B, Shoulson I, Bonilla E and Martin J B 1983 A polymorphic DNA marker genetically linked to Huntington's disease. Nature 306, 234–238
11. Hesse J E, Wieczorek L, Altendorf K, Reicin A S, Dorus E, and Epstein W 1984 Sequence homology between two membrane transport ATPases, the Kdp-ATPase of *Escherichia coli* and the Ca^{2+}-ATPase of sarcoplasmic reticulum. Proc. Natl. Acad. Sci. USA 81, 4746–4750
12. Knecht D A and Dimond R L 1984 Visualization of antigenic proteins on Western blots. Anal. Biochem. 136, 180–184
13. Laemmli U K 1970 Cleavage of structural proteins during the assembly of the head of bacteriophage T_4. Nature 227, 680–685
14. Luben R A, Brazeau, P, Böhlen P and Guillemin R 1982 Monoclonal antibodies to hypothalmic growth hormone-releasing factor with picomoles of antigen. Science 218, 887–889
15. MacLennan D H, Brandl C J, Korczak B and Green N M 1985 Amino-acid sequence of a $Ca^{2+} + Mg^{2+}$-dependent ATPase from rabbit muscle sarcoplasmic reticulum, deduced from its complementary DNA sequence. Nature 316, 696–700
16. Maniatis T, Fritsch E F and Sambrook J 1982 Molecular Cloning. Cold Spring Harbor Laboratory, Cold Spring Harbor, New York, 545 p.
17. Maxam A M and Gilbert W 1977 A new method of sequencing DNA. Proc. Natl. Acad. Sci. USA 74, 560–564
18. Mitchell P 1973 Performance and conservation of osmotic work by proton-coupled porter systems. Bioenergetics 4, 63–91
19. Pick U and Racker E 1979 Inhibition of the (Ca^{2+}) ATPase from sarcoplasmic reticulum by dicyclohexylcarbodiimide: evidence for location of the Ca^{2+} binding site in a hydrophobic region. Biochemistry 18, 108–113
20. Sanger F, Nicklen S and Coulson A R 1977 DNA sequencing with chain-terminating inhibitors. Proc. Natl. Acad. Sci. USA 74, 5463–5467
21. Serrano R 1984 Purification of the proton pumping ATPase from plant plasma membranes. Biochem. Biophys. Res. Commun. 121, 735–740
22. Serrano R 1985 Plasma Membrane ATPase of Plants and Fungi. CRC Press, Florida, 174 p
23. Shull G E, Schwartz A and Lingrel J B 1985 Amino-acid sequence of the catalytic subunit of the ($Na^+ + K^+$) ATPase deduced from a complementary DNA. Nature 316, 691–695
24. Slayman C L and Gradmann D 1975 Electrogenic proton transport in the plasma membrane of Neurospora. Biophysical J. 15, 968–971
25. Slayman C W 1973 The genetic control of membrane transport. Current Topics in Membranes and Transport, Vol 4, p 1–174
26. Soller M and Beckmann J S 1983 Genetic polymorphism in varietal identification and genetic improvement. Theoret. Appl. Genet. 67, 25–33

27 Surowy T K and Sussman M R 1986 Immunological cross-reactivity and inhibitor sensitivities of the plasma membrane [H^+]-ATPase from plants and fungi. Biochim. Biophys. Acta 848, 24–34
28 Sze H and Churchill K A 1981 Mg^{2+}/KCl-ATPase of plant membranes is an electrogenic pump. Proc. Natl. Acad. Sci. USA 78, 5578–5582
29 Thomas P S 1980 Hybridization of denatured RNA and small DNA fragments transferred to nitrocellulose. Proc. Natl. Acad. Sci. USA 77, 5201–5205
30 Towbin H, Staehelin T and Gordon J 1979 Electrophoretic transfer of proteins from polyacrylamide gels to nitrocellulose sheets: procedure and some applications. Proc. Natl. Acad. Sci. USA 76, 4350–4354
31 Uribe E G and Luttge U 1984 Solute transport and the life functions of plants. American Scientist 72, 567–573
32 Vara F and Serrano R 1982 Partial purification and properties of the proton-translocating ATPase of plant plasma membranes. J. Biol. Chem. 257, 12826–12830
33 Vara F and Serrano R 1983 Phosphorylated intermediate of the ATPase of plant plasma membranes. J. Biol. Chem. 258, 5334–5336
34 Walderhaug M O, Post R L, Saccomani G, Leonard R T and Briskin D P 1985 Structural relatedness of three ion-transport adenosine triphosphatases around their active sites of phosphorylation. J. Biol. Chem. 260, 3852–3859
35 Young R A and Davis R W 1983 Efficient isolation of genes by using antibody probes. Proc. Natl. Acad. Sci. USA 80, 1194–1198
36 Young R A and Davis R W 1983 Yeast RNA polymerase II genes: isolation with antibody probes. Science 222, 778–782
37 Young R A and Davis R W 1985 Immunoscreening λgt11 recombinant DNA expression libraries. *In* Genetic Engineering. Vol 7. Eds. J Setlow and A Hollaender. Plenum Press, New York (*In press*)

Control of nutrient concentrations in plant growth media

R.B. COREY and S.M. COMBS
Department of Soil Science, University of Wisconsin-Madison, Madison, WI 53706, USA

Key words Alumina   Concentration   Flowing culture   Ion-exchange resins   Sand culture   Solution culture

Summary   Most nutrient-solution cultures used by plant breeders for screening varieties contain concentrations of P and trace metals far in excess of those found in normal soil solutions. Maintenance of realistic concentrations of these elements requires either that some concentration-buffering medium similar to that found in soil be used or that solutions be replaced before significant depletion occurs, usually by use of flowing cultures. A resin-controlled system for buffering solution pH and concentrations of phosphorus and any cationic nutrients is described. This system is capable of maintaining concentrations of the controlled nutrients within narrow ranges and is relatively inexpensive.
   If the rate at which a nutrient diffuses to the plant root limits uptake in soil systems, solution culture may not be the appropriate screening medium. An equation describing the interacting factors that effect diffusion-controlled uptake is used to show which of these factors a particular screening process includes and which are absent. For nutrients that reversibly adsorb on soil surfaces, a modified sand culture system may simulate the soil better than solution culture. In such a system nutrients are adsorbed on particles of alumina (phosphorus) or on ion-exchange or chelating resins (most soluble cations or anions), and the particles are dispersed throughout a sand matrix. Limitations of these methods are discussed.

Introduction

Morphological and/or physiological adaptations by plants to their natural surroundings have produced a variety of mechanisms for overcoming localized environmental limitations. These plants have become adapted to specific stress conditions and provide plant breeders with breeding material for developing varieties for specific limiting soil conditions. Some plant nutritional problems which cannot be corrected easily or inexpensively by soil amendments may be eliminated by identifying such adaptive characteristics within lines and genetically recombining these in a breeding program[6]. Because large numbers of lines must be screened to identify nutritional adaptations and to determine heritability, this type of breeding program requires fast, inexpensive screening methods. Culture media such as nutrient solutions, soil or low-density solids such as peat, perlite, and vermiculite ('soil-less' soil) are presently used for this purpose. It is important, however, to recognize the limitations these screening methods may impose on extrapolating plant response in a growth chamber or a greenhouse to field performance. For genetic approaches to provide adapted plant varieties for problem soils, screening methods should reflect the soil system as closely as is feasible.

H.W. Gabelman and B.C. Loughman (Eds.), Genetic aspects of plant mineral nutrition.
ISBN-13: 978-94-010-8102-3
© *1987 Martinus Nijhoff Publishers, Dordrecht/Boston/Lancaster.*

The same applies to more closely controlled experiments designed to determine absorption rates and uptake mechanisms. In soils, relatively low but highly buffered concentrations of most plant nutrients are maintained in the soil solution. In traditional non-flowing hydroponic solutions and sand cultures, however, initial concentrations of some nutrients are one to three orders of magnitude higher than in soils, and the systems are not buffered.

In this paper, we shall first use a nutrient uptake model to illustrate the soil and plant factors that control uptake rates in soil systems and to show how these factors differ between soil and solution cultures. Then we shall discuss methods for making rapid screening systems more closely simulate soil conditions.

Nutrient uptake model

The kinetics of nutrient transport and root absorption can be simulated mathematically with a mechanistic model based on diffusion and convection (mass flow) processes[4,18]. Use of a mechanistic model to predict nutrient uptake as opposed to an empirical approach allows better identification of parameters required for accurate prediction of plant uptake. Plant characteristics that enhance uptake of a limiting nutrient or that confer tolerance to toxic conditions may then be selected for inclusion in a breeding program.

A comprehensive transport model describes the flux of ions in the soil solution to root absorbing surfaces by diffusion in response to a concentration gradient and by convection caused primarily by plant transpiration. Diffusion has been found to be the dominant nutrient transport mechanism in soils except for Ca^{2+}, Mg^{2+} and SO_4^{2-}, and most of the subsequent discussion will center on this process.

Soil factors that affect diffusive transport include water content, nutrient concentration in solution, and the ability to resupply absorbed nutrients (buffer power). Important plant factors include root geometry (root radius, presence of root hairs/mycorrhizae) and root uptake physiology (root absorbing power). How these factors interact is shown in Eq. [1], which is a modification of an uptake equation derived by Baldwin et al.[3] that describes the diffusive radial flux of nutrients from an isotropic medium (soil) to a cylindrical sink (plant root), assuming depletion of a cylindrical volume of soil surrounding each segment of root.

$$U = C_{li}b\left(1 - \exp\frac{-2\pi\alpha A_1 r_0 L_v t}{b\left(1 + \frac{\alpha A_1 r_0}{D_1 \theta f}\ln\frac{r_h}{1.65r_0}\right)}\right) \quad (1)$$

Soil factors:
C_{1i} = initial concentration of nutrient in soil solution
b = buffer power
A_1 = fractional area contacting root
θ volumetric water content
D_1 = diffusion coefficient in soil solution
f = conductivity factor related to diffusion pathlength

Plant factor:
U = uptake per unit volume of soil in time, t
α = root absorbing power
r_0 = root radius
r_h = half-distance between roots
L_v = root density

The conductivity factor, f, decreases with a decrease in θ because of greater tortuosity of the diffusion path at lower water contents. The buffer power, b, is equal to the change in concentratin of total labile nutrient per unit change in concentration of that nutrient dissolved in the soil solution. The labile form includes dissolved and reactive adsorbed forms. Soils with high adsorption capacities for specific nutrients generally show high buffer powers for those nutrients. Commonly found ranges in buffer powers for specific nutrients range from less than 1 for nonadsorbed species to more than 1000 for strongly adsorbed species, and generally decrease with increasing saturation of the adsorbing sites with a particular nutrient[4,18].

The root density, L_v, is equal to the length of root per unit volume of soil. The value of L_v is readily determined for roots without root hairs or mycorrhizae, but their presence makes the geometry of the nutrient absorbing system much more difficult to describe quantitatively. The root absorbing power, α, is equal to the uptake flux density divided by the nutrient concentration at the root surface. In some cases, α can be described by a Michaelis-Menten-type plot of flux density ($mol\,cm^{-2}\,s^{-1}$) vs. concentration at the root surface ($mol\,cm^{-3}$). However, this relationship has been shown to depend on the pre-existing nutrient status of the plant[15].

Nutrient availability under greenhouse and field conditions

A common problem with pot cultures, regardless of the culture medium, is that the confined root system has a much higher root density, L_v, than is found under field conditions. This results in faster depletion of the nutrients in a given volume of growth medium than would occur at normal L_v. The problem can be circumvented to some extent by harvesting plants before the roots contact the walls of the containers, which entails either harvesting plants at a very early stage or using very large pots.

Another common problem is greater transpiration per unit length of root and per unit volume of medium from pot cultures than from field-grown plants. This results from greater inputs of advected heat in many growth chamber environments, and from the greater ratio of transpiring surface to root volume in pot cultures. If transpiration per unit length of root is greater in pot cultures, the relative transport by mass flow is larger for plants in pots than for plants in the field.

In addition to these differences that are common to all pot cultures, there are differences associated with specific media.

Solution culture. There are three major differences in addition to those mentioned above, related to nutrient supplying characteristics of solution cultures and soils in the field: (1) Constant stirring of the solution culture limits diffusion as an uptake-controlling process to a thin unstirred layer around the roots and the free space within the cortex, whereas uptake in soil systems is often limited by diffusion within the soil matrix. (2) The buffer power, b, for nutrients in solution is equal to 1 and there is no capacity to replace nutrients absorbed by roots. (3) The buffer power for H^+ is relatively high in soils but very low in solution cultures so that metabolically released H^+ or HCO_3^- causes much greater pH change in the solution system.

Ratios of nutrients in solution cultures are sometimes based on the ratios absorbed by the plants with the absolute concentrations near the upper limit that the plants will tolerate. This maximizes the time that the solution will continue to meet the nutrient requirements of the plant

Table 1. Comparison of nutrient concentrations (ml L^{-1}) in soil solution, culture solutions and concentrations at one-half maximum uptake rate (K_m)

Element	K_m^4	Soil solution[4]	Asher- et al.[2]	Hoagland- Arnon[14]
N	0.4	30	12	252
P	0.1	0.05	0.12-0.78	62
K	1.1	5	10	390
Ca	4	30	10	120
Mg	0.3	30	2.4	48
S	0.5	40	3.2	64
Mn	5	0.3*	0.11	0.5
Zn	0.2	0.02*	0.055	0.05
Cu	0.006	0.10*	0.032	0.02

* Total concentrations. Free-ion concentrations may be an order of magnitude less for Mn and Zn, and two to four orders of magnitude less for Cu.

without change or supplementation. As a result, the concentrations of some nutrients in common nutrient solutions exceed mean soil solution values (C_{li} in Eq. [1]) by as much as two or three orders of magnitude (Table 1). This is the case especially for nutrients with high buffer powers in soils. Obviously, the uptake processes must differ drastically between soil and these solution systems if a given cultivar grows well in either. Therefore, one must question the applicability of solution-derived uptake-rate data to soil systems.

Deficiencies are usually induced in solution cultures either by transferring plants from a complete nutrient solution to one devoid of a specific material or by raising plants from the seedling stage in solutions that contain absolute amounts of a given nutrient insufficient to meet the needs of the plant over the period between solution changes. In the latter case, initial concentrations of nutrients such as phosphorus may be low by solution culture standards, but very high when compared with soil solution values. Therefore, the plant is exposed over a relatively short period of time to a solution concentration that fluctuates from very high to very low by soil solution standards. This contrasts with a deficiency condition in soil in which a low but nearly constant C_{li} is maintained. Because nutrient stress has been shown to increase root absorbing power, α, and high nutrient concentrations following a period of stress have decreased α[9], one would expect plant response to widely fluctuating nutrient levels at the root surface to differ from that obtained with a low but relatively constant concentration.

The flowing culture recirculating system[2,5] provides nutrient solutions with concentrations approximating soil-solution levels and eliminates many of the problems discussed above. However, in order to prevent significant depletion of low-nutrient solutions by actively absorbing plants, high flow rates are necessary. To maintain accurate concentration control, these systems require large volumes of solutions, extensive plumbing and expensive monitoring equipment. The expense involved may make the flowing culture system undesirable for routine screening.

An alternative method for maintaining nutrient concentrations in the soil-solution range is the ion-exchange resin system described later. The resins act as solid phase buffers in hydroponic or sand culture, maintaining low solution levels while allowing control over a wide range of solution compositions. The system is less expensive than flowing culture and control at very low concentrations of P and trace metals is good.

Sand culture. Sand culture introduces a diffusion component to the solution culture system. Because the sand has little tendency to adsorb nutrients from solution, the buffer powers for adsorbable nutrients are very low ($b/\theta = 1$) compared with soils. Because the volumetric solution content, θ, is also low, depletion of nutrients and water is five to ten times faster per unit volume of sand than from an equivalent volume of solution. Therefore solutions must be renewed more frequently than in solution culture.

If used for studying adsorbable ions such as P, rates of diffusive transport in sand cultures will differ from those in soil systems because of the difference in buffer power. Use of activated alumina or resin particles to increase buffer power will be discussed later.

Soilless soil and soil cultures. Soilless soil cultures range in characteristics between sand cultures and soil cultures with water holding characteristics and buffer powers for individual nutrients varying with the materials used. For example, vermiculite has a high buffer power for K, most calcined clay has a high buffer power for P, and peat may have a high buffer power for trace metals. Pot cultures using soil differ from the field systems mainly in the root restriction and transpiration differences described previously. The main problem in using these media for screening is the difficulty in establishing and maintaining specific known levels of plant nutrients.

Control of nutrient concentrations in solution culture

Ion-exchange resin system. The main technical limitations of flowing systems are associated with maintaining very low concentrations of elements such as P and the trace metals. Studies of deficiencies may require concentrations of free ions as low as 10^{-7} to 10^{-8} M for P and below 10^{-11} M for copper. Rapid methods for monitoring these low concentrations are not available at this time.

Ion-exchange resins have a long history of use in plant nutrient studies, particularly in sand cultures[1]. Recently, a combination of resin materials has been used to buffer pH and control activities of P and of cationic nutrients including trace metals in solution cultures[7,8,11]. Four different resin materials were used in this system: (1) a chelating resin with iminodiacetic acid functional groups to control ratios of complexable metals; (2) a weak-acid resin with carboxylic acid functional groups to control pH and contribute to controlling ratios of major cations; and (3) a strong-acid resin with sulfonic acid functional groups to control ratios of major cations; and (4) a strong-acid resin partially saturated with cationic polynuclear Al to provide adsorption sites for control of P concentration[19]. Technical grade resins can be used but they require prior clean-up and conditioning.

Simplified versions of the activity-controlling reactions associated with the specific types of resins used are given below. In these equations, R represents a resin surface site, and K_a, K_{Ca}^{Mg}, K_{Ca}^{K}, and K_p are selectivity coefficients for reactions (2), (4), (6), and (8), respectively. These equations should be considered semi-quantitative because the actual thermodynamic relationships are much more complex.

$$R-COOH \rightleftharpoons R-COO^- + H^+ \tag{2}$$

$$(H^+) = K_a[R-COOH]/[R-COO^-] \tag{3}$$

$$R-Ca + Mg^{2+} \rightleftharpoons R-Mg + Ca^{2+} \tag{4}$$

$$(Ca^{2+})/(Mg^{2+}) = [R-Ca]/[R-Mg]K_{Ca}^{Mg} \tag{5}$$

$$R-Ca_{1/2} + K^+ \rightleftharpoons R-K + 1/2 Ca^{2+} \tag{6}$$

$$(K^+)/(Ca^{2+})^{1/2} = [R-K]/[R-Ca_{1/2}]K_{Ca}^{K} \tag{7}$$

$$R_6 - Al_6(OH)_{12}(H_2O)_{12} + H_2PO_4^- + 1/2 Ca^{2+}$$
$$\rightleftharpoons R_5 - Al_6(OH)_{12}(H_2O)_{10}H_2PO_4 + R-Ca_{1/2} \tag{8}$$

$$(H_2PO_4^-) = (1/K_p)\{R_5 - [Al_6(OH)_{12}(H_2O)_{10}H_2PO_4]\}$$
$$\times [R-Ca_{1/2}]/\{R_6 - [Al_6(OH)_{12}(H_2O)_{12}]\}$$
$$\times (Ca^{2+})^{1/2} \tag{9}$$

The H^+ activity is controlled by the degree of protonation of the carboxyl functional group on the weak acid resin (Eq. (3)). Ratios of various divalent cations in solution are controlled by their ratios on the exchange sites (whether chelating, strong-acid or weak-acid sites) and the appropriate selectivity coefficient (Eq. (5)). Ratios of monovalent cations are governed by a similar mechanism. Ratios of monovalent to divalent ions are governed by Eq. (7), and $H_2PO_4^-$ activity depends on the degree of saturation of the polynuclear Al with phosphate (Eq. (9)).

From the above equations, it is apparent that the activities of H^+ and $H_2PO_4^-$ are largely controlled by the resins, whereas only the *ratios* of the metallic cations are controlled by the resins. Control of absolute concentrations depends on controlling the ionic strength of the solution. This can be achieved by monitoring the electrical conductivity and periodically adjusting it to the desired level by adding appropriate amounts of nutrients in the approximate ratios that they have been absorbed by the plants. There is no need to monitor concentrations of the controlled nutrients or to replace the exact amounts that have been removed because concentrations are very strongly buffered.

The resin system described here exerts no control over anion concentrations or ratios except for $H_2PO_4^-$. Anion concentrations are maintained in the process of adding salts to maintain ionic strength. Anion exchange resins could have been used to control anion ratios, but were not included because they would adsorb anionic chelates used in the studies.

This resin system has been employed successfully for buffering hydroponic solutions used in determining factors that effect the root absorbing power for Cd, Cu, and Zn[7,11]. In these studies, activities of P and of macro- and micronutrient cations were successfully maintained at concentrations normally found in soil solutions. Tomato plants grew as well in the resin system as in 1/2-strength Hoagland's solution[11].

The design of resin-controlled systems can be modified according to the needs of the experiment. The simplest system for screening purposes, if the roots are not to be analyzed and pH above 6.2 is not required, is to allow the resins to float freely in a culture solution with aeration vigorous enough to maintain adequate mixing[11]. If the roots are to be analyzed, the resin must not be allowed to contact them. This can be accomplished by encasing the resins in bags made from a Versapore-800 filter membrane[7,8]. The membrane creates a diffusion barier so the concentration in a culture solution with growing plants is somewhat less than that at equilibrium. Both of these formats have an upper pH limit of about 6.2 because the polynuclear Al necessary for P control is unstable above this pH.

The resin system can also be used in a recirculating culture which is not necessarily limited in pH range. A pump is required to continually recirculate bulk solution from the pot through a resin column to maintain desired pH, P activity and cation ratios. Flow rates must be sufficient to prevent significant nutrient depletion by the roots. Ionic strength can be monitored by electrical conductivity and maintained automatically if desired. Iron, very difficult to buffer by the resin, might be monitored via colorimetry and appropriate levels maintained by addition of an Fe chelate. The resin column can be arranged with each resin type in separate, independently-assembled compartments, allowing for easy treatment changes and resin regeneration. Such a layering arrangement would enable studies to be conducted at pH values greater than 6.2. The recirculated solution could first be passed through a weak-acid resin adjusted to pH 5, optimum for polynuclear Al stability, then in sequence, through the P-control resin, a weak-acid resin adjusted to the desired pH, the strong-acid resin and the chelating resin.

Control of Al and P activities in solution culture

The effects of reactive soil Al on root growth and P availability cause concern in many acid-soil areas[12], and screening for Al tolerance is important[13]. However, Al^{3+} and/or $H_2PO_4^-$ activities in most screening systems are not maintained at realistic soil solution levels.

Ideally, the screening system should allow for independent control of Al and P activities within the limits imposed by precipitation reactions. These limits can be approximated from the following reactions (log K values are for amorphous solid phases[16,17]):

$$Al(OH)_{3(s)} + 3H^+ \rightleftharpoons Al^{3+} + 3H_2O$$
$$\log K = 9.66 \tag{10}$$

$$Al(OH)_2H_2PO_{4(s)} + 2H^+ \rightleftharpoons Al^{3+} + H_2PO_4^- + 2H_2O$$
$$\log K = -1.30 \tag{11}$$

$$Al(OH)_{3(s)} + H_2PO_4^- + H^+ \rightleftharpoons Al(OH)_2H_2PO_{4(s)} + H_2O$$
$$\log K = 10.96 \tag{12}$$

Equation (10) sets the upper limit of Al^{3+} activity at a given pH. Equation (11) sets the upper limit for the product of the Al^{3+} and $H_2PO_4^-$ activities at a given pH when an aluminum phosphate solid phase is present but solid-phase $Al(OH)_3$ is not. Equation (12) sets the upper limit for $H_2PO_4^-$ activity at a given pH when $Al(OH)_3$ (which controls Al^{3+} activity) coexists with aluminum phosphate. Values calculated from these equations will be near the maximum values that can be maintained for short periods in nutrient solutions. Over time, crystalline solid phases may form resulting in lower solubilities.

Below these maximum concentrations, Al^{3+} and $H_2PO_4^-$ activities can be controlled at specific levels with the previously described resin system by varying: (1) the amount of polynuclear Al loaded onto the strong-acid resin; (2) the degree of P saturation of the adsorbed polynuclear Al (Eq. [10]); (3) the pH maintained by the weak-acid resin (Eq. [2]); and (4) the ionic strength of the culture solution.

Control of nutrient concentrations in sand culture

Activated alumina, Al_2O_3, is a porous material that has a high capacity for adsorbed P. This material has been used to buffer P concentrations in sand cultures[10]. In this system, each P-treated alumina granule acts as a point source of P. The rate at which P is delivered to the plant root by diffusion depends on the P concentration at the granule surface and the distance between the granule and the root. Either increasing surface P concentration or decreasing the mean diffusion distance by increasing the number of alumina particles will increase the rate at which P arrives at the plant root.

The equilibrium P concentration supported by the alumina at a given plant uptake rate will be higher than that found in a soil supporting the same uptake rate because the buffer power, b, of the sand between the alumina particle and the root is very low compared with that of a soil. The P-adsorbing surface of a soil is spread much more uniformly throughout the soil matrix and thus P absorbed by the root can be replaced from an adjacent soil particle rather than having to diffuse from a distance alumina particle. Also, some segments of the root system will be closer to alumina particles than other segments so that uptake rate will vary for different segments of the root system. In soils, uptake by individual root segments will be more uniform.

Phosphated alumina appears to be an effective method for establishing P availability levels for screening purposes, but the limitations for interpreting results in terms of soil systems should be kept in mind. Concentration control of other nutrients is no different from the usual sand culture method.

Sand-size resin particles could be used to buffer pH and the cationic nutrients as well as P in sand cultures. This would permit the use of solution concentrations that more realistically match the soil solution, but differences in amount and spatial distribution of buffer power would remain. Systems consisting of resin particles without sand would have buffer powers far in excess of those found in most soils, whereas incorporation of resin particles into a sand matrix would give the same problem of point sources encountered with the alumina system.

Selection of a nutrient-control method

Selection of a method for controlling activities of specific plant nutrients in screening studies depends on many factors including cost, degree of control required, need for a diffusion-controlled system, etc. If a solution culture can be used, the resin system offers a relatively inexpensive means of controlling pH, P, and cationic nutrients, especially if root analyses are not required and the resin can simply be suspended in the solution. The resin system, whether free floating or encased in bags of filter membrane, can give much more precise control than traditional nutrient solution systems and can maintain low concentrations similar to those found in soil solutions.

In studying nutrients that are adsorbed significantly onto surface adsorption sites in soils, a diffusion-limited medium that mimics soil diffusion processes is often desirable. If phosphorus is the only nutrient for which diffusion limitation is desired, dispersing phosphate-treated alumina or polynuclear Al resin particles in a sand culture system will be effective within the limits discussed previously. Cation or anion exchange resins could be used in a similar manner for other nutrients.

For research using solution cultures or in screening studies where cost is not a controlling factor, flowing culture systems can be used. Low concentrations of nutrients can be controlled very accurately, but equipment and instrumentation costs are high. Control might be improved and costs decreased by recycling the flowing nutrient solution through a column containing the mix of resins described previously.

References

1 Arnon D I and Meagher W R 1947 Factors affecting availability of plant nutrients from synthetic ion exchange materials. Soil Sci. 63, 213-221.
2 Asher C J, Ozanne P G and Loneragan J F 1965 A method for controlling the ionic environment of plant roots. Soil Sci. 100, 149 156.
3 Baldwin J P, Nye P H and Tinker P B 1973 Uptake of solutes by multiple root systems from soil. III. A model for calculating the solute uptake by a randomly dispersed root system developing in a finite volume of soil. Plant and soil 38, 621-635.
4 Barber S A 1984 Soil Nutrient Bioavailability. John Wiley & Sons, New York, 398 p.
5 Breeze V G, Canaway R J, Wild A, Hopper M J and Jones L H P 1982 The uptake of phosphorus by plants from flowing nutrient solution. J. Exp. Bot. 33, 183-189.
6 Brown J C, Ambler J E, Chaney R L and Foy C D 1972 Differential responses of plant genotypes to micronutrients. In Micronutrients in Agriculture. Eds. J J Mortvedt, P M Giordano and W R Lindsay. Soil. Sci. Soc. Am., Madison, WI, pp 389 418.
7 Checkai R T, Corey R B and Helmke P A 1981 A method for controlling ionic activities of nutrients in solution culture. Agron. Abstr., p. 82.
8 Checkai R T, Corey R B and Helmke P A 1982 The effects of ionic and complexed Cd on metal uptake by plants. Agron. Abstr., p. 93.
9 Clarkson D T and Scattergood C B 1982 Growth and phosphate transport in barley and tomato plants during the development of, and recovery from, phosphate-stress. J. Exp. Bot. 33, 865 875.
10 Coltman R R, Gerloff G C and Gabelman W H 1982 A sand culture system for simulating plant responses to phosphorus in soil. J. Am. Soc. Hort. Sci. 107, 938 942.
11 Combs S M, Corey R B and Chaney R L 1984 Effect of Cd^{2+}/Zn^{2+} activity ratios on elemental composition of tomato grown in a resin-buffered hydroponic system. Agron. Abstr., p. 201.
12 Foy C D 1974 Effect of aluminum on plant growth. In The Plant Root and Its Environment. Ed. E W Carson. Univ. Press of Virginia, Charlottesville, pp 601 642.
13 Furlani P R and Clark R B 1981 Screening sorghum for aluminum tolerance in nutrient solutions. Agron. J. 73, 587 594.
14 Hoagland D R and Arnon D I 1950 The water culture method for growing plants without soil. Coll. Agric. U C Berkeley, Calif, Agric. Exp. Stn. Circ. 347, p. 1 32.
15 Lauchli A 1984 Mechanisms of nutrient fluxes at membranes of the root surface and their regulation in the whole plant. In Roots, Nutrient and Water Influx and Plant Growth. Eds. S A Barber and D R Bouldin. ASA Special Pub. No. 49, SSSA, CSA, ASA, Madison, WI.
16 Lindsay W L 1979 Chemical Equilibria in Soils. John Wiley and Sons, New York.
17 Lindsay W L, Peech M and Clark J S 1959 Solubility criteria for the existence of variscite in soils. Soil Sci. Soc. Am. Proc. 23, 357 360.
18 Nye P H and Tinker P B 1977 Solute Movement in the Soil-Root System. Univ. of California Press, Berkeley and Los Angeles, 342 p.
19 Robarge W P and Corey R B 1979 Adsorption of phosphate by hydroxy-aluminum species on a cation exchange resin. Soil Sci. Soc. Am. J. 43, 481 487.

Considerations of vesicular-arbuscular mycorrhiza physiology in breeding for enhanced mineral uptake by plants

S.M. SCHWAB
Department of Biology, Eastern Washington University, Cheney, WA 99004, USA

Key words External hyphal growth Mycorrhizal dependency Photosynthate partitioning P uptake Root exudates Vesicular arbuscular mycorrhizae

Summary Because vesicular-arbuscular mycorrhizal (VAM) fungi can increase uptake of P and other immobile minerals, plant breeding programs aimed at improving the efficiency of mineral uptake should take VAM associations into account. The effectiveness of a fungal isolate in enhancing P uptake apparently depends primarily on its ability to form an extensive network of hyphae beyond the root's zone of depletion at a stage in the plant's growth when demand for P is high. Plants which partition relatively little energy to root production are most likely to benefit from VAM associations, and VAM may present less demand for photosynthate than equivalently effective root systems. VAM formation is closely correlated with rates of root exudation. These observations suggest that breeding for less extensive root systems with high rates of root exudation, combined with inoculation with effective strains of VAM fungi would provide for a high ratio of P uptake/photosynthate cost.

Introduction

Symbiotic vesicular-arbuscular mycorrhizal (VAM) associations between certain Zygomycete fungi and crop plants are common in most infertile soils. Exceptions include plants in the families Chenopodiaceae (*e.g.* beets, spinach), Polygonaceae (*e.g.* buckwheat), and Brassicaceae (*e.g.* cabbages, rape), but grains, legumes, solanaceous species, and virtually all perennial crops form VAM. Increases in uptake of P due to VAM formation have been extensively documented[26,27] and zinc, copper, sulfur, and nitrogen also can be absorbed by fungal hyphae and transferred to the host plant[8,14,18,41,42]. With inflows of P into VAM roots sometimes exceeding inflows into noncolonized roots from the same soil by a factor of 3 or more[44,45], any attempt to genetically improve the P uptake ability of a crop plant clearly must consider interactions with VAM fungi. The response of the host plant to VAM depends on the nature of the fungal symbiont and the host, and on interactions with the environment. This review will focus primarily on aspects of the fungus and host that maximize the response of the host to colonization by VAM fungi.

Fungal attributes related to host response

Although the plant breeder is generally more concerned with genetically manipulating the host plant than the fungal endophyte, any well

H.W. Gabelman and B.C. Loughman (Eds.), Genetic aspects of plant mineral nutrition.
ISBN-13: 978-94-010-8102-3
© *1987 Martinus Nijhoff Publishers, Dordrecht/Boston/Lancaster.*

thought out plant breeding program that deals with VAM-mediated phenomena requires an understanding of the biology of the fungal component of the symbiosis as well as the plant component. Only four genera (Glomus, Gigaspora, Acaulospora, and Sclerocystis), containing less than one hundred named species of fungi, form VAM on approximately 200,000 species of flowering plants, many gymnosperms, most ferns and fern allies, and at least some mosses[54]. Although there is obviously little host specificity in regard to ability to colonize roots, VAM are similar to the legume-Rhizobium symbiosis in that infectivity of the microbiol partner is not necessarily correlated with the effectiveness of the symbiosis in enhancing mineral status of the host.

Many VAM researchers argue that the primary, if not sole, effect of VAM on plants is to enhance the uptake of P, and perhaps other immobile minerals, and that all other effects on disease resistance, water relations, and hormone balances are secondary effects of altered P status. Although that argument is vigorously contested by others, this paper will be based on the assumption that characteristics of the fungus that maximize P uptake are the most likely to maximize the host reponse to VAM formation.

Presently there is little evidence that the fungal symbiont possesses unique biochemical properties that allow it to scavenge P that bypasses the uptake abilities of the plant root. Most experimental work with ^{32}P amended soil, or soil amended with insoluble sources of P, shows that VAM fungi extract P from the same pool as is available to roots[38,44]. Comparisons of the kinetics of P uptake did show that VAM tomato (*Lycopersicon esculentum*) roots had a K_m nearly three times lower than non-VAM roots, suggesting that the fungal hyphae did have a higher affinity for P than roots[15]. However, these differences in K_m between colonized and noncolonized roots may not be indicative of a higher affinity for P by the fungus, but rather, reflect the fact that there is greater involvement of inner cortical cells in P-uptake when P is being transported into the root interior than when roots are not colonized by VAM fungi[51]. The possibility that VAM phosphatases release P from organic sources warrants further study[21], but the ability of VAM hyphae to grow beyond the root's zone of mineral depletion and absorb P from soil not exploited by the root system presently seems to best account for the effect of VAM on P uptake[44]. Strains of VAM fungi with a well developed network of external hyphae would therefore be expected to be most effective in enhancing P uptake.

Because measurements of lengths of VAM hyphae in soil are very difficult, few attempts have been made to correlate this parameter with effectiveness of different isolates of VAM fungi. In one study, comparing weight of hyphae of four isolates of VAM fungi retrieved from soil, there was a close correspondence between external fungal biomass and P uptake in onion (*Allium cepa*)[45]. The weight of soil adhering to roots was used as an index of external hyphal development of VAM in another study, and again there was a strong correlation between growth response of citrus and external hyphal growth for six isolates of *Glomus* spp.[23]. In the first study, external hyphal development of the fungi closely paralleled rates of colonization of the root cortex[45]. However, this was not true in the second study[23], and many other investigations have failed to reveal a good correlation between percent of root colonized and growth or mineral content of the host[16,30,36]. Experiments also have shown that external hyphal growth is inhibited by increases in plant P concentrations sooner than internal development is inhibited[3,43,47]. Thus the more easily quantified parameter, colonization of the root cortex, unfortunately does not appear to be a reliable indicator of external hyphal development, especially if measurements are made at only one point in time.

Recent work suggests that even length of external hyphae is not a dependable predictor of fungal efficiency. When a membrane filter — grid intercept technique was used to measure hyphal length, there was no correlation between growth response and hyphal development among four species of VAM fungi on subterranean clover (*Trifolium subterraneum*)[4]. No data on P uptake were reported, but soil had been amended with P to a level adequate for about 60% maximum growth of non VAM clover. Perhaps under these moderately P deficient conditions, VAM-mediated P uptake was not crucial for a growth response. However, it also is likely that neither measurements of external hyphal length nor measurements of colonization of the cortex at only one point in time will provide an adequate picture of the P uptake capabilities of the fungal symbiont. Length of the lag phase and timing of fungal development relative to maximum P demand by the host are not taken into account unless measurements are taken throughout the colonization process. Distribution of hyphae in the soil, while extremely difficult to ascertain, also might be important. Extensive hyphal proliferation within the root's zone of P depletion would be less beneficial for P uptake than a less extensive hyphal network that exploited soil beyond the rhizosphere. Models of mineral uptake that integrate growth rates, and include corrections factors for overlapping depletion zones, such as those derived for roots[46] might be usefully applied to VAM studies to aid in identifying efficient strains of VAM fungi.

Finally, as earlier work with Rhizobium has shown, strains of microbial symbionts that are very efficient at enhancing mineral uptake are of little practical value unless they also can survive and infect their host under field conditions. Considerations of inoculum potential, survival, and competition with less efficient indigenous strains are beyond the scope of this paper but have been recently reviewed elsewhere[1].

Host dependency of VAM

A) Definition of VAM dependency

Even a superficial survey of the literature shows that different plant species vary in the magnitude and direction of their response to formation of VAM[9,24,34]. The term mycorrhizal dependency was coined to describe "the degree to which a plant is dependent on the mycorrhizal condition to produce its maximum growth at a given level of soil fertility" and is usually expressed as the ratio of some measure of growth in mycorrhizal plants to growth in nonmycorrhizal plants[19]. This means of expressing mycorrhizal dependency has no upper limit and the term relative field mycorrhizal dependency (RFMD) has been proposed recently to standardize values between 0 and 1. RFMD is the difference in growth of mycorrhizal and nonmycorrhizal plants divided by the growth of the mycorrhizal plant[39].

Both of the above definitions of mycorrhizal dependency are based on comparing growth of mycorrhizal and nonmycorrhizal plants under a given set of conditions. Values will vary as environmental conditions, particularly mineral supply, change[9,34]. A different approach to determining mycorrhizal dependency is to compare the response of colonized and noncolonized plants to a range of P application. The response curves of the mycorrhizal and nonmycorrhizal treatments can be quantitatively compared using the Mitscherlich equation[2]. Such a comparison enables one to estimate how much applied P is required by nonmycorrhizal plants to achieve the same yield (or P content) as their mycorrhizal counterparts. Although comparisons of response curves are more laborious than comparisons of plant growth under a single set of conditions, the response curve approach allows for comparisons of mycorrhizal dependency that are independent of the supply of P^2.

Regardless of the definition used, it is important to recognize that mycorrhizal dependency is not an immutable characteristic of a plant, but that it varies as environmental conditions and plant age change, and of course depends on the fungal symbiont involved[16,34]. Furthermore, mycorrhizal dependency values will differ for a given set of conditions depending on whether plant growth is measured as total biomass, shoot biomass, fruit or grain yield, height, photosynthetic area, long term survival, or content of a given mineral[30].

One of the first questions a plant breeder needs to address when working with VAM mediated phenomena, is whether it is more desirable to attempt to enhance the plant's response to VAM or to attempt to breed for plants which can approach maximum growth without VAM. Which strategy is more desirable depends on the relative amount of energy partitioned to mineral uptake in species or varieties with low VAM dependency compared to highly VAM dependent varieties colonized by an effective strain of fungus. Although experimental data providing a direct answer to that question are not available, comparisons of the below ground physiology of plants with high and low VAM dependency, as well as the ubiquity of VAM in nature, suggest that dependence on VAM for mineral uptake is frequently more energy efficient than low VAM dependency.

B) Photosynthate partitioning in VAM roots

In VAM dependent species a certain amount of photosynthate must be invested in maintenance of the fungal symbiont. Three separate studies of ^{14}C partitioning on three different host species have shown that between 5 and 15 percent more C was translocated below ground when roots were colonized by VAM fungi than when they were supplied comparable amounts of P from other sources[32,37,50]. There is some evidence that plants compensated for this increased below ground C demand with increased photosynthetic rates but the mechanism for this compensation is not known[37,50] and it is not at all certain that such compensation occurs reliably on a variety of hosts. It is possible that the C demand of VAM over the long term in the field may be quite different than what has been detected with short term labelling experiments with containerized plants. However, the values obtained for C demand do correspond quite well with estimates that in heavily colonized roots the fungal component of VAM comprises about 10 to 20 percent of the total biomass of the root[29,47,53].

Unfortunately, simultaneous studies of P inflows due to colonization of roots by VAM fungi, and studies of photosynthate partitioning to VAM have not been reported. Hence, ratios of P uptake to C drain are not available. However, in onions (*Allium cepa*), VAM portions of the root were calculated to have approximately twice the P uptake rate per cm of root as noncolonized portions of the root[45]. If the C partitioning studies cited above are applicable to onion (one study was done with leek (*Allium porrum*)) then a plant with half its root colonized by VAM would have a roughly 50 percent greater P uptake rate at an investment of about 10 percent more photosynthate than nonmycorrhizal roots.

C) Physiological mechanisms of decreasing VAM dependency

If VAM formation is not considered, mechanistic models predict that the most effective way for a plant to enhance uptake of immobile

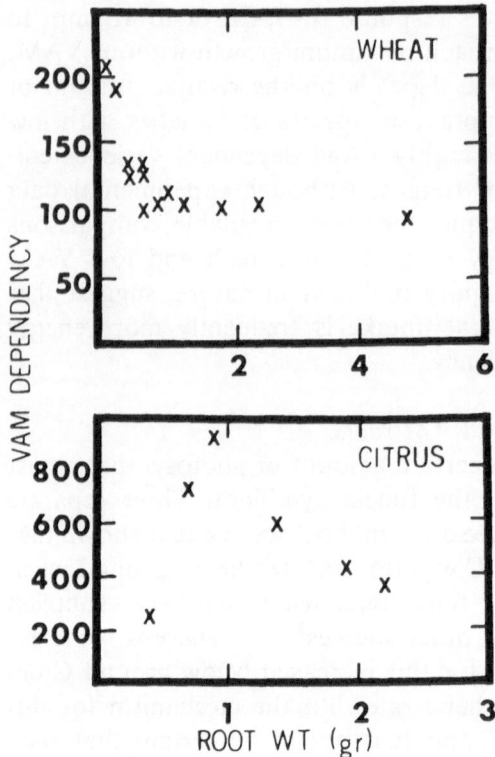

Fig. 1. Relationship to root size to VAM dependency (shoot biomass of VAM plants/shoot biomass of nonVAM plants) of different varieties of wheat and citrus. Redrawn from data in Azcon and Ocampo[9] and Menge et al.[34].

minerals like P is to increase root length[46], and VAM dependency was suggested to be inversely correlated with root length and root hair development many years ago[11,12]. However, this suggestion was based on proportions of root colonized rather than the growth response of the host to colonization, and quantitative measurements of root length in relation to VAM dependency are lacking. VAM dependency was significantly ($p < 0.1$) inversely correlated with root biomass when citrus rootstock cultivars were compared[34] and with root biomass and root/shoot ratio when wheat varieties were compared[9], (Fig. 1), but root biomass is not necessarily a good indication of root length. Using two species that typically do not form VAM at all (*Chenopodium quinona* and *Amaranthus hypochondriacus*) as an extreme example of nondependence on VAM, these species have been found to have up to 50 percent more root mass and more than double the root length, when grown under a range of P availabilities, than moderately VAM dependent tomato (*Lycopersicon esculentum*) plants with similar shoot masses and growth rates (unpubl. data). This limited comparison of mycorrhizal and nonmycorrhizal species suggest that species with low dependence on VAM partition more total biomass to below ground functions as well as increasing the ratio of root length to root biomass.

In addition to enhancing P uptake by increasing root length as a means of reducing VAM dependence, it also may be possible to increase efficiency in P utilization by the plant, or to increase the efficiency of a given amount of root in absorbing P. Regarding the first possibility, there was actually a slight negative correlation between VAM dependence and P concentration in shoot tissue in wheat,[9] and three species of plants that typically do not form VAM all had significantly ($p < 0.05$) higher shoot P concentrations than three noncolonized VAM-dependent species[49]. Thus, more efficient use of P in biomass production, which would result in lower tissue P concentrations, does not appear to be a commonly encountered mechanism for compensating for the absence of P uptake by VAM fungi.

Regarding the possibility of increasing the P uptake ability of a given length of root, some plant species do seem to acidify the rhizosphere and thus solublize P, enabling them to absorb more P than would be predicted based on root length and P availability in the bulk soil. Interestingly, this capability has been best documented for lupins (*Lupinus albus*)[17], rape (*Brassica napus*)[28], and buckwheat (*Fagopyrum esculentum*)[13], all species which are reported to not host VAM fungi. These reports suggest that plants which do not benefit from association with VAM fungi may have a second avenue, other than increased root length, to compensate for the lack of P input by VAM fungi. However, the effectiveness of this mechanism depends on cation/anion balances in soil[13], and roots of two species which do not normally form VAM actually have lower efficiency in P uptake per length of root than noncolonized roots of VAM dependent tomato (unpubl. data). Furthermore, lupins which were particularly efficient in taking up P from soil showed no greater ability to solublize low concentrations of ^{32}P added to soil than species that were much less efficient at P uptake[31]. Hence, even though acidification of the rhizosphere may be a means of enhancing P uptake in some species, the bulk of evidence suggests that plants that achieve maximum P uptake and growth without VAM do so primarily by investing more photosynthate in root production.

Based on the above discussion, selection for greater root length would be the most likely avenue for a plant breeder who wished to reduce VAM dependence to pursue. Although a good estimate of how much additional root length would be required to compensate for the input of VAM is difficult to determine, it almost certainly would be more than the 5 to 15 percent additional photosynthate that VAM require. For example, in one study, the least VAM dependent wheat variety had 42 times more root biomass than the most dependent variety, even though shoot weights and P concentrations were nearly the same in noninoculated plants. Inoculation with VAM fungi doubled the shoot weight and P concentration of the dependent variety but had little effect on the non-dependent variety[9].

Selecting for host response to VAM

Although plants that invest relatively little photosynthate in below ground functions are most likely to show large responses to an effective strain of VAM fungus, selection for a short root system alone obviously will not produce a plant that is efficient in P uptake unless those roots also are capable of being heavily colonized by the symbiotic fungus. The proportion of root colonized by different fungi may not be a good predictor of the effectiveness of any particular isolate of fungus, but it seems reasonable to expect that any given fungus will be more effective on a plant that supports heavy colonization than on a plant that does not. The plant breeder attempting to enhance a plant's response to VAM therefore would want to combine low investment in root production with high ability to support fungal colonization of the root and proliferation of external hyphae.

Because VAM fungi apparently lack saprophytic ability, the rate of transfer of photosynthate from the host to fungus is likely an important factor regulating the growth rate of the fungus. Since the fungus normally does not penetrate the plasma membrane of host cells, host products taken up by the fungus first must be exuded by the host into either the apoplast or the rhizosphere. Several studies have shown that by manipulating environmental variables to increase root exudation, VAM formation also is increased[22,48]. The rate of root exudation is the result of the interaction of many phsiological factors including translocation rates to the root, subcellular partitioning, and cortical cell membrane permeability. Because neither the composition nor the concentration of root exudates always parallels the composition or concentration of soluble material extracted from root cells (Table 1), membrane permeability has been suggested to be particularly important in regulating root exudates and hence VAM formation[22,40,48].

Although extent of VAM formation has been correlated with overall rates of root exudation with a variety of environmental manipulations, an association between VAM and supplies of any single compound or group of compounds has not been found[48,49]. This does not necessarily mean that all components of root exudates are equally important to VAM, but rather may reflect our inability to alter exudation of specific compounds without affecting overall exudation rates. Comparisons of VAM formation in plant varieties that differ markedly in the com-

Table 1. Composition of soluble material exuded from intact roots or extracted from homogenized roots of sudangrass grown at two levels of soil phosphorus*

	Percent of total			
	P deficient		P amended	
	Exudate	Extract	Exudate	Extract
Carbohydrates	12	27	13	26
Amino acids	8	25	9	36
Carboxylic acids	80	48	78	38

* From Schwab et al.[47]

position, but not overall quantity, of root exudates might prove very useful in detecting specific host products that enhance VAM formation.

Despite the fact that VAM formation is consistantly correlated with root exudation when environmental variables are manipulated, there has been little work done to determine if selective breeding for high root exudation rates would enhance VAM formation and host response to VAM. Comparisons of rates of exudation in VAM host and nonhost plant species indicated that genetic variation in exudation rates, as well as environmentally mediated variation, is correlated with VAM formation[49]. However, when different cultivars of wheat were compared, exudation of soluble sugars was only weakly associated with percent of root colonized by VAM fungi, and less closely associated with host response[9].

Other aspects of host-fungus nutrient exchange are only beginning to be explored. VAM fungi seem to produce specific alkaline phosphatases that may be involved in P uptake and transfer, and both phosphatase production and VAM formation are inhibited by high concentrations of P in host tissue[20]. The distribution of ATPase activity in cortical cells is altered in response to penetration of VAM fungi, indicating that the fungus may alter host membrane transport phenomena[33]. The possibility that P and photosynthate exchange are mechanistically linked in a system analogous to the triose-P translocator of chloroplast membranes also has been proposed[52]. The involvement of any of these functions in regulating colonization and response of the host would require a more elaborate system of signals between symbionts than would the scavenging of root exudates passively leaked through somewhat permeable cortical cell membranes. However, the existence, much less the biochemistry and genetics, of such signals remains to be established.

Conclusions

If VAM are to be of practical value to plant breeders and growers it is essential that the response to VAM be predictable. However, it is clearly impractical for plant scientists to test many varieties of a potential host crop against several isolates of VAM fungi, under a broad range of environmental conditions. A theoretical basis for predicting the efficacy of VAM inoculation is needed. Although considerable progress has been made in elucidating the physiology of VAM, there are still major gaps in our understanding of host-fungus interactions.

In order for VAM to be effective in increasing crop yield, VAM must enhance root functions at lower energetic costs to the host than establishing a comparably effective noncolonized root. Therefore, the relative efficiency of roots and fungal hyphae in mineral uptake, expressed in terms of photosynthate investment, under long term field conditions must be determined. Choosing a desirable strain of fungal symbiont will require an ability to determine the photosynthate demands of different strains of fungi balanced against their effectiveness in enhancing P uptake. This will require a much better understanding of the mechanisms of P and C transfer between symbionts, and a close examination of interactions between symbionts at the host-fungus interface.

This review has been based on the assumption that the primary effect of VAM is to enhance P uptake. Much of what has been written probably applies to uptake of other immobile minerals as well, but the effect of VAM on interactions among nutrients has received little experimental or theoretical consideration. The potential for VAM to enhance water relations is reasonably well documented, but it is controversial if VAM modify water uptake by mechanisms independent of P uptake[5,25,35]. There are also suggestions that VAM modify plant hormone balances, but again it has not been clearly established if this effect is primarily related to enhanced mineral uptake or not[6,7,10]. These aspects of VAM symbiosis must be more thoroughly explored before reliable predictions of the effects of VAM on plant growth in the field can be made.

References

1 Abbott L K and Robson A D 1982 The role of vesicular-arbuscular mycorrhizal fungi in agriculture and the selection of fungi for inoculation. Aust. J. Agric. Res. 33, 389–408.
2 Abbott L K and Robson A D 1984 The effect of mycorrhizae on plant growth. In VA Mycorrhizae, Eds. C L Powell and D J Bagyaraj. CRC Press, Boca Raton, Florida. pp 113–129.
3 Abbott L K, Robson A D and De Boer G 1984 The effect of phosphorus on the formation of hyphae in soil by the vesicular-arbuscular mycorrhizal fungus, *Glomus fasciculatum*. New Phytol. 97, 437–446.
4 Abbott L K and Robson A D 1985 Formation of external hyphae in soil by four species of vesicular-arbuscular mycorrhizal fungi. New Phtol. 9, 245–255.
5 Allen M F 1982 Influence of vesicular-arbuscular mycorrhizae on water movement through *Bouteloua gracilis*. New Phytol. 91, 191–196.
6 Allen M F, Moore T S Jr. and Christensen M 1980 Phytohormone changes in *Bouteloua*

gracilis infected by vesicular-arbuscular mycorrhizae. I. Cytokinin increases in the hostplant. Can J.B Bot. 58, 371–374.

7   Allen M F, Moore T S Jr., and Christensen M 1982 Phytohormone changes in *Bouteloua gracilis* infected by vesicular-arbuscular mycorrhizae. II. Altered levels of gibberellin-like substances and abscisic acid in the host plant. Can. J. Bot., 60, 468–471.

8   Ames R N, Reid C P P, Porter L K, and Cambardella C 1983 Hyphal uptake and transport of nitrogen from two ^{15}N-labelled sources by *Glosus mosseae*, a vesicular-arbuscular mycorrhizal fungus. New Phytol. 95, 381–396.

9   Azcon R and Ocampo J A 1981 Factors affecting the vesicular-arbuscular infection and mycorrhizal dependency of thirteen wheat cultivars. New Phytol. 87, 677–685.

10 Barea J M and Axcon-Aguilar C 1982 Production of plant growth-relating substances by the vesicular-arbuscular mycorrhizal fungus, *Glomus mosseae*. Appl. Environ. Microbiol. 43, 810–813.

11 Baylis G T S 1967 Experiments on the ecological significance of phycomycetous mycorrhizals. New Phytol. 66, 231–243.

12 Baylis G T S 1972 Minimum levels of phosphorus for non-mycorrhizal plants. Plant and Soil 36, 233–234.

13 Bekele T, Cino B J, Ehlert P A I, van der Mass A A, and van Diest A 1983 An evaluation of plant-borne factors promoting the solubilization of alkaline rock phosphates. Plant and Soil 75, 361–378.

14 Cooper K M and Tinker P B 1978 Translocation and transfer of nutrients in vesicular-arbuscular mycorrhizas. II. Uptake and translocation of phosphorus, zinc, and sulphur. New Phytol. 81, 43–52.

15 Cress W A, Throneberry G O and Lindsey D L 1979 Kinetics of phosphorus absorption by mycorrhizal and nonmycorrhizal tomato roots. Plant Physiol. 64, 484–487.

16 Furlan V, Fortin J A and Plenchette C 1983 Effects of different vesicular-arbuscular mycorrhizal fungi on growth of *Fraxinus americana*. Can. J. For. Res. 13, 589–593.

17 Gardner W K, Parbery D G and Barber D A 1981 Proteoid root morphology and function in *Lupinus albus*. Plant and Soil 60, 143–147.

18 Gildon A and Tinker P B 1983 Interactions of vesicular-arbuscular mycorrhizal infections and heavy metals in plants. II. The effects of infection on uptake of copper. New Phytol. 95, 263–268.

19 Gerdeman J W 1975 Vesicular-arbuscular mycorrhizae. *In* The Development and Function of Roots. Eds. J G Torrey and D T Clarkson. Academic Press, New York. pp 575–591.

20 Gianinazzi-Pearson V and Gianinazzi S 1978 Enzymatic studies on the metabolism of vesicular-arbuscular mycorrhiza. II. Soluble alkaline phosphate specific to mycorrhizal infection of onion roots. Physiol. Plant. Pathol. 12, 45–53.

21 Gianinazzi-Pearson, Fardeau J C, Asinni S and Gianinazzi S 1981 Sources of additional phosphorus absorbed from soil by vesicular-arbuscular mycorrhizal soy beans. Physiol. Veg. 19, 33–43.

22 Graham J H, Leonard R T and Menge J A 1981 Membrane-mediated decrease in root exudation responsible for phosphorus inhibition of vesicular-arbuscular mycorrhiza formation. Plant Physiol. 68, 548–552.

23 Graham J H, Linderman R G and Menge J A 1982 Development of external hyphae by different isolates of mycorrhizai *Glomus* spp. in relation to root colonization and growth of Troyer citrange. New Phytol. 91, 183–189.

24 Hall I R 1978 Effect of vesicular-arbuscular mycorrhizas on two varieties of maize and one of sweet corn. N.Z.J. Agric. Res. 21, 517–519.

25 Hardie K and Leyton L 1981 The influence of vesicular-arbuscular mycorrhiza on growth and water relations of red clover. I. In phosphate deficient soil. New Phytol. 89, 599–608.

26 Harley J L and Smith S E 1983 Mycorrhizal Symbiosis. Academic Press, New York. pp 77–103.

27 Hayman D S 1983 The physiology of vesicular-arbuscular endomycorrhizal symbiosis. Can. J. Bot. 61, 944–963.

28 Hedley M J White R E and Nye P H 1982 Plant-induced changes in the rhizosphere of rape

(*Brassica napus* var. Emerald) seedlings. III. Changes in *L* value, soil phosphate fractions and phosphate activity. New Phytol. 91, 45–56.
29 Hepper C M 1977 A colorimetric method for estimating vesicular-arbuscular mycorrhizal infection in roots. Soil Biol. Biochem. 9, 15–18.
30 Jensen A 1982 Influence of four vesicular-arbuscular mycorrhizal fungi on nutrient uptake and growth in barley (*Hordeum vulgare*). New Phytol. 90, 45–50.
31 Keay J, Biddiscombe E F and Ozanne P G 1970 The comparative rates of phosphate absorption by eight annual pasteur species. Aust. J. Agric. Res. 21, 33–44.
32 Koch K E and Johnson C R 1984 Photosynthate partitioning in split-root citrus seedlings with mycorrhizal and nonmycorrhizal root systems. Plant Physiol. 75, 26–30.
33 Marx C, Dexheimer J, Gianinazzi-Pearson V and Gianinazzi S 1982 Enzymatic studies on the metabolism of vesicular-arbuscular mycorrhizas. IV. Ultracytoenzymological evidence (ATPase) for active transfer processes in the host-arbuscule interface. New Phytol. 90, 37–43.
34 Menge J A, Johnson E L V and Platt R G 1978 Mycorrhizal dependency of several citrus cultivars under three nutrient regimes. New Phytol. 81, 553–559.
35 Nelson C E and Safir G R 1982 Increased drought tolerance of mycorrhizal onion plants caused by improved phosphorus nutrition. Planta 154, 407–413.
36 Owusu-Bennoah E and Mosse B 1979 Plant growth responses to vesicular-arbuscular mycorrhiza. XI. Field inoculation responses in barley, lucerne, and onion. New Phytol. 83, 671–679.
37 Pang P C and Paul E A 1980 Effects of vesicular-arbuscular mycorrhiza on ^{14}C and ^{15}N distributiuon in nodulated faba beans. Can. J. Soil Sci. 60, 241–250.
38 Pairunan A K, Robson A D and Abbott L K 1980 The effectiveness of vesicular-arbuscular mycorrhizas in increasing growth and phosphorus uptake of subterranean clover from phosphorus sources of different solubilities. New Phytol. 84, 327–338.
39 Plenchette C, Fortin J A and Furlan V 1984 Mycorrhizal dependency of several plant species under field conditions in a soil of moderate P fertility. *In* Proc. 6th N. Amer. conf. on Mycorrhizae. Ed. R Molina. College of Forestry, Oregon St. Univ. Corvallis. p. 401.
40 Ratnayake M, Leonard R T and Menge J A 1978 Root exudation in relation to supply of phosphorus and its possible releveance to mycorrhizal formation. New Phytol. 81, 543–552.
41 Rhodes L H, and Gerdemann J W 1978 Hyphal translocation and uptake of sulphur by vesicular-arbuscular mycorrhizae of onions. Soil Biol. Biochem. 10, 355–360.
42 Rhodes L H and Gerdemann J W 1978 Influence of phosphorus nutrition on sulphur uptake by vesicular-arbuscular mycorrhizae of onions. Soil Biol. Biochem. 10, 361–364.
43 Sanders F E 1975 The effect of foliar applied phosphate on the mycorrhizal infections on onion roots. *In* Endomycorrhizas. Eds. F E Sanders, B Mosse and P B Tinker. Academic Press, New York. pp 261–276.
44 Sanders F E and Tinker P B 1971 Mechanism of absorption of phosphate from soil by *Endogone* mycorrhizas. Nature 233, 278–279.
45 Sanders F E, Tinker P B, Black R L B, and Palmerley S M 1977 The development of endomycorrhizal root systems. I. Spread of infection and growth-promoting effects with four species of vesicular-arbuscular endophyte. New Phytol. 78, 257–268.
46 Schenk M K and Barber S A 1979 Phosphate uptake by corn as affected by soil characteristics and root morphology. Soil Sci. Soc. Am. J. 43, 880–883.
47 Schwab S M, Menge J A and Leonard R T 1983 Comparison of stages of vesicular-arbuscular mycorrhiza formation in sudangrass grown at two levels of phosphorus nutrition. Am. J. Bot. 70, 1225–1232.
48 Schwab S M, Menge J A and Leonard R T 1983 Quantitative and qualitative effects of phosphorus on extracts and exudates of sudangrass roots in relation to vesicular-arbuscular mycorrhiza formation. Plant Physiol. 73, 761–765.
49 Schwab S M, Leonard R T and Menge J A 1984 Quantitative and qualitative comparison of root exudates of mycorrhizal and nonmycorrhizal plant species. Can. J. Bot. 62, 1227–1231.
50 Snellgrove R C, Splittstoesser W E, Stribley D P and Tinker P B 1982 The distribution of carbon and the demand of the fungal symbiont in leek plants with vesicular-arbuscular mycorrhizas. New Phytol. 92, 75–87.
51 Tinker P B and Gildon A 1983 Mycorrhizal fungi and ion uptake. *In* Metals and Micronu-

trients, Uptake and Utilization. Eds. D A Robb and W S Pierpoint. Academic Press, New York. pp 21–32.
52 Tinker P B and Menge J A 1986 Host-fungal transfers in vesicular-arbuscular mycorrhizas. New Phytol. (*In* press).
53 Toth R and Toth D 1982 Quantifying vesicular-arbuscular mycorrhizae using a morphometric technique. Mycologia 74, 182–187.
54 Trappe J M 1982 Synoptic keys to the genera and species of zygomycetous mycorrhizal fungi. Phytopathology 72, 1102–1108.

Progress since the first international symposium: 'Genetic aspects of plant mineral nutrition', Beograd, 1982, and perspectives of future research

M.R. SARIĆ
Institute of Biology, Faculty of Sciences, University of Novi Sad, Yugoslavia

Key words Concentration Efficiency Elements Genotype Mineral nutrition

Summary The paper discusses the problems of genetic aspects of plant mineral nutrition in the light of the results presented at the First and Second Symposia on 'Genetic Aspects of Plant Mineral Nutrition' organized in Beograd in 1982 and Madison in 1985, respectively. On the basis of the results, future directions of research are discussed. The papers deal with the concentration and content of mineral nutrients in different genotypes, physiological and biochemical aspects of the genetic specificity of plant mineral nutrition, relations between plant genotypes and nitrogen fixing micro-organism strains, as well as with some related problems which have been investigated to a lesser extent. Particular attention is paid to papers and problems referring to genetic and breeding research work linked with genetic aspects of plant mineral nutrition as well as the possibilities of developing new cultivars requiring certain soil and mineral nutrition conditions for their cultivation.

Introduction

It is well known that plants require certain chemical elements for the normal maintenance of their physiological and biochemical processes. The available quantities of these elements were often insufficient and this resulted in the application of fertilizers. At present, the application of mineral fertilizers in modern agriculture is a common feature in crop production throughout the world.

In the past, many problems related to mineral nutrition of plants were solved and the results obtained contributed to more efficient utilization of mineral fertilizers. It was found that in general the application of 1 kg of NPK mineral fertilizers increased the grain yield by 10 kg (1:10). However, this ratio varied from 1:8.5 to 1:25.0, depending on the conditions[19]. There is general agreement that 50% of the yield increase in the last few decades is due to the application of mineral fertilizers. However, today we face the problem of obtaining the maximum economically viable yields and particular attention has been paid to abiotic and biotic factors which affect the efficient utilization of applied fertilizers. The results obtained indicate that the genotype requirements for specific elements are different[13]. There are specific relationships between plant genotypes and different strains of nitrogen fixing organisms, an important phenomenon that broadens the problem of the genetic base of plant nutrition. New genotypes, in addition to their improved utilization

H.W. Gabelman and B.C. Loughman (Eds.), Genetic aspects of plant mineral nutrition.
ISBN-13: 978-94-010-8102-3
© *1987 Martinus Nijhoff Publishers, Dordrecht/Boston/Lancaster.*

of energy and atmospheric CO_2 and N_2, should utilize mineral elements more efficiently, thus ensuring maximum and economically viable yields.

The first studies related to the genetic characteristics of mineral nutrition of cultivated plants were published in the twenties and thirties[1,2,6,7,8,9,14,15,16,17,18]. The importance of the problem was summarised in Vavilov's[17,18] citation 'The introduction of chemicals into agriculture up-dates the issue of breeding plants with positive reaction to fertilizer application' and 'geneticists and breeders expect the physiologists to develop the physiology of certain plant species, cultivar physiology and physiological systematics of cultivars'. In this period (1936) a monograph 'Cultivars and Fertilization'[15] was also published.

Much work has been concentrated on genetic aspects of plant mineral nutrition both in theoretical and practical terms, and the results obtained were published in different research journals and special publications[3,4,5,10,20].

A comprehensive review of results, problems and reference literature was presented in papers contained in the Proceedings of the First International Symposium on Genetic Specificity of Plant Mineral Nutrition[11,12]. The papers indicated prospects for future research in the field of genetic aspects of plant mineral nutrition.

On the basis of the results obtained so far, it can be concluded that the mechanism of genetic specificity of plant mineral nutrition integrates a number of morphological and anatomical properties, as well as physiological and biochemical processes, resulting in differences in the uptake and content of mineral elements in genotypes, which specifically reflect on structural and metabolic changes.

The most frequently applied criteria for identifying the existing genetic specificity of plant mineral nutrition in previous studies were the concentration and content of mineral elements in genotypes of different plant species. However, characteristics of the mineral nutrition of certain cultivars are primarily reflected in differences in the uptake of ions, and their transport, distribution and accumulation in certain organs, as well as their re-utilization, efflux, role in metabolism and biomass production. The above mentioned differences were found in plant organs of various age, both with cultivated plants and plants of natural phytocenozes.

Our aim was and still is to solve two main problems: 1. to identify different reactions of existing cultivars, *i.e.* genotypes, to mineral nutrition by applying a number of criteria. 2. to obtain new cultivars, *i.e.* genotypes which can utilize mineral elements more efficiently.

Since the concentration of certain elements in genotypes was the most often used parameter, we are presenting some data indicating the degree of variation of the criteria (Table 1). The results obtained in tackling the first problem should enable the implementation of the second, *i.e.* to find genotypes which could utilize elements more efficiently and economically.

Table 1. The variation of the concentration of mineral elements among the cultivars of the same plant species expressed in percents

Plant species	Number of cultivars or inbreds	Organ	Elements							Authors and years
Spinach	44	Leaves	Cu 63	Fe 86	Ni 93	Mn 69	Zn 60	Mo 59		Karvanek et. al. 1966
Bean	66	Aboveground part	K 42	Na 48	Ca 10	Mg 23				Shea et. al. 1968
Wheat	4	Aboveground part	N 9	P 19	K 21	Ca 14				Sarić et. al. 1969
Soybean	8	Leaves	N 9	P 18	K 29					Hanway et. al. 1971
Maize	43	Aboveground part	Fe 39	Zn 33	Mg 58	Ca 43	Cu 49	Al 41	Mn 73	Clark et. al. 1974
Maize	12	Leaves	P 36	K 34	Ca 30	Mg 39	Fe 16	Zn 71		Broetsch et. al. 1976
Wheat	12	Grains	Ca 54	Mg 59	Fe 67	Zn 46				Nahapetioan et. al. 1976
Vine grapes	5	Leaves	N 16	P 24	K 29	Ca 54	Mg 36			Sarić et. al. 1977
Weeping	11	Plant tops	P 48	Fe 41	Mn 73	Ca 52	Zn 54			Foy et. al. 1977
Maize	10	Grains	N 25	P 35	K 38	Mg 45				Sarić et. al. 1978
Oat	4	Plant tops	P 75	Ca 40	Fe 41					Brown et. al. 1978
Maize	12	Average for root and shoot	N 12	P 20	K 30	Ca 29	Mg 18			Sarić et. al. 1978
Sugarbeet	10	Shoots	N 13	P 26	K 22	Ca 19	Mg 17			Sarić et. al. 1979
Sunflower	22	Average for all organs	N 12	P 36	K 27	Ca 47	Mg 33			Sarić et. al. 1981

Table 1. (Continued)

Plant species	Number of cultivars or inbreds	Organ	Elements									Authors and years
Sorghum	21	Leaves	P 50	K 56	Ca 48	Mg 54	Fe 72	Cu 53	Zn 56	Mn 56	Al 48	Duncan, 1982
Wheat	4	Leaves	N 9	P 6	K 18	Ca 16	Mg 57					Repka, 1982
Variation of minimum and maximum concentration of some elements for 11 plant species			N P K	9-25 6-75 8-56				Ca Mg Fe	10-54 17 58 16 85			

The solution of these two problems should enable improved production of biomass, *i.e.* plant yields in compliance with their adequate mineral nutritional requirements. These problems are linked to research work in the field of physiology, genetics and breeding. One of the main reasons for enhanced interest in solving this problem today is the world energy crisis, *i.e.* the current economic situation involving permanent rise of fertilizer costs and high investments in improving fertility of the types of soils of limiting farming potential.

The main directions of research work were presented at the First and Second International Symposia (Beograd, 1982 and Madison, 1985) 'Genetic Aspects of Plant Mineral Nutrition'. Bearing in mind the aforementioned problems we decided to organise the First International Symposium in order to present the main results and prospects for the future study of genetic aspects of plant mineral nutrition. The Symposium organised in Beograd in 1982 presented the results of intensive research work in the field from its beginning with a limited amount of new research in 58 papers from 20 countries (Table 2). The main problems and fields of research were pointed out in my Introductory paper 'Theoretical and Practical Approaches to the Genetic Specificity of Mineral Nutrition of Plants' (Table 3). The problems encountered in the field are often multi- and interdisciplinary and their solution requires the participation of physiologists, geneticists, biochemists and cytologists.

The Beograd meeting provided an insight into the state of research in the field, as well as consideration of future and new directions of research work. In addition, the Symposium agreed on regular meetings every third year. It was also agreed that the Second International Symposium be organised in Madison in 1985, and we are implementing this decision.

Table 2. Review of sections and number of papers presented at the I International Symposium on Genetic Aspects of Mineral Nutrition of Plants, in Belgrade, 1982

Sections		Number of papers
Introductory paper:	Theoretical and practical approaches to the genetic specificity of mineral nutrition of plants	1
Section I:	Cytological and anatomical changes in different genotypes caused by altered nutrient supply	3
Section II:	Absorption, translocation and accumulation of ions in different genotypes	16
Section III:	The influence of mineral nutrition on physiological and biochemical processes of genotypes	15
Section IV:	The influence of mineral nutrition on yield and quality of different genotypes	9
Section V:	Genetical investigations concerned with selection of genotypes for a more effective use of mineral elements	14
	Total:	58

Table 3. The titles of main problems emphesised in the introductory paper presented at the I International Symposium on Genetic Aspects of Mineral Nutrition of Plants, in Belgrade, 1982

I	The increased efficiency of using both the natural fertility of soils and mineral fertilizers by creating and utilizing suitable cultivars and hybrids
II	The increased efficiency of using mineral nutrients under certain ecological conditions
III	The plant-specific role of microorganisms in enriching soil with nitrogen and soluble forms of other elements
IV	The role of genetic specificity of mineral nutrition in plants in solving the problems of environmental pollution
V	The principles of evaluating the genetic specificity of mineral nutrition in plants
VI	The genotype features influencing uptake of mineral nutrients
VII	Criteria for evaluating the genetic specificity of mineral nutrition in plants
VIII	Methods for selecting genotypes for certain soil types and mineral nutrition

The Second International Symposium is a further development in terms of the programme, problems and results obtained. 80 papers were presented at this Symposium by scientists from 17 countries. Papers are presented for the following sections (Table 4). Particular attention was paid to aluminium, since 13 papers deal with this problem. Special attention was attached to genotypes of sorghum and maize.

At the First Symposium most of the published papers referred to the study of the concentration and content of different minerals in certain organs of various genotypes. The differences obtained in concentration are obvious, although they were closely related to particular plant organs, nutritional conditions, temperatures, *etc.* Studies were often devoted to correlation between ionic concentrations in genotypes sensitive and tolerant to toxic elements. In a number of studies no correlation was found between dry matter production and concentration and content of specific elements in plant tissue.

Certain genotypes with high and low concentrations and different

Table 4. Review of sections and number of papers presented at the II International Symposium on Genetic Aspects of Mineral Nutrition of Plants, in Madison, 1985

Sections		Number of papers
Introductory paper:	Review of Progress to Date in Genetics of Plant Nutrition	1
Section I:	Responses of Wild Plant Ecotypes to Nutrient Deficiency Stress	6
Section II:	Screening Techniques to Detect Nutritional Differences Under Genetic Control	9
Section III:	Tolerances to Salinity and to Metal Toxicities	17
Section IV:	Genotypic Response to Nutrient Deficiency	35
Section V:	Genetic Variation in Microorganism/Host Interactions in Mineral Nutrition	5
Section VI:	Germplasm Resources and Modifications	3
Section VII:	Possibilities and Directions of Future Research	4
	Total:	80

tolerances and efficiencies in the use of particular elements have been identified. However, most of the papers referring to this problem indicate that these properties depend on a number of biotic and abiotic factors. Particularly important is the fact that the genotypes were tolerant only to specific elements. Genotypes were rarely tolerant to several elements, and no genotype had combined tolerance to deficiency or excess of all elements.

Particular attention was paid to the toxic effects of aluminium. Genotypes sensitive to aluminium have a larger uptake of the element in the earlier stages of their development compared to resistant genotypes. In addition, genotypes resistant to aluminium are capable of excreting the aluminium taken up, whereas the sensitive genotypes excrete aluminium with more difficulty. With genotypes more resistant to aluminium, the concentration of Ca and Mg is higher, both in the roots and aerial parts of the plants.

Variation in the concentration and content of certain elements in different genotypes was dependent on the conditions of mineral nutrition and soil properties, as well as other environmental factors (light, temperature, *etc.*).

In several papers presented to the Beograd Symposium there were attempts to determine the existence of genetic specificity of mineral nutrition on the basis of physiological and biochemical parameters. In this connection, study of the activity of enzymes of nitrogen metabolism in different genotypes should be emphasised. The results obtained do not show a consistent relationship between the content of different forms of nitrogen in plants and the activity of nitrate reductase. A considerably smaller number of papers dealt with the activities of enzymes participating in the metabolism of phosphorus in different genotypes. Some indicators in C_3 and C_4 plants were also studied but the results obtained on the concentration of elements and synthesis of organic matter were not consistant. In addition, no correlation between the concentration of selected ions in different genotypes and their photosynthetic activity was found. Future studies should broaden this approach by using specific inhibitors and phytohormones for certain physiological and biochemical processes to aim at possible further explanations of the role of physiological and biochemical processes in assessing the existence of genetic specificity of plant mineral nutrition.

A number of papers presented at the Madison Symposium show that genotypes differ not only in the concentration and total content of certain elements but also in their metabolism. The differences were found both in organs and tissues of certain organs and it was pointed out that genotypes also differ in respect of the synthesis of phosphate carriers.

Results indicating that the genotype distribution of K ions is different even at the subcellular level should be noted as well. Genotypes having a higher concentration of K in their tissues, even if its concentration in the nutrient media is less than their requirements, and still exhibiting K deficiency have been identified. This is due to their inability to mobilise K from the vacuole and transport it to the cytoplasm. The correlations between the anatomical properties of roots, root diameters and properties of xylem elements and Mg uptake were determined. It was also shown that the activities of enzymes of nitrogen metabolism depend both on the properties of inbred lines of C_3 and C_4 plants and on the nitrogen content of the media. Differences between genotypes in terms of their Mo concentrations, as well as correlation between Mo concentrations and nitrate reductase activity were found. Physiological and genetic characteristics of different plant species relating to Zn and Cu uptake were determined by applying specific metabolic inhibitors. Indirect indicators such as the correlation between silica content in corn tissue and resistance to corn borer were also introduced.

In spite of the growing attention paid to the subject of specific relations between plant genotypes and strains of nitrogen fixing microorganisms, contributions at the Beograd Symposium addressing this problem were rather modest. Papers presented at the Madison Symposium show that there is genetic variability in nitrogen fixing components. However, it depends on the plant species and genotype used, as well as the particular strain of *Rhizobium* sp. In some species, nitrogenase activity is primarily conditioned by the size of nodules whereas in others, by their number and weight.

We would like to stress the need for parallel and joint work on creating new genotypes of plants and nitrogen fixing micro-organism strains, in order to obtain the most suitable associations. Future research work should focus on two main problems: 1. explanation of the existence of specific relations between different strains of nitrogen fixing microorganisms and plant genotypes. 2. study of the concentration of other indicators of its metabolism in different genotypes, using two sources of nitrogen, *i.e.* mineral and atmospheric.

The existence of genetic control of inheritance of the concentration of K and Ca was determined. The possibilities of obtaining mutants having different activities of nitrate reductase, as well as of other enzymes which take part in nitrogen metabolism was also proved.

Depending on parental combination, the forms of inheritance may vary. Thus, the concentration of K and P shows an intermedial form of inheritance, whereas N concentration indicates the dominant form, namely that of the better parent, although positive heterosis was also obtained for all three of these elements. The aforementioned results were presented at the First Symposium.

For the Second Symposium we have papers on chromosome changes resulting in genotypes resistant to certain elements *e.g.* Cu, which are extremely important for further work on obtaining genotypes specific for certain nutritional conditions. The work showing the impact of certain genomes and chromosomes on the concentration of a number of elements in wheat should also be mentioned. Finally, we have obtained results indicating that the differences in uptake, distribution and efficiency of P with selected genotypes are genetically controlled, thus showing the particular importance of dominant genes, although additive genes also participate in determining the characters.

Much less work was devoted to investigation of flux-selecting uptake mechanisms and excretion from the roots of different genotypes. These studies could contribute to a better understanding of the role of the plasmalemma and tonoplast in genotypes sensitive and resistant to increased concentrations of certain ions in the nutrient solution. Cation exchange properties could be one of the methods for explaining the sensitivity or resistance to toxic metals in particular.

Metals immobilised in cell walls are not toxic and such genotypes are resistant to these metals. However, plants are more sensitive to metals in the cytoplasm. It is also found that the transport ratio of K/Na is important for shoots and roots of plants. Tolerance of genotypes to salinity increases as this ratio increases and vice versa.

Only a few papers discuss cytological and anatomical changes in genotypes in different conditions of mineral nutrition and, these problems will be important and interesting fields of future research on physiological and biochemical processes, particularly with plants whose economic yield is represented by their roots or tubers (sugar beet, potatoes, *etc.*).

We ask whether it is possible to produce genotypes suitable for certain soil and nutritional conditions. The results obtained so far are encouraging and some fruitful developments can be expected in the future. Particular emphasis should be given to the fact that more attention has been paid to creating cultivars resistant to increased levels of certain elements which, at high concentrations, are toxic to plants (Al, Mn, Na, Cl, Cu, *etc.*). The problem of better utilization of some of the necessary elements, especially N, P, K is almost neglected, *i.e.* the creation of cultivars which could utilize these elements more efficiently.

In natural conditions, the main task is to obtain genotypes for normal soil which would use certain elements more efficiently, particularly in the presence of reduced soil N, P, and K and ensure high production and quality of organic matter. In respect of deficient soils *i.e.* extremely acid or alkaline soils, our task is to obtain cultivars resistant to high or low pH values of the soil solution when there is a deficiency or surplus of certain elements. Usually acid soils have surplus Al and Mn and lack P. Therefore, in this case, cultivars should be resistant — tolerant to potentially toxic quantities of Al and Mn, and should utilise decreased quantities of P more efficiently. Alkaline soils are characterised by higher contents of Na, Cl, Mg and S and cultivars developed for these soils should be tolerant to the elements.

As already pointed out, there are different criteria for assessing the genetic specificity of mineral nutrition. Quantity and quality of synthesised organic matter in a particular genotype under given conditions of mineral nutrition are the most important criteria. All other criteria are of secondary importance, particularly if the practical aspects of the problem are considered.

It seems that investigations dealing with genetic aspects of mineral nutrition should pay more attention to plants as entire systems and try to investigate at least several criteria simultaneously. For example, when investigating the uptake of specific elements by the roots of different genotypes, due consideration should be given to nutrient transport to the shoots, as well as to the concentration, content and synthesis of organic matter. Absorption and transport are not regulated by the same mechanisms. The uptake varies with genotype due to morphological, anatomical and cytological differences in their roots. However, transport could also vary since it is regulated by the properties of conducting vessels, as well as by specific needs of shoots for certain elements. There are hypotheses indicating that the efficiency of uptake depends on the regulation of the relation between uptake and transport of the element. It would be useful to examine whether there are differences in uptake and transport of certain elements by different genotypes when the elements are supplied through the leaves as well as via the roots.

Qualitative differences between genotypes should not be expected in basic metabolism but in the intensity of certain metabolic processes, *i.e.* quantity of specific metabolites. Bearing in mind that certain metabolic processes, which can be performed only in the presence of certain ions, are very specific, we made an attempt to determine some characteristics of basic physiological and biochemical processes for genotype efficient in N and P utilization (Figures 1. and 2.). In future, much more attention should be paid to studies of metabolic sequences. This refers not only to N and P but also to other elements essential for physiological and biochemical processes.

Finally, further research work on the problems of genetic aspects of

GENETIC ASPECTS OF PLANT MINERAL NUTRITION

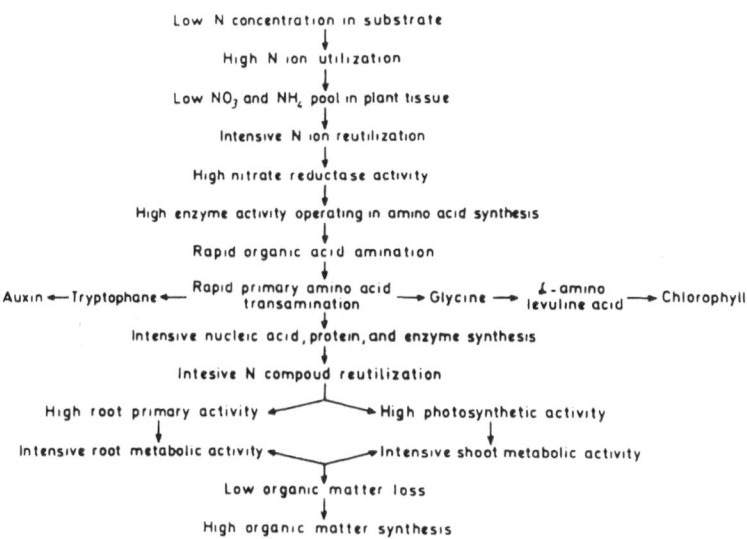

Fig. 1. Some characteristics of basic physiological and biochemical processes for genotypes efficient in nitrogen.

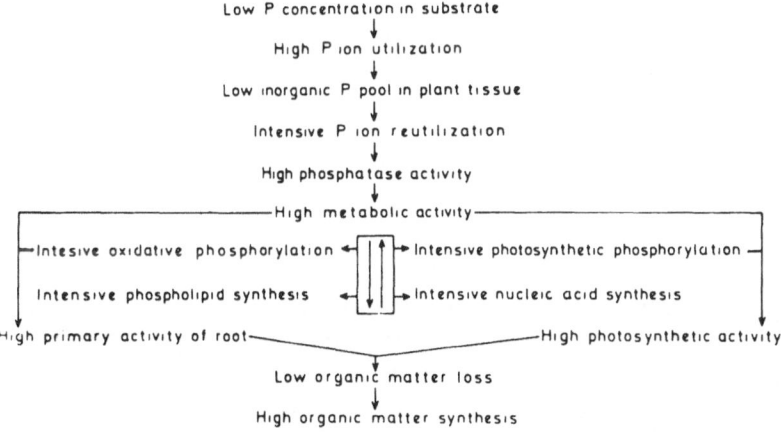

Fig. 2. Some characteristics of basic physiological and biochemical processes for genotypes efficient in phosphorus.

mineral nutrition should take into account the specificity of synthesis of certain metabolites, as well as the role they play. The increased content of certain elements in sugar beet roots negatively affects their quality, *i.e.* makes their processing more difficult. If these elements could be translocated at the end of the vegetative period from roots to leaves, their negative effect on the genotype would be eliminated. In this case, the genetic aspect of mineral nutrition would be assessed on the basis of a translocation indicator, *i.e.* ionic distribution.

Further investigations should be directed in the following areas:
1. Collection of germplasm of wild and cultivated plants and microorganisms and examination of their physiological and biochemical properties that could be used as a model of further research on genetic aspects of mineral nutrition. Particular attention should be paid to genotypes having deficiency or excess of certain elements in their growth media.
2. More detailed study of suitable genotypes on the basis of specific criteria for their further use as models, both in terms of the concentration and content of elements and intensity of reutilization as well as other properties important for solving the specific problems of the genetic base for mineral nutrition.
3. Selection for crossing parental genotypes specific for certain criteria, in order to study the inheritance of the genetic base of mineral nutrition.
4. Development of new cultivars and species with specific mineral requirements, aiming at obtaining maximum yields and quality in adequate soil conditions and with the application of mineral fertilizers.

The problems of genetic aspects of plant nutrition are complex, as is evident from the above discussion, and their solution requires the cooperation of experts from different sciences. The teamwork necessary for this could provide faster and more efficient research and give a more comprehensive insight into the results obtained and newly apparent problems.

Instead of conclusions, problems for further consideration can be stated as follows:

Where in the cell are the genomes responsible for differences in uptake and concentration of specific ions located?

Are the differences in ion concentration of genotypes the result of differences in the level of genomes within species?

Is there a correlation between uptake and content of particular ions and the evolution of development of certain species?

Are there the same trends in ion concentrations in plant tissues and specific subcellular units of genotypes?

Why do the differences in uptake and concentration of elements not have the same trends in all parts of the plant?

Why do genotypes differ in respect to the intensity of uptake and concentration of different elements?

To what extent does the environment affect genotype characteristics contributing to different absorption rates and concentrations of ions and biomass production?

What is the minimum concentration of an element that ensures maximum production of biomass of a particular genotype?

Which mechanisms determine specific relations between plants of a genotype and strains of nitrogen fixing micro-organisms?

How much does the study of the problems concerning genetic aspects of mineral nutrition contribute to our understanding of the mechanism of ion uptake?

Based on the questions mentioned above, the following fundamental problem is set: Which physiological-biochemical mechanisms can be considered as the basic ones to explain the existence of genetic differences in mineral nutrition, *i.e.* more economical utilization of mineral elements by different genotypes?

References

1. Harvey P H 1939 Hereditary variation in plant nutrition. Genetics 24, 437 461.
2. Hoffer G 1926 Some differences in the functioning of self-pollination lines of corn under varying nutritional conditions. J. Am. Soc. Agric. 18, 322–343.
3. IAEA/FAO 1967 Isotopes in plant nutrition and physiology. Proc. Symp. Vienna, September 1966, IAEA Vienna.
4. Jung A G 1978 Crop tolerance to suboptimal land conditions. Am. Agr. Special Publ. 2, 343.
5. Klimaševski E L 1974 Sort i udobrenie AN SSSR Irkutski, 283.
6. Lyness A 1936 Varietal differences in the phosphorus feeding capacity of plants. Plant Physiol. 11, 665 688.
7. Moors C A 1921 The agronomic placement of varieties. J. Am. Soc. Agric. 13.
8. Moors C A 1922 Varieties of corn and their adaptibility to different soils. Univ. Tnn. Agr. Exp. Sta. Bull. 126.
9. Moors C A 1933 The influence of soil productivity on the order of yield in a varietal trial in corn. J. Am. Soc. Agric. 25.
10. Rejmers R E and Klimaševski E L 1969 Fiziologija i biohemija sorata. ANN SSSR Irkutski, 173.
11. Sarić R M 1982 Genetic specificity of mineral nutrition of plants. Serbian Academy of Science and Arts. Scientific assembles XIII No. 3, Beograd, 389.
12. Sarić R M and Loughman B C 1983 Genetic Aspects of Plant Nutrition. Martinus Nijhoff/Dr Junk Publishers Hague, 490.
13. Sarić R M 1984 Genetic improvement of crop yields as related to plant nutrient requirements. 9th World fertil. congress of CIEC, 11—16 june 1984, budapest, Hungary (*In press*).
14. Smith S N 1934 Response of inbred and crosses in maize to variations of nitrogen and phosphorus supplied as nutrients. J. Am. Soc. Agron. 26, 785–804.
15. Sort i udobreni 1936 Moskva izd. VASHNILvip. 1
16. Springfield G and Salter R 1934 Differential response of corn varieties to fertility levels and to seasons. J. Am. Res. 11, 991–1000.
17. Vavilov N I 1932 Genetika na službe socijalističkogo zemljedelia. Soc. rast-vo 4, 19 42
18. Vavilov N I 1934 Osnovnie zadači sovetskoj selekciji rastenii i puti ih osuščestvlenija. Soc. rast-vo 12, 5- 22.
19. Von Peter A 1984 Fertilizers in maximum yield systems. 9th World Fertilizer Congress of CIEC 11 16 june 1984 Budapest Hungary (*In press*).
20. Wright J M 1976 Plant Adaptation to Mineral Stress in Problem Soils. Cornell. Univ. Ithaca New York, 420 p.